ECOTOXICOLOGY

ENVIRONMENTAL SCIENCE AND TECHNOLOGY

A Wiley-Interscience Series of Texts and Monographs

Edited by JERALD L. SCHNOOR, *University of Iowa*
ALEXANDER ZEHNDER, *Swiss Federal Institute for Water Resources and Water Pollution Control*

A complete list of the titles in this series appears at the end of this volume

Series Preface
Environmental Science and Technology

We are in the third decade of the Wiley-Interscience Series of texts and monographs in Environmental Science and Technology. It has a distinguished record of publishing outstanding reference texts on topics in the environmental sciences and engineering technology. Classic books have been published here, graduate students have benefited from the textbooks in this series, and the series has also provided for monographs on new developments in various environmental areas.

As new editors of this Series, we wish to continue the tradition of excellence and to emphasize the interdisciplinary nature of the field of environmental science. We publish texts and monographs in environmental science and technology as it is broadly defined from basic science (biology, chemistry, physics, toxicology) of the environment (air, water, soil) to engineering technology (water and wastewater treatment, air pollution control, solid, soil, and hazardous wastes). The series is dedicated to a scientific description of environmental processes, the prevention of environmental problems, and preservation and remediation technology.

There is a new clarion for the environment. No longer are our pollution problems only local. Rather, the scale has grown to the global level. There is no such place as "upwind" any longer; we are all "downwind" from somebody else in the global environment. We must take care to preserve our resources as never before and to learn how to internalize the cost to prevent environmental degradation into the product that we make. A new "industrial ecology" is emerging that will lessen the impact our way of life has on our surroundings.

In the next 50 years, our population will come close to doubling, and if the developing countries are to improve their standard of living as is needed, we will require a gross world product several times what we currently have. This will create new pressures on the environment, both locally and globally. But there are new opportunities also. The world's people are recognizing the need for sustainable development and leaving a legacy of resources for future generations at least equal to what we had. The goal of this series is to help understand the environment, its functioning, and how problems can be overcome; the series will also provide new insights and new sustainable technologies that will allow us to preserve and hand down an intact environment to future generations.

JERALD L. SCHNOOR AND ALEXANDER J. B. ZEHNDER

Preface

If you know a thing, it is simple.
If it is not simple, you don't know it.

Ecotoxicology aims at characterizing, understanding and predicting deleterious effects of chemicals on biological systems. To this end, it is also concerned with the fate of chemical substances in the environment, with factors that govern the bioavailable portion of compounds and their uptake pathways into organisms, with metabolic activation and deactivation processes as well as interaction mechanisms on the suborganismic level, and with propagation mechanisms of biological effects on higher levels of organization.

Ecotoxicology is a relatively young science, and has a truly applied and interdisciplinary character. Its origin can be considered unique as compared to classical natural sciences. The motivation to undertake ecotoxicological research was based on the recognition that anthropogenic chemicals may affect nontarget organisms, populations and communities in unexpected, extended and in particular hazardous ways. As a result, ecotoxicology has been developed under distinct political and public demands from its beginning.

The need to achieve results of practical relevance for politics and society made the development of ecotoxicology rapid indeed. On the other hand, the theoretical basis was not able to keep pace, and some of the most challenging questions such as long-term effects of xenobiotics on higher levels of biological organization are far from being in the scope of currently available routine work in this field. This situation has also been termed the dilemma of ecotoxicology: The goal to protect the environment against chemical stress requires criteria to detect and evaluate the impact of compounds on natural communities and habitats in the field, whilst current ecotoxicological test methods typically address biological effects of chemical substances on suborganismic entities and organisms, or on artificial communities like microcosms and mesocosms, without yielding an advice for the extrapolation of the results onto ecologically relevant levels of organization and time scales. In this context it should be noted that living systems are by nature subject to stress of many different kinds, and that tracing back effects observed in the field to chemical stressors implies knowledge about natural fluctuations of foodweb characteristics as well as about the impact of nonchemical factors like climate, food abundance and land use.

A prominent and historical example for the complexity of ecotoxicological problems was the observation of population declines of wildlife species like the peregrine falcon (*Falco peregrinus*), the brown pelican (*Pelecanus occidentalis*) and the cormorant (*Phalacrocorax auritus*) in the 1960s and early 1970s in North America. Detailed analyses over several years revealed eggshell thinning as a major factor yielding a significant reduction of the breeding success. This was then traced back to DDE (dichlorodiphenyldichloroethene) residues in the eggs, resulting from metabolic conversion of bioaccumulated DDT (dichloro-diphenyltrichloroethane). Interestingly, a similar decline of the peregrine falcon population in the UK during that time apparently occurred mainly as a result of direct mortality caused by dieldrin, which was used as seed dressing for cereal seeds that were eaten by the bird. Both the effects of DDT and dieldrin on raptorial birds had been completely unexpected, occurred on larger temporal and spatial scales, and were unravelled as late as ca. 15 years after their onset only through extended and specific investigations. Moreover, it may well be that in the UK the in-field susceptibility of wildlife species to dieldrin was in fact enhanced by a parallel but sublethal contamination with DDT. Although the uptake pathways of these persistent organics may look quite understandable from current knowledge about processes of bioaccumulation, their molecular mechanisms of toxicity are still a matter of debate. This example shows how difficult it is to forecast long-term and large-scale effects of chemicals in the environment.

Apart from these and other scientific challenges, public concern about global and principal aspects of environmental pollution has shed new light on the role of ecotoxicology in society. In the 1990s, discussion began to focus on strategies to make the industrial development sustainable in the global sense, and now-adays we realize the need for operational means to foster a corresponding development in both rich and poor countries. Certainly, ecotoxicology will be needed for the finding and setting of suitable environmental quality criteria, but it is also clear that this challenging task of modern societies cannot be solved alone by contributions from the natural sciences. This latter aspect, however, is beyond the scope of the present volume.

As noted above, ecotoxicology is interdisciplinary by nature, and includes a variety of experimental and theoretical methods from related sciences like biology (including ecology), chemistry, toxicology, geology, geography and physics. From this (certainly incomplete) list it is clear that even a comprehens-ive treatment cannot cover all aspects of potential relevance. Whilst most previous books have concentrated on ecotoxicology as a subdiscipline of biol-ogy, the present text discusses the following major areas that are of primary concern for the current understanding and future development of this field: Part 1 presents the ecological fundamentals needed for an understanding of the complexity associated with ecotoxicological effects on population, community and system levels of biological organization. In Part 2, exposure of inorganic and organic chemicals in terrestrial and aquatic environments is discussed in terms of occurrence, transport and distribution as well as transformation and fate. Parts 3 and 4 are concerned with the ultimate goal of ecotoxicology to identify, characterize and predict biological effects of chemicals: In Part 3, bioaccumulation and biological effects on various levels of organization are discussed for terrestrial and aquatic species, and Part 4 contains accounts of current and future test strategies as well as perspectives for an ecotoxicological risk assessment. A more detailed overview of all chapters is given in the following paragraphs, which are structured according to the four parts of the book.

In Part 1 (*Historical Introduction and Ecological Fundamentals*), the historical development of ecotoxicology as a science is reviewed in the introductory chapter by SVEN ERIK JØRGENSEN (*Ecotoxicological Research – Historical Development and Perspectives*). In Chapter 2 (*Ecosystem Principles for Ecotoxicological Analyses*), HELMUT LIETH introduces a definition of ecosystems as consisting of sets of information units that are related to energy and matter, and discusses both ecosystem structures and functions with various examples. Principal types of response of ecosystems towards stress are in the focus of Chapter 3 (*Sensitivity of Ecosystems and Ecotones*) by OTTO FRÄNZLE, where the concepts of stability and resilience of populations and communities as well as of soils as principal regulatory compartments of terrestrial and benthic ecosystems are outlined. Chapter 4 (*Population Dynamics of Plants under Exposure and the Selection of Resistance*) by WILFRIED ERNST presents an analysis of pollution-induced changes in the genetic composition of terrestrial plants, which due to their sedentary character have only limited capacity to compensate loss of biodiversity by immigration of new genotypes. The final chapter of this section (*Community Ecology and Population Interactions in Freshwater Ecosystems*) by BRUNO STREIT emphasizes the need to address natural fluctuations in the community development when analyzing larger-scale ecotoxicological effects.

In Part 2 (*Chemicals in the Environment*), the characteristics of chemical exposure in environmental compartments are discussed in eight chapters, covering inorganic and organic compounds including transformation reactions and fate modelling. Chapter 6 (*Distribution and Biogeochemistry of Inorganic Chemicals in the Environment*) by BERND MARKERT reviews the environmental occurrence and cycling of elements as well as their biological functions for terrestrial plants, which leads to the concept of a biological system of elements. In Chapter 7 (*Speciation of Chemical Elements in the Environment*) by ROLF-DIETER WILKEN, analytical techniques to characterize the speciation of elements are discussed, covering both principal aspects and operational methods for practical applications in environmental sciences.

The following six chapters deal with environmental aspects of organic chemicals. In Chapter 8 (*Multimedia Mass Balance Models of Chemical Distribution and Fate*), DON MACKAY discusses the scientific background and use of multimedia mass balance models for analyzing transport and fate of compounds in the environment on global and regional scales. Hydrolysis, oxidation and reduction as major abiotic transformation pathways for organic chemicals are reviewed by ERIC WEBER in Chapter 9 (*Abiotic Transformation Reactions*), and the role of chemical reductants as well as of electron mediators for redox reactions in anoxic sediments is highlighted. A thorough analysis of the environmental profile of chlorinated xenobiotics is given in Chapters 10 (*Transformation of Chlorinated Xenobiotics in the Environment*) and 11 (*Environmental Fate of Chlorinated Organics*) by HEIDELORE FIEDLER AND CHRISTOPH LAU. The former chapter discusses phase transfer processes like volatilization, sorption and bioconcentration and degradation reactions, and the latter focusses on the occurrence, use patterns and long-range transport phenomena.

Chapter 12 (*Ecochemistry of Toxaphene*) by MEHMET COELHAN AND HARUN PARLAR reviews occurrence, environmental chemistry and ecotoxicological potential of the organochlorine insecticide toxaphene, which has been used in large amounts in the USA for many years and is still of great importance in cotton-growing countries. In the final chapter of this section (*Specimen Banking as an Environmental Surveillance Tool*), ANTONIUS KETTRUP AND PETRA MARTH present the scope and use of environmental specimen banking, which enables retrospective analyses of formerly known and unknown environmental contaminants in the future and with newly developed analytical techniques.

Part 3 (*Bioaccumulation and Biological Effects of Chemicals*) contains a total of eight chapters, discussing bioaccumulation in aquatic and terrestrial species as well as biological effects on various levels of biological organization. Chapter 14 (*Bioaccumulation of Chemicals by Aquatic Organisms*) by DES CONNELL reviews mechanistic and kinetic characteristics of the bioconcentration of chemicals in aquatic organisms as well as its relationship to the lipophilicity of the compounds, including the more complex case of sedimentary infauna. In Chapter 15 (*Metal Bioaccumulation in Freshwater Systems: Experimental Study of the Actions and Interactions between Abiotic and Contamination Factors*) by ALAIN BOUDOU AND coworkers, factors influencing metal bioaccumulation in aquatic organisms are discussed in detail with results from different exposure regimes of cadmium and mercury in indoor microcosms, focussing on differences resulting from different contamination routes (water column or sediment) as well as from different temperatures, pH values and photoperiods.

A process-oriented framework for quantifying different modes of toxic action is presented in Chapter 16 (*Process-oriented Descriptions of Toxic Effects*) by BAS KOOIJMAN, where the Dynamic Energy Budget is used as a concept to establish links between toxic effects on individuals and population parameters like survival, growth and reproduction. The cellular level of biological effects of chemicals as the primary site of action is discussed in Chapter 17 (*Cellular Response Profile to Chemical Stress*) by HELMUT SEGNER AND THOMAS BRAUN-BECK, covering membrane processes, biotransformation, metal homeostasis and various protective mechanisms against chemical stress. Current long-term tests for aquatic organisms used for notification purposes are described and evaluated by HORST PETER AND WOLFGANG HEGER in Chapter 18 (*Long-term Effects of Chemicals in Aquatic Organisms*), and special emphasis is given to a critical discussion of an approach advocated by ECETOC (European Chemical Industry Ecology and Toxicology Centre, Brussels, Belgium) to estimate prolonged toxicity from extrapolation of (more easily available) acute toxicity data with some safety factor.

The impact of soil-borne heavy metals on plants is outlined in Chapter 19 (*Effects of Heavy Metals in Plants at the Cellular and Organismic Level*) by WILFRIED ERNST covering abiotic and biotic factors that affect metal bioavailability as well as plant sensitivity, mechanisms of resistance and implications of metal stress for the life history of individuals. In Chapter 20 (*Assessment of Ecotoxicity on the Population Level Using Demographic Parameters*), NICO VAN STRAALEN AND JAN KAMMENGA review demographic techniques to derive indices of population performance, and introduce a new elasticity index that allows an integrated assessment of toxicant-induced changes in life-cycle parameters. Chapter 21 (*Effects of Pollutants on Soil Invertebrates: Links Between Levels*) by JASON WEEKS discusses approaches to link together biological effects on different levels of organization with soil invertebrates as ecologically relevant targets due to their essential functions in terrestrial ecosystems, and special focus is given to biomarkers as means to detect and integrate ecotoxicological effects.

Ecological risk assessment aims at predicting the likelihood of hazardous effects to occur in the environment, and it requires the integration of knowledge about chemical exposure and biological effects of compounds on the molecular level as well as about mechanisms of propagation of effects through different levels of biological organization. In Part 4 of the book (*Contributions to an Ecological Risk Assessment*), experimental and theoretical techniques in the field of ecotoxicology are discussed that are relevant for risk assessment strategies. Chapter 22 (*Ecotoxic Modes of Action of Chemical Substances*) by GERRIT SCHÜÜRMANN focusses on quantitative structure–activity relationships as a means to identify and characterize the structural disposition of compounds for exerting unspecific and specific modes of toxic action, and discusses descriptors for the physicochemical profile that govern characteristics of exposure and bioavailability. In Chapter 23 (*Endpoints and Thresholds in Ecotoxicology*) by JOHN CAIRNS, analysis of the human society's life support system as both technical and ecological leads to a new concept to establish criteria and thresholds for deleterious effects of toxicants, which is based on the maintenance of ecosystem services that are necessary for a sustainable development. Chapter 24 (*Modeling Ecological Risks of Pesticides: A Review of Available Approaches*) by LARRY BARNTHOUSE reviews the potential of ecological models to characterize and predict adverse effects on organisms, populations and communities in the framework of ecological risk assessment, yielding detailed discussions of age/stage-structured models, individual-based models, metapopulation models, and spatially explicit models.

The next two chapters outline recent developments and future trends of ecotoxicological test methods for terrestrial and aquatic environments. Chapter 25 (*Current and Future Test Strategies in Terrestrial Ecotoxicology*) by REINHARD DEBUS discusses approaches for the hazard assessment of compounds in soils, where only few standardized methods with OECD (Organization for Economic Co-operation and Development, Paris, France) guidelines are available, and emphasizes the need to differentiate between substance-related assessments of soils and evaluations of soil quality in the context of land use scenarios. For bacteria, algae, and invertebrates, a comparative analysis of conventional aquatic tests and alternative microbiotests is given in Chapter 26 (*Alternative Assays for Routine Toxicity Assessments: A Review*) by COLIN JANSSEN, and the validation status and application range of alternative test assays as rapid, simple and low-cost test methods is discussed from the viewpoint of applications in environmental screening programs. The final Chapter 27 (*Legislative Perspective in Ecological Risk Assessment*) by JAN AHLERS AND ROBERT DIDERICH gives a comprehensive outline of a risk assessment scheme based on the ratio between the predicted environmental concentration (PEC) and the predicted no-effect concentration (PNEC) of chemical substances, which is used with some variants by regulatory agencies in most OECD countries and constitutes a framework for the harmonization of legislative procedures on a truly international and transcontinental level.

As outlined above, the presentation of ecotoxicology in this book covers ecological fundamentals needed for an understanding of system-level responses, characteristics and implications of chemical exposure in the environment, and major types of biological effects on terrestrial and aquatic species including approaches to address higher levels of biological organization and strategies to integrate ecotoxicological work in schemes for an ecological risk assessment. An authoritative way to discuss these different aspects would not have been possible without bringing together a large number of distinguished authors, and we would like to express our thanks for their kind cooperation, stimulating input and valuable contributions. There has also been excellent office support from our secretarial staffs, Antje Zschernitz (Leipzig) and Raymonde Figula (Zittau), which is gratefully acknowledged. Moreover, we would also like to express our gratitude to the joint publishers John Wiley & Sons, Inc., and Spektrum Akademischer Verlag for their substantial encouragement and support to undertake this project. Of course we alone remain responsible for the structure of the book, for the selection of its major topics and in particular for all shortcomings. We encourage the readers to inform us about any items of concern, and we hope that the book will contribute to a fruitful discussion and further development of ecotoxicology.

GERRIT SCHÜÜRMANN AND BERND MARKERT

Leipzig and Zittau, May 1997

Contents

3 Sensitivity of Ecosystems and Ecotones

Otto Fränzle

4 Population Dynamics of Plants Under Exposure and the Selection of Resistance

Wilfried H. O. Ernst

PART 4 CONTRIBUTIONS TO AN ECOLOGICAL RISK ASSESSMENT

22 Ecotoxic Modes of Action of Chemical Substances   665
Gerrit Schüürmann

List of Contributors

PD Dr. Jan Ahlers, German Federal Environmental Agency, Risk Assessment of Existing Chemicals (IV 1.2), Mauerstr. 45–52, D-10117 Berlin; Tel.: ++49-30-89033 120, Fax: ++49-30-89033 129.

Dr. Lawrence W. Barnthouse, McLaren Hart Environmental, 109 Jefferson Avenue, Suite D, USA-Oak Ridge, Tennessee 3780; Tel.: ++1-423-482 8978, Fax: ++1-423-482 9473.

Prof. Dr. Alain Boudou, Laboratory of Ecotoxicology, University of Bordeaux I/CNRS, Avenue des Facultés, F-33405 Talence Cedex; Tel.: ++33-5-5684 8808, Fax: ++33-5-5684 8405, email: a.boudou.@ecotox.u-bordeaux.fr.

PD Dr. Thomas Braunbeck, University of Heidelberg, Department of Zoology I, Im Neuenheimer Feld 230, D-69120 Heidelberg; Tel.: ++49-6221-545 668, Fax: ++49-6221-546 161, email: braunbeck@urz.uni-heidelberg.de.

Prof. Dr. John Cairns, Jr., Department of Biology, Virginia Polytechnic and State University, 1020 Derring Hall, USA-Blacksburg, Virginia 24061-0415; Tel.: ++1-540-231 7075, Fax: ++1-540-231 9307, email: cairnsb@vt.edu.

Dr. Mehmet Coelhan, Department of Analytical Chemistry, University of Kassel, D-34109 Kassel; Tel.: ++49-561-804 4319, Fax: ++49-561-804 4480.

Prof. Dr. Des W. Connell, Faculty of Environmental Sciences, Griffith University, Nathan, Qld. 4111, Australia; Tel.: ++61-7-3875 7108, Fax: ++61-7-3875 7459, email: D.Connell@ens.gu.edu.au.

Dr. Reinhard Debus, Fraunhofer Institute for Environmental Chemistry and Ecotoxicology, P.O. Box 1260, D-57377 Schmallenberg; Tel.: ++49-2972-302 255, Fax: ++49-2972-302 319.

Robert Diderich, Ministry of the Environment, Department for the Prevention of Pollution and Risks, 20, avenue de Ségur, F-75302 Paris 07; Tel.: ++33-1-4219 1468, Fax: ++33-1-4219 1544, email: robert.diderich@environnement.gouv.fr.

Prof. Dr. Wilfried H. O. Ernst, Faculty of Biology, Free University, De Boelelaan 1087, NL-1081 HV Amsterdam; Tel.: ++31-20-444 7050, Fax: ++31-20-444 7123.

DR. HEIDELORE FIEDLER*, Chair of Ecological Chemistry and Geochemistry, University of Bayreuth, D-95540 Bayreuth.

PROF. DR. OTTO FRÄNZLE, University of Kiel, Ecology Centre, Schauenburger Str. 112, D-24118 Kiel; Tel.: ++49-431-880 3426, Fax: ++49-431-880 4658.

PD DR. WOLFGANG HEGER, German Federal Environmental Agency, Ecotoxicological Assessment, Bismarckplatz 1, D-14191 Berlin; Tel.: ++49-30-8903 3240, Fax: ++49-30-8903 2285.

DR. BEATRICE INZA, Laboratory of Ecotoxicology, University of Bordeaux I/CNRS, Avenue des Facultés, F-33405 Talence Cedex; Tel.: ++33-5-5684 8808, Fax: ++33-5-5684 8405.

DR. COLIN JANSSEN, University of Ghent, Laboratory for Biological Research in Aquatic Pollution, J. Plateaustraat 22, B-9000 Ghent; Tel.: ++32-9-264 3775, Fax: ++32-9-264 4199, email: colin.janssen@rug.ac.be.

PROF. DR. SVEN ERIK JØRGENSEN, DFH, Institute A, Department of Environmental Chemistry, University Park 2, DK-2100 Copenhagen Ø; Tel.: ++45-3537-0850, Fax: ++45-3537-5744.

DR. JAN E. KAMMENGA, Agricultural University, Department of Nematology, Binnenhaven 10, NL-6709 PD Wageningen; Tel.: ++31-317-482 998.

PROF. DR. ANTONIUS KETTRUP, GSF National Research Centre for Environment and Health, Institute of Ecological Chemistry, Ingolstädter Landstr. 1, D-85764 Oberschleißheim; Tel.: ++49-89-3187 4047, Fax: ++49-89-3187 3371.

PROF. DR. SEBASTIAAN A. L. M. KOOIJMAN, Department of Theoretical Biology, Free University, De Boelelaan 1087, NL-1081 HV Amsterdam; Tel.: ++31-20-4447 130, Fax: ++31-20-4447 123.

CHRISTOPH LAU, Chair of Ecological Chemistry and Geochemistry, University of Bayreuth, D-95440 Bayreuth; Tel.: ++49-921-552 154, Fax: ++49-921-54626.

DR. SYLVIANE LEMAIRE-GONY, Laboratory of Ecotoxicology, University of Bordeaux I/CNRS, Avenue des Facultés, F-33405 Talence Cedex; Tel.: ++33-5-5684 8808, Fax: ++33-5-5684 8405.

PROF. EM. DR. HELMUT LIETH, Institute of Environmental Systems Research, International Ecological Projects, University of Osnabrück, D-49069 Osnabrück, Tel.: ++49-541-969 2547, Fax: ++49-541-969 2570.

PROF. DR. DONALD MACKAY, Trent University, Peterborough, Ontario K9J 7B8, Canada, Tel.: ++1-705-748 1489, Fax: ++1-705-748 1569.

PROF. DR. BERND MARKERT, International Graduate School (IHI) Zittau, Markt 23, D-02763 Zittau; Tel.: ++49-3583-7715 20, Fax: ++49-3583-7715 34, email: novell1.ihi.htw-zittau.de.

* *New address*: Bavarian Institute for Waste Research (Bifa GmbH), Am Mittleren Moos 46a, D-86167 Augsburg; Tel.: ++49-821-7000 198, Fax: ++49-821-7000 100, email: 100143-2665@ compuserve.com.

Dr. Petra Marth, GSF National Research Centre for Environment and Health, Institute of Ecological Chemistry, Ingolstädter Landstr. 1, D-85764 Oberschleißheim; Tel.: ++49-89-3187 4047, Fax: ++49-89-3187 3371.

Dr. Regine Maury-Brachet, Laboratory of Ecotoxicology, University of Bordeaux I/CNRS, Avenue des Facultés, F-33405 Talence Cedex; Tel.: ++33-5-5684 8808, Fax: ++33-5-5684 8405.

Dr. Muriel Odin, Laboratory of Ecotoxicology, University of Bordeaux I/CNRS, Avenue des Facultés, F-33405 Talence Cedex; Tel.: ++33-5-5684 8808, Fax: ++33-5-5684 8405.

Prof. Dr. Harun Parlar, Department of Chemical-Technical Analysis, Technical University of Munich, D-85350 Freising-Weihenstephan; Tel.: ++49-8161-71 3283, Fax: ++49-8161-71 4418.

PD Dr. Horst Peter, German Federal Environmental Agency, Ecotoxicological Assessment, Bismarckplatz 1, D-14191 Berlin; Tel.: ++49-30-8903 3240, Fax: ++49-30-8903 2285.

Dr. Francis Ribeyre, Laboratory of Ecotoxicology, University of Bordeaux I/CNRS, Avenue des Facultés, F-33405 Talence Cedex; Tel.: ++33-5-5684 8808, Fax: ++33-5-5684 8405.

Prof. Dr. Gerrit Schüürmann, Department of Chemical Ecotoxicology, UFZ Centre of Environmental Research, Permoserstr. 15, D-04318 Leipzig; Tel.: ++49-341-235 2309, Fax: ++49-341-235 2401, email: gs@uoe.ufz.de.

PD Dr. Helmut Segner, Department of Chemical Ecotoxicology, UFZ Centre of Environmental Research, Permoserstr. 15, D-04318 Leipzig; Tel.: ++49-341-235 2329, Fax: ++49-341-235 2401.

Prof. Dr. Nico M. van Straalen, Vrije Universiteit (Free University), Department of Ecology and Ecotoxicology, De Boelelaan 1087, NL-1081 NV Amsterdam; Tel.: ++31-20-4447 070, Fax: ++31-20-4447 123.

Prof. Dr. Bruno Streit, Department of Ecology and Evolution, University of Frankfurt, Siesmayerstr. 70, D-60054 Frankfurt; Tel.: ++49-69-798 24711, Fax: ++49-69-798 24820.

Dr. Eric J. Weber, Ecosystems Research Division, National Exposure Research Laboratory, U.S. Environmental Protection Agency, USA-Athens, GA 30605; Tel.: ++1-706-355 8224, Fax: ++1-706-355 8202, email: Weber.Eric@epamail.epa.gov.

Dr. Jason M. Weeks, Institute of Terrestrial Ecology, Monks Wood, Abbots Ripton, GB-Huntingdon, Cambridgeshire PE17 2LS; Tel.: ++44-1487-773 381, Fax: ++44-1487-773 467.

Prof. Dr. Rolf-Dieter Wilken, ESWE–Institute for Water Research and Water Technology GmbH, Söhnleinstr. 158, D-65201 Wiesbaden-Schierstein; Tel.: ++49-611-780 4444, Fax: ++49-611-780 4375, email: wilken@goofy.zdv.uni-mainz.de.

ECOTOXICOLOGY

PART 1

Historical Introduction and Ecological Fundamentals

1

Ecotoxicological Research— Historical Development and Perspectives

Sven Erik Jørgensen (Copenhagen, Denmark)

1.1 SUMMARY

More analyses of ecotoxicological relationships are compulsory for further progress in ecotoxicological research but not sufficient. All the other development tendencies mentioned in this chapter must play an important role also. The full consequence of toxic substances cannot be understood or even described properly without further experience in ecotoxicological modelling and much more research in systems ecology. This will provoke further analyses as the properties of toxic substances and their interactions with the living organisms are important elements in our ecotoxicological models. However, the number of parameters is so great that it will be impossible to perform estimations of all the pertinent parameters within a reasonable time period. Estimation methods will therefore be of great significance in the coming decades and should be developed much further, mainly by gaining experience with as wide a spectrum of estimation pathways as possible.

Environmental management is much more complex today than was foreseen 20–25 years ago. This is mainly due to the unpredicted consequences of the use of about 100,000

Ecotoxicology, Edited by Gerrit Schüürmann and Bernd Markert.
ISBN 0-471-17644-3   © 1998 John Wiley & Sons, Inc. and Spektrum Akademischer Verlag.

chemicals which modern society is producing in such an amount that they may threaten the environment. Fortunately, new tools are facilitating the solving of the recently emerging ecotoxicological problems. Cleaner technology and ecotechnology must work hand in hand with environmental technology if proper solutions to the many serious ecotoxicological problems are to be found.

The challenge to ecotoxicological research in the next century is therefore to intensify the development of the *entire* spectrum of approaches, ideas, tools, and concepts presented in this chapter, which indeed is reflected in the entire volume. Only if the entire spectrum is applied will we have a possibility to solve the extremely important environmental problems associated with the many more or less toxic chemicals that have become an integral part of modern society.

1.2 WHEN DID IT START?

The first green wave started in the second half of the 1960s and focused on a wide range of pollution problems. Among these problems were also ecotoxicological problems: the reduction of birds of prey populations, particularly eagles, due to the biomagnification of dichlorodiphenyltrichloroethane (DDT) and other pesticides through the food chain, unexpected residues of polychlorinated biphenyl (PCB) in seals, and the effects of air pollutants on human health, to mention just a few of the most important problems.

Ecology was a scientific discipline without any roots in the society before this first green wave. If you had asked a man on the street 30 years ago: "What is ecology?" he would not have been able to answer the question, while every schoolchild today will know the meaning of the word "ecology". Most people will even today not know the word "ecotoxicology" but will probably be able to guess that it is a combination of ecology and toxicology, and therefore it must be the science about toxic substances in the environment (nature), i.e., the processes of toxic substances in the environment and their effects on living organisms in nature and on the structure of entire ecosystems.

There is no doubt that Rachel Carson's book *The Silent Spring* has contributed significantly to the focus in the 1960s on pesticides and thereby also on other toxic organic compounds in our environment. Similarly, the debate about and the final discovery of the origin of probably the biggest environmental catastrophe up to now, the Minamata disease in Japan, and later of the itai-itai disease also in Japan, called attention to the role of heavy metals as toxic substances in the environment.

The first green wave initiated in the 1960s was more than just a wave; it was a revolution, in the sense that it changed the basic viewpoints of technology and science. In the 1950s and early 1960s, technology and science were developing very rapidly and the view points were very optimistic: in just a few more decades we will have solved the energy problem for ever—the nuclear energy will be tamed completely—and technology is developing so rapidly that all nations, including developing countries, will have a living standard so high that we cannot imagine today. Technology and science would solve (almost) all the crucial problems of mankind according to the views 40–50 years ago. The situation can best be illustrated by the notion of the "World Equation"—an equation that could explain "everything", meaning that it would represent the very root of all sciences. Scientists were so naive 40–50 years ago that they seriously believed in the possibility of discovering a world equation—one equation that would represent the basis of all physical and chemical forces!

Therefore, the pollution problems emerged like a bomb explosion. These unpleasant side effects of our enormous technological developments came as a surprise, since they were completely overlooked and unexpected. Slowly but surely during the 1960s, 1970s, and 1980s scientists have acknowledged that nature is much more complex than previously thought. Not everything could be reduced to an experiment in the laboratory. Wolfram (1984a,b) talked about irreducible systems, to which most biological systems belong), but required a *synthesis* of many laboratory experiments and/or observations in situ.

Ecotoxicology belongs to one of the new sciences which emerged as a consequence of the adverse effects of technology on complex natural systems. It has been acknowledged in these scientific disciplines that natural (and other) systems are so complex that it is impossible to reach an understanding of all the details of these systems. Like we in nuclear physics have realized that all observations always have an uncertainty caused by our influence on the observations on the small nuclear particles, we have to accept that in the ecological disciplines our description of natural systems and their processes will inevitably have a certain degree of uncertainty due to their enormous complexity.

Nevertheless, ecotoxicological research started in the 1960s to attempt to reveal as many details as possible of the processing of toxic substances in the environment. This research proved to be extremely valuable as it was able to point towards some general rules and classification of the behavior of toxic substances as described in Chapter 3 of this volume. It became, however, clear that the ultimate goals of ecotoxicological research, to determine all processes in nature for all chemicals of interest (there are probably about 100,000 compounds used in such an amount that they could threaten the environment), would be comparable to the fate of Sisyphus. Therefore we have to find an alternative to investigating all of the details of all processes and all chemical compounds. This was one of the focal topics discussed in ecotoxicology in the 1970s and 1980s, and the outcome of this discussion will be presented in the following sections.

1.3 MODELLING AND ECOTOXICOLOGY

When we entered the 1970s, a few but crucial ecotoxicological problems were "isolated and acknowledged". The consequences in the industrialized countries came promptly: the use of DDT was banned, several applications of mercury were banned, and open applications of PCB were banned. Lead-free gasoline was introduced in the late 1970s in the USA. Japan followed shortly after, but Europe hesitated due to the strong vehicle industry in Germany, Italy, France, and Great Britain. Lead-free gasoline was, however, introduced in western European countries later in the 1980s.

At the start of the 1970s, the solution to the pollution problems was still considered a question of "end of the pipe technology". We could solve the pollution problem by introducing another technology, named environmental technology. The USA announced zero discharge in 1985! The pollution problems were just considered another challenge to our steadily growing technologies.

Already in the early 1970s it became clear that we could not solve the problems entirely by end of the pipe technology. We also had to rely on the self-purification processes of nature. The answer was: go modelling. The basic idea behind this new approach is illustrated in Figure 1.1. The emissions caused by our urbanization and technological development have a crucial impact on processes in the environment. It was necessary to know how these processes were influenced, at least partially, and to design a model (i.e., a simplified picture) of the most important environmental processes associated with the focal impact which includes the relationship between the impact and its influence on these processes. The idea was to use this model to find out how much of a reduction in emissions would be necessary to obtain an undisturbed or almost undisturbed environment. A feasible, environmental technology able to reduce the emission sufficiently was selected according to the model results. This approach implied that the self-purification ability of ecosystems could be taken into account in the selection of a proper environmental management strategy and provided a more affordable solution to the pollution problems. In contrast, the approach of "zero discharge" and the use of environmental technology alone would have been prohibitively expensive for all nations including the USA.

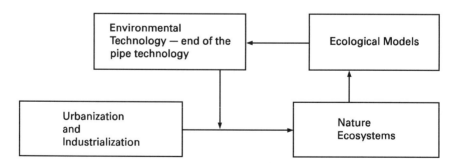

Figure 1.1 The basic idea behind the use of models in the early seventies is shown. The emission by urbanization and industrialization cause an impact on ecosystems, which is symbolized by the arrow between the two boxes. Models based on simplifications are used to quantify this impact. The models are used to select the environmental technology which can reduce the impact of urbanization and industrialization on ecosystems.

The use of environmental management models had already been developed. In the 1920s the well-known Streeter–Phelps model for the oxygen balance in streams was developed, and during the 1960s several amendments to this approach were proposed. However, the wide and fully accepted use of environmental models did not occur until the early 1970s. The first ecotoxicological models emerged in the late 1970s, and the first conference on the development of ecotoxicological models was held in the early 1980s in Copenhagen.

The application of ecotoxicological models in environmental management of emissions of toxic substances, however, quickly became a powerful tool. Two equally applicable approaches were developed in this period: a more ecological approach and a more chemical approach. The ecological approach was mainly inspired by the general application of ecological models in environmental management. It focused on the processes in ecosystems influenced by toxic substances and the distribution of the toxic substances in the ecosystem as a result of these processes. Jørgensen (1983, 1989) gives several illustrative examples of this approach. The chemical approach focuses on the distribution of the chemical of interest in the environment in accordance with the properties of the chemical. The questions are: where would the chemical accumulate? Is it mobile? Is it persistent or biodegradable? Answers to these crucial questions would reveal the general fate of the chemical in nature. Chapters 8 and 10 in this volume present this approach in details.

The two approaches complement each other. The chemical approach allows us to compare easily the fate and thereby the effect of two or more chemicals emitted into the environment. This type of model does not yield a very accurate estimate of the concentrations and the related effects of the focal chemical(s). However, it does indicate which part of the environment is threatened and which chemicals we can or can not allow to be emitted into the environment and at which concentrations. These models are, for instance, extremely useful in our effort to substitute environmentally unacceptable chemicals with other more acceptable chemicals. The ecological approach, in contrast to the chemically oriented models, yields more accurate values of chemical concentrations in the environment, which is of importance in the management decisions associated with the individual ecosystems and with risk assessments in the use of chemicals in certain contexts (see also Chapter 6). These models may be used to determine with reasonable accuracy the amounts of emissions which are permissible from individual sources, but they cannot judge which chemical among a range of possibilities is most safe to use, in contrast to chemical models.

Today, a wide range of models have been developed. They comprise the experience of many case studies where models have been an integrated part of the environmental management decision on the use and emission of toxic substances (Jørgensen et al. 1995). The range of available ecotoxicological models is very wide with respect to chemical compounds, environmental problems, and the ecological components and processes involved. It is likely that more sophisticated models will be developed in the very near future to the benefit of environmental management. Furthermore the two types of ecotoxicological models tend to approach each other in the sense that the chemical models include more and more ecological compartments and processes and the ecological models place more emphasis on the general chemical properties of the compound of interest. Another development which took place, particularly in the 1980s, was the translation of the model results to a risk assessment, which was a more comprehensive way to present the results in a management context. The uncertainty of the model was to a certain extent reflected in the quantification and interpretation of the model results.

1.4 THE TRENDS IN THE 1980s

Three important developments in the 1980s radically changed environmental management including the management of ecotoxicological problems: the important role of diffuse pollution was assessed, the need for use of cleaner technology was emphasized and global pollution problems came into focus in addition to the local and regional ones. How these developments have influenced and changed environmental management strategies is discussed below.

The subject of diffuse pollution, mainly coming from agriculture, was not touched on before the late 1970s. Previously, agriculture, in contrast to industry, was considered to be nature friendly. The intensive use of fertilizers and pesticides, which are used to increase the yield per hectare or per unit of time, has radically changed this image of agriculture. It became clear that environmental technology was not sufficient to solve the pollution problems, because this technology could not cope with diffuse pollution. Pesticides or other toxic organic compounds could be removed from waste water, but the problem associated with contamination and accumulation of persistent compounds in our environment would not be solved if we ignored the enormous amount of more or less toxic compounds emitted as diffuse pollution to the environment by agricultural activities. Therefore we have had to consider new weapons in our effort to abate the pollution problems. How can the non-point pollutants be dealt with? Non-point pollution problems are much more complicated to solve than point pollution problems, provided that we want to maintain the industrialized agriculture in the developed countries. It has therefore been necessary to adapt a multicomponent approach, referred to as ecotechnology or ecological engineering. This involves the employment of many approaches simultaneously, including the use of buffer zones between the natural ecosystems and agriculture, development of a new generation of pesticides, a wider use of biological methods for the control of weeds and herbivorous insects, and development of a completely new ecological agriculture that has banned the use of pesticides and fertilizers.

Cleaner technology is based upon the substitute principle in the sense that if a production method does not include acceptable emissions, it should be replaced by another technology. The development of new-generation pesticides which are less harmful to the environment may be considered as cleaner technology. Formerly, the selection of production methods did not consider pollution problems associated with the production, but rather only which method was easiest to control, involved the lowest costs, and produced the highest quality. The increasing use of "the polluter must pay" principle, introduced for instance through the application of green tax, has provided an economic incentive to consider the environment by introducing new and cleaner production methods, characterized by lower and/or more acceptable emissions than the previously applied methods. This is an extremely important development, since the environment is considered in our selection of technology. The effect on the environment will therefore be a determining factor in our entire industrial and agricultural production in the future. At present we have only seen the tip of the iceberg, but it can be foreseen that our entire society will be changed according to this new view point. The development is expected to be strongly reflected in the emission of toxic substances in particular, simply by the selection of production methods in industry and agriculture which rely on less harmful chemicals. It is not an easy task to determine which chemicals are least harmful. Is a less mobile chemical preferable to a less biodegradable chemical or is the biomagnification the core problem? The chemical models mentioned in Section 1.2 and illustrated in Chapters 8 and 10 are powerful tools which provide a basis for making a proper decision.

The global pollution problems have come more into focus since the beginning of the 1980s. They are mainly related to two atmospheric pollution problems: the green house effect and the reduction of the protective ozone layer. Particularly the latter is strongly related to the ecotoxicological problem complex. Several compounds which are known to react with the ozone layer have been banned in many countries, mainly in the industrialized countries, and it has been agreed internationally to reduce their application according to accepted schedules.

Environmental management has become considerably more complex due to the introduction and acceptance of these three developmental tendencies (ecological engineering, cleaner technology, and global concern). This is illustrated in Figure 1.2. In contrast, Figure 1.1 illustrates the much simpler philosophy behind environmental management 20–25 years ago. This development towards a more and more complex environmental management is, however, not surprising. Environmental problems are very complex, and so it is obvious that they can only be solved by a very wide range of solution tools, which are selected from case to case according to the problem, the character of the compounds in question, and the ecosystem involved. Environmental technology is still a powerful tool, but it has to be seen in relation to other powerful solution models by consideration of the characteristics of the focal problem in the right context.

1.5 ESTIMATION OF ECOTOXICOLOGICAL PROPERTIES

Ecotoxicology as emphasized above deals with very complex systems, and it is not possible to know all the details of all the ecological components and

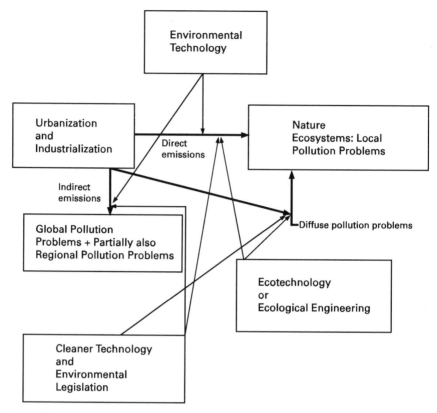

Figure 1.2 The much more complex environmental management of today is illustrated. Bold arrows symbolize impacts and thin arrows ways to control/reduce these impacts. We distinguish local, regional, and global problems from point and non-point pollution problems. Fortunately we have a wide range of tools available to solve the problems: environmental technology, ecotechnology, cleaner technology, and environmental legislation. Ecological models (not shown in this figure, but see Figure 1.1) are widely used to assess ecological risk and select the combinations of tools.

processes involved in a focal ecotoxicological problem. Therefore, we have to use models to provide us with the overview needed to consider at least the most important components and processes. This requires, however, that the properties of the ecological components and the processes and their interactions with the toxic substance of interest are known. This implies that we have to know the pertinent properties of the 100,000 compounds that we are using in our modern society today and the properties related to the interactions of these substances with the wide range of ecological components. If we presume that the more than 5 million species on earth are represented by let us say 25,000 (key) species and that each interaction requires the knowledge of ten parameters (properties), for instance, uptake rate, excretion rate, toxic effect, sublethal effects, and accumulation in various organs, each interaction will easily require at least ten parameters. For each of the 100,000 compounds we would furthermore need to know their physicochemical behavior: their rate of evaporation, solubility in water and fat tissue, acid–base properties, redox properties, photochemical properties, ability to react with air and water, and so on. Let us therefore estimate that our interest in physicochemical processes of the toxic substances will require that we know at least 25 properties for each of the 100,000 compounds of interest. This means that in total we need to know 100,000 $(25,000 \times 10 + 25)$ or more than 25 billions of parameters (properties).

To measure this number of properties is impossible or at least comparable to the work of Sisyphus. In spite of the intensive research performed since the first green wave more than 25 years ago, it has not been possible to measure more than a small percentage of these ecotoxicological properties. The result is that we cannot assess the full environmental consequences for more than a handful of chemicals today. If we continue at the same rate during the next century to assess the environmental properties of chemicals by measurements, we will not even in the year 2100 be able to give all the environmental details for the 100,000 chemicals that we are using today. Since 1982, it has been required in the EU that all environmental properties are known for all new productions of chemical compounds. Thus, it would appear that the chemical industry has launched only very few new products since. This does, however, not change the fact that we only have a very limited knowledge of the 100,000 chemicals that were already in production in 1982. There are only two solutions to this dilemma: we have either to close down the chemical industries or *estimate* the properties of the chemical compounds by the use of various estimation methods. Society has chosen the latter solution.

Research on estimation methods has therefore been intensified during the last decade. Due to this research it is now possible to make reasonable estimations for a wide range of pertinent properties of the chemicals for which measurements are not yet available, There are, as for the models, two methods: a chemical method and an ecological method.

The first method builds on the principle that chemicals of similar structure and with related formulae have similar properties. These are the so-called QSAR and SAR methods. Details of these methods can be found, for instance, in Jørgensen (1989) and Lyman et al. (1994). These methods are under constant development, and the introduction of the concept of molecular connectivity has contributed during the last few years to the significant reduction in uncertainty of parameters estimated by these methods.

The ecological method is based on allometric principles, i.e., that the interaction between an organism and its environment is related to the surface area of the organism, which makes it possible to relate the size of the organism to such important parameters as uptake and excretion rates, biological concentration factor, and ecological magnification factor. It implies that when these rates are known for one organism, they may be estimated for other organisms provided that the sizes of the organisms are known (Jørgensen 1994).

Figures 1.3 and 1.4 give an overview of how a wide range of estimation methods are interconnected. Water solubility and K_{ow}, the partition coefficient

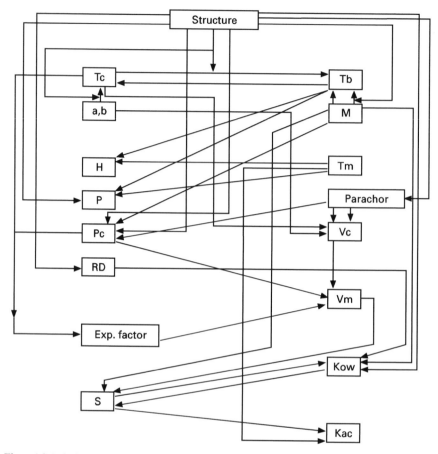

Figure 1.3 Relationships between physical parameters as they are applied in a network of estimation methods. Tc, critical temperature; Tb, boiling point; a and b, constants in Van de Waal's equation; M, molecular weight; H, Henry's constant; Tm, melting point; P, pressure; Pc, critical pressure; Vc, critical volume; RD, molecular refraction index; Vm, molar volume; Kow, octanol–water partition coefficient; s, solubility; Kac, ratio adsorbed on soil with 100% organic carbon in equilibrium with the compound in an aquatic solution.

for water and octanol, are core parameters because they are important for the quantitative description of many processes in the environment, and they are the basis for the estimation of many other parameters, particularly the biological parameters. Figure 1.3 indicates also the use of parameters, e.g., the molecular connectivity, critical data, and molecular refraction coefficient, which are of no use in an environmental context, but which facilitate the estimation of other parameters.

The methods have of course been used on known parameters to assess uncertainty. It is clear from such examinations that the physicochemical parameters can generally be estimated with an uncertainty of 10%–25%, the

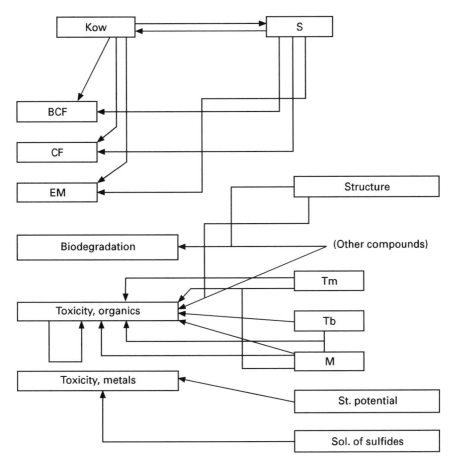

Figure 1.4 Relationships between structure, biological and exotoxicological parameters in a network of estimation methods. Kow, octanol–water partition coefficient; S, solubility; BCF, biological concentration factor; CF, concentration factor; Em, ecological magnification factor; Tm, melting point; Tb, boiling point; M, molecular weight.

biological parameters (bioaccumulation, BCF (Biological Concentration Factor), biodegradation in water and soil) with 20%–50% and the toxicological parameters with 40%–200%. In principle it is possible to estimate the physicochemical parameters from the chemical composition of the compounds of interest (including the important water solubility and K_{OW}) and then to estimate the biological and the toxicological parameters from the estimated physical parameters (Figures 1.3 and 1.4). It is, however, important to make the estimations on the basis of as broad a knowledge as possible. If some of the physicochemical parameters are known, it is important to use this knowledge to obtain as low an uncertainty as possible for the estimated parameters. If for instance the boiling point and Henry's constant are known, the accuracy of an

estimation of the other physicochemical parameters and the biological parameters are improved considerably.

1.6 ECOSYSTEM CONSIDERATIONS

Ecotoxicology focuses on the effects of toxic substances not only at the organism
and population level, but also at the ecosystem level and to an increasing degree.
There has generally been an increasing effort to understand ecosystems at the
system level during the last decade (see for instance Jørgensen 1992, 1997; Hall
1995). Through the research in this field during the last decade it has been
possible to reach an understanding of the hierarchical organization of ecosystems, the importance of the network that binds the ecosystem components
together, and the cycling of mass, energy, and information. Thermodynamics
has been widely applied for a description of the basic properties of ecosystems,
e.g., the buffer capacity associated with change in impacts and the amount of
structure and information stored in the ecosystem.

Biomarkers have been developed to improve the estimation of exposure,
including sublethal exposure, of populations of critical species in ecosystems
(Peakall 1992). They provide increased accuracy in the estimation of impacts
from chronic exposures to defined chemicals in an environment. Whatever their
usefulness, these methods can not be used to assess the impacts of toxic
substances and even chemical mixtures at ecosystem level. It must not in this
context be forgotten that the properties of an ecosystem cannot be equated to
the sum of the properties of its individual components—a widespread and
misleading concept in ecotoxicology. First, many detrimental effects, e.g., impairment of reproductive performance and reduction of growth potential, may
occur at concentrations well below those causing lethality. Second, even if
perfectly understood, the toxicity of a chemical for a specific population is of
little value in characterizing the toxicity that may be manifested in many
ecosystems. Therefore, the current approach must be replaced with examinations of the toxicity of chemicals throughout several ecosystems, which will
require a strong emphasis on basic ecological research. Natural communities
consist of complex assemblies of thousands of species that are in dynamic
equilibrium and that interact with each other and the abiological environment.

Risk assessments of widely used chemicals are often based on more or less
complex models. It is necessary to expand these risk assessments to encompass
the risk of a) reductions in population size and density, b) reduction in diversity
and species richness, c) effects on frequency distribution of species, and d) effects
on the ecological structure of the ecosystem, particularly on a long-term basis.

This expansion of the risk assessment concept to a much wider ecosystem
level has not yet provoked much research. New approaches, new concepts, and
creative ideas are probably needed before a breakthrough in this direction will
occur. The concepts of ecosystem health and ecosystem integrity are probably

the best tools developed up to now. The questions to be answered are: can we define ecological indicators that are able to provide a reasonable assessment of ecosystem health and thereby describe the damage caused by a toxic chemical or a mixture of toxic chemicals at the ecosystem level? Will the vulnerability of ecosystems to other natural impacts also be changed?

1.7 REFERENCES

Hall, C. A. S., Ed., 1995, *Maximum Power: The Ideas and Applications of H. T. Odum*, University Press of Colorado, USA

Jørgensen, S. E., 1983, "Modelling the Distribution and Effect of Toxic Substances in Aquatic Ecosystems," in *Application of Ecological Modelling in Environmental Management, Part A* (S. E. Jørgensen, Ed.), Elsevier, Amsterdam, The Netherlands

Jørgensen, S. E., 1990, *Modelling in Ecotoxicology*, Elsevier, Amsterdam, The Netherlands

Jørgensen, S. E., 1992, *An Introduction of Ecosystem Theories: A Pattern*, Kluwer Academic Publishing, Dordrecht, The Netherlands

Jørgensen, S. E., 1994, *Fundamentals of Ecological Modelling, Developments in Environmental Modelling, 19*, 2nd Edition, Elsevier, Amsterdam, The Netherlands

Jørgensen, S. E., 1997, *An Introduction of Ecosystem Theories: A Pattern*, 2nd Edition, Kluwer Academic Publishing, Dordrecht, The Netherlands

Jørgensen, S. E., B. Halling-Sørensen, and S. N. Nielsen, Eds., 1995, *Handbook of Environmental and Ecological Modelling*, CRC Lewis Publishers, Boca Raton, USA

Lyman, W. J., and D. H. Rosenblat, 1990, *Handbook of Chemical Property Estimation Methods: Environmental Behavior of Organic Compounds*, American Chemical Society, Washington, DC, USA

Peakall, D. B., and J. R. Bart, 1983, "Impacts of Aerial Applications of Insecticides on Forest Birds," *CRC Crit. Rev. in Environ. Control, 13*, 117–165

Wolfram, S., 1984, "Cellular Automata as Models of Complexity," *Nature, 311*, 419–424

Wolfram, S., 1984, "Computer Software in Science and Mathematics," *Sci. Am., 251*, 140–151

2

Ecosystem Principles for Ecotoxicological Analyses

Helmut Lieth (Osnabrück, Germany)

2.1 SUMMARY

This paper compares several definitions of the term ecosystem in their value for ecotoxicological work. The functional aspect of the ecosystem is offered as the most useful definition:

Ecotoxicology, Edited by Gerrit Schüürmann and Bernd Markert.
ISBN 0-471-17644-3   © 1998 John Wiley & Sons, Inc. and Spektrum Akademischer Verlag.

The ecosystem is a system in which energy, material, and information flow with the participation of biota.

Within this definition the ecosystem is discussed in terms of the classical distinction of composition, structure, and function. Structure is outlined in the categories physiognomy, chemistry—elemental as well as compound—genetics, physics, and information. As a bridge between the mainly physicochemically trained ecotoxicologists and ecologists, the physiognomic structures of the main major ecosystem types of the world are shown as informational responses to climatic signals.

The function of ecosystems is discussed with respect to the postulate that the ecosystem

$$\ni = \sum |i|$$

where $|i|$ is understood as sets of noticeable information bits. The noticeable information sets $|i|$ equal specific energy units E_S attached to functional material units M_F:

$$|i| = |E_S + M_F|.$$

The two equations aim at a means to connect information sets of all kinds logically with energy qualities and quantities attached to material quantities. Such a connection may allow us to combine logically socionomic and natural science models and is offered here for further discussion.

The baseline for the biologically available energy, the net primary productivity, is shown as a global pattern using ARC/INFO. Two patterns are compared: one as a function of climate alone and one as a function of climate and soil fertility. These models are referred to as the Rio Model 1995 since they were first shown at the international conference of biogeochemistry in Rio de Janeiro.

2.2 INTRODUCTION

The topic of ecosystems as a scientific discipline is multifaceted. One can divide the treatment of this topic into academic fields such as taxonomy, vegetation pattern, limnology, oceanography, hydrology, and pedology among others. Each field puts a special aspect of the ecosystem in focus. It was, therefore, the primary aim of the first Intecol (International Association of Ecologists, a subdivision of the International Union of Biological Sciences, IUBS) conference in The Hague 1974 to stress unifying concepts of ecosystems (van Dobben and Lowe-McConnell 1975). None of of those treatments of ecosystems covered ecotoxicology as a special field. The speakers at that conference presented the different biological viewpoints of ecosystems and tried to show evolutionary and successional developments, their possible driving force, and the possible final stationary state of a particular ecosystem if such a state can be defined. The history of the earth up to the present time has shown that all parts of the ecosystem, abiotic as well as biotic, have undergone changes abruptly and continuously. We can assume, therefore, that the same will be the case in the future.

The discussion in 1974 was carried out on the basis of earlier hypotheses by, e.g., von Bertalanffy (1951) and Margalef (1968), biologists for whom the biological sector of the ecosystem is the crucial part. The final goal of succession and evolution may be the most elaborate and diverse structure of the biological part of the ecosystem, the biocoenosis. It is postulated that the maximum level of energy flow through the biocoenosis is achieved at a state of homeostasis. At this state the so-called ecological equilibrium between all biota in that biocoenosis is assumed to be realized and this equilibrium needs to be protected.

This concept has been highly stimulating for the analysis of numerous food chains or food webs. The postulate, however, that the highest species diversity would guarantee the stability of the ecosystem has not yet been verified. Lots of details were reported about individual species occupying a niche in the ecosystem, eating and being eaten, and consuming energy and material at each trophic level (see ecological textbooks, e.g., Odum 1971). An overriding principle for the specific structure and function of an ecosystem has not conclusively been offered.

Individual species and ecosystem compositions have undergone drastic changes throughout the earth's history, some due to clearly documented environmental changes and others with seemingly no cause. In spite of that, the assumption is made that the biotic entities at each place are generally in balance and that any change, loss, or alteration of species is catastrophic for the total ecosystem.

This assumption is questionable. The ecosystem is a stationary expression of the response of biota to signals received by environmental forces. We have no proof that the ecosystems we see at present are the ultimate stage of evolution. Species apparently come and go and new ones appear in their place if environmental conditions permit them. What seems to be more stable in ecosystems in general is the chemical composition of plants, animals, and microbes, the erosion and sedimentation patterns of the substrate, the availability of air and water, and the diurnal and seasonal utilization pattern of solar energy input.

A similarly questionable assumption would be that the maximum energy flow through biota is the ultimate goal of ecosystem development. The energy flow through the biota, in all cases analyzed, is only a small portion of the total energy flow at the surface of the earth and is maintained largely through stationary functions of the biological components of the ecosystem in the context of the total energy flow, i.e., evapotranspiration, sensible heat, and biomass turnover, at that point on earth. As far as energy flow patterns are concerned the ecosystem is for all practical purposes the "slave" of environmental forces information in the ecosystem: the abiotic parts of the ecosystem.

Only within the pool of "chemical energy" provided by the plants, i.e., the net primary production (NPP), can the entire ecosystem develop a system of energy diffusion. Because of the fundamental importance of NPP the global pattern is shown in Figures 2.1 and 2.2. The net primary productivity channels only a small fraction of the total solar energy through the ecosystem. A much larger energy percentage is channeled through the evaporation and transpiration of

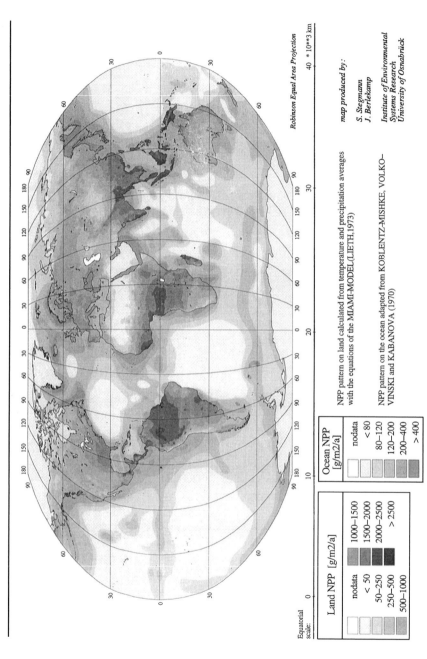

Figure 2.1 Global pattern of net primary productivity — Rio Model 1995. Generated with GIS/ARC INFO.

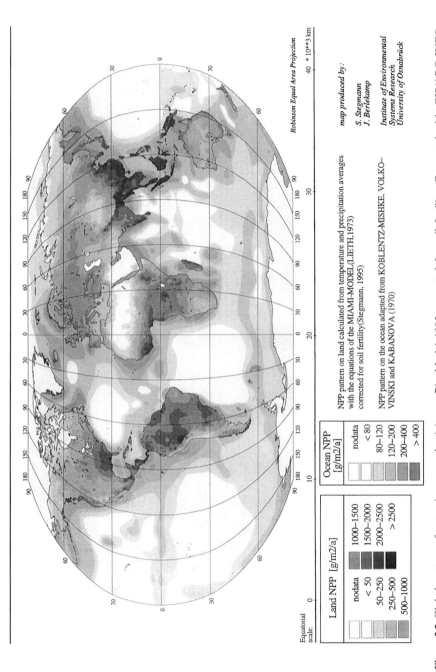

Figure 2.2 Global pattern of net primary productivity — Rio Model 1995, corrected for soil fertility. Generated with GIS/ARC INFO. Figures 2.1 and 2.2 show how the incorporation of additional environmental information results in more exact representations of patterns.

water. If one were to accept the concept of the self design of an ecosystem we would assume that plants have evolved mainly to evaporate water in order to accelerate the solar energy transfer back from the earth to outer space.

No matter what the ultimate driving force of ecosystem evolution may be, the fact remains that environmental constraints shape and limit the ecosystem's functions. Given these abiotic signals which we call constraints the biota develop signals and receptors to control the impact of abiotic forces to some extent.

Environmental forces and sources may be regarded as sets of information to which all ecosystem parts may react. The evolution of biota is, cybernetically speaking, obviously guided by the desire to reach some maximization of homeothermy within the bodies. It is acceptable, therefore, to propose that the entire ecosystem also tries to achieve the best possible homeothermic state as well. For this purpose it is necessary to manipulate the abiotic energy flow in such a way that its impact upon temperature fluctuation for the biota is minimized (see Lieth 1976, 1977). This may be accepted as one of the key aspects for an ecotoxicologist if he wants to assess the influence of adding or subtracting any chemical component from the ecosystem. The main aim of this paper is, therefore, to describe the ecosystem as a basic cybernetic unit, its performance and development being a consequence of the environmental forces controlling its structure and function. For this purpose I shall try to define in the following section the ecosystem including the biological taxa as a manifestation of information systems created by the energy flow on earth.

With regard to natural ecosystems **solar energy** is usually the main driving force; **materials** comprise biomass, soil mass, water, and air, and **information** includes the physical, chemical, and biological components. These components will be detailed further in the following paragraphs.

The stationary state of an ecosystem is determined by the external forces and their fluctuations to which it must adapt. The primary force is the flow of radiant energy onto the earth's surface, through the ecosystem, and from there to outer space. The planetary conditions of the earth on its track around the sun are mainly responsible for quantity, quality, daily fluctuation, and seasonality of the energy influx. Due to the present temperature level on the earth's surface which ranges from about 200 K to 320 K, water exists as solid, liquid, and gas, which makes it convenient for the radiant energy to be either reflected or converted into thermal energy. Ice, water, and water vapor provide the channels through which the main portion of the incoming energy is conducted back to outer space. It leaves the earth's troposphere as radiant energy at a much longer wavelength than the incoming radiation but essentially containing the same energy amount as the incoming energy so that the earth's surface has an average temperature of about 281 K. Individual regions may differ greatly from this average, for example, from an average annual temperature of $-30°C$ in Jacutia/Siberia or Antarctica to $+30°C$ in the Egyptian desert (see Walter and Lieth 1960). The maximum and minimum temperatures in ecosystems may vary much more throughout the day and the season. Within a temperature range of about $100°C$, which exists in many ecosystems throughout the year, all physical and chemical processes supporting the functioning of the ecosystem must occur. The more diverse the physical structure of an ecosystem, the lower the temperature fluctuation may be. Therefore, it appears highly probable that all informational controls in an ecosystem are continuously evolving in order to design the optimal homeothermic stationary ecosystem everywhere on the surface of the earth in response to the ever changing conditions caused by the continuous evolution of the planet earth itself.

2.3 DEFINITION OF AN ECOSYSTEM

Everybody seems to understand what an ecosystem is. If we look closer, however, we find out that this term has several definitions which are quite different in scope with respect to ecotoxicology. A common ecological definition of an ecosystem is:

> A set of species living together in one place in a definable environment.
> D1

This definition requires the existence of fixed boundaries for all species involved. These boundaries, however, are quite different for plants, animals, and microbes. While sets of higher plant species may be countable on several hundred square meters, sets of animal species may cover territories of several square kilometers. Sets of microbes decaying organic litter may be found in areas of less than one square meter. Environmental factors are not uniform over area size differences of such magnitude.

This definition is only acceptable for plant communities and is used here in the context of the habitat concept. For the ecotoxicologist this approach is meaningful if we are concerned with poisonous inputs into ecosystems which may cause changes in the function of individual species, their sudden explosion in development, or their depletion.

An ecosystem may also be defined in terms of functional processes. This type of definition is often used by ecologists and may be based upon the flux of energy, the flux of chemicals, or the interdependence of species. The definition of the ecosystem as an energy flow system based on the thermodynamical principles could read:

> A system through which energy flows in various forms with the participation of biota to a variable degree.
> D2

This definition does not distinguish between biotic and abiotic energy flows. For a thermodynamically consistent energy flow concept this definition is valuable.

Ecologists, however, are mainly interested in the energy flow through biota. Numerous attempts to quantify these flows and to extrapolate basic laws failed because the major part of the energy flowing through an ecosystem is not measurable as biomass.

The approach has some importance for ecotoxicological work. Toxic inputs into the system may change the quantity or quality pattern of the abiotic energy flux (e.g., chlorofluorocarbons, CFC, cause ozone depletion resulting in higher UV radiation at the surface of the ecosystem).

The ecologist normally bases the ecosystem's energy flow on the flux of chemical energy through the food chain:

> The ecosystem is the flux of solar energy and inorganic materials through plant biomass, animal biomass, microbial biomass, and heat and inorganic material.   D3

This is the generally accepted biological trophic level concept (see Odum 1971). It treats the ecosystem as a food chain as shown in Figure 2.3A.

The three definitions D1–D3 given above contain among other principles the assumption of the cycling of elements through the ecosystem. While this is correct in the global context, it is only rudimentarily developed in particular, regionally definable systems. The ecotoxicologist must use this definition cautiously. It is important for the judgment of impacts upon the trophic levels if parts of the system are affected by toxic inputs or by changes in available elemental compositions or chemical compounds.

All definitions given so far lack one important property: the information residing in any ecosystem. The information content in all parts of the system elevates the ecosystem to the level of intelligent systems. Many toxicological implications involve the flow of information as the cause of significant changes in material and energy fluxes in the system. Plants may produce chemicals to protect themselves against animal grazing. Animals may produce toxic chemicals as weapons, humans may produce toxic chemicals to kill each other. Each process is controlled by "bits of information" which flow from one point in the ecosystem to another point. For this reason we must find a way to include the information flow into our definitions as a major characteristic of an ecosystem. This is especially important when we construct computer models to simulate ecosystem performances including human interference (Arbeitsgruppe Systemforschung 1991).

Under such premises we can newly define the term ecosystem for this chapter as follows:

> The ecosystem is a stationary entity of abiotic and biotic compartments, in which energy, material, and information reside and flow in various forms.   D4

The adequate coverage of an ecosystem for an environmental engineer might then go as far as to define the ecosystem as an abstract entity positioned in the energy flow between the sun, the earth's surface, and outer space, which becomes partially attached to chemical material to produce information. If we transfer this sentence into a mathematical conceptual equation we can write

$$i = E + M \qquad (2.1)$$

where **i** stands for information, **E** for energy, and **M** for material.

E comprises radiation as well as chemical energy and may also include other forms of physical energy. In the case of living entities in the ecosystem it includes, besides chemical energy and thermal energy, physical energy needed to conduct water. Since we shall use this definition later on we must describe it in more detail. **Radiant energy** flows in all ecosystems over a wide range of frequencies. The same caloric energy value of different frequency bands may have completely different effects within the ecosystem. The best studied example is the conversion of solar radiation, a mixture between visible and thermal frequency ranges which is converted in plants either into plant biomass or into sensible heat which can be used for the vaporization of liquid water.

M is material in solid, liquid, and gaseous form. The manifestation of **M** in the ecosystem is normally identified as biomass, air, water, soil, or rock, all of which consist of the same set of known chemical elements, albeit in various mixtures, quantities, densities, proportions, and speciation. All of these elements differ widely in acceptance by the biota which is one of the most important properties of environmental material. *The Biological System of the Elements*, recently published by Markert (1994; see his contribution in this volume), shows the affinity of certain chemical elements to those constituting the majority of plant and animal matter: C, O, H, N, S, P.

The chemical elements exist in various species in the biomass and are normally present as part of chemical compounds. The living part of the ecosystem largely consists of compound classes with specific physical and physiological properties, such as carbohydrates, proteins, nucleic acids and so on. The number of known organic chemicals naturally present in ecosystems is steadily increasing. The ecotoxicologist must consider that, additionally, many nonnatural chemical compounds interact with natural compounds and cause unwanted reactions.

The tangible definition of **i** is restricted insofar as we have no way as yet of attaching a measurable weight or energy unit to it. The term information is presently used in many ways with apparently completely different meanings. The bit of information in computer science is the basic yes or no in a computer chip, which is always manufactured in the same configuration. The basic bits in ecosystems, however, are attached to different carriers even down to the level of electron transfers. For an understanding of ecosystems it is important to realize that the majority of biological information is attached to special molecules. The "noticeable information bit" in ecosystems usually contains a bundle of

Environment	Sun	Air	Water	Soil	→ Sensible heat
Flow direction of energy and material	→	⇄	⇄	⇄	⇄
Trophic level class	Primary producers	Consumers	Predators	Scavengers	Decomposers
Main types of biotic taxa	Autotrophic plants and bacteria	Herbivorous animals and plant pathogens	Carnivorous animals and animal pathogens	Carcass-consuming animals	Consumers of dead biomass, bacteria, fungi, small invertebrates
Functional definition: Uptake	Uptake of solar energy for conversion of mineral substance into plant biomass	Uptake of plant material, oxygen, and minerals for conversion into animal biomass	Uptake of live animal biomass, H_2O, and oxygen	Uptake of dead animal biomass, H_2O, and oxygen	Uptake of dead plant and animal biomass, H_2O, and oxygen
Release	H_2O, trace gases, sensible heat, dead biomass	CO_2, H_2O, trace gases, sensible heat, faeces	CO_2, H_2O, sensible heat, faeces		CO_2, H_2O, humus (soil), sensible heat
Contribution to structure	Basic physiognomic structure (Figure 2.5)	Shape of vegetation structure, pattern of taxa	None directly but control of herbivore influence	None	Shape soil structure, influencing abiotic compartments and biotic taxa pattern

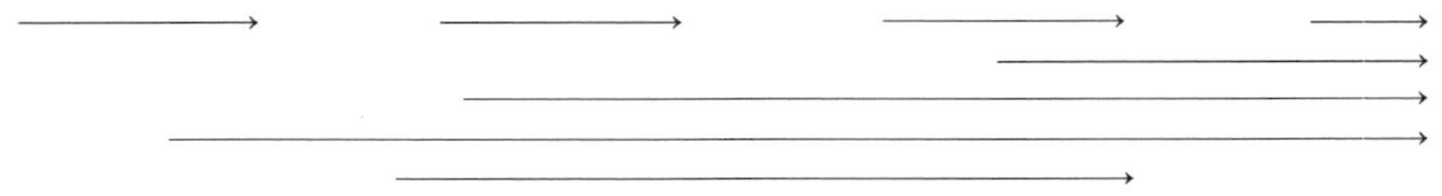

Figure 2.3A The trophic level structure. Caption continued under Figure 2.3B.

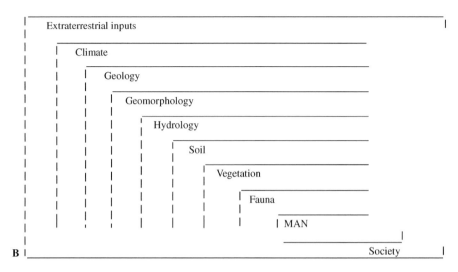

Figure 2.3A,B Descriptions of ecosystem structures. **A** The trophic level structure: food web component classes as commonly used in ecological textbooks (e.g., Odum 1971). The main concern in this concept is the transfer of material and energy through the living compartments of the ecosystems shown in this figure as five trophic level classes. The flow directions of energy and material from the source through the ecosystems are shown by arrows. Each class level lives off one or more of the lower level classes. Each trophic level has a direct connection to the decomposers. The abiotic energy flow as shown in the top line is not considered as such in this concept. The flow of information is not considered at all. The flow of chemicals is not detailed enough to provide ecotoxicological information. **B** Hierarchical rank order model for landscape size ecosystems adapted from Everts et al. (1988). In this concept abiotic and biotic compartments influence each other, as appears to be the case in reality. The hierarchy exists from top to bottom although society can develop counterforces to this hierarchy. This structure is useful for the construction of simulation models which are important for ecotoxicological impact research. The model is based on definition D3. The information concept is not included. The relation between energy content per information set and the number of noticeable information sets per rank is in Figure 2.4. Figures 2.3 and 2.4 together comprise the definition D4.

information bits. Until a single bit of information can be better defined we need to accept the fact, that "a set of information bits" is what we have to deal with in ecosystems analysis. The complete set of chemical, genetic, and structural information makes up the ecosystem so that we may write:

$$\ni = \sum |\mathbf{i}| \tag{2.2}$$

where $\ni$ stands for ecosystem[1] and $|\mathbf{i}|$ for sets of noticeable information attached to matter.

The reason for using this equation in the context of environmental management is that it allows us to generalize the relation within the ecosystem between the living entities as realized sets of information and the abiotic state variables

---

[1] $\ni$ is the Cyrillic equivalent for **E** which is occupied for energy.

and processes. It may also allow us to logically include even social aspects into the ecosystem analysis in the future (Figure 2.3B).

If we accept the description of a "noticeable information bit" in Equation 2.2, we can convert Equation 2.1 to

$$|\mathbf{i}| = |\mathbf{E}_S + \mathbf{M}_F| \tag{2.3}$$

and Equation 2.2 to

$$\ni = \sum |\mathbf{E}_S + \mathbf{M}_F| \tag{2.4}$$

where $\mathbf{E}_S$ = specific energy unit and $\mathbf{M}_F$ = material functional unit.

This may enable us to calculate, after proper specification of $|\mathbf{i}|$, the power of information bits for the energy flow within and through ecosystems. Certain information can cause an enormous flux of energy and material with seemingly no physical energy input. To understand at least the processing of $\mathbf{i}$, we can deduce from many studies of information transfer within ecosystems that $\mathbf{i}$ is attached to a material unit. It is also evident that the relation between $|\mathbf{i}|$, $\mathbf{E}_S$, and $\mathbf{M}_F$ is nonlinear, as suggested in Figure 2.4.

If we accept the fact that differences in $\mathbf{E}$ and $\mathbf{M}$ cause the manifestations of different $|\mathbf{i}|$, then it is equally valid to read Equation 2.3 as: $|\mathbf{i}|$ may equal different material and energy flows where the same total energy flow may be achieved by changing sets of $\mathbf{M}$ to conduct the energy and changing qualities of $\mathbf{E}$ so as to become attached to certain kinds of $\mathbf{M}$ to be conducted (Equation 2.4).

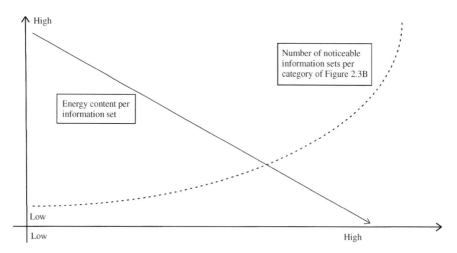

Figure 2.4 The energy content attached to information sets relevant to the ecosystem in the categories of the model shown in Figure 2.3B.

This aspect requires intensive studies if we want to understand the energetic equivalent of information. At present, we can only guess that from the myriads of possible information bits a small portion is possibly noticeable in and by ecosystems where some information bits are used as controls for other bits performing the mass and energy flows. The first estimation of the number of bits making up an ecosystem was given by Patten (1959) as 10^{28}. These bits, however, were defined differently than they are here.

In order to deal logically with an ecosystem we usually start with superficial descriptions of physiognomy, species composition, and environmental forces and sources, and continue with analyses of causal relationships and processes. The final goal is to determine the reason for the evolutionary development of ecosystems including the human species to their present state and their final destination.

2.4 THE STRUCTURE OF ECOSYSTEMS

In order to describe an ecosystem structure and to satisfy our conceptual equation we list and describe some properties of i, E_S, and M_F. This can be done on different levels depending on the size of the entity which we call our ecosystem: global, continental, regional, or landscape unit type of biogeocoenosis. For each of these size classes we consider structural properties in a different way.

The structure of any ecosystem is its stationary aspect. It contains biotic and abiotic elements. The inclusion of the abiotic elements into the total ecosystems concept is becoming more common at present since ecologists are cooperating since 1995 in global change projects (see GCTE 1996). The abiotic elements are soil, water, and air in their individual appearance in various parts of the earth. The biotic elements are divided either into plants, animals, and microbes and parts thereof or into primary producers, consumers, herbivores, predators, scavengers, and decomposers (Figure 2.3A). Each of these groups consists of many species, and each species occupies a certain space and place in the ecosystem. Its particular place is called its niche, which, however, refers mainly to its functional place rather than its structural position.

The biotic structure of the ecosystem is closely controlled by the abiotic components of the system. These structures are called physiognomic aspects, e.g., desert, grassland, woodland, and forest, and are shown in Figure 2.5 and in Table 2.5. Good modern descriptions and presentations of the global pattern of physiognomically defined vegetation units are available in several textbooks of botany, ecology, and zoology (e.g., Archibold 1995; Illies 1971; Kloft 1978; Larcher 1980; Odum 1971; Schultz 1995; Walter 1984; Walter and Breckle 1991; see also Internet citation Olson et al. 1985). Each of these physiognomic classes can be divided in many other units mainly on the basis of their species composition. The physical structure of the ecosystem mainly responds to climatic forces and sources: temperature, wind speed, wave action, precipitation, rain, snow, ice, and drought. In contrast, the species composition generally responds more to edaphic forces and sources: mineral supply, pH, soil, and water and oxygen in the substrate. Since biotic taxa are the product of genetics it is clear that evolution and genetic drift represent important information contained in the species composition.

2.4.1 Physiognomic Structure

In order to demonstrate the strong impact of abiotic signals on biotic responses we show in Figure 2.5 the physiognomy of contrasting ecosystems of the world in a series of 20 photos. For each of the 20 pictures the position on earth where the photo was taken is given and major environmental constraints to which the biota in each system responds are mentioned. In all cases where the vegetation covers the ground completely we find competition among the species. In systems where abiotic constraints limit plant growth we find many adaptations to the specific stress conditions.

Vegetation and fauna are described in numerous other ways in the literature in most languages of the world. The ecotoxicologist will find in most cases the information he needs for local studies. For a global overview maps are available in many geographic textbooks (see e.g., Archibold 1995; Schmithuesen 1976; Schultz 1995; Walter 1984; see also the vegetation type map on Internet by Olsen et al. 1985).

The structure of an ecosystem also contains information about the physical and chemical dimensions of the system in addition to compositional dimensions, organic as well as inorganic. As far as modeling of an ecosystem is concerned all structural features of an ecosystem are regarded as state variables which can perform functions and can receive or send information in and out of the system and between the various state variables of it.

→

Figure 2.5A–U Physiognomic structure of 20 major ecosystem types of the world. Such ecosystem types are called zonal vegetation or vegetation formations in phytogeographical textbooks, or biomes in ecological textbooks, where the coexistence of plants and animals are emphasized. The total ecosystem concept needs to include environmental aspects in the classification as was done by Walter (1984) and Archibold (1995). The photos were selected to show the response of the biotic ecosystem elements to abiotic signals. The majority of plants and animals shown in these pictures are restricted by the specific environmental conditions of the place where they occur. The conditions are referred to as ecological channels (Box 1981). Two important such channels are temperature and precipitation. Their year-round changes can be seen at a glance in so-called climatic diagrams. The climatic diagram corresponding to the ecosystem portrayed is located beside each photo. But before embarking on our journey through selected ecosystems of the world, we should first take a closer look at the information (**A**) provided in the climatic diagrams.

Figure 2.5A Conventions of the climate diagrams used in Figure 2.5B–U (adapted from Ostendorf and Lieth (1982) by S. Riediger; data taken from the GHCN Vose et al. (1995), see Internet Citations). **1** Country name taken from the WMO file. **2** Station name taken from the WMO file. **3** Geographical coordinates, latitude N or S in degrees and minutes, longitude E or W in degrees and minutes, altitude in meters. (Note: If N* is indexed by an asterisk it means that the year in the diagram starts with the month of July.) **4** Length of registration period, first number for temperature, second number for precipitation. **5** Mean annual temperature. **6** Mean annual total sum of precipitation. **7** Baseline with 12 divisions for the months; if N as latitude, starting with January; if S as latitude, starting with July. Climate data are entered at the middle of each month. **8** Ordinate

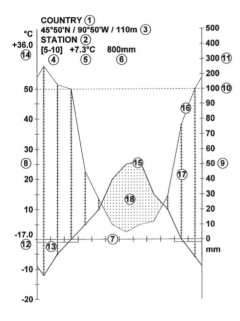

for temperature with 5 degree tick marks. **9** Ordinate for precipitation with 10 mm tick marks; the scales $10° = 20$ mm is the same for all diagrams. **10** 100 mm precipitation line at which **11** shows the marks compressed for 100 mm. **12** Mean temperature of the coldest month in registration period. **13** Months with temperature minima below 0 are blocked. **14** Mean temperature of the warmest month in registration period. **15** Thick line connecting the monthly mean temperature values positioned in the middle for each month. (Note: The astronomic summer is always in the middle of the diagram). **16** Thin line connecting the monthly values for the sum total of precipitation. **17** Vertical lines divided into $1°C = 2$ mm precipitation sections by tick marks when precipitation line exceeds the temperature line indicating humid seasons for the year. **18** Area dotted in distance of $1°C = 2$ mm precipitation if temperature line exceeds precipitation line indicating arid seasons of the year. (Note: A missing number indicates data not available.)

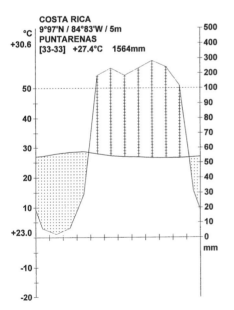

Figure 2.5B The evergreen tropical rainforest grows into 50- to 70-m tall stands of trees of many species from many genera. On fertile soil like here in the Osa peninsula in Costa Rica year-round temperatures of about 27°C and rainfall of up to 3000 mm annually provide conditions for fast growth.

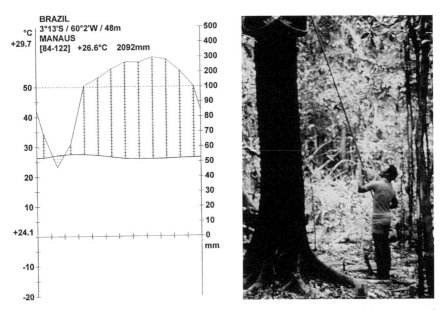

Figure 2.5C In the evergreen tropical rainforest the tree canopy contains most of the green protein-rich herbaceous biomass. A large portion of the animal species lives, therefore, high up in the trees. Insect species crawling up and down the tree trunks are caught with traps mounted at several meters above ground. At this lowland location near the Amazonas river close to Manaus the trees are the only escape when the water level of the river covers the area and only trees reach out of the water. This forest, called varzea, grows up to 40 m, not as tall as the forest outside of the river bottom.

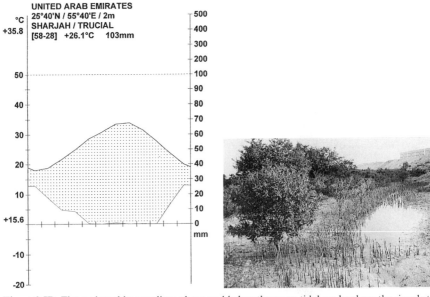

Figure 2.5D Flat and muddy coastlines above and below the mean tidal sea level are the signals to which in the tropical zone mangrove ecosystems respond. The picture was taken at the shore of Abu el Abbyat in Abu Dhabi Emirate. The mangrove ecosystems are among the most productive biological systems on earth, which demonstrates clearly that salinity itself is not a general stress factor. Most of the plants in this system require sodium chloride for proper development. In the temperate zone such locations are occupied by salt marshes, void of trees.

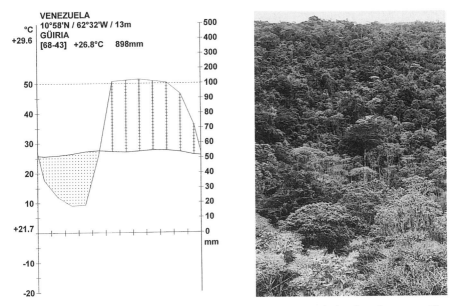

```
VENEZUELA
°C   10°58'N / 62°32'W / 13m          500
+29.6 GÜIRIA                          400
      [68-43]  +26.8°C    898mm       300
                                      200
50                                    100
                                       90
40                                     80
                                       70
30                                     60
                                       50
20                                     40
                                       30
10                                     20
                                       10
+21.7                                   0
                                       mm
-10

-20
```

Figure 2.5E The raingreen tropical forest has a closed green canopy during the rainy season. The trees may be up to 40 m tall at this location near Cumana in Venezuela. With higher rainfall only the top of the largest trees may be deciduous.

Figure 2.5F At nearly the same slope near Cumana as shown in Figure 2.5E during the rainy season, during the dry season the second growth trees look like the summergreen forests in temperate zones of the world in winter, as shown in Figure 2.5G. The canopy at this plot is 15–25 m high.

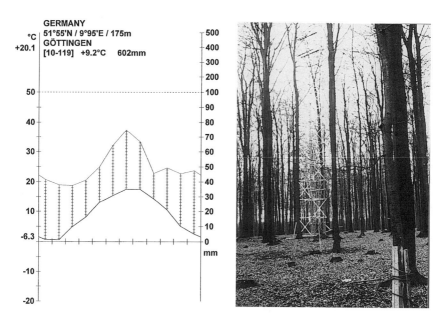

GERMANY
°C | 51°55'N / 9°95'E / 175m
+20.1 | GÖTTINGEN
[10-119] +9.2°C 602mm

Figure 2.5G Summergreen broad-leafed forest covers much of the temperate zones mostly on the continents of the northern hemisphere. The picture was taken at the site of the German International Biological Program near Göttingen where much forest ecology research was done in recent years. The trees grow here up to 40 m tall.

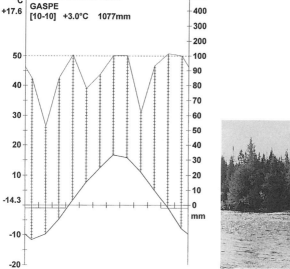

CANADA
°C | 48°77'N / 64°48'W / 33m
+17.6 | GASPE
[10-10] +3.0°C 1077mm

Figure 2.5H The boreal coniferous forest is a relict of the gymnospermeous age several million years ago. The picture was taken at the Gaspé peninsula in Canada. Much of Canada, northern Europe, and Russia is covered by this vegetation type. At several places deciduous broadleaf trees have invaded this territory indicating that angiosperms continue to replace gymnosperms. This forest is usually not more than 30 m tall.

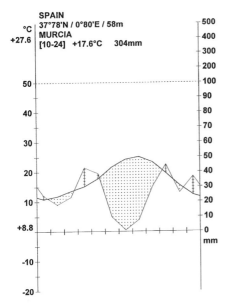

SPAIN
37°78'N / 0°80'E / 58m
MURCIA
[10-24] +17.6°C   304mm

°C
+27.6

+8.8

Figure 2.5I Winterrain, summerdry climate is the abiotic signal to which a forest responds with sclerophyllous leafage. The photo shows a stand of cork oak with a chamaecyparis palm in the foreground in the mountains near Cartagena in Spain. Forests with similar structures are found around the Mediterranean Sea, in California, Chile, South Africa, and southern Australia. Depending on the amount of water available these trees can grow up to 40 m (110 m in some cases).

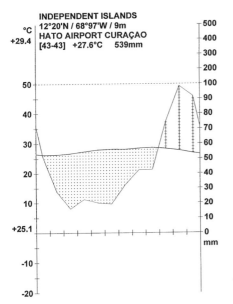

INDEPENDENT ISLANDS
12°20'N / 68°97'W / 9m
HATO AIRPORT CURAÇAO
[43-43] +27.6°C   539mm

°C
+29.4

+25.1

Figure 2.5J In tropical regions with low rainfall various types of dry forests appear. The tree stand is open and less than 10 m tall. Succulence in various forms is one response to the need for water conservation.

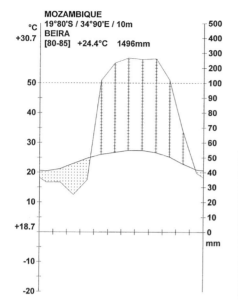

MOZAMBIQUE
19°80'S / 34°90'E / 10m
BEIRA
[80-85]   +24.4°C   1496mm
°C +30.7
+18.7

Figure 2.5K In many of the seasonal rainfall regions in the tropics tall herbaceous stands compete with raingreen forests. The picture was taken near the Zambesi river in Mozambique, where large animal herds such as water buffaloes as shown in the photo or other big animal herds graze. This keeps the tree growth down throughout the year while the grasses start to grow fast after each rain and consume most of the soil water before it can penetrate deeper into the soil, where the tree roots are located.

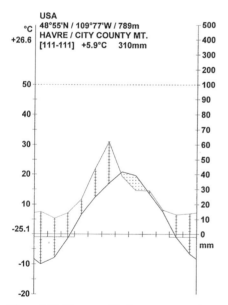

USA
48°55'N / 109°77'W / 789m
HAVRE / CITY COUNTY MT.
[111-111]   +5.9°C   310mm
°C +26.6
-25.1

Figure 2.5L Grassy and herbaceous vegetation occur in the dryer regions of the continents throughout the temperate zones. This picture was taken near the southern border of Montana, USA. It shows grassland invaded by sagebrush in response to heavy grazing by animals. This was formerly the territory of large buffalo herds which could develop in response to the large amount of fodder close to the ground. The region is now used as a cattle range, a system simulating the once naturally occurring buffalo herds.

ICELAND
°C 65°27'N / 13°58'W / 9m
+10.6 HALLORMSTADUR / DALATANGI
[10-29] +3.7°C 656mm

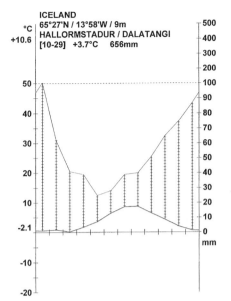

Figure 2.5M The ecosystem type called tundra covers large areas of the northern hemisphere, north of the boreal coniferous forest. Long days during the summer and the action of frost in soil under snow cover during the winter are environmental signals to which the system responds. The picture was taken in the north-eastern part of Iceland. It shows the frost hammocks from the northern direction. The southern slopes of the mounds are covered with small shrubs and low-growing herbaceous plants. The physiognomy of this ecosystem is a result of abiotic and biotic forces. Wet soil expands when freezing and is kept in place by plant roots after thawing.

USA
°C 44°55'N / 110°40'W / 2368m
+16.1 LAKE YELLOWSTONE
[84-86] -0.5°C 501mm

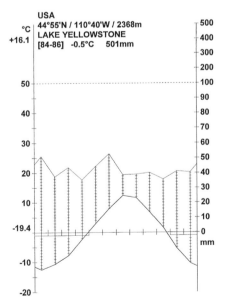

Figure 2.5N Similar tundra-like ecosystems occur at high elevations above the timberline everywhere on earth. In temperate-zone mountains the systems look similar to the northern tundra. The picture was taken on Mount Rainier near Seattle, WA. It shows the alpine grasslands surrounding stunted trees, obviously the timberline. This ecosystem often responds to the mechanical signal of strong winds mixed with icy snow or water probably all year round at this elevation on the top of mountains.

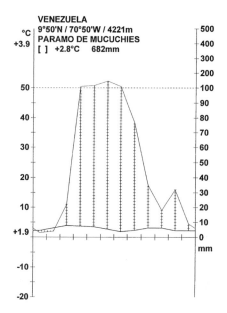

VENEZUELA
9°50'N / 70°50'W / 4221m
PARAMO DE MUCUCHIES
[ ] +2.8°C   682mm

Figure 2.5O Herbaceous ecosystems occur in tropical mountains, at high elevations above the timberline, mixed with tall forbs or short trees such as suffruticose plants. The picture was taken at the Paramo de Mucuchies in the Venezuelan Andes. The system responds to the environmental signal of daily temperature changes, from near freezing point at night to spring-like temperatures at noon, with thick insulating haircoatings as can be seen by the white cover of the *Espeletia* species in the foreground.

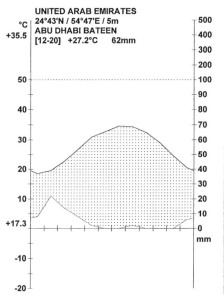

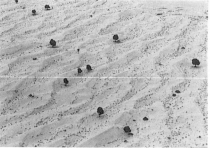

UNITED ARAB EMIRATES
24°43'N / 54°47'E / 5m
ABU DHABI BATEEN
[12-20] +27.2°C   62mm

Figure 2.5P Little water and high temperatures exceeding 50°C near the ground are the environmental signals to which the few plants in the hot desert respond. The picture was taken in the sand desert between Abu Dhabi and Dubai on the Arabian peninsula and shows sand dunes, the result of wind energy, and a few drought-tolerant plants growing between the dunes.

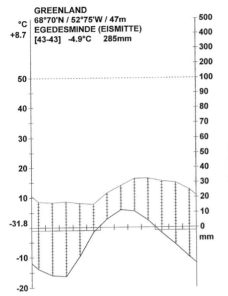

Figure 2.5Q Continuous ice cover and bold rocky faces, often covered with snow, are the environmental signals to which the ice desert must respond. The picture was taken during the month of June from an airplane flying over southern Greenland. Water in this system is mostly ice and therefore unavailable for higher plants.

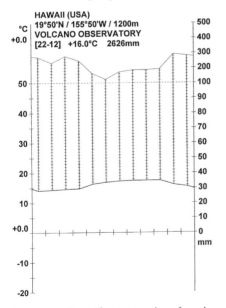

Figure 2.5R Soon after an eruption of a vulcano the hot lava cools down and forms a smooth or rugged surface which does not provide a substrate for living organisms. The picture was taken on Mauna Kea on the island of Hawaii. This environment may contain elements poisonous to most plants which may need to be removed by rainwater before plants and animals can grow extensively. Otherwise this substrate provides sufficient mineral nutrients.

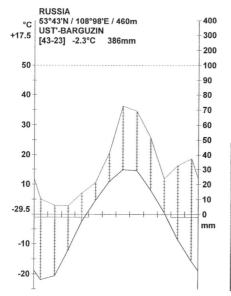

Figure 2.5S Rivers and lakes on land provide signals for special ecosystems. The strength of flowing water and the depth of the water are the signals to which the ecosystems respond. The picture shows the shoreline of Lake Baikal taken during early summer. The lake is at some parts about 1600 m deep — a signal to which parts of the biota readily respond. The shoreline, in contrast, looks much like the shoreline of any other shallow lake in the same climatic environment.

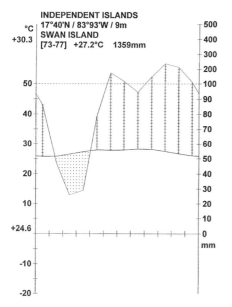

Figure 2.5T An aerial view of a Caribbean Island demonstrates the change of ocean ecosystems from the shore to deep sea. The high salinity, the water temperature, and the strong insolation all year round are the signals to which tidal shoreline vegetation, coral reefs near the shore, and plankton in the open ocean respond.

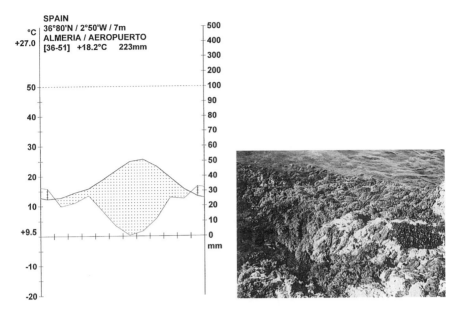

SPAIN
°C 36°80'N / 2°50'W / 7m
+27.0 ALMERIA / AEROPUERTO
[36-51] +18.2°C 223mm

Figure 2.5U Rocky shorelines, heavy wave energy, and tidal inundations are the signals to which many benthic algae respond. The picture was taken near Tossa del Mar in southern Spain. Different algae species in all oceans around the world occupy similar locations and provide the basis for intensive animal growth.

2.4.2 Chemical Structure

The usual ecological treatment of ecosystems is less specific with respect to their chemical structure. The chemistry of ecosystems is nevertheless a key issue for the ecotoxicologist. The chemical structure of the abiotic compartments of ecosystems is usually treated along with topics such as soil, water, and air. For each of these, textbooks and handbooks are available which provide the information needed by the ecotoxicologist.

Less detail is known about the chemical structure of the biotic compartments of the ecosystem. Some elements of the chemical structure have been uncovered by physiological ecologists or nutritionists but a consistent treatment of the chemical structure of ecosystems has received limited attention. The general problem with ecosystem chemistry was described by Bowen (1966), Markert (1993), and Streit (1991), among others. The chemical structure has two major levels: the elemental level and the chemical compound level. For each level we demonstrate in Tables 2.1 and 2.2 some fundamental differences in different ecosystem groups.

2.4.3 Elemental Differences

All naturally occurring chemical elements are present in every plant or animal on earth (Lieth and Markert 1990; Markert and Lieth 1995). They vary, however, significantly in quantity. Lieth and Markert (1990) have started a program called element concentration cadastres in ecosystems. Table 2.1 shows the

Table 2.1 Concentration levels of elements in dry plant tissues[a]

Element	1 Bacteria	2 Plankton, mainly diatoms	3 Brown algae	4 Fungi	5 Bryophytes	6 Ferns	7 Gymnosperms	8 Angiosperms	9 Glycophytes	10 Halophytes	11 Element in earth crust
C	53.80%	22.50%	34.50%	49.40%	45.00%	45.00%	45.00%	45.40%	44.50%	38.8%	320[b]
H	7.40%	4.60%	4.10%	5.50%	5.50%	5.50%	5.50%	5.50%	6.50%	5.6%	0.14%[b]
N	9.60%	3.80%	1.50%	5.10%	2.50%	2.05%	3.20%	3.00%	2.50%	3.5%	46[b]
O	23.00%	44.00%	47.00%	34.00%	45.00%	43.00%	44.00%	41.00%	42.50%	36.5%	47.4%
P	3.00%	0.43%	0.28%	1.40%	0.11%	0.20%	0.29%	0.23%	0.20%	0.17%	0.1%
S	0.53%	0.60%	1.20%	0.40%	0.20%	0.10%	0.11%	0.34%	0.30%	0.2%	0.2%
Si	180	20.00%	0.15%	—	0.20%	0.55%	—	200	0.10%	—	28.2%
Co	—	5	0.7	0.5	0.33	0.8	0.2	0.48	0.2	0.7	25
Cr	—	3.5	1.3	1.5	2	0.8	0.16	0.23	1.5	5	100
Cu	42	200	11	15	7	15	15	14	10	11	55
Fe	250	0.35%	690	130	0.12%	300	130	140	150	614	5.6%
Mn	30	75	53	25	290	250	330	630	200	387	950
Mo	—	1	0.45	1.5	0.7	0.8	0.13	0.9	0.5	25	0.0015
Ni	—	36	3	1.5	2.5	1.5	1.8	2.7	1.5	4.3	75
V	—	5	2	0.67	2.3	0.13	0.69	1.6	0.5	3.5	135
Zn	—	0.26%	150	150	50	77	26	160	50	46	70
Li	—	—	5.4	—	—	—	—	0.1	0.2	—	20
Na	0.46%	0.60%	3.30%	0.15%	0.11%	0.14%	340	0.12%	150	3.9%	2.4%
K	11.50%	—	5.20%	2.23%	0.24%	1.8%	0.63%	1.40%	1.90%	1.6%	2.4%
Rb	—	—	7.4	—	—	—	—	20	50	50	0.09
Cs	—	—	0.067	<0.1	0.2	—	—	0.2	0.2	0.05	0.003
Be	—	—	—	—	—	—	—	<0.1	0.001	—	28
Mg	0.70%	0.32%	0.52%	0.15%	0.30%	0.18%	0.13%	0.32%	0.20%	0.7%	2%
Ca	0.51%	0.80%	1.15%	0.17%	—	0.37%	0.65%	1.80%	1.00%	1.4%	4.2%
Sr	—	260	0.14%	320	15	13	—	26	50	43	0.375
Ba	—	15	31	—	150	8	—	14	40	—	0.425

Table 2.1 (*Continued*)

Element	1 Bacteria	2 Plankton, mainly diatoms	3 Brown algae	4 Fungi	5 Bryophytes	6 Ferns	7 Gymno-sperms	8 Angio-sperms	9 Glycophytes	10 Halophytes	11 Element in earth crust
Ra	—	4×10^{-7}	9×10^{-8}	—	—	—	—	10^{-9}	—	—	—
B	5.5	—	120	5	20	77	63	50	40	96	10
Al	210	0.10%	62	29	0.14%	—	65	550	80	92	8.2%
Ga	—	1.5	0.5	1.5	0.1	0.23	<0.07	0.05	0.1	—	15
In	—	—	—	—	—	—	—	—	0.001	—	0.0001
Tl	—	—	—	—	—	—	—	—	0.05	—	4.5×10^{-7}
Ge	—	—	—	—	—	—	—	—	0.01	—	1.5
Sn	—	35	1.1	5	1	2.3	<0.24	<0.3	0.2	<20	0.002
Pb	—	5	8.4	50	3.3	2.3	1.8	2.7	1	7	0.0125
As	—	—	30	—	—	—	—	0.2	0.1	0.6	1.8
Sb	—	—	—	—	—	—	—	0.06	0.1	0.2	1.4×10^{-4}
Bi	—	—	—	2	1	—	—	0.06	0.01	—	5×10^{-5}
Se	—	—	0.84	—	—	—	—	0.2	0.02	321	4×10^{-6}
Te	—	—	—	—	—	—	—	—	0.05	—	—
Po	—	—	—	—	—	—	—	—	—	—	—
F	—	—	4.5	—	—	—	—	0.50	2	—	625
Cl	0.23%	—	0.47%	1.00%	670	0.60%	—	0.20%	0.20%	7.6%	130
Br	—	—	0.15%	20	5	0.23	—	15	4	500	0.0025
I	—	300	740	—	0.1	0.5	—	0.4	3	—	0.0005
Ag	—	0.25	0.28	0.15	0.1	0.23	0.07	0.06	0.2	0.3	7×10^{-5}
Au	—	—	0.012	—	—	—	—	<0.00045	0.001	<0.0005	4×10^{-6}
Cd	—	0.4	0.4	4	0.1	0.5	0.24	0.64	0.05	1.55	2×10^{-4}
Hg	—	—	0.03	—	<0.3	—	—	0.015	0.1	0.008	8×10^{-5}
Sc	—	—	—	—	—	—	—	0.008	0.02	0.06	22
Y	—	—	—	0.5	0.33	0.77	<0.24	<0.6	0.2	—	0.033

Table 2.1 (*Continued*)

Element	1 Bacteria	2 Plankton, mainly diatoms	3 Brown algae	4 Fungi	5 Bryophytes	6 Ferns	7 Gymnosperms	8 Angiosperms	9 Glycophytes	10 Halophytes	11 Element in earth crust
La	—	—	10	—	3	—	—	0.085	0.2	0.5	0.03
Ce	—	—	—	—	<14	—	—	<34	0.5	0.9	0.06
Pr	—	—	—	—	—	—	—	—	0.05	—	0.086
Nd	—	—	—	—	<6.5	—	—	<24	0.2	—	0.028
Sm	—	—	—	—	—	—	—	—	0.04	—	0.006
Eu	—	—	—	—	—	—	—	0.0055	0.008	0.005	0.0012
Gd	—	—	—	—	—	—	—	0.021	0.04	—	0.0054
Tb	—	—	—	—	—	—	—	—	0.008	—	0.0009
Dy	—	—	—	—	—	—	—	0.0015	0.03	—	0.003
Ho	—	—	—	—	—	—	—	—	0.008	—	0.0012
Er	—	—	—	—	—	—	—	—	0.02	—	0.0028
Tm	—	—	—	—	—	—	—	0.0015	0.004	—	0.0005
Yb	—	—	—	—	0.2	—	—	<0.0015	0.02	—	0.003
Lu	—	—	—	—	—	—	—	—	0.003	—	—
Ac	—	—	—	—	—	—	—	—	—	—	—
Th	—	—	—	0.25	<0.35	—	—	—	0.005	0.07	0.0096
Pa	—	—	—	—	—	—	—	—	—	—	—
U	—	80	12	5	65	—	<0.35	0.038	0.01	0.3	0.0027
Ti	—	20	—	—	—	5.3	0.24	1	5	—	5.7%
Zr	—	—	—	—	0.33	2.3	—	0.64	0.1	—	0.165
Hf	—	—	—	—	—	—	—	—	0.05	—	0.003
Nb	—	—	—	—	—	—	0.3	0.3	0.05	—	0.02
Ta	—	—	—	—	0.3	—	—	—	0.001	—	0.002
W	—	—	0.035	—	—	—	—	0.07	0.2	—	0.0015
Re	—	—	0.014	—	—	—	—	—	—	—	5×10^{-6}

The sequence of elements follows the pattern suggested by Markert (1993).

Sources: Columns 1–8, Lieth (1978); column 9, Markert (1993); column 10, references are listed in Section 2.7.2 of this chapter; column 11, Streit (1991).

[a] Values are given as percentages or in micrograms per gram.

[b] Values are from Schröter et al. (1984).

— Values not available as yet.

differences in elemental concentrations in different plant groups in contrast to their concentrations in the upper 16 km of the earth's crust. Several species groups appear to have specific preferences for certain elements. This is most apparent from the data given in columns 9 and 10. The preference for sodium and chlorine of halophytes is obvious. In addition, many heavy metals are much more highly concentrated in halophytes than in glycophytes, e.g., selenium and cadmium content. One could question the suitability of plant foliage from halophytes for animal fodder.

The differences between the elemental composition of soils and plants are striking. For many elements plants are accumulators, for others, rejectors. In comparing numbers for earth crust and biota one must realize that large differences exist among soil substrates of different geological origin and age.

2.4.4 Differences Between Chemical Compounds

Structural differences between chemical compounds in ecosystems occur in both abiotic and biotic compartments. The differences are so prevalent that they influence major environmental parameters such as pH, chemical compounds, erosion, and weathering of rocks and soil substrate. Many of these properties are described in handbooks of soil science and in soil maps (see FAO/UNESCO 1977). Only one example is mentioned here: the differential vertical movement of silicon dioxide and alumiuum/iron-sesquioxides with different temperatures which leads to the formation of podsols under boreal forests and oxisols under tropical rain forests. The biotic compartments of the ecosystem are capable of changing the top soil's chemical composition drastically.

There are very important differences in chemical composition in the biotic compartments of ecosystems along the trophic structure. The basic difference is shown in Tables 2.2 and 2.3. In these Tables we compare the compound classes usually analyzed to assess food and fodder quality, i.e., carbohydrates, proteins, fats and oils (ether extract), and ash, in plants and animals. The obvious difference is the fact that plants build their structure with carbohydrates while animals build their bodies generally with protein and minerals. However, there are some exceptions. There are plants with an ash content of 20%, e.g., some trees incorporate siliceous material in order to strengthen their trunks. There are animals, especially insects, which maintain their shape with chitin, a structural carbohydrate, when they change from caterpillar stages to pupae and imagines. Further details are available in Lieth (1975, 1978) and in the chemical tables of Ciba Geigy (1977).

One of the main problems in ecotoxicology is the assessment of impacts which man-made chemicals may have on ecosystems, the biotic compartment as well as the abiotic ones. Nature has a great many chemicals which are detrimental to some members of an ecosystem. Chemical elements, poisonous to ecosystem members, were slowly buried in deeper soil layers or down to the ocean floor. Human activities have unearthed these elements or compounds for technical use but to the detriment of ecosystems. This process is continuing at an accelerating pace.

Table 2.2 Chemical differences in primary producers of selected ecosystem types[a]

Ecosystem	Comments	Ash	Structural carbohydrates[b]			Nonstructural carbohydrates	Protein	Ether extract or equivalent	Total (%)
			Cl	Hcl	Lgn.				
Coniferous forest	Biomass, mature stand	0.3	43.5	14.5	30	1.1	1.3	7.7	98.4
	New production	4.2	44.1	8.7	18	15.5	4.0	5.8	100.3
Summergreen deciduous forest	Standing biomass	0.3	46.6	24.0	20	0.8	2.5	1.8	96.0
	New production	4.2	37.0	14.4	12	22.5	6.4	2.8	99.3
Temperate grassland	Above ground	7.6		30.8		50.2	8.7	2.7	100
	Peak biomass	10.0		28.0		48.0	11.0	3.0	100
Salt semideserts	Dry biomass mature plants	21.7		21.7		37.1	11.7	3.3	95.5[c]
Freshwater lake	Plankton	14		18		50	17	1.5	100
	Macrophytes			14–20		43–60	8–19	1.0–2.5	—
	Benthic algae			9–17		36–44	5–18	0.7–2.0	—
	Typha			30–39		38–48	7–12	1.5–3.5	—
	Scirpus	6.5		33		53	7	0.5	100

Source: Lieth (1973, 1975).

[a] Values are given as percentage averages.
[b] Crude fiber; Cl, cellulose; Hcl, hemicellulose; Lgn., lignin.
[c] Rest water up to 100%.
— Data not available in the source.

Table 2.3 Comparison of the chemical composition of trophic levels in a forest ecosystem[a]

Trophic level	Ash	Structural carbo-hydrates	Soluble carbo-hydrates	Protein	Ether extracts
Primary producers	4	75	15–22	4–6	3–6
Consumers	2–28	3–10	10–24	30–75	3–5 (I) 25–35 (V)
Decomposers	5–23	2–11	2–30–60	50–65 (I) 10–40 (P)	4–6 (I) 2–23 (P)

Source: Lieth (1975).
[a] The values given are mean percentages of dry matter.
I, Invertebrates; V, vertebrates; P, fungi.

Equivalent problems are created by human actions with a variety of chemical compounds. Certain plants and animals produce chemicals which are poisonous to other plants and animals. All of these compounds are biodegradable and disintegrate in time into harmless compounds or elements. Human activity, however, has produced chemicals of slow to very slow biodegradation. Such compounds are usually the concern of the ecotoxicologist. Organic compounds are posing the majority of problems since they are produced in such great numbers and quantities. Their addition to the ecosystems of the world is greatly affecting the chemical structure of ecosystems as well as their biological and information structure.

Chemical Compounds with Special Information Content

Some groups of chemical compounds carry special information. These chemicals perform the energy and material transfer and the information exchange. Toxic substances may affect their function. It is, therefore, necessary to discuss these groups briefly. The chemical properties of these compounds are discussed in textbooks of biochemistry and physiology. In the context of this chapter, we can only deal with some specific aspects in the ecosystem. From the potential list of compounds we select: pigments, enzymes, nucleic acid, vitamins, hormones, glycosides, alkaloids, phenolic compounds, and terpenes.

- **Pigments** absorb, conduct, and convert radiant energy into chemical energy, sensible heat, and visible information. Some are very effective at the species level, while others are also effective at the level of the whole ecosystem.
- **Enzymes** control all chemical transfers within and among biotic ecosystem compartments.

- **Nucleic acids** contain all genetic information and control physiological functions.
- **Vitamins** and **hormones** regulate a variety of chemical processes at the species level.
- **Glycosides** and **alkaloids** control interactions and processes among species.
- **Phenolic compounds** play a major role in the conversion of biomass into humus, the organic chemical part of the soil substrates.
- **Terpenes** are important for species interactions and the conversion of biomass energy to sensible heat (fires in dry and semidry ecosystems) at the ecosystem level.

Further information may be found in encyclopedias such as Streit (1991) and Ciba Geigy Tables (1977).

2.4.5 Genetic Structure

The genetic structure in ecosystems is specific for the biotic compartments. The structure is mainly based upon differences in the genetic code which is embedded in nucleic acid compounds.

The genetic structure is manifested

1. at the species level in populations;
2. at the trophic level of species in the ecosystem by the availability of adequate chemical compounds;
3. at the ecosystem level where environmental signals force genetic evolution to create information sets, "species," which can perform its functions within the constraints set by climatic, edaphic, and biotic forces.

Information about these three structural levels are available in textbooks dealing with population genetics and biochemistry of genetics, and in books dealing with nutrition of plants and animals, with ecology in general, and with the habitat concept in ecology.

The genetic structure in ecosystems is considered to be one of the major research problems of ecology. The basic information at the species level is known for a few species. The consequences of genetic structure for species within the trophic level are known to some extent for an even smaller number of species and in some cases for particular races only. The same is true for the consequences in the ecosystem. It is assumed that the genetic structure will yield to the environmental forces while the structure of ecosystems is not taken into consideration. It is very obvious that the genetic structure of ecosystems urgently requires further investigation in order to base current discussions of the ecological impact of gene manipulation on scientifically sound foundations.

2.4.6 Information Structure

The description of information structure in ecosystems is currently only fragmentary. A unifying concept is not yet available. In ecosystem models some aspects are covered by a description of behavior for numerous animals and for some interactions between plant and animal species. The breakdown of signals and responses into bits of information or information sets is still not yet available. We can distinguish so far (selection only):

Abiotic information	Biotic information	Societal information
In the form of environmental quantities for parameter groups from meteorology, hydrology, soil, and extraterrestrial components	Physiological Genetical Sensorial Psychological	Intraspecies Interspecies Family Neighborhood Community Political units

The value of breaking down these categories into basic information bits lies in the potential to combine all bits into one model. This will be especially valuable when we reach the point where energy and mass quantities can be assigned to each "information set". For some types of information bits this will be a difficult task, especially when "virtual" information is involved which may trigger enormous turmoil in biologically controlled systems.

2.4.7 Physical Structure

The earth's surface is seldom a monotonous flat surface. All abiotic forces combined shape any flat places into undulating patterns of various shapes. Over these undulations the vegetation forms an undulating carpet of its own. The physical structure of ecosystems is therefore classified at different levels and categories. For our purpose we distinguish categories such as mountains, hills, dunes, hammocks, valleys, ditches, cracks, solid surfaces, swamps, lakes, creeks, rivers, streams, deltas, deep oceans, shelf, coastal water, and wadden sea. All of these abiotic physical structures have properties important for shaping the physical structure of biota.

The biota displays vertical and horizontal structures. Vertically, we distinguish layers such as subsurface layer, soil layer, ground layer, herbaceous layer, shrub layer, subdominant tree canopy, dominant tree canopy, and aerial layer. In each layer we can distinguish distribution patterns for specimens of individual species. Layering and distribution patterns of plants and animals are important parameters for ecotoxicological work. Many toxic impacts on ecosystems are selective for either whole layers or specific members within a layer. The ecotoxicologist needs to observe structural changes in each category carefully.

The physical structure is one of the prime properties of the ecosystem. It determines the position of energy flow, material transfer, and information exchange. In most standard ecology treatments this aspect is hidden behind species lists, species pattern descriptions, or isolated species function analysis. For ecotoxicological work it is necessary to compile all important parameters of the physical structure of the ecosystem and place the significant transfer elements and processes accordingly in order to localize and forecast probable toxic impacts within an ecosystem. Seldom is a toxicological process noticeable in the entire ecosystem. The so-called early warning of potential risks through increasing toxic levels or through chronic application of low levels is possible in this way: the moss monitoring program in Europe is based on this concept. It has the potential to forecast general detrimental effects when significant correlations between monitoring results and ecosystem inflictions can be established (Markert 1993). Some of these aspects will be discussed later along with ecosystem functions.

2.5 THE FUNCTION OF ECOSYSTEMS

The ecosystems of the world have the following main characteristics in common:

1. the flow of energy;
2. the flow of material;
3. the flow of information;
4. the participation of biota;
5. the participation of liquid water.

The transfer quantities and specific pathways vary widely between ecosystems. Ecosystems are a special case of energy transfer systems which we know to exist in many forms without biota and without liquid water. The participation of biota is the main distinctive characteristic, and the classical description and classification of ecosystems considers mainly the biotic structure and function. This viewpoint may be acceptable for the biologist; however, for the ecotoxicologist it is not acceptable. We can explain the general features of an ecosystem comprehensively for both scientific fields, but more details on the flow of material and water must be included in order to serve adequately the toxicologist. Even more neglected is the information flow in ecosystems. We cover parts of it here because virtual information may greatly influence the application of the results of ecotoxicological work.

2.5.1 Energy Flow

The structure of an ecosystem type can be described in terms of the different possible pathways for energy flow. As mentioned in the introduction solar energy is the main source of energy, which arrives at the outer surface of the atmosphere with a more or less constant level of $2.2 \, \text{cal/cm}^2 = 8.374 \, \text{J cm}^{-2} \, \text{min}^{-1}$ in the wavelength window of 0.5–5 μm. This radiant energy flows through the atmosphere, where it is substantially altered until about 50% of the initial energy hits the soil, water, or biota surface. There it is either reflected, absorbed, and converted into sensible heat or chemical energy or stored in the conversion of liquid water into water vapor and/or used to lift air masses. Depending on local conditions, the different pathways for energy flow conduct different amounts of the energy which, however, leaves eventually quantitatively as radiant energy into outer space mainly in the wavelength window of between 5 and 20 μm.

The energy balance of the earth is such that under present conditions the input of solar energy plus internal energy of the earth equals the energy release from the earth to outer space. The energy balance of an ecosystem on the earth's surface is in most places mainly controlled by the incoming solar radiation. Exceptions are regions of volcanic actions where a continuous supply of hot lava increases the temperature of surface soil and air. Otherwise, the energy balance of an ecosystem can be expressed as

$$I = O + R + A + E + P + S \tag{2.5}$$

where

$I = \text{input}$
$O = \text{output (radiation)}$
$R = \text{reflection}$
$A = \text{advection}$
$E = \text{evaporation}$
$P = \text{photosynthesis (biomass production)}$
$S = \text{temperature change (storage)}.$

E and P may be missing in extreme deserts. P may be missing in ocean deserts.

The flux along these conductors differs in different ecosystem types. Comparative estimates are shown in Table 2.4. The flow values for the energy flowing through biological matter is evident in all ecosystems.

Table 2.4 demonstrates clearly that the energy from the earth's surface is mainly conducted by water evaporation if water is available. It is also striking that photosynthesis conducts only up to 2% in the most productive ecosystem throughout the year. If one compares the percentage of energy flowing through water evaporation with the daily temperature fluctuation it becomes evident

Table 2.4 Possible flux percentages of the total outgoing energy in contrasting ecosystems

	O	R	A (%)	E	P	Total (%)	Approx. day/night temperature changes in °C
Desert	10	60	30	—	—	100	80–100
Tropical forest	5	8	10	75	2.00	100	0–10
Grassland	5	10	25	60	0.50	100.5	10–40
Ocean	5	10	15	70	0.01	100.01	~ 0–10

Source: After Lieth (1976).

that the two are negatively correlated and that evaporation on land increases with vegetation height. This may also suggest that an increase in size of the biota in an ecosystem improves homeothermy which in turn may improve the physiognomic structure of the biocoenoses and is therefore an important feature to be observed in toxicological work. The inputs of solar energy and precipitation as well as relevant temperatures are described by Walter and Lieth (1960) and in the internet file by Vose et al. (1995).

The utilization of energy by the individual species, although low, is of vital importance for the structure of the biocoenosis. Each green plant requires space and light energy for its life. During fast growth, about 70% of the energy received may be used to evaporate water and up to 10% for photosynthesis daily. The remaining radiation energy is converted to either sensible heat or physical force, or is reflected.

The structure of vegetation stands is such that plants of different heights stand together. Bigger trees shade lower trees, and below them are shrubs under which grasses and mosses grow. The radiant energy conversion into plant matter may be compared with the energy conversion of a steam engine where up to 30% of the energy received can be converted to work. The remaining 70% is still available for another plant with a lower energy demand which again uses 30% of the 70%. If we continue such a chain of diminishing energy availability we can try to calculate how many specimens of plants can live under each other within one ecosystem. Simultaneously this chain of diminishing available energy explains the fact that species of different sizes grow together in one ecosystem.

In Figure 2.6 we compare two sets of plant species with different energy utilization efficiency. Each set starts with 100% available energy of which each specimen takes according to its rank 30% or 50% of the energy still available after specimens of higher ranks have taken their share. If we compare the theoretical curves with reality we observe clear similarities. In Figure 2.7A the distribution of biomass across a list of species for several plant communities is compared. The patterns can be explained mainly by the theory above whereby additional forces such as competition, grazing, nutrient availability, trampling, and water consumption may act as modifiers. Figure 2.7A shows the sharp

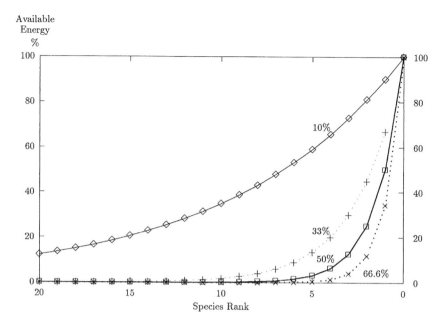

Figure 2.6 Energy utilization model of a hierarchically organized set of plant species in one ecosystem, assuming each species is allowed to use 33% or 50% of the energy not absorbed by species of higher ranks. Ecosystems with high levels of utilizable energy input may have larger numbers of species than ecosystems with lower energy input. This explains differences in species diversity between tropical and boreal ecosystems (see Figure 2.8).

reduction in biomass per species in forest which is evident if one compares the few dominant species with the large number of understory species. Figure 2.7B shows the more equal distribution of biomass among all species in ecosystems stressed by grazing and trampling. The number of total species of biosystems in Figure 2.7A is much larger than those in Figure 2.7B. If our statement above regarding the energy utilization efficiency is correct we would expect to find a higher amount of species per 1000 km^2 in areas with higher net primary productivity.

This argument was analyzed by Malychev (1975). He compared the species numbers in local floras and constructed the pattern shown in Figure 2.8. When we compare this map with the net primary productivity map in Figure 2.1 we find agreement in the tendency of polar/equatorial moist areas and desert/ humid areas. Discrepancies are found in the number of species for climatologically similar areas in Africa vs. America and Asia. It is clear, therefore, that other factors such as abiotic ecosystem compartment diversities and evolutionary age have to be superimposed on the prime pattern provided by the energy utilization efficiency. From experience in the field we know that natural toxicity plays a major role in the impoverishment of species diversity. We may expect,

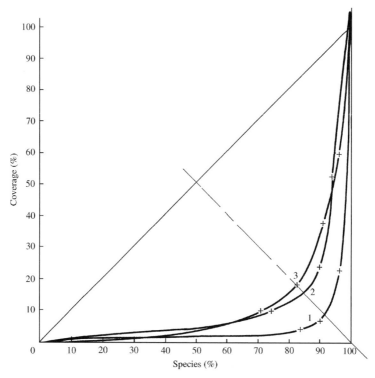

Figure 2.7A Percentage of ground cover (as an indicator for biomass invested in plant species) versus percentage of species contributing to the total ground cover in different ecosystems without undue stress. (After Lieth and Petrall 1972.) Curves of this type are called "wealth distribution" curves. Curve 1 – pine forest in North America in pioneer stage, 2 – pine forest near climax stage, 3 – tropical rain forest in Thailand. The point of inflection for all curves lies near the midpoint projection of the egalitarian distribution of wealth shown as diagonal lines.

therefore, that man-induced toxicity will have the same effect. We have ample proof of this in industrialized areas of the world.

Water Flux as Part of the Energy Flow

Our theoretical example of the energy flow efficiency included the expenditure of water by the system either through transpiration or evaporation. Experiments with lysimeters and irrigated plots have shown that plants use water at a ratio of generally between 300 and 1000 kg per kilogram dry matter produced. This ratio is called water use efficiency or transpiration coefficient. Plants can adapt within certain ranges to abiotic environmental signals, especially to the availability of water and the level of solar input.

The level of responses is controlled evolutionary. It is part of the genetic information which controls body size and structure, physiological processes, leaf

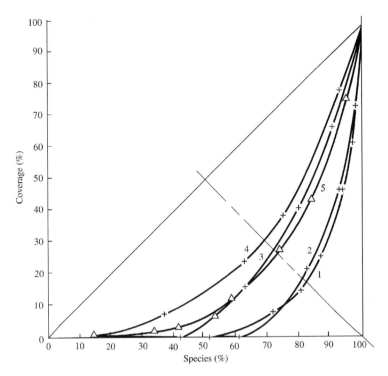

Figure 2.7B "Wealth distribution" pattern for plant species in ecosystems under natural stress. (After Lieth 1975.) Curve 1 – tropical grass land, Ivory Coast, 2 – alpine tundra, Austria, 3 – subtropical dry grass land, Morocco, 4 – arctic tundra, Baffin Island, 1–4 ground cover estimates, 5 – temperate grass land, USA – above ground biomass data. Inflection points disappear under stress because natural regulating forces do not allow high energy utilisation efficiencies for single species. The stress is in grass lands grazing by animals, in tundra ecosystems harsh climate and short vegetation period.

size and water conductivity, as well as root parameters. The propagation and regeneration of plants and animals alike is closely connected to the water flux through the ecosystem.

Besides the direct impact of the water flux upon the anatomy of plants and animals, it appears to have an impact on the total environment in the form of information transfer to the biota. The continuous utilization of water to expand solar energy regulates the temperature range throughout day and night, rainy and sunny seasons, winter and summer.

It appears as if the water flow of the ecosystem in the abiotic and biotic compartments of land ecosystems is such that the temperature range in them remains as small as possible. Water evaporation is used for that purpose which has the effect that the flow of solar energy through the ecosystem in the form of water evaporation is maximized. This fact is clearly demonstrated in Table 2.4.

Figure 2.8 Global floristic richness pattern of vascular plant species averaged for areas of 100,000 km^2 (Malychev 1975).

Temperature stability is most likely sensed by all biota at all levels of information transfer because of the different Michaelis-Menten characteristics in the large sets of enzymes controlling all physiological processes. The contribution of water to the energy flux in aquatic ecosystems is so dominant that the contribution by the biota is almost negligible. If we take the regulation of temperature as a main evolutionary incentive for the development of ecosystem structures it appears logical that biotic structures over large water bodies are as unlikely as biotic structures over climatic deserts on land.

2.5.2 The Flow of Material

The flow of material through ecosystems is one of the main topics of concern in ecotoxicological work. As far as mass flow is concerned, material flow in ecosystems is largely connected to energy flow, especially for the mass flow of water. Within the **soil compartment** the mass flow of liquid and frozen water is not only driven by the immediate solar input but also by geological forces and the general air mass circulation which causes precipitation. It is difficult to calculate the contribution of each force to the abiotic mass flow. Nevertheless the inclusion of the abiotic mass into the total energy/material relation is very important, and new attempts should be made to construct reliable models for the relevant situations on the earth's surface.

The amounts of abiotic air masses flowing through an ecosystem depend on the wind speed. This is important for ecotoxicological work since the effect of toxic substances released into the air by human activity can be estimated on that basis. The exchange of air components between biosphere and atmosphere is very intensive. The biosphere stores large amounts of otherwise airborne elements and compounds at any time. The recent interest in some of the man-made airborne chemicals such as CFC do not directly influence the ecosystem structure and function but instead influence climatic signals which in turn can elicit biospheric responses at several levels.

Within the biotic compartment there is a high correlation between the input of solar radiation, the amount of available water, and the amount of **biomass** produced. The relation is shown in Figure 2.1 and specified in Table 2.5 for different ecosystem types. The overall caloric equivalent of 1 g biomass is 4500 cal. Larger deviations from that figure are possible if high amounts of ash material reduce this figure or if high amounts of oils and terpenes increase the carbon content or the amount of double-bonded carbon. An overview of caloric values of biomass in different ecosystems is given in Table 2.6.

The **chemical differences** of biomass material has crucial consequences for the ecosystem. They determine, for example, whether a forest or grassland is possible or whether a peat bog with mosses or a peat bog with woody overgrowth develops. In Table 2.2 such differences are presented as properties of structurally defined ecosystems. With respect to the functioning of ecosystems these differences determine the longevity of the ecosystem types, which is shown in

Table 2.5 Productivity and standing biomass values for terrestrial vegetation types distinguished in the Osnabrück Biosphere Model

					Net primary production in g/m²			
Ecosystem class, vegetation formation[a]	Area size in 10^3 km³	Mean stand age (years)	Herbaceous portion of biomass	Upper limit	Mean above ground[d]	Max. above ground[d]	Mean total[d]	Max total[d]
1	2	3	4	5	6	7	8	9
1. Tropical formations								
Tropical rainforests	15745	200	0.37	3500	1512 (630)	2924	1897 (706)	2980
Lowland rainforests	—	—	—	2500[c]	—	—	—	—
Mangrove	558	50	0.29	2000[c]	732	—	—	—
Tropical lowland dry forest	6045	80[c]	0.4[c]	1800	624 (325)	1110	752 (340)	1260
Tropical mountain forest	2597	80[c]	0.37[c]	2500	—	—	—	—
Tropical savannah	5546	5	0.98	3500	1465 (747)	2080	—	3340
Tropical paramo woodland	375	10[c]	0.95[c]	1500[c]	—	—	—	—
Tropical herbaceous paramo	50	1[c]	1.0[c]	—	—	—	—	—
2. Subtropical formations								
Subtropical evergreen forests	2821	200	0.37	2800	1458 (537)	1875	1844 (760)	2473
Subtropical mountain forests	—	—	—	2300[c]	588 (407)	875	764 (529)	1138
Subtropical deciduous forests	246	150	0.44	—	—	—	—	—
Subtropical swamp forests	—	—	—	2300[c]	920 (283)	1120	1105 (339)	1345
Subtropical dry woodland	—	—	—	1200	285 (167)	539	506 (297)	981
Xeromorphic formation	4002	20	0.4	—	—	—	—	—
Subtropical xeromorphic shrub formation	—	—	—	800	80	—	—	—
Subtropical halophyte formation	708	5	0.90	—	—	—	—	—
Subtropical savannah	12022	5	0.90	—	—	—	—	—
Subtropical step and grassland	741	1	1.0	—	—	—	—	—
Subtropical and tropical step and grass savannah	—	—	—	3500	1060 (697)	2407	1783 (1116)	3538
Puna step	382	2[c]	1.0[c]	—	—	—	—	—
3. Deserts and semideserts								
Deserts (tropical, subtropical, polar)	9760	5[c]	0.85[c]	—	—	—	—	—
Subtropical desert only	—	—	—	400	83 (49)	168	151 (90)	339[e]
Subtropical semidesert	14559	15[c]	0.85	700	129 (86)	228	235 (160)	417[e]

Standing dry weight biomass in kg/m^2					
Mean above ground[d]	Max. above ground[d]	Mean total[d]	Max. total	Part of Figure 2.5 in which relevant ecosystem is shown	Relevant column[f] for element concentrations shown in Table 2.1
10	11	12	13	14	15

1. Tropical formations

34.10 (10.92)	51.30	38.31 (11.75)	56.75	B	6, 8
—	—	—	—	C	8
6.25	—	11.25	—	D	10
11.65 (5.22)	16.30	14.35 (6.32)	20.00	E/F	8
—	—	—	—	NS	6, 8
1.83 (1.49)	6.13	5.58 (3.22)	9.84	NS	8, 9
—	—	—	—	NS	5, 8
—	—	—	—	O	5, 8

2. Subtropical formations

26.71 (14.40)	37.07	34.07 (14.56)	44.60	NS	6, 8
15.85 (3.19)	18.10	19.80 (3.96)	22.60	NS	6, 8
—	—	—	—	NS	8
33.90 (4.67)	37.20	40.68 (5.62)	44.65	NS	5, 6
4.38 (3.48)	6.40	5.01 (3.37)	9.73	J	NS
—	—	—	—	NS	NS
0.24	—	0.32	—	NS	8, 9
—	—	—	—	NS	10
—	—	—	—	K	8, 9
—	—	—	—	NS	8, 9
0.68 (0.64)	1.97	1.70 (1.29)	4.11[e]	NS	NS
—	—	—	—	NS	5

3. Deserts and semideserts

—	—	—	—	Q	NS
0.36 (0.28)	0.96	0.65 (0.54)	1.87[e]	P	10
0.55 (0.54)	1.31	1.46 (1.29)	2.94	NS	8, 10

Table 2.5 (*Continued*)

					Net primary production in g/m²			
Ecosystem class, vegetation formation[a]	Area size in 10^3 km³	Mean stand age (years)	Herbaceous portion of biomass	Upper limit	Mean above ground[d]	Max. above ground[d]	Mean total[d]	Max total[d]
1	2	3	4	5	6	7	8	9

4. Formation of the temperate zone

Mediterranean sclerophyllous forest	830	100[c]	0.4	1800[c]	750	—	—	—
Mediterranean woodland	3123	15	0.47	1600	481 (210)	850	866[e](378)	1530[e]
Temperate evergreen forest	1086	130	0.29	2300	998 (292)	1658	1282 (388)	2139[e]
Temperate deciduous forest	9360	150	0.38	2300	982 (304)	1670	1167 (347)	1900
Temperate swamp	—	—	—	2300	1229 (592)	1788	1450 (698)	2110[e]
Temperate woodland	3884	25	0.53	—	—	—	—	—
Temperate scrub formation	2992	10[c]	0.85	—	—	—	—	—
Temperate steppes and meadow	8256	1	1.0	—	—	—	—	—
Temperate grasslands and steppes	—	—	—	2300	435 (239)	1131	1300 (742)	3453[e]
Temperate steppes with trees	—	—	—	2000	526	—	—	238
Temperate dry steppes	—	—	—	1000	185 (69)	257	566 (285)	—
Temperate scrub and halophyte steppes	—	—	—	—	—	—	—	—
Cool temperate peat bogs	86	5	0.48	—	—	—	—	—
Temperate heathland and peat bogs	—	—	—	1900	421 (255)	1118	801 (418)	1631

5. Boreal formation

Boreal evergreen needelleaf forest	9632	100	0.34	—	—	—	—	—
Boreal deciduous forests	4886	100	0.38[c]	—	—	—	—	—
Boreal forests	—	—	—	1500	499 (251)	990	617 (274)	1146
Boreal woodland	4108	15	0.6[c]	—	—	—	—	—
Boreal shrub formations	381	10[c]	0.85[c]	—	—	—	—	—
Subpolar shrub formations	—	—	—	1500	509 (318)	1129	620 (395)	1355[e]

6. Arctic–alpine formations

Shrub tundra	5340	10	0.70	—	—	—	—	—
Shrub and lichen tundra and alpine dwarf shrub vegetation	—	—	—	1500	120 (114)	385	349 (345)	1229[e]

Standing dry weight biomass in kg/m²					
Mean above ground[d]	Max. above ground[d]	Mean total[d]	Max. total	Part of Figure 2.5 in which relevant ecosystem is shown	Relevant column[f] for element concentrations shown in Table 2.1
10	11	12	13	14	15

4. Formation of the temperate zone

27.72	—	31.32	—	I	7, 8
2.07 (1.33)	4.87	3.432 (2.73)	8.77[e]	NS	7, 8
67.93 (53.44)	217.00	85.35 (67.32)	273.00[e]	NS	7, 8
27.44 (10.67)	50.00	32.57 (12.43)	57.80	G	8, 9
18.57 (13.66)	31.40	21.64 (15.67)	36.00	NS	5, 9
—	—	—	—	NS	8, 9
—	—	—	—	NS	8, 9
—	—	—	—	L	9
0.32 (0.24)	1.02	2.28 (1.77)	7.87[e]	NS	8, 9
4.38	—	5.78	—	NS	7, 8, 9
0.18	0.37	1.04 (0.78)	2.10	NS	
0.29 (0.22)	0.67	0.77 (0.27)	1.03	NS	10
—	—	—	—	NS	5, 6, 9
1.35 (0.70)	2.51	2.63 (1.25)	4.71[e]	NS	5, 8, 9

5. Boreal formation

—	—	—	—	H	7
—	—	—	—	NS	7, 8
19.34 (10.36)	46.20	25.26 (12.33)	56.65	NS	5, 8
—	—	—	—	NS	
—	—	—	—	NS	8
6.22 (1.97)	8.50	9.30 (3.96)	13.66[e]	NS	8

6. Arctic–alpine formations

—	—	—	—	M/(O)	5, 8
0.68 (0.51)	1.79	1.99 (1.56)	6.53[e]	N	4, 5, 8

Table 2.5 (*Continued*)

Ecosystem class, vegetation formation[a]	Area size in 10^3 km^3	Mean stand age (years)	Herbaceous portion of biomass	Upper limit	Net primary production in g/m^2			
					Mean above ground[d]	Max. above ground[d]	Mean total[d]	Max total[d]
1	2	3	4	5	6	7	8	9
Herbaceous tundra	1070	2	1.0	—	—	—	—	—
Gramineous tundra and alpine meadows	—	—	—	1500	253 (151)	502 (1190)	717 (421)	1502 (3300)
7. Azonal formation[b]								
	1677	5[c]	0.60	3500	1421 (926)	3126	1710 (1078)	3814

Source: Aselmann (1985).

[a] Terminology adapted from Schmithuesen (1976).

[b] Azonal formations are in the majority river bottom vegetation, gallery forests, swampy systems, and other vegetation types outside their coherent range.

[c] Estimation from various publications.

[d] Values in parentheses are standard deviations.

[e] Root values are estimates.

[f] The reference to columns of Table 2.1 refers to the major portion of the biomass. Elements in columns 1 and 4 of Table 2.1 are present in all ecosystems listed here.

—, no information; NS, not shown.

Table 2.5, column 3. The longevity of plants is clearly related to the percentage of structural carbohydrates containing cellulose and lignin. The persistence of the dead biomass is related to the amount of phenolic compounds remaining in the lignin fraction of the decaying litter. For the latter the speed of transfer is largely controlled by the moisture and oxygen level, pH, and the decomposer types possible in this subsystem. Bacterial decay is faster than fungal decay. Bacterial decay occurs in the gut of wood-eating animals and in moist soil. Other animals chew the detritus into small pieces and use the digestible part for themselves but present with their faeces a large surface for bacteria and fungi to utilize the resource. For the development of ecosystem types the longevity of the chemicals in the biomass is important. The turnover speed for biomass is ranked in the following sequence:

fire > herbivores > carnivores > bacteria > fungi > inorganic weathering.

This sequence explains in part the success of the human race on earth: fire is the fastest way to convert biomass back into sensible heat.

Standing dry weight biomass in kg/m^2					
Mean above ground[d]	Max. above ground[d]	Mean total	Max. total	Part of Figure 2.5 in which relevant ecosystem is shown	Relevant column[f] for element concentrations shown in Table 2.1
10	11	12	13	14	15
—	—	—	—	NS	5, 8
0.23 (0.13)	0.49 (0.86)	1.47 (0.80)	3.07[e]	NS	9
		7. Azonal formation[b]			
0.81 (0.74)	2.97	1.75 (1.69)	6.03	S	NS

The Flux of Chemical Elements through Ecosystems

The flux of chemical elements goes along with the transfer of the chemical compounds they form. Their mobility is controlled by their position in gas, water, or solid material, their solubility in water and biomass compounds, and their tolerance by biota. In general, it appears that the biota of an ecosystem screen the surface substrate of the earth for elements which they need, can tolerate, and must reject. Fertile soil reflects in its composition to a large extent the activity of the biota growing on top of it. The rejected elements, such as Na and Cl, are washed out into the ocean, while elements needed such as K and P are conserved. The biological system of elements suggested by Markert (1994) can be analyzed in this context for its flux rate through ecosystems.

The Flux of Chemical Compounds within the Biotic Compartment

Within biotic compartments, chemical compounds flow along the trophic structure of ecosystems. There are large differences between ecosystems with regard to the speed of transfer, the percentage of internally circulating compounds, and the tenure time of some compounds within the ecosystem.

The most significant property of some organic chemicals is their ability to emit, transmit, and receive information. Some of these chemicals are transformed in this action. We shall consider this aspect partly in the discussion of information flow. The correlation of element and chemical compound flow is one of the major problems to be solved before we can approach a better understanding of ecosystem structure and function.

Table 2.6 Average energy values of natural communities, listed in order of increasing value

	Community or species	Location	No. of samples	Leaves (cal/g)	Entire plant (cal/g)	Reference
		Woody communities				
1	Tropical moist forest	Panama	4	3732	ND	Golley (1969)
2	Gallery forest	Panama	4	3879	ND	Golley (1969)
3	Tropical rain forest	Puerto Rico	15	ND	3897	Golley (1961)
4	Premontane tropical forest	Panama	4	4060	ND	Golley (1969)
5	Oak forest	France and Spain	3	ND	47–4900	Lieth (1968)
6	Angiosperm forests	England	8	4759	ND	Ovington and Heitkamp (1960)
7	Scotch pine	England	14	ND	4787	Golley (1961)
8	Poplar	N.W. Territory, Canada	2	4600	4800	Aleksiuk (1970)
9	Willow	N.W. Territory, Canada	2	4600	4800	Aleksiuk (1970)
10	Savanna	Minnesota	Composite	4846	ND	Ovington and Lawrence (1967)
11	Oakwood	Minnesota	Composite	4916	ND	Ovington and Heitkamp (1967)
12	Gymnosperms	England	17	4926	ND	Ovington and Heitkamp (1967)
13	Oak	Minnesota	"Few"	4930	ND	Gorham and Sanger (1967)
14	Alder	N.W. Territory, Canada	2	4800	5130	Aleksiuk (1970)
15	Cedar	Minnesota	"Few"	5250	ND	Gorham and Sanger (1967)
16	Larch	Minnesota	"Few"	5260	ND	Gorham and Sanger (1967)
17	Pine	Minnesota	"Few"	5290	ND	Gorham and Sanger (1967)
18	Alpine shrub	New Hampshire	30	5367	ND	Hadley and Bliss (1964)
19	Alpine shrub	Tirol, Austria	10	5600	5600	Larcher et al. (1973)

Table 2.6 (*Continued*)

Community or species	Location	No. of samples	Leaves (cal/g)	Entire plant (cal/g)	Reference
Herbaceous communities					
1 Perennial grass	Georgia	143	ND	3905	Golley (1961)
2 Prairie grass	Missouri	Composite	4071	ND	Kucera et al. (1967)
3 *Spartina* marsh	Georgia	14	ND	4072	Golley (1961)
4 Desert grass	Utah	24	4080	ND	Cook et al. (1959)
5 Old field herbs	Georgia	35	ND	4177	Golley (1961)
6 Perennial forbs	Michigan	Composite	4315	ND	Weigert and Evans (1964)
7 Perennial grass	Michigan	Composite	4384	ND	Weigert and Evans (1964)
8 Prairie herbs	Minnesota	Composite	4471	ND	Ovington and Lawrence (1967)
9 "Ground flora"	Minnesota	"Few"	4680	ND	Gorham and Sanger (1967)
10 Alpine meadow	New Hampshire	3	ND	4711	Golley (1961)
11 Alpine *Juncus* dwarf heath	New Hampshire	2	ND	4790	Golley (1961)
12 Alpine sedges and herbs	New Hampshire	40	4796	ND	Hadley and Bliss (1964)

Source: Lieth (1978).
ND, no data available.

2.5.3 The Flow of Information

The flow of information in an ecosystem is very diverse. While intensive work is being done on the information flow between biotic compartments and within individual biota, almost no attention has been given to the flow of information between the biotic and abiotic compartments. The relation between these compartments is usually referred to as environmental constraints and energy or material consumes. The fact that all of these interrelations also constitute information is rarely discussed.

If we ever want to understand the energetic equivalent of information bits, we need to answer the question of how information bits are connected to material and energy. Information bits of environmental parameters probably carry enormous amounts of energy and/or material, while the information bits of biotic parameters carry much less energy themselves but may release enormous amounts of energy through connections to societal information bits. For ecotoxicological work it is very important to realize that societal information bits may have a real base in the ecosystem function, but they may also be "virtual" bits simply created by the exchange of information within biotic taxa at all levels.

Biotic information receives the most attention at all levels. Of ecotoxicological concern are the effects of chemicals on all information classes distinguished: physiological, genetical, sensorial, and psychological. The human race is part of the biotic ecosystem compartment and may suffer, therefore, in personal and societal respects as a result of unforeseen information carried by the multitude of chemicals which the ecosystem may constrain naturally or which may be injected into it through the actions of biota including man. The task of solving Equations 2.2 and 2.3 is of high priority. The solution itself is rather important for ecotoxicological work and should receive immediate attention.

2.6 ECOTOXICOLOGY AND ECOSYSTEMS

Toxicology is by definition the study of the impact of chemicals on living systems. Toxicological processes are a concern relevant for all structural and functional levels within and around ecosystems. As a matter of fact they are even a natural part of ecosystem processes. Some are listed in Table 2.7.

The general concept of toxicology usually excludes these facts. We distinguish the production of poison in the skin of a tropical frog as its legitimate defense, whereas if an Indian in Central America uses this substance to poison his arrows the substance is of toxicological concern. The same disparity exists for the acidification of the soil by coniferous trees. In the boreal coniferous forest regions of the world it is considered a natural condition; in forest plantations in other parts of the world it is considered a toxicological process. Given the ambivalence of such widely existing phenomena we can define the toxic substance only as "a chemical substance with a negative impact on biota at the

Table 2.7 Some biotic and abiotic toxicological processes occurring naturally in ecosystems

A: At the species level (all biotic)

Plants may use toxic substances in competing with each other
Plants can use toxic substances against some animals
Animals can use toxic substances to hunt their prey
Fish and amphibians use poisonous secretions for defense
Some algal blooms in oceans and wetlands can produce poisonous chemicals

B: At the ecosystem level

Some forest systems can alter the soil surface in such a manner that many other species cannot germinate (biotic and abiotic)
Eutrophication of water bodies can kill many species by oxygen depletion (biotic)
Salinization of lowlands in semideserts can render such places uninhabitable for plants and animals (abiotic)

wrong place." In the context of an entire ecosystem as defined in definition D4, this is included as noticeable information with impacts on energy and material flow.

Any ecosystem can be influenced by man in several ways, for example:

1. Water pollutants (chemicals) reach rivers as a result of agriculture, housing, and industry.
2. Precipitation chemistry may be altered.
3. Air pollution through industries, housing, and traffic.
4. Changing land use patterns such that dust may be transported from open soil surfaces.
5. Species composition may be changed by adding chemicals to the ecosystem.
6. Species composition may be changed by overexploiting certain species, i.e., plants as well as animals.

Examples 1–4 are intrusions into the environmental part of the ecosystem; examples 5 and 6 are direct impacts on the biotic part of the system. In an aquatic system massive changes in the water quality cause the death of many organisms in the case of poisons, or cause a massive explosion of phytoplankton species to the extent that they cannot be consumed by the zooplankton and are subsequently digested by bacteria. This leads to a decrease in oxygen levels in the layer below the light penetration and influences life in the lower layers of the water.

In terrestrial ecosystems air pollution as well as water pollution from industries usually contain poisonous compounds or elements that may either cause the death of species or poison them so that they are unfit for human consumption. This type of impact on an ecosystem is the most frequently debated subject in which the environmental engineer gets involved. Determination of the cause of the pollution is his obligation in the debate as well as suggesting actions needed for improvement. Land use changes and/or restrictions are often needed to prevent ecosystem intrusions. This is true for mining and industry in order to control erosion, above and below ground water flows. Computer models will be helpful in performing seasonal process analyses in order to predict changes or the chances of remedy successes (Lieth 1990, Arbeitsgruppe Systemforschung 1991).

Species culling by hunting and fishing often leads to ecosystem changes. In such cases the population ecologist is requested to determine the limits which ensure the existence of each population. This aspect is called sustainable harvesting. The environmental engineer must understand this aspect of the ecosystem and must analyze it sufficiently himself or have it done by a specialist of the particular species group endangered in the respective system. The task becomes more difficult when the number of important species increases. It is important, however, to distinguish between changes occurring through selective culling of species and those which are due to the impact of chemicals added.

In any ecotoxicological debate the naturalists and ecologists will defend the position that each species is

- important in its own right;
- important in its position of the food web;
- important for the total system;
- important for genetic evolution;
- important for mankind.

The environmental engineer can and must respond to this position. Simultaneously, the ecotoxicologist needs to consider the three major functional properties of the ecosystem; energy, material and information flow, and their level of accumulation achieved in any given ecosystem. This chapter has outlined guidelines for this task.

Acknowledgements. I wish to thank Ms Rebecca Smith and Mr Jens Varnskühler (International Graduate School (IHI) Zittau) for preparing the final version of the manuscript and Ms Marina Moschenko, Ms Andrea Noel, and Mr Uwe Menzel (University of Osnabrück, Institute for Environmental Systems Research) for their contributions to the manuscript. Without their help this chapter would not have been completed in this form in time.

2.7 REFERENCES

2.7.1 Literature

Aleksiuk, M., 1970, "The Seasonal Food Regime of Arctic Beavers," *Ecology, 51*, 267–270

Arbeitsgruppe Systemforschung, 1991, "Nitratversickerung im Kreis Vechta: Simulation und ihr Praxisbezug," *Berichte aus der ökologischen Forschung*, Vol. 3, Jülich Forschungszentrum GmbH, Jülich, Germany

Archibold, O. W., 1995, *Ecology of World Vegetation*, Chapman and Hall, London, UK

Aselmann, I., 1985, *Zur Beziehung von Netto-Primärproduktion und Biomasse*. Naturforschende Gesellschaft, Emden, Germany

v. Bertalanffy, L., 1951, *Theoretische Biologie*, Vol. 2, Franke, Bern, Switzerland

Bowen, H. J. M., 1966, *Trace Elements in Biochemistry*, Academic Press, London, UK

Box, E., 1981, *Macroclimate and Plant Forms: An Introduction to Predictive Modeling in Phytogeography*, Dr. W. Junk, The Hague, The Netherlands

v. Dobben, W. H., and R. H. Lowe-McConnell, Eds., 1975, *Unifying Concepts in Ecology*, Dr. W. Junk, The Hague, The Netherlands

Ciba Geigy, 1977, *Wissenschaftliche Tabellen*, Vol. 4, Ciba Geigy, Basel, Switzerland

Cook, C. W., L. A. Stoddert, and L. E. Harris, 1959, "The Chemical Content in Various Portions of the Current Growth of Salt Desert Shrubs and Grasses During Winter," *Ecology*, 40, 644–650

Everts, H. H., A. P. Grootjans, and N. P. J. de Vries, 1988, "Distribution of Marsh Plants as Guidelines for Geohydrological Research," *Colloques phytosociologiques*, 16, 271–292

FAO/UNESCO, *Soil Map of the World*, Vol. 10, UNESCO, Paris, France

Gaussen, H., 1954, Théories et Classification des Climates et Microclimates, *8 Congrès International de Botanique*, Sections 7 et 8, Paris, France, pp. 125–130

GCTE, 1996, "GCTE's New Focus 4: Global Change and Ecological Complexity. GCTE News 9: 3," *CSIRO Division of Wildlife and Ecology*, ACT, Lyneham, Australia

Golley, F. B., 1961, "Energy Values of Ecological Materials," *Ecology*, 42, 581–584

Golley, F. B., 1969, "Caloric Value of Wet Tropical Forest Vegetation," *Ecology*, 50, 517–519

Gorham, E., and J. Sanger, 1967, "Caloric Values of Organic Material in Woodland, Swamp, and Lake Soils," *Ecology*, 48, 492–494

Hadley, E. B., and L. C. Bliss, 1964, "Energy Relationships of Alpine Plants on Mt. Washington, New Hampshire," *Ecol. Monogr.*, 34, 334–357

Illies, J., 1971, *Einführung in die Tiergeographie*, Fischer, Stuttgart, Germany

Jordan, C. F., 1971, "A World Pattern in Plant Energetics," *American Scientist*, 59(4), 425–433

Kloft, W. J., 1978, *Ökologie der Tiere*, Ulmer, Stuttgart, Germany

Koblentz-Mishke, O. I., V. V. Volkovinski, and Y. G. Kabanova, 1970, "New Data on a Magnitude of Primary Production in the Oceans," *Dokl. Akad. Nauk USSR, Ser. Biol.*, 183(5), 1189–1192

Kucera, C. L., R. C. Dahlman, and M. R. Koelling, 1967, "Total Net Productivity and Turnover on an Energy Basis for Tallgrass Prairie," *Ecology*, 48, 536–541

Larcher, W., 1980, *Ökologie der Pflanzen*, 3rd Edn., Ulmer, Stuttgart, Germany

Lieth, H., 1968, "The Measurement of Calorific Values of Biological Material and the Determination of Ecological Efficiency," in *Functioning of Terrestrial Ecosystems at the Primary Production Level. Proc. Copenhagen Symposium*, UNESCO, 233–241

Lieth, H., 1972, "Über die Primärproduktion der Pflanzendecke der Erde," *Angew. Botanik, 46,* 1–37

Lieth, H., 1973, *Chemical Differences in Contrasting Ecosystems and Their Trophic Levels. An Exploration of a New Viewpoint in Systems Ecology,* US IBP EDF Biome Memoreport, 73–76

Lieth, H., 1975, "Some Prospects Beyond Production Measurement: Comparative Analysis of Some Biomass Properties on the Ecosystem Level," in *Primary Productivity of the Biosphere* (H. Lieth, and R. H. Whittaker, Eds.), Springer, Heidelberg, New York

Lieth, H., 1976, "Biophysikalische Fragestellungen in der Ökologie und Umweltforschung, Part 2: Extremalprinzipien in Ökosystemen," *Rad. Environm. Biophysics, 13,* 337–351

Lieth, H., 1977, "Energy Flow and Efficiency. Differences in Plants and Plant Communities," in *Applications of Calorimetry in Life Sciences* (I. Lamprecht, and B. Scharschmidt, Eds.), de Gruyter, Berlin, New York, pp. 325–336

Lieth, H., 1978, "Patterns of Primary Productivity in the Biosphere," *Benchmark Papers in Ecology, 8,* Hutchinson and Ross, Stroudsburg PA, Dowden

Lieth, H. 1990, "Entwicklung und Ziele der Systemökologie." *Z. Angew. Umweltf. 3(4),* 373–393

Lieth, H., and P. Petrall, P., 1972, *Papers on Productivity and Succession in Ecosystems,* US – Deciduous Forest Biome Memo Report 72-10, UNC Dept. of Botany, Chapel Hill, USA, 51 pp

Lieth, H., and R. H. Whittaker, Eds., 1975, *Primary Productivity of the Biosphere.* Springer, Heidelberg, New York

Lieth, H., and B. Markert, 1990, *Element Concentration Cadasters in Ecosystems,* VCH Publishers, Weinheim, New York

Malyshev, L. I., 1975, "The Quantitative Analysis of Flora: Spatial Diversity, Level of Specific Richness, and Representativity of Sampling Areas," *Botanicheskij Zhurnal, 60,* 1537–1550

Markert, B., Ed., 1993, *Plants as Biomonitors,* VCH Publishers, Weinheim, New York

Markert, B., 1994, "The Biological System of the Elements (BSE) for Terrestrial Plants (Glycophytes)," *Sci. Total Environ., 155,* 221–228

Markert, B., and H. Lieth, 1995, "Instrumental Multielement Analysis in Plant Materials—A Modern Method in Environmental Chemistry and Tropical Systems Research," Abstract. *Biosphere and Atmospheric Changes. 12th International Symposium on Environmental Biogeochemistry,* 3–8 September 1995, Rio de Janeiro, Brazil

Margalef, R., 1968, *Perspectives in Ecological Theory,* University of Chicago Press, Chicago, London

Odum, E. P., 1971, *Fundamentals of Ecology,* 3rd Edn., Saunders, Philadelphia, London, Toronto

Ostendorf, B., and H. Lieth, 1982, "The Computer Drawn Climate Diagrams," *Selected Climatic Data for a Global Set of Standard Stations* (M. J. Müller, Ed.), Task for Vegetation Science 5, Dr. W. Junk, The Hague, Boston, London

Ovington, J. D., and D. Heitkamp, 1960, "The Accumulation of Energy in Forest Plantations in Britain," *Ecology, 48,* 639–646

Ovington, J. D., and D. B. Lawrence, 1967, "Comparative Chlorophyll and Energy Studies of Prairie, Savanna, Oak Forest, and Maize Field Ecosystem," *Ecology, 48,* 515–524.

Patten, B. C., 1959, "An Introduction to the Cybernetics of the Ecosystem. The Trophic-Dynamic Aspect," *Ecology*, *40*, 221–231

Schmithuesen, J., 1976, *Atlas zur Biogeographie*, Bibliographisches Institut, Mannheim, Germany

Schröter, W., K.-H. Lautenschläger, and H. Wibrak, 1984, *Taschenbuch der Chemie*, Verlag Harry Deutsch, Thun, Frankfurt/M., Germany

Schultz, J., 1995, *The Ecozones of the World*, Springer, Heidelberg, New York

Streit, B., 1991, *Lexikon Ökotoxikologie*, VCH Publishers, Weinheim, Germany

Walter, H., 1955, "Klimadiagramme als Mittel zur Beurteilung der Klimaverhältnisse für ökologische, vegetationskundliche und landwirtschaftliche Zwecke," *Berichte der Deutschen Botanischen Gesellschaft*, *68*, 331–344

Walter, H., 1958, Klimatypen dargestellt durch Klimadiagramme, *Geographisches Taschenbuch 1958/59*, M. Steiner, Wiesbaden, Germany, pp. 540–543

Walter, H., 1984, *Vegetation und Klimazonen*, 5th Edn., Ulmer, Stuttgart, Germany

Walter, H., and H. Lieth, 1960, *Climate Diagram World Atlas*, Fischer, Jena, Germany

Walter, H., and S.-W. Breckle, 1991, *Ökologische Grundlagen in globaler Sicht*, Fischer, Stuttgart, Germany

Weigert, R. G., and F. C. Evans, 1964, "Primary Production and Disappearance of Dead Vegetation on an Old Field in South-Eastern Michigan," *Ecology*, *45*, 49–63

2.7.2 Literature List for Element Concentrations in Halophytes (Table 2.1, Column 10)

Ayoub, A. T., 1994, "Some Features of Salt Tolerance in Senna (*Cassia acutifolia*), in Sudan," in *Task for Vegetation Science 32* (V. R. Squires, and A. T. Ayoub, Eds.), Kluwer Academic Publishers, Dordrecht, Boston, London, pp. 297–301

Badger, K. S., and I. A. Ungar, 1990, "Effects of Soil Salinity on Growth and Ion Content of the Inland Halophyte *Hordeum jubatum*," *Bot. Gaz.*, *151*(3), 314–421

Bayoumi, M. T., and H. M. El Shaer, 1994, "Impact of Halophytes on Animal Health and Nutrition," in *Task for Vegetation Science 32* (V. R. Squires, and A. T. Ayoub, Eds.), Kluwer Academic Publishers, Dordrecht, Boston, London, pp. 267–272

Böer, B., 1991, *Vegetationskartierung auf den Inseln Abu Dhabi und Rafiq mit Erfassung wichtiger ökologischer Parameter*, MS Thesis, Universität Osnabrück, Germany

Brevedan, R. E., O. A. F. Fernandez, and C. B. Villamil, 1994, "Halophytes as a Resource for Livestock Husbandry in South America," in *Task for Vegetation Science 32* (V. R. Squires, and A. T. Ayoub, Eds.), Kluwer Academic Publishers, Dordrecht, Boston, London, pp. 175–199

El Shaer, H. M., and E. A. Gihad, 1994, "Halophytes as Animal Feeds in Egyptian Deserts," in *Task for Vegetation Science 32* (V. R. Squires, and A. T. Ayoub, Eds.), Kluwer Academic Publishers, Dordrecht, Boston, London, UK, pp. 281–284

Flowers, T. J., and D. Dalmond, 1992, "Protein Synthesis in Halophytes: The Influence of Potassium, Sodium and Magnesium In Vitro," *Plant Soil*, *146*, 153–161

Gheler Arce, E., 1992, *Mineralstoffgehalte von Salzböden und Halophyten des bolivianischen Altiplano*, PhD Thesis, Universität Göttingen, Germany

Gihad, E. A., and H. M. El Shaer, 1994, "Utilization of Halophytes by Livestock on Rangelands: Problems and Prospects," in *Task for Vegetation Science 32* (V. R. Squires, and A. T. Ayoub, Eds.), Kluwer Academic Publishers, Dordrecht, Boston, London, pp. 77–96

Glenn, E. P., R. S. Swingle, J. J. Riley, C. U. Mota, M. C. Watson, and V. R. Squires, 1994, "North American Halophytes: Potential Use in Animal Husbandry," in *Task for Vegetation Science 32* (V. R. Squires, and A. T. Ayoub, Eds.), Kluwer Academic Publishers, Dordrecht, Boston, London, pp. 165–174

Jayasekera, R. L. 1988, *Growth Characteristics and Uptake of Minerals of the Two Mangrove Species Rhizophora mangle L. and Rhizophora mucronata LAMK. under Different Environmental Conditions*, Ph.D. Thesis, Universität Osnabrück, Germany

Jin, Q., 1994, "*Alhagi sparsifolia*: A Potentially Utilizable Forage in Saline Soil," in *Task for Vegetation Science 32* (V. R. Squires, and A. T. Ayoub, Eds.), Kluwer Academic Publishers, Dordrecht, Boston, London, pp. 285–288

Koocheki, A., and M. N. Mohalati, 1994, "Feed Value of Some Halophytic Range Plants of Arid Regions of Iran," in *Task for Vegetation Science 32* (V. R. Squires, and A. T. Ayoub, Eds.), Kluwer Academic Publishers, Dordrecht, Boston, London, pp. 249–253

Kürschner, H., 1983, "Vegetationsanalytische Untersuchungen an Halophytenfluren Zentralanatoliens (Türkei)," *Beihefte zum Tübinger Atlas des Vorderen Orients: Reihe A, Naturwiss., Nr 11*, Dr. Ludwig Reichert Verlag, Wiesbaden, Germany

Lieth, A. F., 1994, *Use of Seawater for Growth and Productivity of Halophytes in the Gulf Region*, MS Thesis, Faculty of Science of the United Arab Emirates University

Lieth, H., R. Jayasekera, and B. Markert, 1987, "The Application of Element Concentration Cadastres to Different Ecological and Ecophysiological Problems," in *Proceedings of the Third Italo-Hungarian Symposium on Spectrochemistry* (S. Caroli, G. Rossi, E. Sabioni, and K. Zimmer, Eds.), Commission of the European Communities Joint Research Center Ispra, 8–12 June, 1987, pp. 71–86

Markert, B., and R. Jayasekera, 1987, "Elemental Composition of Different Plant Species," *J. Plant Nutrition, 10*(7), 783–794

Miyamoto, S., E. P. Glenn, and N. T. Singh, 1994, "Utilization of Halophytic Plants for Fodder Production with Brackish Water in Subtropic Deserts," in *Task for Vegetation Science 32* (V. R. Squires, and A. T. Ayoub, Eds.), Kluwer Academic Publishers, Dordrecht, Boston, London, pp. 43–75

O'Leary, J. W., and E. P. Glenn, 1994, "Global Distribution and Potential for Halophytes," in *Task for Vegetation Science 32* (V. R. Squires, and A. T. Ayoub, Eds.), Kluwer Academic Publishers, Dordrecht, Boston, London, pp. 7–17

Retana, J., D. R. Parker, C. Amrhein, and A. L. Plage, 1993, "Growth and Trace Element Concentrations of Five Plant Species Grown in a Highly Saline Soil. *J. Environm. Qual., 22*, 805–811

Stichting Marine Cultures Oosterschelde, 1992–1995, "Saline Crops. A Contribution to the Diversification of the Production of Vegetable Crops by Research on the Cultivation Methods and Selection of Halophytes," *SMCO–NIOO–UG–VUB–IBET, Annual Reports, 1992–1995*, Foundation for Marine Cultures Oosterschelde, Wilhelminadrop, The Netherlands

Swingle, R. S., E. P. Glenn, and J. J. Riley, 1994, "Halophytes in Mixed Feeds for Livestock," in *Task for Vegetation Science 32* (V. R. Squires, and A. T. Ayoub, Eds.), Kluwer Academic Publishers, Dordrecht, Boston, London, pp. 97–100

Wilson, A. D., 1994, "Halophytic Shrubs in Semi-arid Regions of Australia: Value for Grazing and Land Stabilization," in *Task for Vegetation Science 32* (V. R. Squires, and A. T. Ayoub, Eds.), Kluwer Academic Publishers, Dordrecht, Boston, London, pp. 101–113

Zhang, L.-Y., X. Yang, and Z. Yun, 1994, "Halophytes and Halophytic Plant Communities in Inner-Asia," in *Task for Vegetation Science 32* (V. R. Squires, and A. T. Ayoub, Eds.), Kluwer Academic Publishers, Dordrecht, Boston, London, pp. 115–122

2.7.3 Internet Citations

As of 1996 a variety of ecologically relevant data are available through Internet. The following data bases were used:

For additional evaluation of vegetation types a 0.5×0.5 degree global land vegetation cover by **Olsen, J. S., J. A. Watts, and L. J. Allison, 1985, "Major world ecosystem complexes ranked by carbon in live vegetation. Code: ftp://cdias.esd.ornl.gov/pub/ndp 017/***

For the comparison of climatic signals with ecosystem types on land, climatic data are available through historical climatology network: GHCN version 1 and 2 by **Vose, R. S., T. C. Peterson, and R. L. Schmoyer, 1995 ff.** under the code: **ftp://cdiaac.esd.ornl.gov/pub/ndp 041**

3

Sensitivity of Ecosystems and Ecotones

Otto Fränzle (*Kiel, Germany*)

3.1 SUMMARY

In ecology the sensitivity of biota or ecosystems can be appropriately defined in terms of population, community, and soil stabilities. Generalized theoretical considerations lead to empirically relevant demographic and nondemographic stability measures. These are derived from sufficiently realistic models of community or systems behavior and corresponding field tests on various levels of generality. Soil sensitivity is, in an ecotoxicological

Ecotoxicology, Edited by Gerrit Schüürmann and Bernd Markert.
ISBN 0-471-17644-3   © 1998 John Wiley & Sons, Inc. and Spektrum Akademischer Verlag.

respect, defined in terms of essential ecological functions, i.e., the regulatory, site (habitat), and productive functions of the structurally complex soil cover. Terrestrial, lotic, lentic, and wetland ecotones play an essential role in coupling ecosystems, and consequently their stability and resilience properties merit particular attention. In aquatic ecosystems stress is usually defined as a measurable change of the state of organisms, as induced by an environmental change, which renders the individual, population, or community more vulnerable to further adverse impacts. In both aquatic and terrestrial ecosystems K-selected populations with their high competitive capacity, high inherent survivorship but low reproductive rates are normally resistant to disturbances, but once perturbed they have little possibilities to recover, i.e., low resilience. The r-selected populations, by contrast, have less resistance but distinctly higher resilience. This inverse relationship between resistance and resilience is of particular importance in an ecotoxicological context.

3.2 INTRODUCTION

In ecology the sensitivity of biota or ecosystems to disturbances may be defined in terms of population, community, and soil stability. Questions of persistence and the probability of extinction are of particular importance in this connection, and hence these measures are also frequently defined as aspects of stability. Stability denotes the ability of a system to return to an equilibrium state after a limited disturbance, while persistence or buffering capacity is an indication of resilience in the sense defined by Holling (1976), i.e., the capacity to absorb changes of state and driving variables and parameters. The more rapidly a system returns to equilibrium and with the least fluctuations, the more stable it is; this means, in terms of thermodynamics, that stability is coupled with a (relative) minimum of entropy production (Fränzle 1993; Prigogine 1976). Further differentiations lead to various notions of community stability and nondemographic stability measures.

Two questions are basic in this context: What makes certain species, populations, communities, or ecosystems vulnerable to particular kinds of change or impact? How far do ecosystems respond as entities, and to which extent, in contrast, is their response no more than the combination of the individual responses and diverse strategies of their component species?

3.3 POPULATION AND COMMUNITY STABILITY

3.3.1 Different Notions of Stability

A distinction may be made between local and global stability. *Local stability* or stability in the vicinity of an equilibrium point describes the tendency of a community to return to its original state when subjected to a minor perturbation. *Global stability* describes this tendency when the community is subjected to a major disturbance. The graphical visualization of solutions of the equations of population dynamics for an m-species community may be represented on some m-dimensional surface, where each point marks a set of populations. In this visualization equilibrium situations are in principle characterized by those points where the surface is flat; equating the configuration with that of a landscape, this means in customary geographical notation on hilltops and in valley bottoms. The hilltop equilibria, however, are obviously labile, i.e., unable to survive the smallest displacement; only the valley bottoms are stable configurations.

Looking beyond the realm of linearized stability in the immediate neighborhood of equilibrium points, for which straightforward mathematical tools exist, the situation becomes more complicated. The appropriate representation of the *global stability* of a system implies recourse to nonlinear equations of population biology. Again in geographical terms, such a global analysis aims at comprehending the stability of an entire landscape, and not only just that of the immediate vicinity of equilibrium points. This involves an appropriate recognition of the fact that real-world environments are uncertain and stochastic, which means that the corresponding environmental parameters in the model equations exhibit random fluctuations (Begon et al. 1990; Gigon 1983; van der Maarel 1976).

This leads to a yet more general meaning of stability, termed *structural stability*, which refers to the qualitative effects upon solutions of the model equations which are due to gradual variations in the model parameters. Thus, a system may be considered structurally stable if these solutions change in a continuous manner. Conversely, a system is structurally unstable if gradual changes in the system parameters, e.g., alterations in site factors of a community, produce qualitatively discontinuous effects (cf. Fränzle 1993; Jørgensen 1990; van der Maarel 1976).

3.3.2 Demographic Measures of Stability

Single Populations

Beginning with the behavior of a single population $N(t)$, where N is the number of individuals (or sometimes their biomass) at time t, population growth can be characterized by the familiar logistic

$$dN/dt = rN(1 - N/K) \tag{3.1}$$

K may designate the carrying capacity of the environment, as determined by food, space, predators, etc.; r is the intrinsic growth rate, free from environmental constraints. The specific form of Equation 3.1 is not to be taken seriously: it is rather representative of a wide class of population equations with density-dependent, nonlinear regulatory mechanisms (cf. May 1975, 1981). In any such dynamical system there is a "characteristic return time," T_R, which indicates the time the population needs to return to equilibrium, following a disturbance. It ensues from Equation 3.1 that $T_R = 1/r$.

In terms of the deliberately oversimplified notions of r and K selection (MacArthur 1972), these features permit some generalizations of the life history strategies of species. Organisms of a K-selected species "see" their environment as relatively stable and predictable; consequently, the population is usually around its globally stable equilibrium value, $N^* = K$. Emphasis is on *population statics*, which means that K strategists tend to be good competitors and to have fewer offspring but invest more time and energy in raising them. Conversely, an r-selected organism "sees" its environment as relatively unstable and unpredictable, is usually at low population values, but undergoes episodes of boom and bust under favorable conditions. Thus the evolutionary pressures here are for opportunism, i.e., for large r to exploit the transient good times, producing many offspring, few of which can expect to mature. Hence, the emphasis is on *population dynamics*. In reality there is no such clear-cut distinction between r- and K-selection; both are no more than the opposite ends of a continuum.

In the real world regulatory effects necessarily operate with a certain time delay, the characteristic magnitude of which may be denoted by T. Their incorporation into Equation 3.1 leads to a generalization of the logistic

$$dN/dt = rN[1 - N(t - T)/K] \tag{3.2}$$

The dynamical behavior of this equation depends on the relative magnitude of the two time scales, T and $T_R = 1/r$. If the time delay T is short in comparison to the characteristic return time T_R, disturbances will be damped monotonically back to the above equilibrium point $N^* = K$. As T approaches T_R, there is a tendency for the inherent regulatory mechanisms to produce overshoot and overcompensation, i.e., an oscillatory return to the equilibrium point. Finally, as T becomes significantly larger than T_R, the pattern of overcompensation leads to self-sustaining stable limit cycles. The amplitude and period of the oscillations of population density $N(t)$ are determined uniquely by the parameters in Equation 3.2.

Difference equations admit of more realistic complications than the above differential equations. First, they more appropriately describe the fact that for many plant and animal species generations do not overlap. Second, they show that the regular pattern of stable cycles can give way to apparently chaotic fluctuations, if the nonlinearities are sufficiently severe. For further variations on the theme of relative time scales the reader is referred to May (1981).

Here it may suffice to consider some consequences for the response of a population under variable environmental conditions in terms of stability and resilience. In an unpredictable environment, r selection is advantageous to recover from bad times and to exploit the good ones. But largish r, and consequently short T_R, condemns the population to overcompensation which is unfavorable to population regulation, thus exacerbating the degree of environmental unpredictability. Conversely, relatively small r values imply a long response time; thus the population may better average over environmental variations, but with the disadvantage of increased sensitivity, i.e., only slow recovery from disturbances.

Two Populations

An extension of the preceding model helps to elucidate some basic features of two populations interacting with variable intensity as prey-predator, competitors, or mutualists. The classical model for a highly simplified prey-predator system are the Lotka-Volterra differential equations which were modified later to more realistically include density dependence and to allow for effects such as saturation in the predator's capacity to respond to increasing prey densities (May 1981).

More complicated model variations allow for predators adaptation to increasing prey numbers. In vertebrate populations such a functional response is stabilizing at low prey densities, but destabilizing at high densities. If the environmental carrying capacity for a prey, K, is distinctly higher than the equilibrium prey density, the stabilizing elements contributed to the system dynamics by the prey density dependence is likely to be weak. Stability studies by May (1975) have further shown a propensity toward stable cycles when the growth rate of the prey population exceeds that of the predators.

Modelling the competition aspect in the behavior of a two-population biocenosis leads to extensions of the above single-species logistic equation:

$$dN_1/dt = r_1 N_1 \left[1 - (N_1 + \alpha_{12} N_2)/K_1 \right] \tag{3.3}$$

$$dN_2/dt = r_2 N_2 \left[1 - (N_2 + \alpha_{21} N_1)/K_2 \right] \tag{3.4}$$

K_1 and K_2 here denote the carrying capacities of the environment, as perceived by the species 1 and 2, respectively; r_1 and r_2 are the corresponding intrinsic growth rates; α_{12} is a competition coefficient measuring the extent to which species 2 presses upon the resources exploited by species 1; α_{21} is the corresponding coefficient for the effect of species 1 on species 2 (May 1981).

Characterizing sensitivity in terms of stability the solution of Equations 3.3 and 3.4 indicates the possibility of a stable equilibrium point, if intraspecific competition is stronger than interspecific competition (i.e., $\alpha_{12} \alpha_{21} < 1$). If, however, interspecific competition is stronger than intraspecific ($\alpha_{12} \alpha_{21} > 1$), no stable coexistence of both is possible. Also in the case when the two species use the resources in an identical manner (i.e., $\alpha_{12} = \alpha_{21} = \alpha_1$; $K_1 = K_2$), a coexistence is not possible. Other models lead to similar conclusions (Nunney 1980).

In the case of host–parasitoid interaction, the searching behavior of the parasitoid in a spatially and temporally inhomogeneous environment is particulary important. Wiegand and Wissel (1994) developed a patch-dynamics model which shows that a slight desynchronization between host and parasitoid populations leads to coexistence while perfect synchronization causes instability. In spatial terms the model confirms the empirical findings that stability increases with increasing aggregation of parasitoids. If hosts stay in the patch where they hatch, this leads to inherent fluctuations and strongly chaotic behavior. Coexistence becomes possible if the system is able to smooth the strong host fluctuations in the individual patches. This implies that parasitoids avoid patches with low host density and that the time spent by a parasitoid for searching appropriate patches is relatively short.

A consequence of these theoretical reflections is the "competitive exclusion principle" which says that no two species can stably coexist provided they make their livings in identical ways or, in other terms, they cannot occupy the same niche. A closer scrutiny of these statements leads to the semantic question of identicalness and to the pragmatic one of which are the limits to niche overlap and the limits to similarity among competing species.

In comparison with two-species prey-predator and competitive systems, mutualistically interacting populations have received distinctly less attention. "Minimally realistic models" (May 1981) for two such mutualists must allow for saturation in the magnitude of at least one of the reciprocal benefits. The resulting equilibrium situation is punctiform and tends to be less stable insofar as perturbations are damped more slowly than in two-population systems without mutualistic interaction (Hirsch and Smale 1974; Whittaker 1975).

Real-world examples of mutualistic systems which have evolved to the point where at least one of the partners finds the other necessary for its existence, are plants and specifically adapted seed dispersers or ants and acacias. The long-term persistence of such systems will be favored by stable and predictable environments such as those provided by the tropical lowland rainforest (Temple 1977; Hill and Blackmore 1980; Bristow 1981).

Community Behavior

Multispecies generalizations of the preceding considerations lead to models of community behavior. They are, in mathematical terms, necessarily much more complicated if formulated as systems of simultaneous differential equations, since already 3-dimensional systems frequently display a rich dynamical complexity appropriately reflected in strange attractors. Therefore this section is limited to a brief overview of existing models, followed by a more detailed description of the structure of real communities and resilience conditions.

Biodiversity and Stability of Model Communities. An important issue of theoretical ecology is how biodiversity relates to community structure and which community-level properties emerge from the disparate interactions of organisms, populations, and site qualities. Possible examples of such "emergent properties" include trophic and guild structures, stability, resilience, and successional stages. A generalized offshoot of the concept of emergent properties is the notion that biological systems are hierarchically organized, with new properties at each level of the hierarchy (O'Neill et al. 1989; Allen and Starr 1982; Solbrig and Nicolis 1991). According to this view, diversity is better understood if ecosystems or biota are decomposed hierarchically so that each process can be viewed as a stabilizing or disruptive factor at each level in a hierarchy of temporal and spatial scales.

During the 1950s and 1960s, the "conventional wisdom" in ecology (Begon et al. 1990) was that increased complexity within or diversity of a community leads to increased stability (Elton 1958; MacArthur 1955). May (1972), however, came to contrasting conclusions by means of model food webs comprising a number of interacting species. The term β_{ij} was used to measure the effect of species j's density on species i's rate of increase. Thus β_{ij} would be zero when there was no effect, while both β_{ij} and β_{ji} would be negative for two competing species, and β_{ij} would be positive and β_{ji} negative for a predator (i) and its prey (j). Setting all self-regulatory terms (β_{ii}, β_{jj}, etc.) in his randomly connected networks at -1, May distributed all other β-values at random, including a certain number of zeros. Thus the cybernetic webs serving as ecosystem models could be described by three parameters: S, the number of species; C, the connectance of the web, i.e., the fraction of all possible pairs of species interacting directly ($\beta_{ij} \neq 0$); and $\bar{\beta}$, the average interaction strength, i.e., the average of the non-zero β values, disregarding sign.

The comparative analysis of these networks showed that they were only likely to be stable, i.e., the populations thus represented would return to equilibrium after a small disturbance if:

$$\bar{\beta}(SC)^{1/2} < 1 \qquad (3.5)$$

In other words, May's model suggests, like others, that "too rich a web connectance or too large an average interaction strength leads to instability. The larger the number of species, the more pronounced the effect." This inference is clearly in contradiction to the above conventional wisdom, but it indicates that there is no general, unavoidable connection between complexity or diversity and community stability (cf. Wissel 1981). Another question is how far May's result is an artifact arising out of the particular characteristics of the model and the interpretative techniques applied (Begon et al. 1990; Jørgensen 1990), and to which extent it can be corroborated by stability-oriented diversity analyses in the field (Section 3.3.3).

The picture also alters if, instead of focussing on local stability and correspondingly minor disturbances, larger perturbations are considered or "species-deletion stability," which is particularly interesting from an ecotoxicological point of view. Under the assumption of the deletion of one species of a community owing to a persistent impact, the system is said to be species-deletion stable if all of the remaining species are retained at locally stable equilibria. In the simple six-species community of Figure 3.1, containing two top and two intermediate predators and two basal species (either plants or categories of dead organic matter), interaction strengths were defined at random, while connectance was

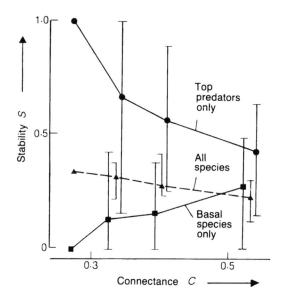

Figure 3.1 Species-deletion stability plotted against connectance (C) for six-species models. Results are for systems where (i) only top predators are deleted, (ii) only basal species are deleted, and (iii) all species are deleted in turn. The bars show the range in species-deletion stability for all models of a given connectance. After Begon et al. (1990).

varied systematically; thus complexity could be equated with connectance. Again, stability generally decreased with increasing diversity; but this trend was reversed when basal species or elements were eliminated. However, within the simulated small subset of stable communities, resilience actually increased with complexity.

Other theoretical approaches to the stability problem are due to McMurtrie (1975), Pimm (1982), and Ulanowicz (1986), who established criteria for the stability of ecosystems under certain restrictive conditions, such as zero-sum games. In particular, they have shown that ecosystems can become unstable against weak perturbations when the complexity increases beyond a certain value, although the transition is not always sharp (Cohen and Newman 1984). Deriving sensitivity measures from community descriptors Prahl-Wostl and Ulanowicz (1993) found "ascendency" (A) to be an appropriate index that includes both system size and structure

$$A = \sum_{j=0}^{n} \sum_{i=1}^{n+1} T_{ji} W_{ji} \qquad (3.6)$$

where T_{ji} denotes the physical magnitude of flow summed over time, and W_{ji}, sensitivity of A to a minute change in T_{ji}. Thus, a flow is chracterized by high structural weights if it is specific for a particular network function of the system considered.

The results of model simulations prove the usefulness of this approach in order to define the importance of a species for the dynamics of a community or an ecosystem (Pahl-Wostl 1994). Irrespective of a rather low annually averaged biomass, a species may be crucial during a certain period of the year, and its decrease (due to external or internal disturbance) has a pronounced effect on the whole system because it cannot be compensated by other species.

The application of whole system descriptors in simulation models was also advocated by others. Jørgensen (1992) used exergy as a goal function in models to account for changes in species composition. Bachas and Huberman (1987) related the complexity of hierarchical structures, as measured by their diversity (Ceccato and Huberman 1988), to dynamical behavior. The result suggests that there might be a quantifiable relation between the diversity of a hierarchical structure and its stability. Furthermore, results of Doreian (1986) and Briand and Cohen (1987) suggest that the effective dimensionality of the space in which the interactions occur, and the overlaps in food webs, contribute to the nature of the structural representation of the community. Hence, it may be concluded that their stability will likewise be affected. This is in agreement with the results of a comparative model analysis of unstructured cooperative interactions between arbitrary species on the one hand and ecosystems with pyramidal organization on the other.

Returning to the original problem set out by May (1972), this author derived a condition of stability for more general systems as a function of species diversity and the strength of interactions. The stability of such systems is determined by the behavior of the largest eigenvalue of matrices governing the response of the system to small perturbations. As a result May and coworkers showed in the case of nonhierarchical organizations how the removal of a zero-sum game condition can lead to a further reduction in stability. Examining hierarchical ecologies, they demonstrate that these hierarchical structures are intrinsically more stable than unstructured ones (Hogg and Huberman 1989).

The conflicting results amongst the models developed so far must be considered in the light of model structure, in particular the number and composition of elements, i.e., species simulated. In comparison to real communities with hundreds or thousands of species (Urban et al. 1987) and highly variable degrees of interaction, most models are extremely "impoverished" in species. Furthermore, the models quite often refer to randomly constructed communities, while real communities and ecosystems are far from randomly constructed and normally display a complex hierarchical structure (Fränzle 1993; Solbrig and Nicolis 1991). Unstable communities are liable to collapse when they experience environmental conditions which reveal their instability; but the range and predictability of conditions may vary markedly from place to place. Under stable and predictable site conditions, a community will only experience limited fluctuations, and thus even a dynamically fragile one may still persist. By contrast, in a variable and largely unpredictable environment, only dynamically robust communities are likely to persist.

Stability-oriented Biodiversity Analyses of Real Communities. Following May's approach studies have been performed on the general importance of dynamic stability constraints by examining the relationship between S, C, and β in real communities. It follows from inequality (Equation 3.5) that increases in S will lead to decreased stability unless there are compensatory decreases in C or β. Since data on interaction strengths for whole communities are unavailable, it is usually assumed that β is constant. In such a case, communities with more species would only retain stability if there is a corresponding decline in average connectance, which means that the product SC should remain approximately constant.

Rejmanek and Stary (1979) studied geographically distinct plant-aphid-parasitoid communities in central Europe and found a decrease in C with an increase in S, with SC lying between 2 and 6. Evaluating literature, Briand (1983) analyzed 40 terrestrial, freshwater, and marine food webs and calculated a single value for connectance on the basis of both predator-prey and presumed competitive interactions, i.e., C_{max}. Once again, connectance decreased in an approximately hyberbolic way with species number, and SC was about 7.

McNaughton's field test (1977) of May's inference, carried out in 17 grassland stands in Tanzania's Serengeti National Park, showed "that both average interaction strength and connectance declined as species richness of the grassland increased. The correlation was somewhat stronger for interaction strength than for connectance." Hence, it may be surmised, assuming the grassland communities analyzed are representative biocenoses, that species-poor stands are likely to be characterized by strong interactions among these species, while species interacting with many others do so only relatively weakly. Diffuse competition obviously increases with the number of species. In addition, McNaughton's findings suggest that communities have a block-like or patchy organization consisting of species interacting among themselves but only little with species in other blocks. An analogous situation is found in tropical rainforests, which is described later.

McNaughton (1977) also tested the prediction that complex communities are less likely to return to their state prior to perturbation. In the first experiment, the perturbation resulted from addition of plant nutrients, while in the second, it involved the action of grazing buffaloes. The results summarized in Table 3.1 indicate that each perturbation significantly reduced the diversity of the species-rich but not the species-poor community.

Van den Bergh (1981) studied the dynamics of partly grazed and partly hayed perennial grassland near Wageningen. The main species were: *Festuca rubra, Agrostis (tenuis* and *stolonifera), Anthoxantum odoratum, Holcus lanatus, Rumex acetosa, Trisetum flavescens, Alopecurus pratensis,* and *Lolium perenne.* The diversity analysis showed, in terms of frequency percentages, in addition to very rapid fluctuations (from year to year), long-lasting fluctuations (about 10 years) and comparatively slow trends (over 20 years). Fluctuations were much more pronounced on the hayed plots than on the grazed ones. Higher fertility on the P, K plots stimulates productive grasses with the result that less productive

Table 3.1 Influence of nutrient addition on species richness, equitability ($H/\ln S$), and diversity in two fields; and influence of grazing by african buffalo on species diversity of two vegetation plots

	Control plots	Experimental plots	Statistical significance
	Nutrient addition		
Species richness per G.5-m² plot			
Species-poor plot	20.8	22.5	n.s.*
Species-rich plot	31.0	30.8	n.s.
Equitability			
Species-poor plot	0.660	0.615	n.s.
Species-rich plot	0.793	0.740	$p < 0.05$
Diversity			
Species-poor plot	2.001	1.915	n.s.
Species-rich plot	2.722	2.532	$p < 0.05$
	Grazing		
Species diversity			
Species-poor plot	1.069	1.357	n.s.
Species-rich plot	1.783	1.302	$p < 0.005$

Source: After McNaughton (1977).
*n.s. not significant.

species such as *Agrostis* and *Anthoxanthum* are crowded out. The most obvious difference between the grazed and the hayed *P, K* plots is the increase of *Lolium* on the former and *Dactylis* on the latter. The most dynamic picture is obtained from the *N, P, K* plots, especially from the hayed ones.

It ensues from part of these analyses and studies on the species diversity of tropical rainforest ecosystems in particular that a climax community is more likely to have both low fluctuations in composition (i.e., high stability) and low resilience, the more homogeneous its environment in space and time is. Figure 3.2 summarizes the diversity and abundance spectra of numerous neo- and paleotropical rainforest communities in relation to soil properties (Fränzle 1994a).

The diagram clearly shows that in most cases rainforest stands on highly nutrient depleted ferralsols, acrisols, and podzols of tropical lowlands attain diversity indices above 90% of the theoretical maximum. However, a further differentiation according to local nutrient status and water budget is not possible with the data available. Nevertheless stand 38 (Borneo) is of particular interest since it is a representative of a forest which developed on a tropical bog; its exceptionally low index value allows the conjecture that its nutrient supply has fallen below a critical threshold.

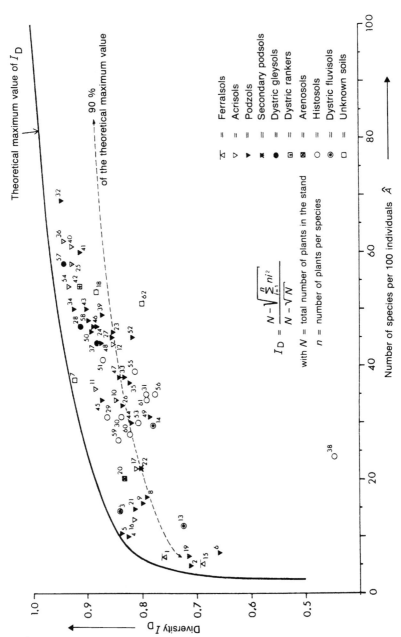

Figure 3.2 Diversity and abundance spectra of neo- and paleotropical rainforest communities.

This specific correlation pattern is in obvious contrast to the first of Thienemann's (1954) ecological principles which were first propounded in 1920 but repeatedly discovered and put forward as new ideas: The greater the diversity of site conditions, the larger is the number of species which make up the biotic community. In fact this "normal" diversity/nutrient relationship holds only for relatively young (postglacial) ectropical ecosystems; for tropical systems developed during periods of hundred thousands of years under conditions of (relative) climatic and geomorphic stability the inverse relationship is valid for thermodynamic reasons (Fränzle 1994a). Since a full account cannot be given in the present sensitivity-related context, the essential evolutionary feedback mechanisms are enumerated which, in variable combination, most effectively contribute to reduce the entropy production of the rainforest biomes and to simultaneously increase their entropy export into the environment. Since, for elementary thermodynamic reasons, stability of an open system is coupled with a (relative) minimum of total entropy (cf. Prigogine 1976), this state of biome evolution is equivalent to a (relative or overall) maximum of system stability.

The first of these adaptive mechanisms developed by the rainforest stands with their mosaic-like structure owing to the patchiness of rejuvenation processes is a dense root mat in intimate contact with litter and with root tips growing upwards into the fallen litter. Mycorrhizal associations benefit their host plants by effectively increasing the absorptive surface of the root system, thus providing for increased uptake of nutrients. The dense fabric of predominantly fine roots also plays an important part in exchange and adsorption of nutrients from throughfall water. In addition, the structure of the foliage favors the use of nutrients by a long active life, the retransport of certain nutrients before leaf shedding, and a high polyphenol content and coriaceous nature which both reduce herbivory. Another factor of high importance is the multi-layered structure of the forest and the activities of epiphytes and microorganisms on much of the exposed surfaces which together form highly efficient filtering systems scavenging nutrients from rainwater. Finally, biological nitrogen fixation in the root–humus–soil interface appears to play an important role in the nutrient budget (Fittkau 1991).

Even in the simplest cells the normal metabolic pathways imply several thousand complex chemical reactions which must be coordinated by means of an extremely sophisticated functional network. This means that hierarchical order in both a functional and spatiotemporal respect constitutes a further and most powerful negentropic and consequently stabilizing factor.

The effectiveness of these negentropic processes is further enhanced by very efficient entropy fluxes related to the transpiration and nocturnal respiration of plants. The molal entropy of H_2O increases from 63 $J\,mol^{-1}\,K^{-1}$ (liquid) to 189 $J\,mol^{-1}\,K^{-1}$ (gas) in the course of evaporation, and CO_2 has a molal entropy of no less than 214 $J\,mol^{-1}\,K^{-1}$. Consequently, also the reverse process, the photosynthetic fixation of CO_2, is of comparable importance for the negentropy balance of the system (Fränzle 1994a).

In terms of May's stability criterion (Equation 3.5) these findings would indicate a very low degree of connectance and average interaction strength between plants of the same species. This is clearly corroborated by the exceedingly high diversity values of Figure 3.2, showing that often the average density of arborescent species is in the order of magnitude of one specimen/hectare only. With regard to nutrient uptake from the soil this means a well-balanced, low-grade intra- and interspecific competition involving a marked selection of K strategies and the formation of numerous ecological niches. Therefore, it seems more likely that the highly complex rainforest communities with their relatively constant environments are more susceptible to outside, unnatural disturbances than the simpler, more robust communities of ectropical regions.

The example of Vietnam's inland forests is illuminating in this respect, since they still bear the scars of a 10-year herbicide spraying programme by the US Air Force during the Vietnam war. According to official US figures, some 10.3% of these forests, in addition to 36.1% of mangrove forests, 3% of cultivated land, and 5% of other land were sprayed. Thus some 90,000 tons of herbicides and antiplant chemicals were used between 1962 and 1971 (Hay 1983). Most of the spraying was with Agent Orange (i.e., a 1:1 mixture of 2,4-D and 2,4,5-trichlorophenoxy-acetic acid), Agent White (2,4-D and picloram), and Agent Blue (cacodylic acid). Some 20 years after the spraying ended in 1971, there has been little regrowth in areas sprayed three or four times. The fertility of the soils is rapidly further reduced if forests are replaced with grassland or bamboo, as has happened in much of the sprayed region. In particular, there has been a loss of minerals and nitrogen and in some cases a fall in soil pH, all of which inhibit a diversified recolonization.

Defoliation has clearly affected the faunal assemblages, not only in the areas totally defoliated and subsequently converted into grassland, but also in those regions less frequently sprayed where the plants in the upper reaches of the forest were damaged. In both areas, the number of animals and birds has dropped dramatically, and animals are also at considerable risk in some of the forest areas which have been isolated from the main body by herbicide application. Furthermore there is concern that Agent Orange contained a high concentration of the highly toxic contaminant 2,3,7,8-tetrachlorodibenzo-p-dioxin, which is both teratogenic and carcinogenic in animals, and also toxic to humans.

3.3.3 Nondemographic Measures of Stability

It ensues from the introductory remarks on stability and resilience that community and ecosystem stability can also be viewed from perspectives other than demographic ones. The highly different aspects of functioning, e.g., primary productivity, nutrient cycles, and energy and water balance, and composite aspects of structure such as standing-crop biomass appear to be particularly important (Begon et al. 1990).

McNaughton's (1977) studies on Serengeti grassland communities showed that species-rich associations responded to perturbations with a significant drop in species diversity, while species-poor communities did not. In terms of *primary productivity* however, grazing as a natural perturbation reduced the standing-crop biomass in the species-poor grassland much more than that of species-rich communities. Similarly, several studies have revealed a marked influence of *food web* structure on community resilience in response to perturbations of energy and nutrient supplies. O'Neill (1976) considered a three-compartment system consisting of active (photoautotrophic) plant tissue, heterotrophs, and dead organic matter, and found that the rate of change in standing crop depends on the energy transfer between these compartments. With regard to the autotrophic compartments it depends on the input variable net primary productivity and two output variables (fraction consumed by heterotrophs and fraction lost to soil by litterfall) in addition to a great variety of translocation processes. The rate of change of the heterotrophic compartment depends on two inputs (consumption of living plant biomass and dead organic matter) and two outputs (respiratory heat loss and defecation). Finally, as Figure 3.3 shows, the energetic exchange rate of soil organic matter depends on two inputs (litterfall and defecation) and two outputs (physical transport into neighboring ecosystems and consumption by heterotrophs).

On the basis of this simple configuration, and inserting real data from communities representing a pond, a spring and tundra, tropical forest, temperate deciduous forest, and a salt marsh, O'Neill (1976) subjected the models of these communities to a standard perturbation which consisted in a 10% decrease in the initial standing-crop vegetation. The rates of recovery toward equilibrium thus monitored and plotted as a function of the *energy input* per unit standing crop of living tissue are presented in Figure 3.4.

It shows that the flux of energy through the different subsystems has an important influence on community resilience; the higher it is per unit standing

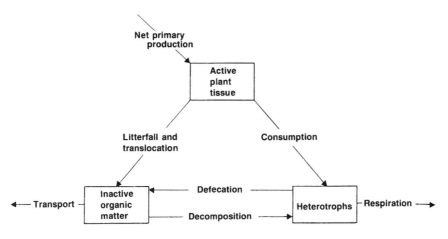

Figure 3.3 Three-compartment model of an ecosystem with tropical energy transfers. After O'Neill (1976).

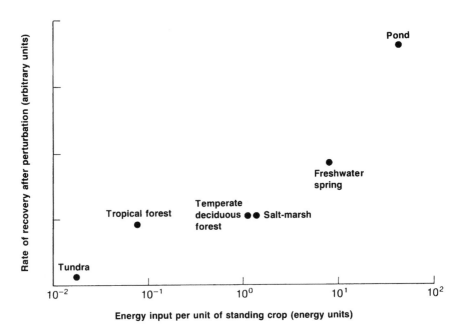

Figure 3.4 The rate of recovery (index of resilience) after perturbation (as a function of energy input per unit standing crop) for models of six contrasting communities. The pond community was most resilient to perturbation, the tundra was least resilient. After O'Neill (1976).

crop, the more quickly the effects of perturbations will be compensated. This, in turn, seems to depend in part on the relative importance of heterotrophs in the system considered.

The pond as the most resilient system has a heterotrophic biomass 5.4 times that of autotrophs, reflecting the high turnover rate of phytoplankton; the least resilient tundra has a heterotroph/autotroph ratio of only 0.004. De Angelis (1980) came to analogous conclusions in terms of residence times of various nutrients in different compartments of woodland communities. Thus, a unit of nitrogen exists in soil for an average of 109 years, is taken up into forest biomass for 88 years, remains in the litter for up to 5 years, and passes through the detrivore compartment in a few days. In contrast to nitrogen, calcium is a less tightly cycled element whose residence times in soil and biomass only amount to 32 and 8 years, respectively.

More detailed long-term investigations in the forest and agroecosystems of the Bornhöved study area (Schleswig-Holstein, Germany) corroborate these findings. Here, the nitrogen deposition amounted to 19.9 and 18.2 kg ha^{-1} a^{-1} in 1989 and 1990, respectively, with an ammonia fraction of 47% (60%) and 23% (7%) of organic N compounds, while nitrate deposition yielded 6.1 and 5.9 kg ha^{-1} a^{-1}.

The nitrogen balance of a field is summarized in Table 3.2.

Table 3.2 Nitrogen balance of a field in the Bornhöved lake district

Process	Amount of nitrogen measured $(kg\,ha^{-1}\,a^{-1})$	
	1989	1990
Initial N content	51[a]	38
Total N input	384	222
NO_3N	30	14
NH_4N	118	64
N_{org}	236	144.5
Mineralization	221[b]	215[b]
Plant uptake	206	121
N_x leakage	110	129
Other N losses[c]	7	7
Residual N_{min} content	39	4.5
Balance	+59	+70

[a] Deduced value. [b] Net mineralization. [c] Denitrification and volatization of ammonia.

The wet deposition of nitrates via canopy flow in a neighboring beech stand (*Asperulo-Fagetum*) of relatively uniform age structure (100 years) amounted to 30.3 kg ha^{-1} a^{-1} in 1989 and 37.5 kg ha^{-1} a^{-1} in 1990, again with a major ammonia fraction. Comparably high were the deposition rates in the soils of a neighboring alder stand (*Carici elongatae-Alnetum*), where an additional nitrogen fixation is due to the symbiotic activity of *Frankia*.

With regard to internal nitrogen fluxes the beech and alder stands differ considerably. In the former case litterfall transported 78 kg N per hectare and year to the soil surface, where the chemical and microbial transformation processes induce seasonal changes in N_{min} content as illustrated in Figure 3.5.

In the light of the site characteristics of this stand (Dystric Cambisol) the figures are indicative of a rather limited N export from the community, which is further corroborated by measurements of the N fluxes in the related seepage water. In the distinctly more resilient alder stand the nitrogen input due to litterfall amounted to 109.4 kg ha^{-1} a^{-1} in 1989 and 104.5 kg ha^{-1} a^{-1} in 1990. In comparison with the above *Asperulo Fagetum* microbial degradation of the litter is very high in the *Alnetum*, and within 18 months almost all of the leaf tissue is degraded.

Werner and Stickan (1983) investigated the biodiversity of a pasture (*Lolio-Cynosuretum*) and the net photosynthesis of the dominant species of this community as indicators of chemical stress. The effects of the test chemicals pentachlorophenol (PCP) and atrazine (A) were determined by means of comparing treated and untreated plots. Changes in the indices of similarity (i.e., Sørensen index and percentage similarity) and the dominance structure show

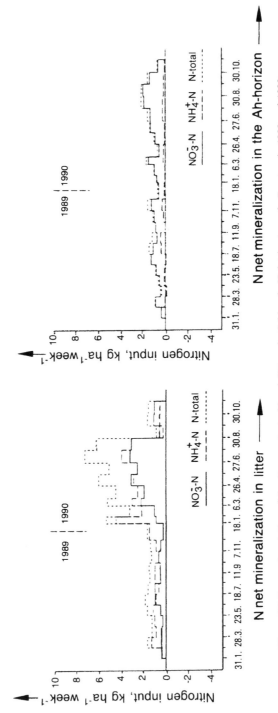

Figure 3.5 Seasonal variation of the N_{min} content of beech litter of a Schleswig-Holstein *Asperulo-Fagetum* in 1989 and 1990.

that PCP causes a disappearance of less dominant species (*Taraxacum officinale, Veronica chamaedrys, Cerastium holosteoides, Medicago lupulina, Bellis perennis*), while A affects more the dominant ones (*Trifolium repens, Achillea millefolium*). Also the analysis of photosynthetic activity displayed characteristic injury dynamics. After application of PCP at a concentration of 0.5 g/m^2 the CO_2 gas exchange decreased in the first three or four days; thereafter a short-term recovery was observed. Different species showed different sensitivities; the same applies to different phenological phases.

3.4 SENSITIVITY OF THE SOIL SYSTEM

Soil is one of the principal regulatory compartments of all terrestrial and benthic ecosystems. Analogous to the preceding stability and resilience-related definitions of sensitivity, the sensitivity of soils to physical and chemical impact may be defined as character and velocity of sequential change in soil properties after a disturbance.

In view of the position of soil in ecosystems its sensitivity should be defined in terms of essential soil functions. In the present ecotoxicological context the ecological functions, i.e., the regulatory, site (habitat), and productive functions, appear more important than others which are therefore simply summarized in Figure 3.6, where arrows indicate structural and functional interrelationships.

The *regulatory function* is an expression of different cascading properties of a soil relating to fluxes of matter and energy (Fränzle 1993). *Filtering capacity* denotes the retention quality with regard to particulate or colloidal matter, while *buffering* means retention of dissolved substances due to chemical or physicochemical processes such as sorption, ion exchange, ionization, hydrolysis, oxidation-reduction, complexation, precipitation, and bioaccumulation. The *transformation capacity* indicates the soil potential for microbial and chemical transformation processes in relation to both natural compounds and xenobiotics.

The *habitat function* describes the physiologically relevant qualities of soil in relation to microbial communities, vegetation, and faunal assemblages. By analogy the *productive function* of soil indicates the potential for growing food and fodder plants; consequently positive or negative effects on this type of function are primarily assessed in economic rather than in ecological terms.

3.4.1 Sensitivity-Related Soil Qualities and Processes

Pedological processes related to regulatory and habitat functions constitute important boundary conditions for energetic and material flux processes in soil and largely define the resultant balances. Soil flora and fauna account for maximally 90% of the transformative capacity of soil; therefore, it follows that soil biota are particularly sensitive and worth protecting (Blume 1992; Fränzle

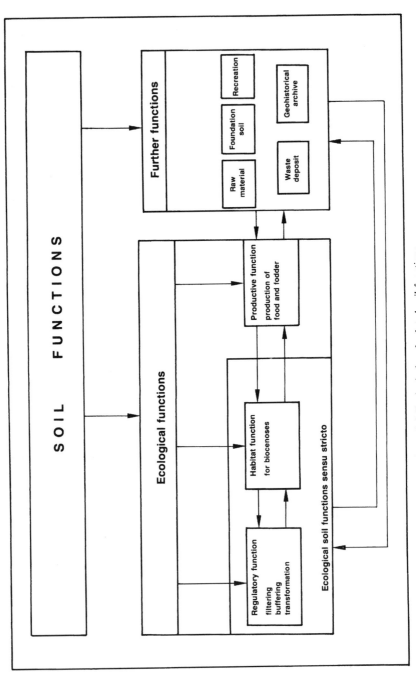

Figure 3.6 Ecological and related soil functions.

1993; cf. Chapter 21). Table 3.3 provides a summary of relevant interrelationships between ecological soil functions, pedological processes, and boundary conditions or controlling variables.

It ensues from Table 3.3 that soil sensitivity to chemical impact is a highly variable property which can only be defined with a resonable amount of practical accuracy when related to the ecological soil functions, pedological processes, and agricultural or silvicultural use patterns. The importance of compound element speciation is derived from the fact that different species will exhibit different mobilities and reactivities in soils and sediments, different availabilities for plant uptake, and different toxicities for organisms.

Mathematical modelling of these processes is a field of current intensive work, and the variety of models has increased considerably during the last two decades. In general, soil/groundwater modelling concepts refer to point source pollution and can be categorized into (1) unsaturated soil, (2) saturated soil and subsoil (groundwater), (3) geochemical, and (4) ranking. The first two categories are designed to describe the behavior of dissolved organic and inorganic chemicals or immiscible fluids for nonaqueous phase compounds; these concepts are comparable in their physicochemical approach and mathematics. The third enters into chemistry and speciation modelling, while the fourth follows a screening approach (Bonazountas 1987; Fränzle 1993; Mattheß 1990).

In view of the high-grade spatial variability of the soil cover and related vegetation complexes the application of such models poses problems. They result from the fact that models are generally calibrated and validated on the basis of a more or less punctiform input data set only, and, therefore, the appropriate areal extrapolation is faced with difficulties.

An adequate solution therefore involves the combination of the model with a geographic information system which maintains the topologic structure of the environment-specific climatologic, hydrologic, and soil data as input to the model and to validate model output. Examples of this assessment approach at both the large-scale and small-scale levels are given in Fränzle (1993) and Reiche (1991).

3.4.2 Ecological Soil Functions as Sensitivity Measures

Using the ecological functions as indicator variables for chemical stress involves the regionalized determination of geogenic and anthropogenic background concentrations of potentially hazardous chemicals and the assessment of different degrees of adverse chemical effects on soils and related biota (Fränzle et al. 1993). Table 3.4 summarizes the essentials of such an evaluation of stress tolerance of soils as ecosystem compartments.

Thus, the preventive threshold value C_p characterizes chemical concentration levels not affecting any essential ecological soil function. The C_{eff} value is indicative of first and low-level adverse effects and is defined by means of ecotoxicological test procedures and a determination of the physicochemical behavior of xenobiotics in soil. By contrast the C_{dis} value indicates long-lasting or permanent disturbances of essential biotic and abiotic soil functions.

Table 3.3 Ecological soil functions related to soil properties and processes

	Regulatory functions		Habitat functions	Productive function
Partial function	*Buffering*	*Filtering*	*Transformation*	
			Substrate for soil, biota, plants, and animals	Economy-oriented production of useful plants
			Transformative and genetic potential of flora and fauna	Increased productivity
			Production of biomass	Sustainable production
Processes	Adsorption, desorption	Mechanical retention of solid and colloidal substances in dependence on pore spectrum	Microbial transformation	Production-related adequate nutrient, water, air, and energy balance
	Precipitation, crystallization			
	Neutralization		Biochemical and abiotic chemical processes	Production-related adequate nutrient, water, air, and energy balance
			Nutrient, water and energy fluxes	
	Biological uptake			

Table 3.3 (Continued)

	Regulatory functions	Habitat functions	Productive function
Controlling variables	Organic, matter (content, composition)	Density and composition of microbial communities	Nutrient content (N, P, K, S, Ca, Mg, …)
	Clay (content, composition)		Base saturation
	Granulometry pore-size spectrum	Soil moisture	
	pH value		C/N ratio
	Size of particulate xenobiotic	Aeration, redox potential	
	Redox potential		
	Organic matter	Temperature	Field capacity
	Soil moisture	pH value	
	Soil temperature	Content and composition of soil organic matter	Water balance
	Hydraulic conductivity		
	Microbial activity		Pore spectrum
	Concentration and chemical character of substances in soil solution[1]		
	Intensity of seepage and groundwater flow		

Source: Fränzle (1993).

[1] Vapor pressure, volatility, water solubility, n-octanol/water partition coefficient.

Table 3.4 Complex ecological threshold values for soil quality characterization

C_p Natural background "clean soil"	C_{eff} Range of tolerable disturbance; no further chemical stress	C_{dis} Ranges of permanent heavy disturbances; changes in land use; hazard assessment and restoration indicated
No reduction of ecological soil functions	First low-level adverse effects on ecological soil functions	Long-lasting or permanent disturbances of essential ecological functions
Multifunctionality: optimum range of ecological soil functions		
No detrimental fluxes of chemicals	No detrimental fluxes of chemicals	Detrimental fluxes of hazardous chemicals into neighboring ecosystems
Optimum conservation of species	Almost complete conservation of species (HC$_5$)	High-grade species deletion (HC$_{50}$)
Site-specific quasi-natural climax communities		
Any form of ecologically indicated land use possible	Gradually increasing reduction of soil functions	Possibilities for land use reduced
		Biotic and abiotic soil functions reduced

Among the extrapolation methods for determining concentrations which give rise to ecologically relevant effects the van Straalen and Denneman (1989) approach appears most appropriate. It defines a hazard concentration (HC_p) which ensures the protection of a certain number of soil-living animals. Thus the approach is explicitly based on resilience considerations using "no observed effect concentrations" (NOECs) as reference values. The hazard assessment involves four steps:

(I) Determination of at least 5 NOECs.

(II) Standardization of NOEC values with regard to a standard soil from at least three taxonomic groups according to the following equation

$$NOEC' = NOEC\ (L, H) \frac{R(25, 10)}{R(L, H)} \tag{3.7}$$

where

NOEC′	=	NOEC of standard soil
NOEC (L, H)	=	experimental NOEC with given clay (L) and organic matter content (H)
R (25, 10)	=	reference value of standard soil with 25% clay and 10% organic matter
R (L, H)	=	reference value on the basis of the clay and organic matter content of the test soil.

(III) Determination of a safety factor T in relation to the different sensitivity of indicator species:

$$T = \exp\left[\frac{3S_m d_m}{\pi^2} \ln\left(\frac{1 - \delta_1}{\delta_2}\right)\right] \tag{3.8}$$

where

S_m	=	standard deviation of ln NOECs
d_m	=	factor dependent on the number of test species
δ_1	=	proportion of nonprotected species, related to defined hazard concentration (HC_p)
δ_2	=	probability to overestimate HC_p
T	=	safety factor for NOEC average.

(IV) Based on this safety factor (T) the hazard concentration HC_p is defined as

$$HC_p = \frac{\exp(x_m)}{T} = \frac{\overline{NOEC}}{T} \tag{3.9}$$

where

m	= number of test species
x_m	= mean value of ln NOECs
$\overline{\text{NOEC}}$	= geometric mean value of NOECs
s_m	= standard deviation of ln NOECs.

The exemplary application of Equation 3.9 to Cd and a test system of *Dendrobaena rubida, Lumbricus rubellus, Eisenia foetida, Helix aspera, Porcellio scaber, Platynothrus peltife*, and *Orchsella cincta* would yield a HC_5 (i.e., $\delta_1 = 0.05$) of 0.16 mg Cd per kilogram soil. In comparison, the present A level of the Dutch "Leidraad Bodemsanering" is equivalent to 0.80 mg Cd per kilogram standard soil.

As long as the C_{eff} value, defined by means of the above procedure, is not exceeded, there will be no essential reduction in ecological soil functions, and consequently the long-term multifunctionality of soil is sustained. In contrast to this threshold the C_{dis} value indicates a distinctly higher concentration of pollutants in soil which exerts long-lasting or irreversible effects on biotic and regulative soil functions. For assessment purposes basically the same approach is foreseen as for the definition of C_{eff} values, i.e., a multispecies determination of HC_p with a p around 50. Under these circumstances also neighboring ecosystems may be affected due to export of primary pollutants or their metabolites. Thus, the C_{dis} value characterizes a threshold which necessitates protective or rehabilitation measures. In dependence on both quality and intensity of land use these may be quite different, including, e.g., limitations of crop production or reduction of fertilizer and pesticide application. In terms of sensitivity this means that a soil is the more sensitive with regard to essential ecological functions, the lower the above threshold values are.

3.5 ECOTONES

3.5.1 Ecotone Typology

Terrestrial and Lotic Ecotones

In view of the ecological importance of ecotones three broad classes of ecotones should be distinguished. *Terrestrial ecotones* may be pragmatically defined as discontinuities between ecosystems. Consequently, the most obvious ecotones in terrestrial ecosystem complexes involve transitions between communities dominated by different life forms. Large-scale examples are transitions from tundra to taiga along thermal gradients or transitions from prairie to forest along continental moisture gradients. A frequent small-scale type is the transition from field to woodlot in agricultural landscapes.

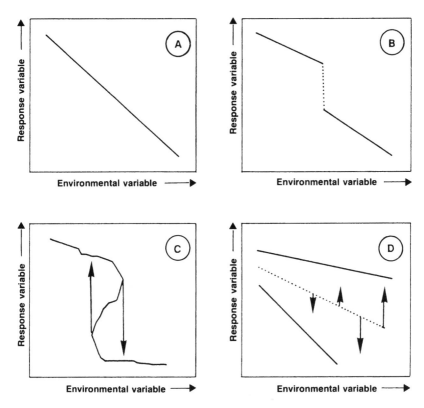

Figure 3.7 Ecosystem response to different types of environmental gradients. (A) Case in which the response of the ecosystem is smooth with respect to the environmental variable. (B) Case in which the response is discontinuous. (C) Case in which the response is folded and results in hysteresis. (D) Case in which the system can have multiple responses. Modified after Naiman and Décamps (1988).

Ecotones are characterized by the nature of physiologically relevant gradients as Figure 3.7 shows. Temporally and spatially variable gradients induce the development of species-poor communities such as the *Agropyro-Rumicion crispi*. In contrast, sharp and stable gradients lead to species-rich communities, e.g., those belonging to the syntagmatic class of the *Trifolio-Geranieta sanguinei* (Westhoff 1974).

The characterization of terrestrial ecotones is highly dependent on contrasts between the different elements defining a landscape. The degree of contrast is necessarily a function of the variables used by the observer to categorize the landscape elements (Fränzle et al. 1987; Urban et al. 1987). Thus, a more precise and generally applicable method to define ecotones is based on areally valid measurements of ecologically relevant parameters and a subsequent multivariate cluster analysis. The result is operationally homogeneous spatial units (patches) separated by distinct boundaries, whose location and geometry depend on the variables used to characterize the primary measuring points or elementary spatial units, respectively, the fusion level of clustering adopted, and finally the scale selected (Fränzle 1994b).

Lotic ecotones can be defined, in an analogous way, as fluvial boundaries. This definition indicates that resource patches are separated by both longitudinal (i.e., upstream/downstream) and lateral (landward) ecotones which operate over highly various temporal and spatial scales. This type of ecotone appears to be highly sensitive to landscape changes caused by physical and biotic disturbances. Examples of direct influences of disturbances are removal of riparian vegetation by floods or landslides and edaphic or vegetative modifications by animals and, in particular, humans. Indirect influences, such as changes in concentration gradients of dissolved chemical compounds and changes in pathways of chemical reactions, may result from changes in edaphic and vegetative properties following direct impacts (Blume 1992; Fränzle 1993; Stigliani 1988; Stockey 1994; Wissmar et al. 1990).

Lentic and Wetland Ecotones

Lentic ecotones, i.e., boundaries between water bodies (lakes, ponds, reservoirs) and adjacent terrestrial patches, also play an important role in coupling ecosystems. With their steep gradients and typically great heterogeneity lentic ecotones can contribute considerably to mitigating unfavorable changes in aquatic ecosystems. Furthermore, they regulate the landscape mosaic by controlling energy and nutrient flow between adjacent patches in the ecosystems (Kluge and Fränzle 1992; Pieczinska 1990). *Wetlands* are defined as lands transitional between terrestrial and aquatic ecosystems where the water table is at or near the surface or the land is seasonally covered by shallow water. Therefore they support, at least periodically, hydrophytes, and their substrate is classified predominantly as an undrained hydric soil (Schleuß 1992). A clear-cut ecological distinction separates *tidal* and *inland wetlands* which may be subdivided further (Mitsch and Gosselink 1986).

Wetlands have internal and external boundaries separating distinct vegetation patches (Holland et al. 1990; Scholle 1991). They can be categorized into various types according to the flows which dominate in the tidal and inland environments. In both tidal and inland wetlands vertical and horizontal transfers occur across a series of surficial and lateral boundaries. Flows across surficial boundaries, e.g., graded sediment beds or soil horizons, include transfer from aerobic to anaerobic soils, from soils to surficial vegetation and litter and vice versa, and from open water to the atmosphere. Transfers across lateral boundaries include material flows from the upland to the wetland (upland/wetland ecotones) or from the wetland into adjacent open water (wetland/open water ecotones), from groundwater aquifers into soils, or across vegetation patches with each community dominated by different species (wetland/wetland ecotones).

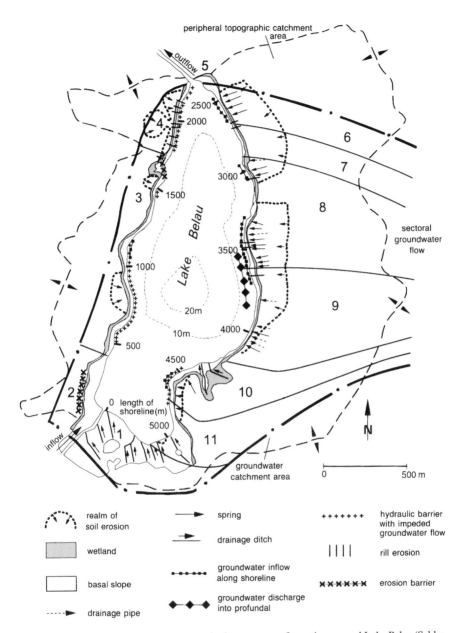

Figure 3.8 Groundwater flow and hydrogeologic ecotone configuration around Lake Belau (Schleswig-Holstein).

Figures 3.8 and 3.9 illustrate a combined lentic/wetland/upland ecotone situation in the Bornhöved Lakes District (Schleswig-Holstein).

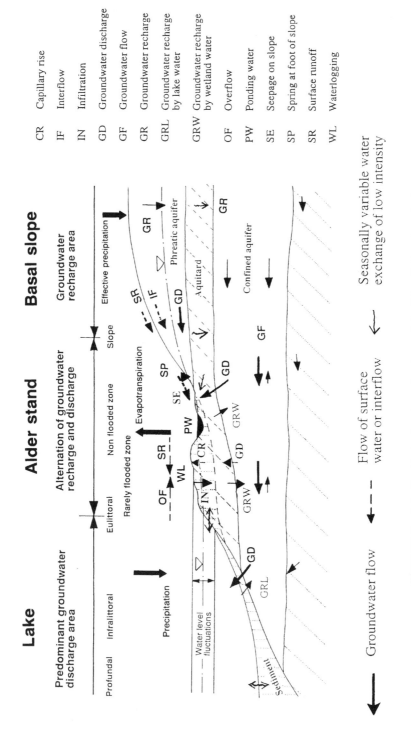

Figure 3.9 Lentic/wetland/upland ecotone situation at Lake Belau (Schleswig-Holstein).

The figures show that lentic ecotones form unusually differentiated habitats. Various microenvironments are formed within the shore zone of the water body, and similar microenvironments occur in certain sections of water bodies contrasting in size, depth, productivity, and catchment area (Schernewski 1992). Lentic ecotones have rich plant and animal communities characterized by a shifting proportion of terrestrial and aquatic organisms.

3.5.2 Sensitivity of Ecotones

Similar to communities and ecosystems, ecotones also differ in physical stability or resistance (resilience) to changes caused by disturbances. In contrast to lotic ecotones, adjacent *terrestrial* ecosystems and *ecotones* may be viewed as occupying distinctly more stable positions in a landscape where the intervals of natural disturbances due to geomorphic processes are normally longer than the time for patch adjustment to them.

Distinguishing reactions, recovery, and persistence phases, and the recurrence intervals as characteristic parameters of system response to physical disturbance (Chorley et al. 1984), recovery, and disturbance recurrence intervals are of particular ecological importance. The ratio of the recovery to disturbance recurrence intervals (R/D) suggests differences in recovery or sensitivity characteristics of both ecotones and ecosystems. For unstable systems, the R/D ratios exceed 1. Typically this is the case for predominantly transient systems such as many *lotic ecotones* (Wissmar et al. 1990). Normally, deposition and erosion interact at varying intensities and frequencies with the riparian vegetation so that the communities are kept in early stages of succession. A decrease in seasonal flooding of the riparian vegetation results in changes via several possible successional pathways, which may lead to the formation of purely terrestrial communities when the system is no longer influenced by flooding.

Basically the same applies to lentic ecotones. Furthermore the above example of Lake Belau (Figure 3.8) illustrates a high-grade selectivity of ecotone reactions to different types of stress. In ecotone sections characterized by high amounts of water transfer (Figure 3.10), communities react distinctly more sensitive to water stress than to chemical impact via groundwater (Fränzle and Bobrowski 1983; Hemprich 1991). Vice versa, chemical stress is more pronounced in the drier ecotone sections around the lake.

Examples of this type support the assertion of Naiman et al. (1988) that a paradox of modelling interfaces predicts when and where ecotone parameters become unpredictable relative to adjacent homogeneous sites. Chaos theory may be useful in dealing with this uncertainty since ecotones are not simple averages of adjacent systems; the spatiotemporal pattern of environmental factors and the interactions between patches are as important in determining ecotone dynamics on different scales as are the characteristics of the adjacent patches.

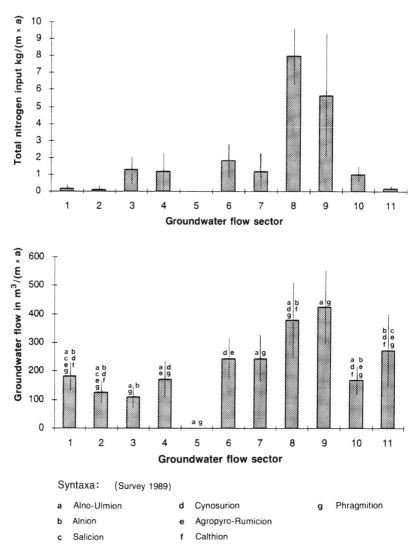

Figure 3.10 Water flow and nitrogen transport across the lentic ecotones around Lake Belau (Schleswig-Holstein), related to the distribution of phytosociological syntaxa.

3.6 SENSITIVITY OF AQUATIC ECOSYSTEMS

Sensitivity of aquatic ecosystems to disturbances depends on the type and size of the respective system and on the nature of disturbance or stressors. Thus, stress is usually defined as a measurable alteration of the state of organisms as induced by an environmental change, which renders the individual, population, or community more vulnerable to further environmental change.

3.6.1 Reactions of Aquatic Ecosystems to Stress

Like terrestrial ecosystems, aquatic ecosystems under stress also undergo changes in both structure and function. Changes in structure are manifested by changes in the composition of populations of different species in the ecosystem. Changes in function are represented by differences in the organic matter productivity of the respective ecosystem and in the rates of release and utilization of different gases and minerals. We can therefore speak of the sensitivity of ecosystem structure and sensitivity of ecosystem function.

There is no uniformity of opinion as to whether the structure or function of ecosystems in general is more sensitive to various stressors. Several researchers feel that functional variables, especially those that are substrate-limited, will always be less sensitive than structural measures because there is functional redundancy in the community. They suggest that any loss of functional capacity by one organism is immediately compensated by increased activity of another organism. Other scientists feel that functional capacity can be affected before compensatory mechanisms operate. This is especially likely when those compensatory mechanisms themselves are adversely affected by the stress or when they operate on a more lengthy time scale relative to the functional measure (Cairns and Niederlehner 1993).

Different reactions of aquatic ecosystems to stress according to Cairns and Niederlehner (1993) are as follows:

1. Community respiration increases.
2. Productivity/respiration becomes unbalanced.
3. Productivity/biomass ratio increases as energy is diverted from growth and reproduction into acclimatation and compensation.
4. Importance of auxiliary energy increases (import becomes necessary).
5. Export of primary productivity increases.
6. Nutrient turnover and nutrient loss increases.
7. One-way transport increases and internal cycling decreases.
8. Lifespan decreases, turnover of organisms increases.
9. Trophic dynamics shifts, food chains shorten, functional diversity declines.
10. Efficiency of resources use decreases.
11. Condition declines.

In the direction of slowly and rapidly reacting ecosystems, at opposite ends of the scale are pelagic lake and reservoir communities composed of very short-living (a few hours up to a few days) species which react to any stress rapidly by their appearance or disappearance, and terrestrial forest ecosystems with very long-living tree species which are unable of rapid presence/absence types of reactions. Benthic communities of rivers and fish populations are intermediate: their life span is measurable in months up to years (cf. Chapter 5).

In aquatic ecosystems, many reports have shown a greater relative sensitivity for structural than functional variables. For example, Schindler (1987), during his extensive comparative studies of Canadian Shield lakes, found no significant changes in decomposition or nutrient cycling in whole lakes treated with acid, but species composition of phytoplankton was among the earliest indicators of change. Crossey et al. (1988) found in impaired rivers that measures of production and respiration were more variable than the macroinvertebrate composition. Crumby et al. (1990) studied the biological reaction of the Roaring River in Tennessee to stress caused by various constructions around the river and by inadequate agricultural practices in the watershed. Species composition changes were reflected in a general decline in numbers of intolerant species and a simultaneous increase of tolerant species.

However, many reports of the greater sensitivity of functional variables of aquatic ecosystems also exist. Rodgers et al. (1980) found that process rate changes were more sensitive than biomass or chlorophyll concentration in detecting the effects of diverse chemicals on the periphyton in artificial streams. When dealing with enrichment, functional measures are often a good warning indicator. Uhlmann et al. (1978) and Gnauck (1982) summarized structural and functional changes in aquatic ecosystems and gave examples of experiments equivalent to those of van Voris et al. (1980). Figure 3.11 compares the degree of variability (expressed by a relative index of instability) during a 50-day experiment with an experimental purification pond for several structural and a few functional variables.

Definitely, the highest instability equivalent to the highest sensitivity is seen (with the exception of the alga *Scenedesmus obliquus*) for the biomass of individual species of organisms (index values between 0.85 and 1.45). In contrast, the functional variables such as oxygen concentration, primary production, turbidity, and the summarizing biomass of functional groups of organisms such as zooplankton or phytoplankton or chlorophyll-a levels have an index close to and below 0.5. The least sensitive appeared to be the organic carbon elimination capacity of the pond which characterizes, from the human point of view, its most important function (Straškraba 1995).

Once the stress has ceased, two possibilities exist: either reversible changes are induced or the changes prove irreversible. If the disturbance is sufficiently regular and of near-natural origin, then components of the ecosystem may adapt and eventually require disturbance to maintain a normal, resilient system, e.g., tidal pools require daily exchange with the sea. Another example is periodic pools, which are inhabited by animals adapted to desiccation. Their characteristic fauna will disappear when water is permanently present because they are inferior in competition with longer-lived species not needing desiccation.

For chemical stress, the recovery of aquatic ecosystems depends mainly on the degree of accumulation of the respective chemical in the environment and the rate of flow. This conclusion can be corroborated by comparing the findings in lakes with those in rivers. The recovery of eutrophic or highly polluted lakes after the sources of organic pollution and phosphorus causing lake

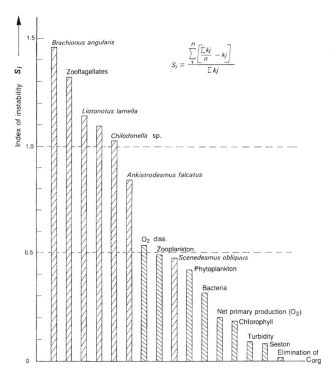

Figure 3.11 Comparison of the degree of variability, expressed in terms of an instability index, during a 50-day experiment with an artificial purification pond. Moidfied after Uhlmann et al. (1978).

eutrophication have been reduced is very slow. Recovery takes up to 10–12 years. The retarded reaction is due to enormous quantities of decomposable organic matter and phosphorus stored in the bottom mud. The oxygen at the bottom is consumed during the decomposition of organic matter, and in such conditions phosphorus is continuously released. This phenomenon is called internal P load, which indicates that when the external load ceases, an internal one plays a major role in continuing eutrophication. On the other side of the spectrum are rivers with high flushing. For rivers, Yount and Niemi (1990) studied the recovery of rivers after a stress and found that an almost full recovery of rivers may take place within about 2 years.

3.6.2 Sensitivity to Different Stressors

The sensitivity of marine and freshwater fish worldwide to recent changes in the environment was determined by examining changed species diversity (Moyle and Leidy 1992). The authors found that the numbers of extinct, endangered, and threatened species and species of special concern represent 21%–69% of the total fauna in different regions of the world, with the exception of 9% only in Latin America. Highest numbers were found in California and South Africa, low figures were found in Iran, Australia and Sri Lanka, but also in Arkansas. They recognized the major causes of this situation to be the extraction of water for various human uses, physical alteration of fish habitats (mainly by channelization, construction of weirs and reservoirs, siltation from mass erosion, and degradation of wetlands), and pollution (mainly municipal and industrial point-source pollution, agricultural pollution, and acid rain).

The sensitivity of aquatic ecosystems to different specific stressors depends on the type of ecosystem and the agent. For chemical agents the degree of interference with the structure and metabolism of organisms is important (cf. Chapter 18). However, even such components as structural macroelements of organisms may be harmful once their concentrations in the environment exceeds certain levels. This is, e.g., the case with plant nutrients.

The sensitivity of aquatic ecosystems to ionizing radiation is not well known. However, it may be expected that some reactions parallel to those observed for terrestrial ecosystems will be valid. The latter were studied in detail by Woodwell (1967). He demonstrated that in a mixed forest/shrub/grassland exposed to different levels of ionizing radiation the diversity of species is more sensitive than organic production. The sensitivity increases with the size of organisms. The reason is that increased size puts greater demand on the photosynthetic mechanism for maintenance, leaving less for growth and repair. Variation in sensitivity was also observed to correlate with chromosome size: plants with large chromosomes were more sensitive than those with smaller chromosomes.

Sensitivity to communal organic pollution is well known for both stream and lake ecosystems; this is the classical type of pollution studied by progressively quantitative methods since the beginning of this century. From the pollution point of view the organic matter is not characterized as a chemical species, but instead in a summarizing way, namely the total or the easily decomposable or refractory matter. The difference between the two categories is arbitrary: easily degradable matter is matter which is decomposed by the natural population of microorganisms within five days' incubation. In streams, systematic changes of the composition of benthos are observed under the effect of certain amounts of organic pollution. This is the basis for the bioindicator approach to detect the degree of organic pollution. However, geographic differences in the benthic fauna of different regions, dependence on stream characteristics such as size, rate of flow, and bottom type, as well as the combination of pollution types complicate the evaluation (cf. Chapters 14 and 18).

3.7 APPRAISAL

The analysis of real communities and the comparative assessment of ecological models show that real communities are not those homogeneous and temporally invariant systems described by simple Lotka–Volterra equations and exemplified by most laboratory microcosms. Population dynamics of real-world ecosystems are spatially distributed in general, and associated with a marked temporal variation, providing a multitude of ways in which the probability of coexistence is enhanced and biodiversity increased. Coexistence under stochastic, non-equilibrium conditions as described by patch-dynamics models at different scales can be just as strong and stable as that occurring under a deterministic, niche-differentiation model.

In a single patch species extinctions can occur as a result of competitive exclusion, overexploitation, and other destabilizing interspecific interactions or they may be due to environmental instability, e.g., unpredictable disturbances or changes in conditions. De Angelis and Waterhouse (1987) and Remmert (1987) have emphasized that the macroscale integration of unstable patches which are out of phase with each other can lead to persistent species-rich communities. In this connection it is worth remembering that soil conditions are largely controlled by pH and redox systems. Since both are nonlinear interrelated buffering systems, soil stability with regard to sorption and mobilization processes is liable to unpredictable, chaotic changes.

An important parallel seems to exist between the properties of a community and the properties of its component populations which will be subject to a (relatively) high degree of K selection in stable environments; r selection predominates in variable environments. K-selected populations with their high competitive capacity, high inherent survivorship, but low reproductive rates are normally resistant to disturbances. However, once perturbed they have little possibility to recover, i.e., low resilience. The r-selected populations, by contrast, have less resistance but distinctly higher resilience. This inverse relationship between resistance and resilience is of particular importance in the ecotoxicological context.

It ensues from the above stability considerations that food chains should be shorter in "fluctuating" environments, since only the most resilient food webs would be expected to persist, and short food chains are most resilient. An analysis of Briand's findings (1983), however, shows that further critical tests of these logical ideas are required, since the average maximal food chain lengths for 40 communities from unstable and "constant" environments did not differ significantly.

In conclusion, it may be said that most communities are probably organized by a temporally and spatially varying mixture of "forces," namely competition, predation, and disturbances, with competition and predation being presumably less important in more disturbed environments. Consequently, there is no such thing as a single stability or sensitivity measure for a community. Stability varies with the aspect of the community or ecosystem under study and the nature of the disturbance. Without pertinent semantic specifications relating to defined temporal and spatial scales (duration and dimensionality of observations), indicator variables and the nature of perturbation, statements about systems stability and sensitivity would be meaningless in a general ecological and a more specific ecotoxicological context.

3.8 REFERENCES

Allen, T. H. F., and T. B. Starr, 1982, *Hierarchy*, Chicago University Press, Chicago, USA

Bachas, C. P., and B. A. Huberman, 1987, "Complexity and Ultradiffusion," *J. Phys. A.*, *20*, 4995–5014

BBA (Biologische Bundesanstalt für Land- und Forstwirtschaft), 1990, *Bestimmung der Reproduktionsleistung von Folsomia candida (WILLEM) in künstlichem Boden (Draft)*, Braunschweig, Germany

Begon, M., J. L. Harper, and C. R. Townsend, 1990, *Ecology: Individuals, Populations and Communities*, Blackwell, Boston, Melbourne

Blume, H.-P. (Ed.), 1992, *Handbuch des Bodenschutzes*, Ecomed, Landsberg/Lech, Germany

Bonazountas, M., 1987, "Chemical Fate Modelling in Soil Systems: A State-Of-The-Art Review," in *Scientific Basis for Soil Protection in the European Community* (H. Barth, P. 'Hermite, (Eds.), Elsevier, London, New York, pp. 487–566

Briand, F., 1983, "Environmental Control of Food Web Structure," *Ecology*, *64*, 253–263

Briand, F., and J. E. Cohen, 1987, "Environmental Correlates of Food Chain Length," *Science*, *238*, 956–960

Bristow, C. M., 1981, "The Role of Mutualists in Structuring the Honeydew-Producing Guild on *Veronia noveboracensis*," *Ph.D. Thesis*, Princeton University, USA

Cairns Jr., J., and B. R. Niederlehner, 1993, "Ecological Function and Resilience: Neglected Criteria for Environmental Impact Assessment and Ecological Risk Analysis," *Environ. Professional*, *15*, 116–124

Ceccato, H. A., and B. A. Huberman, 1988, "The Complexity of Hierarchical Systems," *Physica Scr.*, *37*, 145–150

Chorley, R. J., S. A. Schumm, and D. E. Sudgen, 1984, *Geomorphology*, Methuen, London, UK

Cohen, J. E., and C. M. Newman, 1984, "The Stability of Large Random Matrices and Their Products," *Ann. Probab.*, *12*, 283–310

Crossey, M. J., and T. W. LaPoint, 1988, "A Comparison of Periphyton Community Structural and Functional Responses to Heavy Metals," *Hydrobiologia*, *162*, 109–121

Crumby, W. D., M.-A. Webb, F. J. Bulow, and H. J. Cathey, 1990, "Changes in Biotic Integrity of a River in North-Central Tennessee," *Trans. Am. Fish. Soc.*, *119*, 885–893

de Angelis, D. L., 1980, "Energy Flow, Nutrient Cycling and Ecosystem Resilience," *Ecology*, *61*, 764–771

de Angelis, D. L., and J. C. Waterhouse, 1987, "Equilibrium and Nonequilibrium Concepts in Ecological Models," *Ecological Monographs*, *57*, 1–21

Doreian, P., "Analyzing Overlaps in Food Webs", *J. Soc. Biol. Struct.*, 1986, 9, 115–139

Elton, C., 1958, *The Ecology of Invasion by Animals and Plants*, Methuen, London, UK

Fittkau, E. J., 1991, "Tropische Regenwälder — Ökologische Zusammenhänge", *Bensberger Protokolle*. Vol. 66, Thomas-Morus-Akademie Bensberg, Germany, pp. 27–63

Fränzle, O., 1993, *Contaminants in Terrestrial Environments*, Springer, Berlin, Heidelberg, Budapest

Fränzle, O., 1994a, "Thermodynamic Aspects of Species Diversity in Tropical and Ectropical Plant Communities," *Ecol. Modelling, 75/76,* 63–70

Fränzle, O., 1994b, "Modellierung des Chemikalienverhaltens in terrestrischen Ökosystemen auf unterschiedlichen Raum-Zeit-Skalen," in *Bewertung des ökologischen Gefährdungspotentials von Chemikalien* (E. Bayer, H. Behret, Eds.), *GDCh-Monographien, 1,* 45–90

Fränzle, O., and U. Bobrowski, 1983, "Untersuchungen zur ökologischen Aussagefähigkeit floristisch definierter Vegetationseinheiten," *Verh. Ges. Ökol., 11,* 101–109

Fränzle, O., D. Kuhnt, G. Kuhnt, R. Zölitz, et al., 1987, "Auswahl der Hauptforschungsräume für das Ökosystemforschungsprogramm der Bundesrepublik Deutschland," *Forschungsbericht 101 04 043/02 im Umweltforschungsplan des BMU,* University of Kiel, Germany

Gigon, A., 1983, "Über das biologische Gleichgewicht und seine Beziehungen zur ökologischen Stabilität," *Ber. Geobot. Inst. Eidg. Techn. Hochsch.,* Stift Rübel Zürich, *50,* 149–177

Gnauck, A., 1982, "Strukturelle und funktionelle Änderungen in aquatischen Ökosystemen," *Kongreß- und Tagungsberichte Martin-Luther-Universität Halle-Wittenberg,* Germany, pp. 335–344

Hill, M. G., and P. J. M. Blackmore, 1980, "Interactions Between Ants and Coccid, *Icerya seychellarum,* on Aldabra Atoll," *Oecologia, 45,* 360–365

Hirsch, M. W., and S. Smale, 1974, *Differential Equations, Dynamical Systems, and Linear Algebra,* Academic Press, New York, USA

Hogg, T., B. A. Huberman, and J. M. McGlade, 1989, "The Stability of Ecosystems," *Proc. R. Soc. B., 237,* 43–51

Holland, M. M., D. F. Whigham, and B. Gopal, 1990, "The Characteristics of Wetland Ecotones," in *The Ecology and Management of Aquatic-Terrestrial Ecotones* (R. J. Naiman, H. Décamps, Eds.), UNESCO, Paris, Parthenon, Park Ridge, pp. 171–198

Holling, C. S., 1976, "Resilience and Stability of Ecosystems," in *Evolution and Consciousness* (E. Jantsch, and C. H. Waddington, Eds.), Addison–Wesley, Reading, Mass., USA, pp. 73–92

Jørgensen, S. E., 1990, *Modelling in Ecotoxicology,* Elsevier, Amsterdam, The Netherlands

Jørgensen, S. E., 1992, "Development of Models Able to Account for Changes in Species Composition," *Ecol. Modelling, 62,* 195–208

Kluge, W., and O. Fränzle, 1992, "Einfluß von terrestrisch-aquatischen Ökotonen auf Wasser- und Stoffaustausch zwischen Umland und See," *Verh. Ges. Ökol., 21,* 401–407

MacArthur, R. H., 1995, "Fluctuations of Animal Populations and a Measure of Community Stability," *Ecology, 36,* 533–536

MacArthur, R. H., 1972, *Geographical Ecology,* Harper and Row, New York, USA

Mattheß, G., 1990, *Die Beschaffenheit des Grundwassers,* Borntraeger, Stuttgart, Berlin, Germany

May, R. M., 1972, "Will a Large Complex System be Stable?" *Nature, 238,* 413–414

May, R. M., 1975, *Stability and Complexity in Model Ecosystems,* Princeton University Press, Princeton

May, R. M., 1981, *Theoretical Ecology,* Blackwell Scientific Publications, Oxford, UK

McMurtrie, R. E., 1975, "Determinants of Stability of Large Randomly Connected Systems," *J. Theor. Biol., 50,* 1–11

McNaughton, S. J., 1977, "Diversity and Stability of Ecological Communities: A Comment on the Role of Empiricism in Ecology," *Am. Nat., 111*, 515–525

Mitsch, W. J., and J. G. Gosselink, 1986, *Wetlands*, Van Nostrand Reinhold, New York, USA

Moyle, P. B., and R. A. Leidy, 1992, "Loss of Biodiversity in Aquatic Ecosystems: Evidence from Fish Faunas," in *From Conservation Biology: The Theory and Practice of Nature, Conservation, Preservation and Management* (P. S. Fielder, and K. J. Subodh, Eds.), Chapman and Hall, New York, USA, pp. 127–169

Naiman, R. J., H. Décamps, J. Pastor, and C. A. J. Johnston, 1988, "The Potential Importance of Boundaries to Fluvial Ecosystems," *J. North Am. Benthol. Soc., 7*, 289–306

Nunney, L., 1980, "The Stability of Complex Model Ecosystems," *Am. Nat., 115*, 639–649

O'Neill, R. V., 1976, "Ecosystem Persistence and Heterotrophic Regulation," *Ecology, 57*, 1244–1253

O'Neill, R. V., A. R. Johnson, and A. W. King, 1989, "A Hierarchical Framework for the Analysis of Scale," *Landscape Ecol., 3*, 193–205

Pahl-Wostl, C., 1994, Sensitivity Analysis of Ecosystem Dynamics Based on Macroscopic Community Descriptors: a Simulation Study," *Ecol. Modelling, 75/76*, 51–61

Pahl-Wostl, C., and R. Ulanowicz, 1993, "Quantification of Species as Functional Units Within an Ecological Network," *Ecol. Modelling, 66*, 65–79

Pieczinska, E., 1990, "Lentic Aquatic-Terrestrial Ecotones: Their Structure, Functions, and Importance," in *The Ecology and Management of Aquatic Terrestrial Ecotones* (R. J. Naiman, and H. Décamps, Eds.), UNESCO, Paris, Parthenon, Park Ridge, pp. 103–140

Pimm, S. L., 1982, *Food Webs*, Chapman and Hall, London, UK

Prigogine, I., 1976, "Order through Fluctuation: Self-organization and Social System," in *Evolution and Consciousness* (E. Jantsch, and C. H. Waddington, Eds.), Addison-Wesley, Reading, Mass., USA, pp. 93–133

Reiche, E. W., 1991, "Entwicklung, Validierung und Anwendung eines Modellsystems zur Beschreibung und flächenhaften Bilanzierung der Wasser- und Stickstoffdynamik in Böden," *Kieler Geogr. Schriften, 79*

Rejmanek, M., and P. Stary, 1979, "Connectance in Real Biotic Communities and Critical Values for Stability in Model Ecosystems," *Nature, 280*, 311–313

Remmert, H., 1987, "Sukzessionen im Klimax-System," *Verh. Ges. Ökol. XVI*, 27–34

Rodgers Jr., J. H., K. L. Dickson, and J. Cairns, *American Society for Testing and Materials*, (R. G. Wetzel, Ed.), Philadelphia, USA, pp. 142–167

Schernewski, G., 1992, "Raumzeitliche Prozesse und Strukturen im Wasserkörper des Belauer Sees," *EcoSys-Beiträge zur Ökosystemforschung*, Suppl. Vol. 1, Kiel, Germany

Schindler, D. W., 1987, "Detecting Ecosystem Responses to Anthropogenic Stress," *Can. J. Fish. Aquat. Sci., 44*, Suppl. 1, 6–25

Schleuß, U., 1992, "Böden und Bodenschaften einer norddeutschen Moränenlandschaft," *EcoSys Suppl. 2*, Kiel, Germany

Scholle, D., 1991, "Vegetationskundliche Untersuchungen im Raum Bornhöved und deren Auswertung mit Hilfe eines Geographischen Informationssystems (GIS)," MSc Thesis, Saarbrücken, Germany

Solbrig, O. T., and G, Nicolis, 1991, *Perspectives in Biological Complexity*, IUBS, Paris, France

Stigliani, W. M., 1988, "Changes in Valued "Capacities" of Soil and Sediments as Indicators of Nonlinear and Time-Delayed Environmental Effects," *IIASA Ecoscript, 35*, International Institute for Applied Systems Analysis, Laxenburg, Austria

Stockey, A., 1994, "Etablierung, Sukzession und Diversität von Bachufervegetation," *Bielefelder Ökologische Beiträge, 7*

Straškraba, M., 1995, "Nutrient Cycles and Productivity of Terrestrial and Aquatic Ecosystems," in *Ullmann's Encyclopedia of Industrial Chemistry*, VCH-Verlagsgesellschaft, Weinheim, Germany, pp. 40–54

Temple, S. A., 1977, "Plant-Animal Mutualism: Coevolution with Dodo Leads to Near Extinction of Plants," *Science, 197*, 885–886

Uhlmann, D., H. Mihan, and A. Gnauck, 1978, "Schwankungen des Sauerstoffhaushalts und der biologischen Struktur extrem nährstoffreicher Gewässer unter gleichbleibenden Umweltbedingungen (Modellversuche)," *Acta Hydrochim. Hydrobiol., 6*, 421–444

Ulanowicz, R. E., 1986, *Growth and Development: Ecosystems Phenomenology*, Springer, New York, USA

Urban, D. L., R. V. O'Neill, and H. H. Shugart, 1987, "Landscape Ecology," *BioScience, 37*, 119–127

van den Bergh, J. P., 1981, "Interaction Between Plants and Population Dynamics," *Verh. Ges., Ökol., 11*, 155–163

van der Maarel, E., 1976, "On the Establishment of Plant Community Boundaries," *Ber. Dtsch. Bot. Ges., 89*, 415–443

van Straalen, N. M., and C. A. J. Denneman, 1989, "Ecotoxicological Evaluation of Soil Quality Criteria," *Ecotox. Environ. Saf., 18*, 241–251

van Voris, P., R. V. O'Neill, W. R. Emanuel, and H. H. Shugart, 1980, "Functional Complexity and Ecosystem Stability," *Ecology, 6*, 1352–1360

Werner, and W. Stickau, 1983, "Veränderungen der Artenzusammensetzung und Photosyntheseleistung als Anzeiger für die chemische Belastung auf Grünlandökosysteme," *Verh. Ges. Ökol., 11*, 463–477

Westhoff, V., 1974 "Stufen und Formen von Vegetationsgrenzen und ihre methodische Annäherung," in *Tatsachen und Probleme der Grenzen in der Vegetation* (W. H. Sommer, W. H., and R. Tüxen, Eds.), Cramer, Lehre, Germany, pp. 45–64

Whittaker, J. B., 1979, "Invertebrate Grazing, Competition, and Plant Dynamics," *Symp. Br. Ecol. Soc., 20*, 207–222

Wiegand, T., and C. W. Wissel, 1994, "Host-Parasitoid Models in Temporally and Spatially Varying Environment," *Ecol. Modelling, 75/76*, 161–170

Wissel, C., 1981, "Lassen sich ökologische Instabilitäten vorhersagen?," *Verh. Ges. Ökol., 11*, 143–152

Wissmar, R. C., and F. J. Swanson, 1990, *The Ecology and Management of Aquatic-Terrestrial Ecotones* (R. J. Naiman, and H. Décamps, Eds.), Paris, UNESCO, Parthenon, Park Ridge, pp. 171–198

Woodwell, G. M., 1967, "Radiation and the Patterns of Nature," *Science, 156*, 3774, 461–470

Yount, J. D., and G. J. Niemi, 1990, "Recovery of Lotic Communities and Ecosystems from Disturbance: Theory and Applications," *Environ. Managem., 14* (5)

4

Population Dynamics of Plants Under Exposure and the Selection of Resistance

Wilfried H. O. Ernst (Amsterdam, The Netherlands)

4.1 SUMMARY

The quantitative (Pearl-Verhulst function) and qualitative (Hardy-Weinberg equilibrium) principles of an analysis of plant populations are elaborated in relation to the impact of high concentrations of chemical elements and organic compounds. As soon as plants cannot react to an imbalance of an element or organic compound by acclimation, populations are negatively affected in their quantity and the genetic composition will be changed in favor of resistant genotypes. The quantitative changes enable the establishment of ecological effect concentration and can be described by an equation with a linear or sigmoid shape. The reaction to an injury may not be immediate and can be delayed by decades either due to the buffer and/or complexing capacity of the soil or due to the life history and/or rooting pattern of the species. Qualitative changes in a population will impair the ecosystem, losing initially its stability and structure. At very high and long

Ecotoxicology, Edited by Gerrit Schüürmann and Bernd Markert.
ISBN 0-471-17644-3   © 1998 John Wiley & Sons, Inc. and Spektrum Akademischer Verlag.

exposures an ecosystem will become impoverished, but the remaining low biodiversity may be sufficient to establish a new ecological equilibrium.

4.2 INTRODUCTION

A change in environmental quality can have two consequences for plant populations: (1) the aimed application of chemicals as pesticides to control populations of those organisms which hamper the productivity of other desired plant species, e.g., weeds in agricultural fields, i.e., the intended extinction of populations and species; (2) the exposure to primarily unintended emissions of potentially toxic levels of chemicals, e.g., gaseous pollutants from industrial processes (SO_2) and biological industries (NH_3), particulates such as heavy metals by smelters, and radionuclides by power stations (Chernobyl accident). These emissions can affect all organisms including agricultural, horticultural, and silvicultural crops.

Generally, each plant population, except highly selected crop species, consists of a number of different genotypes. As soon as the genotype diversity of a population is very high, differentiation can be found between and within populations in relation to nearly every environmental factor (Antonovics 1984; Bradshaw 1984). All of these genotypes, however, differ from each other in the potential, timing, and intensity of the response to the changed environmental condition(s). These responses may be ranked from very sensitive to very resistant. Because of the sedentary character of plants—in contrast to a lot of animal species—the immigration of new genotypes into a local population is very restricted. Therefore, the various responses of the different genotypes will affect the genetic structure of a population and determine the survival and viability of a population in the long term.

4.3 QUANTITATIVE PRINCIPLES OF POPULATION DYNAMICS IN PLANTS

In population dynamics only four components determine the number of individuals in a population: birth (B), death (D), immigration (I), and emigration (E). The number (N) in a population at time $t + 1$ is related to that at time t by the equation

$$N_{t+1} = N_t + B - D + I - E \tag{4.1}$$

The change in numbers give

$$\Delta N = B - D + I - E \tag{4.2}$$

An exposure to a surplus of a chemical substance will generally increase the mortality rate. The population declines in numbers if the rate of birth and/or immigration does not compensate the loss by death and emigration. The strong decrease in lichens during increased exposure to enhanced concentrations of SO_2 in industrial areas is a well-documented example of population dynamics under exposure (Ferry et al. 1973; Guderian 1985). After strong reduction of SO_2 emissions from 1968 onwards in the Netherlands and from the mid 1970s onwards in other western European countries the lichen populations are now recovering. However, the process is slow because all increases are primarily due to the immigration of vegetative propagules from less-polluted sites.

Equation 4.2 fits well for populations with nonoverlapping generations, as usually realized by annual plant species with short-living seed banks. In all other annual and perennial plant species the occurrence of overlapping generations is the rule. The various age classes in such populations differ in their demand for space and resources, resulting in different competitive abilities.

The relation between population size, sources, and conditions can be described by the logistic (Verhulst–Pearl) equation

$$\mathrm{d}N/\mathrm{d}t = r^* N^* [(K - N)/K] \tag{4.3}$$

where r is the intrinsic rate of increase, N the number of individuals, and K the carrying capacity supporting the number of individuals in a particular environment. The different demands of sources and space by the various plant species will influence the population dynamics of species to different degrees. Exposure to chemical compounds which can be used in plant metabolism, such as nitrogen and sulfur, may have other consequences for the carrying capacity and the rate of population increase than chemicals without any metabolic stimulus. Sulfur enrichment of the soil by dry and wet deposition (Wookley and Ineson 1991) has increased the carrying capacity of sulfur-demanding plant species belonging to the family *Brassicaceae*. Members of this plant family have enormously enlarged their area during the last 40 years in the Netherlands and the western part of Germany (Ernst 1993a), in contrast to the decrease in SO_2-sensitive species by nearly one-third.

For the implementation of the logistic growth curve (Equation 4.3) another aspect of plants has to be considered, i.e., the plasticity of their performance. The mean plant dry weight of each member of a population (W) is related to the density of surviving plants per unit (N) by the equation (Watkinson 1984)

$$W = W_m^* (1 + aN)^{-b} \tag{4.4}$$

where W_m is the mean yield of an isolated plant, a the area required to achieve W_m, b the efficiency of resource utilization, and N, the number of individuals per study area (which may represent the whole population). If $b = 1$, the total yield per unit area becomes independent of plant density. This situation is defined as the "law of constant yield" (Shinokazi and Kira 1956). This law is only defined for annual plant species. The plasticity and genetic diversity among and between

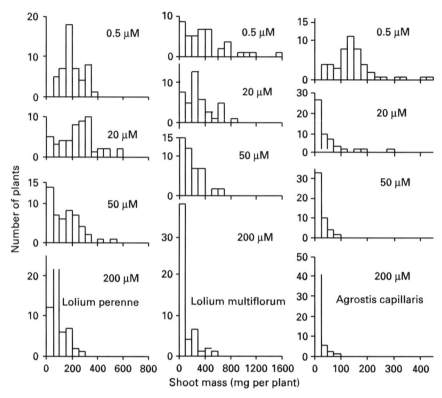

Figure 4.1 Impact of increasing concentrations of copper (µM Cu in nutrient solution) on shoot dry mass of *Lolium perenne*, *Lolium multiflorum*, and *Agrostis capillaris* after 7 weeks of exposure. After Lolkema (1985).

plant species makes it difficult to find a real mean yield of an individual plant in a field population and to distinguish this W_m from the plant mass at low chronic exposure. In experimental research exposure to increasing concentration of copper diminished strongly the plant dry mass in a species-specific manner (Figure 4.1); the smallest individuals of several grass and herb species were found at the highest copper concentration (Lolkema 1985; Ernst 1993b).

All of the considerations formalized in Equations 4.1 to 4.4 are designed for populations with nonoverlapping generations. The theory of population dynamics of overlapping generations is still in its infancy and will not help to facilitate the judgement of population dynamics of wild plant species under exposure.

4.4 QUALITATIVE PRINCIPLES OF POPULATION DYNAMICS

Quantitative population dynamics treats all members of a population as equal. Wild plant populations, however, are characterized by a high degree of genetic diversity, even those under long-term exposure (Verkleij et al. 1985). Genotypes can differ in morphological traits such as hairiness and flower color as in *Digitalis purpurea* (Ernst 1985), flower morphs as in *Senecio vulgaris* (Hull 1974), and timing of germination and flowering as in winter and summer population of *Senecio sylvestris* (Ernst 1989). All of these genotypes can be identified in the field, and changes in genotype frequency can be followed over time (Ernst 1987, 1993b). The identification of all other genotypes in a population demands more laborious techniques such as the analysis of isozymes (Loveless and Hamrick 1984), multiple restriction fragment length polymorphism (RFLP) analysis (Mastuura and Fujita 1995), and random amplified polymorphic DNA analysis (RAPD; Peakall et al. 1995). As shown for the dioecious buffalo grass *Buchloe dactyloides* each technique has its own advantages and disadvantages: RADPs revealed 14% greater variation among populations of different geographic regions compared to allozymes, but 12% less variation among individuals within populations (Peakall et al. 1995).

Due to the genetic variability of plant populations the logistic growth equation (Equation 4.3) has to be modified to

$$dN/dt = r_i^* N_i^* [(K_i - N_i)/K_i] \tag{4.5}$$

where the subscript i indicates the genotypes in a population. Each genotype has its own demands for space and resources. Up to now such an analysis has not been carried out for any field population.

In a large random-mating population with no selection, mutation, or migration the gene frequency and the genotype frequency follows the Hardy–Weinberg principle, i.e., the frequency (p, q) of two alleles (A, a) remains constant over generations, because

$$p + q = 1 \tag{4.6}$$

For a gene with two alleles A and a the genotypic frequency in the next generation results in

$$p^2 (AA) + 2pq (Aa) + q^2 (aa) = 1 \tag{4.7}$$

Under exposure to chemical substances not all genotypes will show the same fitness. Therefore, the selection factor s has to be introduced in the calculation. Assuming that the directional selection is against the homozygous recessive individuals (aa), then the relative fitness w of these individuals is $1 - s$, a number greater than zero. When $s = 1$, no individual will survive. In a Mendelian population the proportion AA remains p^2, the proportion Aa remains $2pq$, and the proportion aa becomes $q^2(1 - s)$.

When the selection pressure is high (s approaching 1) a strong and often rapid change in the genetic structure will be the consequence. Famous examples are the high selective forces of increased concentrations of heavy metals in the vicinity of smelters (Bradshaw 1976; Ernst 1976; Dueck et al. 1984), SO_2 in the vicinity of coal-fired industries (Bell and Mudd 1976; Ernst et al. 1985), and the treatment of arable fields with the herbicides triazine and chlorsulfuron (Holt et al. 1981; Hall and Devine 1990). The copper-resistant genotypes in a population of the perennial grass *Agrostis capillaris* growing in the vicinity of a copper refinery increased from 6.3% after 4 years of exposure to 27.6% after 14 years and to 40% after 70 years (Bradshaw 1976).

Most plant populations are regularly under selection pressure, not only due to exposure to acute and chronic concentrations of chemical substances from anthropogenic sources, but also due to natural processes such as changing temperature and atmospheric physics and chemistry, drought and flooding, pathogens, and herbivores. In wild plant populations under chronic exposure it is difficult to decide upon the most selective force which changes the genetic structure. An example for this difficulty is the hairiness of the perennial herb *Silene dioica* in serpentine and normal grassland (Westerbergh and Nyberg 1995). Selective grazing by two *Arion* slugs obviously counteracted the flow of the recessive allele determining glabrousness from serpentine into normal grassland populations, because the slugs were rare in the serpentine grassland.

4.5 REACTION PATTERN TO EXPOSURE—THE DOSE

Five hundred years ago the fundamentals of toxicology were established by Paracelsus (1493–1541): "Sola dosis facit venum." (It is only the dose that makes the poison.) The toxicological response level of individuals to exposure is calculated as the dose of a chemical substance which causes death of part of or the whole population. The lethal dose (LD) is 100% (LD_{100}) when all organisms die. Toxicological tests take mostly a LD_{50}; this means that 50% of the (experimental) population die after a certain period of exposure, often 48 h, 96 h, and 60 days. All admissions of chemical compounds are based on this testing protocol.

In ecotoxicology a great refinement of this protocol is necessary to ensure the maintenance of biodiversity and population integrity. Therefore, ecological effect parameters are introduced, which can be handled in field research. The highest concentration of a chemical substance in the environment without any effect is defined as "no effect concentration" (NEC). It is difficult to ensure no effect. Therefore, NEC is replaced by the "no observed effect concentration" (NOEC), which opens the possibility to focus on ecologically relevant changes in biological processes. The other extreme of a response to exposure is a complete effect, the "effect concentration" of 100% (EC_{100}). At EC_{100} the metabolic process within an organism is out of function, but it may be restored after transfer to a clean environment. Therefore EC_{100} is not identical to LD_{100}.

Only when the ecological parameter is the survival of the population, may EC_{100} be identical to LD_{100}. This very essential difference between the EC and LD concepts will be highlighted by an example. A toxic concentration of a metal hampers all root growth, resulting in an EC_{100}, but not in an LD_{100}. After transfer of the plant into a clean environment, root growth can be initiated by undifferentiated cells near the shoot and the plant can restart growth; thus it has survived despite a temporary exposure to an EC_{100}. This testing procedure is well known for the establishment of the metal resistance in plants, the so-called improved rooting test (Schat and Ten Bookum 1992).

In the case of EC, the scale can be subdivided to more sensitive subunits, EC_{10}, EC_{25}, EC_{50}, and EC_{90}. An EC_{10} value may be difficult to establish in a real field population due to the high genetic variability and phenotypic plasticity in contrast to experimental testing of well-defined genotypes and clones.

The dose–effect relationship is quite different for chemical elements or compounds which are necessary for plant growth at low concentrations. In the case of an essential element the dose–response relationship shows a bimodality (Figure 4.2). At a very low supply of an essential chemical element the plant suffers from a shortage which may result in bad performance or even death due to starvation. When the chemical element concentration surpasses the optimum for plant growth it causes adverse effects, finally resulting in death due to toxicity ($EC_{100} = LD_{100}$). The interest of ecotoxicology is primarily not at the shortage site, but rather at the surplus site of the plant (Ernst 1994). Nonessential elements and compounds do not cause a deficiency or shortage. Therefore, the measurement of ecotoxicological parameters starts at the optimum of the plant response, the NOEC (Figure 4.3a). The response of a plant exposed to a surplus of a chemical substance can vary. An increasing concentration of the subsance (log c) may result in a linear dose–effect relationship. Each atom or

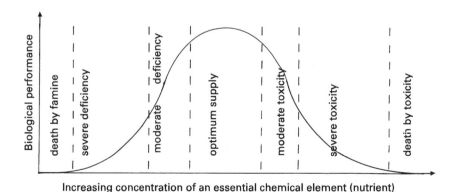

Figure 4.2 Dose–response relationship of a chemical element which is necessary for plant growth. After Ernst (1996).

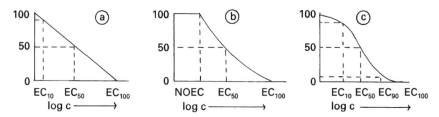

Figure 4.3 Three types of dose–response relationships: (a) linear, (b) hyperbolic with a range of no observable effect concentration (NOEC), (c) sigmoidal. EC, effect concentration. After Ernst (1994).

molecule of a substance is effective so that the number of affected cells, tissues, or individuals increases exponentially (Figure 4.3b). An example is the linear decrease in P uptake by the perennial grass *Holcus lanatus* with increasing external arsenate concentration and high external P supply (Meharg and Macnair 1990). In this case the calculated EC_{50} value is at 8 mM external As (Ernst 1994).

A chemical substance may not immediately affect an organism at a relatively low concentration (NOEC) because the organism can detoxify the substance by metabolisation, complexation, compartmentation, or excretion (Figure 4.3b). The NOEC and EC values may depend on the processes being analyzed, as can be observed by analyzing various biological processes in a plant exposed to cadmium (Figure 4.4). Root elongation and in vitro activity of the enzyme nitrate reductase in Cd-sensitive plants of the perennial herb *Silene vulgaris* have a NOEC of 0.8 µM Cd after 3 days and 15 min exposure, respectively. Cd concentrations in the roots are measurable by standard atomic absorption procedures at 0.3 µM external Cd. The synthesis of phytochelatins, a stress peptide synthesized only upon metal exposure (De Knecht et al. 1994; see Chapter 19), occurred at 0.15 µM external Cd, i.e., a factor of 5 lower than the NOEC for root elongation growth.

For most biological processes the parameter selected may decide upon the NOEC. In an ecological context the effect of exposure on health and survival will be relevant at the individual level, that on reproduction and competitive ability at the population level. Biomarkers as early warning systems (Ernst and Peterson 1994) may be helpful for an evaluation of chronic effects, i.e., long-term exposure at low levels. The importance of the registration of low EC concentration is currently being realized in view of the observation of fertility losses in man, fish, and other animals after long-term exposure to low levels of estrogen-like substances (Colborn et al. 1993; Sharpe and Skakkebaek 1993) and growth stimulation of plants (Shore et al. 1992).

Instead of linearity or hyperbolic response function the dose–effect relationship may have a sigmoid shape (Figure 4.3c). The ecotoxicological consequence of such a response is an increased EC_{10} value in comparison with the linear function and a diminished effect concentration in the range between EC_{50} and

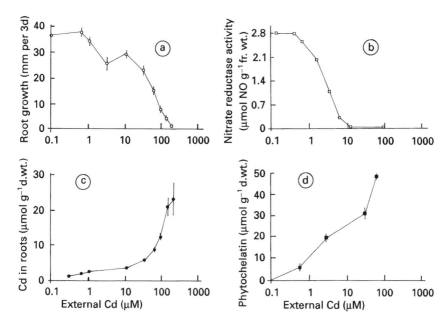

Figure 4.4 The response of plants of *Silene vulgaris* to increasing concentrations of cadmium. Measured parameters are: (a) root elongation growth, (b) in vitro activity of nitrate reductase, (c) Cd concentration in roots, (d) phytochelatin production. Data base from De Knecht et al. (1994).

EC_{90}. An example of a sigmoid response curve is the pronounced decrease in P uptake by *H. lanatus* with increasing As concentration, but now, in contrast to the earlier example, at low external P supply (Meharg and Macnair 1990). This exposure results in an EC_{50} value of 0.25 mM external Cd, i.e., a factor of 32 lower than at the same external Cd concentration but at higher P supply (Ernst 1994).

4.6 EXPOSURE CONCENTRATION AND EXPOSURE TIME

The ecotoxicological effects of a substance does not depend only on the exposure concentration (c) but also on the exposure time (t). This relation can be expressed by the equation

$$c \times t = \alpha \tag{4.8}$$

The constant α depends on the parameter which is measured and the life history of an organism. By considering the logarithimic concentration effects Equation 4.8 can be transformed into

$$\log c = \log \alpha - \log t \tag{4.9}$$

Both parameters (c, t) can be presented as log–log diagrams (Figure 4.5). The ideal situation will result in a straight line with a slope of -1. Both equations (4.8 and 4.9) will fit if the substances are not metabolized by plants. When a substance is incorporated into plant metabolites, e.g., NH_3 into amino acids, then the exposure time will go to infinity at low exposure concentrations (Figure 4.6). In such a situation a species-specific threshold level (NOEC) will disturb the

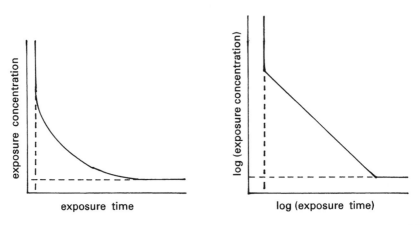

Figure 4.5 The relation between exposure time (t) and exposure concentration (c) on a linear (left) and logarithmic scale (right).

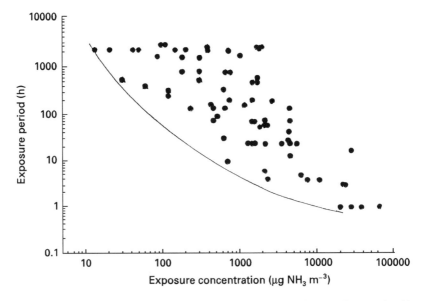

Figure 4.6 Exposure period and exposure concentration interaction of many plant species. Above the solid curve ammonia has adverse effects. Modified after Van der Eerden (1992).

formalized relation between concentration and time. At extreme high exposure concentrations a similar problem will arise. When a high concentration of the substance disrupts the biomembrane there is a lower time limit or threshold of reaction time.

After a short exposure to a high concentration immediate effects may occur, defined as the acute exposure. Toxicity tests are concentrated on the measurement of high concentration at short exposure times so that the acute toxicity can be judged. In ecotoxicology the long-term effects of relatively low exposure concentrations are of great interest due to the longevity of organisms, being more than a hundred years for several grass-, herb-, and tree species. This is called the chronic exposure or chronic toxicity of a chemical substance. A low concentration may not have a measurable effect on the short term, but only on the long-term, as mentioned above for the estrogen-like substances. Chronic toxicity can result in delayed effects.

In the case of plants the various plant parts may be confronted with the same pollutant at different times and at different concentrations in life history. This situation is the case especially if the chemical substance has an inherent persistency as that of heavy metals. They cannot be broken down because they are basic chemical elements. Injury by air pollutants of plant leaves was recognized more than a century ago (Haselhoff and Lindau 1903); research activities have been enhanced during the past 40 years (Guderian 1977; Mudd and Kozlowski 1975; Alscher and Wellburn 1994). Interest in the impact of air pollutants on the plant as a whole, however, has emerged only in recent times, e.g., translocation of photosynthates from shoot to root and mycorrhization under exposure (Termorshuizen and Schaffers 1987). These processes are difficult to observe and are regarded as NOEC. The relation of forest decrease to increased soil acidification (Van Breemen et al. 1982) is a typical example of a delayed exposure.

The relation between exposure time and exposure concentration does not consider changes in the genetic structure of plant populations during long-term exposure. Therefore Equation 4.8 is not applicable to such a situation. It roughly fits plant populations in situations where the pollutant is not constantly present, as shown in Figure 4.6 for ammonia (Van der Eerden 1992).

4.7 EXPOSURE TIME AND LIFE HISTORY

In plants certain life stages have a very short exposure time by nature. Pollen grains and stigmas of flowers are very short living. Therefore the exposure concentration has to be very high for an ecological effect. Observations of infertility due to exposure of pollen grains and stigmas to environmental pollution are scarce. NOEC values after pollen germination for exposure to heavy metals were in the vicinity of 50 μM Cu and more than 100 μM Zn (Searcey and Mulcahy 1985a,b).

Seed development takes place over weeks in a lot of herbs and grasses, over months in sedges, and over years in coniferous trees. Ecotoxicological effects are caused by the translocation of pollutants from vegetative plant parts to the seeds. After the Chernobyl catastrophe the seeds of Scotch pine trees (Coniferae) absorbed a dose of radionuclides varying from 0.4 to 12 Gy. The frequency of mutation was on average ten times higher than the mutation rate of nonexposed tree populations (Kalchenko et al. 1990).

In most plant species germination of seeds is a rapid process when constraints of primary and/or secondary dormancy are removed. The young root (radicula) is immediately exposed to all chemical substances dissolved in the soil solution or in the water. The rapidly dividing cells are very sensitive to exposure (Davies et al. 1995). They die off at inappropriate exposure concentrations so that no cell division and root growth can occur (Lolkema 1985). This process is often camouflaged because root death occurs some weeks earlier than the cotyledons disappear, suggesting a longer time for a NOEC.

In deep-rooting plants the root system may be protected against substances for decades or even centuries, whereas the shoot can be immediately affected by aerial deposition and an effect on the roots near the soil surface will follow within months or years. As long as the EC does not severely affect the lifespan and production of leaves the plant can produce seeds. Whereas plant species with only shallow roots may disappear from the ecosystem or will be selected for highly adapted (resistant) genotypes, the individuals of the deep-rooting species can persist without any acclimation or adaptation up to the time that they die due to senescence. When the offspring of these deep-rooting plants germinate, the seedling will be exposed to a high concentration of the pollutant (acute toxicity) and will have great difficulties to survive. The difference between the zinc resistance of the shallow-rooting grass *A. capillaris* and that of the deep-rooting grass *Molinia coerulea* in a soil contaminated by metal fallout is an example of delayed chronic exposure (Ernst 1996).

4.8 CONCLUSION

When plants cannot counteract exposure to chemicals by acclimation (phenotypic plasticity), the chemical substances will affect population dynamics. The presence of genetic diversity is the absolute requirement for adaptation of plants to anthropogenic environmental change. Many plant species have insufficient genetic variability in this respect. At low genetic variability and low exposure concentration the mortality will not be sufficient to eliminate all individuals within a local population. At high exposure concentration, however, the probability of causing extinction of the population could increase. In contrast, high genetic variability in a population offers the opportunity for survival of the least-injured, best-adapted genotypes which can recruit a new genetic structure in the surviving population. Not all plant species and populations have the same potential to withstand the selection pressure; the speed of the environmental change has to be in line with the speed of population response. The genus *Agrostis* has an enormous potential for the selection of resistance to heavy metals, independent of whether the smelter was constructed in a formerly metal-clean environment such as at Prescot (Bradshaw 1979) and at Budel-Dorplein (Dueck et al. 1984) or in the vicinity of former ore outcrops as at Harlingerode (Ernst 1974). Within 5 years after exposure to metal pollution a population of *A. capillaris* consisted of zinc-resistant individuals; thus, this species responded quickly to the environmental change. The potential of *Agrostis* species to evolve resistance against pesticides has not been detected. On the other hand, certain agricultural weeds such as *Senecio vulgaris* can respond to intensive application of herbicides by developing herbicide resistance (Holt et al. 1981), but they have obviously a low potential for development of metal resistance. The speed of the selection of herbicide resistance is less than that of metal resistance. One of the reasons for different evolution patterns may be the stability and repair of biomembranes, although a taxonomic affinity of species does not guarantee a prediction of their evolutionary potential. Ecosystems with plant populations having low evolutionary potential will lose their biodiversity under exposure. The highly metal-resistant grass "savannas" around smelter sites are an example of this decrease in biodiversity.

4.9 REFERENCES

Alscher, R. G., and A. R. Wellburn, Ed., 1994, *Plant Responses to Gaseous Environment*, Chapman and Hall, London, UK

Antonovics, J., 1984, "Genetic Variation Within Populations," in *Perspectives of Plant Population Ecology* (R. Dirzo, and J. Sarukhan, Eds.), Sinauer Associates Publishers, Sunderland, USA, pp. 229–241

Bell, J. N. B., and C. H. Mudd, 1976, "Sulphur Dioxide Resistance in Plants: a Case Study of *Lolium perenne*," in *Effects of Air Pollutants* (T. A. Mansfield, Ed.), Cambridge University Press, Cambridge, UK, pp. 87–103

Bosbach, K., H. Hurka, and R. Haase, 1982, "The Soil Seed Bank of *Capsella bursa-pastoris* (Cruciferae): Its Influence on Population Variability," *Flora, 172*, 47–56

Bradshaw, A. D., 1976, "Pollution and Evolution," in *Effects of Air Pollutants* (T. A. Mansfield, Ed.), Cambridge University Press, Cambridge, UK, pp. 135–149

Bradshaw, A. D., "Ecological Significance of Genetic Variation Between Populations," in *Perspectives of Plant Population Ecology* (R. Dirzo, and J. Sarukhan, Eds.), Sinauer Associates Publishers, Sunderland, USA, pp. 213–228

Colburn, T., F.S. vom Saal, and A. M. Soto, 1993, "Developmental Effects of Endocrine-Disrupting Chemicals in Wildlife and Humans," *Environ. Health Perspect., 101*, 378–384

Davies, K. L., M. S. Davies, and D. Francis, 1995, "The Effects of Zinc on Cell Viability and on Mitochondrial Structure in Contrasting Cultivars of *Festuca rubra* L.—A Rapid Test for Zinc Tolerance," *Environ. Pollut., 88*, 109–113

De Knecht, J. A., M. van Dillen, P. L. M. Koevoets, H. Schat, J. A. C. Verkleij, and W. H. O. Ernst, 1994, "Phytochelatins in Cadmium-Sensitive and Cadmium-Tolerant *Silene vulgaris*," *Plant Physiol., 104*, 255–261

Dueck, T. A., W. H. O. Ernst, J. Faber and F. Pasman, 1984, "Heavy Metal Immission and Genetic Constitution of Plant Populations in the Vicinity of Two Metal Emission Sources," *Angew. Bot., 58*, 47–59

Ernst, W. H. O., 1974, *Schwermetallvegetation der Erde*, G. Fischer Verlag, Stuttgart, Germany

Ernst, W.H.O., 1976, "Physiological and Biochemical Aspects of Metal Tolerance," in *Effects of Air Pollutants* (T. A. Mansfield, Ed.), Cambridge University Press, Cambridge, UK, pp. 115–133

Ernst, W. H. O., 1987, "Scarcity of Flower Colour Polymorphism in Populations of Digitalis purpurea," *Flora, 179*, 231–239

Ernst, W. H. O., 1989, "Selection of Winter and Summer Annual Life Forms in Populations of *Senecio sylvaticus* L.," *Flora, 182*, 221–231

Ernst, W. H. O., 1993a, "Ecological Aspects of Sulfur in Higher Plants: The Impact of SO₂ and the Evolution of the Biosynthesis of Organic Sulfur Compounds on Populations and Ecosystems," in: *Sulfur Nutrition and Assimilation in Higher Plants* (L. J. De Kok, I. Stulen, H. Rennenberg, C. Brunold, and W. E. Rauser, Eds.), SPB Academic Publishing, The Hague, The Netherlands, pp. 295–313

Ernst, W. H. O., 1993b, "Population Dynamics, Evolution and Environmental Stress," in *Plant Adaptation to Environmental Stress* (L. Fowden, T. Mansfield, and J. Stoddart, Eds.), Chapman and Hall, London, UK, pp. 19–44

Ernst, W. H. O., 1994, "Wirkungen erhöhter Bodengehalte an Arsen, Blei und Cadmium auf Pflanzen," in *Beurteilung von Schwermetallen in Böden von Ballungsgebieten: Arsen, Blei und Cadmium*, Dechema, Frankfurt/Main, Germany, pp. 319–355

Ernst, W. H. O., 1996, "Stress durch Schwermetalle," in *Stress bei Pflanzen* (C. Brunold, R. Brändle, and A. Rüegsegger, Eds.). UTB Ulmer, Stuttgart, Germany

Ernst, W. H. O., A. E. G. Tonneijck, and F. J. M. Pasman, 1985, "Ecotypic Response of *Silene cucubalus* to Air Pollutants (SO₂, O₃)," *J. Plant Physiol., 118*, 439–450

Ernst, W. H. O., and P. J. Peterson, 1994, "The Role of Biomarkers in Environmental Assessment. (4). Terrestrial plants," *Ecotoxicology, 3*, 180–192

Ferry, B. W., M. S. Baddeley, and D. L. Hawksworth, Eds, *Air Pollution and Lichens*, Athlone Press, London, UK

Guderian, R., 1985, *Air Pollution by Phytochemical Oxidants*, Springer Verlag, Berlin, Germany

Hall, L. M., and M. D. Devine, 1990, "Cross-Resistance of a Chlorsulfuron Resistant Biotype of *Stellaria media* to a Triazolopyrimidine Herbicide," *Plant Physiol., 93*, 962–966

Haselhoff, E., and G. Lindau, 1903, *Die Beschädigung der Vegetation durch Rauch*, Borntraeger Verlag, Berlin, Germany

Holt, J. S., J. J. Sternler, and S. R. Radosevich, 1981, "Differential Light Responses of Photosynthesis by Triazine Resistant and Triazine Susceptible *Senecio vulgaris*," *Plant Physiol., 67*, 744–748

Hull, P., 1974, "Self-Fertilisation and Distribution of the Radiate Form of *Senecio vulgaris* L. in Central Scotland," *Watsonia, 10*, 65–75

Kalchenko, V. A., J. S. Fedotov, and N. A. Arkhipov, 1990, "Dynamics of Mutation Rate in the Progeny of Chernobyl Population of *Pinus sylvestris*," in *Biological and Ecological Aspects of After effects of the Accident on the Chernobyl AEP*, Moscow, Russia, 158 pp

Lolkema, P. C., 1985, *Copper Resistance in Higher Plants*. Doctorate Thesis, Vrije Universiteit, Amsterdam, The Netherlands

Loveless, M. D., and J. L. Hamrick, 1984, "Ecological Determinants of Genetic Structure in Plant Populations. *Annu. Rev. Ecol. Systemat., 15*, 65–85

Mastuura, S., and Y. Fujita, 1995, "RFLPs in Japanese Cucumber Varieties," *Breed. Sci., 45*, 91–95

Meharg, A. A., and M. R. Macnair, 1990, "An Altered Phosphate Uptake System in Arsenate-Tolerant *Holcus lanatus* L., *New Phytol., 116*, 29–35

Mudd, J. B., and T. T. Kozlowski, Eds., 1975, *Responses of Plants to Air Pollution*, Academic Press, London, UK

Peakall, R., P. E. Smouse, and D. E. Huff, 1995, "Evolutionary Implications of Allozyme and RADP Variation in Diploid Populations of Dioecious Buffalo Grass *Buchloe dactyloides*," *Mol. Ecol., 4*, 135–147

Schat, H., and W. M. Ten Bookum, 1992, "Genetic Control of Copper Tolerance in *Silene vulgaris*," *Heredity, 68*, 219–229

Searcey, K. B., and D. L. Mulcahy, 1985a, "The Parallel Expression of Metal Tolerance in Pollen and Sporophytes of *Silene dioica* (L.) Clairv., *S. alba* (Mill.) *Krause* and *Mimulus guttatus* DC.," *Theor. Appl. Genet., 69*, 597–602

Searcey, K. B., and D. L. Mulcahy, 1985b, "Pollen Selection and the Gametophytic Expression of Metal Tolerance in *Silene dioica* (Caryopylllaceae) and *Mimulus guttatus* (Scrophulariaceae)," *Am. J. Bot., 72*, 1700–1706

Sharpe, R. M., and N. E. Skakkebaek, 1993, "Are Estrogens Involved in Falling Sperm Counts and Disorders of the Male Reproductive Tract," *Lancet, 3412*, 1393–1395

Shinokazi, K., and T. Kira, 1956, "Intraspecific Competition Among Higher Plants. VII. Logistic Theory of the C-D Effect," *Journal of the Institute of Polytechnic, Osaka City University Ser. D, 7*, 35–72

Shore, L. S., Y. Kapulnik, B. Ben-Dor, Y.Fridman, S. Wininger, and M. Shemesh, 1992, "Effects of Estrone and 17 β-Estradiol on Vegetative growth of *Medicago sativa*," *Physiol. Pl., 84*, 217–222

Termorshuizen, A. J., and A. P. Schaffers, 1987, "Occurrence of Carpophores of Ectomycorrhizal Fungi in Selected Stands of *Pinus sylvestris* in the Netherlands in Relation to Stand Vitality and Air Pollution," *Plant Soil, 104*, 209–217

Van Breemen, N., P. A. Burrough, E. J. Veldhorst, H. F. Van Dobben, T. De Wit, T. B. Ridder, and F. F. R. Reijnders, 1982, "Soil Acidification from Atmospheric Ammonium Sulphate in Forest Canopy Throughfall," *Nature, 299*, 548–550

Van der Eerden, L. J. M. (1992) *Fertilizing Effects of Atmospheric Ammonia on Semi-natural Vegetation*, Doctorate Thesis, Vrije Universiteit, Amsterdam

Verkleij, J. A. C., W. B. Bast-Cramer, and H. Levering, 1985, "Effects of Heavy Metal Stress on the Genetic Structure of Populations of *Silene cucubalus*," in *Structure and Functioning of Plant Populations* (J. Haeck, J. W. Woldendorp, Eds.), North-Holland/Elsevier, Amsterdam, The Netherlands, pp. 355–365

Watkinson, A. R., 1984, "Yield–Density Relationship: The Influence of Resource Availability on Growth and Self-Thinning in Populations of *Vulpia fasciculata*," *Ann. Bot.*, *53*, 469–482

Westerbergh, A. and A. B. Nyberg, 1995, "Selective Grazing of Hairless *Silene dioica* Plants by Land Gastropods," *Oikos*, *73*, 289–298

Wookey, P. A., and P. I. Ineson, 1991, "Chemical Changes in Decomposing Forest Litter in Response to Atmospheric Sulphur Dioxide," *J. Soil Sci.*, *42*, 615–628

5

Community Ecology and Population Interactions in Freshwater Systems

Bruno Streit (*Frankfurt, Germany*)

5.1 SUMMARY

Effects of pollutants on the biota are observed at various levels of organization, from enzymes to individuals to multispecies communities. Ecotoxicological studies of biological communities are frequently performed on a mesocosm scale. In order to interpret results of toxicity realistically, fundamentals of community ecology have to be applied at a similar depth as those of toxicology and environmental chemistry. An inherent problem with ecotoxicological analyses of communities is noise, i.e., the observation of unpredictable fluctuations in number, biomass, or activity of the various organisms under

Ecotoxicology, Edited by Gerrit Schüürmann and Bernd Markert.
ISBN 0-471-17644-3   © 1998 John Wiley & Sons, Inc. and Spektrum Akademischer Verlag.

consideration. We thus feel that a thorough understanding of community ecology and the working of population interactions is urgently needed for the ecotoxicologist. The goal of the present chapter is to provide the reader with some basic knowledge in this field of ecology, especially with respect to its importance for toxicity test assessments.

5.2 TOXICANTS AND COMMUNITIES

Pollutants can cause effects within ecosystems at suborganismic to superorganismic levels: they can influence molecules, tissues, organs, individuals, populations, and whole communities. In the same way that a system is generally considered to represent more than the sum of the elements, it is an equally fundamental insight that the toxic responses of individuals, taken from a biological community and studied in isolation, cannot predict the response of the more complex systems, such as dynamical population systems, interacting biological communities, or even whole ecosystems. Ecotoxicological tests have to be performed, therefore, at different levels of organization, from molecules to ecosystems.

We will start with two definitions:

◆ A *population* in ecology is a group of individuals of the same species, which, in the ideal case, represents a single reproductive community. Within a given aquatic system, such as a lake, the terms population and species are often used synonymously, as the total of individuals of a species within the ecosystem is usually considered to represent a population. Note that groups of individuals of a given local species sometimes do not represent true populations, but single or multiple "clones" with no genetic diversity within clones. This is observed, e.g., in the extensive "populations" of the common reed in littoral zones of lakes. Also parthenogenetically reproducing daphnids or rotifers are mixtures of clones rather than true populations (e.g., Schwenk and Spaak 1995).*

◆ A *biological community* or "biocenosis" refers to a group of interacting populations (species) in a more or less well-defined ecosystem. This ecosystem may have either sharp boundaries, such as the banks and the inlet and outlet of a lake, or loosely defined boundaries, such as the upper and lower "end" of a mountain stream or a lowland river system. Major biocenoses of

---

* Note that for ecotoxicological test experiments the term "population" is sometimes misused to refer to a number of individuals of the same species that constitute a single test "population" within a vessel. They are of the same size and age (e.g., a group of *Daphnia* or of *Tubifex*) but often do not represent a population in the biological–ecological sense. Depending on the species and the experimental design, they may represent a group of unrelated individuals, or a single clone. Similarly, "communities" of multispecies experiments do not represent biological communities (but rather species assemblages) if they do not consist of long-term interacting populations.

freshwater ecosystems are the planktonic community, the benthic community, the community of lotic river environments, and the communities of subterranean waters. The science that studies factors and interactions influencing the composition and dynamics of biological communities is called "community ecology."

Whereas toxic effects on suborganismic to organismic levels can be easily measured in small-scale laboratory bioassays, effects on populations and communities are more difficult to test; the full set of possible interactions in nature have to be taken into consideration. There are several reasons why the interpretation of pollutant effects on natural or artificial multispecies communities is difficult:

1. Every natural community is unique with respect to its numerical species composition. Results cannot necessarily be directly transferred from one community to the other.
2. A species may be genetically different at various localities or may be conditioned (physiologically influenced) distinctly by local environmental influences of the different aquatic systems.
3. Sublethal effects may be significantly measurable in communities only after a long time.
4. Observed ecological effects, such as an increase in algal biomass as a consequence of ultraviolet-B radiation, may be the result of altered biological interactions, e.g., a more severe damage of the consumer population (in this case chironomid larvae) than of the algae (Bothwell et al. 1994).
5. A community will typically be variable in time with respect to its species and individual composition even in the absence of pollutants. The reason for this can lie in abiotic factors (e.g., change in nutrient supply or temperature regime), in biotic factors (e.g., numerical changes of coexisting species), or even in intrinsic factors (demographically induced stochasticity, genetical changes).

As a consequence, it may be impossible to relate any natural community response unequivocally to the effect of a certain pollutant. Nevertheless, studies of toxic effects on biological communities in the form of multispecies tests are frequently considered important for at least three reasons: (1) they have the potential of revealing combined and interacting effects on natural complex entities most reliably; (2) they will automatically include members of the biological community which differ in their sensitivity and, thus, often include very sensitive forms; (3) they have, therefore, the potential of a sensitive and realistic analysis of a whole community and ecosystem.

Single-species tests, on the other hand, are considered to have their own advantages: species are usually chosen on the basis of being easy to handle, of representing an important part of the ecosystem, or of being sensitive and therefore having the potential of an "early warning" indicator. Single-species tests are much easier to perform and are often cheaper, and their results are easier to interpret than multispecies tests, both statistically and theoretically.

5.3 ABIOTIC AND BIOTIC FACTORS INFLUENCING COMMUNITIES

How do biological communities get established in nature? How constant or fluctuating are species compositions and numerical relationships over time, and how do they respond to environmental pollution stress? In the evaluation of pollution effects at the community and ecosytem levels, a thorough understanding of two sets of factors influencing the survival and fecundity of species is important in order to allow a realistic interpretation of published test results: (1) the direct influences of various abiotic factors, such as temperature, nutrients, and toxicants, and (2) interactions between species. Are biological communities governed rather by a set of abiotic factors or by natural biotic interactions?

Abiotic factors provide species with a set of functional variables such as adequate sunlight energy, temperature ranges, and nutrient concentrations over time, and—in the case of pollutants—may cause direct physiological toxicity. Scholars focusing on the analysis of abiotic factors as determinants of the structure and interactions of communities have collected a huge amount of data on local environmental variables. Variables frequently studied include various elemental concentrations in the water (e.g., O_2, Ca^{2+}, phosphorous), pH, water depth, turbidity, temperature, solar radiation, and stream current. Since biological species are mostly adapted to a set of habitats through evolutionary processes, and not to just a single habitat with a well-defined set of parameters, environmental variables are usually tolerated within a certain range, below and above which the fitness will significantly diminish (i.e., less progenies are produced) or become reduced to zero. It is a general observation in ecology that the more dominant the extremes of one or more physical variables become, the more intensively will they affect the community, e.g., too low a pH value, too high a heavy metal concentration, too long or severe a flood period. For instance, at low calcium concentrations, many biological species with a calciferous skeleton will naturally disappear, such as the majority of molluscs. So this factor will exert a severe impact on the community. A similar but even stronger effect would result from a reduction in pH, either caused naturally or by human activities. As long as the chemical and physical variables are kept within an average frame, however, the community will largely depend on and react to biological interactions, such as food supply and the presence of competitors or predators.

Scholars focusing on *biological interactions* have always been faced with the great problem of the inherent complexity of any community and the fact that every component interacts with every other component to some degree.

A certain critical simplification of analyses and assumptions are therefore a prerequisite for a successful study. Probably the first to publish a diagram of a food web was the North American animal ecologist V.E. Shelford in 1913. The first to grasp the full theoretical significance and importance of food webs was his younger British colleague C. Elton, as described in his book *Animal Ecology* in 1927. He also gave us a modern definition of the ecological niche and coined the term food chain.

Today's concepts of food webs concentrate on their specific patterns and the factors important for their structuring. Patterns of food webs define positional relationships between species that occur more often than expected by chance. This sort of interest in patterns of food webs began nearly two decades ago with Cohen's book *Food Webs and Niche Space* (1978) and was continued with Pimm's book *Food Webs* (1982). Later studies were done by Briand and Cohen (1987), Cohen (1989), and Lawton (1989). More recent reviews concentrating on the specific situation of freshwater systems are, e.g., those by Strong (1992), Hildrew (1992), Wooton (1994), Streit (1995), and various contributions in Streit et al. (1997).

Any biocenosis can be "explained" on the basis of ruling abiotic and biotic factors and on the basis of functional system parameters, e.g., photosynthesis, productivity, respiration, or food web structure. Abiotic factors are by no means a separate category of influences, but are frequently intimately related to biotic ones, in that abiotic factors may change individual life histories of certain species, which themselves will affect survival of other species by means of altered biological interactions and, therefore, also the food web structure. We thus come to three more definitions important for an in-depth treatment of the topic:

◆ A *food chain* is a sequence of organisms in a trophic system (i.e., an energy transfer system of a biological community). In this system, some organisms, representing the higher trophic level, typically feed upon certain others, representing the lower trophic level.

◆ A *food web* is a trophic system composed of interconnected food chains. In natural systems, most so-called food chains are rather food webs when studied in detail. Yet, a few dominating food chain structures are typically observed even within complex food web communities, so that it is frequently acceptable to talk of food chains even in the case of an overall food web structure.

◆ *Life history* in ecology comprises those features or traits of an organism that influence survival and fecundity. Among others, they include reproductive rate, age at maturity, and reproductive risk, most of which are affectable also by environmental pollutants. These traits are usually considered to be under evolutionary selection.

We now have some basic conceptual as well as terminological instruments for a more in-depth treatment of aspects of community ecology and to relate recent results and concepts in this field to those of ecotoxicology in the following sections.

5.4 WHY CONCENTRATE ON FRESHWATER SYSTEMS IN THIS CHAPTER?

Each of the major ecological systems—freshwater, marine, and terrestrial—needs a specific approach for a detailed study and treatment. The present chapter will concentrate on biological communities of freshwater ecosystems. To focus on freshwater systems more profoundly than on others can be justified for various reasons.

Many tests in ecotoxicology are performed by using limnic prokaryotic and eukaryotic unicellular organisms, freshwater vascular plants, or freshwater animals. They are used because of their importance for testing freshwater as a drinking water resource for humans. They are also used because of the relative simplicity of test procedures and the ease of establishing well-defined environmental variables, such as homogeneous temperatures. Freshwater organisms have therefore been considered "model systems" for the study of fundamental aspects of general population biology (e.g., Streit and Städler 1995) and community ecology (e.g., Carpenter 1988). They are in many respects better understood in their underlying regulating mechanisms than many terrestrial or marine communities, and research on them has a long tradition (e.g., Streit 1995).

The major principles found by studying freshwater ecological systems can nevertheless be applied also to the other ecological environments. This chapter will first present some traditional concepts on the community in freshwater systems and then proceed to current ideas on the community ecology of freshwater ecosystems and try to explain which factors are important in regulating sizes and densities of natural populations.

5.5 THE BIOLOGICAL COMMUNITY: SUPERORGANISM OR LOOSE SPECIES ASSEMBLAGE?

For practical reasons it is convenient to subdivide the continuum of physical appearances of freshwater environments and their communities into a few major types. In lakes and other nonflowing freshwater systems, the two major communities are the pelagic communities (plankton and nekton) and benthic or benthonic communities (benthon) on or near the bottom of lakes. In flowing water systems, a distinction between lotic (fast-flowing) and lenitic (calm) environments is frequently made. Here, the benthic communities are generally more prominent than the pelagic ones, which are even absent in small and fast-flowing rivers (Figure 5.1). Examples of minor types of communities are the pleuston and

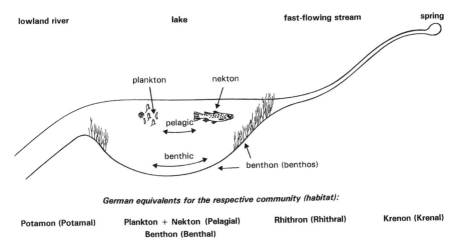

German equivalents for the respective community (habitat):

Potamon (Potamal)   Plankton + Nekton (Pelagial)   Rhithron (Rhithral)   Krenon (Krenal)
                              Benthon (Benthal)

Figure 5.1 Diagrammatic representation of the major freshwater communities in a spring, mountain stream, lake, and lowland river. Note that the ending '-al' refers to the respective environment, whereas the ending '-on' refers to the respective community (benthon and benthos are synonyms, but 'benthon' is preferred). The English equivalents for the German terms 'Potamon' and 'Potamal', 'Rhithron' and 'Rhithral', and 'Krenon' and 'Krenal' are currently unusual. Also note that the term 'habitat' is rather infrequently used in aquatic systems, in contrast to terrestrial ones.

neuston communities at the surface of lenitic areas of aquatic systems. From a functional point of view, the various subcommunities are interconnected in most aquatic systems to form the local biological community.

Freshwater communities are frequently composed of hundreds to thousands of animal, plant, fungal, protist, and bacterial species, which constitute the biological community as a whole. In the past, communities and ecosystems have been considered as a "superorganism" or a "community organism" by various researchers. The reason was that they seemingly exhibit features analogous to organisms, such as juvenile stages (e.g., a newly created glacial lake with few organisms), maturation (e.g., a sequence of increasing nutrient supply as the lake gets older and shallower), and aging with the final stage of individual death (e.g., the final stage of a lake being a swamp or finally a forest ecosystem). The high complexity and (assumed) stability of mature ecosystems and communities, in this case of nutrient-poor to nutrient-rich lakes, was considered to be linked to a high information content of the systems, and hence increased feedback loops, resulting in a high level of self-organization (e.g., Margalef 1968; Odum 1969). The comparison with individual organisms is a very rough analogy, however, as ecosystems provide many features and characteristics different from individuals.

There have been strong counter-positions to this approach, such as the "individualistic concept" of Gleason (1926), a plant ecologist who emphasized the independent reactions of populations to environmental factors. However, this concept too was realistic only to some extent. For instance, the process of colonization of a newly formed habitat (e.g., a basin filled with water) initially shows this kind of behavior. The more species it contains and the more their populations reach the carrying capacity of the system, the more important interactions between species become. These interactions include intra- and interspecific competition, predator–prey interactions and parasite–host relationships, chemical interferences, aspects of genetic heterogeneity, selection and hybridization processes, and the increasing impact of invading species from other geographical areas. The latter include new competitors (e.g., exotic plants competing with local ones), predators (e.g., locally introduced fish species feeding on amphibia), or parasites (e.g., the fungus species introduced from North America that caused the near-extinction of the European crayfish, *Astacus astacus*). Population interactions will thus have to be an integral part of any theory of biological communities in freshwater systems. The population biological interaction approach represents an alternative to the more mechanically operating energy- and nutrient-based approach. Population interactions can be studied from aspects of demography and mathematics, from population genetics and coevolution, and from detailed experiments and observations of manipulated complex communities. (For some recent general considerations on aspects of holistic vs analytical approaches see, e.g., Rigler and Peters, 1995, and Calow 1995).

The following part will exemplify the problems of determinism and unpredictable fluctuations in biological communities. It was shown mathematically that increased complexity does not necessarily result in increased stability against external disturbances and stress factors, including pollutants, and that the respective analogy between the properties and stabilities of a mechanical web with those of a so-called food web is unsuitable (see May 1972, 1973, 1974). Mathematically, nonlinear interrelationships between populations (representing the "elements" of the system under consideration) may easily result in unpredictable dynamics of population systems in space and time. May showed that external factors influence ecological subsystems with totally unpredictable consequences for the whole system. Further, genetic processes, such as mutations, hybridizations, and other kinds of genetic rearrangements, can lead to new modes of interactions and are practically unpredictable.

But not all aspects of ecosystems are unpredictable. Some quantities and qualities are highly predictable, such as energy flow and certain relationships between trophic levels, if certain environmental information is available. Also, we can predict with a more or less high probability the predominance of certain species under a given set of environmental parameters (known abiotic factors, known set of species available), at least for some time period. But a deterministic (i.e., 100% accurate) prognosis will never be possible, and the longer the period for prognosis is, the less predictable are statements on numerical compositions of species assemblages.

A long-term study of *Daphnia* dynamics in a North American lake may illustrate this kind of problem. In Oneida Lake, a shallow and nutrient-rich lake on the Ontario Lake Plain of New York, structural changes in the fish community were matched by shifts in species composition of the zooplankton community. *Daphnia galeata* and *Daphnia retrocurva* were the most abundant species from 1964 to 1969; *Daphnia pulex* dominated throughout the 1970s, but *D. galeata* reemerged to become the predominant daphnid after 1981 (Figure 5.2). Disappearance and reappearance of *D. galeata* was associated with changes in abundance of the youngest year class of yellow perch and other zooplankton-feeding fish (Mills and Forney 1988). Although retrospective explanations are possible to some degree, predictions would remain relatively unprecise for future situations and are possible only on the basis of probabilities.

5.6 CLASSIFICATIONS AND CHARACTERIZATIONS OF FRESHWATER COMMUNITIES

Freshwater ecosystems and their communities have been classified many times and from various perspectives in the past. Classifications were based either on (1) *phenomenological* characteristics, such as certain chemical properties or the occurrence or absence of certain species, or according to (2) *functional* and *dynamical characteristics*, such as primary productivity (low and high productive ecosystems) or the energy source for the ecosystem (autotrophy- vs heterotrophy-based ecosystems). Due to the highly variable nature of freshwater ecosystems, a great number of classification systems were created, some of which have been in use for a long time, whereas others have been abandoned or are of historical interest only.

The two types of classifications cannot be sharply separated from each other, as dynamical models often are based on phenomenological characterizations. A classic example of the latter would be the distinction between lakes of different oxygen concentration curves within their water bodies. It was only after the inclusion of photosynthesis and respiration rates of higher and lower parts of the water body into the model and the accumulation of knowledge about water exchange between the various parts of the lake, that a functional model could replace the static one. We will use this example in the following as the first of various types of classifications and characterizations:

1. A traditional approach to lake typology was based on the observation of early hydrobiologists that certain chemical characteristics, especially oxygen concentration, correlated with the occurrence or absence of certain invertebrate species, which therefore were considered to be indicative of the respective chemical (and some physical) characteristics of the lake. This kind of approach was conceived independently during the 1910s by the Swede E. Naumann, who coined the terms "oligotrophic" and "eutrophic" for lakes with few and extensive nutrients, respectively, and the German A. Thienemann, who coined the (now only historical) terms of "Chironomus lake" and "Tanytarsus lake." Oligotrophic and eutrophic lakes could be

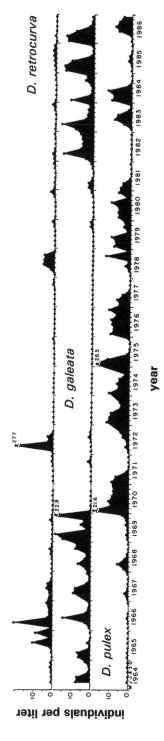

Figure 5.2 Irregular changes in mean density of three *Daphnia* species (*D. retrocurva, D. galeata, D. pulex*) in May–October in Oneida Lake since 1964. This alkaline lake is situated some 20 km northwest of Syracuse, New York, at 112 m above sea level. Its mean depth is only 6.8 m and it contains nearly 60 fish species and 19 zooplankton species. From Mills and Forney (1988).

described by a whole set of characteristics (Thienemann 1925). Oligotrophic lakes exhibited a low nutrient supply for algae and for subsequent trophic levels and, thus, a low overall productivity, a high oxygen concentration from the surface down to the bottom, and typically the presence of larvae of the midge genus *Tanytarsus* in the lake sediment. Eutrophic lakes exhibited a higher nutrient supply, a higher productivity, a reduced or even vanishing oxygen concentration towards the lake bottom, and larvae of the midge *Chironomus* as a characteristic indicator organism.

2. Other typologies, forwarded by fish biologists, used various fish species to characterize lakes, e.g., species of the salmon family (Salmonidae), the white fish (Coregonidae), or the carp family (Cyprinidae).

3. Running water systems were classified according to hydraulic properties or according to the degree of pollution. In running water systems, it has become common to correlate organic pollution by biochemically degradable material with the local biocenosis. Thus, polluted areas could be categorized by the presence or absence of a set of characteristic species of invertebrates, algae, and microorganisms, which thus represented indicator species (Kolkwitz and Marsson 1902; Steinmann and Surbeck 1918). The resulting categories were those of saprobic zones, from oligosaprobic to polysaprobic. They were later largely replaced by "water qualities," which were defined also on the basis of direct chemical measurements (in Germany typically represented by different colors and denoted as "Gewässergüteklassen").

4. A different typology for communities in running water systems is based on hydraulic properties and certain biological–ecological features of the community as a direct response and adaptation of the various species to these properties. The velocity of flowing and the extent of turbulence, the average depth, and the development of the temperature regime all change unidirectionally from mountainous to lowland areas. The major subcommunities are the communities of the spring region (the corresponding biotope in German is called the "Krenal"), fast-flowing stream communities (in German called "Rhithral"), and communities of lowland rivers (in German called "Potamal"). The environment of the icy waters of glacial areas, allowing only very few species to survive due to extremely low temperature and poor nutrient supply, is called the "Kryal" in German. Each of these parts has its own set of characteristic organisms with morphological, physiological, or ethological properties corresponding, mainly as a result of adaptation, to the physical characteristics of their environment.

5. A purely numerical approach for both running and nonrunning water systems is the characterization of the respective communities according to low or high species richness. This kind of approach had at least two bases: (1) the observation that in heavily polluted areas only one or a few species dominate numerically (e.g., tubificid worms); (2) the idea that ecosystems with a low species richness could be considered "immature" and "on the way" of reestablishing. The information content of these systems, due to fewer

feedback loops with other species, is lower than that of undisturbed systems. An index based on information theory therefore seemed to be suitable to express the degree of stress exerted on the respective system. The most widely used index is the Shannon–Weaver index (Shannon and Weaver 1949), but indices based on other mathematical models are all in use. Although the Shannon–Weaver index does not provide any real data on "information content" of the communities, it is used because it is easy to calculate: in order to calculate this index, it is even not absolutely necessary to determine single species correctly; they may just be named "species A", "species B", etc. In certain multispecies toxicity tests, species richness is a frequently used toxicological "end point."

6. Categorizations based on the role of each species in the ecosystem represent a functional approach. The question of why there are so many species in an ecosystem can in part be answered by the various functions and microenvironments the community has to cope with. The major subdivisions are oxic and (continuously or temporarily) anoxic environments. Among the animal community, some representatives are adapted to crack mechanically firm substrates by adequate biting mechanisms; others are adapted to degrade chemical compounds enzymatically, which are otherwise highly stable, such as lignin or many of the anthropogenic compounds. Further, there are organisms adapted either to live in sediment, on rocks or vascular plants, or in the pelagic environment, either passively or actively swimming. The function of a species within a working ecosystem is called its *ecological niche*. Species that are functionally similar, though not identical, are said to form a guild (see Section 5.7). Natural aquatic systems provide a great many ecological niches, which is usually considered favorable for system stability against anthropogenic influences.

7. Another functional approach is energetical analysis, whereby low- and high-productive aquatic systems can be distinguished. Whereas two-level interaction studies have been performed many times, multilevel interactions have been rarely studied; this is a consequence of the complex matter and the often long-term fluctuations of multispecies systems. In a very general sense, low productivity is caused by low nutrient supply, high productivity by high nutrient supply (see also point 1; adequate techniques to measure productivity are based on ^{14}C incorporation by algae within exposed bottles containing ambient water samples). The energetical approach allows estimates of the productivity of overall consumer populations, such as fish. Yet, it may fail to predict specifically which species will be the dominant one, as the numerical success of single species not only depends on food supply, but also on biological interactions. These include so-called exploitative and apparent competition, direct and indirect predator influences, parasitic impact, and various kinds of indirect mutualism, chemical interferences, and so on.

8. Recent functional approaches are therefore largely based on population biological interactions. They include aspects of energy flow and specific

functions in the sense of ecophysiological properties of the species, highlight all kinds of interactions between the various populations of complex food webs, and are flexible in the use of various theoretical perspectives. They use dynamic interaction models as well as various kinds of more complex community system models. For instance, they focus on the nature and flexibility of the trophic position of a species within a guild and the way they alter resource partition as a consequence of changing nutrient supply or variable predator influences. A special version of the population biological interaction approach is to look at communities from the perspective of island biogeography theory (MacArthur and Wilson 1967). The origin of this theory lies in the observation that isolated habitats, such as oceanic islands, will mostly reveal a certain percentage of species invading the system per time unit from outside, whereas a certain percentage of species is disappearing due to local extinction. This model is applicable for many lake or river systems.

In the following sections we will concentrate on the latter three functional approaches (i.e., points 6–8).

5.7 MANY NICHES FOR MANY SPECIES

The nutritional, and thus the energetic basis of freshwater communities consists of either aquatic primary producers such as algae, or of preformed organic material such as decaying vascular plant material, derived from within the water body or from adjacent trees, or of organic pollution introduced into the water system. A usual distinction, therefore, is between ecosystems based on local primary producers (autotrophy-based ecosystems), and detritus-based ecosystems (heterotrophy-based ecosystems).

The first type (*autotrophy-based ecosystems*) is typical for intact ecosystems with an equilibrium between primary production and decomposition rates within the same water body. In nutrient-based ecosystems with a considerable growth of algae, grazers form the first link of the food chain. In lakes these may be planktonic herbivores, in rivers they are mostly represented by benthic herbivores. Macrophytes (higher vascular plants) and various forms of large algae are usually eaten only after decay and so provide the energy source for the second type of ecosystem.

The second type of ecosystem (*heterotrophy-based ecosystems*) can occur in various forms: if preformed material such as leaves or branches are important energetical sources, shredders will act as primary consumers of this detritus-based food chain. If dissolved organic compounds, such as carbohydrates or organic acids, are important energetical sources, various kinds of microorganisms, bacteria, and fungi will function as (heterotrophic) primary producers. Their biomass will be transferred into the detritivore biomass, like that of filter-feeding simuliid larvae or tubificid worms; from a trophical standpoint they act as primary consumers in this case. To some extent, small organic particles and flocks, e.g., fecal pellets, will constitute the energetical basis for sediment eaters and filterers. It is especially the adhering microbial biomass that contains a high nutritional value for them. In all ecosystem types, there exists an intimate interrelation between bacteria, fungi, algae, aquatic vascular plants, and the animal components of the ecosystem.

As has become clear from the above-mentioned interrelationships, many heterotrophic species in freshwater environments take up food from different trophic levels and can thus be termed omnivores. This is true for the zooplankton as well as the zoobenthon communities. In a certain sense, it is more convenient therefore to classify consumers according to other criteria: it is the physicochemical state in which a specific nutritional organic compound occurs in the environment that will actually determine the consumer type feeding upon it. Solved carbohydrates are usually taken up effectively by bacteria or fungi. If the carbohydrates constitute solid particles (e.g., of cellulose), however, they will be taken up by organisms adapted to the uptake of detrital particles, potentially preceded by adequate mechanical breakdown. Further, if the food particles are small and suspended in the water they will be taken up by other species than if they are sedimented. The diversity of food sources is therefore the basis for the co-occurrence of a multitude of functional organismic groups in any natural habitat. If any major type is completely absent from a system for any reason, this may result in an accumulation of the respective organic material (accumulation of debris in the absence of adequate detritivores, explosive algal growth in the absence of grazers, etc.).

We can distinguish at least four major categories of *feeding niches*, based on mechanical food uptake mechanisms for animals. These do, however, not directly correspond to the major categories of the food chain, i.e., herbivores (feeding on living algae or higher plants), carnivores (feeding on animals), and detritivores (feeding on dead particles, usually with adhering microorganisms). The four categories are (based in part on Cummins (1974) and Streit (1995)):

1. Shredders: they chew, mine, or bore large particles such as leaves, stems, and branches, which are mostly dead. Examples are the crustacean species *Gammarus* and *Asellus* as well as several insect larvae. Shredders are typically detritivores.

2. Grazers, including scrapers: they graze and scrape the periphyton off other surfaces. Examples are many snails and mayfly larvae. Also herbivore filterers are often called grazers, especially the herbivorous zooplankton species. Grazers are either primarily herbivores (in environments rich in suspended algae) or omnivores (e.g., when grazing unselectively on surfaces).

3. Filterers: these encompass passive filterers (or collectors; e.g., simuliids, using the stream current for food particle supply) and active filterers (or gatherers; e.g., mussels and many fish). They filter particulate matter, alive or dead, from the water or from the sediment. Examples are daphnids and mussels, but also

tubificids and planktivorous fish. Filterers are either herbivores (e.g., daphnids), carnivores (e.g., fish), or detritivores (e.g., simuliids, grazing on detrital particles in polluted streams).

4. Engulfers and piercers: engulfers eat whole prey items (e.g., many fish that feed upon small invertebrates), piercers pierce the prey and suck fluids out (this mechanism leads towards a parasitic mode of feeding, e.g., in leeches). Engulfers and piercers are either true carnivores, parasites, or parasitoids (a special parasite-like group within insects), but may also feed upon algae (e.g., *Chaetogaster* species).

The problem of niche separation and niche overlap is discussed in more detail below.

The reason why the phytoplankton community is said to represent a guild (as explained in Section 5.6, point 6) is because all algal species have basically similar requirements for their species-specific set of nutrients, which is a consequence of the universal composition of the major constituents of living cells. Yet the requirements differ for at least two reasons: first, some algal species need special elements that others do not need at all or not to the same extent, e.g., silicium for the skeleton of the diatoms. Second, their metabolic capacities for uptake vary in that, e.g., some species need a high external concentration to build up their body biomass; for instance, with regard to phosphorous this is true for euglenids, which therefore are predominantly found in phosphorous-rich environments. Other differences between algae are found with respect to the presence of certain vitamins (e.g., cobalamin) in the ambient water, since they cannot synthesize them, as is the case with humans.

Also many freshwater zooplankton species resemble each other with regard to preferred food particle sizes. The reason lies in the linear dimensions of their uptake mechanisms (e.g., size of filter-feeding setae in daphnids), which cause a partial niche overlapping. The observation that the phyto- and zooplankton communities represent guilds and nevertheless allow numerous species to co-exist in nearly all lakes has been called the paradox of the plankton (Hutchinson 1961). It is now explained in part as a consequence of species-specific uptake mechanisms, fluctuations in abiotic ambient conditions and in complex biological interactions, as discussed more thoroughly in Sections 5.8 and 5.9.

The degree of niche overlap between related species is exemplified by the herbivore zooplankton guild. The species of this grazing community, consisting primarily of small crustacean species, actually prefer slightly different food items in terms of size (e.g., maximum length of diatoms), quality (e.g., excluding cyanobacteria), and quantity (different minimum concentrations needed to allow a positive individual energy balance). The minimum concentration needed to sustain a viable population is called the threshold concentration of the respective species. Comparisons between different species of daphnids have revealed that under steady-state and low-mortality conditions, the threshold concentration lies between 0.01 and 0.04 mg C per liter for *Daphnia* and

threshold food concentration (μg C per liter)

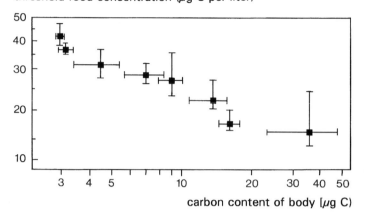

carbon content of body (μg C)

Figure 5.3 Threshold food concentrations for various daphnid species (genera *Daphnia* and *Ceriodaphnia*) taken as the intercepts of the regression lines with the zero growth level plotted against species-specific body size expressed as body carbon of 6-day-old animals grown at high food concentrations (green algae). Adapted from Gliwicz (1990).

Ceriodaphnia species of different body sizes (*Daphnia magna* being the largest, *Ceriodaphnia reticulata*, the smallest species; Gliwicz 1990; Figure 5.3). For planktonic rotifers, the corresponding figures range from 0.03 mg C per liter in *Keratella cochlearis* to 0.4–0.5 mg C per liter in *Keratella crassa* and *Brachionus rubens*; thus, compared to daphnids, the range for rotifers is shifted toward more eutrophic conditions (Walz 1995).

Grazers among the zooplankton are not purely herbivorous (algae consuming), but also omnivorous; they feed on a variety of food particles of adequate size, including various heterotrophic forms such as flagellates or even detrital particles. Bacteria are taken up passively, as they are frequently attached to detrital particles. Nevertheless, various zooplankton species are capable of actively rejecting individual items, e.g., cyanobacteria, that are usually of low nutritional value.

5.8 TWO-LEVEL TROPHIC INTERACTIONS

Populations of freshwater communities are maintained by the energy flow through the ecosystem. In the pelagic communities, simultaneous processes of build-up and degradation of organic material are observable with usually a surplus of production in the epilimnion that is transferred by sedimentation into deeper zones of the lake, where degradation processes become predominant or even exclusive.

The size and dynamics of populations in natural aquatic systems is roughly determined by two factors: the supply of nutrients, which directly favor algal growth and indirectly all other trophic levels, and the impact of potential predators on trophic levels below them. The two opposing effects on the numerical status of any population are called *bottom-up* and *top-down* effects. A bottom-up effect in a pelagic environment means that a high concentration of nutrients increases primary productivity and subsequent energy transfer from algae to small zooplankton, from them to larger zooplankton, then to planktivorous (often small) fish, and finally to piscivirous fish or birds. The top carnivores can lead to depression effects on consumer populations by feeding pressures. Should top carnivores become abundant and exert a strong pressure on planktivorous fish, these may decline in numbers despite a possibly extensive supply of zooplankton (top-down effect; Figure 5.4). The zooplankters may actually become very abundant as their direct predators become scarce and they may consume extensively on the algae, which may thus be dramatically reduced in numbers (known as zooplankton 'grazing effects'). These multitrophic interactions with differential impacts on the various trophic groups is called the trophic cascade.

Both effects—top-down and bottom-up—can operate at most levels of the food chain. The population densities are also influenced by processes such as competition, host–parasite interactions, and abiotic factors. The questions of how effective energy transfer is, what transfer efficiencies mean for maximum

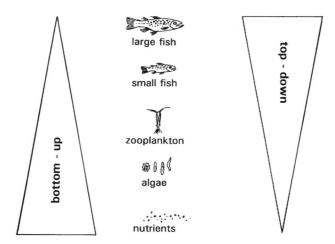

Figure 5.4 Bottom-up vs top-down effects. Populations that represent major trophic levels are regulated either by nutrient (and energy) supply at the basis of the energetic pyramid, or by the predator pressure of populations one or more trophic levels higher. The first is called 'bottom-up effect', the second, 'top-down effect'. The relative influence of the two directions of regulation is usually variable. Both effects can be manipulated, e.g., by increased fertility, or by introduction of top carnivores into the system. Influence of toxicants can profoundly alter actual regulation interactions.

food chain length and for food web complexity, and what the potential impact of pollutants on these energetic interrelationships may be are discussed below.

The number of trophic levels possible within a biological community is determined by the set of ecological efficiencies of energy transfers between the members of the community. These include the transfer efficiency of ingested food into production of the respective population. The gross production efficiency (P/C, "production to consumption [= ingestion] rate") is composed of two independent efficiencies: the assimilation efficiency (A/C, "assimilation [or absorption] to consumption rate") and the net production efficiency (P/A, "production to assimilation rate"). The lower the assimilation efficiency, the more the organism eliminates nonused energy, largely in the form of feces. The lower the net production efficiency the more the organism uses for respiration and other nonproductive purposes (e.g., sliming, fugitive behavior).

Assimilation efficiency (A/I) will naturally differ between different food types, since, e.g., animal tissues are generally more digestible than plant tissues. Also, feeding on easily degradable fresh algal or microbial items results in higher percentages of food absorption compared with the feeding of poorly degradable material, such as washed-out leaves or branches. Extensive surveys on assimilation efficiencies in benthic invertebrates have shown that we can distinguish different functional niche groups (as characterized in Section 5.7), which exhibit also different assimilation efficiencies (Figure 5.5, left part).

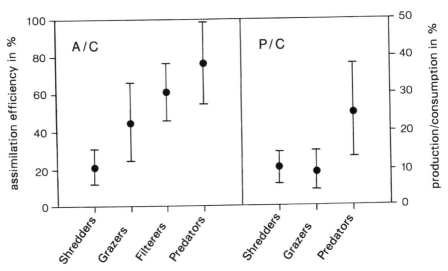

Figure 5.5 Energetic parameters for benthic freshwater invertebrates with different feeding mechanisms. The left part of the figure represents the assimilation/consumption efficiency (A/C), indicating the effectiveness of the digestion process; the right part shows the production/consumption efficiency (P/C), indicating the effectiveness of tissue production from consumed (= ingested) food. From Streit (1995)).

Assimilation efficiency differs for different compounds of the food particles. Usually, protein constituents are assimilated to a high percentage, cellulose to a much lower percentage or not at all (usually only with the symbiotic interaction of microorganisms within the gut system). Similar differences are observed for the various metals, depending also on speciation (oxidation states, chelate structures, etc.). Organic environmental pollutants are absorbed at highly variable efficiencies from nearly 0% to nearly 100%. In aquatic ecotoxicology, assimilation efficiency with regard to pollutants is usually called either absorption efficiency or dietary or oral bioavailability.

Net production efficiencies (P/A) are generally lower in warm-blooded than in cold-blooded organisms. They are also low in animals with complex life cycles or with major biological requirements other than growth and reproduction, e.g., territorial behavior or extensive migrations. They typically range from 1% to 5% in aquatic bird and mammal species and are frequently around to be 10% in long-lived ectotherms such as fish or molluscs, and between about 10% and 40% in short-lived herbivorous invertebrates. Temporarily, the values may even exceed 50%, as reported from temporary growth phases in fast-growing zooplankton species and in some carnivorous invertebrates. For overviews and more detailed examples, see, e.g., McNeill and Lawton (1970), Humphries (1979), and Streit (1975, 1976, 1995).

In experiments it is frequently easier (and for ecological and ecotoxicological purposes more adequate) to combine A/C and P/A to the single net production efficiency P/C. Both A/C and P/A seem to be variable even within single groups of benthic freshwater invertebrates, as the mechanisms for food gathering obviously operate at highly different efficiencies. Figure 5.5 (right part) provides some examples of these combined P/C values for the groups mentioned above.

Efficiency values vary a lot under various biological and ecological conditions and also under various kinds of stress situations. The latter include natural ones, such as predator avoidance behavior, and anthropogenic ones, such as thermal or chemical pollution. Thermal pollution may directly influence biochemical transfer efficiencies, as enzymes may be forced to operate outside optimal thermal ranges and thus with reduced biochemical efficiencies. Also chemical pollution has definite effects on production efficiencies, as shown, e.g., by applying herbicides to experimental conditions (Streit and Peter 1978). The reason may only in part lie in the surplus energy needed for transformation of the xenobiotics; the by far more important part of additional energy uptake seems to be caused by complex organismic reactions and deviations from normal behavior, such as extensive sliming or hyperactivity.

In addition to the energetic efficiency parameters presented, other parameters and activity rates are in use for community and ecosystem characterizations. For instance, rates of ingestion or production based on mean biomass of individuals are usually termed "specific rates" (e.g., specific ingestion rate, specific assimilation rate, specific growth rate) of the respective individuals or populations. Their use allows easy comparison of energy transfer and elemental turnover rates for communities and ecosystems. The most widespread are specific production rates of individuals and populations, for which extensive tables have been compiled (Zaika 1973). Other important parameters include production to biomass (P/B) ratios and production to respiration (P/R) ratios, both based either on populations or whole ecosystems.

The meaning of the various efficiencies and ratios for practical purposes varies. Whereas in fish ponds a high production rate (not respiration rate) is desired, in sewage ponds, a high respiration (and thereby potentially a high degradation rate of biochemically degradable material) is the ultimate goal, not a high production rate. Pollutants can affect both production and respiration, although an increase in respiration rates is much more common.

5.9 MULTILEVEL TROPHIC INTERACTIONS

Species assemblages, population densities, and other aspects of community patterns are based primarily on the energy supply and predator pressures, i.e., bottom-up und top-down effects, as explained in the previous section. But these factors are not the only ones that determine the dynamical system behavior of biological communities. There are to some extent fluctuating and variable structural components that can rearrange under the influence of external factors. These include (1) variable abiotic factors, (2) changes for stochastic reasons (as outlined in Section 5.5), (3) human impacts such as fishing or release of new species, and (4) seminatural invasions of exotic species.

The majority of our knowledge of complex interactions in communities comes from pelagic studies in lakes, focusing on plankton and fish, rather than from benthic communities. The major reason for this may lie in the fact that pelagic environments are relatively homogeneous, easy to describe, sample, and manipulate, and that their communities are frequently less speciose than those of the benthic environments. In the latter, the patchiness is an inherent characteristic of the habitat and makes analysis more difficult. Few multilevel trophic studies have really focused on these benthic environments; some recent ones are those from Brönmark (1992), Rosemond et al. (1993), Wooster (1994), and Schlosser (1995). We will therefore give examples of multilevel interactions from the plankton community and from isolated tank experiments and microbial systems below.

Seasonal shifts in numerical species compositions and size distributions of zooplankton communities have been observed in many temperate lake systems. A frequent kind of shift is from the predominance of rather large zooplankton species early in spring to smaller ones as the summer progresses. Explanations have been found in seasonal changes of adequate algal food sources and in various forms of interactions. Grazing efficiencies of zooplankton are usually dependent on the algal species and predator pressures by planktivorous fish species. The latter may be the juveniles of larger species (which often feed on zooplankton and naturally occur in highest densities after their hatching period) or they may be special planktivorous fish species. Since planktivorous fish take up preferentially larger zooplankton prey, which they can visually localize more easily, smaller zooplankton taxa have been found to be selected for as a consequence of this kind of feeding pressure.

This observation was first published by Brooks and Dodson (1965) and has become known as the "size-efficiency hypothesis." It was originally based on observations of plankton communities in Crystal Lake, Connecticut (USA), after the introduction of a planktivorous fish species (Figure 5.6). The lake had not contained planktivorous fishes before. The hypothesis is based on the combined effects of competition between two competing zooplankton "subguilds" (large and small zooplankters) and the selective influence of predatory fishes. The authors believed that in the absence of planktivorous fishes, all zooplankters competed for the 1- to 15-μm particulate matter in the water and that the larger zooplankters were more successful. Small animals were thus thought to be excluded by starvation if large ones were present. Fish, however, and also amphibians, where occurring, select larger Crustacean plankton, such as cladocerans and adult copepods. Depending on the intensity of predation, smaller zooplankters, such as small species of crustaceans, and rotifers, may be able to coexist or even replace large forms where predation is intense.

Later studies forced the hypothesis to be modified. They failed to demonstrate any greater feeding efficiency of large zooplankton species compared with smaller ones. The predominance of larger species when vertebrate predation on

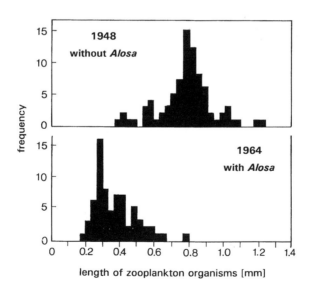

Figure 5.6 Size distribution of the mainly crustacean zooplankton of Crystal Lake, Connecticut (USA), before (1942) and after (1964) introduction of a planktivorous fish, *Alosa aestivalis*. The effect of the fish has been to replace a community of large species by one comprised of smaller species (according to Brooks and Dodson (1965).

them is relaxed seemed to rest on size-selective predation of the smaller zooplankton species by carnivorous plankters in the same way as the predominance of the small forms depends on vertebrate predation of the larger ones (Dodson 1974). A critical review on the topic was provided by Hall et al. in 1976.

So far, we have confined ourselves to biological interactions within the animal components of communities. However, dynamic interactions, which are beyond direct energetical influences, are observed also between primary producers, various consumer levels, and the microbial community. An example of complex relationships between nutrients (and consequently primary production) and consumer densities was presented by Leibold and Wilbur (1992). The authors found that zooplankton biomass in mesocosm tank experiments could either increase or diminish after the addition of nutrients, depending on food web structure. For instance, phytoplankton concentration increased in experiments with either tadpoles (frog larvae of the genus *Rana*) alone or *Rana* and *Daphnia* together, but when *Daphnia* alone or neither of the two grazers were present, phytoplankton concentration decreased. The exact cause for these complex interactions with the other zooplankton representatives remains to be determined. From a mathematical standpoint, Abrams (1993) has shown that it can be possible for increased nutrients to decrease population sizes at all levels when there are more than two species per level. This seems highly paradoxical, but similar phenomena in networks of interconnected elements have been shown in other systems, including technical ones. For instance, increasing the number of roads in a traffic network can decrease transit speed for all travelers (Cohen et al. 1990).

Such experiments and theoretical considerations illustrate that changes in nutrient supply can alter energy flow, food web structures, and biological interactions dramatically and that they can operate at complexity levels and system sizes used typically for multispecies toxicity tests.

Other authors have focused attention on the question of whether natural population fluctuations depend on the length of a food chain, and whether food specialists (feeding on only a few selected prey species) lead to more or less population fluctuations compared with nonspecialists (especially omnivores). Such a study, using a small-scale experimental design with bacteria and various species of protozoans, was performed by Lawler and Morin (1993). They found that population fluctuations and extinctions both increased with increasing food chain length and that omnivorous protozoans endured fluctuations in prey abundance better than specialists feeding on fewer types of prey.

5.10 NONTROPHIC INTERFERENCES

Besides trophic interactions, other kinds of interferences between species may vastly influence community structures, two of which are mentioned here: chemical interferences and the influence of species invasions, caused mainly through anthropogenic activities, which have strongly altered many original community structures and will further alter them.

Chemical interactions can be encountered in two different forms:

1. Some species simply harbor harmful chemical compounds within their body cells, so-called toxins, and are thus unpalatable to others. The compounds are usually stored within the individuals, but can also be released, e.g., by algae that form water blooms. Well-known examples among the primary producers are many cyanobacteria, which are of poor food quality or even unpalatable to many aquatic herbivores. Examples from the animal kingdom can be found especially in various aquatic coleopteran and hemipteran insects (Dytiscidae, Gyrinidae, Notonectidae). These insects have special glands to store distasteful secretions (e.g., Scrimshaw and Kerfoot 1987).

2. A second type of chemical substance is released by some species and influences growth, fecundity, or survival of other species, thus exerting a severe control over local species compositions. The classic study with freshwater species on this topic focused on interactions between various rotifers, especially *Brachionus* and *Asplanchna* species (e.g., Gilbert 1967), in which the predatory *Asplanchna* releases an organic compound altering morphological characteristics of the *Brachionus* species and thus influencing interactions. Besides rotifers, also the morphology of some daphnids is influenced by certain substances, especially by inducing the formation of a helmet. A more recent study has shown that the final size of *Daphnia* species may be influenced by the presence of certain vertebrate or invertebrate predators (e.g., Stibor and Ling 1994): *Daphnia hyalina* individuals that had been transferred into "conditioned water" (previously in contact with predatory fishes) allocated less energy into growth and thus remained smaller. Invertebrate predators, such as the planktonic dipteran larvae of *Chaoborus*, tend to elicit the opposite allocation shift. Such a physiological and life history response can be considered adaptive. The reduction in body size seems to be an adequate avoidance mechanism against fish predators, whereas the alternative strategy might be adaptive in defense against the relatively small-sized *Chaoborus* larvae, which catch daphnids and other prey items individually. There exist further complex interactions between helmet forming in some daphnids (frequently considered as a defense mechanism against planktivorous fish), vertical migration, and phenotypic plasticity (De Meesters et al. 1995).

An aspect of increasing importance for freshwater ecosystems is the *colonization by species from distant areas*, which have been accidentally released by man, passively transported by vessels, immigrated by means of canal systems, or actively introduced for fisheries or biomanipulations. In many major river and lake systems with a high vessel traffic, such as the Rhine river system in Europe or the Great Lakes system in North America, and in systems with a heavy

number of exotic species

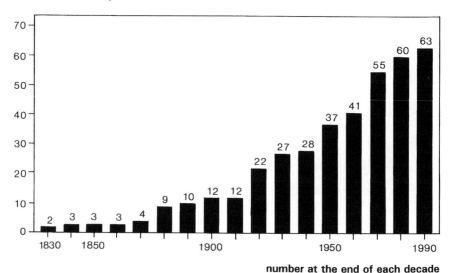

number at the end of each decade

Figure 5.7 Increase in neozoan species in the Rhine shipway. Adapted from Kinzelbach (1995).

impact by introduced fish species, as in New Zealand or even the great East African lakes, exotic species have changed communities fundamentally. Introduced species may again be replaced after some time by new invaders, but usually not by the original local species (Figure 5.7).

For instance, zebra mussel (*Dreissena polymorpha*) invasions were recorded already in their 19th century in central and western European rivers and lakes, and their numbers considerably increased in the 1970s. In North America, this ponto-caspian species invaded the Great Lakes system in 1985/86 and immediately became a pest, partly covering firm surfaces with densities of several hundred thousand individuals per square meter. In Europe, however, especially in the Rhine river system, where the same serious development occurred some one to two decades earlier, populations started to decline again in the 1980s due to several factors: predator pressures by birds such as coots (*Fulica atra*) have increased, and filter-feeding competitors, such as the amphipod species *Corophium curvispinum* (another invader from the Ponto-Caspian), have become dominant. The latter species began to cover rocks as well as zebra mussels with densities up to roughly 100 000 per square meter in the late 1980s (e.g., van der Velde et al. 1994).

Gammarus species composition has also undergone dramatic changes in the Rhine river. Indigenous representatives of this genus were widely used as important test organisms for some standard toxicity tests. Further, their presence or absence in natural waters was frequently considered as indicative of certain water quality standards, for which they are no longer useful in these rivers. The species *Gammarus tigrinus*, transferred from North America to Central Europe, has largely replaced the autochthonous species *Gammarus fossarum* and *Gammarus pulex*. The reasons for its colonization success may lie in its higher tolerance of temperature, salinity, and selected chemicals, such as organophosphorous insecticides (Streit and Kuhn 1994). Recently (i.e., since the middle of the 1990s, the invasion of the large ponto-caspian species *Dikerogammarus haemobaphes* has begun to reduce population densities of *G. tigrinus* substantially (e.g., Kinzelbach 1995). This species probably entered the Rhine system through the Rhine–Main–Donau canal.

5.11 CONCLUSIONS: HOW SHOULD TOXIC EFFECTS ON COMMUNITIES BE INTERPRETED?

Toxicity tests are designed for many levels of complexity and will be described in much more detail in later chapters of this book. Multispecies tests range from small, two-component systems using predator–prey or competition outcome, or the chemical fate as toxicological end points, to whole ecosystem manipulations that focus directly on properties thought to be of value for protection in natural systems. The most widely used systems are the following categories:

- Small multispecies tests with few components include simple predator–prey or competitive interaction tests, typically with unicellular species. They do not include interactions between chemicals and abiotic substances or complex interactions between various species; they are typically performed with small and short-living species.
- Midsized microcosm tests include laboratory-scaled pond and stream systems, often in about the 50–100 liter size category. Taxonomic groups tested may include various species of phyto- and zooplankton, benthon, and macrophytes. They are typically performed with rather small species, or species easy to handle within the microcosms.
- Large-scale test systems comprise artificial pond mesocosms. Pond mesocosms vary in size, shape, and substrate. Also stream systems have been used to evaluate various chemicals. This test size may represent the optimal compromise between whole ecosystem studies and single-species studies; the individual effects causing the complex reaction of the system, however, can usually not be disentangled.
- The ultimate multispecies toxicity tests are occasional experimental manipulations of whole ecosystems. The most famous studies of this kind have been performed in Canada. For practical, ethical, and regulatory reasons these experiments are rarely done with natural systems anymore.

Toxicological end points often include taxonomic richness and similarity, and aspects of energy flow. From an information theoretical standpoint, the variability in the composition of a community (or any other end point of a toxicity test or monitoring) is equivalent to noise. The increased noise in multispecies tests has been one of several reasons (e.g., of Kooijman 1985) to argue that they are not more sensitive than conventional tests, whereas others argue that for end points such as species richness and species composition the signal/noise ratio is uniformly high and comparable in sensitivity to end points in very simple tests (Cairns and Cherry 1993). It seems, however, that the practicability and information content obtained from multispecies tests depends largely on the choice of the specific multispecies test design. The need for a scientific basis of environmental protection and management has led to the development of increasingly sophisticated approaches and models that address the responses to stress of community and ecosystem structure and function. As pointed out in Section 5.2, the various test systems will probably all have their specific advantages and disadvantages, and combinations of various test systems and scales may give the optimum in terms of realistic evaluations of pollutant impacts on ecosystems.

Finding solutions to the environmental problems requires the integration of experimental and theoretical approaches in pure ecotoxicology as well as in pure ecology. Traditionally, experts of either field often neglect vast parts of theory in the fields that is not primarily theirs. The interweaving of the two fields will hopefully allow the development of more adequate models as well as biomonitoring and bioassay systems in the future. It is the interplay between pure and applied science, i.e., of theory and practice, which will allow progress of a responsible survey and management of aquatic ecosystems.

5.12 OUTLOOK

I hope to have shown that biological communities are not static entities on which we can put some simple quality label. They are rather highly dynamical systems with frequently unpredictable outcomes in response to environmental variables. Stochastic effects can occur even in very simple interacting systems, and long-term trends in community change can occur for a variety of biological rather than pollution-induced reasons, including colonization of invading species or genetic shifts of populations.

What are the current and future research perspectives in the field of freshwater community ecology with an emphasis on ecotoxicology? Some major topics may be the following: (1) more detailed studies on the interrelationship between nutrients, phototrophic production, heterotrophic microbial production, and consumer ingestion rates are needed. Studies on abiotic and biotic factors should be connected, and also the various aspects of biological interactions (between bacteria, fungi, protists, plants, and animals) should be studied in coherent research projects. The results may allow a better understanding of the dynamic coexistence of populations. (2) Studies will have to focus on chemical effects on individual energy allocation and life history traits of interacting populations and species. (3) The use of nonequilibrium models in experiments may prove fruitful for future studies. Empirical studies will have to take into account features of chaos theory when discussing heterogeneity in space and time. Spatial heterogeneity, genetic heterogeneity, genetic exchange at various levels (e.g., introgression), parasitism (parasitized vs nonparasitized populations), and processes of selection and species turnover as the consequence of invading species on a world-wide scale will be important for an understanding of global community changes. (4) From a methodological standpoint, successful studies will have to make extensive use of experiments with various levels of complexity and size, such as microcosm experiments in the laboratory, mesocosm studies, and field studies. Ideally these studies should be conducted as combined research programs of (a) theoretical and pure ecologial studies and (b) applications for the optimization of toxicity tests.

5.13 REFERENCES

Abrams, P. A., 1993, "Effect of Increased Productivity on the Abundances of Trophic Levels," *Am. Nat.*, *141*, 351–371

Bothwell, M. L., D. M. J. Sherbot, and C. M. Pllock, 1994, "Ecosystem Response to Solar ultraviolet-B Radiation: Influence of Trophic-Level Interactions," *Science*, *265*, 97–100

Briand, F., and J. E. Cohen, 1987, "Environmental Correlates of Food Chain Length," *Science*, *238*, 956–960

Brönmark, C., S. P. Klosiewski, and R. A. Stein, 1992, "Indirect Effects of Predation in a Freshwater, Benthic Food Chain," *Ecology*, *73*, 1662–1674

Brooks, J. L., and S. I. Dodson, 1965, "Predation, Body Size and Composition of Plankton," *Science*, *150*, 28–35

Cairns Jr., J., and D. S. Cherry, 1993, "Freshwater Multi-species Test Systems," *Handbook of Ecotoxicology. Vol. I* (P. Calow, Ed.), Blackwell Science, Oxford, UK, pp. 101–116

Calow, P., 1995, "Book Review on F. H. Rigler and R. H. Peters: Science and Limnology," *J. Anim. Ecol.*, *64*, 791–792

Carpenter, S. R., 1988, *Complex Interactions in Lake Communities*, Springer, New York, USA

Cohen, J. E., 1978, *Food Webs and Niche Space*, Princeton University Press, Princeton, USA

Cohen, J. E., 1989, "Food Webs and Community Structure," in (J. Roughgarden, R. M. May, and S. A. Levin, Eds.), Perspectives in Ecological Theory, Princeton University Press, Princeton, USA

Cohen, J. E., F. Briand, and C. M. Newman, 1990, *Community Food Webs: Data and Theory*, Springer, Berlin, Germany

Cummins, K. W., 1974, "Structure and Function of Stream Ecosystems, *BioScience*, *24*, 631–641

De Meesters, L., L. J. Weider, and R. Tollrian, 1995, "Alternative Antipredator Defences and Genetic Polymorphism in a Pelagic Predator–Prey System," *Nature*, *378*, 483–485

Dodson, S. I., 1974, "Zooplankton Competition and Predation: An Experimental Test of the Size–Efficiency Hypothesis," *Ecology, 55*, 605–613

Elton, C., 1927, *Animal Ecology*, Sidgwick and Jackson, London, UK

Gilbert, J. J., 1967, "*Asplanchna* and Posterolateral Spine Production in *Brachionus calyciflorus*," *Arch. Hydrobiol., 64*, 1–62

Gleason, H. A., 1926, "The Individualistic Concept of the Plant Association," *Bull. Torrey Bot. Club, 53*, 7–26

Gliwicz, Z. M., 1990, "Food Thresholds and Body Size in Cladocerans," *Nature, 343*, 638–640

Hall, D. J., S. T. Threlkeld, C. W. Burns, and P. H. Crawley, 1976, "The Size and Efficiency Hypothesis and the Size Structure of Zooplankton Communities ," *Annu. Rev. Evol. Systemat., 7*, 177–208

Hildrew, A. G., 1992, "Food Webs and Species Interactions," in *The Rivers Handbook. Hydrological and Ecological Principles* (P. Calow, and G. E. Petts, Eds.), Blackwell, Oxford, UK, pp. 309–330

Hutchinson, G. E., 1961, "The Paradox of the Plankton," *Am. Nat., 95*, 137–145

Kinzelbach, R., 1995, "Neozoans in European Waters—Exemplifying the Worldwide Process of Invasion and Species Mixing," *Experientia, 51*, 526–538

Kolkwitz, R., and M. Marsson, 1902, "Grundsätzliches für die biologische Beurteilung des Wassers nach seiner Flora und Fauna," *Mitt. K. Prüfanst. Wasservers. Abwasserbes. Berlin-Dahlem, 1*, 33–72

Lawler, S. P., and P. J. Morin, 1993, "Food Web Architecture and Population Dynamics in Laboratory Microcosms of Protists," *Am. Nat., 141*, 675–686

Lawton, J. H., 1989, "Food Webs," in *Ecological Concepts* (J. M. Cherrett, Ed.), Blackwell, Oxford, UK, pp. 43–78

Leibold, M. A., and H. M. Wilbur, 1992, "Interactions Between Food-Web Structure and Nutrients on Pond Organisms," *Nature, 360*, 341–343

MacArthur, R. H., and E. O. Wilson, 1967, *The Theory of Island Biogeography*, Princeton University Press, Princeton, USA

Margalef, R., 1968, *Perspectives in Ecological Theory*, University of Chicago Press, Chicago, USA

May, R. M., 1972, "Will a Large Complex System be Stable?" *Nature, 238*, 413–414

May, R. M., 1973, *Stability and Complexity in Model Ecosystems*, Princeton University Press, USA

May, R. M., 1974, "Biological Populations with Nonoverlapping Generations: Stable Points, Stable Cycles, and Chaos," *Science, 186*, 645–647

McNeill, S., and J. H. Lawton, 1970, "Annual Production and Respiration in Animal Populations," *Nature, 225*, 472–474

Mills, E. L., and J. L. Forney, 1988, "Trophic Dynamics and Development of Freshwater Pelagic Food Webs," in *Complex Interactions in Lake Communities* (S. R. Carpenter, Ed.), Springer, New York, USA

Odum, E. P., 1969, "The Strategy of Ecosystem Development," *Science, 164*, 262–270

Pimm, S. L., 1982, *Food Webs*, Chapman and Hall, London, UK

Rigler, F. H., and R. H. Peters, 1995, *Science and Limnology. Excellence in Ecology, 6*, Ecology Institute, Nordbünte, Germany

Rosemond, A. D., P. J. Mulholland, and J. W. Elwood, 1993, "Top-Down and Bottom-Up Control of Stream Periphyton: Effects of Nutrients and Herbivores," *Ecology, 74*, 1264–1280

Schlosser, I. J., 1995, "Dispersal, Boundary Processes, and Trophic-Level Interactions in Streams Adjacent to Beaver Ponds," *Ecology, 76*, 908–925

Schwenk, K., P. Spaak, 1995, "Evolutionary and Ecological Consequences of Inter-specific Hybridization in Cladocerans," *Experientia, 51*, 465–481

Shannon, E. E., and W. Weaver, 1949, *The Mathematical Theory of Communication.* University of Illinois Press, Urbana, Illinois, USA

Shelford, V. E., 1913, *Animal Communities in Temperate America as Illustrated in the Chicago Region. A Study in Animal Ecology.* Bulletin of The Geographical Society of Chicago 5, Reprint by Arno Press, New York, USA, 1977

Stibor, H., and J. Ling, 1994, "Predator-Induced Phenotypic Variations in the Pattern of Growth and Reproduction in *Daphnia hyalina* (Crustacea: Cladocera), *Funct. Ecol., 8*, 97–101

Steinmann, P., and G. Surbeck, 1918, *Die Wirkung organischer Verunreinigungen auf die Fauna schweizerischer fliessender Gewässer*, Schweizerisches Departement des Innern, Bern, Switzerland

Streit, B., 1975, "Experimentelle Untersuchungen zum Stoffhaushalt von *Ancylus fluviatilis* (Gastropoda-Basommatophora). 1. Ingestion, Assimilation, Wachstum und Eiablage," *Arch. Hydrobiol. Suppl., 47*, 458–514

Streit, B., 1976, "Energy Flow in Four Different Field Populations of *Ancylus fluviatilis* (Gastropoda-Basommatophora)," *Oecologia (Berl.), 22*, 261–273

Streit, B., 1995, "Energy Flow and Community Structure in Freshwater Ecosystems," *Experientia, 51*, 425–436

Streit, B., and K. Kuhn, 1994, "Effects of Organophosphorous Insecticides on Autochthonous and Introduced *Gammarus* Species," *Water Sci. Technol., 29*, 233–240

Streit, B., and H.-M. Peter, 1978, "Long-Term Effects of Atrazine to Selected Freshwater Invertebrates," *Arch. Hydrobiol. Suppl., 55*, 62–77

Streit, B., and T. Städler, 1995, "Freshwater Invertebrates as Model Systems in Population Ecology and Genetics," *Experientia, 51*, 423–424

Streit, B., T. Städler, and C. M. Lively, 1997, *Evolutionary Ecology of Freshwater Animals*, Birkhäuser, Basel and Boston, in press

Strong, D. R., 1992, "Are Trophic Cascades all Wet? Differentiation and Donor-Control in Speciose Ecosystems," *Ecology, 73*, 747–754

Thienemann, A., 1925, *Die Binnengewässer Mitteleuropas. Die Binnengewässer Band 1*, E. Schweizerbart'sche Verlagsbuchhandlung, Stuttgart, Germany

van der Velde, G., B. G. P. Paffen, F. W. B. van den Brink, A. bij de Vaate, and H. A. Jenner, 1994, "Decline of Zebra Mussel Populations in the Rhine," *Naturwissenschaften, 81*, 32–34

Wooster, D., 1994, "Predator Impacts on Stream Benthic Prey," *Oecologia, 99*, 7–15

Wooton, J. T., 1994, "The Nature and Consequences of Indirect Effects on Ecological comunities," *Annu. Rev. Ecol. Syst., 25*, 443–466

Walz, N., 1995, "Rotifer Populations in Plankton Communities: Energetics and Life History Strategies," *Experientia, 51*, 437–453

Zaika, V. E., 1973, *Specific Production of Aquatic Invertebrates*, Wiley, New York, USA

PART 2

Chemicals in the Environment

6

Distribution and Biogeochemistry of Inorganic Chemicals in the Environment

Bernd Markert (Zittau, Germany)

6.1 SUMMARY

This chapter gives an overview of the occurrence and distribution of inorganic chemical substances in various environmental compartments and the rates of flow between these.

Ecotoxicology, Edited by Gerrit Schüürmann and Bernd Markert.
ISBN 0-471-17644-3   © 1998 John Wiley & Sons, Inc. and Spektrum Akademischer Verlag.

Special emphasis is placed on naturally occurring substances and the influence of anthropogenic activities on their geochemical distribution. The significance of chemical elements in living organisms is considered in greater detail, and aspects of essentiality, the biological function of individual elements in plant and animal organisms, the uptake form of individual elements by plants and their interelemental correlations are discussed. Typical data are given for emissions of chemical elements into the environment by man, special attention being given to the cycles and residence times of various inorganic substances in the environment. In some cases these indicate a hazard to the biocoenoses from emissions to which they are not adapted either quantitatively (release of large amounts of naturally occurring elements) or qualitatively (release of elements that do not occur in nature at all, or are very rare). Two examples that show to varying degrees the incongruity of essentiality and possible pollutant effects serve to illustrate these aspects. Nitrogen occurs widely in the ecosystem as a macronutrient, but in certain types of bonds it may be considered an environmental pollutant. Tin, likewise, is an essential trace element for at least some groups of organisms; but its organic compounds, in particular, are among the most toxic substances created biogenically or produced by man and released into the environment. Examples are given of corresponding effects in the animal and plant kingdoms. The chapter finally outlines possible remedial and revitalizing measures for soils contaminated with heavy metals, which as a field will acquire much greater significance in the future.

6.2 OCCURRENCE OF CHEMICAL ELEMENTS IN THE ENVIRONMENT

Current studies of the *global* cycles of carbon, sulfur, and phosphorus and studies of trace gases in the atmosphere show how important it is to know about the worldwide circulation of these substances between the atmosphere, biosphere, hydrosphere, and geosphere. Similar attention must be given to work on *regional* changes in the inorganic chemical world (Adriano 1986, 1992, 1994; Adriano et al. 1994; Bloemen et al. 1995; Brooks 1993; Duvigneaud and Denayer De Smet 1973; Ernst 1974, 1993; Fiedler and Rösler 1993; Fränzle 1993, 1994; Herpin et al. 1995; Keune et al. 1991; Lacerda and Salomons 1991; Likens et al. 1977; Markert 1993a, 1995; McGrath and Loveland 1992; Moldan and Cerny 1994; Salomons and Förstner 1984; Thornton 1988). In ecosystems the paths and whereabouts of individual inorganic substances may be influenced in a specific manner by organismic activity, for example, the selective uptake and accumulation of elements (Alloway 1993; Baker 1981; Bruns et al. 1995; Davies 1992; Ernst and Joossee Van Damme 1983; Kinzl 1982; Lepp 1992; Matschullat and Müller 1994; Nagel and Loskill 1991; Oehlmann et al. 1995; Oehlmann and Markert 1996; Parlar and Angerhöfer 1991; Rossbach et al. 1992; Wilken 1992; Wolterbeek et al. 1995). Substances occurring together may have a positive and/or negative effect on their transportation or accumulation within the organism or community of organisms. We must assume that there are interactions between organismic activity and the flow rates and flow patterns of individual inorganic substances through the components and compartments of an ecosystem that can only be interpreted correctly after an analysis of both inorganic and organic substance patterns that are as complete as possible (Fortescue 1980; Jayasekera 1993 and 1994; Kovacs et al. 1993 and 1994; Lieth and Markert 1990; Markert 1992, 1993b, 1996; Roth 1992; Sansoni 1987). There is no doubt that for most substances and ecosystems the nature and extent of these mutual influences depends on abiotic factors such as weather conditions. The flow rates and flow patterns therefore vary in accordance with these factors (Farago 1994; Iyengar 1989; Markert 1994a; Sansoni and Iyengar 1978; Streit and Stumm 1993; Ure and Davidson 1995).

By and large the earth's crust may be regarded as a natural reservoir for all chemical elements found in the biosphere (Bowen 1979; Hamilton 1979; Ittekkot et al. 1990; Kabata-Pendias and Pendias 1992; Merian 1991; Streit 1994a, b). Over 99% of the total mass of the earth's crust is made up of only eight of the 88 elements that occur naturally. It consists of 46.4% oxygen, 28.15% silicon, 8.23% aluminium, 5.63% iron, 4.15% calcium, 2.36% sodium, 2.33% magnesium, and 2.09% potassium. Of these eight most common elements in the earth's crust, oxygen is the only nonmetal. Together, the remaining 80 elements of the periodic system account for less than 1% of the total.

On average, 85%–90% of the fresh weight of living plant organs, i.e., those with an active metabolism, consists of water. The dry matter of the bodies of plants is made up primarily of the following elements: carbon (44.5%); oxygen (42.5%); hydrogen (6.5%); nitrogen (2.5%); phosphorus (0.2%); sulfur (0.3%); and the alkaline or alkaline earth metals potassium (1.9%), calcium (1.0%), and magnesium (0.2%). This means that in contrast to the earth's crust the greater part of organic life is formed from nonmetals. Since they occur in large quantities in living organisms the above nine elements are termed macroelements. In addition, there are so-called microelements. These include Na, Cl, Si, V, Cr, Mo, Mn, Fe, Co, Ni, Cu, Zn, B, Sn, Se, F, and I. However, this traditional classification as macro- or microelements, based solely on the physical quantity of an element in an organism, has had to be modified considerably in modern plant, animal, and human physiology. For example, the list of macroelements has had to be extended for certain groups of organisms. Silicon must be regarded as a macroelement in diatoms and horse-tails. Moreover, element-specific and organism-specific accumulations frequently occur as a result of particular locational and genetical constellations. Sodium, bromine, and chlorine are accumulated by many halophytes; copper, nickel, zinc, lead, cadmium, and other heavy metals are taken up in large quantities by metallophytes.

Both macro- and microelements are nutrients that are necessary for the growth and normal development of an organism; their function cannot be taken on by any other element. In other words, they are essential. This is why macro- and microelements are also called macro- and micronutrients (Table 6.1). Of the microelements listed above, V, Ni, Sn, Se, F, and I are so far only known to have a physiological function in animal organisms; the element B has only been proved to have a function in plants. Appendix A.6.1 gives an overview of the occurrence and significance of all chemical elements.

Table 6.1 Classification of the chemical elements according to their physiological aspects

Structural elements:	C, H, O, N, P, S, Si, Ca
Electrolytic elements:	K, Na, Ca, Cl, Mg
Enzymatic elements:	V^a, Cr, Mo, Mn, Fe, Co, Ni^a, Cu, Zn, B^b, Sn^a, Se^a, F^a, I^a, Mg

So far without a biological function:

1st main group: Li, Rb, Cs, (Fr)	2nd main group: Be, Sr, Ba, Ra
3rd main group: Al, Ga, In, Tl	4th main group: Ge, Pb
5th main group: As, Sb, Bi	6th main group: Te, Po
7th main group: Br, (At)	
8th main group: He, Ne, Ar, Kr, Xe, Rn	

1st subgroup: Sc, Y	2nd subgroup: Ti, Zr, Hf
3rd subgroup: Tb, Ta	4th subgroup: W
5th subgroup: (Tc), Re	6th subgroup: Ru, Os
7th subgroup: Rh, Ir	8th subgroup: Pd, Pt
9th subgroup: Ag, Au	10th subgroup: Cd, Hg

Lanthanides: La, Ce, Pr, Nd, (Pm), Sm, Eu, Gd, Tb, Dy, Ho, Er, Tm, Yb, Lu
Actinides: Ac, Th, Pa, U, (Np), (Pu), (Am), (Cm), (Bk), (Cf), (Es), (Fm), (Md), (No), (Lr)

Source: From Markert (1992).

( ): Elements that do not occur naturally, based on Sansoni and Iyengar (1978).
[a]Essential significance so far determined for animal organisms only.
[b]Essential significance so far determined for plant organisms only.

In addition to the macro- and micronutrients described above, living organisms also contain a number of other chemical elements that have become accessible to chemical analysis since the advent of sensitive instrumental techniques. It is safe to say that about 60–70 elements can now be determined quantitatively in animal and plant organisms. It will probably be possible to measure quantities of the remaining 20 elements within the next few years, when suitable analytical methods become available.

It is interesting to plot the molar masses of individual element concentrations in plant organisms against the molar masses of the mean element concentrations in the earth's crust (Figure 6.1). Most of the element concentrations are to be found along the dividing line between the two subsets of concentrations. On the one hand this shows that the extraterrestrial origin of the elements is reflected in both the earth's crust and living biomass, but it also demonstrates that some elements have acquired a special significance during the history of evolution and most particularly within the biological life of today. These are the elements C, H, O, K, Ca, Si, Na, Fe, P, S, N, Mn, B, Zn, Cu, Ni, Cr, Co, Cl, V, F, Rb, Sr, Ba, Ti, and Al contained in group I in Figure 6.1. All of them except the last five have an essential function at least in certain groups of organisms, and even these last five elements—Rb, Sr, Ba, Ti, and Al—must be assumed to have an essential

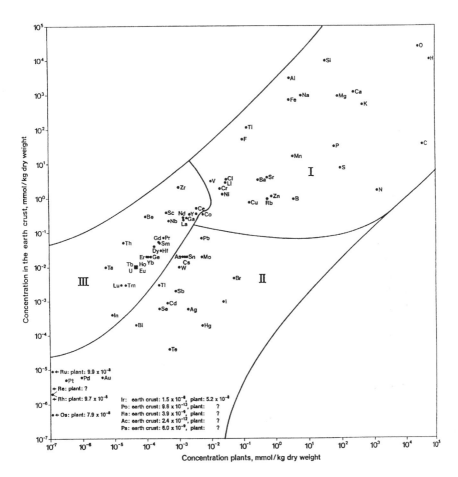

Figure 6.1 Mean concentrations of individual chemical elements in plants plotted against the mean concentrations of elements in the earth's crust (all values in mmol/kg dry weight). From Markert (1996).

function although we cannot define it accurately at present. Group II contains elements that have an essential function in some cases (I, Mo, Se, and Sn) but are also highly toxic, usually even at low concentrations. This applies especially to the heavy metals Pb, Cd, As, Ti, Hg, etc.. Group III consists of elements that have not exchanged their passive role in the earth's crust for an active part in organismic life in the course of evolution. In particular, group III contains the lanthanoids and platinum metals. The notion, developed from the history of evolution, that certain chemical elements were selected to form the basis of organismic life led to the establishment of the first biological system of the elements (BSE).

6.3 THE SIGNIFICANCE OF CHEMICAL ELEMENTS IN LIVING ORGANISMS

Biological processes on the molecular level are frequently based on physical and chemical conditions whose fundamental systematics on the chemical side in the periodic system of the elements (PSE) were determined by Mendeleyev and Meyer in 1869. However, these physical and chemical regularities are frequently modified in biological systems. The reason for this is the adaptation of all organismic life to the aqueous environment. For example, in many biochemical processes ions have to pass through cell membranes. It is remarkable that according to the position in the PSE the larger K^+ ion ($\emptyset$: 0.133 nm) passes through the cell membrane more easily than the smaller Na^+ ion ($\emptyset$: 0.95 nm). The explanation for this is to be found in the reversal of the size ratio of the two ions in the hydrated state ($\emptyset$ K^+_{hydr}: 0.340 nm; $\emptyset$ Na^+_{hydr}: 0.480 nm). Even if transport phenomena of this nature can be explained relatively easily by the classic tools of chemical knowledge, the demands made on the possibilities of physicochemical interpretation become increasingly difficult to satisfy with the growing degree of complexity of physiological processes. The demands made on the physicochemical approach to physiological processes become more burdensome if, for example, a biochemically substantiated explanation of the effect of individual substances or groups of substances on various groups of organisms is involved. Effect research of this type includes questions of both the essentiality and also the toxicity of individual substances or groups of substances. The PSE does not permit any approach which would enable the functional essentiality or the chronic or acute toxicity of individual elements or their compounds to be derived from their position in the PSE.

This becomes particularly clear whenever we search for common physiological features among elements within one group of the periodic system of the elements. What, for example, do the elements of the 4th main group (carbon (C), silicon (Si), germanium (Ge), tin (Sn), and lead (Pb)) have in common from a physiological point of view? Carbon (C) largely occurs in living systems as a structural element in macromolecules, silicon (Si) is considered to be a "strengthening" element in a few special groups of organisms (diatoms, grasses, and some others), germanium (Ge) has as yet no known physiological function, although in the form of spiro-germanium (4,4-dialkyl-4-germacyclohexanone and 8,8-dialkyl-8-germaazaspiro (4,5) decane) it is used medically as an antitumor agent. Tin is an important component of various enzyme systems for all vertebrates, and at present lead appears only to have a toxic effect on the systems of all living organisms.

Even in a consideration of diagonal relationships within the periodic system, similarities and relations between individual elements are only apparent on a superficial level. For instance, it is striking for the elements of the 4th to 7th main group (C, P, Se, and I) that all of these elements have an essential significance. A consideration of the elements of the 3rd to 6th main group (B, Si, As, Te) reveals that these elements only have an essential physiological function for special groups of organisms. In the case of boron (B) this is true of plants; with silicon (Si), as already mentioned above, this holds for some specific groups of organisms; while arsenic (As) is regarded as a growth promoter for domestic animals (e.g., As deficiency leads to cardiac death in the third generation of goats), and tellurium (Te) has no recognized physiological function as yet. As far as we know, the elements arranged in the 3rd to 6th main group (Al, Ge, Sb, Po) do not have any essential function at all. On the contrary, it is assumed that these elements have a toxic effect on living organisms even at low concentrations. That is to say, in biological and medical research, the arrangement of the elements in the chemical periodic system does not provide any clear answers to questions about the toxicity and essentiality of individual elements or their forms of bonds.

The periodic system of the elements has long been criticized by chemists since it displays a wide range of confusing irregularities and the PSE basically only appears periodic upon superficial consideration. Allen (1992) recently put forward the question of what the common physicochemical features between carbon, silicon, tin, and lead could be since they are shown to be related to element group 4, or between boron (B) and aluminium (Al) in element group 3, whereas, for example, boron and silicon, on the one hand, and aluminium and beryllium, on the other, have many common features. Allen comes to the conclusion that something fundamental is missing from the PSE as our basic instrument for organizing chemical phenomena. It contains no information on the energetic states of the atoms, without which no satisfactory description of the structure of matter can be given. Even if all the graphical tricks of the trade are applied, it is only possible to represent the PSE in two dimensions. The inconsistencies of the system can only be solved by introducing a further, third dimension. To this end, Allen defined the concept of configurational energy, a mean function of the electron energy of an atom. Such energies can be determined by spectroscopic methods from the ionization potentials of the valance electrons. Allen conjectures that the configuration energy is the long sought after, missing third dimension of the PSE. Its suitability for predicting the properties of as yet unknown chemical compounds can be applied as a touchstone.

Due to the inadequate chemical description of the biological and medical relations of the chemical elements in the PSE, a biological system of the elements (BSE) is proposed below, as shown in Figure 6.2. The compilation of this system of the elements is based on considerations arising in particular from data on the multielement analysis of terrestrial plants. The data were compiled from standard reference materials from the National Institute of Standards and Technology (NIST, Gaithersburg, MD, US) and highly accurate research samples. The data pool for plant reference materials is composed of certified element contents for citrus leaves (NIST SRM 1572), tomato leaves (NIST SRM 1573), pine needles (NIST SRM 1575), and Bowen's kale. The data pool was supplemented by multielement data spectra from the following plants: leaves of *Betula pendula*, needles of *Pinus sylvestris*, leaves of *Vaccinium vitis-idaea*, leaves of *Vaccinium*

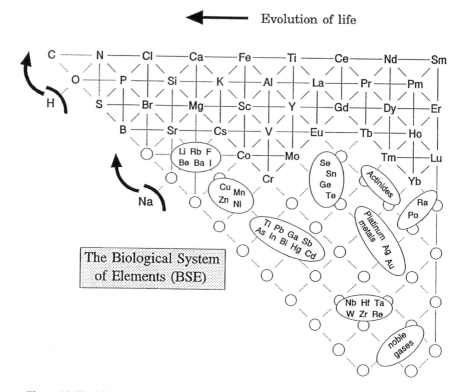

Figure 6.2 The biological system of the elements. For details, see text. From Markert 1994b.

myrtillus, the aboveground parts of *Deschampsia flexuosa*, *Molinia caerulea*, and *Polytrichum formosum*, and different *Sphagnum* species. All of the plants were gathered in the course of the 1987 vegetation period on Grasmoor near Osnabrück (Markert 1993). Only data from terrestrial plants of freshwater systems (glycophytes) were used; no results of typical accumulator plants or halophytes enter the data base. The following criteria are of significance: interelemental correlations, biological function of individual elements, and uptake form of individual elements by the plant.

6.3.1 Interelemental Correlations

Garten (1976) discovered high interelemental correlations for the elements P, N, K, Ca, and Mg for 54 species growing in the field, resulting independently of the site of the respective plant species studied, which are expressed in a high positive correlation according to the linear correlation of the individual measured values. The high correlation coefficients for N and P were attributed to the high association of the two elements, especially during protein biosynthesis, while for

Ca and Mg they were explained on the basis of common enzyme activators during various metabolic processes (Garten 1976). The results of multielement analyses in the 1980s permitted the element spectrum to be extended for correlation analyses of this type. The result of these correlations are given by Markert (1993) and cannot be discussed in detail here. However, the correlation data represented an important criterion in compiling the BSE.

It was striking that of the alkaline metals, in particular the element potassium (K), and of the alkaline earth metals Ca and Mg, and to a certain extent also Sr, display both high correlations to each other and also to the macronutrients N and P. It is interesting that the correlation tendency drops considerably in the sequence Ca, Mg, Sr, and Ba, although the ionic radii of the hydrated elements do not display any great differences (Ca^{2+}: 0.6 nm; Mg^{2+}: 0.8 nm; Sr^{2+}: 0.5 nm; Ba^{2+}: 0.5 nm).

The readiness of potassium to form high positive correlations with almost all macroelements (K/Ca: 0.7545; K/N: 0.8370; K/Mg: 0.7928; K/P: 0.7768) and with the halogens (K/Br: 0.8684; K/Cl: 0.6904) is most probably an indication of its outstanding role as an electrolytic element in plant metabolism. If metals are involved then the biochemical function of the electrolytes is largely determined by their tendency to complexation. The alkaline metals have hardly any tendency towards complex formation, the alkaline earth metals a moderate tendency and the transition metals, a strong tendency. This property also determines the form in which these elements are transported in the plant organism. With respect to their function, sodium and potassium ions mainly function as transporters of charges, magnesium and calcium as stabilizers of organic structures and information transmitters, and the transition metals, in combination with proteins, as catalysts. Together with large organic anions, chloride ions ensure the maintenance of electroneutrality in the case of the intercompartmental displacement of charge carriers. High positive interactions were established for the two halogens, Br and Cl, both with each other ($r = 0.7587$) and with K, N, and P. In this connection, the tendency of Br to form positive correlations was even greater than that of Cl: Br/K (0.8684), Cl/K (0.6904); Br/P (0.9276), Cl/P (0.6572); Br/N (0.8081); Cl/N (0.2540). The high correlations of the halogens with K could involve a typical charge antagonism between K^+ ions and halogenide ions.

The high correlation tendency of the elements Al, Fe, Sc, and La, which as a rule is greater than $r = +0.9$, can be attributed to the trivalent charge state of the cations of these elements and the very similar radii of the hydrated ions Al^{3+}, Fe^{3+}, and La^{3+} of 0.9 nm. This must also be the reason for the correlations between the lanthanide elements and for those between the lanthanide elements and Al, Fe, and Sc. The high correlations between boron and P ($r = +0.7917$) and N ($r = +0.8121$) confirm that boron can be regarded as an essential element for plants. The high correlation of P and N at $r = +0.8352$ confirms the high correlation coefficient reported by Garten (1976; $r = +0.84$) and the high degree of association which the two elements display, especially during protein biosynthesis.

6.3.2 The Biological Function of Individual Elements

Since the significance of an element in the living organism does not depend on the amount contained in the organism, a systematic division according to physiological and biochemical aspects seems meaningful. Those elements (C, H, O, N, P, S, Si, and Ca) which participate in the constitution of the functional molecular structural elements of the cell metabolism (proteins, lipids, carbohydrates, nucleic acids, etc.) or display a direct supporting or strengthening character (calcium, silicon) are termed structural elements. Nitrogen and sulfur are biochemically integrated into the carbon chain, i.e., after reduction of their generally high oxidation state (nitrate or sulfate) they are firmly bound to the organic substance. In contrast, phosphorus, boron, and silicon are present in their highest oxidation state and are not reduced; indeed they rather tend towards ester formation with OH groups of the most varied molecules, particularly the sugars. All structural elements are in the top left of the BSE. They seem to be the elements which during evolution have developed via the photochemical process from a silicon dioxide matrix towards the basic organic matrix as the "building blocks" of life. As the former structure-forming element on earth, silicon is now only present as a structural element totally (e.g., diatoms) or partially (e.g., Sphenopsida) in lower plants.

The hydrogen atom represents an exception within the structural elements, as also in the chemical system of the elements. Although it is contained in almost all structure-forming macromolecules, it additionally has two further decisive functions to fulfill in the living organism. It provides the reduction equivalents in many redox processes (usually via the system of nicotinamide adenine dinucleotide or nicotinamide adenine dinucleotide phosphate or flavin adenine dinucleotide), and in the form of the H_3O^+ ion it is also responsible for the pH value conditions in the cell body. Hydrogen was therefore not rigidly fixed in the BSE, but its special position was rather accommodated in a more or less dynamic manner.

In contrast, the so-called electrolytic elements (K, Na, Ca, Cl, and Mg) are required for the constitution of specific physiological potentials and are important for maintaining defined osmolytic conditions in cell metabolism (Table 6.1). The element calcium may thus occur simultaneously as a structural element and an electrolytic element. In the BSE the electrolytic elements are positioned directly adjacent to the structural elements since a compartmentalization in the cell environment was only possible by the formation of membrane systems due to the constitution of structure-forming units, and a constitution of potential differences was only able to result due to this type of compartmentalization of the cell environment. On the basis of the high correlation coefficients to their directly adjacent elements in the BSE, Br, Cs, and Sr can probably also be included among the essential electrolytic elements. Sodium is an exception here. Sodium can probably be regarded as the "information" element, at least in rapidly progressing reactions (for example in the conduction of excitation at neurons), since potential reversals frequently occur by changing the permeability for Na^+ ions at the cell membranes, thus making the information flow possible at all. The high affinity of Na for organic structures is clearly expressed in the high correlation coefficients with the individual nonmetals (Na/B: 0.72; Na/Br: 0.78; Na/N: 0.76; Na/P: 0.81; Na/S: 0.97).

A number of chemical elements, above all metallic ions, exercise a catalytic function in cell metabolism as metal complex compounds. These elements are termed enzymatic elements (Table 6.1). They include, for example, V, Cr, Mo, Mn, Fe, Co, Ni, Cu, Zn, Sn, and Se. Due to the current imprecise data, it has not yet been possible to offer a satisfactory solution to the conclusive position of these elements in the BSE, so that the physiologically related elements still have to be combined into groups. However, it may be assumed that elements such as Ge and Te, which are grouped together with the elements Se and Sn in the BSE, will be assigned an essential function in the future.

6.3.3 The Uptake Form of Individual Elements by the Plant

The relationship between the nutrient supply in the substrate (soil, nutrient solution, atmosphere) and the activity in the plant is not linear over a wide range, but rather describes an optimum curve. Ingestion is adequately described in quantitative terms by the intensity and extent of uptake up to a certain point in time. In the case of a defined nutrient, uptake by the plant depends on the supply of the nutrient in the uptake medium and on its availability. The plant does not generally have any positive influence on the supply, although it does on the material and spatial availability of the nutrients. Nutrient availability may thus be altered in a material respect for example by influencing the pH value of the soil solution (excretion of H_3O^+ or HCO_3^- ions by the root), by the release of organic acids with a chelating effect by the root, by the participation of microorganisms (mycorrhiza), and also by influencing the redox potential in the soil by H_3O^+ and O_2 excretion at the root surface.

The most readily available elements are present in the form of ions or as a soluble organic complex in the soil solution. The most poorly available elements are those which are firmly bound to the soil structure, e.g., as secondary constituents in the crystal structure of primary minerals. In the BSE those elements primarily taken up in the form of neutral molecules are arranged at the top left. They include carbon and oxygen (uptake form mainly CO_2) and hydrogen (uptake in the form of H_2O). However, sulfur and nitrogen can also be taken up as neutral molecules in the form of SO_2 and NO_2, particularly if they occur as deficiency elements in the primary nutrient medium, the soil. Otherwise in the BSE, nitrogen and sulfur form a transition to those elements which are primarily available to the plant organism in an anionic form. They occur here as a function of the pH value and redox conditions mainly in the form of HSO_4^-, SO_4^{2-}, or NO_3^-. The elements chlorine (as Cl^-), bromine (as Br^-), and boron (as $B(OH)_4^-$) are similarly taken up anionically. All other elements, if they do not occur in a chelated form as a complex, are mainly incorporated cationically and are therefore located on the right of the BSE. This includes, in particular, the alkaline and alkaline earth elements as well as most of the heavy metals.

On the basis of similar ionic radii and charge conditions of the individual elements and element compounds, each element may in principle be of significance for the physiology of an organism. In this regard it does not matter in the first instance whether these chemical elements have a direct promoting or inhibiting effect on structural or enzyme systems, or whether by their presence they compete indirectly with other reactive substances. However, from their position in the BSE it can be concluded that a direct physiological essentiality will have to be assigned to elements such as Br, Sr, Cs, Ge, and Te in the future, whereas an exclusively toxic function will continue to be allocated to elements such as Tl, Pb, Ga, Sb, In, Bi, Hg, and Cd; alternatively, the latter may only be able to intervene indirectly in the physiology of the living organism. Further data from trace element research is required in order to complete the BSE and to verify statements derived from it.

6.4 THE EMISSION OF CHEMICAL ELEMENTS INTO THE ENVIRONMENT

Discussion of the insidious poisoning of our environment with heavy metals and other inorganic substances and compounds has drawn attention to the importance of ecotoxicology. As far back as 1988, Nriagu and Pacyna prepared initial estimates of worldwide pollution of the environment with heavy metals. A comprehensive stocktaking of worldwide emissions of metals in the most important environmental fields showed that the activities of man have become the decisive factor in the global cycle.

An analysis of coal and lignite, for example, showed that these are now the principle cause of pollution of the atmosphere by mercury, molybdenum, and selenium and also arsenic, chromium, manganese, antimony, and thallium. The burning of petroleum contributes the greatest emissions of vanadium and nickel. Nonferrous metalworking is the largest source of lead (alongside petrol combustion in motor vehicles) and also of arsenic, cadmium, copper, and zinc. Nriagu and Pacyna estimate that an average of 7,800 t arsenic, 1,000 t cadmium, 19,000 t copper, 516,000 t manganese, 19,000 t lead, 66,000 t vanadium, 46,000 t zinc, and 6,000 t mercury are emitted into the atmosphere each year.

The main causes of water pollution are household waste (arsenic, chromium, copper, manganese, and nickel), coal-fired power stations (arsenic, mercury, and selenium), iron and steel production (chromium, molybdenum, zinc, and antimony), and metal smelters (cadmium, nickel, antimony, and selenium). The quantities emitted each year are astronomical. The estimate for manganese is between 100,000 and 400,000 t, for lead between 100,000 and 180,000 t, and for cadmium between 2,000 and 17,000 t. Of these metals, 25% are thought to enter rivers and lakes; soils are also heavily polluted with metals. If the calculated quantities were to be spread evenly over the entire cultivated area of the earth the results would be between 1 g/ha for cadmium and antimony, 50 g for lead, copper, and chromium, and 65 g for manganese and zinc.

Elements that once occurred only in extremely small concentrations in nature are now increasingly entering our environment. The installation of platinum-based catalytic converters in motor vehicles, for example, is expected to result in additional emissions of ruthenium, rhodium, palladium, osmium, and indium, which occur as companions to platinum, as well as greater amounts of platinum itself. The lanthanoids, another group of elements little examined so far, are likely to acquire greater significance in ecotoxicological research because of their increased emissions. Some of the lanthanoids already have an established technological use. Samarium, praseodymium, and cerium are used as components of alloys for magnetic materials; lanthanum and yttrium are constituents of the new high-temperature superconductors. Some of the lanthanoid elements act as catalysts in chemical reactions and are required for manufacturing laser crystals. In Appendix A.6.2 an attempt has been made to calculate the annual production of all naturally occurring elements, with the exception of the noble gases, for the year 2000 (Markert 1992).

6.5 CYCLES AND RESIDENCE TIMES OF CHEMICAL ELEMENTS IN THE ENVIRONMENT

Nearly all chemical substances are involved in so-called cycles. They move between the atmosphere, hydrosphere, lithosphere, and biosphere to varying extents and at different speeds. Such material cycles are a basic requirement of the life processes known to take place on the earth today (Bliefert 1994). As described earlier, all living organisms require elements such as phosphorus, sulfur, calcium, potassium, and magnesium, in addition to hydrogen, oxygen, carbon, and nitrogen. Not least biological processes cause these elements to accumulate in the earth's crust, sometimes in large amounts, in the form of coal or calcium carbonate in coral reefs. Table 6.2 gives an initial estimate of the binding of chemical elements in the vegetable biomass as a whole. Moreover, some organisms can accumulate comparatively rare elements far beyond the normal levels. In such cases we speak of hyperaccumulators.

Cycles of materials and elements can be shown in the form of models. The atmosphere and rivers are the important areas of the environment for the transportation of most substances. Both also function as reservoirs, although the rivers are only of secondary importance for most compounds. It is the soil and the oceans that have the greatest significance as reservoirs.

As a rule it is difficult to estimate the quantities of substances present in a large environmental compartment and often more difficult still to make a reasonably reliable assessment of the quantities that move from one compartment to another in the course of a year. The published data are therefore rough

Table 6.2 Total element content in the world plant biomass in t

Ac	?	Hf	9.2×10^4	Rb	9.2×10^7
Ag	3.682×10^5	Hg	1.841×10^5	Re	?
Al	1.47×10^8	Ho	1.472×10^4	Rh	1.84×10^1
As	1.841×10^5	I	5.523×10^6	Ru	1.84×10^1
Au	1.841×10^3	In	1.841×10^3	S	5.523×10^{10}
B	7.3640×10^7	Ir	1.841×10^2	Sb	1.841×10^5
Ba	7.3640×10^7	K	3.497×10^{10}	Sc	3.682×10^4
Be	1.841×10^3	La	3.682×10^5	Se	3.682×10^4
Bi	1.841×10^4	Li	3.682×10^5	Si	1.841×10^9
Br	7.364×10^6	Lu	5.523×10^3	Sm	7.364×10^4
C	8.19×10^{11}	Mg	3.682×10^5	Sn	3.682×10^5
Ca	1.841×10^{10}	Mn	3.682×10^8	Sr	9.2×10^7
Cd	9.2×10^4	Mo	9.2×10^5	Ta	1.841×10^3
Ce	9.2×10^5	N	4.602×10^{10}	Tb	1.472×10^4
Cl	3.682×10^9	Na	2.76×10^8	Te	9.2×10^4
Co	3.682×10^5	Nb	9.2×10^4	Th	9.2×10^3
Cr	2.7615×10^6	Nd	3.682×10^5	Tl	9.2×10^4
Cs	3.682×10^5	Ni	2.76×10^6	Ti	9.2×10^6
Cu	1.841×10^7	O	7.824×10^{11}	Tm	7.364×10^3
Dy	5.523×10^4	Os	2.7615×10^1	U	1.841×10^4
Er	3.682×10^4	P	3.682×10^{10}	V	9.2×10^3
Eu	1.472×10^4	Pa	?	W	3.682×10^5
F	3.682×10^6	Pb	1.841×10^6	Y	3.682×10^5
Fe	2.76×10^8	Pd	1.841×10^2	Yb	3.682×10^4
Ga	1.841×10^4	Po	?	Zn	9.2×10^7
Gd	7.364×10^4	Pr	9.2×10^4	Zr	1.841×10^5
Ge	1.841×10^4	Pt	9.2×10^1		
H	1.196×10^{11}	Ra	?		

estimates and usually vary from one author to another. Table 6.3 uses six selected elements as examples of the quantities moving among the various compartments. These figures already give an impression of how different the paths of pollutants in the environment can be. In the case of carbon, for instance, the two cycles between the soil and the atmosphere (AS and SA) are almost balanced, the comparatively small difference of 0.07×10^{14} kg/a between AS and SA being responsible for the accumulation of carbon in the atmosphere (Bliefert 1994). In the case of phosphorus the ocean is the main sink; this is shown by the wide gap between OA and AO and the high value for RO. The copper and lead emissions are for the most part carried by the rivers into the sea, where they accumulate in the sediments.

The elements or compounds whose cycles are yet to be investigated have a specific (mean) life or residence time in the various compartments of the environment. For many substances in the atmosphere this is in the order of a year, exceptions being highly reactive substances such as OH and NO_2 and

Table 6.3 Interchange of some important elements among the four environmental compartments atmosphere (A), soil (S), oceans (O), and rivers (R)

Element		AS	SA	RO	OA	AO
O (as H_2O)	in 10^{17} kg/a	1.0	0.7	0.3	3.7	3.4
C	in 10^{14} kg/a	1.2	1.27	< 0.01	0.1	0.1
S	in 10^{14} kg/a	0.7	1.6	2.1	1.6	2.6
P	in 10^9 kg/a	3.2	4.3	19	0.3	1.4
Pb	in 10^8 kg/a	3.2	4.7	7.8	< 0.01	1.4
Cu	in 10^7 kg/a	6.2	7.1	632	< 0.01	1.3

Source: From Bliefert (1994).

highly inert substances such as N_2 and CF_4. The residence times for substances in the hydrosphere and pedosphere must be considered two to three magnitudes greater, i.e., hundreds or even thousands of years, and in the lithosphere they are in the region of millions of years and longer. The short residence times of many substances in the atmosphere are one important reason for giving much scope to discussion of this storage and transport medium in environmental chemistry.

In addition to the reactivity of a substance there are other factors that may determine its residence time in the same environment. For example, the level of carbon dioxide in the atmosphere can easily be changed by the burning of fossil fuels because of the small amount present. On the other hand, the much greater proportion of oxygen remains virtually unchanged in spite of the O_2 consumption involved in combustion. Global cycles yield qualitative information on shifts in worldwide mass transport processes brought about by man. However, natural mass exchange mainly takes place on a regional scale. To illustrate this the two following examples—a macronutrient and a micronutrient—will be examined with regard to both their life-promoting and toxicological features.

6.5.1 Nitrogen as an Example of a Macronutrient and "Environmental Pollutant"

The only minerals containing significant amounts of nitrogen are saltpeter (KNO_3) and Chile saltpeter ($NaNO_3$). Otherwise, little nitrogen is to be found in the earth's crust or in minerals. In its elemental form, as N_2, nitrogen occurs mainly in the atmosphere. In the environment it undergoes various microbial transformations. Denitrification is the conversion of nitrates into free nitrogen by bacteria, with nitrogen monoxide and nitrous oxide as by-products. The bacteria responsible for the process are facultative anaerobes. They use the oxygen of nitrates as a hydrogen acceptor in breaking down organic nutrients (denitrification), a property that is exploited for removing nitrogen in sewage treatment. Large amounts of N_2O are produced by denitrification in the soils of the tropical rain forests. Strong irrigation of the soil and lack of air promote this activity. Conversely, nitrifying bacteria are able to oxidize NH_4^+ to NO_3^-. In sewage treatment these aerobic bacteria are also used to produce nitrate—later converted to nitrogen by denitrifying bacteria—from ammonium nitrogen. In agriculture this process leads to depletion of the nitrogen pool, as nitrate is more easily leached out than ammonium.

All reactions in which atmospheric nitrogen, which is chemically very inactive because of its stable triple bond, is converted into reactive or biologically available compounds with hydrogen, carbon, and/or oxygen are termed nitrogen fixation. These reactions are very important because of their essentiality for plant, animal, and human life. Fire, lightning, and the like may split the triple N–N bond, forming NO_x. The combustion of petrol and other fossil fuels also produces NO_x from N_2. These are examples of chemical nitrogen fixation. The quantities produced in this way in densely populated areas may be considerable, but at the global level this form of nitrogen fixation is of no great significance.

An important process is industrial nitrogen fixation. The chemical industry synthesizes ammonia from N_2 and H_2 by the Haber process; much of this ammonia is then processed further to make nitrogen fertilizers. As NH_4^+ and above all as NO_3^-, the "fixed" nitrogen is converted into proteins by terrestrial and aquatic plants. Conversely, N_2 is released into the atmosphere by decaying organisms, oxidation, denitrifying bacteria, and other mechanisms. Another part of the cycle consists of the uptake of plants as food by animals; animals bind nitrogen in protein and then die and decay, thus releasing nitrogen again through denitrification.

Greatly increased anthropogenic emissions of nitrogen-containing compounds have raised the inputs of nitrogen into terrestrial and aquatic ecosystems far above the natural level. For example, the maximum loads for nitrogen calculated by the "critical loads/critical levels" method are exceeded—sometimes considerably—by present emissions in many parts of Germany. The magnitude of nitrogen inputs has already led to harmful accumulations in ecosystems.

The eutrophicating and acidifying effect of atmospheric nitrogen deposition disrupts and endangers forest ecosystems. The eutrophication of natural areas of poor soil (moors, dry meadows, and heaths) and modification or reduction of the spectrum of species to the point of irreversible habitat alterations are further consequences of nitrogen deposition. Man-made excesses of nitrogen impair the quality of water bodies in general: increased nitrate levels in ground water and drinking water may endanger health, while high ammonium and nitrate concentrations in surface waters have a toxic effect on aquatic organisms. Both contribute to the eutrophication of continental coastal waters especially, with the familiar results (growth of algae). Emissions of laughing gas from soils and water bodies are partly responsible for the destruction of the ozone shield and the anthropogenically induced greenhouse effect. Nitrogen oxide emissions contribute, as precursor substances, to the formation of tropospheric ozone (photochemical smog). Tropospheric ozone is a greenhouse gas and can have toxic effects on living organisms even at low concentrations.

The main causes of inputs of nitrogen into the biosphere are agriculture, transport, energy conversion, and heating, but also human nutrition with the resulting problems of sewage disposal and management of residues and waste.

According to the German Federal Office of the Environment (UBA (Umwelt-bundesamt) 1994), the best chances of reducing emissions in Germany are in the field of agriculture. The efficient production, storage, and application of organic fertilizers alone would reduce overall nitrogen emissions by about 12%. Restricting the amount of stock that may be kept on a specific area, aiming at an average density of 0.6 dung units per hectare of agricultural land throughout Germany (at present it is 0.91 units/ha), would reduce the overall emissions by just under 15%. A change in eating habits (half animal and half vegetable protein) combined with a reduction in agricultural exports would lead to a reduction of about 12%. Basic interdisciplinary research results in the field of intensive farming and the leaching out of nitrates have been compiled by the University of Osnabrück following a project lasting several years and are summarized in the form of a simulation model (Forschungszentrum Jülich GmbH 1991; Kramer 1990).

According to the German Federal Office of the Environment (UBA 1994) the potential reduction of emissions in the transport sector is comparatively small. Replacing old vehicles by cars with multifunctional catalytic converters (all to be replaced by the year 2005) may reduce nitrogen emissions by about 6% (of the total). Speed limits, including safety-conscious, "defensive" driving, and increased use of public transport accompanied by minimization of traffic (within about three years) would allow a reduction of about 3% in each case.

In the power production/heating sector, all the possibilities of reducing emissions account for about 3% of the total at most. In the field of human nutrition, adjustment to the World Health Organization standard (in respect of protein and energy requirements, for example) would reduce nitrogen emissions by around 7% and nitrogen elimination in sewage treatment plants would lead to an overall reduction of approximately 3%.

6.5.2 Tin as an Example of a Microelement That is Both Essential and Toxic

With ten natural and two artificial isotopes, tin (Sn) has a greater number of known isotopes than any other element. Tin makes up about 0.0035% of the earth's crust, which means that this heavy metal has a similar geochemical significance to that of cobalt, yttrium, and cerium. Cassiterite (tin stone, SnO_2) and stannite (tin pyrites, Cu_2FeSnS_4) are the most important tin ores. Tin is one of the elements that occur mainly in the crystal structures of rock-forming silicates and oxides in the earth's crust; accordingly, the heavy metal must be regarded as a trace element in these minerals (Wedepohl 1991). Lignite and coal contain relatively large amounts of tin, an average of 1.0 and 2.6 mg/kg, respectively (Yudovich et al. 1972; Brumsack et al. 1984), whereas crude oil, with 10 µg/kg, contains comparatively little (Bertine and Goldberg 1971). In its compounds the heavy metal occurs in the oxidation states +2 and +4, both valences being about equally common. About 70% of the world's reserves of tin are owned by the countries Australia, Bolivia, Indonesia, Malaysia, Nigeria, Thailand, and Zaire, which joined to form the Association of Tin Producing Countries (ATPC) in 1983. In open-cast mining, ores with a tin content of only 0.1% are considered workable; under ground, ores with an Sn content of 0.3% or more are mined (Falbe and Regitz 1992).

In 1989, worldwide Sn production was 231,800 t from smelting and an additional 34,700 t from recycling of waste and scrap metal. Annual production is expected to reach the 300,000 t level by the year 2000 (Kabata-Pendias and Pendias 1992). Most of the tin produced is used for making tinplate for the food preserving industry (40%), solder (30%), synthesis of organotin compounds (15%), bearing metals, bronzes and other tin alloys, and float glass (for production quantities and uses of organotin compounds see below). In particular the Sn–Cu alloy known as bronze is widely used and has achieved great significance in art and history. It was in the Bronze Age (3500 to 3200 B.C.) that weapons, jewellery, and tools were first made from metal to any significant extent.

Inorganic tin is only very slightly toxic, and since the Middle Ages the metal has used for making plates, jugs, and drinking vessels. If tin is ingested, only a very small proportion (<5%) is absorbed by the body and has a systemic effect. Industrial dust may be so heavily contaminated with tin that grazing cattle are endangered when it is deposited on the pasture plants. In dogs, an oral dose of 200–300 mg $SnCl_4$/kg body weight is lethal. In contrast to the very slight or absent toxicity of the inorganic Sn compounds, the toxicity of the organotin compounds is very high; some are indeed used as antitumour drugs and cytostatic agents (Gielen 1987). It was long debated whether tin is also an essential microelement. In contrast to bacteria, fungi, algae, and higher plants, in which Sn does not appear to be necessary, tests on vertebrates showed that an Sn deficiency results in poor appetite, hair loss, and acne (Schwarz et al. 1970); today there is discussion as to whether Sn as a constituent of gastrin has a function in the digestive system of vertebrates (Kieffer 1991). Because of the extensive use of canned foods over the past decades, no cases of Sn deficiency in man have become known. The ADI (acceptable daily intake) is stated to be 2 mg Sn/kg body weight. Absorption of large quantities of tin may inhibit certain enzymes, especially the succinic dehydrogenase of the citric acid cycle and acid phosphatase.

The wide range of applications of tin is leading to an increasing contamination of all compartments of the environment; however relatively little reliable analytical data is available on this heavy metal compared with others. This is also reflected in the small number of certified standard reference materials for tin. According to Bennett (1981), the average concentrations are 30 ng Sn/m^3 for unpolluted air, 40 ng/L for water, and 200 μg/kg for food. Tin concentrations of 40–700 mg/kg dry matter are now found in sewage sludge, while various authors state the maximum acceptable level in arable soils to be 50 mg Sn/kg dry weight. Unpolluted soils contain an average of 1–20 mg Sn/kg dry weight (Markert 1993a). Under reductive conditions the mobility of Sn in soils is comparable to that of Cd, Cr, Mn, and Sr and thus very high for a heavy metal (for an overview see Kabata-Pendias and Pendias 1992). Commercial fertilizers, which account for the largest input of substances into agricultural land in Central Europe alongside applied sewage sludge, atmospheric inputs, and residues from harvesting, contain an average of 0.8–20 mg Sn/kg dry matter. Markert (1993a) gives average concentrations of 0.8–7 mg Sn/kg dry weight for plant material. Levels of 60 mg/kg dry weight or more are considered raised or even potentially toxic (Kabata-Pendias and Pendias 1992).

Alkylation of heavy metals generally results in compounds that are more toxic than the initial inorganic substances. It has been proved that the elements As, Hg, Se, Te, and also Sn undergo microbial methylation under environmental conditions. But while the formation of biogenic methyltin compounds and their possible danger to ecosystems has long been the subject of intensive research in the aquatic sector (cf., for example, Tugurul et al. 1983; Hamaski et al. 1991), not much progress has yet been made with risk assessment in the terrestrial environment. Moreover, recent investigations have shown that, for example, alkyl lead and methyltin compounds as methylating agents are capable of transferring their organic constituents to other metals, giving rise to further organometallic compounds (Kabata-Pendias and Pendias 1992).

Organotin compounds have been produced since the middle of the last century, but their industrial use did not start until 100 years later, when their biocidal effect was discovered. After oral intake their rates of absorption are much higher (10%–90%) than in the case of inorganic Sn salts ($< 5\%$). The degree of toxicity of the compounds depends basically on the length of the alkyl chain. The aryl compounds, for instance, are scarcely more hazardous than inorganic Sn, but the alkyltin compounds of the butyl and phenyl series are very highly toxic. The biocide tributyltin (TBT), especially, is one of the most poisonous substances that have ever been produced and released into the environment (World Health Organisation 1980; Goldberg 1986; Müller et al. 1989; Stewart et al. 1992). Because of its great efficacy as a biocide, this organometallic compound is used for numerous applications, including anti-fouling paints, preservatives for wood and other materials, textiles, sealing and casting compounds (e.g., polyurethane foams), as a stabilizer for plastics, in paints and adhesives, and also in mineral products such as insulating materials (Blunden et al. 1984; Hall and Pinkney 1985). TBT is further used in agriculture and horticulture as a biocide to combat fungi, bacteria, ants, insects, molluscs and rodents. Annual production of organotin compounds has risen steadily from a few tons in 1950 to at least 35,000 tons in 1985, 5,000 tons of this being TBT (Goldberg 1986; Maguire 1987).

TBT is usually produced from oxide, fluoride, sulfide, chloride, or acetate and dissolves in water, forming a hydrated cation (TBT$^+$). With an octanol/water distribution coefficient (P_{ow}) of 1600 to 1700 the compound is extremely lipophilic (Maguire 1987) and therefore dissolves only very slightly in artificial seawater with saturation values of 3–10 mg/L (Blunden et al. 1984). For this reason TBT is very quickly and effectively taken up and accumulated by organisms and sediments. The TBT levels found in the soft bodies of marine Prosobranchiata may be anything up to 300,000 times higher than in the surrounding seawater. Some organs even have bioconcentration factors of up to 1.8×10^6 (Oehlmann et al. 1993; Oehlmann 1994).

Tributyltin compounds damage the organism at several levels. It is known that TBT not only acts as a general metabolic toxin (Krajnc et al. 1984; Vos et al. 1984) and cytotoxin (Wester et al. 1990; Arkawa 1991; Jensen et al. 1991) but also damages membranes (Snoeij et al. 1986; Saxena 1987; Zucker et al. 1988; Antonenko 1990), disrupting the respiratory chain in the mitochondria and photosynthesis (Aldridge and Street 1964; Saxena 1987; Powers and Beavis 1991). The teratogenic effect of this class of compounds has also been described a number of times (Weis and Kim 1988; Walker et al. 1989; Fent and Meier 1992; Yonemoto et al. 1993). Furthermore, there is definite proof that TBT and other organotin compounds have a neurotoxic (Fendt and Meier 1992), carcinogenic (Wester and Canton 1987; Walker et al. 1989), and mutagenic (Davis et al. 1987) effect, not only in man.

Environmental concentrations of the various organotin compounds in limnic and marine waters have been determined in several countries since suitable measuring techniques became available in the late 1970s. The highest TBT level measured in freshwater was 3,000 ng/L in the yacht harbour of St. Clair Lake in Canada (Hall and Pinkney 1985). In various rivers, lakes, and inland waters in the USA and Germany the maximum concentrations were around 1,600 ng TBT/L (for an overview see Hall and Pinkney 1985). Also in the marine environment, similar TBT levels were found in the water. Even in France, where there have been restrictions on the use of TBT since 1982, high environmental concentrations of the biocide are still to be found. Alzieu et al. (1990, 1991) detected concentrations of between 2 and 833 ng TBT/L on the French Atlantic and Mediterranean coasts. In Brittany the values are only between < 2.5 and 80 ng TBT/L, since the tidal range there usually results in a complete exchange of the water even in highly polluted areas (Oehlmann et al. 1993).

TBT accumulation in sediments is especially important and usually occurs in magnitude of 10^3, which means that harbor sediments may be polluted with anything up to several mg/kg of the biocide (Kram et al. 1989). This constitutes a particular hazard to animal and plant species that live near the ground or in the sediment. In the UK the clam *Scrobicularia plana* was found to be an endangered species because of the high accumulation of TBT in the sediment (Langston and Burt 1991).

The various animal and plant groups show very different reactions to pollution with TBT. While it takes very high concentrations to kill some algae, crustaceans and fish are much more sensitive. The organisms most sensitive to TBT are molluscs such as snails and mussels. One of the reasons for these differences is the greatly varying ability of the groups of organisms to excrete the TBT they take up, and especially their ability to break it down into less toxic compounds. It is known that TBT is broken down into inorganic tin by a cytochrome-P-450-dependent multifunctional oxygenase system (MFO) via dibutyl tin and monobutyl tin in the microsomes of the mesenteron gland of marine invertebrates and the liver of vertebrates (Lee 1986; Stegeman and Kloepper-Sams 1987; Livingstone et al. 1989; Fent and Stegeman 1991). However, the MFO content and, thus, the ability to break down TBT is far lower in molluscs than in crustaceans and vertebrates (Lee 1986). This is why snails and mussels are found to have much higher TBT levels in their tissues than other invertebrates and fish at the same environmental concentrations. The bio-concentration factors (BCFs) determined are between 1.2×10^3 and 3.0×10^4 in the case of algae (Rexrode 1987). Of the invertebrates, annelids and crustaceans have relatively low BCFs at 3.4×10^3 and 4.4×10^3, respectively (Langston et al. 1987; Evans and Laughlin 1984), while molluscs and especially predatory Prosobranchiata have the highest. The BCFs determined were between 1.5×10^3 and 3.0×10^3 in mussels and oysters (Langston and Burt 1991), between 3.8×10^3 and 4.4×10^3 in herbivorous snails such as *Littorina littorea* (Langston et al. 1987; Matthiessen et al. 1991), and between 1.25×10^4 and 2.6×10^5 in carnivorous species (Bryan et al. 1986; Oehlmann et al. 1992). This shows that the position in the food chain is one of the factors that determine body burdens of TBT and, thus, indirectly, the sensitivity of a species to TBT, since predatory species feed on prey whose TBT levels are usually well above those of the plant material that serves as food for the herbivorous species.

The sand-shrimp *Crangon crangon* is killed as an adult within 96 h at TBT concentrations of 41 µg/L, but for the larvae a concentration of 2 µg/L is lethal (Hall and Pinkney 1985). Similar ontogenetic differences are also found with other crustaceans (e.g., *Gammarus oceanicus*, *Metamysidopsis elongata*), Polychaeta (e.g., *Neanthes arenaceodentata*), fish (e.g., *Cyprinodon variegatus*, *Solea solea*), and mussels (e.g., *Mercenaria mercenaria*, *Crassostrea gigas*). Chronically toxic TBT concentrations are naturally below the levels determined in acute tests, as the organisms are exposed to the pollutant for a much longer time. With values of less than 1 µg/L for *Gammarus oceanicus* (Laughlin et al. 1984), crustaceans have a relatively high threshold limit value for growth inhibition. In fish this effective concentration is much lower, at approximately 200 ng TBT/L for *Salmo gairdneri* (Seinen et al. 1981), and is only exceeded by the high degree of sensitivity of the mussels. In *Mytilus edulis* further growth is inhibited from about 60 ng TBT/L (Thain and Waldock 1986; Salazar and Salazar 1987), in adult *Crassostrea gigas* from 200 ng TBT/L, and in the larvae of this species, from as little as 5 ng/L (Hall and Pinkney 1985; Nell and Chvojka 1992). In the diatom *Skeletonema costatum*, photosynthesis is so effectively inhibited from as little as 100 ng TBT/L that no further growth is possible. The threshold concentrations for other planktonic algae are considerably higher (Beaumont and Newman 1986). The comparison shows that although certain systematic trends in TBT sensitivity can be detected, these are on no account to be regarded as hard and fast rules. The increased TBT sensitivity of young animals and larvae can also be demonstrated in respect of chronic effects. The NOEC (no observable effect concentration) for growth inhibition in the Pacific oyster is less than 5 ng TBT/L; in the edible mussel, which is also found in large areas of the northern and western Baltic Sea, it is less than 50 ng TBT/L (Dixon and Prosser 1985). Such TBT concentrations are considerably exceeded in many European coastal regions.

Other chronic TBT effects have been observed in fish, crabs, and starfish. At the 800-ng-TBT/L level degenerative changes occur in the gills, eyes, skin, and skeletal muscles of fish and result in increased mortality (Fent and Meier 1992). In crabs it is known that from about 500 ng TBT/L deformities occur in regenerated extremities such as pincers and running legs, and ecdysis is delayed (Weis and Kim 1988). Because of their deformed pincers and mouth parts the animals are no longer able to take in food. In starfish the regeneration of damaged or lost arms is suppressed at levels as low as 100 ng TBT/L (Walsh et al. 1986).

In the late 1970s shell deformities and failure of the larvae to develop into adults were first detected in the Pacific oyster *Crassostrea gigas*, which could not be explained by seasonal or climatic factors. In wide areas of Europe the oyster population broke down, causing considerable economic losses (Alzieu et al. 1989). The restrictions on the use of TBT that were introduced in France in 1982 were intended to bring TBT contamination of the coastal waters below the NOEC of 50 ng/L for shell abnormalities. Pollution levels fell noticeably and oyster populations recovered locally (Alzieu et al. 1987), but even today the TBT concentrations are above this NOEC along large sections of the French coast (Alzieu et al. 1990, 1991).

By far the most sensitive bioindicator system for TBT is based on the pseudohermaphroditism or imposex phenomenon (Jenner 1979 and Smith 1981, respectively) in marine Prosobranchiata. In these heteroecious species the females develop parts of the male reproductive system, usually a penis and/or a vas deferens (sperm duct), when exposed to TBT. Imposex has now been described in well over 100 marine species from all parts of the world (cf. Fioroni et al. 1991). As the imposex phenomenon develops, the size of the penis and the extent of the vas deferens steadily increase without at first impairing the reproductive ability of the females. In the final stage the oviduct is either blocked by hyperplastic vas deferens tissue or the ontogenetic closure of the initially open female genital tract fails to take place, so that intact egg capsules can no longer be deposited or produced. The result is sterility (Gibbs et al. 1990; Oehlmann et al. 1992; Oehlmann 1994). In all of the species that have been investigated in detail the occurrence of imposex in the vicinity of harbors and marinas shows that this phenomenon is an appropriate means of monitoring the effects of TBT. Near such sources of TBT emissions the imposex intensities (vas deferens sequence index or average penis length of females) found in the populations are much greater than elsewhere (Figure 6.3). The specific threshold values for sterility in females are 2 ng TBT as Sn/L for the species *Nucella lapillus* and 8 ng TBT as Sn/L for *Ocenebra erinacea*. In the highly sensitive species *Ocinebrina aciculata*, sterile specimens are already to be found at levels below 1.5 ng TBT as Sn/L.

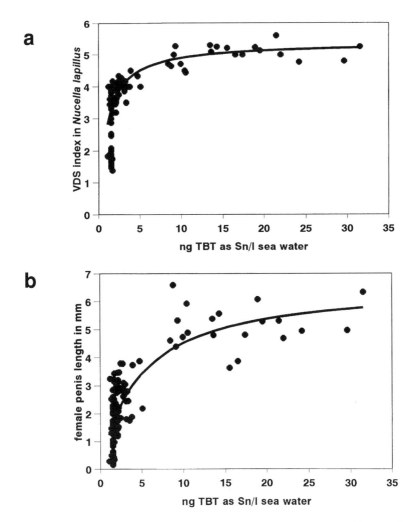

Figure 6.3 *Nucella lapillus*. Relationship between aqueous TBT concentrations and imposex intensities, measured as VDS (vas deferens sequence) index (a) and average female penis length (b). (a): y $= (5.38 \cdot x) - (0.974 + x)$; $n = 131$ populations analyzed; $r = 0.678$; $p < 0.0005$. (b): y $= (6.63 \cdot x)$ $-(4.67 + x)$; $n = 131$ populations analyzed; $r = 0.831$; $p < 0.0005$. From Oehlmann (1994).

6.6 REMEDIAL MEASURES FOR SOILS CONTAMINATED WITH HEAVY METALS

Where soils are polluted with ecotoxic substances and/or substances toxic to man, precautions have to be taken to prevent harm to both human beings and the environment as far as this is possible (Frankenberger 1992; Förstner 1994). The nature and extent of such precautions depends on the pollutants and—in

Germany, at least—the actual or planned use of the site (Eikmann et al. 1989; Franzius 1994; Weber and Neumaier 1993). If it is intended to use the site as a car park, the remedial action does not have to be as thorough as it would in the case of a children's playground.

Increased heavy metal concentrations in soils can usually be traced back to anthropogenic sources. Above all, these include inadequately safeguarded abandoned waste disposal sites, spoil from ore and coal mining, and point inputs from the metalworking industry (Ernst and Joose-van Damme 1983; Weber and Neumaier 1993). However, high heavy metal levels may also have geogenic origins. The situation always becomes very complex when inputs from human activities are added to an existing geogenic load as is the case in the Freiberg area of Saxony (Markus 1995).

For cleaning up such sites a number of different methods are already available or currently being developed and tested (Alef 1994; Weber and Neumaier 1993; Wille 1993). Table 6.4 gives a summary of decontamination and safeguarding techniques suitable for treating soils and ground waters polluted with heavy metals. By "safeguarding techniques" we mean methods that reduce the hazard to man and the environment from the contaminated soil but leave this contamination where it is.

Excavation and dumping of the polluted material without previous treatment is one of the simplest procedures of all. It was the method of choice when remedial work on abandoned hazardous sites began in many parts of the world, as it has the advantage that the contaminated soil is removed from the dangerous site and deposited in a safe and sanitary manner in a landfill. In this case the pollutants are not treated. When the extent of the remedial work on polluted sites became known the method was usually abandoned, as it takes up a great amount of landfill space (Wille 1993).

Table 6.4 Decontamination and safeguarding techniques for abandoned sites polluted with heavy metals

Technique	Where applied	Status
Excavation and dumping	Off site	Introduced
Immobilization	In situ/off site	Introduced
Vitrification	Off site	Introduced
Covering (with unpolluted earth)	In situ	Introduced
Encapsulation	In situ	Introduced
Washing	In situ/on site	Introduced
Extraction	In situ/on site	Introduced
Steam purification	On site/off site	Introduced
Electromigration, electroosmosis	In situ	Used in isolated cases
Phytoremediation	In situ	At the development stage
Bioleaching	In situ/on site	At the development stage

In *immobilization* the pollutants are converted by chemicals added to the soil (Wille 1993). Heavy metals are usually bound in the form of sulfides. In most cases additional ground engineering measures are necessary, especially when heavy metals are to be immobilized (Weber and Neumaier 1993). As an in situ procedure this method is fairly complex, as the whole of the contaminated area has to be brought into contact with the reagents. As an off site procedure it can also be used for pretreating contaminated soils that are later to be dumped.

Vitrification is a method of pretreating soils that are to be dumped. The sealing of pollutants in glass is generally regarded as a permanent immobilization measure (Weber and Neumaier 1993). The technique is energy-intensive because of the high temperatures required for vitrification.

Covering the contaminated area with unpolluted earth is an extremely simple method that is only acceptable and useful in exceptional cases. It can prevent the dispersal of contaminants that are insoluble in water through the air and their direct contact with biota (Wille 1993).

Encapsulation of polluted sites is carried out with vertical subterranean curtains or sheet piling ending in an aquiclude or artificial sealing layer (e.g., an injected grout floor). In addition, the surface is usually covered with plastic sheeting and sealed as a protection against precipitation and escaping gases. Of the techniques for safeguarding alone, encapsulation is widely regarded as the most effective (Weber and Neumaier 1993).

Washing generally means treatment of the soil with water or aqueous solutions. In its simplest form it consists of wet mechanical separation of the fine and coarse particles; as a rule, the highly polluted fine material is removed to a landfill while the coarse material is put back in place. This can be done in situ or on site. More complex processes consist of pretreatment in mills and sieves followed by separation of the pollutants from the soil particles. The soil is swirled with water or dilute acid, sometimes under pressure (Weber and Neumaier 1993; Wille 1993). The heavy metals are precipitated out of the contaminated water in the form of hydroxides. The resulting sludge is dewatered and usually deposited in a landfill; after primary treatment the water can be taken to a sewage works. The purified earth can be put back in place, but the original soil structure is lost and it is biologically dead. Washing is one of the most common methods of purifying soils contaminated with heavy metals (Weber and Neumaier 1993; Wille 1993).

In *extraction* the soils are treated with organic solvents (Weber and Neumaier 1993). This process is only used to combat heavy metal contamination if organic pollutants are also present and all of the contaminants are to be extracted at the same time.

The process and results of *steam purification* correspond largely to those of high-pressure washing with water, but as a rule it is carried out without additives to the water (Weber and Neumaier 1993; Wille 1993).

In *electromigration* and *electroosmosis*, electrodes are inserted into the contaminated soil to create a d.c. voltage field. Within this field, charged particles move to the electrodes, which are surrounded with filter materials and a current of water. They accumulate in the filter devices and are pumped off into collecting tanks. The contaminated liquid is treated as in the washing process. The method can be used under sealed-off areas and on cohesive soils, but at present it is still at the trial stage (Wille 1993).

Phytoremediation is the decontamination of soils polluted with heavy metals by means of hyperaccumulating plants (Alef 1994). The contaminated plants are reaped and burnt. The ash, which is heavily polluted with heavy metals, is treated or deposited in a landfill. This is a highly environment-friendly technique in which the structure and nutrient content of the soil remains unimpaired; it is currently being tested on a large scale by several study groups. One of the difficulties (as yet) is the choice of suitable plants that display both high uptake rates into their overground parts and sufficiently fast growth (Banks et al. 1994; Bücking and Heyser 1994).

In *bioleaching* the heavy metals are released from the soil into the surrounding or interstitial water through the activity of microorganisms (Haddadin et al. 1995; Kapoor et al. 1995). The polluted water is collected and treated. This method can be carried out both in bioreactors and in situ. Like photoremediation it results in a largely undisturbed soil whose multifunctionality has been restored after the treatment. The technique has been used for some time in ore leaching. Its use in remedial work on abandoned hazardous sites is being investigated at a few individual research establishments.

Acknowledgements. I wish to thank Dr. Gernot Kayser and Dr. Jörg Oehlmann (both of IHI Zittau) for their valuable assistance in preparing this article. My thanks also to Mrs. Marion Braase (Buchholz, near Hamburg) for the English translation.

6.7 REFERENCES

Adriano, D. C., 1986, *Trace Elements in the Terrestrial Environment*, Springer, Berlin, Heidelberg, New York

Adriano, D. C., Ed., 1992, *Biogeochemistry of Trace Metals*, Lewis Publishers, Boca Raton, Florida, USA

Adriano, D. C., 1994, *Global Center for Environmental Restoration (GCER). A Proposal to Establish an International Center for Soil Remediation*, Draft 11/04/1994, Aiken South Carolina 29802, USA

Adriano, D. C., Z. S. Chen, and S. S. Yang, 1994, "Biogeochemistry of Trace Elements," *Sci. Technol. Let.*, Northwood, A special issue of *Environmental Geochemistry and Health*, Vol. 16

Aldridge, W. N., and B. W. Street, 1964, "Oxidative Phosphorylation: Biochemical Effects and Properties of Trialkyltins," *Biochem. J.*, *91*, 287–297

Alef, K., 1994, *Biologische Bodensanierung*, VCH, Weinheim, Germany

Allen, L. C., 1992, "Extension and Completion of the Periodic Table," *J. Am. Chem. Soc.*, *114*, 1510–1514

Alloway, B. J., Ed., 1993, *Heavy Metals in Soils*, Blackie Academic & Professional, London, Glasgow, New York, Tokyo, Melbourne, Madras

Alzieu, C., G. Barbier, and J. Sanjuan, 1987, "Évolution des Teneurs en Cuivre des Huîtres du Bassin d'Arcachon: Influence de la Législation sur les Peintures Antisalissures," *Oceanol. Acta*, *10*, 463–468

Alzieu, C., M. Héral, and J.-P. Dreno, 1989, "Les Peintures Marines Antisalissures et Leur Impact sur L'ostréiculture," *Equinoxe*, *24*, 22–31

Alzieu, C., P. Michel, J. Sanjuan, and B. Averty, 1990, "Tributyltin Levels in French Mediterranean Coastal Waters," *Appl. Organometal. Chem.*, *4*, 55–61

Alzieu, C., P. Michel, I. Tolosa, E. Bacci, L. D. Mee, and J. W. Readman, 1991, "Organotin Compounds in the Mediterranean: a Continuing Cause of Concern," *Mar. Environ. Res.*, *32*, 261–270

Antonenko, Y. N., 1990, "Electrically Silent Anion Transport Through Bilayer Lipid Membrane Induced by Tributyltin and Triethyllead," *J. Membr. Biol.*, *113*, 109–113

Arakawa, Y., 1991, "Biochemical Activity of Organotin Compounds and their Mechanisms," *J. Pharmacobio-Dyn.*, *14*, 131

Baker, A. J. M., 1981, "Accumulators and Excluders—Strategies in the reponse of plants to heavy metals," *J. Plant Nutr.*, *3*, 643–654

Banks, M.-K., A. P. Schwab, G. R. Fleming, and B. A. Hetrick, 1994, "Effects of Plants and Soil Microflora on Leaching of Zinc from Mine Tailings," *Chemosphere*, *29*, 1691–1699

Beaumont, A. R., and P. B. Newman, 1986, "Low Levels of Tributyl Tin Reduce Growth of Marine Micro-algae," *Mar. Pollut. Bull.*, *17*, 457–461

Bennett, B. G., 1981, "Exposure Commitment Assessments of Environmental Pollutants. Summary Exposure Assessments for Mercury, Nickel, Tin," *MARC Report No. 25*, *Vol. 1*, Monitoring and Assessment Research Centre, London

Bertine, K. K., and E. D. Goldberg, 1971, "Fossil Fuel Combustion and the Major Sedimentary Cycle," *Science*, *173*, 233–235

Bliefert, C., 1994, *Umweltchemie*, VCH, Weinheim, New York, Tokyo

Bloemen, M. L., B. Markert, and H. Lieth, 1995, "The distribution of Cd, Cu, Pb, and Zn in topsoils of Osnabrück in relation to land use," *Sci. Total Environ.*, *166*, 137–148

Blunden, S. J., L. A. Hobbs, and P. J. Smith, 1984, "The Environmental Chemistry of Organotin Compounds," *Environ. Chem.*, *3*, 49–77

Bodek, I., W. J. Lyman, W. F. Reehl, and D. H. Rosenblatt, Eds., 1988, *Environmental Inorganic Chemistry*, Pergamon Press, New York, USA

Bowen, H. J. M., 1979, *Environmental Chemistry of the Elements*, Academic Press, London, UK

Brooks, R. R., 1993, "Geobotanical and Biogeochemical Methods for Detecting Mineralization and Pollution from Heavy Metals in Oceania, Asia, and The Americas," in *Plants as Biomonitors, Indicators for Heavy Metals in the Terrestrial Environment* (B. Markert, Ed.), VCH, Weinheim, New York, Tokyo, pp. 127–149

Brumsack, H. J., H. Heinrichs, and H. Lange, 1984, "West German Coal Power Plants as Sources of Potentially Toxic Emissions," *Environ. Technol. Lett.*, *5*, 7–22

Bruns, I., A. Siebert, R. Baumbach, J. Miersch, D. Günther, B. Markert, and G. J. Krauss, 1995, "Analysis of Heavy Metals and Sulphur-Rich Compounds in the Water Moss *Fontinalis antipyretica* L. ex Hedw.," *Fresenius J. Anal. Chem., 353*, 101–104

Bryan, G. W., P. E. Gibbs, L. G. Hummerstone, and G. R. Burt, 1986, "The Decline of the Gastropod *Nucella lapillus* Around South-West England; Evidence for the Effect of Tributyltin from Antifouling Paints," *J. Mar. Biol. Ass. U.K. 66*, 611–640

Bücking, H., and W. Heyser, 1994, "The Effect of Ectomycorrhizal Fungi on Zn Uptake and Distribution in Seedlings of *Pinus sylvestris L.*," *Plant Soil, 167*, 203–212

Champ, M. A., and F. L. Lowenstein, 1987, "TBT: The Dilemma of High-Technology Antifouling Paints," *Oceanus, 30*, 69–77

Davies, B. E., 1992, "Trace Metals in the Environment, Retrospect and Prospect," in *Biogeochemistry of Trace Metals* (D. C. Adriano, Ed.), Lewis Publishers, Boca Raton, Florida, USA, pp. 1–17

Davis, A., R. Barale, G. Brun, R. Forster, T. Günther, H. Hautefeuille, C. A. van der Heijden, A. G. A. C. Knaap, R. Krowke, T. Kuroki, N. Loprieno, C. Malaveille, H. J. Merker, M. Monaco, P. Mosesso, D. Neubert, H. Norppa, M. Sorsa, E. Vogel, C. E. Voogd, M. Umeda, and H. Bartsch, 1987, "Evaluation of the Genetic and Embryotoxic Effects of Bis(tri-*n*-butyltin)oxide (TBTO), a Broad-Spectrum Pesticide, in Multiple In Vivo and In Vitro Short-Term Tests," *Mutat. Res., 188*, 65–95

Dixon, D. R., and H. Prosser, 1986, "An Investigation of the Genotoxic Effects of an Organotin Antifouling Compound (Bis[tributyltin]oxid) on the Chromosomes of the Edible Mussel, *Mytilus edulis*," *Aquat. Toxicol., 8*, 185–195

Duvigneaud, P., and S. Denaeyer-De Smet, 1973, "Biological Cycling of Minerals in Temperate Deciduous Forests," in *Analysis of Temperate Forest Ecosystems, Ecological Studies 1* (D. E. Reichle, Ed.), Springer, Berlin, Heidelberg, New York

Eikmann, T., S. Michels, T. Krieger, and H. J. Einbrodt, 1989, "Umwelthygienische Grundlagen und Problematik der Richtwertfestsetzung für Schadstoffe in Böden," *Forum Städtehygiene, 40*, 333–337

Ernst, W. H. O., 1974, *Schwermetallvegetation der Erde*, Gustav Fischer Verlag, Stuttgart, Germany

Ernst, W. H. O., 1993, "Geobotanical and Biogeochemical Prospecting and Analysis of Heavy Metal Deposits in Europe and Africa," in *Plants as Biomonitors, Indicators for Heavy Metals in the Terrestrial Environment* (B. Markert, Ed.), VCH, Weinheim, New York, Tokyo, pp. 107–126

Ernst, W. H. O., and E. N. G. Joossee Van Damme, 1983, *Umweltbelastung durch Mineralstoffe*, Gustav Fischer Verlag, Stuttgart, Germany

Evans, D. W., and R. B. Laughlin, 1984, "Accumulation of Bis(tributyltin) Oxide by the Mud Crab, *Rhithropanopeus harrisii*," *Chemosphere, 13*, 213–219

Falbe, J., and M. Regitz, Ed., 1992, *Römpp-Chemie-Lexikon*, 9th Edn., Thieme, Stuttgart, New York

Farago, M. E., Ed., 1994, *Plants and the Chemical Elements*, VCH, Weinheim, New York, Tokyo

Federal Office of the Environment—see UBA

Fent, K., and W. Meier, 1992, "Tributyltin-Induced Effects on Early Life Stages of Minnows *Phoxinus phoxinus*," *Arch. Environ. Contam. Toxicol., 22*, 428–438

Fent, K., and J. J. Stegeman, 1991, "Effects of Tributyltin Chloride In Vitro on the Hepatic Microsomal Monooxygenase System in the Fish *Stenotomus chrysops*," *Aquat. Toxicol.*, *20*, 159–168

Fiedler, H. J., and H. J. Rösler, 1993, *Spurenelemente in der Umwelt*, Gustav Fischer Verlag, Jena, Stuttgart, Germany

Fioroni, P., J. Oehlmann, and E. Stroben, 1991, "The Pseudohermaphroditism of Prosobranchs; Morphological Aspects," *Zool. Anz, 226*, 1–26

Forschungszentrum Jülich GmbH, Ed., 1991, *Intensivlandwirtschaft und Nitratbelastung, Berichte aus der ökologischen Forschung*, Band 3, Jülich, Germany

Fortescue, J. A. C., 1980, *Environmental Geochemistry*, Springer, Berlin, Heidelberg, New York

Förstner, U., 1994, *Umweltschutztechnik*, Springer, Berlin, Heidelberg, New York

Franzius, V., Ed., 1994, *Sanierung kontaminierter Standorte*, Erich Schmidt Verlag, Berlin, Germany

Fränzle, O., 1993, *Contaminants in Terrestrial Environments*, Springer, Berlin, Heidelberg, New York

Fränzle, O., 1994, "Representative Soil Sampling," in *Environmental Sampling for Trace Analysis* (B. Markert, Ed.), VCH, Weinheim, New York, Tokyo, 305–320

Frankenberger, W. T., Jr., and U. Karlson, 1992, "Dissipation of Soil Selenium by Microbiological Volatilization," in *Biogeochemistry of Trace Metals* (D. C. Adriano, Ed.), Lewis Publishers, Boca Raton, Florida, USA, 365–381

Garten, C. T., 1976, "Correlations Between Concentrations of Elements in Plants," *Nature, 261*, 686–688

Gibbs, P. E., G. W. Bryan, P. L. Pascoe, and G. R. Burt, 1990, "Reproductive Abnormalities in Female *Ocenebra erinacea* (Gastropoda) Resulting from Tributyltin-Induced Imposex," *J. Mar. Biol. Ass. U.K., 70*, 639–656

Gielen, M. F., 1987, *Metal-Based Anti-tumour Drugs*, Freund Publishers, London, UK

Goldberg, E. D. 1986, "TBT: An Environmental Dilemma," *Environment, 28*, 17–20; 41–44

Golley, F. B., T. Richardson, and R. G. Clements, 1978, "Element Concentrations in Tropical Forests and Soils in Northwestern Columbia," *Biotropica, 10*, 144–151

Haddadin, J., C. Dagot, and M. Fick, 1995, "Models of Bacterial Leaching," *Enzyme Microb. Technol., 17*, 290–305

Hall, L. W., and A. E. Pinkney 1985, "Acute and Sublethal Effects of Organotin Compounds in Aquatic Biota: An Interpretative Literature Evaluation," *Crit. Rev. Toxicol., 14*, 159–209

Hamasaki, T., H. Nagase, T. Sato, H. Kito, and Y. Ose, 1991, "Production of Methyltin Compounds Related to Possible Conditions in the Environment," *Appl. Organomet. Chem., 5*, 83–90

Hamilton, E. I., 1979, *The Chemical Elements and Man*, John Wiley & Sons, New York, USA

Herpin, U., B. Markert, U. Siewers, and H. Lieth, 1995, *Monitoring der Schwermetallbelastung in der Bundesrepublik Deutschland mit Hilfe von Moosanalysen*, Forschungsbericht 108 02087, UBA-FB, 94–125

Ittekkot, V., S. Kempe, W. Michaelis, and A. Spitzy, Eds., 1990, *Facets of Modern Biogeochemistry*, Springer, Berlin, Heidelberg, New York

Iyengar, G. V., 1989, *Elemental Analysis of Biological Systems*, CRC Press, Boca Raton, USA

Jayasekera, R., 1993, "Concentrations of Selected Heavy Metals Indifferent Compartments of a Mountain Rain Forest Ecosystem in Sri Lanka," in *Plants as Biomonitors, Indicators for Heavy Metals in the Terrestrial Environment* (B. Markert, Ed.), VCH, Weinheim, New York, Tokyo, pp. 613–622

Jayasekera, R., 1994, "Sampling of Tropical Terrestrial Plants with Particular Reference in the Determination of Trace Elements," in *Environmental Sampling for Trace Analysis* (B. Markert, Ed.), VCH, Weinheim, New York, Tokyo, pp. 443–448

Jenner, M. G., 1979, "Pseudohermaphroditism in *Ilyanassa obsoleta* (Mollusca: Neogastropoda)," *Science, 205*, 1407–1409

Jensen, K. G., A. Önfelt, M. Wallin, V. Lidums, and O. Andersen, 1991, "Effects of Organotin Compounds in Mitosis, Spindle Structure, Toxicity and In Vitro Microtubule Assembly," *Mutagenesis, 6*, 409–416

Kabata-Pendias, A., and K. Pendias, 1992, *Trace Elements in Soils and Plants*, CRC Press, Boca Raton, Florida, USA

Kapoor, A., T. Viraraghavan, A. T. Hanson, and Z. Samani, 1995, "Remediation of Chromium-Containing Soils by Heap Leaching: Column Study," *J. Environ. Eng., 121*, 366–367

Keune, H., A. B. Murray, and H. Benking, 1991, "Harmonization of Environmental Measurement," *GeoJournal, 23*, 249–255

Kieffer, E., 1991, "Metals as Essential Trace Elements for Plants, Animals, and Humans," in *Metals and Their Compounds in the Environment. Occurrence, Analysis and Biological Relevance* (E. Merian, Ed.), VCH Publishers, Weinheim, New York, Basel, Cambridge, pp. 481–489

Kinzl, H., Ed., 1982, *Pflanzenökologie und Mineralstoffwechsel*, Verlag Eugen Ulmer, Stuttgart, Germany

Kovacs, M., K. Penksza, G. Turcsanyi, L. Kaszab, and P. Szoke, 1993, "Multielement-Analysen der Arten eines Waldsteppen-Waldes in Ungarn," *Phytocoenologia, 23*, 257–267

Kovacs, M., K. Penksza, G. Turcsanyi, L. Kaszab, S. Toth, and P. Szoke, 1994, "Comparative Investigation of the Distribution of Chemical Elements in an Aceri Tatarico-Quercetum Plant Community and in Stands of Cultivated Plants," in *Environmental Sampling for Trace Analysis* (B. Markert, Ed.), VCH, Weinheim, New York, Tokyo, pp. 435–442

Krajnc, E. I., P. W. Wester, J. G. Loeber, F. X. R. Van Leeuwen, J. G. Vos, H. M. A. G. Vaessen, and C. A. Van der Heijden, 1984, "Toxicity of Bis(tri-n-butyltin)oxide in the Rat. I. Short-Term Effects on General Parameters and on the Endocrine and Lymphoid Systems," *Toxicol. Appl. Pharmacol., 75*, 363–386

Kram, M. L., P. M. Stang, and P. F. Seligman, 1989, "Absorption and Desorption of Tributyltin in Sediments of San Diego Bay and Pearl Harbor," *Appl. Organometal. Chem., 3*, 523–536

Kramer, M., 1990, "Ein einzelbetrieblich basiertes Simulationsmodell der regionalen Agrarstrukturentwicklung," in *Schriftenreihe zur Angewandten Systemforschung* (H. Lieth, B. Meyer, T. Witte, Eds.), Band 1, LIT-Verlag, Münster, Germany

Lacerda, L. D., and W. Salomons, 1991, *Mercury in the Amazon: a Chemical Time Bomb?*, Dutch Ministry of Housing/Physical Planning and Environments, Haren, The Netherlands

Langston, W. J., and G. R. Burt, 1991, "Bioavailability and Effects of Sediment-Bound TBT in Deposit-Feeding Clams, *Scrobicularia plana*," *Mar. Environ. Res.*, 32, 61–77

Langston, W. J., G. R. Burt, and Z. Mingjiang, 1987, "Tin and Organotin in Water, Sediments, and Benthic Organisms from Poole Harbour," *Mar. Pollut. Bull.*, 18, 634–639

Laughlin, R. B., K. Nordlund, and O. Lindén, 1984, "Long-Term Effects of Tributyltin Compounds on the Baltic Amphipod, *Gammarus oceanicus*," *Mar. Environ. Res.*, 12, 243–272

Lee, R. F., 1986, "Metabolism of Bis(tributyltin)oxide by Estuarine Animals," *Oceans '86. Conference Record*, 4, 1182–1188

Lepp, N. W., 1992, "Uptake and Accumulation of Metals in Bacteria and Fungi," in *Biogeochemistry of Trace Metals* (D. C. Adriano, Ed.), Lewis Publishers, Boca Raton, Florida, pp. 277–298

Lieth, H., and B. Markert, (Eds.), 1990, *Element Concentration Cadasters in Ecosystems*, VCH, Weinheim, New York, Tokyo

Likens, G. E., F. H. Bormann, R. S. Pierce, J. S. Eaton, and N. M. Johnson, 1977, *Biogeochemistry of a Forested Ecosystem*, Springer, Berlin, Heidelberg, New York

Livingstone, D. R., M. A. Kirchin, and A. Wiseman, 1989, "Cytochrome *P*-450 and Oxidative Metabolism in Molluscs," *Xenobiotica*, 19, 1041–1062

Maguire, R. J., 1987, "Environmental Aspects of Tributyltin," *Appl. Organometal. Chem.*, 1, 475–498

Markert, B., 1992, "Presence and Significance of Naturally Occurring Chemical Elements of the Periodic System in the Plant Organism and Consequences for Future Investigations on Inorganic Environmental Chemistry in Ecosystems," *Vegetatio*, 103, 1–30

Markert, B., Ed., 1993a, *Plants as Biomonitors. Indicators for Heavy Metals in the Terrestrial Environment*, VCH, Weinheim, New York, Tokyo

Markert, B., 1993b, *Instrumentelle Multielementanalyse von Pflanzenproben*, VCH, Weinheim, New York, Tokyo

Markert, B., Ed., 1994a, *Environmental Sampling for Trace Analysis*, VCH, Weinheim, New York, Tokyo

Markert, B., Ed., 1994b, "The Biological System of the Elements (BSE) for Terrestrial Plants (Glycophytes)," *Sci. Total Environ.*, 155, 221–228

Markert, B., 1995, "Quality Assurance of Plant Sampling and Storage," in *Quality Assurance in Environmental Monitoring for Trace Element Determination* (P. Quevauviller, Ed.), VCH, Weinheim, New York, Tokyo, 215–244

Markert, B., 1996, *Instrumental Element and Multi-element Analysis of Plant Samples— Methods and Applications*, John Wiley & Sons, New York, USA, 2nd Revised Edn., 296 pp.

Markus, K., 1995, "Bodenbelastungsgebiete im Freistaat Sachsen," *Vortrag Sächsische Bodenschutztage*, 20–21 September 1995, Dresden

Matschullat, J., and G. Müller, Eds., 1994, *Geowissenschaften und Umwelt*, Springer, Berlin, Heidelberg, New York

Matthiessen, P., R. Waldock, J. E. Thain, S. Milton, and S. Scorpe-Howe, 1991, "Changes in Periwinkle (*Littorina littorea*) Population Following the Ban on TBT-Based Antifoulings on Small Boats," *Int. Council Expl. Sea. Mar. Environ. Qual. Com.*, CM 1991/E:5 (o.S.)

McGrath, S. P., and P. J. Loveland, 1992, *The Soil Geochemical Atlas of England and Wales*, Blackie Academic & Professional, London, Glasgow, New York, Tokyo, Melbourne, Madras

Merian, E., Ed., 1991, *Metals and Their Compounds in the Environment*, VCH, Weinheim, New York, Tokyo

Moldan, B., and J. Cerny, Eds., 1994, *Biogeochemistry of Small Catchments*, John Wiley & Sons, Chicester, New York, Brisbane, Toronto, Singapore

Müller, M. D., L. Renberg, and G. Rippen, 1989, "Tributyltin in the Environment—Sources, Fate and Determination: An Assessment of Present Status and Research Needs," *Chemosphere*, *18*, 2015–2042

Nagel, R., and R. Loskill, Eds., 1991, *Bioaccumulation in Aquatic Systems*, VCH, Weinheim, New York, Tokyo

Nell, J. A., and R. Chvojka, 1992, "The Effect of Bis-tributyltin Oxide (TBTO) and Copper on the Growth of Juvenile Sydney Rock Oysters *Saccostrea commercialis* (IREDALE and ROUGHLEY) and Pacific Oysters *Crassostrea gigas* THUNBERG," *Sci. Total Environ.*, *125*, 193–201

Nriagu, J. O., and J. M. Pacyna, 1988, "Quantitative Assessment of Worldwide Contamination of Air, Water and Soils by Trace Metals," *Nature*, *333*, 134–139

Oehlmann, J., 1994, *Imposex bei Muriciden (Gastropoda, Prosobranchia), eine ökotoxikologische Untersuchung zu TBT-Effekten*, Cuvillier, Göttingen, Germany

Oehlmann, J., E. Stroben, and P. Fioroni, 1992, "The Rough Tingle *Ocenebra erinacea* (Gastropoda: Muricidae): an Exhibitor of Imposex in Comparison to *Nucella lapillus*," *Helgoländer Meeresunters*, *46*, 311–328

Oehlmann, J., E. Stroben, and P. Fioroni, 1993, "Fréquence et Degré D'expression du Pseudohermaphrodisme Chez Quelques Prosobranches Sténoglosses des Côtes Françaises (Surtout de la Baie de Morlaix et de la Manche). 2. Situation Jusqu'au Printemps de 1992," *Cah. Biol. Mar.*, *34*, 343–362

Oehlmann, J., E. Stroben, U. Schulte-Oehlmann, B. Bauer, P. Fiorini, and B. Markert, 1996, "Tributyltin Biomonitoring Using Brosobranchs as Sentinel Organisms," *Fresenius J. Anal. Chem.*, *354*, 540–545

Oehlmann, J., and B. Markert, 1996, *Humantoxikologie*, Wissenschaftliche Verlagsgesellschaft, Stuttgart, Germany, 260 pp

Parlar, H., and D. Angerhöfer, 1991, *Chemische Ökotoxikologie*, Springer, Berlin, Heidelberg, New York

Powers, M. F., and A. D. Beavis, 1991, "Triorganotins Inhibit the Mitochondrial Inner Membrane Anion Channel," *J. Biol. Chem.*, *266*, 17250–17256

Rexrode, M., 1987, "Ecotoxicity of Tributyltin," *Oceans '87, Conference Record*, *4*, 1443–1455

Rossbach, M., J. D. Schladot, and P. Ostapczuk, 1992, *Specimen Banking*, Springer, Berlin, Heidelberg, New York

Roth, M., 1992, "Metals in Invertebrate Animals of a Forest Ecosystem," in *Biogeochemistry of Trace Metals* (D. C. Adriano, Ed.), Lewis Publishers, Boca Raton, Florida, USA, 299–328

Salomons, W., and U. Förstner, 1984, *Metals in the Hydrocycle*, Springer, Berlin, Heidelberg, New York

Salazar, M. H., and S. M. Salazar, 1987, "Tributyltin Effects on Juvenile Mussel Growth," *Oceans '87*, Conference *Record, 4*, 1504–1510

Sansoni, B., 1987, "Multielement Analysis for Environmental Characterization," *Pure Appl. Chem.*, 579–610

Sansoni, B., and V. Iyengar, 1978, *Sampling and Sample Preparation Methods for the Analysis of Trace Elements in Biological Materials*, Forschungszentrum Jülich, Germany, Spezieller Bericht des Forschungszentrums, Jülich, 13

Saxena, A. K., 1987, "Organotin Compounds: Toxicology and Biomedicinal Applications," *Appl. Organometal. Chem., 1*, 39–56

Schwarz, K., D. B. Milne, and E. Vinyard, 1970. "Growth Effects of Tin Compounds in Rats Maintained in Trace Element-Controlled Atmosphere," *Biochem. Biophys. Res. Commun., 40*, 22–25

Seinen, W., T. Helder, H. Vernij, A. Penninks, and P. Leeuwangh, 1981, "Short Term Toxicity of Tri-*n*-butyltinchloride in Rainbow Trout (*Salmo gairdneri* RICHARDSON) Yolk Sac Fry," *Sci. Total Environ., 19*, 155–166

Smith, B. S., 1981, "Tributyltin Compounds Induce Male Characteristics on Female Mud Snails *Nassarius obsoletus* = *Ilyanassa obsoleta*," *J. Appl. Toxicol., 1*, 141–144

Snoeij, N. J., P. M. Punt, A. H. Penninks, and W. Seinen, 1986, "Effects of Tri-*n*-butyltin Chloride on Energy Metabolism, Macromolecular Synthesis, Precursor Uptake and Cyclic AMP Production in Isolated Rat Thymocytes," *Biochim. Biophys. Acta, 852*, 234–243

Stegeman, J. J., and P. J. Kloepper-Sams, 1987, "Cytochrome *P*-450 Isozymes and Monooxygenase Activity in Aquatic Animals," *Environ. Health Perspect., 71*, 87–95

Stewart, C., S. J. de Mora, M. R. L. Jones, and M. C. Miller, 1992, "Imposex in New Zealand Neogastropods," *Mar. Pollut. Bull., 24*, 204–209

Streit, B., 1994a, *Lexikon Ökotoxikologie*, 2nd Edn., VCH, Weinheim, New York, Tokyo

Streit, B., 1994b, *Ökologie, Meyers Forum*, B. I.-Taschenbuchverlag, München, Leipzig, Wien, Zürich

Streit, B., and W. Stumm, 1993, "Chemical Properties of Metals and the Process of Bioaccumulation in Terrestrial Plants," in *Plants as Biomonitors, Indicators for Heavy Metals in the Terrestrial Environment*, VCH, Weinheim, New York, Tokyo, pp. 31–62

Thain, J. E., and M. J. Waldock, 1986, "The Impact of Tributyl Tin (TBT) Antifouling Paints on Molluscan Fisheries," *Wat. Sci. Tech., 18*, 193–202

Thornton, I., Ed., 1988, *Geochemistry and Health*, Science Review Ltd., Northwood

Tugurul, S., T. I. Balkas, and E. D. Goldberg, 1983, "Methyltins in the Marine Environment," *Mar. Pollut. Bull., 14*, 297–303

UBA (Umweltbundesamt), Ed., 1994, *Jahresbericht*, Berlin, pp. 231–234

Ure, A. M., and C. M. Davidson, Eds., 1995, *Chemical Speciation in the Environment*, Chapman & Hall, Glasgow, UK

Vos, J. G., A. De Klerk, E. I. Krajnc, W. Kruizinga, B. Van Ommen, and J. Rozing, 1984, "Toxicity of Bis(tri-*n*-butyltin)oxide in the Rat. II. Suppression of Thymus-Dependent Immune Responses and of Parameters of Nonspecific Resistance After Short-Term Exposure," *Toxicol. Appl. Pharmacol., 75*, 387–408

Walker, W. W., C. S. Heard, K. Lotz, T. F. Lytle, W. E. Hawkins, C. S. Barnes, D. H. Barnes, and R. M. Overstreet, 1989, "Tumorigenic, Growth, Reproductive, and Developmental Effects in Medaka Exposed to Bis(tri-*n*-butyltin) Oxide," *Oceans '89, Conference Record*, *2*, 516–524

Walsh, G. E., L. L. McLaughlin, M. K. Louie, C. H. Deans, and E. M. Lores, 1986, "Inhibition of Arm Regeneration by *Ophioderma brevispina* (Echinodermata, Ophiuroidea) by Tributyltin Oxide and Triphenyltin Oxide. *Ecotoxicol. Environ. Safety*, *12*, 95–100

Weber, H. H., and H. Neumaier, Eds., 1993, *Altlasten—Erkennen, Bewerten, Sanieren*, 2nd Edn., Springer, Berlin, Germany

Wedepohl, K. H., 1991, "The Composition of the Upper Earth's Crust and the Natural Cycles of Selected Metals. Metals in Natural Raw Materials. Natural Resources," in *Metals and Their Compounds in the Environment. Occurrence, Analysis and Biological Relevance* (E. Merian, Ed.), VCH Publishers, Weinheim, New York, Basel, Cambridge pp. 3–17

Weis, J. S., and K. Kim, 1988, "Tributyltin is a Teratogen in Producing Deformities in Limbs of the Fiddler Crab, *Uca pugilator*," *Arch. Environ. Contam. Toxicol.*, *17*, 583–587

Wester, P. W., and J. H. Canton, 1987, "Histopathological Study of *Poecilia reticulata* (guppy) After Long-Term Exposure to Bis(tri-*n*-butyltin)oxide (TBTO) and Di-*n*-butyltindichloride (DBTC)," *Aquat. Toxicol.*, *10*, 143–165

Wester, P. W., J. H. Canton, A. A. J. Van Iersel, E. I. Krajnc, and H. A. M. G. Vaessen, 1990, "The Toxicity of Bis(tri-*n*-butyltin)oxide (TBTO) and Di-*n*-butyltindichloride (DBTC) in the Small Fish Species *Oryzias latipes* (medaka) and *Poecilia reticulata* (guppy)," *Aquat. Toxicol.*, *16*, 53–72

Wille, F., 1993, *Bodensanierungsverfahren*, Vogel Buchverlag, Würzburg, Germany

Wilken, R. D., 1992, "Mercury Analysis: A Special Example of Species Analysis," *Fresenius J. Anal. Chem.*, *342*, 795–802

Wolterbeek, H. T., P. Kuik T. G. Verburg, U. Herpin, B. Markert, and L. Thöni, 1995, "Moss Interspecies Comparisons in Trace Element Concentrations," *Environ. Monit. and Assessm.*, *35*, 263–286

World Health Organization, Ed., 1980, *Tin and Organotin Compounds: A Preliminary Review*, Environmental Health Criteria, Vol. 15, Geneva, 109 pp

Yonemoto, J., H. Shiraishi, and Y. Soma, 1993, "In Vitro Assessment of Teratogenic Potential of Organotin Compounds Using Rat Embryo Limb Bud Cell Cultures," *Toxicol. Lett.*, *66*, 183–191

Yudovich, Y. E., A. A. Korycheva, A. S. Obrucknikov, and Y. V. Stepanov, 1972, "Mean Trace-Element Contents in Coal," *Geochem. Int.*, *9*, 712–720

Zucker, R. M., K. H. Elstein, R. E. Easterling, H. P. Ting-Beall, J. W. Allis, and E. J. Massaro, 1988, "Effects of Tributyltin on Biomembranes: Alteration of Flow Cytometric Parameters and Inhibition of Sodium, Potassium-ATPase Two-Dimensional Crystallization," *Toxicol. Appl. Pharmacol.*, *96*, 393–403

Appendix A.6.1 Essentiality, Occurrence, Toxicity, and Uptake Form of Individual Elements in the Environment. The uptake of many metals in the form of chelate complexes was not taken into consideration. Explanations of abbreviations are given on page 214.

	Essentiality	Occurrence (mg/kg dry weight)	Toxicity	Form of uptake
Ac	Bac Alg Fun HPl An — — — — —	S: ? P: ? R: ?	Toxic	Unknown

A: 15–1000 Bq/kg in plants grown on thorium-rich soils (Bowen 1979)

Ag	Bac Alg Fun HPl An — — — — —	S: 0.02–0.09 P: 0.06-0.3 R: 0.2	Toxic M: 60 mg/d	Ag^+ $AgCl_2^-$

A: Lycoperdales, *Eriogonum ovalifolium*
Sp: Used medically as an antibacterial ointment for burns (especially for *Pseudomonas aeruginosa*) and in dental fillings. Interaction with Cu and Se in metabolism

Al	Bac Alg Fun HPl An — — — — —	S: 71000 P: 90–530 R: 80	Toxic* Pl: 0.1–30 mg/l	Al^{3+} $Al(OH)_4^-$ $Al(OH)_3$

A: Diapensiaceae, Ericaceae, Melastomaceae, Symplocaceae, Theaceae, Orites excelsa
F: Essentiality discussed for ferns, possibly activation of some dehydrogenases and dehydrogenase enzymes
Sp: *Aluminum is toxic for plants and fish. Possibly a component factor in novel forest damage since Al is made more readily available by acid deposition. Possible connection with Alzheimer's disease

As	Bac Alg Fun HPl An* — — — — — ?	S: 0.1–20 P: 0.01–1.5 R: 0.1	Toxic** Pl: 0.02–7.5mg/l M: 5-50 mg/d	$HAsO_4^{2-}$ $H_2AsO_4^-$

F: As is essential for red algae.
De: *As deficiency causes a reduction in growth and affects reproduction in vertebrates. As deficiency leads to cardiac death in the third generation for goats.
Sp: **Toxicity increases from As via As(V) to As(III)

Au	Bac Alg Fun HPl An — — — — —	S: 0.001-0.002 P: 0.01–0.04 R: 0.001	Slightly toxic*	$Au(OH)_3$ $AuCl_2$

A: Possibly *Tectona grandi*
Sp: Au is administered in organic compounds as a medicinal treatment for arthritis. *Au(III) is more toxic than Au(I)

Appendix A.6.1 (*Continued*)

	Essentiality	Occurrence (mg/kg dry weight)	Toxicity	Form of uptake
B	Bac Alg Fun HPl An ? + − + −	S: 5–80 P: 30–75 R: 40	Pl: 1–5 mg/l M: 4 g/d	$B(OH)_3$ $B(OH)_4^-$

A: Cruciferaceae and Leguminosae
F: B is of significance for cell division, possibly also involved in glycometabolism and sugar transport. It participates in flavonoid and nucleic acid synthesis and in cell wall construction, and it stimulates N fixation by bacteria
De: Deficiency symptoms are known worldwide (disturbance in growth, restricted root branching, phloem necrosis, fructification disturbances)
Sp: Borates serve for water softening in detergents which involves dangers for groundwater and irrigation water

| Ba | Bac Alg Fun HPl An
− − − − − | S: 500

P: 10–100
R: 40 | Relatively
harmless
Pl: 500 mg/l
M: 200 mg/dT | Ba^{2+} |

A: *Bertholletia excelsa, plankton; Chaetoceros curvisetus* and *Rhizosolenia calcaravis*
Sp: A possible essentiality discussed for mammals

| Be | Bac Alg Fun HPl An
− − − − − − | S: 0.1–5
P: 0.001–0.4
R: 0.001 | Toxic*
Pl: 0.5 mg/l | $BeOH^+$ |
| A: | *Vaccinium myrtillus* and *Vicia sylvatica* | | | |

Sp: *Both Be metals and Be compounds are considered to be allergenic. The carcinogenicity of Be has been demonstrated for several animal species

| Bi | Bac Alg Fun HPl An
− − − − − | S: 0.2
P: 60 ppb
R: 0.01 | Toxic
Pl: 27 mg/l
Rat: 160 mg/d | BiO^+
$Bi(OH)_2^+$ |

| Br | Bac Alg Fun HPl An
− ± − − ? | S: 1–10
M: 15
R: 4 | Pl: 15–600 mg/l
M: 3 g/d | Br^-
BrO^-
HBrO |

A: Some red and brown algae, some Porifera, many corals, a few molluscs (Aplysia and Muricidae), fungi (Amanita ssp)
F: Essentiality for mammals under discussion
Sp.: Sources of bromide emissions are antiknock agents, fumigants for preservation purposes, insecticides, and flame-proofing agents

Appendix A.6.1 *(Continued)*

	Essentiality	Occurrence (mg/kg dry weight)	Toxicity	Form of uptake
C	Bac Alg Fun HPl An + + + + +	S: Variable* P: 45% R: 44.5%	Many toxic compounds, CO_2**	CO_2 HCO_3^-

F: Basic structural element of all organic compounds (sugar, fats, proteins, etc.)
Sp: *Very variable, depending on soil type and soil horizon
**CO_2 and CH_4 cause the greenhouse effect; the average CO_2 content of the atmosphere is 340 ppm

Ca	Bac Alg Fun HPl An + + − + +	S: 0.1%–1.2% P: 1% R: 1%	Very slightly toxic	Ca^{2+} $CaOH^+$

A: Some red algae, shells of many invertebrates, bone
F: Structural constituent of cell walls, constituent of bone, physiological regulation function, enzyme activator, electrochemical function
De: Deficiency in plants causes disturbed division growth (small cells), drying of leaf tips, leaf deformation, restricted root growth

Cd	Bac Alg Fun HPl An* − − − − ?	S: 0.01–3 P: 0.03–0.5 R: 0.05	Toxic** Pl: 0.2–9 mg/l M: 3–330 mg/d	Cd^{2+} $CdOH^+$

A: Agaricus and other fungi, molluscs (*Ostrea* spp.)
Sp: *Goats fed dry feed with 15 µg Cd/kg and rats fed on a low Cd diet grew more slowly than control animals with 300 µg/kg feed. The goats with a low Cd diet had difficulty in conceiving. **Itai-Itai disease (Japan): the Cd content of river water from a silver mine used to irrigate rice paddies led to increased Cd content in the rice and thus to skeleton deformation and spontaneous fractures in man. Released from metallurgical plants, towns, refuse incineration plants, cigarette smoke, mineral fertilizers and sewage sludge

Ce	Bac Alg Fun HPl An − − − − −	S: 50 P: 0.25–0.55 R: 0.5	Slightly toxic	Ce^{3+} $CeOH^{2+}$

A: *Carya* spp.
Sp: See under La

Appendix A.6.1 (*Continued*)

		Essentiality	Occurrence (mg/kg dry weight)	Toxicity	Form of uptake
Cl		Bac Alg Fun Hpl An ± + − + +	S: 100* P: 0.2%–2% R: 0.2%	Relatively nontoxic**	Cl^-

A: Halophytes such as Chenopodiaceae, Frankeniaceae, Plumbaginaceae; mangroves such as Rhizophoraceae and Verbenaceae
F: Osmolytic function, enzyme activation
De: Wilting and root thickening
Sp.: *Very high concentrations in arid and semiarid soils
**Cl_2 and a large number of organochlorine compounds are highly toxic. Apart from the natural emission potential from the ocean, its use in deicing salt for roads can be regarded as a potential danger for soils and waters. Chlorine is a raw material for the production of solvents and is an active substance in bleaches. It is also used for the stabilization and purification of water

		Essentiality	Occurrence	Toxicity	Form of uptake
Co		Bac Alg Fun Hpl* An ± ± − ? +	S: 1–40 P: 0.02–0.5 R: 0.2	Weakly toxic Pl: 0.1–3 mg/l M: 500 mg/d	Co^{2+} $CoCO_3$

A: *Clethra barbinervis, Crotalaria cobaticola, Nyssa sylvatica*
F: Part of vitamin B_{12}, enzymatic
De: Anemia, vitamin B_{12} deficiency, disturbance of nucleic acid synthesis
Sp: Whether Co is essential for higher plants remains unclear; nevertheless it is undoubtedly necessary for the N_2-fixation system of Rhizobium bacteria living in a state of symbiosis with the Leguminosae. The addition of slight quantities of Co often leads to improved yields

		Essentiality	Occurrence	Toxicity	Form of uptake
Cr		Bac Alg Fun Hpl An − − − − +	S: 2–100 P: 0.2–1 R: 1.5	Toxic* Pl: 1 mg Cr(VI)/l M: 3 g/d	$Cr(OH)_3$ CrO_4^{2-}

A: *Leptospermum scoparium, Pimelia suteri*
F: Insulin intensification, glucose tolerance function
De: Diabetes, increased serum lipids
Sp: *Cr(VI) is about 1000 times more toxic than Cr(III) (basically only Cr(VI) is capable of passing through the cell membrane). In cells Cr(III) is preferentially found in the walls and, Cr(VI) in the cell sap, whereas the concentration in the mitochondria and the cell nuclei is comparatively low

		Essentiality	Occurrence	Toxicity	Form of uptake
Cs		Bac Alg Fun Hpl An − − − − −	S: 1–20 P: 0.03–0.44 R: 0.2	Relatively harmless	Cs^+

Sp: Cs^{134} and Cs^{137} are released during nuclear fission

Appendix A.6.1 (*Continued*)

	Essentiality	Occurrence (mg/kg dry weight)	Toxicity	Form of uptake
Cu	Bac Alg Fun HPl An + + + + +	S: 1–80 P: 2–20 mg/l R: 10	Toxic* Pl: 0.5–8 mg/l M: 250 mg/d	$CuOH^+$ $CuCO_3$

A: *Aeolanthus biformifolus, Becium homblei, Cryptosepalum maraviense, Elsholtzia haichowensis, Gypsophila patrinii, Lychnis alpina, Polycarpaea spirostylis, Silene dioica, Silene vulgaris, Triumfetta welwitschi, Uapaca* spp., *Veronica glaberrima*
F: Energy metabolism, N metabolism, oxidizing systems, elastin cross-linkage, catalytic function in many redox reactions
De: e.g., grey speck disease of cereals, drying of leaf tips, wilting, spot chlorosis of young leaves, blocking of oxidation (respiration), anemia, changes in bone formation
Sp: *Released into drinking water from copper pipes

	Essentiality	Occurrence (mg/kg dry weight)	Toxicity	Form of uptake
Dy	Bac Alg Fun HPl An – – – – –	S: 5 P: 0.025–0.05 R: 0.03	Slightly toxic	Dy^{3+} $DyOH^{2+}$

A: *Carya* spp.
Sp: See under La

	Essentiality	Occurrence (mg/kg dry weight)	Toxicity	Form of uptake
Er	Bac Alg Fun HPl An – – – – –	S: 2 P: 0.015–0.030 R: 0.02	Slightly toxic	Er^{3+} $ErOH^{2+}$

A: *Carya* spp.
Sp: See under La

	Essentiality	Occurrence (mg/kg dry weight)	Toxicity	Form of uptake
Eu	Bac Alg Fun HPl An – – – – –	S: 1 P: 0.005 – 0.015 R: 0.008	Slightly toxic	Eu^{3+} $EuOH^{2+}$

A: *Carya* spp.
Sp: See under La

	Essentiality	Occurrence (mg/kg dry weight)	Toxicity	Form of uptake
F	Bac Alg Fun HPl An – – – – +	S: 10–1000 P: 2–20 R: 2	Toxic* Pl: 5 mg/l M: 20 mg/d	F^- HF

A: *Acacia georginae, Dichapetalum* spp., *Gastrolobium grandiflorum, Porifera Dysidea crawshayi*, human teeth
F: In mammals fluorine strengthens the teeth during development. An excess of fluorine can lead to fluorosis as is known from grazing animals in North Africa
Sp: *Fluorine is released from the ceramic, cement, and brick-making industry. The resulting dusts and gases containing fluoride can lead to agricultural and forestry damage

Appendix A.6.1 (*Continued*)

	Essentiality	Occurrence (mg/kg dry weight)	Toxicity	Form of uptake
Fe	Bac Alg Fun HPl An + + + + +	S: 0.7%–42% P: 5–200 R: 150	Toxic Pl: 10–200 mg/l M: 200 mg/d	Fe^{2+} $Fe(OH)_2{}^+$

A: Iron bacteria, lichen *Acarospora smaragdula*, red blood corpuscles
F: Chlorophyll synthesis, many iron enzymes, ferritin as the storage and transport form, hemoglobin
De: Straw-colored intercostal chloroses or whitening of young leaves, apical buds suppressed, anemia, growth reduction, hemolysis

Ga	Bac Alg Fun HPl An – – – – –	S: 0.1–10 P: 0.01–0.23 R: 0.1	Rat: 10 mg/d	$Ga(OH)_4^-$

Sp: Used for antitumorigenic purposes; various medical applications

Gd	Bac Alg Fun HPl An – – – – –	S: 4 P: 0.03–0.06 R: 0.04	Slightly toxic	Gd^{3+} $GdOH^{2+}$

A: *Carya* spp.
Sp: See under La

Ge	Bac Alg Fun HPl An – – – – –	S: 1 P: 1–2.4 R: 0.01	Toxic (apart from GeH_4)	$Ge(OH)_4$

Sp: Germanium organic compounds are used as chemotherapeutical agents for bacteria. Spiro-germanium (4.4-dialkyl-4-germacyclo-hexanone) and 8.8-dialkyl-8-germaazaspiro(4.5)decane is used for antitumorigenic purposes

H	Bac Alg Fun HPl An + + + + +	S: Variable P: 4.1%–7.2% R: 6.5%	As D_2O	H_2O

F: Participates in the structure of a large number of organic compounds and supplies reduction equivalents in physiological processes

Hf	Bac Alg Fun HPl An – – – – –	S: 6 P: 0.001–1 R: 0.05	Toxic	Unknown

Appendix A.6.1 (*Continued*)

		Essentiality	Occurrence (mg/kg dry weight)	Toxicity	Form of uptake

Hg Bac Alg Fun HPl An
— — — — —

S: 0.01–1   Toxic* $Hg(OH)_2$
P: 0.005–0.2   M: 0.4 mg/d   HgOHCl
R: 0.1
A: *Minuartia setacea, Betula papyrifera*
Sp: *Toxicity increases from elemental mercury to ionic mercury to organomercury compounds. Amalgam dental fillings may lead to allergies. Minamata disease: a disease which occurred from 1953–1960 in Minamata Bay, Japan. Water containing methyl mercury contaminated the food fish. In humans, the disease took the form of serious kidney damage and damage to the immunological system also leading to fatalities

Ho Bac Alg Fun HPl An
— — — — —

S: 0.6   Slightly toxic Ho^{3+}
P: 0.005–0.015 $HoOH^{2+}$
R: 0.008
A: *Carya* spp.
Sp: See under La

I Bac Alg Fun HPl An
— ± — — +

S: 1–5   Relative nontoxic* I^-
P: 0.07–10   Pl: 1 mg/l IO_3^-
R: 3   M: 2 mg/d
A: Red and brown algae, many Porifera, some Coelenterata, *Feijoa sellowiana*
F: Necessary for the function of the thyroid gland hormone thyroxine
De: Goiter formation, cretinism
Sp: *Various iodine isotopes are released during nuclear weapons tests and reactor accidents which may become dangerous due to the accumulation of iodine in the thyroid glands of mammals and humans

In Bac Alg Fun HPl An
— — — — —

S: 0.2–0.5   Rat: 200 mg/d $In(OH)_4^-$
P: 0.0005–0.002
R: 0.001

Ir Bac Alg Fun HPl An
— — — — —

S: ?
P: ?   Unknown   Unknown
R: 0.00001
Sp: See under Pt

K Bac Alg Fun HPl An
+ + + + +

S: 0.2% − 2.2%   Relatively harmless K^+
P: 0.5% − 3.4%
R: 1.9%
F: Electrochemical and catalytical function; enzyme activation
De: Deficiency disturbs the water balance (drying of leaf tips) and causes leaf curling (wilting) on older leaves and root rot

Appendix A.6.1 (*Continued*)

	Essentiality	Occurrence (mg/kg dry weight)	Toxicity	Form of uptake
La	Bac Alg Fun Hpl An − − − − −	S: 40 P: 0.15–0.25 R: 0.2	Slightly toxic Rat: 720 mg/d	La^{3+} $LaOH^{2+}$

A: *Carya* spp.
Sp: The lanthanides are not considered to be essential and are only slightly toxic. In environmental specimens they obey the Harkins rule which says that a lanthanide with an odd atomic number occurs in a lower concentration than the directly adjacent lanthanide element with an even atomic number

	Essentiality	Occurrence (mg/kg dry weight)	Toxicity	Form of uptake
Li	Bac Alg Fun Hpl An − − − − ?	S: 1–100 P: 0.01–3.1 R: 0.2	Slightly toxic Pl: 30 mg/l M: 200 mg/d	Li^+

A: Solanaceae in arid climates
Sp: Lithium compounds are used to treat manic depressives. Possible essentiality for mammals under discussion

	Essentiality	Occurrence (mg/kg dry weight)	Toxicity	Form of uptake
Lu	Bac Alg Fun Hpl An − − − − −	S: 0.4 P: 0.0025–0.005 R: 0.003	Slightly toxic	Lu^{3+} $LuOH^{2+}$

A: *Carya* spp.
Sp: See under La

	Essentiality	Occurrence (mg/kg dry weight)	Toxicity	Form of uptake
Mg	Bac Alg Fun Hpl An + + + + +	S: 500–5000* P: 1000–9000 R: 2000	Very slightly toxic	Mg^{2+} $MgOH^+$

A: Marine algae accumulate 6000–20000 mg/kg dry weight
F: Enzyme component; important for chlorophyll structure; electrochemical and catalytic functions
De: Deficiency causes stunted growth and intercostal chlorosis on older leaves
Sp: *Higher Mg contents in the soil arise if rock contains $MgCO_3$

	Essentiality	Occurrence (mg/kg dry weight)	Toxicity	Form of uptake
Mn	Bac Alg Fun Hpl An + + + + +	S: 20–3000 P: 1–700 R: 200	Slightly toxic Pl: 1–100 mg/l Rat: 10–20 mg/d	Mn^{2+}

A: Ericaceae and Theaceae, *Diatomeae coscinodiscus*, Porifera *Terpios zeteki*, Annelida *hermione*, Ascidiae *Didemnum* and *Halocynthia*
F: Nucleic acid synthesis; photolysis of water during the light reaction of photosynthesis; stabilizes chloroplast structure; necessary for the metabolism of mucopolysaccharides and the activity of superoxide dismutase, arginase, pyruvate carboxylase, malate enzyme
De: Chloroses and necroses on young leaves, defoliation, bone deformation, anemia, reduced growth

Appendx A.6.1 (*Continued*)

	Essentiality	Occurrence (mg/kg dry weight)	Toxicity	Form of uptake
Mo	Bac Alg Fun HPl An + + + + +	S: 0.2–5 P: 0.03–5 R: 0.5	Slightly toxic Pl: 0.5–2 mg/l Rat: 5 mg/d	MoO_4^{2-}

A: *Grindelia fastigiata*
F: N fixation; P metabolism; Fe absorption and translocation; xanthine oxidase and sulfoxidase activity
De: Disturbed growth and shoot deformation, discoloration of leaf edges, disturbance of fatty acid formation from carbohydrates
Sp: Leguminosae require about three times as much Mo as other spermatophytes, since N fixation by the symbiotic rhizobia requires Mo

	Essentiality	Occurrence (mg/kg dry weight)	Toxicity	Form of uptake
N	Bac Alg Fun HPl An + + + + +	S: 2000 P: 1.2%–7.5% R: 2.5%	Ecotoxic	NO_3^- NH_4^+ N_2

F: Important for the structure of many organic compounds and for many metabolic and physiological functions
De: Stunted growth or dwarfism, bulky growth and skleromorphosis, premature yellowing of older leaves
Sp: Simple nitrogen compounds today represent an extensive ecotoxicological problem, e.g., the nitrate problem associated with large-scale livestock farming, NO_2 emissions, and N_2O as a greenhouse gas in the atmosphere

	Essentiality	Occurrence (mg/kg dry weight)	Toxicity	Form of uptake
Na	Bac Alg Fun HPl An ± ± − ± +	S: variable P: 35–1000 R: 150	Relatively harmless	Na^+

A: Halophytes; some Chenopodiaceae, Frankeniaceae, and Plumbaginaceae; mangroves such as Avicennia, Bruguiera, and Rhizophora
F: Electrochemical function; enzyme activation

	Essentiality	Occurrence (mg/kg dry weight)	Toxicity	Form of uptake
Nb	Bac Alg Fun HPl An − − − − −	S: 10 P: 0.28 R: 0.05	Slightly toxic	

A: Ascidian: *Molgula manhattensis*

	Essentiality	Occurrence (mg/kg dry weight)	Toxicity	Form of uptake
Nd	Bac Alg Fun HPl An − − − − −	S: 35 P: 0.1–0.25 R: 0.2	Slightly toxic	Nd^{3+} $NdOH^{2+}$

A: *Carya* spp.
Sp: See under La

Appendix A.6.1 (*Continued*)

	Essentiality	Occurrence (mg/kg dry weight)	Toxicity	Form of uptake
Ni	Bac Alg Fun HPl* An ± ± − ± +	S: 2–50 P: 0.4–4 R: 1.5	Toxic ** Pl: 0.5–2 mg/l Rat: 50 mg/d	Ni^{2+}

A: *Alyssum bertolinii* and *Alyssum murale*, *Dicoma* spp., *Homalium* spp., *Hybanthus floribundus*, *Pimelia suteri*, *Planchonella* spp., *Psychotria* spp., *Rinorea bengalensis*, *Sebertia* spp., Poriferae *Dysidea*
F: Interaction with iron resorption
De: Growth reduction
Sp: *For some plants and microorganisms Ni is a component of urease and therefore essential. **$Ni(CO)_4$ is a highly toxic industrial product

O	Bac* Alg Fun HPl An ± + + + +	S: 49% P: 40–44% R: 40.5%	Toxic in the form of O_3 and peroxide	O_2 CO_2

F: Important for the structure of many organic compounds provides oxidation equivalents in metabolism
Sp: *Lethal for obligate anaerobic microorganisms

Os	Bac Alg Fun HPl An − − − − −	S: ? P: ? R: 0.000015	Very toxic in the form of OsO_4	

Sp: See under Pt

P	Bac Alg Fun HPl An + + + + +	S: 200–800 P: 120–30000 R: 2000	Ecotoxic*	HPO_4^{2-} $H_2PO_4^-$

F: Forms hydroxyapatite in mammals, necessary for bone and teeth formation; necessary for energy metabolism, phosphorylation, DNA and ATP structure
De: Disturbance of reproductive processes (flowering inhibition), bulky growth, dry tips in the case of conifer needles
Sp: *Eutrophication of waters; phosphate esters which enter the water as insecticides have a toxic effect on much aquatic life, phosphates are relatively harmless; phosphoric acid esters and PH_3 are very toxic

Pa	Bac Alg Fun HPl An − − − − −	S: ? P: ? R: ?	Unknown	Unknown

Appendix A.6.1 (*Continued*)

	Essentiality	Occurrence (mg/kg dry weight)	Toxicity	Form of uptake
Pb	Bac Alg Fun HPl An* — — — — ?	S: 0.1-200 P: 0.1–5 R: 1	Toxic Pl: 3–20 mg/l M: 1 mg/d	$PbCO_3$

A: *Amorpha canescens, Minuartia verna,* lichen *Stereocaulan pileatum*
F: *Essentiality for vertebrates under discussion. Pb enters into the environment (currently to a decreasing extent) particularly from the use of tetraethyl lead as an antiknock agent for petrol engines

	Essentiality	Occurrence (mg/kg dry weight)	Toxicity	Form of uptake
Pd	Bac Alg Fun HPl An — — — — —	S: ? P: ? R: 0.0001	Toxic	Unknown

Sp: See under Pt

	Essentiality	Occurrence (mg/kg dry weight)	Toxicity	Form of uptake
Po	Bac Alg Fun HPl An — — — — —	S: 8–220 Bq/kg P: 8–12 Bq/kg R: ?	Highly toxic for vertebrates	Unknown

	Essentiality	Occurrence (mg/kg dry weight)	Toxicity	Form of uptake
Pr	Bac Alg Fun HPl An — — — — — —	S: 3–12 P: 0.03–0.06 R: 0.05	Slightly toxic	Pr^{3+} $PrOH^{2+}$

A: *Carya* spp.
Sp: See under La

	Essentiality	Occurrence (mg/kg dry weight)	Toxicity	Form of uptake
Pt	Bac Alg Fun HPl An — — — — —	S: ? P: ? R: 0.00005	Slightly toxic	Unknown

Sp: 0.9–2.3 g Pt is contained in three-way catalytic converters. Pt metal and its salts cause allergic contact eczema. Cis-dichloroplatinum(II) complexes are used in cancer therapy. Pt and its family enter into the environment due to natural wear from exhaust catalytic converters in cars. Together with aluminum oxide as the carrier material, these compounds are released into the air in a highly dispersive metallic form as suspended matter and subsequently deposited. The consequences of this are still unclear, particularly since the concentrations of platinum metal naturally occurring in the environment are very low and often cannot be determined exactly

	Essentiality	Occurrence (mg/kg dry weight)	Toxicity	Form of uptake
Ra	Bac Alg Fun HPl An — — — — —	S: ? P: 0.03–1.6 ppt R: ?	Radioactive*	Ra^{2+}

A: *Bertholletia excelsa*
Sp: *Similar chemical behavior to Ba and Ca and is therefore incorporated into the bone substance; 10–20 μg is sufficient to cause bone-marrow depression and myelosarcomas

Appendix A.6.1 (*Continued*)

	Essentiality	Occurrence (mg/kg dry weight)	Toxicity	Form of uptake
Rb	Bac Alg Fun HPl An − − − − −	S: 10–100 P: 1–50 R: 50	Slightly toxic Rat: 10 mg/d	Rb^+
	Sp: Similar to K; may replace K at bonding locations, but not in its physiological effect			
Re	Bac Alg Fun HPl An − − − − −	S: ? P: ? R: ?	?	ReO_4^-
Rh	Bac Alg Fun HPl An − − − − −	S: ? P: ? R: 0.00001	?	Unknown
	Sp: See under Pt			
Ru	Bac Alg Fun HPl An − − − − −	S: ? P: ? R: 0.00001	?	Unknown
	Sp: See under Pt			
S	Bac Alg Fun HPL An + + + + +	S: 200–2000* P: 600–10000 R: 3000	Ecotoxic**	SO_4^{2-} HSO_4^-
	A: Individual plants of the Cruciferae, *Alium* spp., sulfur bacteria, vertebrate hair, feathers F: Constitutent of amino acids (cysteine and methionine), coenzymes, acid mucopolysaccharides, and sulfuric acid esters De: Very similar to N deficiency; intercostal chloroses of young leaves; premature yellowing of leaves and needles Sp: *Considerably higher content in gypsum soils. **Contribution of anthropogenic SO_2 emissions to "new" forest damage and soil acidification			
Sb	Bac Alg Fun HPl An − − − − −	S: 0.01–1 P: 0.1–200 ppb R: 0.1	Toxic* M: 100 mg/d Rat: 10–75 mg/d,	$Sb(OH)_6^-$
	Sp: *Sb(III) is more toxic than Sb(V)			
Sc	Bac Alg Fun HPl An − − − − −	S: 0.5–45 P: 0.01–0.2 R: 0.02	Slightly toxic	$Sc(OH)_3$

Appendix A.6.1 (*Continued*)

	Essentiality	Occurrence (mg/kg dry weight)	Toxicity	Form of uptake
Se	Bac Alg Fun HPl An − − − ± +	S: 0.01 P: 0.01–2 R: 0.02	Toxic* Pl: 1–2 mg Se(IV)/l M: 5 mg/d Rat: 1–2 mg/d,	SeO_3^{2-}

A: Fungus *Boletus edulis*; individual species from the families of the Compositae, Lecythidaceae, Leguminosae (e.g., Astragalus), and Rubiaceae
F: Component of glutathione peroxidase
De: Deficiency causes lipid peroxidation, endemic cardiomyopathy and hemolysis in animals and man
Sp: *Selenites and selenates are very toxic. The toxicity of As, Hg, Cd, Tl, and NO_3^- is reduced if Se is taken up at the same time. The toxic effect results from the replacement of sulfur by Se in amino acids

	Essentiality	Occurrence	Toxicity	Form of uptake
Si	Bac Alg Fun HPl An + − − ± +	S: 33%* P: 200–8000 R: 1000	Physically, e.g., in the form of asbestos	$Si(OH)_4$

A: Diatoms, radiolarians and siliceous sponges, Sphenopsida, Cyperaceae, Gramineae, Juncaceae
F: Structural component of siliceous skeletons; calcification
De: Growth disturbances, bone deformation
Sp: *Clearly lower contents in particular soils, e.g., limestone. Organic Si compounds may be highly toxic. Some Si compounds such as Si halogen compounds are corrosive

	Essentiality	Occurrence	Toxicity	Form of uptake
Sm	Bac Alg Fun HPl An − − − − −	S: 4.5 P: 0.02–0.04 R: 0.04	Slightly toxic	Sm^{3+} $SmOH^{2+}$

A: *Carya* spp.
Sp: See under La

	Essentiality	Occurrence	Toxicity	Form of uptake
Sn	Bac Alg Fun HPl An* − − − − ?	S: 1–20 P: 0.8–7 R: 0.2	Very slightly M: 2 g/d	$SnO(OH)_3^-$

A: *Silene vulgaris*; Porifera *Terpios zeteki*
F: *Sn may possibly be essential for vertebrates
De: Growth disturbances occur in vertebrates; furthermore digestive enzymes are not secreted; imposex by TBT (tributyltin)
Sp: Organotin compounds (e.g., triphenyltin) are used as fungicides, insecticides, and bactericides

Appendix A.6.1 (*Continued*)

	Essentiality	Occurrence (mg/kg dry weight)	Toxicity	Form of uptake
Sr	Bac Alg Fun Hpl An	S: 20–3500	In the form	Sr^{2+}
	— — — — —	P: 3–400	of Sr^{90}*	
		R: 50		
	A: Protozoa Acanthometra, brown algae			
	Sp: Sr seems to be essential for some organisms. However, this requires further investigation. *Sr^{90} is a decay product from nuclear explosions and as a consequence of its similarity to Ca is incorporated into bone substance.			
Ta	Bac Alg Fun Hpl An	S: 0.5–4	Slightly toxic	Unknown
	— — — — —	P: <0.001	Rat: 300 mg/d	
		R: 0.001		
	A: Ascidia *Stylea plicata*			
Tb	Bac Alg Fun Hpl An	S: 0.7	Slightly toxic	Tb^{3+}
	— — — — —	P: 0.005–0.015		$TbOH^{2+}$
		R: 0.008		
	A: *Carya* spp.			
	Sp: See under La			
Te	Bac Alg Fun Hpl An	S: ?	Toxic	$HTeO_3^-$
	— — — — —	P: 0.01–0.35	Pl: 6 mg/l	
		R: 0.05	Rat: 1–9 mg/d	
Th	Bac Alg Fun Hpl An	S: 9	Toxic	$Th(OH)_4^{4-n}$
	— — — — —	P: 0.03–1.3		
		R: 0.005		
Tl	Bac Alg Fun Hpl An	S: 0.01–0.5	Toxic	Tl^+
	— — — — —	P: 0.03–0.3	Pl: 1 mg/l	
		R: 0.05	M: 600 mg/d	
	Sp: Particularly high environmental concentrations in the vicinity of cement factories			
Ti	Bac Alg Fun Hpl An*	S: 1500–5000**	Very slightly	$Ti(OH)_4$
	— — — — —	P: 0.02–56	Toxic***	
		R: 5		
	F: *Although Ti is not regarded as essential for plants it may nevertheless play a positive role in cereal growth and for N_2 fixation by Leguminosae.			
	** 15% in the upper layers of lateritic soils. The measured Ti content during plant analysis may be used as an indicator of contamination by soil particles.			
	***Due to its grain size of 20 μm, the titanium dioxide pigment is classified as dust pollution			

Appendix A.6.1 (*Continued*)

	Essentiality	Occurrence (mg/kg dry weight)	Toxicity	Form of uptake
Tm	Bac Alg Fun HPl An − − − − −	S: 0.6 P: 0.0025–0.005 R: 0.004	Slightly toxic	Tm^{3+} $TmOH^{2+}$
	A: *Carya* spp. Sp: See under La			
U	Bac Alg Fun HPl An − − − − −	S: 0.01–1 P: 0.005–0.06 R: 0.01	Highly toxic Rat: 36 mg/d	$UO_2(CO_3)_3^{4-}$
	A: *Coprosma arborea, Uncinia leptostachya*, some corals			
V	BaC Alg Fun HPl An − ± − ± ±	S: 10–100 P: 0.001–10 R: 0.5	Toxic P: 10–40 mg/l Rat: 0.25 mg/d	$H_2VO_4^-$ HVO_4^{2-}
	A: Fungi: *Amanita muscaria, Astragalus confertiflorus*; ascidians F: Inhibition of chloresterol synthesis De: Growth reduction, changes in lipid metabolism, fertility disturbances Sp: Some sea cucumbers, in isolated cases molluscs and generally ascidians contain high V concentrations. Some species of ascidians have a vanadium protein complex in their blood and other species, a pyrrole complex in the green blood corpuscles, the vanadocytes. The general function of vanadium is unclear			
W	Bac Alg Fun HPl An − − − − −	S: 1.5 P: 0.0005–0.15 R: 0.2	Slightly toxic Pl: 10 mg/l Rat: 30–50 mg/d	WO_4^{2-}
	A: *Pinus cembra* Sp: The physiological effect of tungsten is as an antagonist to molybdenum where the replacement of Mo by W in enzymes (e.g., xanthine oxidase) generally leads to a drop in activity			
Y	Bac Alg Fun HPl An − − − − −	S: 40 P: 0.15–0.77 R: 0.2	Slightly toxic	$Y(OH)_3$
	A: Poriferae *Melithoca* spp., *Carya* spp.			
Yb	Bac Alg Fun HPl An − − − − −	S: 3 P: 0.015–0.030 R: 0.02	Slightly toxic	Yb^{3+} $YbOH^{2+}$
	A: *Carya* spp. Sp: See under La			

Appendix A.6.1 (*Continued*)

	Essentiality	Occurrence (mg/kg dry weight)	Toxicity	Form of uptake
Zn	Bac Alg Fun HPl An + + + + +	S: 3–300 P: 15–150 R: 50	Toxic Pl: 60–400 mg/l M: 150–600 mg/l	Zn^{2+} $ZnOH^+$ $ZnCO_3$

A: *Armeria maritima* subsp. *halleri, Minuartia verna, Silene vulgaris, Thlaspi alpestre, Viola tricolor* var. *calaminaria*

F: Necessary for chlorophyll formation; enzyme activator; necessary for energy metabolism (dehydrogenases), protein degradation, formation of growth substance (IES), and transcription

De: Growth inhibition, whitish green discoloration of older leaves, fructification disturbances

Sp: Toxic symptoms in humans are sexual immaturity, skin lesions, and grey hair. Soils rich in Zn are termed "calamine soils" and support "calamine flora"

| Zr | Bac Alg Fun HPl An
−   −   −   −   − | S: 1–300
P: 0.3–2
R: 0.1 | Slightly toxic
Rat: 250 mg/d | $Zr(OH)_4^{4-n}$ |

Source: From Markert (1992), modified.

Bac, essentiality for bacteria; Alg, essentiality for algae; Fun, essentiality for fungi; HPl, essentiality for higher plants; An, essentiality for animals; +, essential; −, as yet no essential significance; ±, essentiality only demonstrated for certain species; ?, essentiality under discussion, when ? referes to essentiality; S, average contents in soils; P, average contents in plants; R, average content in the reference plant (after Markert 1991); Pl, average toxicity concentration for plants; M, average toxicity concentration for man; Rat, average toxicity concentration for rats; l, lethal dose; d, daily intake; F, examples of the element function; A, accumulator organisms; De, deficiency symptoms in case of insufficient supply of the element; Sp, special features of the element; ?, no information available, if ? does not refer to essentiality.

Appendix A.6.2 Estimated Annual Production of Individual Elements in the Year 2000 (Unless Otherwise Specified) and Examples of Their Technical Application.

Symbol	Estimated annual production in the year 2000 (in 1000 t)	Examples of technical application
Ac	No data available	Ac is used as a radioactive source to generate alpha radiation
Ag	12	Ag is used for photographic material, electric controllers and conductors, coins, medals, jewellery, silverware, alkali batteries, mirrors, catalysts, hard alloy and silver-plated objects
Al	60000	Aluminum is used, for example, in producing sheet metal, wires, and alloys. Al salts are used in sewage plants to precipitate phosphate
As	51	Metallic arsenic is used as an alloying element to increase hardness. Copper arsenide, $Cu(AsO_2)_2$, is effective as an insecticide and fungicide
Au	2	Gold is applied in electroplating, electronics, and jewellery production
B	1000 (1974)	Boron is a component of alloys. Boron nitrides (in the diamond-like modification) are important abrasives. Boron compounds are used in the glass, ceramic, and enamel industries (particularly borax and boric acid) and also serve as fertilizers and pesticides
Ba	5000	Barium sulfate is used in X-ray diagnostics as a contrast medium; barium carbonate is used for waste water purification and as a rat poison
Be	15	Be is used for X-ray windows. Alloys of Be with Cu, Al, Ni, Co, and Fe increase hardness and temperature- and corrosion-stability. It is also found in clock springs, surgical instruments, electrical engineering (as electric insulators), and in aerospace applications
Bi	6	Bi preparations (bismuth oxide chloride and other salts) are used for cosmetic articles and as soluble salts for pharmaceuticals. Bi is used for easily fusible alloys and also in silver mirrors as well as in battery cathodes, semiconductors, and catalysts

Appendix A.6.2 (*Continued*)

Symbol	Estimated annual production in the year 2000 (in 1000 t)	Examples of technical application
Br	35	Bromium and its compounds are applied as anti-knock agents, fumigants, preservatives, insecticides, flameproofing agents, and in the production of pesticides, dyes, pharmaceuticals, and photochemicals
C	6×10^6	Carbon in its elemental form (diamond or graphite) or in the form of its compounds is used in a large number of technical processes, particularly as an energy carrier
Cd	20	Cd is used in the production of Ni/Cd batteries, for corrosion control, for pigments, and as a plastic stabilizer
Ce	0.3 (1979)	Cerium oxides are important constituents of self-cleaning ovens; the element also serves as a glass polishing agent (see also La)
Cl	81000 (1979)	Cl is a basic material for the production of solvents and is an active substance in bleaches. It is used for sterilizing and conditioning water (chlorination)
Co	30	Co is utilized for hard alloys, to harden tungsten carbide, and as a catalyst. It is also a constituent of glasses, pottery, and blue and green pigments
Cr	3750	Cr is used in metallurgy to prevent rust and as a basic material for the production of paints. It is also used for catalysts, in tanneries, and for impregnating wood
Cs	0.03	Used in photocells and as a solid rocket propellant ($CsBH_4$). CsCl is used in electric bulbs and lamps, Cs_2CO_3 as a catalyst and for the production of cathode material
Cu	12000	50% of Cu applications are in electrical engineering. Cu is also used for alloys, water mains, roofing, household goods, and coins. Cu^{2+} in the form of copper sulfate is used in agriculture as an additive to green fodder in cases of Cu deficiency, in various compositions as a fungicide or bactericide, and as an algicide and molluscicide in water

Appendix A.6.2 (*Continued*)

Symbol	Estimated annual production in the year 2000 (in 1000 t)	Examples of technical application
Dy	No data available	Dy is occasionally used in nuclear engineering (see also La)
Er	No data available	Er is used in alloys, in nuclear engineering, and to color glasses and enamel (see also La)
Eu	No data available	Eu serves as a neutron absorber, an activator in scintillation crystals, and a material in lasers and color television picture tubes (see also La)
F	3500	Fluorine is released from the aluminum, ceramic, cement, and brick-making industry. The resulting dusts and gases containing fluoride may cause the fluoride content of soils in the vicinity of such industrial areas to rise dramatically, which leads to agricultural and forest damage
Fe	1×10^6	Fe is used as a construction material and for many special purposes. Relatively small quantities of Fe oxides and Fe salts are used as paint pigments or to precipitate water impurities in the so-called third purification stage, and thus end up in the discharge flow
Ga	0.015 (1974)	Gallium is a by-product of aluminum production. It is primarily applied in the semiconductor industry. Gallium arsenide can be found in solar cells and in various telecommunication and supercomputing
Gd	No data available	Gd is used for high-temperature alloys, superconductors, magnets, and electronic components (see also La)
Ge	0.2	Ge has applications in optical components such as lenses, prisms, and windows in infrared spectroscopy. Ge is also used as a catalyst, an alloying element, and in semiconductor technology
H	No data available	Hydrogen in its elemental form or in the form of its compounds is utilized for a wide range of technical processes
Hf	0.0013 (1974)	Hafnium is used for control rods in nuclear reactors, as a filling for flash bulbs, and for alloys

Appendix A.6.2 (*Continued*)

Symbol	Estimated annual production in the year 2000 (in 1000 t)	Examples of technical application
Hg	14	Component of scientific instruments (thermometers, barometers), electrical equipment, and dental fillings
Ho	No data available	(see under La)
I	7 (1974)	Iodine is used in the chemical industry (e.g., as a catalyst and stabilizer, and in the paint industry) and in photography and medicine (tincture of iodine, X-ray contrast medium, iodine tablets)
In	0.06	Used for alloys, in the semiconductor industry, for special coatings, and for fusion threads. $InCl_3$ is used in the manufacture of fluorescent lamps
Ir	No data available	Ir is utilized for hard platinum alloys, for contacts and fountain pen nibs
K	18500 (1979)	Used for alloys and organic synthesis. Soluble potassium salts are used as a fertilizer
La	0.3 (1979)	Used in alloys and as an additive in intensive light sources. Lanthanum oxide is added to glasses as a stabilizer to combat base influences. The lanthanides are used in industry as catalysts, for mineral oil cracking, as luminescent material for color TV sets, and as additives for Hg and fluorescent lamps. They are also employed to improve light spectra, for special glasses, permanent magnets and as control rods for nuclear fuel rods
Li	33	Used in alloys (e.g., in the production of wheel bearings). It is a component of electrodes, lubricating greases, and reactor coolant. LiH is also used as a reactor fuel and for drying, condensation, and reduction agents. LiOH is used as an "air purifier" and $LiClO_4$ as rocket fuel
Lu	No data available	Lu is used as a catalyst (see also La)
Mg	242	Constituent of many alloys and fertilizers
Mn	18000	Mn is used in steel production to bind oxygen and sulfur, and also in alloys and batteries. Of the organic compounds, the fungicide manganese-ethylene-bis-dithiocarbamate and the antiknock agent methylcyclopentadienyl-manganese-tricarbonyl should be mentioned

Appendix A.6.2 (*Continued*)

Symbol	Estimated annual production in the year 2000 (in 1000 t)	Examples of technical application
Mo	130	Molybdenum is used in steel production and is a component of pigments, catalysts, lubricants, and flameproofing agents
N	120×10^3 (1979)	N_2 serves as an inert protective gas for welding and is employed in semiconductor production, for deep freezing of food, to displace air from partially filled fuel tanks, and as a propellant in aerosol sprays and fire extinguishers. Liquid N_2 is an important coolant. In large-scale manufacturing N is an important raw material for the synthesis of compounds containing nitrogen (N oxides, amides, cyanides, nitrides etc.)
Na	53000 (1979)	Used to produce antiknock agents in metallurgy, for gas-discharge lamps, in fast breeders, and solar power stations as a coolant; sodium salts, used as fertilizers (e.g., sodium nitrate and sodium molybdate)
Nb	5 (1979)	Nb is used for alloys. It also serves as a construction material in space capsules, for welding stainless steels, and as a material for fuel rod claddings
Nd	No data available	Nd is employed to color glass and enamel and as a laser material (see also La)
Ni	1500	Ni is used in many Ni alloys (used in kitchen appliances, coins, jewelry, turbines, etc.), and as Ni/Cd accumulators and in catalysts. Nickel tetracarbonyl occurs as an intermediate product in nickel purification and is also used for production processes
O	23×10^6	Oxygen in its elemental form or in the form of its compounds is used for a wide range of technical processes
Os	No data available	Os is used for very hard alloys, for pen nibs, bearings, contacts, and microscopy stains
P	11200 (1979)	Most technically produced phosphorus products are used for phosphoric acids and phosphates (e.g., for detergents). Further uses of P are as a component of copper phosphide, friction surfaces on match boxes (red P), military incendiaries, and semiconductors

Appendix A.6.2 (*Continued*)

Symbol	Estimated annual production in the year 2000 (in 1000 t)	Examples of technical application
Pa	No data available	No information
Pb	5000	Pb is used in the accumulator industry, for cable coatings, die castings, and antiknock agents
Pd	No data available	Used as a catalyst, in dental prostheses and for jewelry alloys
Po	No data available	Radioactive, therefore no general application
Pr	No data available	Applied in electrodes for arc lamps (see also La)
Pt	0.2 (all Pt metals) (1991)	Used as a catalyst and for crucibles, electrodes, and articles of jewelry. In the form of the cis-dichloro-platinum(II) complex it is used in medicine for cancer therapy
Ra	No data available	Previously used in the clock-making industry (dial illumination) and in radiotherapy; today used together with beryllium as a source of high-energy neutrons
Rb	0.003	Rb is utilized in semiconductor technology and as a material for photocathodes; rubidium carbonate is used to manufacture special glasses
Re	0.014	Re is used for heating filaments, thermocouples, fountain pen nibs, filaments in flashlights, and as a catalyst
Rh	No data available	Rh is used for various alloys and, for example, is processed into catalysts, heating spirals, and thermocouples. It is also used in jewelry
Ru	No data available	Ru serves as a catalyst and is also used for alloys
S	120×10^3 (1979)	Most is processed in sulfuric acid production; a small fraction is used in elemental form for vulcanization and the manufacture of matches, fungicides, paints, gunpowder, and medical preparations
Sb	100	Used in the semiconductor industry as an alloy component, in rubber additives, in pigments, and in paints
Sc	No data available	Sc is released during uranium smelting. It is used in nuclear engineering and as an additive for light sources

Appendix A.6.2 (*Continued*)

Symbol	Estimated annual production in the year 2000 (in 1000 t)	Examples of technical application
Se	2	Selenium dioxide is used in the glass industry and for electroplating; selenites are also used in the gas industry and as a feed additive. Cadmium selenide is utilized in semiconductor production
Si	380×10^3 (1979)	In its elemental form Si is almost exclusively used as a semiconductor. Si compounds are major constituents of glass, porcelain, earthenware, and cement. SiC is a crystalline solid of great hardness and strength
Sm	No data available	Sm has applications as a neutron absorber in nuclear reactors, in permanent magnets, and also as a catalyst (see also La)
Sn	300	Sn is used as a coating for sheet iron (corrosion protection, manufacture of tin cans). Organotin compounds (e.g., triphenyltin) serve as fungicides, insecticides, and bactericides, and are also used as PVC and PCB thermal stabilizers
Sr	120	Used to refine alloys
Ta	0.5 (1979)	Ta is utilized for electric capacitors and for lining chemical reactors. It is a component of steel alloys. Cutting tools often contain tantalum carbide
Tb	No data available	Tb is used in lasers, fluorescent materials, and in high-temperature fuel cells (see also La)
Te	0.25	Tellurium is added to steel, lead alloys, etc. It is also used in photography and medicine
Th	0.7 (1984)	^{232}Th is used in breeder reactors as a fertile material to produce ^{233}U. ThO_2 is used in the production of crucibles, for heating conductors, and as a catalyst for organic syntheses. ThC_2 (thorium carbide) is utilized as a nuclear fuel in nuclear power stations
Tl	0.03 (1984)	Used for alloys, low-temperature thermometers, in electronics, and for special glasses; Tl_2SO_4 is used as rat poison
Ti	1800	Ti is used for titanium dioxide pigments in the production of oil-based paints, plastics, rubber, paper, ceramics, fibers, printing inks, cosmetics, and foodstuffs

Appendix A.6.2 (*Continued*)

Symbol	Estimated annual production in the year 2000 (in 1000 t)	Examples of technical application
Tm	No data available	No information available
U	250	Uranium is primarily employed as a fuel in nuclear reactors and to breed plutonium and other trans-uranic elements. ^{235}U serves as bomb material and as an additive to natural uranium for nuclear fuel rods
V	35	V is mainly used for steel production. Vanadium pentoxide serves as a catalyst in technical processes
W	47	Tungsten is used for the production of hard metals, electrodes, coiled filaments, heating elements, and as a contact material for electric switches
Y	No data available	Yttrium is used for the red components in the picture tubes of color TV sets and for alloys. In the form of barium yttrium cuprates it is suitable for high-temperature superconductors
Yb	No data available	Ytterbium is used as an alloying element (see also La)
Zn	11000	In its metallic form Zn is mainly used as corrosion protection for electroplating in iron and steel production. Zinc oxide is used, for example, for catalysts and pigments; zinc bacitracin is used as a growth-promoting agent in pig and poultry breeding. Various zinc salts serve as insecticides and fungicides
Zr	500	Zr is a material used in aerospace and reactor technology. Zr compounds are also used to impregnate textiles, in leather tanning, and in the glass and ceramics industry. ZrO_2 is used as an abrasive, a white pigment for porcelain, and for fireproof apparatus

Source: Modified from Markert (1992).

7

Speciation of Chemical Elements in the Environment

Rolf-Dieter Wilken (Wiesbaden, Germany)

7.1 SUMMARY

Speciation is a new island in the ocean of analytical sciences. It describes the main properties of a compound in terms of chemical bonding between its atoms or molecules and is useful for the understanding of transport behavior, toxicity, and decontamination. There are two different definitions of species to be regarded: classically and operationally defined species. Whereas the classically defined species analysis is understood and powerful enough to measure concentrations down to ecologically important concentrations, operationally defined species are a mixture of classically defined species which are important for transport behavior on the small scale of membrane penetration or on the larger scale of transport in a river or in a soil column. Examples of research on these

Ecotoxicology, Edited by Gerrit Schüürmann and Bernd Markert.
ISBN 0-471-17644-3   © 1998 John Wiley & Sons, Inc. and Spektrum Akademischer Verlag.

different species are given in this chapter. Analytical methods for species determination are also described. The determinations of spatial structures which could be a link between the different classically and operationally defined species are still lacking.

This chapter focuses on mercury, the element with the widest range of differentiation in species behavior: it includes soluble and insoluble, volatile and solid, and toxic and harmless compounds.

7.2 INTRODUCTION

In the past, most chemical analyses sought to determine the total content of metals, whereas the determination of organic molecules nowadays is carried out by examining both structure and behavior in reactivity. A total differentiation between chemical forms of metals and elements is necessary in order to predict transport behavior, to affect patterns of toxicity, and to develop remediation strategies in the case of contamination.

For species analysis, therefore, the whole classical analytical process must be revised. Speciation requires new analytical strategies for the determination of bonding and a new philosophy of quality management in determination. Samples cannot be stabilized by preservatives: these change the speciation of elements in a given matrix. Care must be taken to ensure that the bioactivity in samples does not affect speciation. The analytical tools for speciation analysis are a derivatization of "hyphenated methods," normally a combination of a chromatographic process with an atomic-specific detector or a molecular identification after separation.

7.3 WHAT DOES "SPECIATION" MEAN?

The question of what "speciation" means is often asked. The answer could be, as the IUPAC (International Union of Pure and Applied Chemistry) defines: "Speciation is the process yielding evidence of the atomic or molecular form of an analyte." In this chapter the term "speciation" is used in its extended meaning: binding forms of elements, exactly definable or only operationally defined. An example of this definition is given in Table 7.1.

7.4 SPECIATION OF ELEMENTS

Most of the elements known are able to form species, even the noble gases under special circumstances. The type of speciation normally depends on the oxidation state and chemical surroundings of the central atom, which may be ionic or covalent.

There are many elements which form species in the environment. Much research has been done on the Cr^{3+} and Cr^{6+} species because it is well known

Table 7.1 Parent species, selected matrix species, and one of the most frequently occurring parent species of the analytical species of methylmercury(II)

Parent species	Matrix	Matrix species	Analytical species
CH_3Hg^+	Air	CH_3HgL^a	
	Water	$CH_3Hg(OH)$	$\Rightarrow CH_3HgCl$
	Biological tissue	CH_3Hg-S-protein	
	Sediment	CH_3Hg-humics	

Source: After Bernhard et al. (1986).
[a]L, Weak ligand.

that Cr^{6+} is carcinogenic. Organic compounds of Hg, Sn, or Pb are more toxic than the inorganic ones; this is in contrast to the arsenic compounds. Much work on speciation is done in the field of bonding to humic substances. The difficulty with these agglomerates or compounds is that they are poorly defined; however, they have a major influence on transport behavior in a natural ecosystem.

In speciation analyses two different definitions of speciation can be found. From the chemist's point of view it is clear that there are well-defined molecules, normally with a metal atom which should be considered. This is not the view of an ecologist, in whose field transport behavior and toxicity are the most important points. These latter traits often do not depend on certain identifiable molecules, but rather are determined by a group of different molecules with similar behavior. Hence, two different approaches to species identification have been devised and are in use; these are outlined in the following sections.

7.4.1 Classically Defined Species

Covalently bound elements are considered as species in its classical meaning. Such elements include arsenic, cadmium, chromium, copper, nickel, antimony, selenium, tin, and lead. Mercury is an element with special properties; it forms not only insoluble (HgS) but also soluble (Hg^{++}) and volatile compounds (Hg^0, dimethylmercury). The organic species of mercury are mostly more toxic than the inorganic ones.

The determination of such well-defined species is done generally by the scheme shown in Figure 7.1, and consists of a separation procedure followed by element-specific detection. It is still a major problem in the analysis of metal species to get the undisturbed species into the detection system, especially when the species are in a special balance with the matrix and other compounds.

7.4.2 Operationally Defined Species

Species can also be described by their behavior during extraction or chemical processes. This could be demonstrated with operationally defined mercury

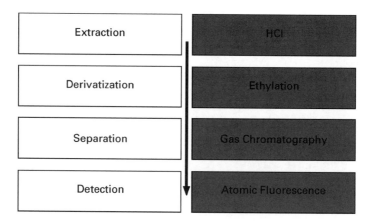

Figure 7.1 Principle of species analysis, as used in Hg species analysis.

species in river systems. They may be described as "volatile" when they can be purged by air or gases. For the chemist these operationally defined species mainly include Hg^0 and dimethylmercury. If a mercury species can be reduced by $SnCl_2$ to Hg^0, it is described as being "reactive". Such species include Hg^0, Hg^{2+}, and mercury which can easily be detached from surfaces.

On examination of the transport behavior of mercury through a soil passage, the importance of operationally defined species becomes clearer: in a first approach it is necessary to know the water soluble amounts of mercury. An example of such an analysis is given in Figure 7.2.

In this example the mercury species were extracted by mercury free water (10:1, 1 hour) in 5 separated slices of the soil column. The extract was filtered through a 0.45 μm glass fiber filter first and afterwards by ultrafiltration steps. In the whole profile the main transport of mercury in depth occurs by mercury complexes with a molecular weight > 1000 MW. Regarding also sulfur and DOC it can be concluded that there is no direct correlation between all these parameters, but between the molecular weight and the mercury, sulfur and DOC content.

Another determination of operationally defined compound classes is possible by electrochemical methods: therefore, in some cases solvated or complexed species are separately determined before or after an oxidizing digestion. Distinctions between species groups can be made by membrane separation, of which the experiment shown in Figure 7.2 is an example. Another example is the penetration of methylmercury, and not ionic mercury, through certain pore-free membranes (Wilken and Hintelmann 1990).

7.4.3 Dynamics of Species

Some species are very stable; others are more labile. This depends on the strength of the bonding to the surroundings. Figure 7.3 shows how rapidly

Ultrafiltration

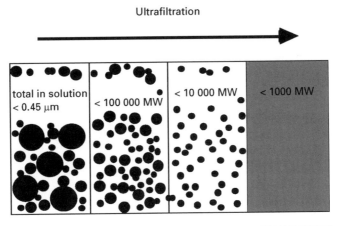

Profile depth [cm]	Hg concentration [µg/L]			
Organic surface	486	242	50	6
1–3	506	353	108	19
3–20	348	176	129	9
30–70	34	n.a.	14	11
70–150	38	n.a.	26	7

Profile depth [cm]	DOC and S concentration [mg/L]							
	DOC	S	DOC	S	DOC	S	DOC	S
Organic surface	n.a.	n.a.	n.a.	n.a.	n.a.	n.a.	n.a.	n.a.
1–3	84	1.8	9 1	2.0	9 0	2.2	49	1.3
3–20	n.a.	n.a.	n.a.	n.a.	n.a.	n.a.	n.a.	n.a.
30–70	72	2.2	n.a.	n.a.	68	2.4	63	1.6
70–150	94	2.2	n.a.	n.a.	84	2.0	48	1.6

n.a., not analyzed; DOC, Dissolved Organic Carbon.

Figure 7.2 Operationally defined species with mercury compounds as an example. Hg penetration through a soil column depends on Hg-DOC (Dissolved Organic Carbon) and sulfur bonds in bigger complexes.

species of mercury can change from ionic mercury to volatile species under environmental conditions with the influence of bacteria (Ebinghaus et al. 1994). Methylmercury, which in analytical processes is very stable, is degraded or formed depending on oxygen concentration and bacterial activity. Another example is the evaporation of mercury as dimethylmercury from contaminated floodplain soils after addition of water, as shown in Figure 7.4. This reaction is explained by Craig and Moreton (1983) through the equation: $2MeHg^+ + S^{2-} \Rightarrow (MeHg)_2S \Rightarrow HgS + Me_2Hg$, so that water under anoxic conditions is

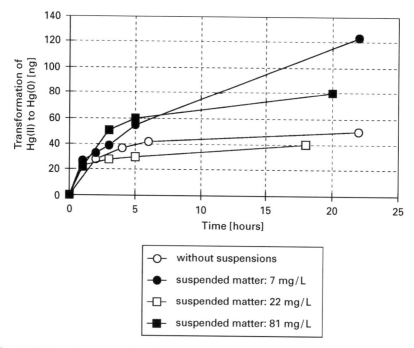

Figure 7.3 Dynamic processes of species: conversion of ionic mercury to elemental mercury in Elbe river water. 750 ng $HgCl_2$ was added; anaerobic conditions prevailed.

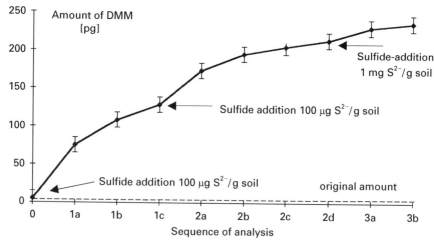

Figure 7.4 Development of dimethylmercury (DMM) from a contaminated floodplain soil.

necessary for this reaction. Other changes are slower, but can also influence the species concentrations. Hence, the matrix can change mercury species added to Elbe floodplain soil to other mercury species. This is shown in Figure 7.5 (Wallschläger et al. 1995). In this experiment an extraction procedure was used

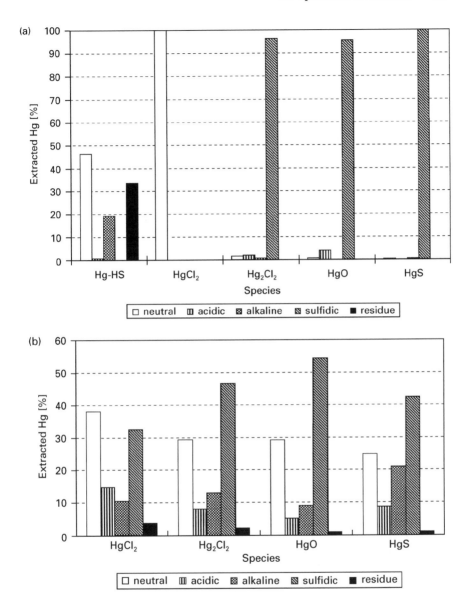

Figure 7.5 Change of mercury species added to an Elbe river floodplain soil in about 50 h, demonstrated by their solubility. Extraction steps: water, acid (HNO₃, pH 2), alkaline water (KOH, 1 M), sulfidic water (Na₂S + KOH), and residue (HNO₃ digestion). (a) The species added (Hg-HS: Hg with humic substances); (b) the species mixed with floodplain soil after 50 h.

which extracts the soil or chemicals in the consecutive steps: water, acid (HNO₃, pH 2), alkaline water (KOH, 1 M), sulfidic water (Na₂S + KOH), and residue (HNO₃ digestion). This is also an example of the influence of the matrix on the species present in a given environment.

7.5 ANALYTICAL TOOLS FOR SPECIES DETERMINATION

In most cases of species determination a separation of the different species can be achieved by chromatographic separation techniques. A derivatization of the different compounds should lead to thermal stability, an improved separation in gas chromatography (GC), and a better detectability in the detection system. An overview of the principle steps of species analysis is given in Figure 7.1.

Often, a derivatization should be performed before separation. This is done for improvement of volatilization in the case of a GC separation or to achieve better conditions for HPLC separation. For many metal organyles often the alkylation for GC-AAS (gas chromatography coupled to an atomic-absorption spectrometer) separation/detection (Wilken et al. 1994) is performed (Figure 7.6). In the case of difficult mercury species, derivatization by thioethanol, liquid separation, and detection by atomic fluorescence is suggested (Wilken 1992). Methods of species determination have been described in recently published books (Ure and Davidson 1995; van Leeuwen and Buffle 1993; Kramer and Allen 1991; Broekaert et al. 1990). A promising new method is the coupling of GC with inductively coupled plasma and mass spectroscopy (ICP/MS), by which the best determination limits for metal species are reported (Prange and Jantzen 1995a). This method is shown in Figure 7.7.

For mercury separation in a GC system an ethylation is performed (Bloom 1989); the detection is preferably performed using the atomic fluorescence

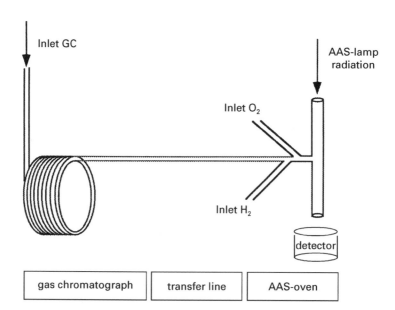

Figure 7.6 Gas chromatography and atomic absorption coupled for element-specific determination of species.

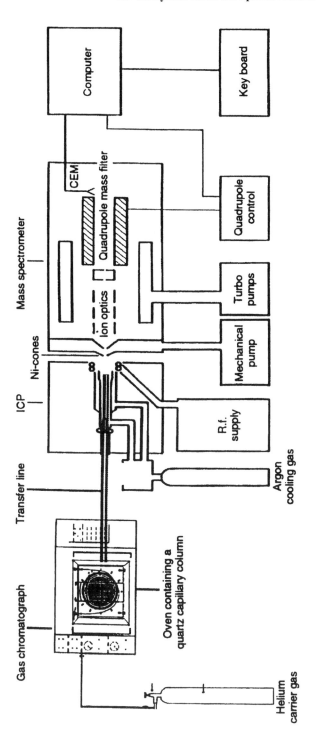

Figure 7.7 Coupling of gas chromatography with inductively coupled plasma and mass spectroscopy (R.f., radio frequency; CEM, channel electron multiplier).

method afterwards. A new combination for the determination of many metallic species is the GC-ICP/MS combination. The ethylation is done by ethylmagnesium chloride; the detection limits are in the range of 100 fg for tetraethyllead, 120 fg for diethylmercury, and 50 fg for tetrabutyltin. The special development is the interface for the coupling of the GC system to the ICP torch, which must permit transport of the compounds without irreversible adsorption or destruction in the transfer line to the detection system. The inner surface and the heating of the transfer line are the special features of this combination (Prange and Jantzen 1995b). The results obtained for the defined species of tin, mercury, and lead are given in Figure 7.8.

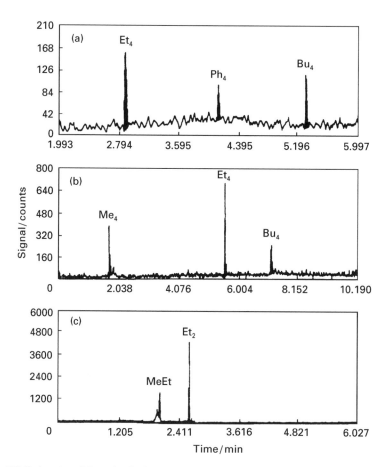

Figure 7.8 Estimation of detection limits (LOD) of the GC-ICP/MS coupling for the determination of Sn (a), Pb (b), and Hg (c) compounds (Prange and Jantzen 1995). Added amounts: 150 fg Sn as Bu_4Sn, LOD (3 s) 50 fg Sn; 1000 fg Pb as Et_4Pb, LOD (3 s) 100 fg Pb; 10^4 fg Hg as Et_2Hg, LOD (3 s) 120 fg Hg.

7.6 FUTURE ASPECTS

One of the main challenges for future species research is the determination of matrix species as defined in Table 7.1. Whereas analytical methods for classically defined species are well developed and sufficient for the evaluation of their impact in the ecosystem, the matrix species, responsible for the behavior of the species, are not yet well understood. This includes also the dynamics of species changing in the environment. How is a certain compound bound in a natural surface water system? How rapidly does this change? What are the influencing parameters? What is the transport vehicle of a parent species through a river system, or, on another scale, through a membrane? These questions lead to the question of remediation of contaminated sites, where man can influence the degradation of toxic species to prevent further hazards to the ecosystem.

7.6.1 Measurement Needed

The analysis of "species of species," the operationally defined group of species, or as named in Table 7.1, the matrix species, is the new challenge for the analytical chemist. It is a challenge because the chemist wants clear results; operationally defined compounds cannot be so well defined. The humic substances, for example, play a substantial role as a matrix species. They are not well defined because of their complex polymer structure but are important for many transport processes as illustrated, for example, in Figure 7.2. A new approach would therefore be to develop an exact and reliable definition of separation and behavior of such operationally defined compounds. This is not easy because most of the factors influencing a species in its concentration and composition are not well known.

A promising approach is, therefore, the elaboration of separation steps and the determination of relevant parameters of the composition and structure of these species. The next step must be the development from a one- and two-dimensional chemical analysis of species to the third dimension, i.e., the development of methods for the analysis of the spatial structure of species. The problems with this kind of analysis are obvious: the concentrations are normally very low and the species very labile when transferred to a different matrix for the analytical procedure.

7.6.2 Remediation Approaches

The identification of harmful species in the ecosystem is one task: to abolish the threat is another. It is simple but often too expensive to clean up a contaminated site by excavation and deposition of the material in a closed system, e.g., underground storage. Under certain conditions, methods of extraction or biological remediation can be used. In some cases, and mercury decontamination is an example, thermal methods are possible (Wilken et al. 1995). Optimization of such decontamination procedures with respect to energy consumption depends on the species involved. In the case of mercury species, Hg^0 evaporates around 100°C, whereas HgS needs temperatures above 350°C (Windmöller et al. 1995; Bombach et al. 1994). It is therefore one of the aims of species research to reduce all mercury species to Hg^0 to get optimal conditions for the thermal process. This philosophy can also be developed for extraction methods, changing the species composition in such a way that extractable species are dominant or in equilibrium with other species so that the extraction is successful.

7.7 OUTLOOK

Species analysis is an important tool to

- predict behavior of chemicals in the ecosystem and
- develop remediation measures in the case of a contamination.

An attempt to close the link between chemical analysis and the behavior of compounds in the environment by species analysis is therefore a promising approach.

7.8 REFERENCES

Bernhard, M., F. E. Brinckman, and K. J. Irgolic, 1986, "Why Speciation," in *The Importance of Chemical "Speciation" in Environmental Processes*, (M. Bernhard, F. E. Brinckman, P. J. Sadler, Eds.), Dahlem Konferenzen, Springer, Berlin, Germany, 1986, pp. 7–14

Bloom, N., 1989, "Determination of Picogram Levels of Methylmercury by Aqueous Phase Ethylation, Followed by Cryogenic Gas Chromatography with Cold Vapour Atomic Fluorescence Detection," *Can. J. Fish Aquat. Sci.*, *46*, 1131–1140

Bombach, G., K. Bombach, and W. Klemm, 1994, "Speciation of Mercury in Soils and Sediments by Thermal Evaporation and Cold Vapor Atomic Absorption," *Fresenius J. Anal. Chem.*, *350*, 18–20

Broekaert, J. A. C., S. Gücer, and F. Adams, *Metal Speciation in the Environment*, NATO ASI Series, Ecological Sciences, Vol. 23, Springer, Berlin, Heidelberg, New York

Craig, P. J., and P. A. Moreton, 1983, "Total Mercury, Methyl Mercury and Sulphide in River Carron Sediments," *Mar. Poll. Bull.*, *14*, 408–411

Ebinghaus, R., H. Hintelmann, and R.-D. Wilken, 1994, "Mercury Cycling in Surface Waters and in the Atmosphere – Species Analysis for the Investigation of Transformation – and Transport Properties of Mercury," *Fresenius J. Anal. Chem.*, *350*, 21–29

Kramer, J. R., and H. E. Allen, (Eds.), 1991, *Metal Speciation: Theory, Analysis and Application*, Lewis Publishers, Chelsea, USA

Leeuwen, van, H. P., and J. Buffle, 1993, *Environmental Particles*, Vol. II, Louis Publishers, Jersey, 1993

Prange, A., and E. Jantzen, 1995, "Determination of Organometallic Species by GC-ICP-MS," *J. Anal. At. Spectr.*, *10*, 105–109

Prange, A., and E. Jantzen, 1995, "Determination of Organometallic Species by Gas Chromatography Inductively Coupled Plasma Mass Spectrometry," *J. Anal. At. Spectr.*, *10*, 105

Ure, A. M., and C. M. Davidson, 1995, *Chemical Speciation in the Environment*, Blackie Academic Professional, London, UK

Wallschläger, D., H. Hintelmann, R. D. Evans, and R.-D. Wilken, 1995, "Volatilization of Dimethylmercury and Elemental Mercury from River Elbe Floodplain Soils," *Water Air Soil Pollut.*, *80*, 1325–1329

Wilken, R.-D., 1992, "Mercury Analysis, a Special Example of Species Analysis," *Fresenius J. Anal. Chem.*, *342*, 795–801

Wilken, R.-D., H. Hintelmann, 1990, Analysis of Mercury-Species in Sediments, in *Metal Speciation in the Environment*, (J. A. C. Broekaert, S. Gücer, and F. Adams, Eds.), NATO ASI Series, Vol. G 23, Springer, Berlin, Germany, 1990, pp. 339–359

Wilken, R.-D., J. Kuballa, E. Jantzen, 1994, "Organotins: Their Analysis and Assessment in the Elbe River System, Northern Germany," *Fresenius J. Anal. Chem.*, *350*, 77–84

Wilken, R.-D., M. Hempel, and I. Richter-Poltiz, 1995, "Mercury Contamination and Decontamination," *Int. Conf. Heavy Metals in the Environment*, Proceedings, CEP Consultants, Edinburgh, Hamburg, 1995, Vol. 2, 42–51

Windmöller, C. C., R.-D. Wilken, and W. de F. Jardim, 1995, "Mercury Speciation in Contaminated Soils by Thermal Release Analysis," *Water Air Soil Pollut.*, *89*, 399–416

8

Multimedia Mass Balance Models of Chemical Distribution and Fate

Don Mackay (Peterborough, Canada)

8.1 SUMMARY

The role of multimedia mass balance models for expressing quantitatively the sources, transport, and transformation of chemicals in the environment is described. These models usually treat the environment as a set of connected well-mixed or homogeneous compartments into which a chemical is introduced under various steady-state or dynamic conditions, and to which a selection of simplifying assumptions is applied. The

Ecotoxicology, Edited by Gerrit Schüürmann and Bernd Markert.
ISBN 0-471-17644-3   © 1998 John Wiley & Sons, Inc. and Spektrum Akademischer Verlag.

environments may be evaluative (hypothetical) or real. An example is presented of the deduced environmental fate of phenanthrene in a series of increasingly complex evaluative environments. Various applications of multimedia models for specific purposes such as risk assessment of new and existing chemicals, priority setting among groups of chemicals, and design of monitoring programs are described.

"I often say that when you can measure what you are speaking about and express it in numbers you know something about it; but when you cannot measure it, when you cannot express it in numbers, your knowledge is of a meagre and unsatisfactory kind; it may be the beginning of knowledge, but you have scarcely, in your thoughts, advanced to the stage of science ... "   —Lord Kelvin

8.2 INTRODUCTION

The science of ecotoxicology is regarded by many as originating with the publication of Rachel Carson's *Silent Spring* in 1962. This book catalyzed a widespread realization that chemicals discharged into our finite environment can migrate between air, water, soils, and sediments, and into biota by pathways which are often surprising and cause unexpected and undesirable effects on organisms such as birds. The advent of gas chromatographic analysis and its ability to measure contaminants at concentrations below 1 mg/L or one part per million resulted in contaminants being identified in remarkably high concentrations in diverse media such as fish, birds, wildlife, and humans, often far from obvious sources. Contaminants were found in Arctic and Antarctic environments, previously believed to be pristine by virtue of their remoteness from sources. Gradually a *qualitative* picture emerged of sources and pathways, causes and effects. In recent decades the focus has been on *quantifying* these processes and developing a predictive capability; indeed much of this text is devoted to reviewing our present knowledge of how chemicals move into and affect ecosystems and how these effects may be assessed.

The challenging task of translating qualitative observations into quantitative statements is usually accomplished by some form of mathematical model in which equations describe the mass balance and relevant processes incorporating appropriate parameter values. The equations are then applied or solved to test if the model assertions agree with observations, and the findings are interpreted. Among the earliest environmental models are those of oxygen depletion in rivers, such models being generally applied to what is now regarded as a single phase or medium. There has been impressive progress in developing models describing the dispersion of chemicals in the atmosphere, in lakes and estuaries, and in soils, usually for agrochemical purposes. These models usually treat loss from the medium in question as permanent; for example, a chemical which evaporates from a lake does not return by absorption or deposition. Such assumptions permit the modeler to focus greater detail on the medium in question, allowing for segmentation into several connected compartments, and enabling spatial differences in concentration to be described by analytical expressions.

Multimedia models seek to provide a broader and more complete picture of chemical fate by including all media to and from which the chemical may migrate. All significant transport and transformation pathways are included and a total mass balance is established including all possible fates. The merit of this approach is that unexpected pathways such as those identified in *Silent Spring* should be revealed. The primary media of accumulation are determined, as are the most important processes, the approximate persistence of the chemical, and in some cases order-of-magnitude estimates of concentration. A picture emerges of the general behavioral characteristics of the chemical which is valuable as a contribution to more effective assessment and management to ensure an absence of adverse ecotoxicological effects.

In this chapter the mathematical principles which form the foundation of mass balance models are outlined and the media to which they are applied are described. Various modelling approaches are discussed and selected current models are described. The input and output of a typical multimedia model are presented to provide the reader with a concrete example of the nature of these models. This is followed by an account of applications including a discussion of several issues facing the model developer and user.

8.3 MATHEMATICAL BASIS

Figure 8.1 illustrates part of a multimedia model in which the primary compartment, medium, or "box" is water, which for illustrative purposes is in contact with air, bottom sediment, and fish. A key assumption is that the water is homogeneous or well mixed; thus, the concentration of the chemical, C_W mol/m^3, is spatially constant. A volume V_W m^3 can be assigned; thus, the amount of chemical present in this box of water is $V_W C_W$ or M_W mol. There may be direct inputs to the water by discharge from industrial, municipal, or other sources, and in some cases by formation of the substance at a total rate E mol/h. Normally there is inflow and outflow of water at a rate of say G m^3/h; thus the residence time of water in the system is V_W/G hours. The inflowing water may contain a chemical at a concentration C_I mol/m^3; thus, the input rate of chemical by advection is GC_I mol/h. The outflowing water will contain a chemical at the prevailing concentration C_W; thus, the outflow rate is GC_W mol/h. This is termed the "continuous stirred tank reactor" or CSTR assumption, since it is widely used in chemical engineering analyses of chemical reactors.

There may be inputs by transport from other media, namely air, sediment, and fish at rates J_{AW}, J_{SW}, and J_{FW} mol/h and corresponding transport losses

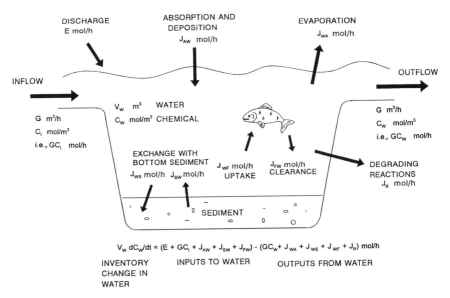

Figure 8.1 The mass balance concept applied to a single-medium "box" of well-mixed water.

from the water at rates J_{WA}, J_{WS}, and J_{WF} mol/h. Note that the order of the subscripts denotes the direction of transport. Finally there may be reaction or transformation of the substance by processes such as biodegradation or photolysis at a total J_R mol/h.

The axiomatic statement of mass balance takes the form:

$$\text{Rate of inventory change} = \text{input rate} - \text{output rate}$$

or

$$V_W \, dC_W/dt = (E + GC_I + J_{AW} + J_{SW} + J_{FW})$$
$$- (GC_W + J_{WA} + J_{WS} + J_{WF} + J_R) \qquad (8.1)$$

It is usual to express all of the output rates as a function of C_W; for example, J_R could be $VC_W k_R$ where k_R is a first order rate constant for reaction (h^{-1}). If the input terms are known, this differential equation can then be solved to give a dynamic or unsteady state solution for C_W as a function of time. The solution may be analytical or by numerical integration.

A simple and very convenient expression is obtained if the left-hand side can be set to zero implying that conditions are steady state or constant with time. The equation becomes algebraic and C_W can be deduced directly. Such situations exist or are approximated when there is prolonged constant discharge of a chemical to the water and consequently a build-up of constant conditions. Such situations may be approximated in the real environment if discharges are steady, environmental conditions change little with time, and if the chemical is persistent, i.e., it is degraded relatively slowly so that its residence time is long. This residence time can be deduced as $M_W/$(total input or output rate) h where M_W or $V_W C_W$ is the constant inventory in the water.

In practice, when modelling multimedia systems, differential or algebraic equations are written for each medium (e.g., air, sediment) and a set of simultaneous mass balance equations must be solved. Terms such as air to water transfer J_{AW} are expressed as a function of the concentration in air; thus, if there are N media, a total of N unknown concentrations is obtained and a solution is possible. The challenge is to define the discharge rates and the various transport and transformation terms as a function of concentration for the chemical of interest. This requires information on the properties of the chemical, for example, its reactivity, transport, and partitioning tendencies, and of the environment, such as volumes, areas, compositions, and temperature, and discharge or "loading" data. The availability or accuracy of this information usually limits the accuracy of the model.

8.4 ENVIRONMENTAL MEDIA

Examination of the fate of a number of chemicals with diverse properties suggests that to account properly for all significant processes requires that most of the media in Table 8.1 be included. Aerosol particles are not important for chloroform but they are critically important for PCBs (polychlorinated biphenyls). The dilemma is that some ten to 15 media must be included, rendering the mathematics complex. Provision of all of the required input data is very demanding and expensive.

The response to this dilemma is to apply Occam's razor or the principle of parsimony and include only the necessary complexities, or in this context reduce the number of equations to a minimum by lumping media together, by applying the steady state assumption where possible to render differential equations algebraic, and by applying partitioning equilibrium assumptions where applicable. For example, in Figure 8.1 the quantity of a chemical in the fish is probably very small compared to that in the water, so it may be possible to ignore the fish medium. Often the fish can be assumed to be in a steady state with respect to the water; hence, C_F can be expressed in terms of C_W using a bioaccumulation factor K_B or C_F/C_W, and the amount in the fish M_F can be added to M_W to give an amount in the bulk phase of water plus fish. An even more restrictive approach is to assume that thermodynamic equilibrium exists between fish and water; thus, the concentration in fish C_F is $K_{FW}C_W$ where K_{FW} is an equilibrium partition coefficient or bioconcentration factor. The amount in the bulk water plus fish phase thus becomes:

$$V_W C_W + V_F C_F = V_W C_W + V_F K_{FW} C_W = C_W(V_W + V_F K_{FW}) \quad (8.2)$$

Table 8.1 Environmental media

Primary media	Secondary media
Air	Aerosols, rain, snow
Water	Suspended solids, biota, e.g., fish
Soils	Mineral matter, organic matter,
(possibly several	pore water, air, vegetation
layers or types)	
Sediments	Sediment solids, pore water, biota
(possibly several layers)	
Other media	Mammals, birds, groundwater

This "equilibrium assumption" approach is particularly useful and can be applied when the time for attainment of equilibrium is short compared with the time for transport and transformation from the bulk phase, i.e., the residence time of the chemical in the phase. It is useful to apply this equilibrium assumption to air, aerosols, and rain, to the solids, water, and air in the soil matrix, to water, fish, and suspended matter in the water column, and to solids, water, and biota in the bottom sediment phase. More problematic is how vegetation which is in contact with air and soil should be treated.

Soils and bottom sediments present a special problem because they are usually heterogeneous and it is not obvious what vertical depth should be selected. It may be necessary to segment these media into two or more layers. This can give concentrations more representative of reality, but it introduces the requirement to define layer–to–layer transport rates. Most models treat the four bulk or primary media, air, water, soil, and sediment as illustrated in Table 8.1.

8.5 LEVELS OF COMPLEXITY

As models have been developed and applied, a terminology has evolved for describing models of differing complexity. A fuller account of these definitions is given by Cowan et al. (1995) and Mackay (1991).

Level I models are merely models of the equilibrium distribution of a fixed quantity of a chemical between the media, i.e., there are no inputs or outputs and it is assumed that the chemical is conserved. This is useful for gaining an initial impression of partitioning tendencies, and for obtaining order-of-magnitude estimates of relative concentrations.

Level II models treat a steady state input–output situation and they also assume intermedia equilibrium; thus, there is no need to specify the intermedia transport rates J_{ij}. Total input rates to the entire system are defined and output rates are defined for reactions in all media and advective flows in air and water. A steady-state assumption is usually applied, but unsteady state versions are possible. This level is used to gain an appreciation of the relative rates of processes of loss by reaction and advection, and to give an estimate of overall chemical residence time or persistence.

The preferred model is Level III which quantifies the rates of all intermedia transport processes and thus treats the more realistic case of nonequilibrium between media. It describes steady-state conditions, i.e., the constant conditions which apply after prolonged steady discharges of a chemical. The discharges must be specified individually for each medium. The solution is thus of a set of simultaneous algebraic equations which can be accomplished by algebra or by matrix manipulation. This level gives rates of all transport and transformation processes.

Finally, Level IV models are similar to Level III models but describe unsteady-state conditions and can yield information on how long the response time may be for concentrations to build up or decay following changes in discharge rate. The solution is of a set of simultaneous ordinary differential equations. This level is usually not required because the response time of the system can be estimated from the chemical residence times in Level III.

Cowan et al. (1995) described the characteristics of four such models. Examples of such models are also given by Mackay et al. (1991) and Mackay and Paterson (1991). It should be emphasized that this terminology is applicable to all models, not just fugacity models.

Modelers can use some discretion in setting up and solving these equations. Some prefer computer spreadsheets, while others write programs in conventional programming code. Output data may be purely tabular, or a more attractive graphical output may be used. It seems likely that with the continued development of computer software there will emerge a variety of approaches, with the most useful thriving and the less useful falling into deserved obscurity.

8.6 MODEL STRUCTURE AND VARIABLES

The one area in which there are major differences between programs is in the selection of algebraic variables used to define the quantities of a chemical, and from them the rates of processes of transport and transformation.

The *conventional concentration* approach as described earlier uses C_W as the key parameter describing the presence of a chemical in the water column. Units of mol/m^3 or g/m^3 or some other system of units may be used, but adherence to the SI system is now preferred. All process rates are expressed as a function of C_W. Reaction is usually a VCk product where V is volume and k is a first order rate constant, and advection is a GC product as illustrated earlier. Diffusion may be expressed as KAC or DAC/Y where K is a mass transfer coefficient which can be viewed as diffusivity D divided by a diffusion path length Y, and A is area. A disadvantage of the conventional approach is that the relative significance of disparate processes cannot be readily compared because of the variety of terms used to define rates.

This problem can be remedied by expressing all process rates using first order *rate constants* $k_i(\text{h}^{-1})$; thus, all rates are $k_i M$ mol/h if M has units of moles. The definition of k is achieved by writing the conventional rate expression in mol/h and then dividing by CV or M. For example, the rate constant for evaporation from water by diffusion becomes $KAC/(CV)$ or KA/V or K/H, where H is V/A, the depth of the water column. The advantage of this approach is that all processes are similarly expressed and the total rate of loss is $M\sum k_i$ where $\sum k_i$ is the sum of all the rate constants. The relative significance of diverse processes can be readily identified, the faster (larger) rate constants being most important. Mackay et al. (1994) have used this approach to describe steady and unsteady state conditions in multimedia lake systems.

The third approach is to use the equilibrium criterion of fugacity as advocated by Mackay (1991). A complete description of this approach is beyond the scope of this chapter, but it proves to be particularly useful for multimedia calculations. Concentration C mol/m^3 is related to fugacity f (Pa) by the expression:

$$C = Zf \tag{8.3}$$

where Z is a capacity term, usually termed a Z value with units of mol/m^3.Pa which is specific to the chemical, medium, and temperature. Fugacity is essentially a corrected partial pressure or escaping tendency; thus, when two media (1 and 2) are in equilibrium the fugacities of the chemical f_1 and f_2 are equal and

$$C_1/C_2 = Z_1 f_1/Z_2 f_2 = Z_1/Z_2 = K_{12} \tag{8.4}$$

where K_{12} is an equilibrium partition coefficient. Each Z is "half" a partition coefficient and represents the tendency of the chemical to be absorbed into that phase. High concentrations occur when Z is large. Recipes for estimating Z for organic chemicals in a variety of media are given by Mackay (1991).

Rates of transport and transformation are deduced as Df where D is a rate parameter with units of mol/h.Pa which is similar in principle to a rate constant. As with the rate constant approach the total rate is $f\sum D$ and diverse processes can be readily compared. Indeed it can be shown that k_i is $D_i/(VZ)$.

Different modelers and groups thus adopt different approaches when compiling multimedia mass balance models, but in most cases these models can be shown to be similar or even identical algebraically. Modelers strive to incorporate the latest knowledge of the phenomena of equilibrium partitioning, intermedia transport, and transformation processes into appropriate mathematical expressions, and where necessary using correlations for parameters such as partition coefficients and transfer coefficients. Differences between models thus tend to reflect differences of opinion on the number of media which should be included and how they are connected, the detailed expressions used, and the parameter estimation procedures. Regardless of the approach there is a strong case for ensuring that the models are well documented and transparent, i.e., that they contain no hidden assumptions and can if necessary be subjected to detailed peer review. Further, the modeler has an obligation to present the results not just as a single quantity such as a concentration of 5 μg/g but as a range or distribution, for example a 90% probability that the value lies between 2 and 15 μg/g. This avoids giving an accidental impression of undeserved accuracy.

8.7 EXISTING MODELS

A considerable number of models exists in various stages of development, use, and neglect and with various spheres of application. There is no simple system for discriminating between multimedia and other models because a fairly continuous spectrum exists from comprehensive multimedia to several media to single media models. Mackay (1995) has reviewed many of these models. It is impossible to discuss more than a small fraction of the models available; thus the aim is to convey here an impression of the range of currently available models.

No account of these models would be complete without an acknowledgement of the foresight of the early model developers, notably Baughman, Lassiter, and Burns who pioneered the US EPA's EXAMS model which is primarily a water quality or aquatic model, but which set a standard for subsequent models in terms of scientific content, transparency, and accessibility. EXAMS was an early "evaluative" model in which the fate of the chemical was estimated in a purely hypothetical or ideal environment with dimensions and properties defined by the modeler rather than by nature. The use of evaluative models has been particularly valuable for elucidating the behavioral characteristics of chemicals separately from the dependence of that behavior on the properties of the environment. The "fugacity" models evolved from the early evaluative model conceived by Baughman and Lassiter (1978).

In general then, there are two types of models: *evaluative models*, which treat chemical fate in hypothetical environments, and *models of real systems* which purport to describe chemical fate in real regions such as a river or a nation. Some models contain data on several regions and an evaluative environment may be included. Evaluative models cannot be validated, whereas limited validation is possible for real models. It must be appreciated, however, that a model may give satisfactory predictions for one chemical (suggesting that it is valid) but it may fail for another. Some researchers hold the view that validation in its strictest sense is an impossible goal because examples of invalidity can always be found (Oreskes et al. 1994). Perhaps the goal should be to demonstrate that the model is sufficiently accurate for a defined group of chemicals so that it can play a useful and reliable scientific and regulatory role.

8.7.1 Fugacity Models

The earliest evaluative models which employed fugacity as a means of expressing interphase equilibrium treated an area of 1 km^2 of which 70% was water (Mackay 1979). More recent evaluative models treat larger areas of some

100 000 km^2 (Mackay et al. 1991) and have been used as a basis for depicting chemical fate in a series of handbooks as described later. The "real" CHEM-CAN model of chemical fate in Canada divides that country into 24 regions and includes the effect of temperature (Mackay et al. 1995a). CHEMFRANCE is a similar model which can be applied to regions of France (Devillers 1995). HAZCHEM (ECETOC 1993) was developed by an industry group in the Netherlands using the generic model of Mackay et al. (1991) as a basis but modifying the method of solution. CalTOX was developed by McKone (1993) to address the fate of chemicals originating from waste sites in California. It contains multiple soil layers and a comprehensive treatment of human exposure. More recent fugacity models include the EQC (Equilibrium Criterion) models which treat not only conventional organic contaminants but also involatile substances, such as metals, and sparingly soluble and even insoluble substances such as polymers.

8.7.2 Nonfugacity Models

Among the early multimedia models were MEPAS by Droppo et al. (1989). An evaluative model compiled by Cohen et al. (1990) gives more detailed treatment of atmospheric deposition and fate in soils. SimpleBOX is an eight-compartment model developed in and applicable to the Netherlands (van de Meent 1993).

8.7.3 Round Robins

An interesting and illuminating exercise was sponsored by SETAC in 1994 in which four models (CalTOX, CHEMCAN, HAZCHEM, and SimpleBOX) were compared by running them for a similar environment and selected chemicals with a common set of properties. Initial comparison results showed surprisingly large differences which were identified as being attributable to three primary causes: (i) differences in interpretation of input data, (ii) use of different concentration units, e.g., wet or dry weight concentration bases, and (iii) differences in the selection of intermedia transport parameters such as mass transfer coefficients. In a subsequent analysis all four models were adjusted to eliminate such differences, and they were then found to give essentially identical results. They are thus structurally similar and the primary differences arise from different methods of describing the same phenomena. Full details of this exercise are given in a report by Cowan et al. (1995).

Any discussion of environmental fate models is incomplete without acknowledgement of the fact that extensive efforts have been devoted to modelling radionuclides. The broad community of ecotoxicologists which includes chemists, physicists, and biologists who are concerned with chemical substances in the environment, have very little contact with the scientific community which addresses the fate and effects of radionuclides. Following the initial use of nuclear weapons in 1945 and the later extensive weapons testing and development of nuclear power plants, considerable efforts were devoted to investigating and modelling the fate of radioactive materials, especially as they enter food chains. The BIOMOVS program has involved a "round robin" comparison of multimedia models of radionuclide fate. It is regrettable that this insularity persists since both chemical and radionuclide modelling groups could benefit from the experience of the other.

8.8 AN ILLUSTRATION OF A MULTIMEDIA MODEL

To illustrate the concepts described in this chapter an example is given of a multimedia model of phenanthrene in an evaluative or generic environment. A complete description of the model structure, the working equations, and justification for the chemical properties is beyond the scope of this chapter. The example is taken directly from a published model which has been developed to illustrate chemical fate in a series of *Illustrated Handbooks of Physical Chemical Properties and Environmental Fate for Organic Chemicals* published in four volumes. Volume 4 includes a disk containing a spreadsheet and BASIC programs of the model, and each volume contains an explanation of the model structure (Mackay et al. 1995).

Briefly, the environment comprises an area of $100\,000$ km^2, about the size of England or Ohio, and consists of four primary compartments: air, soil, water (which covers 10% of the area), and bottom sediment. The chemical selected (phenanthrene) is convenient because it is fairly hydrophobic and persistent and has a tendency to partition into all media in significant quantities. The selected properties of phenanthrene are given in Table 8.2.

8.8.1 Level I

The results of the Level I calculation illustrated in Figure 8.2 suggests that if $100\,000$ kg (100 tonnes) are introduced into the $100\,000$ km^2 environment, 1.85% will partition into air at a concentration of 18.5 ng/m^3. The water will

Table 8.2 Properties of phenanthrene at 25 °C used in the evaluative multimedia model (Mackay et al. 1992)

Molecular mass	178.2 g/mol
Melting point	101°C
Vapor pressure	0.020 Pa
Solubility in water	1.10 g/m^3
Log octanol–water partition coefficient	4.57
Half-life in air	55 h
Half-life in water	550 h
Half-life in soil	5500 h
Half-life in sediment	17 000 h

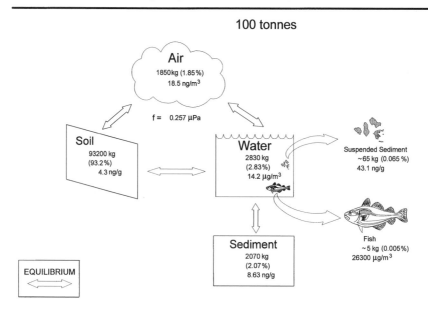

Figure 8.2 Illustration of a Level I evaluation of the equilibrium partitioning of 100000 kg of phenanthrene.

contain 2.83% at a low concentration of 14 μg/m³. Soils will contain 93% of the phenanthrene at 4.3 ng/g, and sediments, about 2% at 8.6 ng/g. These soil and sediment values would barely be detectable as a result of the moderate tendency of phenanthrene to be sorbed to organic matter in these media. There is evidence of bioconcentration with a fish concentration of 26 ng/g, i.e., a bioconcentration factor of 1860. The prevailing fugacity is 2.6×10^{-7} Pa, i.e., 0.26 μPa. It is clearly important that the sorption characteristics of this substance be accurately known because most of it is sorbed to soil or sediments.

8.8.2 Level II

The Level II calculation illustrated in Figure 8.3 includes estimated reaction half-lives of 55 h in air, 550 h in water, 5500 h in soil, and 17 000 h in sediment. These half-lives represent log means of a range rather than precise numbers. For example 55 h is the approximate log mean of the range 30–100 h. No reaction is included for suspended sediment or fish. The steady input of 1000 kg/h results in an overall fugacity of 4.3×10^{-6} Pa or 4.3 μPa, which is about 17 times the Level I value. The concentrations and amounts in each medium are thus about 17 times the Level I values. The relative mass distributions in Levels I and II are identical. The primary loss mechanisms are reaction and advection in air, which

Fugacity Model Level II   100 000 km²

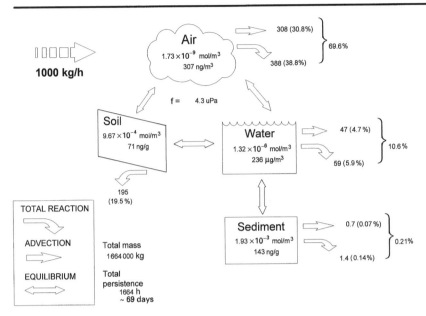

Figure 8.3 Illustration of a Level II evaluation of the steady-state, equilibrium condition arising from the discharge of 1000 kg/h of phenanthrene.

account for about 70% of the input. Most of the remainder is lost by reaction in soil. The water and sediment loss processes are unimportant, largely because so little of the chemical is present in these media, and because of the slower reaction rates. Since there is an inventory in the system of 1 664 000 kg, the overall residence time is 1664 h or 69 days.

The key reactions occur in air and soil; thus, a need for accurate half-lives in these media is indicated.

8.8.3 Level III

The Level III results are shown in a series of four figures illustrating the behavior of phenanthrene when discharged to air, water, and soil individually, and in combination.

The first figure (Figure 8.4) describes the condition if 1000 kg/h is emitted into air. The media are no longer at equilibrium, having fugacities of: air, 6.0 μPa; water, 0.66 μPa; soil, 0.11 μPa; and sediment, 1.2 μPa. The result is similar to the Level II calculation with 43 200 kg in air, 7470 kg in water, 41 500 kg in soil, and 9940 kg in sediment. It can be concluded that phenanthrene discharged to the atmosphere has limited potential to enter other media. The rate of transfer from

Fugacity Model Level III 100 000 km²

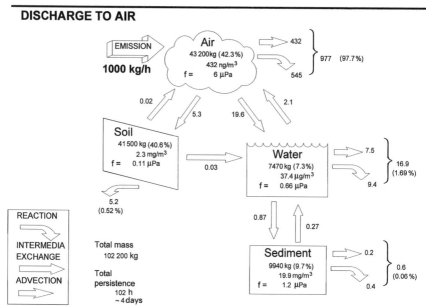

Figure 8.4 Illustration of a Level III evaluation of the steady-state, nonequilibrium condition arising from the discharge of 1000 kg/h of phenanthrene to air.

air to water is only about 20 kg/h, i.e., 2% of the input, but over time an appreciable quantity of about 17 400 kg builds up in the water and sediment. The overall residence time is 102 h and is controlled by the reaction and advection losses from air which total 977 kg/h, i.e., they remove 97.7% of the input.

If 1000 kg/h of phenanthrene is discharged to water, as in Figure 8.5, there is predictably a much higher concentration in water (by a factor of 51). There is reaction of 481 kg/h in water, advective outflow of 382 kg/h, and transfer to air of 108 kg/h, with a small net loss of 31 kg/h to sediment. The amount in the water is 382 000 kg; thus, the residence time in the water is 382 h and the overall environmental residence time is a longer 899 h largely because of the large amount in the sediment (508 000 kg) with its long residence time of 16 400 h. The key processes are thus reaction in water (half-life 550 h), evaporation (half-life 2500 h), and advective outflow (residence time 1000 h). Clearly, competition between reaction and evaporation in the water determines the overall fate. Of the phenanthrene discharged, 42% is now found in the water, and the concentration is fairly high, namely 1.9×10^{-3} g/m³ or 1900 ng/L.

Figure 8.6 shows the fate if discharge is into soil. The amount in soil is 7.86×10^6 kg, reflecting a 7860-h residence time. The rate of reaction in soil is

Fugacity Model Level III 100 000 km^2

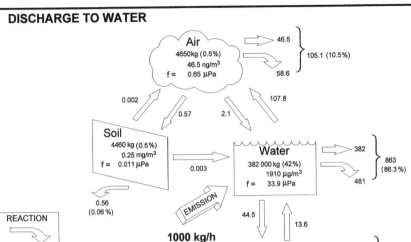

DISCHARGE TO WATER

Figure 8.5 Illustration of a Level III evaluation of the steady-state, nonequilibrium condition arising from the discharge of 1000 kg/h of phenanthrene to water.

990 kg/h; loss mechanisms are transfer to water by run-off at a rate of 6.2 kg/h, with a relatively minor loss (3.9 kg/h) to air by evaporation. The soil concentration of 0.44 g/m^3 is controlled almost entirely by the rate at which the phenanthrene reacts.

The net result is that phenanthrene behaves very differently when discharged to the three media. If discharged to air it reacts and advects rapidly with a residence time of 102 h or about 4 days, with slow transport to soil or water; however, there is an appreciable build up of phenanthrene in soil. If discharged to water it reacts and advects with an overall residence time of 899 h or 5 weeks. If discharged to soil it mostly reacts, with an overall residence time of about 7860 h or nearly a year.

The final scenario, shown in Figure 8.7, is a combination of discharges: 600 kg/h to air, 300 kg/h to water, and 100 kg/h to soil. Inspection shows that the concentrations, amounts, and rates of transport and transformation are simply linearly added fractions of the three single-media discharge scenarios.

In this multimedia discharge scenario the overall residence time is 1117 h, which can be viewed as the sum of 60% of the air residence time, 30% of the water residence time, and 10% of the overall soil residence time. The overall amount in the environment of 1.17 × 10^6 kg is thus largely controlled by the discharges to soil.

Fugacity Model Level III 100 000 km²

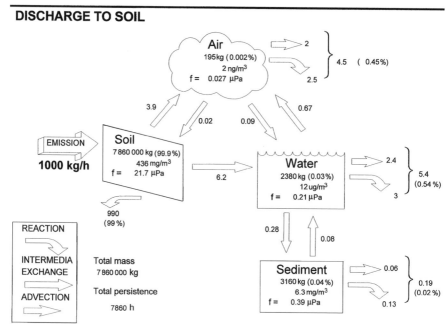

Figure 8.6 Illustration of a Level III evaluation of the steady-state, nonequilibrium condition arising from the discharge of 1000 kg/h of phenanthrene to soil.

These single-media behavior profiles in combination with the fourth give a comprehensive illustration and explanation of the environmental fate characteristics of phenanthrene. They show the relative importance of the various intermedia transport processes and how levels in various media arise from discharges to other media.

For example, if there is discharge to water it can be expected that considerable sediment contamination will result. Soil can become contaminated from discharge to air. The evaluation essentially translates physicochemical data into environmental fate information. The general fate characteristics as elucidated by the generic environment are believed to apply at least approximately to other real environments. Such evaluations should, in most cases, identify the most important physicochemical properties, reactions, and intermedia transport parameters. Sensitivity analyses can be conducted. With a knowledge of the key parameters, more effort can be devoted to obtaining more accurate values.

A Level IV calculation could be completed to show how long it would take for these steady-state concentrations to build up or decay. This is particularly important when it is suspected that soils and sediments may slowly build up high concentrations from which they will take a long time to recover.

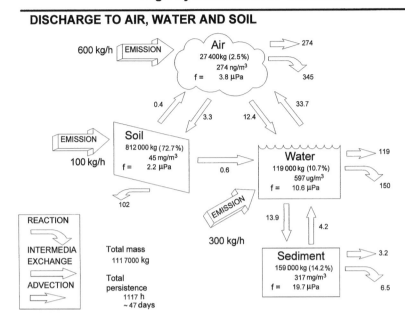

Figure 8.7 Illustration of a Level III evaluation of the steady-state, nonequilibrium condition arising from the discharge of phenanthrene to air, water, and soil.

8.9 APPLICATIONS OF MULTIMEDIA MODELS

A recent report sponsored by the Society for Environmental Toxicology and Chemistry (SETAC) comprehensively reviewed the application of multimedia models (Cowan et al. 1995). The following is largely derived from that report, to which the reader is referred for a more detailed treatment.

The primary application of multimedia models has been to the *risk assessment of new and existing chemicals*, usually as part of mandated regulatory proced- ures. Most industrially advanced countries have legislation requiring the assess- ment of risk associated with selected high priority chemicals, and chemicals which are new to commerce and are produced or imported in substantial quantities. Attempts have been made to harmonize the procedures for chemical data acquisition and the assessment processes by OECD and within the Euro- pean Union, but to date no real standardization has yet been achieved (EEC 1993). Whereas insights into chemical fate and likely concentration levels in various media can be obtained by analysis of environmental samples for existing chemicals, no such insights are possible for "new" chemicals. Such information can only be obtained by running multimedia models in a predictive mode. The incentive for advance assessment is particularly obvious when a new chemical is proposed as a more benign substitute for an existing chemical, such as DDT or PCBs, which has proved to be environmentally unacceptable. The aim is to avoid repeating the past mistakes which were exposed in *Silent Spring*.

The specific information which can be gained includes:

- the media in which primary accumulation is expected, e.g., air or soil
- whether or not bioaccumulation is likely
- the key processes affecting fate, e.g., evaporation from water or biodegradation in soils
- the likely order of magnitude of environmental concentrations, e.g., 0.1–10 µg/L in water or 30–500 ng/g in fish
- how current or future local discharges (e.g., 100 t/y) compare with advective inputs in air or water entering the region as a result of chemical use "upstream" in neighboring regions
- the likely time scale during which concentrations are expected to build up to or recover from steady levels, e.g., 1 month or 5 years.

It is noteworthy that different jurisdictions have different requirements for input or notification data. Some require experimental determination while others are satisfied with estimates based on structure–activity relationships. Often the assessment system is tiered with increasingly stringent requirements and evaluations being applied to chemicals of most concern. Regrettably, despite the obvious advantage of sharing assessment tasks and results internationally (since benzene is equally bioaccumulative and toxic in Taiwan and Turkey) each jurisdiction seems determined to carry out its own assessment. It is likely that a set of somewhat comparable multimedia models will evolve with differing input requirement and outputs as dictated by legal legislation. It is hoped that some form of continuing model intercalibration can be fostered and that the flexibility to include new modelling approaches will be encouraged.

A second application is to chemical *ranking, scoring,* or *priority setting.* Since there are some 60 000–100 000 chemicals in commerce, it is obviously not feasible, nor is it necessary, to conduct detailed assessments for all chemicals. A simple model may be adequate to demonstrate low or high priority using information on properties, persistence, and toxicity. Early scoring systems assigned points for factors such as toxicity, bioaccumulation, and persistence; however, such approaches are ultimately subjective and suspect because of the need to weight such factors which are so different in nature.

A third, more specific application is to the assessment of *risk from hazardous waste sites.* Notable in this regard is the CalTOX model by McKone (1993) which includes estimates of human exposure by a variety of routes including inhalation, diet, and dermal absorption. Such model applications are particularly useful when a jurisdiction must set priorities among a large number of sites with a diversity of wastes and containment characteristics.

Models such as CalTOX explicitly treat *exposure by all indirect routes* which may not be treated by single-media models. For example, if hexachlorobenzene is discharged to water it may be thought that most exposure will be from ingestion of water. In reality it is probable that consumption of contaminated fish and inhalation of evaporated chemical are more important. A multimedia model coupled to estimates of exposure to all affected media can highlight the significant, and often surprising routes. A classic example is exposure to volatile organic chemicals present in potable water during washing and showering.

Models can also assist the *optimization of chemical testing* requirements. Concentrations as deduced by models obviously depend on numerous input parameters such as vapor pressure, but the sensitivity of results to input parameters can differ greatly. Often when a parameter has a value outside a certain range, e.g., above or below 0.1–100 Pa, its actual value is of little consequence because it is very volatile or very involatile. Since accurate experimental determinations can be expensive it is invaluable to have information on the required level of accuracy, which is a function of inherent sensitivity of the model output to these input data. Models can identify and quantify these sensitivities.

Models can also assist in the *design of monitoring programs* by showing which media will probably experience concentrations above nondetection levels. Regrettably many expensive monitoring programs merely generate large quantities of "nondetects". A prior modelling exercise can often reveal the likely futility of certain analyses so that efforts can be focussed on where they will be most useful. A classic case is the use of fish as biomonitors of water contamination.

Other more recent and exploratory applications of multimedia models are to the estimation of *global dispersion* of chemicals (Wania and Mackay 1993, 1995; Mackay and Wania 1995), testing the *coherence of environmental quality criteria*, i.e., whether or not criteria developed for benzene, air, and water are realistic given that these media are in contact and influence each other (van de Meent and De Bruijn 1995), and as part of the process of *life cycle assessment* (Guinee and Heijungs 1993).

In summary, it is clear that by providing a method by which estimates of chemical emission rates, chemical properties, and environmental conditions can be combined to deduce even approximate concentrations, models can contribute to more effective environmental management in a variety of contexts. It is hoped that in the coming years there will be more successful demonstrations of the utility of these models which will encourage their continued and expanded development and use.

8.10 CONCLUSIONS

In this chapter the aim has been to show how multimedia mass balance models are structured and can be used to reveal quantitatively the fate of chemicals in the environment. Concentrations can be predicted and compared with observations, at least for existing chemicals, leading to a degree of validation. Successful demonstrations of model applications lend credibility to the use of models in a more predictive mode. This is especially important for "new" chemicals in which models are the only predictive tool available. Because the environment is complex and chemicals vary so greatly in properties, the human mind is incapable of processing the diversity of "input data" to deduce environmental fate. It needs the assistance of the mass balance models as a tool towards quantitative understanding.

The multimedia mass balance model should thus be viewed as one tool among the many tools used by the ecotoxicologist who strives to understand natural and anthropogenic chemicals present in our environment and ensure that the well-being of humans and our fellow creatures is not adversely affected by their presence.

8.11 REFERENCES

Baughman, G. L., and R. R. Lassiter, 1978, in *Estimating the Hazard of Chemical Substances to Aquatic Life* (J. Cairns Jr., K. G. Dickson, and A. W. Maki, Eds.), American Society Testing and Materials Tech. Pub. 657, Philadelphia, PA, USA

Carson, R., 1962, *Silent Spring*, Houghton Mifflin, Boston, MA, USA

Cohen, Y., W. Tsai, S. L. Chetty, and G. Mayer, 1990, "Dynamic Partitioning of Organic Chemicals in Regional Environments: A Multimedia Screening-Level Modeling Approach," *Environ. Sci. Technol.*, *24*, 1549–1558

Cowan, C. E., D. Mackay, T. C. J. Feijtel, D. van de Meent, A. Di Guardo, J. Davies, and N. Mackay, 1995, *The Multimedia Fate Model: A Vital Tool for Predicting the Fate of Chemicals*, SETAC Press, Pensacola, FL, USA

Devillers, J., S. Bintein, and W. Karcher, 1995, "CHEMFRANCE: A Regional Level III Fugacity Model Applied to France," *Chemosphere*, *3*, 457–476

Droppo, J. G., D. L. Strenge, J. W. Buck, B. L. Hoopes, R. D. Brockhaus, M. B. Walter, and G. Whelan, 1989, *Multimedia Environmental Pollutant Assessment System (MEPAS) Application Guidance*, Vols 1 and 2, Pacific Northwest Laboratories, USA

EEC, 1993, "Technical Guidance Document in Support of the Risk Assessment Directive (93/67/EEC) for New Substances Notified in Accordance with the Requirements of Council Directive 67/548/EEC," Brussels, Belgium

ECETOC, 1993, *Environmental Hazard Assessment of Substances*, Technical Report No. 51, January 1993, European Centre for Ecotoxicology and Toxicology of Chemicals, Brussels, Belgium

Guinee, J., and R. Heijungs, 1993, "A Proposal for the Classification of Toxic Substances Within the Framework of Life Cycle Assessment of Products," *Chemosphere*, *26*, 1925–1944

Mackay, D., 1979, "Finding Fugacity Feasible," *Environ. Sci. Technol.*, *13*, 1218–1223

Mackay, D., 1991, *Multimedia Environmental Fate Models: The Fugacity Approach*, Lewis Publications, Chelsea, MI, USA

Mackay, D., and S. Paterson, 1991, "Evaluating the Multimedia Fate of Organic Chemicals: A Level III Fugacity Model," *Environ. Sci. Technol.*, *25*, 427–436

Mackay, D., W. Y. Shiu, K. C. Ma, 1991, *Illustrated Handbook of Physical Chemical Properties for Organic Chemicals*, Vol. I, Lewis Publications, Boca Raton, FL, USA

Mackay, D., W. Y. Shiu, and K. C. Ma, 1992, *Illustrated Handbook of Physical-Chemical Properties and Environmental Fate for Organic Chemicals*, Vol. II, CRC Press, Boca Raton, FL, USA

Mackay, D., 1995, "Fate Modeling", *Fundamentals of Aquatic Toxicology*, Chapter 17 (G. M. Rand, Ed.), Taylor and Francis, New York, USA

Mackay D., 1994, "Fate Models" in *Handbook of Ecotoxicology* (P. Calow, Ed.), Vol. 2, Blackwell Scientific Publications, Oxford, UK, pp. 348–367

Mackay, D., S. Sang, P. Vlahos, M. Diamond, F. Gobas, and D. Dolan, 1994, "A Rate Constant Model of Chemical Dynamics in a Lake Ecosystem: PCBs in Lake Ontario," *J. Great Lakes Res.*, *20* (4), 625–642

Mackay, D., and F. Wania, 1995, "Transport of Contaminants to the Arctic: Partitioning, Processes and Models," *Sci. Total Environ.*, *160/161*, 25–38

Mackay, D., W. Y. Shiu, and K. C. Ma, 1995, *Illustrated Handbook of Physical–Chemical Properties and Environmental Fate for Organic Chemicals*, Vol. IV, CRC Press, Boca Raton, FL, USA

Mackay, D., S. Paterson, D. D. Tam, A. Di Guardo, and D. Kane, 1995a, *ChemCAN: A Regional Level III Fugacity Model for Assessing Chemical Fate in Canada*, Report to Health Canada, Ottawa, Canada

McKone, T. E., 1993, *CalTOX, A Multimedia Total-Exposure Model for Hazardous Wastes Sites Part II: The Dynamic Multimedia Transport and Transformation Model*, Report prepared for the State of California, Department Toxic Substances Control by the Lawrence Livermore National Laboratory No. UCRL-CR-111456PtII, Livermore, CA, USA

Oreskes, N., K. Shrader-Frechette, and K. Belitz, 1994, "Verification, Validation and Confirmation of Numerical Models in the Earth Sciences," *Science*, *263*, 641–646

van de Meent, D., 1993, *SimpleBOX: a Generic Multi-Media Fate Evaluation Model*, RIVM Report No. 6727200001, Bilthoven, The Netherlands

van de Meent, D., and J. H. M. de Bruijn, 1995, "A Modeling Procedure to Evaluate the Coherence of Independent Derived Environmental Quality Objectives for Air, Water and Soil," *Environ. Toxicol. Chem.*, *14*, 177

Wania, F., and D. Mackay, 1993, "Modelling the Global Distribution of Toxaphene: A Discussion of Feasibility and Desirability," *Chemosphere*, *27*, 2079–2094

Wania, F. and D. Mackay, 1995, "A Global Distribution Model for Persistent Organic Chemicals," *Sci. Total Environ.*, *160/161*, 211–232

9

Abiotic Transformation Reactions

Eric J. Weber (*Athens, USA*)

9.1 SUMMARY

This work provides an introduction to the major abiotic transformation pathways of organic chemicals in aquatic ecosystems. The types of functional groups that are susceptible to abiotic hydrolysis and redox reactions and reaction mechanisms for their transformation are presented. The reaction mechanisms are used as a framework for discussing factors that affect reaction rates and product distributions. Although our understanding of hydrolysis pathways are quite advanced in comparison to other transformation processes, prediction of hydrolysis rates still requires the extrapolation of kinetics data that have been measured in the laboratory to transformations that occur in natural aquatic ecosystems. Our limited understanding of reaction mechanisms for redox reactions is currently a barrier to the prediction of absolute reduction rates, and of the manner in which reaction rates will vary from one environmental system to another.

Ecotoxicology, Edited by Gerrit Schüürmann and Bernd Markert.
ISBN 0-471-17644-3   © 1998 John Wiley & Sons, Inc. and Spektrum Akademischer Verlag.

9.2 INTRODUCTION

The purpose of this chapter is to provide the reader with a framework for assessing the environmental fate of organic chemicals in aquatic ecosystems with respect to abiotic transformation reactions. In making such an assessment, the primary questions that are usually addressed include (1) will a given transformation reaction occur, (2) what will be the timescale of that reaction process, and (3) what reaction products will result from the transformation process? Once it has been determined that the chemical of interest will be degraded, the most difficult task becomes putting a time frame on the transformation process. Will the transformation process occur on a timescale of hours, days, weeks, months, or years? A transformation process occurring with a half-life on the order of months can be a significant process for a groundwater contaminant that may have a residence time of many years. In contrast, a long half-life for hydrolysis for a chemical that is readily removed from a water body by a physical process such as volatilization will not be significant.

The major reaction pathways that are discussed here include hydrolysis, and oxidative and reductive transformations. Reaction mechanisms for these processes are discussed in some detail. An aspect that each of these transformation processes have in common is that they involve reaction of the organic chemical of interest with a chemical species; however, the chemical species may be the direct result of a microbial reaction. As a result, the distinction between abiotic and biological transformation pathways often becomes blurred. For example, the reduction of nitroaromatics by iron-bearing minerals has been observed in laboratory studies (Heijman et al. 1995). The overall rate of reduction, however, was found to be dependent on microbial reduction of Fe(III) to Fe(II).

The first step in determining whether a chemical will degrade is to identify whether the chemical contains a functional group that is susceptible to abiotic transformation reactions. A functional group is defined as an atom or group of atoms that defines the structure of a particular family of organic compounds and, at the same time, determines their physical and chemical properties. Functional groups provide a "handle" or reactive site where transformations can occur. The reactivity of functional groups arises from their electron-rich or electron-poor properties. As part of this discussion, we present the major functional groups that are susceptible to hydrolysis, oxidation, and reduction under environmental conditions. Once the functional groups have been identified, the more challenging task is to determine how substituent groups will affect the reactivity of a functional group. Neighboring groups may act to stabilize the functional group or activate it toward transformation.

9.3 HYDROLYSIS

Hydrolysis is defined as the reaction of water with a substrate resulting in the cleavage of a covalent bond and formation of a new covalent bond with oxygen at the reaction center (the atom that is bonded to the leaving group; Larson and Weber 1994). The major types of functional groups that are susceptible to hydrolysis in aquatic ecosystems are summarized in Figure 9.1. Even within one class of chemicals, hydrolysis half-lives may vary by several orders of magnitude. Although our understanding of hydrolysis pathways are quite advanced in comparison with other transformation processes, prediction of hydrolysis rates still requires the extrapolation of kinetics data that have been measured in the laboratory to transformations that occur in natural aquatic ecosystems. A unique aspect concerning the prediction of hydrolysis rates of organic chemicals is that, in the majority of cases, we can ignore the environmental system of interest except for one parameter—hydrogen ion activity (pH). In the following discussion we examine how pH affects hydrolysis rates.

9.3.1 Hydrolysis Kinetics

The rate term for hydrolysis is described by:

$$d[RX]/dt = k_{hyd}[RX] = k_a[H^+][RX] + k_n[RX] + k_b[OH^-][RX] \qquad (9.1)$$

where RX is the concentration of the hydrolyzable compound; k_{hyd} is the observed or measured hydrolysis rate constant; and k_a, k_n, and k_b are the rate constants for the acid-catalyzed, neutral, and base-catalyzed processes, respectively. Assuming that the individual rate processes for the acid, neutral, and base hydrolyses obey first-order kinetics with respect to the hydrolyzable chemical, RX, we can write the following equation for k_{hyd}:

$$k_{hyd} = k_a[H^+] + k_n + k_b[OH^-] \qquad (9.2)$$

From the equilibrium term for the ionization of water, K_w:

$$K_w = [OH^-][H^+] = 1 \times 10^{-14} \qquad (9.3)$$

it is possible to substitute for $[OH^-]$ in Equation 9.2 giving:

$$k_{hyd} = k_a[H^+] + k_n + k_b(K_w/[H^+]) \qquad (9.4)$$

Because k_{hyd} is a pseudo-first-order rate constant at a fixed pH, the half-life for hydrolysis can be calculated from k_{hyd}:

$$t_{1/2} = \ln 2/k_{hyd} \qquad (9.5)$$

It is apparent from Equation 9.4 that the overall rate constant for hydrolysis, k_{hyd}, depends on pH and the magnitude of the rate constants for the individual

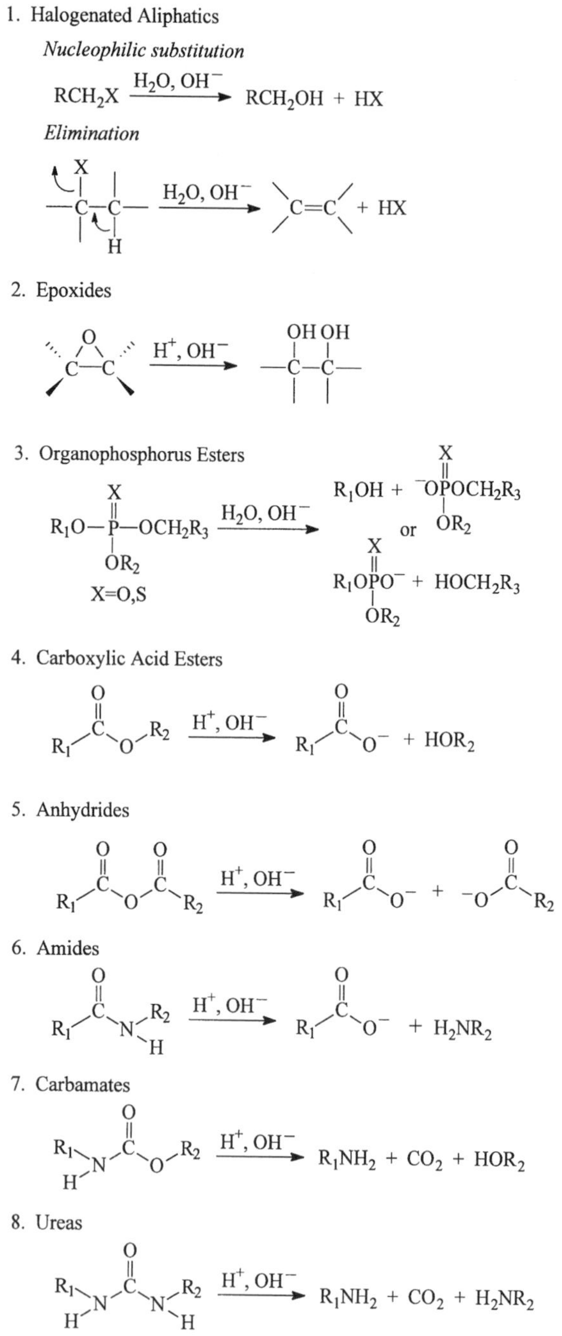

1. Halogenated Aliphatics

   Nucleophilic substitution

   $$RCH_2X \xrightarrow{\text{H}_2\text{O, OH}^-} RCH_2OH + HX$$

   Elimination

   $$-\overset{|}{\underset{|}{C}}-\overset{X}{\underset{H}{\overset{|}{C}}}- \xrightarrow{\text{H}_2\text{O, OH}^-} \diagdown C=C\diagup + HX$$

2. Epoxides

   $$\overset{O}{\underset{C-C}{\diagup\diagdown}} \xrightarrow{\text{H}^+,\,\text{OH}^-} \overset{\text{OH OH}}{-\underset{|}{\overset{|}{C}}-\underset{|}{\overset{|}{C}}-}$$

3. Organophosphorus Esters

   $$R_1O-\overset{X}{\underset{OR_2}{\overset{\|}{P}}}-OCH_2R_3 \xrightarrow{\text{H}_2\text{O, OH}^-} \begin{array}{l} R_1OH + \ ^-O\overset{X}{\overset{\|}{P}}OCH_2R_3 \\ \text{or}\quad OR_2 \\ R_1O\overset{X}{\overset{\|}{P}}O^- + HOCH_2R_3 \\ \quad OR_2 \end{array}$$

   $X=O,S$

4. Carboxylic Acid Esters

   $$R_1-\overset{O}{\overset{\|}{C}}-O-R_2 \xrightarrow{\text{H}^+,\,\text{OH}^-} R_1-\overset{O}{\overset{\|}{C}}-O^- + HOR_2$$

5. Anhydrides

   $$R_1-\overset{O}{\overset{\|}{C}}-O-\overset{O}{\overset{\|}{C}}-R_2 \xrightarrow{\text{H}^+,\,\text{OH}^-} R_1-\overset{O}{\overset{\|}{C}}-O^- + \ ^-O-\overset{O}{\overset{\|}{C}}-R_2$$

6. Amides

   $$R_1-\overset{O}{\overset{\|}{C}}-\underset{H}{\overset{}{N}}-R_2 \xrightarrow{\text{H}^+,\,\text{OH}^-} R_1-\overset{O}{\overset{\|}{C}}-O^- + H_2NR_2$$

7. Carbamates

   $$\underset{H}{\overset{R_1}{N}}-\overset{O}{\overset{\|}{C}}-O-R_2 \xrightarrow{\text{H}^+,\,\text{OH}^-} R_1NH_2 + CO_2 + HOR_2$$

8. Ureas

   $$\underset{H}{\overset{R_1}{N}}-\overset{O}{\overset{\|}{C}}-\underset{H}{\overset{R_2}{N}} \xrightarrow{\text{H}^+,\,\text{OH}^-} R_1NH_2 + CO_2 + H_2NR_2$$

Figure 9.1 Examples of hydrolyzable functional groups.

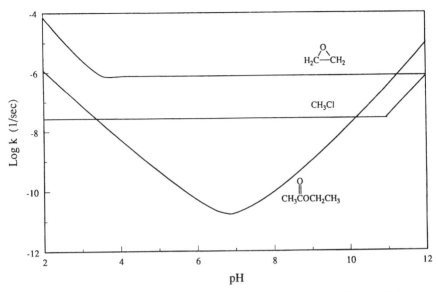

Figure 9.2 pH-rate profiles for the hydrolysis of ethylene oxide, methyl chloride, and ethyl acetate.

processes. Determining the pH dependence for the hydrolysis kinetics of a particular chemical requires measuring hydrolysis rate constants over a wide range of pH. Plots of $\log k_{hyd}$ versus pH are very useful for determining the contribution of the acid, neutral, and base terms for hydrolysis of the compound of interest at a specific pH. Figure 9.2 illustrates the dependence of $\log k_{hyd}$ on pH for several environmental chemicals of interest. These data demonstrate that the relationship between hydrolysis kinetics and pH is dependent on the nature of the hydrolyzable functional group. It is apparent from these data that only in the case of ethyl acetate will acid- and base-catalyzed hydrolysis contribute to hydrolysis at pH 7. For many chemicals, however, acid- and base-catalyzed hydrolysis may be the dominant process at pH 7.

9.3.2 Mechanisms of Hydrolysis

Hydrolysis occurs through two basic mechanisms; nucleophilic addition–elimination and nucleophilic substitution. By examining these mechanisms in some detail, it is possible to gain an understanding of how the relationship between hydrolysis kinetics and pH is dependent on the nature of the hydrolyzable functional group. The addition–elimination mechanism for hydrolysis is a two-step process involving nucleophilic addition at the acyl group to give a tetrahedral intermediate and elimination of the leaving group as shown below.

$$\underset{\text{O}}{\overset{\text{O}}{\underset{\|}{R-C-X}}} + Y: \rightleftarrows R-\underset{\underset{Y}{|}}{\overset{\overset{O^-}{|}}{C}}-X$$

$$R-\underset{\underset{Y}{|}}{\overset{\overset{O^-}{|}}{C}}-X \;\rightleftharpoons\; R-\overset{\overset{O}{\|}}{C}-Y \;+\; X:$$

The classes of chemicals that hydrolyze through the addition–elimination mechanism are those containing acyl groups [RC(O)] such as carboxylic acid derivatives (e.g., lactones, anhydrides, and amides) and carbonic acid derivatives (e.g., carbonates, carbamates, and ureas). Acid-catalyzed hydrolysis occurs because protonation of the carbonyl oxygen makes the carbon atom more susceptible to nucleophilic attack by H_2O by increasing the partial positive charge on carbon. The mechanism for the acid-catalyzed hydrolysis of a carboxylic acid ester is shown below. A tetrahedral intermediate is formed that may revert back to starting materials or eliminate the alkoxy group to give the hydrolysis products.

$$R_1-\overset{\overset{O}{\|}}{C}-OR_2 \;+\; H^+ \;\rightleftharpoons\; R_1-\overset{\overset{\oplus OH}{\|}}{C}-OR_2$$

$$R_1-\overset{\overset{\oplus OH}{\|}}{C}-OR_2 \;+\; H_2O \;\rightleftharpoons\; R_1-\underset{\underset{\oplus OH_2}{|}}{\overset{\overset{OH}{|}}{C}}-OR_2$$

$$R_1-\underset{\underset{\oplus OH_2}{|}}{\overset{\overset{OH}{|}}{C}}-OR_2 \;\overset{-H^+}{\rightleftharpoons}\; R_1-\underset{\underset{HO\;\;H}{|}}{\overset{\overset{OH}{|}}{C}}-\overset{\oplus}{O}R_2$$

$$R_1-\underset{\underset{\overset{|}{O}\diagdown H}{\overset{|}{\oplus}}}{\overset{\overset{HO\;\;H}{|}}{C}}-OR_2 \;\rightleftharpoons\; R_1-\overset{\overset{O}{\|}}{C}-OH \;+\; R_2OH \;+\; H^+$$

Base catalysis occurs because OH^- is a much stronger nucleophile than H_2O (approximately by a factor of 10^3). The base-catalyzed mechanism for the hydrolysis of a carboxylic acid ester is shown below.

$$R_1-\overset{\overset{O}{\|}}{C}-OR_2 \;+\; OH^- \;\rightleftharpoons\; R_1-\underset{\underset{OH}{|}}{\overset{\overset{O^-}{|}}{C}}-OR_2$$

$$R_1-\underset{\underset{OH}{|}}{\overset{\overset{O^-}{|}}{C}}-OR_2 \;\rightleftharpoons\; R_1-\overset{\overset{O}{\|}}{C}-OH \;+\; R_2O^-$$

$$R_1-\overset{\overset{O}{\|}}{C}-OH \;+\; R_2O^- \;\rightleftharpoons\; R_1-\overset{\overset{O}{\|}}{C}-O^- \;+\; R_2OH$$

The reaction mechanisms for nucleophilic substitution reactions are typically described by the limiting cases of the ionization mechanism (S_N1, substitution nucleophilic unimolecular) and the direct displacement mechanism (S_N2, substitution nucleophilic bimolecular) (Gleave et al. 1935). The S_N1 and S_N2 mechanisms describe the extremes in nucleophilic substitution reactions. Usually a combination of these reaction mechanisms occur simultaneously. The S_N1 mechanism proceeds by a rate-determining heterolytic dissociation of the substrate to an sp^2 hybridized carbocation, R^+ (commonly referred to as a carbonium ion), and the leaving group X^-:

$$RX \rightarrow [R^{\delta^+} \text{---} X^{\delta^-}] \rightarrow R^+ + X^- \qquad (9.6)$$

where $[R^{\delta^+} \text{---} X^{\delta^-}]$ represents a transition state, an energy maximum in the dissociation of RX to R^+ and X^- ions. The S_N2 mechanism proceeds through a transition state in which bond breaking and bond making occur simultaneously. In general, hydrolysis reactions occurring through S_N1 mechanisms are more sensitive to electronic effects, whereas those occurring through S_N2 mechanisms are more sensitive to steric effects.

Classes of chemicals that hydrolyze through direct nucleophilic substitution include halogenated aliphatics, epoxides, and organophosphorous compounds. Each of these classes are subject to base catalysis due to the fact that OH^- is a much stronger nucleophile than H_2O. However epoxides and organophosphorous compounds are susceptible to acid-catalyzed hydrolysis, but halogenated aliphatics are not (see the example of methyl chloride in Figure 9.2). Halogenated aliphatics are simply not strong enough bases for protonation to occur. As with chemicals containing acyl groups, epoxides can be protonated at the oxygen atom under acidic conditions. The oxygen–carbon bond is weakened and nucleophilic attack of H_2O occurs at the least sterically hindered carbon.

An obvious consequence of these mechanisms is that any substituent of a molecule that blocks the approach of the attacking nucleophile or decreases the partial positive charge at the reactive center will decrease hydrolysis rates. In contrast, substituents that are electron withdrawing in nature and increase the partial positive charge at the reaction center will enhance nucleophilic attack

Table 9.1 Hydrolysis of aliphatic esters at pH 7 and 25°C

R_1	R_2	$k_a[H^+](s^{-1})$	$k_n(s^{-1})$	$k_b[OH^-](s^{-1})$	$k_{hyd}(s^{-1})$	$t_{1/2}$
Me	Et	1.1×10^{-11}	1.5×10^{-10}	1.1×10^{-8}	1.1×10^{-8}	2.0 y
$ClCH_2$	Me	8.5×10^{-12}	2.1×10^{-7}	1.4×10^{-5}	1.4×10^{-5}	14 h
Cl_2CH	Me	2.3×10^{-11}	1.5×10^{-5}	2.8×10^{-4}	3.0×10^{-4}	38 m
Cl_3C	Me	n.a.	$\geq 7.7 \times 10^{-4}$	n.a.	$\geq 7.7 \times 10^{-4}$	≤ 15 m
Me	i-Pr	6.0×10^{-12}	n.a.	2.6×10^{-9}	2.6×10^{-9}	8.4 y
Me	t-Bu	1.3×10^{-11}	n.a.	1.5×10^{-10}	1.6×10^{-10}	140 y

Source: Data taken from Mabey and Mill (1978).
n.a., Data not available; y, years; m, months.

and increase hydrolysis rates. These effects of substituent groups on hydrolysis rates for a series of aliphatic esters are illustrated in Table 9.1. The addition of electron-withdrawing chloride substituents to the R_1 group drastically increases the neutral and base-catalyzed hydrolysis rate constants, whereas increasing the steric bulk of R_2 (t-Bu $>$ i-Pr $>$ Et) significantly decreases the base-catalyzed hydrolysis rate constant.

9.4 REDOX REACTIONS

The oxidation–reduction or redox reaction is a process involving the transfer of electrons from one reactant to another. Oxidation involves the loss of electrons and reduction, the gain of electrons. The tendency of substances to gain or lose electrons varies greatly. Oxidizing agents possess a strong affinity for electrons and cause other substances to be oxidized by abstracting electrons from them. In the process, the oxidizing agent accepts electrons and is thereby reduced. Reducing agents readily give up electrons and thereby cause some other species to be reduced. A consequence of this electron transfer is the oxidation of the reducing agent. Separation of a redox reaction into its component parts (i.e., into half-reactions) is a convenient way of indicating clearly the species that gains electrons and the one that loses them. For example, the overall reaction for the reduction of nitrobenzene by ferrous iron is obtained by combining the half-reaction for the oxidation of iron(II)

$$6Fe^{2+} \rightleftharpoons 6Fe^{3+} + 6e^- \tag{9.7}$$

with the half-reaction the reduction of nitrobenzene

$$PhNO_2 + 6e^- + 6H^+ \rightleftharpoons PhNH_2 + 2H_2O \tag{9.8}$$

to give

$$PhNO_2 + 6Fe^{2+} + 6H^+ \rightleftharpoons PhNH_2 + 6Fe^{3+} + 2H_2O \qquad (9.9)$$

It is clear from this analysis that the reduction of a nitro group to an amine requires the transfer of six electrons. It will become apparent from the following discussion that much of the difficulty in predicting reactions rates for redox reactions in environmental systems is the difficulty in identifying the half-reaction for naturally occurring oxidants and reductants. This is especially true for the reductive transformation of organic pollutants.

9.4.1 Oxidation

Although oxidation is the dominant removal process for most organic chemicals in the atmosphere, oxidation in aquatic ecosystems is important for a relatively small class of organic chemicals. Almost all oxidative transformations are kinetically second-order reactions in which the reaction rate is proportional to the concentrations of both the oxidizing agent, Ox, and the substrate of interest, P:

$$- dA/dt = k_{ox}[Ox][P] \qquad (9.10)$$

Oxidants in aquatic ecosystems include ground-state species such as oxygen (O_2), metal ions, and mineral oxides, activated oxygen species such as singlet oxygen ($^1O_2$), and radical species including hydroxyl radicals (OH•) and organic peroxyl radicals (ROO•).

Ground-State Oxidants

Oxygen is the most abundant oxidizing agent in the water column; however, oxidation of organic chemicals by O_2 alone is generally quite slow. This is attributed to the fact that O_2 is fundamentally a diradical species; it has a triplet configuration with one electron localized on each of the oxygen atoms. A quantum mechanical spin barrier exists for the reaction of O_2 with organic chemicals that do not have functional groups with unpaired electrons. Consequently, only readily oxidizable compounds such as phenols, anilines, and mercaptans that readily form free radicals react with O_2 at rates fast enough to be significant.

Probably of greater environmental significance in aquatic ecosystems is oxidation catalyzed by metal ions and mineral oxides. For example, recent studies suggest that oxidation of phenols in aquatic systems contaminated with hexavalent chromium (Cr(VI)) is an important reaction process. Kinetic studies have demonstrated that chromium-mediated oxidation of phenols can occur with a half-life ranging from minutes to months (Elovitz and Fish 1994). The accepted mechanism for the oxidation of phenols by Cr(VI) involves the rapid initial formation of a chromate-phenol ester as verified by spectroscopy (Elovitz

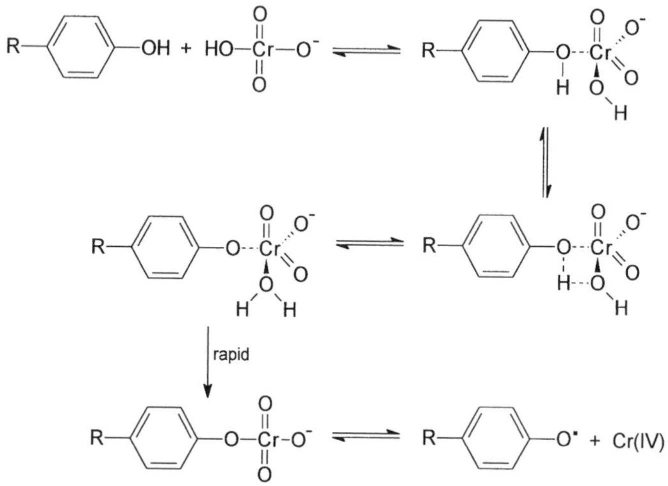

Figure 9.3 Reaction mechanism for the oxidation of phenols by Cr(VI).

and Fish 1995; Figure 9.3). The subsequent rate-limiting ester decomposition proceeds via innersphere electron transfer. At pH values above 5, unimolecular decomposition of the ester occurs, possibly via a homolytic cleavage of the chromium–oxygen bond, leading to the formation of Cr(V) and phenoxy radical products. The resulting phenoxy radicals undergo further oxidation, ortho coupling and polymerization. For example, oxidation of 2,4-dimethylphenol by Cr(VI) results in the formation of a quinone and radical coupling products as illustrated here:

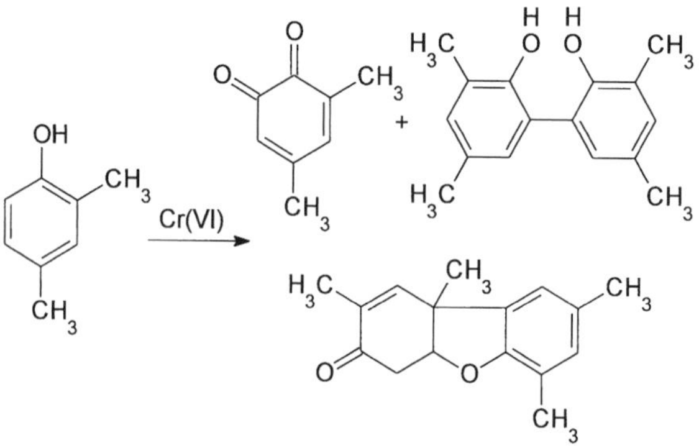

Mineral oxides containing iron(III), manganese(III), manganese(IV), and other transition metals have been demonstrated in laboratory studies to be potential oxidizing agents for organic pollutants in aquatic ecosystems. These processes are probably most important in groundwater environments. Because the solubilities of these higher valent states of iron and manganese are extremely low, oxidation of the organic chemical must take place at the mineral/water interface. Consequently, the overall rates for oxidation depend upon rates and extent of adsorption, as well as rates of electron transfer and subsequent reactions. Classes of organic chemicals that are known to be susceptible to oxidation by mineral surfaces include phenols and aromatic amines. Rates of oxidation generally decrease as Hammett constants of ring substituents become more positive, reflecting trends in basicity, nucleophilicity, and half-wave potentials of substituted phenols and anilines (Stone 1987).

A reaction mechanism for the oxidation of chlorophenols by manganese(III/IV) oxide surfaces is presented below (Ulrich and Stone 1989).

(1) Surface complex formation:

$$> Mn^{3+}\text{-}OH + ArOH \rightleftharpoons Mn^{3+}\text{-}OAR + H_2O$$

(2) Electron transfer:

$$> Mn^{3+}\text{-}OAr \rightleftharpoons (> Mn^{3+}, \bullet OAR)$$

(3) Release of phenoxy radical:

$$(> Mn^{3+}, \bullet OAR) + H_2O \rightleftharpoons Mn^{2+}\text{-}OH_2 + \bullet OAR$$

(4) Release of reduced Mn(II):

$$Mn^{2+}\text{-}OH_2 \rightleftharpoons Mn^{2+}(+ \text{free underlying site})$$

(5) Coupling and further oxidation:

$$ARO\bullet \rightarrow \text{quinones, dimers, and polymeric oxidation products.}$$

Singlet Oxygen

Singlet oxygen is the first electronically excited state of molecular oxygen. It is formed from sensitizers that absorb light and transfer the energy to dissolved, ground-state triplet oxygen. The primary sensitizer for the formation of $^1O_2$ in natural aquatic ecosystems is dissolved organic matter (DOM). It has been estimated that the steady-state concentration of $^1O_2$ at the surface of a eutropic freshwater body under noontime, midsummer conditions is approximately 2×10^{-13} M and the average concentration of $^1O_2$ for the first meter of such a water column is estimated to be 4×10^{-14} M (Haag and Hoigne 1986). Because $^1O_2$ is a moderately reactive electrophile it can play a significant role, even at these low concentrations, in the environmental fate of a variety of electron-rich organic compounds including alkenes that are substituted with

$$\text{Initiation} \begin{cases} A{-}B \longrightarrow A\cdot + B\cdot \\ A\cdot + RH \longrightarrow R\cdot + AH \end{cases}$$

$$\text{Propagation} \begin{cases} R\cdot + O_2 \longrightarrow ROO\cdot \\ ROO\cdot + RH \longrightarrow ROOH + R\cdot \end{cases}$$

$$\text{Termination} \begin{cases} 2\,ROO\cdot \longrightarrow ROO{-}OOR \\ ROO\cdot + R\cdot \longrightarrow ROOR \\ 2\,R\cdot \longrightarrow R{-}R \end{cases}$$

Figure 9.4 Reaction mechanism for radical reactions.

electron-donating groups, phenols, anilines, and mercaptans. Dienes such as furfuryl alcohol (FFA) are frequently used as trapping agents for $^1O_2$ determination. Reaction of $^1O_2$ with FFA occurs by a Diels-Alder reaction as shown below (Haag et al. 1984).

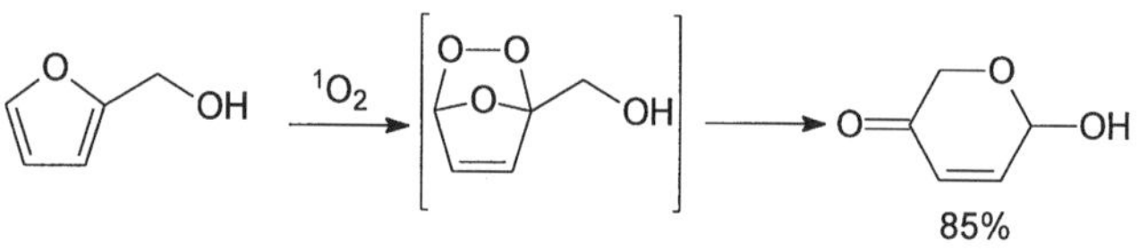

Radical Oxidants

Free radical oxidation occurs through a series of reactions referred to as an initiation step, propagation, and a subsequent termination step as illustrated in Figure 9.4.

The propagation reactions occur by 4 basic processes (Hutzinger 1980), which are outlined as follows:

(1) H atom transfer

R = Alkyl or H

(2) Addition to double bonds

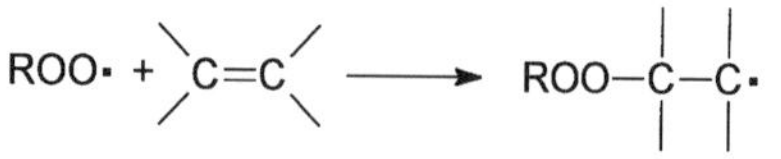

R = Alkyl or H

(3) HO• addition to aromatics

$$HO\cdot + \text{(benzene)} \longrightarrow \text{(cyclohexadienyl radical with HO, H)}$$

(4) $RO_2\cdot$ transfer of O atoms to nucleophilic species

$$ROO\cdot + NO \longrightarrow RO\cdot + NO_2^-$$

The reactivities of peroxyl radicals (ROO•) and the hydroxyl radical (HO•) range over 12 orders of magnitude. These photoactivated species exhibit very short lives (in the order of picoseconds to milliseconds) and low steady-state concentrations (approximately 10^{-18} M for HO• and 10^{-10} M for ROO•) due to rapid quenching by water, DOM, and other trace constituents. Because the concentration of HO• in natural waters is so low, HO• is of negligible importance compared with ROO•. In the atmosphere the opposite is true; HO• is the only radical oxidant of importance. The hydroxyl radical is an extremely potent and nonselective oxidizing agent. Although it reacts with many organic chemicals with a second-order rate constant of approximately 6×10^9 $M^{-1}s^{-1}$, at the very low steady-state concentration of HO• typically found in natural waters, a pseudo-first-order rate constant of 1.8×10^{-8} s^{-1} is obtained, which corresponds to a half-life of 446 days (Schwarzenbach et al. 1993). In nitrate-rich waters, half-lives for organic chemicals may be decreased by a factor of 100 because of HO• production from the photolysis of nitrate (Zepp et al. 1987).

Formation of peroxyl radicals occurs from the reaction of photochemically excited chromophores of DOM with ground-state oxygen. At typical concentrations of DOM found in natural waters (5 mg C L^{-1}), peroxyl radical species are not efficiently scavenged by DOM; thus, they are available for reaction with easily oxidizable organic chemicals including phenols, mercaptans, and anilines. The dominant reaction of ROO• with phenols is abstraction of the phenolic hydrogen as shown below for 4-methylphenol. As expected, electron-donating groups (e.g., alkyl groups) will accelerate this process. For example, the estimated half-life for phenol for reaction with ROO• in lake water at 47°N latitude is 200 days (Faust and Hoigne 1987). For 4-methylphenol, the estimated half-life was shortened to 16 days.

$$H_3C-\text{(phenyl)}-OH + ROO\cdot \longrightarrow H_3C-\text{(phenyl)}-O\cdot + ROOH$$

9.4.2 Reduction

Of the abiotic transformation processes discussed in this chapter, reductive transformations are probably the most poorly understood. This cannot be attributed to a lack of effort on the behalf of environmental scientists and engineers, as reductive transformations has been an intense area of research in recent years. Much of the impetus for this work results from the fact that unlike other transformation process such as hydrolysis, reductive transformations often result in the formation of reaction products that are of more concern with respect to their ecological impact than the parent compound (Larson and Weber 1994). Furthermore, organic chemicals that were previously thought to be stable in aquatic ecosystems because they did not contain hydrolyzable functional groups (e.g., nitroaromatics and aromatic azo compounds) have been shown in laboratory studies to undergo facile reduction in anoxic systems (Macalady et al. 1986). Figure 9.5 summarizes the functional groups that have been identified as being susceptible to reduction in anoxic environments.

Of the classes of chemicals listed in Figure 9.5, the nitroaromatics and the aromatic azo compounds are probably the most thoroughly investigated. This can be attributed to their common occurrence in agrochemicals, munitions, dyestuffs, and other important synthetic chemicals. In reducing environments, such as anoxic sediments, the reduction of nitroaromatics can occur with half-lives in the order of minutes to hours. Fate models that do not take into account the potential for the reductive transformation of these types of chemicals can greatly overestimate their concentrations in aquatic ecosystems. Furthermore, and perhaps more importantly, the potential for the formation of aromatic amines, which are generally considered a hazardous class of chemicals, would be ignored.

Disperse Blue 79, which is one of the largest volume dyes currently on the market, is an example of an environmentally significant chemical that has been shown in laboratory studies to undergo both aromatic nitro and azo reduction in anoxic sediment water systems (Weber and Adams 1995). The results of kinetic and product studies of Disperse Blue 79 are consistent with a reaction pathway in which reductive cleavage of the azo linkage of Disperse Blue 79 occurs initially, resulting in the formation of 2-bromo-4,6-dinitroaniline (BDNA) and the *N,N*-disubstituted 1,4-diaminobenzene (Figure 9.6). Subsequently, facile reduction of the 2-nitro group of BDNA occurs to give the 1,2-diaminobenzene, which in turn is slowly reduced to the triaminobenzene. The *N,N*-disubstituted 1,4-diaminobenzene undergoes a cyclization reaction through nucleophilic attack of the unsubstituted amino group on the adjacent amide group to give the benzimidazole.

Reduction Kinetics

Although our understanding of reductive transformations in the environment has progressed to the point that we can identify the types of functional groups that will be susceptible to reduction, our limited understanding of reaction mechanisms for such processes is currently a barrier to the prediction of absolute reduction rates and of the manner in which reaction rates will vary from one environmental system to another. The obvious question remains

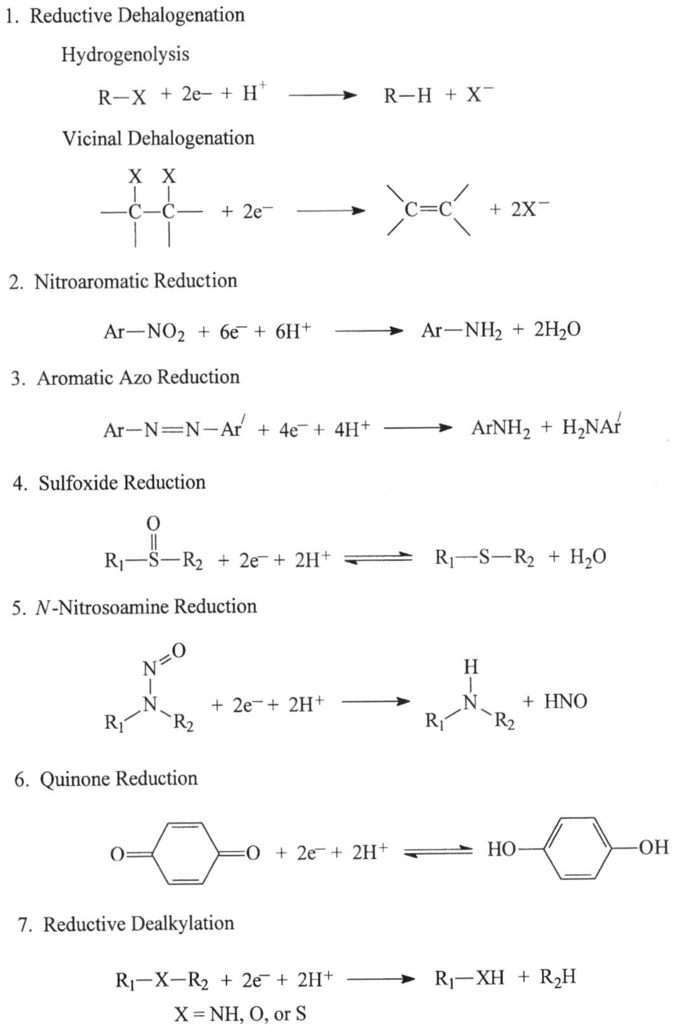

1. Reductive Dehalogenation

Hydrogenolysis

$$R-X + 2e^- + H^+ \longrightarrow R-H + X^-$$

Vicinal Dehalogenation

$$\underset{\substack{| \\ }}{\overset{\substack{X \\ |}}{-C}}-\underset{\substack{| \\ }}{\overset{\substack{X \\ |}}{C}}- + 2e^- \longrightarrow \underset{/}{\overset{\backslash}{C}}=\underset{\backslash}{\overset{/}{C}} + 2X^-$$

2. Nitroaromatic Reduction

$$Ar-NO_2 + 6e^- + 6H^+ \longrightarrow Ar-NH_2 + 2H_2O$$

3. Aromatic Azo Reduction

$$Ar-N{=}N-Ar' + 4e^- + 4H^+ \longrightarrow ArNH_2 + H_2NAr'$$

4. Sulfoxide Reduction

$$R_1-\overset{\overset{\displaystyle O}{\|}}{S}-R_2 + 2e^- + 2H^+ \rightleftharpoons R_1-S-R_2 + H_2O$$

5. N-Nitrosoamine Reduction

$$\underset{R_1 \diagup \quad \diagdown R_2}{\overset{\overset{\displaystyle N^{\nearrow O}}{|}}{N}} + 2e^- + 2H^+ \longrightarrow \underset{R_1 \diagup \quad \diagdown R_2}{\overset{\overset{\displaystyle H}{|}}{N}} + HNO$$

6. Quinone Reduction

$$O{=}\!\!\left\langle\!\!\bigcirc\!\!\right\rangle\!\!{=}O + 2e^- + 2H^+ \rightleftharpoons HO{-}\!\!\left\langle\!\!\bigcirc\!\!\right\rangle\!\!{-}OH$$

7. Reductive Dealkylation

$$R_1-X-R_2 + 2e^- + 2H^+ \longrightarrow R_1-XH + R_2H$$
$$X = NH, O, \text{ or } S$$

Figure 9.5 Reductive transformations known to occur in natural reducing environments.

"what are the sources of electrons in natural reducing environments?" The cumulative knowledge in this area suggests that naturally occurring reductants are a complex array of species ranging from chemical or "abiotic" reagents such as iron-bearing minerals, sulfide, and natural organic matter, through extra-cellular biochemical reducing agents such as iron porphyrins, corrinoids, and redox active enzymes, to biological systems such as microbial populations (Ulrich and Stone 1989; Macalady et al. 1986; Tratnyek and Wolfe 1990). Accordingly, the rate term for the reductive transformation of organic pollutants

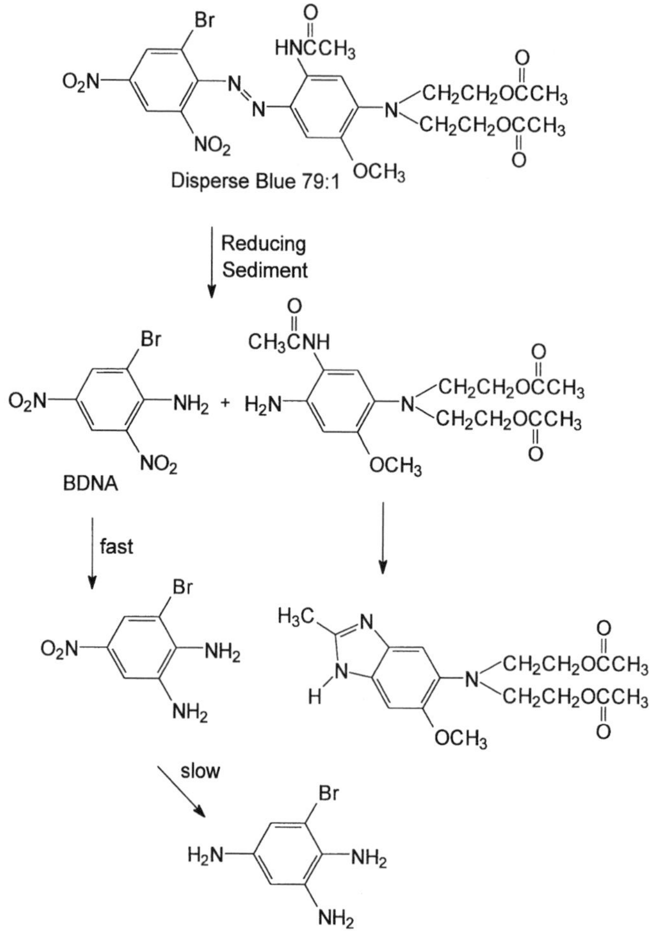

Figure 9.6 Reaction pathway for the reduction of Disperse Blue 79 in an anoxic sediment–water system.

in anoxic sediments and aquifer materials can be written as:

$$- d[P]/dt = k_{chem}[P][R_{chem}] + k_{biochem}[P][R_{biochem}] + k_{bio}[P][R_{bio}] \quad (9.11)$$

where P is pollutant; k_{chem}, $k_{biochem}$, and k_{bio} are the second-order rate constants for the chemical, biochemical, and biological reduction processes, respectively; and R_{chem}, $R_{biochem}$, and R_{bio} are chemical, biochemical, and biological reductants, respectively. Of course, each of the rate terms for chemical, biochemical, and biological reduction can be broken down further to represent the multitude of reductants that are likely to be present in natural systems.

Clearly, the contribution of each of the terms described for the overall rate constant for reduction will be dependent on the class of the organic chemical in question and the reducing system of interest. For example, a growing body of evidence suggests that the reduction of nitroaromatics and aromatic azos in anoxic sediments and aquifers occurs primarily by chemical or abiotic processes (Larson and Weber 1994; Weber and Adams 1995; Sanders and Wolfe 1985). These reductive transformations are characterized by the lack of observation of a lag phase and by insensitivity to the addition of biological inhibitors. The situation for halogenated organics is quite different, though some of the highly halogenated aliphatics, such as hexachloroethane, undergo rapid abiotic reduction in anoxic pond sediments (Jafvert and Wolfe 1987). In contrast, lower halogenated aliphatics, such as 1,2-dichloroethane and vinyl chloride, as well as the halogenated aromatics (i.e., polychlorinated biphenyls (PCBs) chlorinated phenols, anilines, etc.), require viable microbial communities for reductive dechlorination to occur (Vogel et al. 1987). A scenario emerges that indicates that electron transfer between these less reactive substrates and the array of chemical reductants present in anoxic systems is not thermodynamically feasible and that the enzyme systems of microorganisms are necessary to serve as catalysts in order for electron transfer to occur.

Naturally Occurring Chemical Reductants

Few investigations have focused on identifying and quantifying the important chemical reductants in anoxic sediments and aquifers. The list of chemical species that have been suggested as reductants in these systems includes iron- and sulfur-bearing minerals, sulfide, hydroquinones, reduced porphyrins, and natural organic matter (Larson and Weber 1994). Very recent work suggests that the iron-bearing minerals play a dominant role in the abiotic reduction of organic pollutants in anoxic systems. Iron-bearing minerals are known to be a dominant constituent of many soils, sediments, and aquifers (Tan 1982; Stumm 1992). From a thermodynamic point of view, it is readily apparent that surface-bound Fe(II) and structural Fe(II) are much stronger reductants than solution phase Fe(II) (Haderlein and Schwarzenbach 1995). The very low reduction potentials of Fe(II)/Fe(III) redox couples associated with mineral oxides are quite remarkable. For example, the couple Fe_2SiO_4 (fayalite)–Fe_3O_4 (magnetite) has a reduction potential similar to that for the reduction of water to hydrogen gas (Stumm 1992). The significant lowering of the reduction potential of the iron redox couple results from the formation of a surface–Fe(III) complex that is more stable than the surface–Fe(II) complex.

Only recently has the ability of iron-bearing minerals to reduce organic chemicals been demonstrated in the laboratory. For example, the reduction of substituted nitrobenzenes in mineral suspensions treated with Fe(II) has been observed by Klausen et al. (1995). These investigators concluded that the reductant in these systems was Fe(II) adsorbed to the surface of the iron-bearing mineral. Reduction of nitrobenzenes by either Fe(II) or magnetite separately was quite slow. The combination of Fe(II) and magnetite resulted in the facile reduction of nitrobenzenes. These results are in direct support of the experimental observations made by Heijmann et al. (1995) of the reduction of nitrobenzenes in laboratory columns containing aquifer material. It had been proposed that reduction occurred primarily by reaction with surface-bound iron species. The surface-bound Fe(II) species were thought to be a direct result of microbial-mediated reduction of surface-bound Fe(III) species. By measuring reduction rates for a series of monosubstituted nitrobenzenes, the investigators were able to conclude that the rate-determining step was not electron transfer from surface-bound Fe(II), but regeneration of the reactive Fe(II) sites by iron-reducing bacteria.

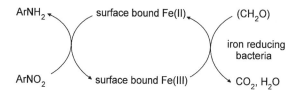

Iron-bearing mineral oxides have also been shown to effect the reductive dechlorination of halogenated aliphatics. Hexachloroethane and carbon tetrachloride were readily reduced in aqueous systems containing biotite, vermiculite, pyrite, or marcasite (Kriegman-King and Reinhard 1991; Kriegman-King and Reinhard 1992). The addition of sulfide to the biotite and vermiculite reaction systems was found to enhance rates of electron transfer significantly. It was proposed that sulfide may behave as a catalyst to regenerate Fe(II) sites on the mineral surfaces. This scenario is quite similar to that proposed by Klausen et al. (1995), except that the electron source for reduction of the iron oxide is supplied by a chemical rather than by microbes.

Role of Electron-Transfer Mediators

In the previous discussion concerning reductive transformations in mineral oxide systems, it was generally assumed that reduction of the organic substrate occurred at the surface of the mineral oxide. In more complex natural systems such as anoxic sediments and aquifers, naturally occurring chemical and biochemical species may be present that act as electron shuttles, moving electrons from the reducing surface of the mineral oxide to the organic pollutants in solution. A reaction mechanism involving electron-transfer mediators has been invoked to account for the extremely fast reaction kinetics observed for such compounds as methyl parathion in reducing sediments:

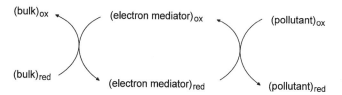

In this scenario, the bulk reductant rapidly reduces an electron carrier or mediator, which in turn transfers electrons to the pollutant of interest. The oxidized electron-transfer mediator is then rapidly reduced again by the bulk reductant, which enables the redox cycle to continue. Support for electron transfer by this type of mechanism derives from laboratory studies of "clean" systems in which the reduction kinetics for a chemical of interest is measured in the presence of a bulk reductant (typically sulfide) with and without the addition of an electron-transfer mediator. Generally, the reduction of nitroaromatics and halogenated organics by bulk reductants is very slow. The addition of electron-transfer mediators are often found to greatly accelerate reduction rates (Schwarzenbach et al. 1990).

Several chemical substances have been proposed as electron-transfer mediators, including natural organic matter (Dunnivant et al. 1992; Curtis and Reinhard 1994), iron porphyrins (Baxter 1990; Klecka and Gonsior 1984), corrinoids (Krone et al. 1989; Assaf-Anid et al. 1994), and bacterial transition-metal coenzymes such as vitamin B_{12} and hematin (Assaf-Anid and Hayes 1994; Gantzer and Wackett 1991; Chiu and Reinhard 1995). Extracellular enzymes have also been proposed to play significant roles as electron-transfer mediators in aquatic ecosystems (Price and Morel 1990).

Although numerous laboratory studies have established the plausibility of electron transfer by electron-transfer mediators, this concept has not been demonstrated in natural reducing systems. A process involving electron transfer by electron-transfer mediators provides a plausible explanation for an apparent dichotomy that has been well documented by several investigators in our laboratory (Sanders and Wholfe 1985; Weber and Wolfe 1987). Reduction rates for nitroaromatics, aromatic azos, and halogenated ethanes in anoxic pond sediments have been found to increase with sediment concentration, yet increased partitioning to the sediment inhibits reduction. These observations are consistent with a scenario in which the sediment provides a source of electron-transfer mediators. Of course, as sediment concentration increases so would the concentration of electron-transfer mediators, and hence, the overall rate of reduction would increase. Partitioning to the sediment would render the organic chemical unavailable for reaction with the electron-transfer mediators. In a similar manner, the partitioning of hydrophobic organic pollutants such as PAHs (polyaromatic hydrocarbons) and PCBs makes them less available for biological degradation (Harms and Zehnder 1995).

9.4.3 Quantitative Structure–Activity Relationships for Redox Reactions

Quantitative structure–activity relationship (QSAR) analysis can be a powerful tool for predicting reaction rates of unstudied chemicals and the development of mechanistic models. Although it is not possible to predict the absolute rates for the reduction of organic chemicals in environmental systems, recent laboratory studies have established relationships between substrates and relative reactivities (Tratnyek et al. 1991). The success of these studies has resulted from the use of one-electron reduction potentials, which is a measure of the substrate's "willingness" to accept electrons. Although the reductive transformation of organic functional groups is often described as occurring through the transfer of electron pairs (Figure 9.5), the accepted theory is that electrons are transferred one by one (Semenov 1958). The general mechanism for one-electron transfer, as described by Schwarzenbach et al. (1993), is:

$$P + R \rightleftharpoons (PR) \rightarrow [PR \leftrightarrow P^- R^{\cdot+}] \rightarrow P^{\cdot-} R^{\cdot+} \leftrightarrow P^{\cdot-} + R^{\cdot} \qquad (9.12)$$

reactants   precursor   transition   successor   products
             complex   state   complex

where P is an organic pollutant (electron acceptor) and R is a reductant (electron donor). The precursor complex, (PR), describes the electron coupling prior to the one-electron transfer, which occurs in the transition state. After electron transfer, a successor complex is proposed that subsequently dissociates to provide the radical ions. The concept of the one-electron transfer scheme provides a common feature to the seemingly unrelated reductive transformation pathways presented in Figure 9.5. Each of these processes occurs initially by transfer of a single electron in the rate-determining step. In each case a radical anion is formed that is more susceptible to reduction than the parent compound.

To date, QSAR analyses that incorporate the use of one-electron potentials have been limited to redox reactions in model systems (Tratnyek and Hoigne 1991). For example, excellent correlations have been observed between second-order rate constants for the reduction of substituted nitrobenzenes in sulfide/quinone model systems and their standard one-electron reduction potentials (Schwarzenbach and Stierli 1990). Based on this correlation, the investigators were able to conclude that the overall rate of reduction was limited only by the electron-transfer step. In contrast to the quinone-based model system, the correlation between reduction rates and one-electron reduction potentials suggested that the reaction rate was not being controlled by the electron-transfer step but by formation of the precursor complex. The very good correlations often found between rate constants for a redox reaction of closely related organic chemicals and their one-electron redox potentials suggests that such correlations may be observed in more complex natural systems.

9.5 CONCLUSIONS

The environmental fate of a large number of organic chemicals is controlled by abiotic transformation processes. With respect to hydrolytic reactions, our understanding of the relationship between hydrolysis kinetics and pH of most hydrolyzable functional groups is quite good. Hydrolysis kinetic data generated in the laboratory can often be successfully extrapolated to natural aquatic ecosystems. Of course, in any such extrapolation the possibility exists that components of the natural system, such as dissolved organic matter, metal ions, salts, and surfaces, will significantly affect reaction rates and product distributions. The challenge remains to identify those classes of chemicals whose hydrolysis rates will be altered and to quantify these effects. The task of developing predictive models for redox reactions is clearly more daunting. The complexity of environmental systems and the inability thus far to identify the reductants may provide insurmountable barriers to the development of predictive models. Our accumulated knowledge of redox reactions in natural systems suggests that natural surfaces (e.g., sediments, soils, mineral oxides) play crucial roles in these processes. Clearly, the ability to understand and thus predict reaction rates for redox reactions in environmental systems will in part be dependent on our ability to delineate the role that natural surfaces play in the overall process.

9.6 REFERENCES

Assaf-Anid, N., K. F. Hayes, T. M. Vogel, 1994, "Reductive Dechlorination of Carbon Tetrachloride by Cobalamin(II) in the Presence of Dithiothreitol: Mechanistic Study, Effect of Redox Potential and pH," *Environ. Sci. Technol., 28,* 246–252

Baxter, R. M., 1990, "Reductive Dechlorination of Certain Chlorinated Organic Compounds by Reduced Hematin Compared with their Behavior in the Environment," *Chemosphere, 21,* 451–458

Chiu, P.-C., and M. Reinhard, 1995, "Metallocoenzyme-Mediated Reductive Transformation of Carbon Tetrachloride in Titanium(III) Citrate Aqueous Solution," *Environ. Sci. Technol., 29,* 595–603

Curtis, G. P., and Reinhard, M., 1994, "Reductive Dehalogenation of Hexachloroethane, Carbon Tetrachloride, and Bromoform by Anthrahydroquinone Disulfonate and Humic Acid," *Environ. Sci. Technol., 28,* 2393–2401

Dunnivant, F. M., R. P. Schwarzenbach, and D. L. Macalady, 1992, "Reduction of Substituted Nitrobenzenes in Aqueous Solutions Containing Natural Organic Matter," *Environ. Sci. Technol., 26,* 2133–2141

Elovitz, M. S., and W. Fish, 1994, "Redox Interactions of Cr(VI) and Substituted Phenols: Kinetic investigation," *Environ. Sci. Technol., 28,* 2161–2169

Elovitz, M. S., and W. Fish, 1995, "Redox Interactions of Cr(VI) and Substituted Phenols: Products and Mechanism," *Environ. Sci. Technol., 29,* 1933–1943

Faust, B. C., and J. Hoigne, 1987, "Sensitized Photo-Oxidation of Alkylphenols by Fulvic Acid and Natural Water," *Environ. Sci. Technol., 21,* 957–964

Gantzer, C. J., and L. P. Wackett, 1991, "Reductive Dechlorination Catalyzed by Bacterial Transition-Metal Coenzymes," *Environ. Sci. Technol., 25,* 715–722

Gleave, J. L., H. D. Hughes, and C. K. Ingold, 1935, "Mechanism of Substitution at a Saturated Carbon Atom," *J. Chem. Soc.,* 236

Haag, W. R. and J. Hoigne, 1986, "Singlet Oxygen in Surface Waters. 3. Photochemical Formation and Steady-State Concentrations in Various Types of Waters," *Environ. Sci. Technol., 20,* 341–348

Haag, W. R., J. Hoigne, E. Gassmann, and A. M. Braun, 1984, "Singlet Oxygen in Surface Waters–Part I: Furfuryl Alcohol as a Trapping Agent," *Chemosphere, 13*, 631–640

Haderlein, S. B., and R. P. Schwarzenbach, 1995, in *Biodegradation of Nitroaromatic Compounds*, J. C. Spain, Ed, Plenum Press, New York, USA, pp. 199–225

Harms, H. and A. J. B. Zehnder, 1995, "Bioavailability of sorbed 3-chlorodibenzofuran," *Appl. Environ. Microbiol.,* 27–33

Heijman, C. G., E. Grieder, C. Holliger, and R. P. Schwarzenbach, 1995, "Reductior of Nitroaromatic Compounds Coupled to Microbial Iron Reduction in Laboratory Aquifer Columns," *Environ. Sci. Technol., 29*, 775–783

Jafvert, C. T., and N.L. Wolfe, 1987, "Degradation of Selected Halogenated Ethanes in Anoxic Sediment–Water Systems, *Environ. Toxicol. Chem., 6*, 827–837

Klausen, J., S. P. Trober, S. B. Haderlein, and R. P. Schwarzenbach, 1995, "Reduction of Substituted Nitrobenzenes by Fe(II) in Aqueous Mineral Suspensions," *Environ. Sci. Technol., 29*, 2396–2404

Klecka, G. M., and S. J. Gonsior, 1984, "Reductive Dechlorination of Chlorinated Methanes and Ethanes by Reduced Iron(II)Porphyrins," *Chemosphere, 13*, 391–402

Kriegman-King, M. R., and M. Reinhard, 1991, *Organic Substances and Sediments in Water* (R. Baker, Ed.), Lewis Publishers, Chelsea, MI, USA, pp. 349–364

Kriegman-King, M. R., and M. Reinhard, 1992, "Transformation of Carbon Tetrachloride in the Presence of Sulfide, Biotite and Vermiculite," *Environ. Sci. Technol., 26*, 2198–2206

Krone, U. E., R. K. Thauer, and H. P. C. Hogenkamp, 1989, "Reductive Dehalogenation of Chlorinated C1-Hydrocarbons Mediated by Corrinoids," *Biochemistry, 28*, 4908–4914

Larson, R. A., and E. J. Weber, 1994, *Reaction Mechanisms in Environmental Organic Chemistry*, Lewis Publishers, Chelsea, MI, USA

Mabey, W. R., and T. Mill, 1978, "Critical Reviews of Hydrolysis of Organic Compounds in Water Under Environmental Conditions," *J. Phys. Chem. Ref. Data, 7*, 383–415

Macalady, D. L., P. G. Tratnyek, and T. J. Grundl, 1986, "Abiotic Reductions Reactions of Anthropogenic Organic Chemicals in Anaerobic Systems: A Critical Review," *J. Contam. Hydrol., 1*, 1–28

Mill, T., 1980, in *The Handbook of Environmental Chemistry., Vol 2, Part A: Reactions and Processes* (O. Hutzinger, Ed.), Springer, Berlin, Germany, pp. 77–105

Price, N. M., and F. M. M. Morel, 1990, in *Aquatic Chemical Kinetics* (W. Stumm, Ed.), Wiley, New York, USA, pp. 199–233

Sanders, P., and N. L. Wolfe, 1984, *190th National Meeting of the American Chemical Society, Chicago, IL, 1985, Abstracts*. American Chemical Society, Washington, DC, USA

Schwarzenbach, R. P., R. Stierli, K. Lanz, and J. Zeyer, 1990, "Quinone and Iron Porphyrin Mediated Reduction of Nitroaromatic Compounds in Homogeneous Aqueous Solution," *Environ. Sci. Technol., 24*, 1566–1574

Schwarzenbach, R. P., P. M. Gschwend, and D. M. Imboden, 1993, in *Environmental Organic Chemistry*. Wiley, New York, USA

Semenov, N. N., 1958, in *Some Problems in Chemical Kinetics and Reactivity*, Princeton University Press, Princeton, NJ, USA

Stone, A. T., 1987, "Reductive Dissolution of Manganese(III/IV) Oxides by Substituted Phenols," *Environ. Sci. Technol.*, *21*, 979–988

Stumm, W., 1992, *Chemistry of the Solid-Water Interface*, Wiley, New York, USA

Tan, K. H., 1982, *Principles of Soil Chemistry*, Marcel Dekker, New York, USA

Tratnyek, P. G., J. Hoigne, J. Zeyer, and R. P. Schwarzenbach, 1991, "QSAR Analyses of Oxidation and Reduction Rates of Environmental Organic Pollutants in Model Systems," *Sci. Total. Environ.*, *109/110*, 327–341

Tratnyek, P. G., and N. L. Wolfe, 1990, "Characterization of the Reducing Properties of Anaerobic Sediment Slurries Using Redox Indicators," *Environ. Toxicol. Chem.*, *9*, 289–295

Ulrich, H. -J., and A. T. Stone, 1989, "Oxidation of Chlorophenols Adsorbed to Manganese Oxide Surfaces," *Environ. Sci. Technol.*, *23*, 421–428

Vogel, T. M., C. S. Criddle, and P. L. McCarty, 1987, "Transformations of Halogenated Aliphatic Compounds," *Environ. Sci. Technol.*, *21*, 722–736

Weber, E. J., and R. L. Adams, 1995, "Chemical- and Sediment-Mediated Reduction of the Azo Dye Disperse Blue 79," *Environ. Sci. Technol.*, *29*, 1163–1170

Weber, E. J., and N. L. Wolfe, 1987, "Kinetic Studies of the Reduction of Aromatic Azo Compounds in Anaerobic Sediment/Water Systems," *Environ. Toxicol. Chem.*, *6*, 911–919

Zepp, R. G., J. Hoigne, and H. Bader, 1987, "Nitrate-Induced Photooxidation of Trace Organic Chemicals in Water," *Environ. Sci. Technol.*, *21*, 443–450

10

Transformation of Chlorinated Xenobiotics in the Environment

Heidelore Fiedler and Christoph Lau (Bayreuth, Germany)

10.1 SUMMARY

This chapter gives an overview of transport and transformation processes that chlorinated organics may undergo under environmental conditions. In addition, some basic

Ecotoxicology, Edited by Gerrit Schüürmann and Bernd Markert.
ISBN 0-471-17644-3   © 1998 John Wiley & Sons, Inc. and Spektrum Akademischer Verlag.

definitions and terms used to describe the fate of chemicals in the environment are given. Important processes that occur in the environment are volatilization, sorptive processes, so-called third phase phenomena, biotic and abiotic degradation processes, and bioconcentration. Special emphasis is placed on biological transformation processes. Although most information is obtained from laboratory experiments, results from field studies are included whenever available.

10.2 DEFINITIONS AND BASICS OF ENVIRONMENTAL PROCESSES

10.2.1 Volatilization

Volatilization is the process in which a chemical is transferred from a condensed phase (liquid or solid) into a gaseous phase. The transfer of a chemical from or to gaseous environmental phases, such as air, is governed by its vapor pressure. The vapor pressure of a compound depends on the environmental temperature. Vapor pressures of organochlorines under standard conditions ($T = 25°C$) vary over many orders of magnitude, from nearly 1, e.g., methylene chloride, to down to 10^{-12} atm for polychlorinated biphenyls (PCB) and DDT (Dichlorodiphenyl trichloroethane). Even for compounds with such low vapor pressures, volatilization might be an important environmental process due to the compounds' extremely low solubilities in water. Combined with the resulting high aqueous fugacities, these compounds thus still partition to a considerable extent into the atmosphere (Schwarzenbach et al. 1993). Some physicochemical constants for organochlorines are given in Table 10.1.

According to Suntio et al. (1988a) volatilization behavior of chemicals can be defined according to Henry's Law Constant (K_H) and classified as follows:

- Compounds with a K_H value larger than 100 $Pa\,mol^{-1}\,m^{-3}$ ($= 0.001$ atm) volatilizes rapidly from water; the diffusional resistance is the controlling

Table 10.1 Physicochemical constants for some semivolatile compounds

Compound	$-\log c_w$	$\log K_{OW}$	$\log K_H$
Hexachlorobenzene (HCB)	7.69	5.5	0.18
Benzo[a]pyrene (BaP)	7.31	6.5	-2.24
2,2′,5,5′-TetraCB (PCB 52)	7.06	6.18	-0.54
DecaCB (PCB 209)	10.55	8.23	-1.73
2,3,7,8-Cl$_4$DD	10.3	6.64	-1.3
p,p′-DDT	7.85	6.36	-2.02
Lindane (γ-HCH)	4.59	3.78	-2.49

Source: Schwarzenbach et al. (1993).

K_{OW}, Octanol-water partition coefficient at 25°C [$(mol\,L^{-1}$ octanol) $(mol\,L^{-1}$ water)$^{-1}$]; c_w, aqueous solubility at 25°C ($mol\,L^{-1}$); K_H, Henry's Law Constant at 25°C ($L\,atm^{-1}\,mol^{-1}$).

factor of the partitioning process. Volatilization may take place before biodegradation processes are possible. Examples include most chlorinated short-chain hydrocarbons with a K_H ranging between 40 and 2290 Pa mol^{-1} m^{-3} (= 0.0003–0.001 atm) at 20°C (Gossett 1987; Tse et al. 1992).

- For compounds with a K_H in the range 100–25 Pa mol^{-1} m^{-3}, volatilization will be appreciably slower due to stronger diffusional resistance. Many chlorobenzenes fall into this category, exhibiting a K_H of 40–190 Pa mol^{-1} m^{-3} (Ten Hulscher et al. 1992).

- Compounds with a K_H value below 25 Pa mol^{-1} m^{-3} (= 0.0003 atm) have a very low volatilization rate. Competing processes, such as sorption to solids or colloidal material, might prevail. In the case of a K_H value below 1 Pa mol^{-1} m^{-3} volatilization from bulk aqueous phase is insignificant. Examples include most PCB (K_H: 1–35 Pa mol^{-1} m^{-3}; Brunner et al. 1990; Dunnivant et al. 1988; and Murphy et al. 1987). Some PCDD (polychlorinated dibenzo-p-dioxins) have K_H values as low as 0.1–4.8 Pa mol^{-1} m^{-3} (Shiu et al. 1988).

These general rules, however, were obtained under standard conditions and, thus, are only of limited value for partitioning processes in natural waters (Suntio et al. 1988b). Since volatilization is sensitive to temperature variations, it may exhibit diurnal as well as seasonal and/or geographical trends. Such temperature dependency of K_H has been investigated for halogenated aliphatics (Tse et al. 1992; Gossett, 1987), α- and γ-hexachlorocyclohexane (α- and γ-HCH; Kucklick et al. 1991), as well as for PCB and chlorinated benzenes (Ten Hulscher et al. 1992). For PCB, a doubling in K_H per 10°C increase in temperature was observed. Moreover, the wind speed also influences volatilization in natural waters or outdoor treatment facilities. Results of wind tunnel experiments suggest that under field conditions mass transfer coefficients will be lower than those measured in the laboratory (Mackay and Yeun 1983). In addition, the presence of other dissolved chemicals such as electrolytes, solvents, detergents, and dissolved organic matter affect the activity of a chemical in aqueous solution and, thus, their apparent solubility and vapor pressure.

Gschwend and Schwarzenbach (1992) also underline the importance of dissolved natural molecules for environmental processes. The effect of dissolved organic carbon on air–water partitioning of hydrophobic compounds was demonstrated by Hassett and coworkers (Yin and Hassett 1989; Hassett and Milicic 1985; Jota and Hasset 1991). Finally, the partitioning process may compete with sorption as a second pathway for compounds leaving the aqueous phase.

10.2.2 Sorptive and Desorptive Processes

Sorption to solids as well as to dissolved organic carbon may be an important pathway in the fate of chlorinated organic chemicals in aqueous environments.

This is particularly true for chemicals exhibiting a relatively low K_H. Adsorption occurs when attractive forces cause a chemical to accumulate at the surface of a solid phase. Absorption occurs when a chemical is incorporated into another; there is no clear boundary between the two substances, and usually the absorbing medium increases its volume as a result of the process. Surfaces of minerals are considered important for sorption of organic contaminants if the organic carbon content in the solution is low (Haderlein and Schwarzenbach 1993).

There are two major thermodynamic theories dealing with a mathematical description of sorption processes. Both sorption models assume the sorbent to be homogeneous with respect to the sorptive sites, a prerequisite which is not likely to be fulfilled by natural sorbents. Both models also predict linear sorption behavior as long as only a small fraction of the sorption sites is occupied by the sorbing molecules.

A compound's sorption to solids as well as its biological uptake is inversely correlated with its aqueous solubility. The organic-carbon-normalized sorption coefficients of a chemical correlate well with the hydrophobicity parameter K_{ow}; no surface saturation behavior is visible (i.e., the sorption isotherms are linear).

For equilibrium reactions also desorption of contaminants is important in order to make the compound available for biological degradation processes. A significant correlation between the desorption rate constant and the sorption coefficient can be observed for nonionic organic compounds. In desorption experiments, small amounts (0.4%–1.0%) of tetrachloroethene and dibromo-chloropropane were found to irreversibly sorb to solids of low organic carbon content (0.19%; Pignatello 1991). The correlation between sorption coefficient and desorption rate constant for nonpolar organic compounds also applies to chlorophenols. Lee et al. (1991) measured sorption coefficients and desorption rate constants for ionized as well as for nonionized chlorophenols. For both groups of species, sorption coefficients were found to increase with increasing degree of chlorination. When ionized, a larger portion is readily desorbed from the solids as in the neutral state.

In reality, however, equilibrium is not always reached. Ball and Roberts (1991) studied the sorption behavior of trichloroethylene (TCE) and tetrachloro-benzene with solids of low organic carbon content (0.02%). Contact times of up to more than 100 days were necessary to obtain equilibrium. Based on correlations between sorption coefficient and organic matter content, sorption was about a magnitude larger than predicted. Thus, either the compounds sorb to the mineral phases, or equilibrium was not achieved in previous experiments.

The environmental relevance of sorptive processes has been described in cases of agricultural surface runoff which is a major contributor of fairly water soluble organochlorines, such as the triazine herbicides or alachlor, in surface waters (Squillace and Thurman 1992). Klaine et al. (1988) found that a total of 1.5% of the atrazine applied to an agricultural field was lost by surface runoff.

Sorption of chlorinated chemicals to field sediments were determined for:

- Chlorinated phenols (Xie et al. 1986; Leuenberger et al. 1985; Paasivirta et al. 1990)
- PCB (Brannon et al. 1991; Eisenreich et al. 1989; Nondek and Frolikowa 1991)
- Chlorinated catechols and guaiacols (Paasivirta et al. 1990; Remberger et al. 1988; Remberger et al. 1993)
- Chlorinated dibenzo-*p*-dioxins, dibenzofurans, and related compounds (Brusseau et al. 1990; Czuczwa and Hites 1984)
- Chlorinated hydrocarbon pesticides (Eisenreich et al. 1989) and AOX (adsorbable organic halogens) (Kostiuk et al. 1990).

The importance of desorption for the bioavailability of contaminants is under-lined by Speitel et al. (1989). Neither 2,4-dichlorophenol nor pentachlorophenol (PCP) were significantly degraded when adsorbed to activated carbon (Speitel et al. 1989). Desorption from solids appeared to be the limiting factor in the degradation of 2,4-dichlorophenol sorbed to activated carbon and of α-hexa-chlorocyclohexane (α-HCH) sorbed to calcareous soil. The rate of degradation of PCP sorbed to activated carbon was limited by slow microbial kinetics (Speitel et al. 1989). Particle diffusion models have been derived for describing the effect of desorption on the biodegradation of organic chemicals in solids–water systems (Rijnaarts et al. 1990; Speitel et al. 1989; Scow and Hutson 1992).

The "third phase" hypothesis has been put forward to account for sorption–desorption anomalies and the effect of particle concentration on sorption coefficients (Schrap and Opperhuizen 1991). Both effects cannot be explained by simple partitioning of hydrophobic compounds between the aqueous and the solid phase. Expanding on the partition model by a third phase, however, can account for such effects. Resendes et al. (1992) support the third phase hypothe-sis particularly for the interaction between chlorinated aromatic compounds, water, sediment, and dissolved organic carbon.

10.2.3 Bioconcentration

Substances of low biological degradability tend to accumulate throughout trophic levels of the food net. Definitions of the terms bioconcentration, bio-accumulation, and biomagnification according to de Wit (1992) are as follows:

Bioconcentration is mainly used in the discussion of fish or other gill-breath-ing aquatic organisms. Bioconcentration can be defined as the direct, nondietary uptake of chemicals from water, e.g., in fish through the gills (see also Opper-huizen 1986). It tends to level off at higher concentrations. A bioconcentration factor (BCF) can be defined as the ratio of concentration in the organism, c_{org}, to that in water, c_{water}:

$$BCF = c_{org}/c_{water}$$

Bioconcentration of chlorinated benzenes, biphenyls, and phenylbenzoylureas has also been observed in aquatic plants and phytoplankton under laboratory conditions (Wolf et al. 1991; Gobas et al. 1991; Yu-yun et al. 1993). In contrast to bioconcentration in fish, the bioconcentration factor in plants did not level off (Gobas et al. 1991). Nevertheless, the partitioning coefficient $\log K_{OW}$ very accurately predicted the lipid-based bioconcentration factor, implying that the partition hypothesis is also applicable to plants. This hypothesis also provides more evidence that bioconcentration in fish is largely influenced by fish physiology.

Bioaccumulation describes the uptake of chemicals by food. A bioaccumulation factor (BAF) is expressed as the ratio of concentration in lipid weight in the organism, $c_{lip.,org}$, to that in its food, $c_{lip.,food}$:

$$BAF = c_{lip.,org}/c_{lip.,food}$$

Biomagnification is used for bioaccumulation on higher trophic levels in a food web. Analogously, a biomagnification factor (BMF) is defined as the ratio of concentration in lipid weight in the predator, $c_{lip.,predator}$, to that in its prey, $c_{lip.,prey}$:

$$BMF = c_{lip.,predator}/c_{lip.,prey}$$

Biomagnification of polychlorinated hydrocarbons in the Great Lakes has been demonstrated by, for example, Oliver and Niimi (1988). Applying the toxic equivalency factors (TEF) for PCB (Safe 1990) it has been demonstrated that to a large extent the biological potency of polychlorinated hydrocarbon extracts from Great Lakes biota can be attributed to a few *non-ortho*-substituted PCB congeners (Smith et al. 1990).

10.3 BIOLOGICAL DEGRADATION

Biological degradation, or biodegradation, can generally be defined as biological transformation of organic compounds.

10.3.1 General Considerations

Biodegradation changes the original molecular structure of a compound and ultimately yields the inorganic products carbon dioxide, water, sulfite, ammonia, or phosphate (aerobic biodegradation) or methane and sulfide (anaerobic biodegradation). In the case of only small structural transformations, the term "biotransformation" may be used instead. Metabolites of biodegradation processes can be more recalcitrant and/or toxic than the parent compounds. They often exhibit different partitioning characteristics in the environment to those of the parent compounds.

10.3.2 Environmental Factors Influencing Biodegradation Rates

Ideally, an organism utilizes the chlorinated substrate as the sole source of both cell carbon and energy. The compounds may be transformed metabolically, when serving as carbon and energy sources, or cometabolically, in which case the organism does not gain energy or carbon. This idealized situation, however, hardly exists in natural systems, where usually a range of natural nonchlorinated substrates are available in high concentration and easily accessible for microbial growth. The substrates may originate, e.g., from plant material deposited in the aquatic and sediment phases.

The *concentration of the substrate* is significant from several points of view:

- Chlorinated compounds may be toxic even to the metabolizing organism; exposure to low concentrations might be tolerable.
- The substrate concentration might determine the metabolic pathway and influence the rate of metabolic reactions. For example, at low substrate concentrations (100 μg/L) 3,4,5-trichloroguaiacol is quantitatively transformed by cell suspensions of a *Rhodococcus* sp. only to the corresponding veratrole, whereas at high concentrations (10 mg/L) a number of products are produced, including 1,2,3-trichloro-4,5,6-trimethoxybenzene (Allard et al. 1985; Neilson 1990).
- For the metabolism of structurally diverse compounds including 2,4-D, there seems to exist a threshold concentration. Below this threshold concentration, rates of biodegradation are extremely low or nonexistent (Hoover et al. 1986) because transforming enzymes are not induced to activity.

Also sediment affects the biodegradation as sediment and dissolved organic carbon are potential sources of the carbon needed by the dechlorinating microorganisms (Gibson and Suflita 1990). Bioavailability (which is usually correlated to the mobility) of the substrate can be limited when it is bound to sediments or other matrices. These bound residues may not be bioavailable even if they are recoverable using standard extraction methods.

Pritchard et al. (1987) showed slow degradation of *p*-chlorophenol in estuarine water samples. Addition of detrial sediment, however, resulted in immediate and rapid degradation. Also, addition of sterile sediment or sand resulted in approximately four to six times more CO_2 development than observed in the original water sample. Densities of *p*-chlorophenol-degrading bacteria associated with the detrial sediment were 100 times greater than those enumerated in water.

Other factors that affect biodegradation rates *in situ* are: temperature, pH value, redox potential, sulfate and nitrate concentrations, O_2 concentration, biomass concentration, and interspecies competition in the consortia. Spatial availability of these compounds may be influenced by aeration and the depth of the reactor basin. Laboratory observations regarding the metabolism of single substrates by clearly defined microbial species do not reflect the real outdoors situation with mixed substrates and mixed cultures of microorganisms.

Thus, one needs to be careful in applying results from controlled laboratory experiments to field situations. Conditions in soils, water, or other environmental compartments are generally far from being ideal for supporting microbial degradation activity. For this reason a compound also should only be termed recalcitrant in combination with the particular process parameters and/or environmental conditions at which no (bio)degradation was observed.

10.3.3 Chemical Factors Influencing Biodegradation Rates

Microorganisms obtain energy from a wide variety of electron donors and acceptors, depending on the prevailing redox conditions. Aerobic metabolism dominates if a sufficient amount of oxygen is present. When oxygen is depleted, however, other electron acceptors are used, such as nitrate, sulfate, and carbon dioxide. Transformation of chlorinated compounds is affected by the availability of the different electron acceptors and the resulting redox conditions. In addition transformation processes are influenced by the reactivity of enzymes and coenzymes associated with the degrading microorganisms.

The critical step in the degradation of chlorinated compounds is the cleavage of the carbon–halogen bond. Two distinct strategies are employed by bacteria for degrading chlorinated aromatic compounds: the chlorine substituent is removed from the aromatic ring either as an initial degradation step or after ring cleavage from a chlorinated aliphatic intermediate. Direct elimination of chlorine substituents from the aromatic ring occurs by replacement either through hydroxyl groups (hydrolytical or oxygenolytical displacement) or through hydrogen atoms (reductive dechlorination).

Algae and fungi, for example, mediate the oxidation of aromatic hydrocarbons but cannot cleave aromatic rings. In bacteria, degradation of aromatic hydrocarbons is carried out by a dioxygenase system, whereas in fungi and higher organisms, a monooxygenase system is used.

10.3.4 General Rules for Degradability of Chlorinated Organics

The extent of degradation depends on the chemical group to be degraded. The degradability of a compound is very much controlled by the position and number of chlorine substituents—less chlorinated compounds are degraded slower than the polychlorinated ones—and the overall recalcitrance of the chemical class a compound belongs to.

Aliphatics are biodegraded according to the following general rules (Korte 1987, 1992):

- Aliphatic compounds are increasingly recalcitrant against degradation with increasing chain length.
- Branching increases recalcitrance.
- Alkenes are easier degraded than alkanes, and alkanes easier than aromatic compounds.

Qualitative observations on the influence of the molecules' structure on their biodegradability are (Hauk and Schramm 1990):

- Functional groups, such as hydroxyl or carboxylate groups of benzene rings, are rapidly degraded. This is in agreement with the general observation that water-soluble compounds are degraded faster than water-insoluble chemicals.
- Halogen, nitro, and sulfonate substituents inhibit biodegradation.
- *Ortho* substitution of aromatic compounds yields slightly more biodegradable compounds than *para* substitution. *Meta*-substituted compounds are the least biodegradable.

If degradation data of an organic compound are not available, structural analogies to other compounds may help to evaluate its biodegradation potential. However, applying quantitative structure–activity relationship analyses (QSAR) to biodegradation is limited to structural classes of compounds which have already been investigated. QSAR have successfully been applied to single-step transformation reactions (Paris and Wolfe 1987), but extension to the multistep processes involved in biodegradation presents a much greater challenge.

10.3.5 Aerobic Environments and Aerobic Respiration

Aerobic Biodegradation

Conventional aerobic biodegradation involves the oxidation of organic chemicals used as carbon and energy sources for biological growth. Typically, the major oxidized product is carbon dioxide, whereas water is produced by oxygen reduction. The extent of biodegradation can be correlated with bacterial oxygen consumption.

For degradation under aerobic conditions, the following generalizations apply:

- The rate of biodegradation increases with decreasing lipophilicity (Banerjee et al. 1984).

● Higher chlorinated organics are less readily biodegraded than lower chlorinated organic compounds (Banerjee et al. 1984). Short-chain compounds are most readily degraded at chlorine levels below 60%. All 1-monosubstituted alkanes and many dichloroalkanes have been found to be aerobically degradable (Zitomer and Speece 1993). Complete breakdown was observed for a long-chain chlorinated paraffin within 8 weeks (Madeley and Birtley 1989). Also, monosubstituted chloroanilines are more easily degraded than polysubstituted ones (Pearson 1982).

The ability of aerobic microorganisms to oxidize a range of chlorinated paraffins also depends upon their previous acclimatization to the particular environmental conditions, the hydrocarbon chain length, and its degree of chlorination. Microorganisms previously conditioned to specific chlorinated paraffins show a greater ability to degrade the compounds than normal sewage treatment organisms.

Conventionally cultured aerobic bacteria efficiently degrade aromatic compounds which are anaerobically recalcitrant. These microorganisms often produce mixed function oxidase enzymes that initiate aromatic ring cleavage (Table 10.2). Aerobic degradation processes are limited in the case of recalcitrant, highly chlorinated chemicals, such as hexachlorobenzene, tri- and tetrachloroethylene, and tetrachloromethane. These chemicals are degraded at appreciable rates only under anaerobic conditions.

Exceptions to these generalizations are highly chlorinated short-chain compounds, such as trichloroethylene, 1,1,1-trichloroethane, and chloroform. They are biotransformed under aerobic conditions if methane, phenol, or toluene is provided as a primary source of carbon and energy for biological growth (Zitomer and Speece 1993). However, these reactions are cometabolic rather than conventional.

Cometabolic Aerobic Biodegradation

Microbial cometabolism, also termed gratuitous metabolism (Young and Hägg-blom 1991), is defined as the transformation of a non-growth substrate in the obligate presence of a growth substrate or another transformable compound (Janke and Ihn 1990). Cometabolic reactions may aid to prevent dead-end situations, where no metabolic pathway for a substance is available or no induction of a metabolic reaction takes place. It often has been found in mixed cultures of microorganisms.

Most reported cometabolic reactions involve the aerobic transformation of one- and two-carbon chlorinated aliphatics if methane is provided as a primary substrate. This process is catalyzed by a methane monooxygenase (MMO) enzyme produced by bacteria that employ methane as a primary substrate (methanotrophic bacteria). In addition, the MMO enzyme catalyzes transformations of *cis*- and *trans*-1,2-dichloroethylene, vinyl chloride, chloroform, and

Table 10.2 Oxygenases involved in the degradation of chlorinated aliphatic compounds

Organisms/inducer (enzyme)	Substrates	References
Nitrosomonas europaea/ ammonia (ammonia monooxygenase)	Dichloromethane, trichloromethane, 1,1,2- and 1,1,1-trichloroethane, monochloroethylene, dichloroethylenes, trichloroethylene, 1,2,3-trichloropropane	Vanelli et al. (1990)
Nitrosolobus multiformis/ ammonia (ammonia monooxygenase)	1,2-dichloropropane	Rasche et al. (1990)
Pseudomonas cepacia/ toluene, phenol (monooxygenase)	Trichloroethylene	Nelson et al. (1987)
Pseudomonas putida, *E. coli*/toluene (dioxygenase)	Trichloroethylene	Zylstra et al. (1989)
Mycobacterium ssp./ propane (propane monooxygenase)	Monochloroethylene, 1,1-dichloroethylene, *cis*- and *trans*-1,2-dichloroethylene, trichloroethylene	Wackett et al. (1989)
Pseudomonas putida/ camphor (cytochrome P450)	1,1,2-Trichloroethane	Castro and Belser (1990)
Methylosinus trichosporium/methane (methane monooxygenase)	Dichloromethane, trichloromethane, chloroform, 1,1-dichloroethane, 1,2-dichloroethane, 1,1-dichloroethylene, *cis*- and *trans*-1,2-dichloroethylene, 1,2-dichloropropane, *trans*-1,3-dichloropropylene, trichloroethylene	Oldenhuis et al. (1989)
Alcaligenes eutrophus	Trichloroethylene, phenol, 2,4-D	Harker and Kim (1990)

dichloromethane. In contrast, tetrachloroethylene and tetrachloromethane appear resistant to the enzyme (Oldenhuis et al. 1989).

Other reports of cometabolic oxidations describe bacteria that use propane, toluene, phenol, cresol, dichlorophenoxyacetic acid, and ammonia as primary substrates (Wackett et al. 1989; Nelson et al. 1987; Harker and Kim 1990; Vanelli et al. 1990).

Reactions

Hydrolytic Replacement of Chlorine Atoms by Hydroxyl Groups. The metabolism of chloroalkanoates and chloroalkanes involves replacement of the chlorine atom by hydroxyl groups. It is mediated by a series of dehalogenases. The dechlorinated products then enter central catabolic pathways. They are finally metabolized through tricarboxylic acid or glyoxylate metabolism, for example, in the dechlorination of dichloromethane by *Hyphomicrobium* sp. (Neilson 1990). Major metabolites of dichloromethane are carbon monoxide and carbon dioxide.

Xanthobacter autotrophicus is able to utilize a number of halogenated short-chain hydrocarbons, for example, dichloroethane (Neilson 1990) and halogenated carboxylic acids, as the sole carbon source for growth. The organism produces two different dehalogenases, one of which specifically degrades halogenated alkanes, whereas the other degrades halogenated carboxylic acids (Figure 10.1; Janssen et al. 1985).

Direct replacement of chlorine substituents with hydroxyl groups in aromatic chlorocompounds before ring cleavage has been observed in 4-chlorobenzoate (Neilson et al. 1991; Marks et al. 1984) and pentachlorophenol (Häggblom 1990). This reaction is mediated by *Rhodococcus chlorophenolicus* and *Actinomycetes* (Neilson et al. 1991).

R. chlorophenolicus preferably degrades polychlorinated phenols. The bacterium hydroxylates chlorophenols at position 4 whether or not this position is substituted by chlorine. It even metabolizes PCP and several tetrachlorophenols. 2,3,4,6- and 2,3,5,6-tetrachlorophenol are degraded almost completely, while 2,3,4,5-tetrachlorophenol and the trichlorophenols are degraded in part at concentrations of 2 μM and 10 μM. However, *R. chlorophenolicus* seems not to degrade dichlorophenols in significant amounts (Apajalahti and Salkinoja-Salonen 1986; Häggblom et al. 1989).

Strains of *R. chlorophenolicus* carry out the degradation of several chlorinated guaiacols. Degradation of higher chlorinated congeners is analogous to PCP degradation: the 4-chloro substituents in tetrachloroguaiacol and 3,4,5-trichloro-2,6-dimethoxyphenol are displaced by hydroxyl groups. Like chlorinated phenols (Table 10.3), chloroguaiacols with fewer chlorine atoms are more slowly metabolized (Table 10.4; Häggblom et al. 1988a).

Reductive Elimination of Chlorine Substituents by Hydrogen Atoms (Before Ring Cleavage). Chlorine may be reductively replaced by hydrogen. This

$$ClCH_2\text{-}CH_2Cl \rightarrow ClCH_2\text{-}CH_2OH \rightarrow ClCH_2\text{-}CHO \rightarrow ClCH_2\text{-}CO_2H \rightarrow$$

$$HOCH_2\text{-}CO_2H$$

Figure 10.1 Degradation of 1,2-dichloroethane by *Xanthobacter autotrophicus* (Janssen et al. 1985).

Table 10.3 Reduction of chlorophenols (%) after incubation of chlorophenols (2 µM) with cells of *Rhodococcus chlorophenolicus* for 14 days

Chlorophenol	% Reduction
Pentachloro-	100
2,3,4,5-Tetrachloro-	100
2,3,5,6-Tetrachloro-	100
2,4,6-Trichloro-	94
2,4,5-Trichloro-	70
3,4,5-Trichloro-	25
2,6-Dichloro-	81
2,3-Dichloro-	52
2,4-Dichloro-	15

Source: Häggblom et al. (1989).

Table 10.4 Reduction of chloroguaiacols (%) after incubation of 5 µM chloroguaiacols with cells of *Rhodococcus chlorophenolicus* for 14 days

Chloroguaiacol	% Reduction
3,5-Dichloro-	100
3,4-Dichloro-	17
4,5-Dichloro-	5
3,4,6-Trichloro-	100
4,5,6-Trichloro-	16
Tetrachloro-	100

Source: Häggblom et al. (1988a)

pathway was observed in the metabolism of 2,4-D and 2,4,5-trichlorophenoxyacetic acid (2,4,5-T; Neilson et al. 1991). Also the degradation of PCP by *R. chlorophenolicus* might in part be due to reductive elimination of chlorine (Neilson 1990). Tetrachlorohydrochinone, the first intermediate in PCP degradation is degraded through a hydrolytic dechlorination and three reductive dechlorinations producing 1,2,4-trihydroxybenzene. All chlorine substituents are thus removed before ring cleavage (Häggblom 1990).

Direct Elimination of the Chlorine Substituent through Double-Bond Formation (Before Ring Cleavage). This mechanism has been found to represent the degradation of DDT by *Aerobacter aerogenes*, a facultatively anaerobic bacterium (Neilson 1990). γ-HCH (lindane) is transformed to pentachlorocyclohexene

(Neilson 1990), β-HCH to tetrachlorocyclohexene (Fiedler et al. 1993); successive dechlorination yields chlorinated benzenes (Fiedler et al. 1993). Only aerobic microorganisms show growth with lindane as the sole carbon and energy source (Bachmann et al. 1988).

Hydroxylation or Dioxygenation of the Aromatic Ring (Ring Cleavage). Ring cleavage may occur either between the atoms bearing the hydroxy groups (*ortho, endo,* or 1:2 cleavage) or between one of the carbon atoms bearing a hydroxy group and the adjacent nonhydroxylated atom (*meta, exo,* or 2:3 cleavage). Dechlorination of the intermediate after cleavage produces acetate or succinate, which then may enter the catabolic degradation pathway. Chloride appears to be spontaneously eliminated after isomerases and reductases have acted to destabilize the carbon halogen bond.

The type of cleavage mechanism is critical to the cell, since operation of the 2:3 cleavage produces metabolites which may toxify certain degradation enzymes. This is the case, for example, for enzymes facilitating the cleavage of 3-chlorocatechol (Bartels et al. 1984; Klecka and Gibson 1981). Such reactions are called "lethal synthesis".

In the case of polycyclic aromatic compounds (such as PCB or naphthalenes) the ring with fewer chlorine substituents is attacked first (Neilson et al. 1990). However, ring cleavage is inhibited by steric hindrance (Dieter 1990). Thus, aromatic compounds appear relatively stable. PCB degradation usually proceeds via dioxygenase attacks between carbon 2 and 3 or carbon 3 and 4, resulting in the formation of chlorobenzoic acid, chloroacetophenone, and other degradation products (Sijm 1992; Young and Häggblom 1991). Other bacterial genera of PCB-degraders include *Alcaligenes, Pseudomonas, Actinebacter, Bacillus, Acetobacter,* and *Klebsiella.*

Aerobic degradation of mono- and dichlorinated phenols and benzoates generally proceeds *via* hydroxylation or dioxygenation. This degradation pathway has been demonstrated with several genera of bacteria, including several strains of *Pseudomonas* sp., *Rhodococcus* sp., and *Nocardia* sp., with one species each of *Achromobacter, Bacillus,* as well as with *Mycobacterium coeliacum.* Key metabolites in the degradation of chlorinated aromatics are chlorocatechols, produced by hydroxylation of chlorophenols and by dioxygenation of chlorinated aromatic hydrocarbons, chlorinated benzoates, and chloroanilines (Neilson et al 1990). Dechlorination occurs after ring fission (Neilson 1990; Young and Häggblom 1991).

Ortho Methylation. Ortho methylation (*o* methylation) of aromatic compounds is an important transformation reaction, since methylation increases the lipophilicity of the compound and thus the potential for bioaccumulation. Moreover, some of the resulting metabolites resist further degradation and have 10–100 times higher bioconcentration factors than their precursors (Winter et al. 1991).

O methylation is found in aquatic as well as in terrestrial environments (Neilson et al. 1991), e.g., for 2,4-D and 2,4,5-T (Neilson 1990). *O* methylation rates strongly depend on the number and positions of the substituents. Replacement of the *ortho*-chlorine substituents of polychlorophenols with one (guaiacols) or two methoxy groups (syringols) apparently decreases the rate of degradation (Häggblom 1990).

Bacteria facilitating *o* methylation belong to the genera *Acinetobacter*, *Pseudomonas*, *Rhodococcus*, and *Mycobacterium*. Under aerobic or micro-aerophilic conditions, they have been observed to *o*-methylate chlorinated phenols and phenol derivatives (guaiacols, syringols, catechols, hydroquinones, phenoxy phenols; Neilson et al. 1990; Häggblom 1990). Several species of the genus *Rhodococcus* were observed to *o*-methylate chlorinated phenols, independent of whether or not the strain had been previously exposed to a chlorinated aromatic compound (Häggblom et al. 1988a). Different strains of *Rhodococcus* preferred as substrates for *o*-methylation compounds in which the hydroxyl group is flanked by two chlorine substituents. Chlorophenols having one or no *ortho* chlorines were only slowly methylated (Häggblom et al. 1988b). This is of importance in so far as *Rhodococci* are widespread species in nature and may therefore play an important role in both the degradation and transformation of chlorophenols in the environment.

Other Mechanisms. In some cases the degradation mechanism has not yet been identified. According to Aust (1993), it has been shown that the white-rot fungi, which degrade lignin in wood, also are capable of degrading a wide range of otherwise very recalcitrant organic pollutants. Extracellular mechanisms exist to detoxify toxic chemicals in a way that resistance to toxicity is provided. Mechanisms also provide for further oxidation of already highly oxidized chemicals, including DDT and chlorinated phenols. Extracellular peroxidases (lignin peroxidases) can catalyze either oxidations or reductions. White-rot fungi produce oxalate which reacts with the veratryl alcohol cation radical to produce the anion radical of oxalate. The oxalate anion radical is an excellent reductant, capable of, e.g., reductive dechlorination of carbon tetrachloride. Several mechanisms for reductions and oxidations exist. The fungus thus has unique capabilities making it suitable for the bioremediation of many different pollutants, including complex, toxic mixtures in various matrices.

10.3.6 Anaerobic Environments and Anaerobic Respiration

Anaerobic Biodegradation

Anaerobic degradation is an important process in the natural environment because chlorinated organics partition into anaerobic environmental compartments such as sediments (which tend to be anaerobic after a few centimeters below the interphase with water and near discharges with organic loads), anaerobic water bodies, waste dumps, anaerobic sites in soils, etc. Anaerobic methods of waste water treatment are considered environmentally safe, since few toxic chemicals can be stripped into the ambient air.

Unlike in aerobic degradation, in anaerobic degradation the chlorinated molecule is used as a direct source of electrons. It reductively degrades the compounds and eventually may lead to the conversion of chloro-organic compounds into methane, carbon dioxide, and other inorganic products. Compared to aerobic degradation rates, anaerobic degradation rates often are much slower (Neilson et al. 1991).

Evidence suggests that reductive dechlorination is not simply an electrophilic or nucleophilic substitution (Dolfing and Tiedje 1991). This is based on the observation that the effects of aryl substitution on chlorobenzoate did not correlate with Hammett substituent constants. The findings indicate that biologically catalyzed dechlorination reactions may be more complicated than most abiotic substitutions.

Whereas in studies with aerobic organisms pure cultures have been used, experimental difficulties in isolating pure cultures of anaerobic bacteria dictated reliance on metabolically stable consortia. In the most simplistic case, the consortium of bacteria is composed of

- Acidogenic bacteria, which transform complex organic compounds into acetate, carbon dioxide, and hydrogen
- Methanogenic bacteria, which convert the intermediates into methane.

The presence of sulfate may inhibit dechlorination of aromatic compounds in anaerobic aquifer material, possibly because of competition for hydrogen (Gibson and Suflita 1990). Similarly, the addition of sulfate to methanogenic freshwater sediment cultures slowed down the rate of 2,4-dichlorophenol (2,4-DCP) dechlorination (Kohring et al. 1989). In marine environments, sulfate reduction is the major electron sink during anaerobic degradation of organic matter. It accounts for more than 50% of the mineralization of organic matter. The marine environment is a rich source of biologically produced chlorinated compounds. Presumably, bacteria capable of aerobic or anaerobic dechlorination could evolve in such habitats (Häggblom and Young 1990).

Reactions

The primary reaction process for anaerobic dechlorination of chlorinated organic compounds is reductive dechlorination. Mostly cometabolic reaction chains may lead to complete mineralization of the organochlorine compounds. Reductive dechlorination usually has been observed in combination with methanogenesis, a reaction whereby under anaerobic conditions methane is produced.

Reductive Dechlorination and Methanogenesis. Methanogenesis is the anaerobic production of methane by microbial degradation. In the degradation of chlorinated organic compounds, it is usually combined with reductive dechlorination. The short-chain degradation products are further reduced to methane. Methanogenic conditions are characterized by an extremely low redox potential ($< -300\,\text{mV}$). They support and generate dechlorinating activity (Young and Häggblom 1991).

Reductive dechlorination is the successive shedding of chlorine atoms under reduced, anaerobic conditions and is a common initial step in the biodegradation of chlorinated organics (Zitomer and Speece 1993; Holliger et al. 1989). The dechlorination reactions are usually catalyzed biologically. It is assumed that reductive dechlorination will not yield energy to the degrading organism; therefore, the degradation process is considered cometabolic (Shelton and Tiedje 1984).

Theoretical methane and carbon dioxide production may be calculated from reaction stoichiometry. The obtained quantity is referred to as the theoretical gas production. Actual gas production, however, usually is considerably lower than theoretically obtained values. This is due to bacterial synthesis and incomplete or no conversion of some compound. For example, measurements of the gas produced by anaerobic consortia degrading various organic chemicals determined that many chlorophenols did not support anaerobic gas production (Battersby and Wilson 1989). It was concluded that chlorosubstituents inhibited anaerobic gas formation, whereas carboxyl and hydroxyl groups resulted in enhanced gas formation.

Investigators have attempted to categorize bacteria that carry out reductive dechlorination. It is most important to distinguish between strict anaerobes, such as methanogens, and facultative anaerobes. Both types of bacteria have been observed to catalyze reductive dechlorination reactions (Mohn and Tiedje 1992).

Reductive dechlorination is relatively rapid for chemicals with a high number of chlorine substituents, such as PCB, hexachlorobenzene, tetra-, trichloroethene, tetrachloromethane, and 1,1,1-trichloroethane when compared with their less-chlorinated homologues. The resulting products after dechlorination usually are more susceptible to a hydrolytic and oxidative process and less susceptible to further reduction. Hence, less-chlorinated compounds tend to persist longer in anaerobic environments than highly chlorinated compounds. However, if the environment becomes aerobic, lower chlorinated chemicals might be degraded by aerobic bacteria.

One of the benefits of reductive dechlorination is that highly chlorinated, often toxic compounds may be dechlorinated, yielding less-chlorinated compounds. These are generally less toxic and more amenable to further aerobic and anaerobic biodegradation (Bouwer and McCarty 1983). However, due to their increased water solubility, they might become more mobile in the environment, as is the case for less-chlorinated phenols in contaminated soils.

Reductive dechlorination has been studied in various habitats. It has not been observed in sterile sediment or in acclimated cultures incubated aerobically (Suflita et al. 1982). It appears to primarily occur together with methanogenesis.

It is found less frequently when sulfate- or nitrate-reducing conditions predominate (Sharak-Genthner et al. 1989).

Compounds subject to the most significant dechlorination processes usually do not transform under conventional aerobic or anaerobic conditions. This category includes insecticides, PCB, chlorinated benzenes, tetrachloroethylene, and tetrachloromethane. Other aromatic compounds that undergo reductive dechlorination are chlorophenols, chloroguaiacols, chloroveratroles, and chlorocatechols. Examples of compounds which undergo dechlorination are given below.

Chlorinated aliphatics may be reductively dechlorinated under anaerobic conditions. It was found that 0.75 mg/L of tetrachloroethylene can be sequentially dechlorinated to ethylene in enrichment cultures that produce methane (Freedman and Gossett 1991). In addition, the same methanogenic culture dechlorinated 91 mg/L of tetrachlorethylene to ethylene. However, methanogenesis ceased as the vinyl chloride conversion to ethylene increased. Thus, methanogenesis might not be a necessary condition for the reduction of tetrachloroethylene or vinyl chloride. Nevertheless, methanogenic bacteria may still play an essential role in these processes (DiStefano et al. 1991).

In a study by Freedman and Gossett (1991), a methanogenic consortium degraded dichloromethane. The predominant pathway for degradation was oxidation to carbon dioxide. Thus, the methane produced resulted from carbon dioxide reduction, but not from direct reduction of dichloromethane. Selective inhibition of methanogens did not affect the rate of degradation. It was concluded that acetogenic bacteria accomplished the degradation. Freedman et al. (1989) also observed dechlorination of tetrachloroethylene.

In a methanogenic process, chlorinated alkanes, such as trichloromethane, tetrachloromethane and tetrachloroethane, have been found to be degraded by *Acetobacterium woodii* and *Clostridium* sp. (Gälli and McCarthy 1989; Egli et al. 1988). The dichloromethane formed in these experiments appeared stable towards further degradation. In addition, tetrachloromethane was shown to undergo reductive dechlorination, whereas chloroform and dichloromethane were identified as intermediates; major products were carbon dioxide and acetate (Egli et al. 1988).

Other reports indicate that 1,1,1-trichloroethane, chloroform, and tetrachloromethane are reductively dechlorinated by anaerobic, acetogenic bacteria (Gälli and McCarthy 1989). Reductive dechlorination of 1,1,2-trichloroethane was found for *Pseudomonas putida* (Neilson et al. 1991). In a fixed-bed column filled with a mixture (3:1) of anaerobic sediment from the Rhine river and anaerobic granular sludge, tetrachloroethene was degraded to ethane.

No chlorinated compounds remained in the presence of lactate as an electron donor (De Bruin et al. 1992). A partial mineralization of tetrachloroethene (24% mineralization) was observed in a continuous-flow, fixed-film column (Vogel and McCarthy 1985). Trichloroethylene was the major intermediate formed, but traces of dichloroethylene isomers and vinyl chloride were also found. It was hypothesized that some of the vinyl chloride was transformed via an alcohol and an aldehyde to CO_2. Anaerobic degradation of TCE yields the toxic metabolite vinyl chloride (Neilson et al. 1991). Vinyl chloride can be further reduced to ethylene when methanol is provided as electron donor (Freedman and Gossett 1989).

Chlorinated cyclohexane degradation under methanogenic conditions yields monochlorobenzene (Holliger et al. 1989). *Clostridium rectum* degrades γ-HCH through tetrachlorocyclohexane and chlorobenzene (successive dechlorination; Neilson et al. 1990; Dieter 1990); γ-HCH (lindane) may be transformed through microbial activity to the more persistent α- and β-HCH. In soils, considerable lindane degradation proceeds only under anaerobic conditions, while aerobic degradation can be neglected (Niedermaier and Zech 1990).

Likewise, in surface waters and sediments anaerobic degradation rates exceed those under aerobic conditions (Fiedler et al. 1993). α-HCH was found to be bioconverted under methanogenic conditions, but with a very long "lag phase". Degradation started after about 30 days at a rate of 13 mg/kg soil per day. About 85% of the initial α-HCH was converted to monochlorobenzene, 3,5-dichlorophenol, and a trichlorophenol isomer, possibly 2,4,5-trichlorophenol. No significant bioconversion of α-HCH was observed under both denitrifying and sulfate-reducing conditions (Bachmann et al. 1988).

Chlorinated benzenes are dechlorinated by *Staphylococcus epidermidis* (Tsuchiya and Yamaha 1984). 1,2,4-trichlorobenzene was converted to dichlorobenzene, which was further converted to monochlorobenzene. This reaction only occurred with H_2 in the gas phase.

2,4,5-T, in enrichments of 3-chlorobenzoate, was either dechlorinated at the *para* position or first converted to 2,4,5-trichlorophenol and then dechlorinated at the *ortho* positions (Holliger et al. 1989). *Chlorobenzoates* are solely dechlorinated in the *meta* position. Total dechlorination of chlorobenzoate was necessary before the compound could be further mineralized to CH_4 and CO_2 (Suflita et al. 1982).

Hexachlorobenzene dechlorinates to tri- and dichlorobenzenes in anaerobic sewage sludge, and the chlorine is sequentially removed from the aromatic ring. Hexachlorobenzene reduces to pentachlorobenzene, then to 1,2,3,5-tetrachlorobenzene, and finally to 1,3,5-trichlorobenzene (Fathepure et al. 1988). All three isomers of trichlorobenzene have been reductively dechlorinated to monochlorobenzene via dichlorobenzenes in anaerobic sediment from the Rhine River (Bosma et al. 1988). Dechlorinating activity could only be maintained when an electron donor such as lactate, ethanol, and hydrogen was added.

Chlorophenols, chloroguaiacols, chloroveratroles, and *chlorocatechols*: anaerobic degradation of chlorophenols only proceeds under methanogenic conditions (Holliger et al. 1989). Methanogenic degradation of 2,6-dichlorophenol from surface layer sediments via reductive *ortho* dechlorination yielded 2-monochlorophenol and phenol (Nowak and Janke 1990). Dechlorination first takes place in an *ortho* position, then in *para* and finally in *meta* position. These compounds were reductively dechlorinated in an upflow anaerobic sludge blanket reactor, with glucose, methanol, and acetate employed as primary substrates (Woods et al. 1989). Complete mineralization did not occur; however, the chlorinated compounds were transformed to less-chlorinated homologues. Chlorocatechols from previous demethylation can undergo dechlorination (Neilson et al. 1990). The dechlorination reaction with chlorocatechols shows a high specificity in terms of the number and positions of chlorine substituents (Allard 1991).

An interesting phenomenon, finally, is the effect of the interactions between different classes of compounds on degradation. They allow for conditioning of consortia to certain compounds, but also may inhibit degradation of others. For example, relatively easily degradable chloroguaiacols retarded the degradation of PCP, while poorly degradable chloroguaiacols and -syringols had little or no effect on the rate of PCP degradation (Häggblom 1990). Many of the PCP-degrading strains have a wide substrate specificity for polychlorinated phenols, but poorly degrade mono- and dichlorophenols (Häggblom 1990).

Ortho Demethylation. *Ortho* demethylation is a reaction where the methyl group in an *ortho* position of an organic aromatic compound is eliminated. It has been found for chlorinated guaiacols and veratroles which transform into their corresponding catechols (Neilson et al. 1991, Neilson et al. 1990). *Ortho* demethylation is viewed as one of the reasons for the failure to recover chlorinated guaiacols and veratroles in sediment samples.

Other Degradation Pathways. In a number of cases the converting bacterium has been identified, but the exact metabolic pathway for the degradation of organochlorines is not yet known. However, some observations have been made, which are listed here as follows:

- Chlorolignins (complex phenyl-group-containing compounds) are degraded by a white-rot fungus (*Phanerochaete chrysosporium*) to some extent (Neilson et al. 1991; Eriksson and Kolar 1985). White-rot fungi excrete extracellular enzymes which efficiently degrade high molecular weight chlorolignins.
- *P. chrysosporium* dechlorinates bleach plant effluent by converting the organically bound chlorine into inorganic chloride (Prasad and Joyce 1992).
- Chlorinated diterpenoids, such as abietic acids ($C_{20}H_{30}O_2$), can be degraded by the fungus *Mortierella isabellina* through hydroxylation and oxidation reactions (Neilson et al. 1991).
- Buturon (a phenyl-urea herbicide) yields small metabolites such as chloroaniline and methoxy chloroaniline for several years after application (Scheunert 1991).
- For the fungicide quintozen (pentachloronitrobenzene), 28 metabolites containing chlorine atoms were found 8 years after application, e.g., chlorophenols, chloroanisols, chlorobenzenes, and chloroanilines (Scheunert 1991).

10.3.7 Sequential Biodegradation

Since strong inhibition has been shown in some cases of reductive dechlorination by other chlorinated chemicals (Suflita et al. 1983) or by chemical contaminations (Mikesell and Boyd 1986), the composition of waste (water) governs whether or not reductive dechlorination occurs. A change in waste composition or variations in sequences of aerobic or anaerobic process steps might allow full use of the potential of anaerobic bacteria for reductive dechlorination in waste and wastewater treatment.

Thus, occasionally, anaerobic–aerobic sequences are more successful in reducing toxicity than aerobically activated sludge alone. Such a sequence may be capable of mineralizing otherwise recalcitrant compounds. For example, tetrachloroethylene and tetrachloromethane may be mineralized by a sequential anaerobic–aerobic process. Initial anaerobic stages may accomplish reductive dechlorination, producing TCE and chloroform. Subsequent aerobic, methanotrophic stages may convert TCE and chloroform to carbon dioxide and water. Alternatively, anaerobic reductive dechlorination may produce vinyl chloride and chloromethane, which may be degraded in conventional aerobic processes if volatilization losses are minimized (Zitomer and Speece 1993).

In some situations, the primary function of aerobic stages in sequential anaerobic–aerobic systems has been efficient in reducing BOD (biological oxygen demand). Anaerobic stages may aid in the transformation of specific organic chemicals. For example, toxic chlorinated compounds of spent pulp bleaching liquors have successfully been detoxified in an anaerobic fluidized bed reactor followed by an aerobic trickling filter (Kringstad and Lindström 1984). Toxicity of highly chlorinated bleaching wastewater from the pulp and paper industry has been reduced by this process. Examples of sequential biodegradation are as follows:

- Chlorinated alkanes: tetrachloroethylene and chloroform have been degraded in a two-stage biofilm reactor consisting of an anaerobic column followed by a conventional aerobic column (Fathepure and Vogel 1991). Reductive dechlorination occurred in the anaerobic column, and trichlorinated and dichlorinated products were formed. In the aerobic column, the less-chlorinated intermediates were substantially transformed into carbon dioxide and nonvolatile products. The two-stage process resulted in 61%, 49%, and 23% mineralization of chloroform, tetrachloroethylene, and hexachlorobenzene, respectively. Dechlorination was most extensive when acetate served as the primary substrate; it occurred to a lesser extent when glucose and methanol served as primary substrates.
- Effluent from a kraft softwood bleach plant was treated by a two-stage process: white-rot fungus followed by a mixed anaerobic population in an anaerobic reactor. This treatment degrades and dechlorinates both high and low molecular weight chlorinated organic compounds (Prasad and Joyce 1992). The AOX removal efficiency in this sequential biological treatment was 66%.

10.3.8 Degradation Under Environmental Conditions

Many degradation processes have been observed in laboratories, where conditions considerably differ from conditions in the outdoor "natural" environment, i.e., the aquatic systems of streams, lakes, and the sea. In those systems, microorganisms are provided with a number of substances which are much better available and accessible than the organochlorines to be degraded.

The concentration of the chemical to be degraded is important. For example, laboratory results suggest that reductive dechlorination of PCB is less effective at concentrations below 200 ppm. However, PCB are designated as hazardous at much lower concentrations than those used in laboratory studies. Availability from soils or sediment of newly added PCB will be different from availability of PCB that have aged in their natural environment and may be tightly bound (Young and Häggblom 1991).

Besides substrate differences, a difference in reaction temperature often may facilitate other reactions than those observed under laboratory conditions. Also, sunlight and the corresponding algae productivity, and accordingly oxygen conditions influence the physicochemical milieu in natural environments. Finally, due to different physicochemical conditions, the consortia of microorganisms in the natural environment may differ considerably from those investigated in the laboratory.

Biodegradation of chlorinated compounds in natural systems may be facilitated by the application of nonchlorinated analogues. This has been demonstrated with, for example, chlorinated anilines (You and Bartha 1982) and PCB (Brunner et al. 1985). In other cases, organisms degrading chlorinated substrates do not necessarily degrade the nonchlorinated analogues: organisms degrading 2,6-dichlorotoluene (Vandenberg et al. 1981), PCP (Stanlake and Finn 1982), and 4-chlorobenzoate (Barton and Crawford 1988) appeared to be unable to degrade the nonchlorinated compounds.

For biological treatment it is important to choose strains of microorganisms —whether they are natural or genetically engineered—which are well adapted to the particular environmental conditions if breakdown of the target pollutants is to be successful (McClure et al. 1991). Details of degradation processes under environmental conditions for several compounds are given below.

Chlorinated aliphatics: aerobic biodegradation of dichloromethane in soils was found to have half-lives ranging from 1 to about 200 days at concentrations ranging from 0.1 to 5.0 ppm. Anaerobic degradation had longer acclimation periods but higher subsequent rates of degradation. The degradation rates were dependent on soil type, substrate concentration, and redox potential of the soil. Preexposure to dichloromethane increased degradation in subsequent exposures (Davis and Madsen 1991).

Chlorinated phenols: degradation of monochlorinated phenols and 2,4-dichlorophenol (0.1 mM) under anaerobic (sulfidogenic) conditions in a sediment under saline and freshwater conditions took about 120–200 days for complete removal, with a lag period of 50–100 days. After acclimatization the degradation rate considerably increased (Häggblom and Young 1990). The biodegradation of PCP in outdoor, aquatic environments was examined with man-made channels. The channels were dosed continuously with various concentrations of PCP. The biodegradation became significant about 3 weeks after the initial dose. It eventually became the primary mechanism of removal, accounting for a 26%–46% (dose-dependent) decline in initial PCP concentration. Total bacterial numbers in the channels were not affected significantly by PCP concentrations (Pignatello et al. 1983).

Edgehill and Finn (1983) showed that soil was more rapidly cleared of PCP if inoculated with specialized *Arthrobacter* cells. A soil contaminated with a high concentration of 2,4,5-T was repeatedly inoculated with *Pseudomonas cepacia*, which is capable of utilizing 2,4,5-T as the sole source of carbon and energy (Kilbane et al. 1983). The contaminated soil samples showed more than 90% degradation after six treatments with the strain, while the concentration of the contaminant remained unchanged in the uninoculated soil.

In anaerobic sediments monochlorophenols were degraded more readily than monochlorobenzoates. Within the chlorophenols the relative order of degradability was *ortho > meta > para* (Sharak-Genthner et al. 1989).

Chloroanilines: degradation of dichloroaniline (2,4- and 3,4-) to monochloroanilines in anaerobic pond sediments was completed in about 8 weeks. Reductive dechlorination of 3,4-dichloroaniline started after a lag period of 3 weeks. Cross-acclimation could be observed for dechlorination of dichloroanilines and dichlorophenols. Anaerobic bacteria in pond sediment acclimated to dechlorinate dichlorophenol (2,4- or 3,4-) and rapidly dechlorinated dichloroaniline (2,4- or 3,4-) without any time lag (Struijs and Rogers 1989). Cross-acclimation among the monochlorinated phenols and the corresponding monochlorinated anilines did not occur in these studies. Apparently, fundamental differences exist between enzymatic machineries that reductively dechlorinate mono- and dichlorophenols.

Chlorobenzenes and *chlorobenzoates*: unlike chlorophenols, degradability of chlorobenzoates was *meta > ortho > para* (Sharak-Genthner et al. 1989). All three isomers of trichlorobenzene were reductively dechlorinated to monochlorobenzene via dichlorobenzenes in anaerobic sediment from the Rhine River (Bosma et al. 1988). Dechlorinating activity could only be maintained if an electron donor such as lactate, ethanol, or hydrogen was added.

10.4 NONBIOLOGICAL DEGRADATION PROCESSES

10.4.1 Photochemical Degradation

Numerous transformation reactions are photochemically mediated, such as dechlorination, oxidation, and ring cleavage (Neilson et al. 1991). Photochemical degradation is an abiotic degradation process for organic chemicals and is based on absorption of electromagnetic radiation. For the sake of clarity, it is presented here separately from other chemical degradation mechanisms.

Photochemical degradation is limited to the compartments atmosphere, water, and surfaces of vegetation and soil. In the atmosphere of northern Europe significant photochemical reactions in the fate of chemicals is limited to the summer months (Neilson et al. 1991). In aqueous environments, color and clarity of the water governs photolysis rates. In eutrophic waters, light penetration will only be a few meters at maximum. For example, pulp mill effluents, especially high molecular weight chlorination products, show strong light absorption so that light penetration is limited to the surface layer. Therefore, photolysis as well as photosynthesis are inhibited (Kordsachia 1985).

Photolysis works by two different mechanisms, i.e., direct and sensitized photolysis.

- *Direct photolysis* of chemicals often involves and incorporates molecular oxygen. Usually, the initial step is the formation of free radicals. The reaction proceeds after absorption of light by the pollutant itself.

- *Sensitized photolysis* is initiated through light absorption by a second chemical that subsequently reacts with the pollutant or mediates the degradation reaction for the pollutant. An example is the excitation of oxygen that yields singlet oxygen, which can react with other compounds.

The primary products of photolytic degradation of chloro-organics are carbon dioxide (CO_2) and hydrochloric acid (HCl). They have been found as degradation products for the following chemicals after exposure to wavelengths greater than 290 nm (for comparison: UV-B light has wavelengths in the range 280–320 nm; Korte 1987, 1992):

- Hexachlorobutadiene
- Dichloropropene
- Dichloroethene
- Tetrachloroethene
- Dichlorofluoromethane
- Trichlorofluoromethane
- Chlorinated benzenes

Tetrachloroethylene is transformed photolytically to trichloroacetic acid (TCA), which is also used as a herbicide. Atmospheric TCE degradation yields dichloroacetic acid, and atmospheric radical reactions also degrade TCE to TCA (Frank et al. 1991).

The rate of photolysis is a function of the quantum yield of the reaction, the absorption cross section of the molecule, and the solar flux in the absorption spectrum of the chemical. A number of chlorinated chemicals change their absorption spectrum to shorter wavelengths when they are sorbed to active inorganic surfaces (e.g., PCP, tetrachlorobiphenyl, DDT, DDE). By these means photodegradation is enhanced (Ell et al. 1990) and photochemical reactions which usually would not occur under stratospheric conditions are possible. For PCP, this so-called bathochrome shift in the absorption frequency changes the absorption maximum from 290 nm to 319 nm. Complete photodegradation of PCP in the atmosphere can be expected within 5–7 days (Korte 1987, 1992).

10.4.2 Chemical Degradation

Chemical degradation, as opposed to biological degradation, proceeds without any enzymatic catalysts. Therefore, chemical degradation rates often are slow compared with biological degradation rates, especially for recalcitrant compounds such as chlorinated organics. Nevertheless, for environmental compartments of negligible biological activity, such as groundwater or the atmosphere, chemical degradation may be the only sink for chlorinated organics.

10.4.3 Hydrolysis

Hydrolysis—or the more general term solvolysis—is a chemical degradation reaction that cleaves covalent bonds while incorporating H_2O into a molecule. In aquatic systems, many organic chemicals readily hydrolyze. For a number of pesticides, hydrolysis is a detoxification mechanism. Heptachlor, for example, undergoes hydrolysis under environmental conditions (Korte 1987, 1992) in a dechlorination reaction.

If the pH regime of an aquatic system is known, the rate of hydrolytic degradation can be reasonably well predicted (Baughman and Burns 1980). Alkaline conditions, as present in seawater with a pH above 7, make organic compounds more susceptible to hydrolysis. In some cases their half-lives, i.e., the time it takes for half of the original amount of the chemical to be degraded, are drastically decreased (Jeffers et al. 1989).

Abiotic hydrolysis rates were examined for a number of chlorinated compounds of one- to three-carbon atoms. For standard conditions (25°C, pH = 7) half-lives due to hydrolysis were calculated as 0.004–2.2 years for chlorinated propanes, 0.4–140 years for chlorinated ethanes, and 106–1010 years for chlorinated ethylenes. In the natural environment, a combination of steric and energetic effects slows down hydrolysis of persistent compounds (Jeffers et al. 1989).

Under anaerobic conditions, at a pH of about 7, the abiotic dechlorination of 1,2-dichloroethane proceeds primarily via reaction with water and sulfide. In the presence of sulfide, the half-life of 1,2-dichloroethane under environmental conditions (pH = 7, 15°C) in water is 23 years. For sulfide-free water, it is even longer, by approximately one order of magnitude (Barbash and Reinhard 1989). For PCB under environmental conditions hydrolysis is negligible (Fiedler et al. 1995).

10.4.4 Oxidation

Oxidation is a chemical degradation process often initiated by reactive oxygen species, such as free radicals, singlet oxygen ($^1O_2$), peroxides, or ozone (O_3). Oxygen molecules hardly are directly involved in the reaction. They merely act as an ultimate electron sink in most environmental oxidation reactions (Mill 1990). Important oxidants in the troposphere are OH radicals and ozone. In fog droplets, oxidation facilitated by hydroxy radicals might be a significant removal process for chlorinated aromatic compounds (Sedlak and Andren 1991a, 1991b).

Ozone reacts with halogenated double bonds, as found in DDE, chloroethene, and other compounds. The rate of reaction decreases with increasing degree of chlorination. Steric effects also play an important role in oxidation reactions. Singlet oxygen does not react with chlorinated double bonds or double bonds with steric hindrance (Korte 1987, 1992).

Chlorinated organics in the troposphere—in the form of gaseous compounds—are transformed into radicals when reacting with hydroxy radicals. Chlorine atoms resulting from these reactions destroy ozone molecules (Graedel 1980). In addition to well-known chlorinated compounds, such as chlorofluorocarbon, stratospheric chlorine is also naturally supplied to some degree from methyl chloride produced in marine environments. These compounds add to the so-called greenhouse effect, which, however, is not the subject of this chapter.

Oxidation reactions in soils can be catalyzed either by enzymes (i.e., biotic degradation) or metal oxides (i.e., chemical degradation; Korte, 1987, 1992).

10.5 REFERENCES

Allard, A. S., 1991, "Dechlorination of Chlorocatechols by Stable Enrichment Cultures of Anaerobic Bacteria," *Appl. Environ. Microbiol.*, *57*, 77–84

Allard, A.-S., M. Remberger, and A. H. Neilson, 1985, "Bacterial O-Methylation of Chloroguaiacols: Effect of Substrate Concentration, Cell Density and Growth Conditions," *Appl. Environ. Microbiol.*, *49*, 279–288

Apajalahti, J. H. A., and M. S. Salkinoja-Salonen, 1986, "Degradation of Polychlorinated Phenols by *Rhodococcus Chlorophenolicus*," *Appl. Microbiol. Biotechnol.*, *25*, 62–67

Aust, S. D., 1993, "Biodegradation of Halogenated Organic Pollutants," Presentation on the *Dioxin '93 – 13th Int. Symp. on Chlorinated Dioxins and Related Compounds*, Vienna, Austria, 20–24, September 1993; *Environ. Sci. Pollut. Res.*, *1*, 61

Bachmann, A., P. Walet, P. Wijnen, W. De Bruin, J. L. M. Huntjens, W. Roelofsen, and A. J. B. Zehnder, 1988, "Biodegradation of α- and β-Hexachlorocyclohexane in a Soil Slurry under Different Redox Conditions," *Appl. Environ. Microbiol.*, *54*, 143–149

Ball, W. P., and P. V. Roberts, 1991, "Long-Term Sorption of Halogenated Organic Chemicals by Aquifer Material. 1. Equilibrium," *Environ. Sci. Technol.*, *25*, 1223–1237

Banerjee, S., R. H. Sugatt, and D. P. V. O'Grady, 1984, "A Simple Method for Determining Bioconcentration Parameters of Hydrophobic Compounds," *Environ. Sci. Technol.*, *18*, 79–81

Barbash, J. E., and M. Reinhard, 1989, "Abiotic Dehalogenation of 1,2-Dichloroethane and 1,2-Dibromomethane in Aqueous Solution Containing Hydrogen Sulfide," *Environ. Sci. Technol.*, *23*, 1349–1357

Bartels, I., H.-J. Knackmus, and W. Reineke, 1984, "Suicide Inactivation of Catechol 2,3-Dioxygenase from *Pseudomonas putida* by 3-Halocatechols," *Appl. Environ. Microbiol.*, *47*, 500–505

Barton, M. R., and R. L. Crawford, 1988, "Novel Biotransformation of 4-Chlorobiphenyl by a *Pseudomonas* sp.," *Appl. Environ. Microbiol.*, *54*, 594–595

Battersby N. S., and V. Wilson, 1989, "Survey of the Anaerobic Biodegradation Potential of Organic Chemicals in Digesting Sludge," *Appl. Environ. Microbiol.*, *55*, 433–439

Baughman, G. L., and L. A. Burns, 1980, "Transport and Transformation of Chemicals," in *Handbook of Environmental Chemistry 2A* (O. Hutzinger, Ed.), Springer, Heidelberg, Germany, pp. 1–18

Bosma, T. N. P., J. R. van der Meer, G. Schraa, M. E. Tros, and A. J. B. Zehnder, 1988, "Reductive Dechlorination of All Trichloro- and Dichlorobenzene Isomers," *FEMS Microbiol. Ecol.*, *53*, 223–229

Bouwer, E. J., and P. L. McCarthy, 1983, "Transformations of Halogenated Organic Compounds under Denitrification Conditions," *Appl. Environ. Microbiol.*, *45*, 1295–1299

Brannon, J. M., T. E. Myers, D. Gunnison, and C. B. Price, 1991, "Nonconstant Polychlorinated Biphenyl Partitioning in New Bedford Harbor Sediment during Sequential Batch Leaching," *Environ. Sci. Technol.*, *25*, 1082–1087

Brunner, S., E. Hornung, H. Santl, E. Wolff, O. Piringer, J. Altschuh, and R. Brüggemann, 1990, "Henry's Law Constants for Polychlorinated Biphenyls: Experimental Determination and Structure–Property Relationships," *Environ. Sci. Technol.*, *24*, 1751–1754

Brusseau, M. L., R. E. Jessup, and P. S. C. Rao, 1990, "Sorption Kinetics of Organic Chemicals: Evaluation of Gas-Purge and Miscible-Displacement Techniques," *Environ. Sci. Technol.*, *24*, 727–735

Czuczwa, J. M., and R. A. Hites, 1984, "Environmental Fate of Combustion-Generated Polychlorinated Dioxins and Furans," *Environ. Sci. Technol.*, *18*, 444–450

Davis, J. W., and S. S. Madsen, 1991, "The Biodegradation of Methylene Chloride in Soils," *Environ. Tox. Chem.*, *10*, 463–474

De Bruin, W. P., M. J. J. Kotterman, M. A. Posthumus, G. Schraa, and A. J. B. Zehnder, 1992, "Complete Biological Reductive Transformation of Tetrachloroethene to Ethane," *Appl. Environ. Microbiol.*, *58*, 1996–2000

de Wit C., 1993, The Toxicology Forum, Cosponsored by the Bundesminister für Umwelt, Naturschutz und Reaktorsicherheit, Senator für Stadtentwicklung und Umweltschutz von Berlin, and Minister für Umwelt, Naturschutz und Raumordnung des Landes Brandenburg, pp. 354 and 359, Berlin, Germany, 9–11 November 1992

Dieter, H. H., 1990, "Halogen-Organische Verbindungen," *UWSF-Z. Umweltchem. Ökotox.*, *2*, 220–225

DiStefano, T. D., J. M. Gossett, and S. H. Zinder, 1991, "Reductive Dechlorination of High Concentrations of Tetrachloroethene to Ethene by an Anaerobic Enrichment Culture in the Absence of Methanogenesis," *Appl. Environ. Microbiol.*, *57*, 2287–2292

Dolfing, J., and J. M. Tiedje, 1991, "Influence of Substituents on Reductive Dehalogenation of 3-Chlorobenzoate Analogs," *Appl. Environ. Microbiol.*, *57*, 820–824

Dunnivant, F. M., J. T. Coates, and A. W. Elzerman, 1988, "Experimentally Determined Henry's Law Constants for 17 Polychlorobiphenyl Congeners," *Environ. Sci. Technol.*, *22*, 448–452

Edgehill, R. U., and R. K. Finn, 1983, "Microbial Treatment of Soil to Remove Pentachlorophenol," *Appl. Environ. Microbiol.*, *45*, 1122–1125

Egli, C., T. Tschan, R. Scholtz, A. M. Cook, and T. Leisinger, 1988, "Transformation of Tetrachloromethane to Dichloromethane and Carbon Dioxide by *Acetobacterium woodii*," *Appl. Environ. Microbiol.*, *54*, 2819–2824

Eisenreich, S. J., P. D. Capel, J. A. Robbins, and R. Bourbonniere, 1989, "Accumulation and Diagenesis of Chlorinated Hydrocarbons in Lacustrine Sediments," *Environ. Sci. Technol.*, *23*, 1116–1126

Ell, R., H. Fiedler, and O. Hutzinger, 1990, *Toxic and Persistent Organic Compounds, Distribution and Health Effects in Humans: Tetrachloroethene*, Review, Chair of Ecological Chemistry and Geochemistry, University of Bayreuth, Germany

Eriksson, K. E., and M. C. Kolar, 1985, "Microbial Degradation of Chlorolignins," *Environ. Sci. Technol.*, *19*, 1085–1089

Fathepure, B. Z., J. M. Tiedje, and S. A. Boyd, 1988, "Reductive Dechlorination of Hexachlorobenzene to Tri- and Dichlorobenzenes in Anaerobic Sewage Sludge," *Appl. Environ. Microbiol.*, *54*, 327–330.

Fiedler, H., C. Lau, S. Schulz, C. Wagner, O. Hutzinger, and K.T. von der Trenck, 1995, Stoffbericht Polychlorierte Biphenyle (PCB). Materialien zur Altlastenbearbeitung *16/95*. Landesanstalt für Umweltschutz Baden-Württemberg (Ed.), Karlsruhe, Germany

Fiedler, H., M. Hub, and O. Hutzinger, 1993, Stoffbericht Hexachlorcyclohexan (HCH). *Handbuch Altlasten – Texte und Berichte zur Altlastenbearbeitung 9/93*, Landesanstalt für Umweltschutz Baden-Württemberg (Ed.), Karlsruhe, Germany

Frank, H., W. Frank, and M. Gey, 1991, "C_1- und C_2-Halogenkohlenwasserstoffe," *UWSF-Z. Umweltchem. Ökotox.*, *3*, 167–175

Freedman, D. L., and J. M. Gossett, 1991, "Biodegradation of Dichloromethane and Its Utilization as a Growth Substrate under Methanogenic Conditions," *Appl. Environ. Microbiol.*, *57*, 2847–2857

Freedman, D. L., and J. M. Gossett, 1989, "Biological Reductive Dechlorination of Tetrachloroethylene and Trichloroethylene to Ethylene under Methanogenic Conditions," *Appl. Environ. Microbiol.*, *55*, 2144–2151

Gälli, R., and P. L. McCarthy, 1989, "Biotransformation of 1,1,1-Trichloroethane, Trichloromethane and Tetrachloromethane by a *Clostridium* sp," *Appl. Environ. Microbiol.*, *55*, 837–844

Gibson, S. A., and J. M. Suflita, 1990, "Anaerobic Biodegradation of 2,4,5-Trichlorophenoxacetic Acid in Samples from Methanogenic Aquifer: Stimulation by Short-Chain Organic Acids and Alcohols," *Appl. Environ. Microbiol.*, *56*, 1825–1832

Gobas, F. A. P. C., E. J. McNeil, L. Lovett-Doust, and G. D. Haffner, 1991, "Bioconcentration of Chlorinated Aromatic Hydrocarbons in Aquatic Macrophytes," *Environ. Sci. Technol.*, *25*, 924–929

Gosset J. M. (1987), "Measurement of Henry's Law Constants for C_1 and C_2 Chlorinated Hydrocarbons," *Environ. Sci. Technol.*, *21*, 202–208

Graedel, T. E., 1980, "Atmospheric Photochemistry," in *Environmental Chemistry* (O. Hutzinger, Ed.), *2A*, Springer, Heidelberg, Germany, pp. 107–144

Gschwend, P. M., and R. P. Schwarzenbach, 1992, "Physical Chemistry of Organic Compounds in the Marine Environment," *Mar. Chem.*, *39*, 187–207

Haderlein, S., and R. P. Schwarzenbach, 1993, "Adsorption of Substituted Nitrobenzenes and Nitrophenols to Mineral Surfaces," *Environ. Sci. Technol.*, *27*, 316–326

Häggblom, M., 1990, "Mechanisms of Bacterial Degradation and Transformation of Chlorinated Monoaromatic Compounds," *J. Basic Microbiol.*, *30*, 115–141

Häggblom, M. M., and L. Y. Young, 1990, "Chlorophenol Degradation Coupled to Sulfate Reduction," *Appl. Environ. Microbiol.*, *56*, 3255–3260

Häggblom, M. M., D. Janke, and M. S. Salkinoja-Salonen, 1989, "Transformations of Chlorinated Phenolic Compounds in the Genus Rhodococcus," *Microb. Ecol.*, *18*, 147–159

Häggblom, M., J. H. A. Apajalahti, and M. S. Salkinoja-Salonen, 1988a, "Hydroxylation and Dechlorination of Chlorinated Guaiacols and Syringols by *Rhodococcus chlorophenolicus*," *Appl. Environ. Microbiol.*, *54*, 683–687

Häggblom, M., L. J. Nohynek, and M. S. Salkinoja-Salonen, 1988b, "Degradation and O-methylation of Polychlorinated Phenolic Compounds by Strains of *Rhodococcus and Mycobacterium*," *Appl. Environ. Microbiol.*, *54*, 3043–3052

Harker, A. R., and Y. K. Kim, 1990, "Trichloroethylene Degradation by Two Independent Aromatic-Degrading Pathways in *Alcaligenes eutrophus*," *Appl. Environ. Microbiol.*, *56*, 1179–1181

Hassett, J. P., and E. Milicic, 1985, "Determination of Equilibrium and Rate Constants for Binding of a Polychlorinated Biphenyl Congener By Dissolved Humic Substances," *Environ Sci. Technol.*, *19*, 638–643

Hauk, A., and K. W. Schramm, 1990, "Molecular Modeling in Environmental Science," *Toxicol. Environ. Chem.*, *26*, 45–54

Holliger, C., A. J. M. Stams, and A. J. B. Zehnder, 1989, "Anaerobic Degradation of Recalcitrant Compounds," *Anaerobic Dig. Proc. Int. Symp.*, 211–224

Hoover, D. G., G. E. Borgonovi, S. H. Jones, and M. Alexander, 1986, "Anomalies in Mineralization of Low Concentrations of Organic Compounds in Lake Water and Sewage," *Appl. Environ. Microbiol.*, *51*, 226–232

Janke, D., and W. Ihn, 1990, "Bacterial Cometabolism of Chlorinated Phenols and Anilines: Biotechnological Implications of the Phenomenon," *Organohalogen Compd.*, *1*, Eco-Informa Press, Bayreuth, Germany, pp. 355–358

Janssen, D. B., A. Scheper, L. Dijkhuizen, and B. Witholt, 1985, "Degradation of Halogenated Aliphatic Compounds by *Xanthobacter autotrophicus*," *Appl. Environ. Microbiol.*, *49*, 673–677

Jeffers, J., L. M. Ward, L. M. Woytowitch, and N. L. Wolfe, 1989, "Homogeneous Hydrolysis Rate Constants for Selected Chlorinated Methanes, Ethanes, Ethenes, and Propanes," *Environ. Sci. Technol.*, *23*, 965–969

Jota, M. A. T., and J. P. Hassett, 1991, "Effects of Environmental Variables on Binding of a PCB Congener by Dissolved Humic Substances," *Environ. Toxicol. Chem.*, *10*, 483–491

Kilbane, J. J., D. K. Chatterjee, and A. M. Chakrabarty, 1983, "Detoxification of 2,4,5-Trichlorophenoxyacetic Acid from Contaminated Soil by *Pseudomonas cepacia*," *Appl. Environ. Microbiol.*, *45*, 1697–1700

Klaine, S. J., M. L. Hinman, K. R. Sauser, J. R. Martin, and L. W. Moore, 1988, "Characterization of Agricultural Nonpoint Pollution: Pesticide Migration in a West Tennessee Watershed," *Environ. Toxicol. Chem.*, *7*, 609–614

Klecka, G. M., and D. T. Gibson, 1981, "Inhibition of Catechol 2,3-Dioxygenase from *Pseudomonas putida* by 3-Chlorocatechol," *Appl. Environ. Microbiol.*, *41*, 1159–1165

Kohring, G.-W., J. E. Rogers, J. Wiegel, 1989, "Anaerobic Biodegradation of 2,4-Dichlorophenol in Freshwater Lake Sediments at Different Temperatures," *Appl. Environ. Microbiol.*, *55*, 348–353.

Kordsachia, O., 1985, "Organische Chlorverbindungen in Bleichabwässern," *Papier*, *39*, 207–213

Korte, F., 1987 and 1992, *Lehrbuch der ökologischen Chemie*, 2nd and 3rd Edition, Thieme Verlag, Stuttgart, Germany

Kostiuk, E., G. L. Amy, and E. Aprahamian, 1990, "The Partitioning and Removal of Organic Halide Across an Aerated Stabilization Basin (ASB): Treating Kraft Mill Wastewater," *Environ. Eng.*, 1990, 865–872

Kringstad, K. P., and K. Lindström, 1984, "Spent Liquors from Pulp Bleaching," *Environ. Sci. Technol.*, *18*, 236A–248A

Kucklick, J. R., D. A. Hinckley, and T. F. Bidleman, 1991, "Determination of Henry's Law Constants for Hexachlorocyclohexanes in Distilled Water and Artificial Seawater as a Function of Temperature," *Mar. Chem.*, *34*, 197–209

Lee, L. S., P. S. C. Rao, M. L. Brusseau, 1991, "Nonequilibrium Sorption and Transport of Neutral and Ionized Chlorophenols," *Environ. Sci. Technol.*, *25*, 722–729

Leuenberger, C., W. Giger, R. Coney, J. W. Graydon, and E. Molnar-Kubica, 1985, "Persistent Chemicals in Pulp Mill Effluents," *Water Res.*, *19*, 885–894

Mackay, D., and A. T. K. Yeun, 1983, "Mass Transfer Coefficient Correlations for Volatilization of Organic Solutes from Water," *Environ. Sci. Technol.*, *17*, 211–217

Madeley, J. R., and R. D. N. Birtley, 1980, "Chlorinated Paraffins and the Environment," *Environ. Sci. Technol.*, *14*, 1215–1221

Marks, T. S., A. E. W. Smith, and A. V. Quirk, 1984, "Degradation of 4-chlorobenzoic Acid by *Arthrobacter* sp.," *Appl. Environ. Microbiol.*, *48*, 1020–1025

McClure, N. C., A. J. Weightman, and J. C. Fry, 1989, "Survival of *Psedomonas putida* UWC1 Containing Cloned Catabolic Genes in a Model Activated-Sludge Unit," *Appl. Environ. Microbiol.*, *55*, 2627–2634

Mikesell, M. D., and S. A. Boyd, 1986, "Complete Reductive Dechlorination and Mineralization of Pentachlorophenol by Anaerobic Microorganisms," *Appl. Environ. Microbiol.*, *52*, 861–865

Mill, T., 1980, "Chemical and Photo Oxidation," in *Environmental Chemistry 2A* (O. Hutzinger, Ed.), Springer, Heidelberg, Germany, pp. 77–106

Mohn, W. W., and J. M. Tiedje, 1991, "Evidence for Chemiosmotic Coupling of Reductive Dechlorination and ATP Synthesis in *Desulfomonile tiedjei*," *Arch. Microbiol.*, *157*, 1–6

Murphy, T. J., M. D. Mullin, and J. A. Meyer, 1987, "Equilibration of Polychlorinated Biphenyls and Toxaphene with Air and Water," *Environ Sci. Technol.*, *21*, 155–162

Neilson, A. H., 1990, "A Review: The Biodegradation of Halogenated Organic Compounds," *J. Appl. Bacteriol.*, *69*, 445–470

Neilson, A. H., A.-S. Allard, P.-A. Hynning, and M. Remberger, 1991, "Distribution, Fate and Persistence of Organochlorine Compounds Formed during Production of Bleached Pulp," *Toxicol. Environ. Chem.*, *30*, 3–41

Neilson, A. H., A.-S. Allard, P.-A. Hynning, M. Remberger, and T. Viktor, 1990, "The Environmental Fate of Chlorophenolic Constituents of Bleachery Effluents," *Tappi J.*, *73*, 239–247

Nelson, M. J. K., S. O. Montgomery, W. R. Mahaffey, and P. H. Pritchard, 1987, "Biodegradation of Trichloroethylene and Involvement of an Aromatic Biodegradation Pathway," *Appl. Environ. Microbiol.*, *53*, 949–954

Niedermaier, M., and W. Zech, 1990, "Lindan im Waldhumus," *UWSF-Z. Umweltchem. Ökotox.*, *2*, 6–9

Nondek, L., and N. Frolikova, 1991, "Polychlorinated Biphenyls in the Hydrosphere of Czechloslovakia," *Chemosphere*, *23*, 269–280

Nowak, J., and D. Janke, 1990, "Reductive Dechlorination of 2,6-Dichlorophenol in an Anaerobic Digester," in *DIOXIN '90*, Vol. 1, Short Papers – Toxicology-Environment, Food, Exposure-Risk, Eco-Informa-Press, Bayreuth, Germany, pp. 363–366

Oldenhuis, R., R. L. J. M. Vink, D. B. Janssen, and B. Witholt, 1989, "Degradation of Chlorinated Aliphatic Hydrocarbons by *Methylosinus trichosporium* Expressing Soluble Methane Monooxygenase," *Appl. Environ. Microbiol.*, *55*, 2819–2826

Oliver, B. G., and A. J. Niimi, 1988, "Trophodynamic Analysis of Polychlorinated Biphenyl Congeners and Other Chlorinated Hydrocarbons in the Lake Ontario Ecosystem," *Environ. Sci. Technol.*, *22*, 388–397

Opperhuizen, A., 1986, "Bioconcentration in Fish and Other Distribution Processes of Hydrophobic Chemicals in Aqueous Environments," *Ph.D. Thesis*, University of Amsterdam, The Netherlands

Paasivirta, J., H. Hakala, T. Otollinen, J. Särkkä, L. Welling, R. Paukku, and R. Lammi, 1990, "Organic Chlorine Compounds in Lake Sediments. III. Chlorohydrocarbons, Free and Bound Chlorophenols," *Chemosphere*, *20*, 1355–1370

Paris, D. F., and N. L. Wolfe, 1987, "Relationship between Properties of a Series of Anilines and Their Transformation by Bacteria," *Appl. Environ. Microbiol.*, *53*, 911–916

Pearson, C. R., 1982, "Halogenated Aromatics," in *Environmental Chemistry* 3 B (O. Hutzinger, Ed.), Springer, Heidelberg, Germany, pp. 89–116

Pignatello, J. J., 1991, "Desorption of Tetrachloroethene and 1,2-Dibromo-3-Chloropropane from Aquifer Sediments," *Environ. Toxicol. Chem.*, *10*, 1399–1404

Prasad, D.Y., and T. W. Joyce, 1992, "Removal of Chlorinated Organics from Kraft Softwood Bleach Plant Effluent by Sequential Biological Treatment Using White-Rot Fungus and an Anaerobic Reactor," in *46th Purdue Industrial Waste Conf. Proc.*, Lewis, Chelsea, USA

Pritchard, P. H., E. J. O'Neill, C. M. Spain, and D. G. Ahearn, 1987, "Physical and Biological Parameters that Determine the Fate of *p*-Chlorophenol in Laboratory Test Systems," *Appl. Environ. Microbiol.*, *53*, 1833–1838

Remberger, M., P.-A. Hynning, and A. Neilson, 1993, "Release of Chlorocatechols from a Contaminated Sediment," *Environ. Sci. Technol.*, *27*, 158–164

Remberger, M., P. A. Hymming, and A. H. Neilson, 1988, "Comparison of Procedures for Recovering Chloroguaiacols and Chlorocatechols from Contaminated Sediments," *Environ. Toxicol. Chem.*, *7*, 795–805

Resendes, J., W. Y. Shiu, and D. Mackay, 1992, "Sensing the Fugacity of Hydrophobic Organic Chemicals in Aqueous Systems," *Environ. Sci. Technol.*, *26*, 2281–2387

Rijnaarts, H. H. M., A. Bachmann, J. C. Jumelet, and A. J. B. Zehnder, 1990, "Effect of Desorption and Intraparticle Mass Transfer on the Aerobic Biomineralization of α-Hexachlorocyclohexan in Contaminated Calcareous Soil," *Environ. Sci. Technol.*, *24*, 1349–1354

Safe, S., 1990, "Polychlorinated Biphenyls (PCBs), Dibenzo-*p*-dioxins (PCDDs), Dibenzofurans (PCDFs), and Related Compounds: Environmental and Mechanistic Considerations which Support the Development of Toxic Equivalency Factors (TEFs)," *CRC Crit. Rev. Toxicol.*, *21*, 51–88

Scheunert, I., 1991, "Langzeitverhalten von Chemikalien im Boden," *UWSF-Z. Umweltchem. Ökotox.*, *1*, 28–32

Schrap, S. M., and A. Opperhuizen, 1991, "On the Contradiction Between Experimental Sorption Data and the Sorption Partitioning Model," *Chemosphere*, *24*, 1259–1282

Schwarzenbach, R. P., P. M. Gschwend, and D. M. Imboden, 1993, *Environmental Organic Chemistry*, Wiley, New York, Chichester, Brisbane, Toronto, Singapore

Schwarzenbach, R. P., W. Giger, E. Hoehn, and J. K. Schneider, 1983, "Behavior of Organic Compounds during Infiltration of River Water to Groundwater. Field Studies," *Environ. Sci. Technol.*, *17*, 472–479

Scow, K. M., and J. Hutson, 1992, "Effect of Diffusion and Sorption on the Kinetics of Biodegradation: Theoretical Considerations," *Am. J. Soil Sci. Soc.*, *56*, 119–127

Sedlak, D. L., and A. W. Andren, 1991a, "Oxidation of Chlorobenzene with Fenton's Reagent," *Environ. Sci. Technol.*, *25*, 777–782

Sedlak, D. L., and A. W. Andren, 1991b, "Aqueous Phase Oxidation of Polychlorinated Biphenyls by Hydroxy Radicals," *Environ. Sci. Technol.*, *25*, 1419–1427

Sharak-Genthner, B. R., W. A. Price, and P. H. Prichard, 1989, "Anaerobic Degradation of Chloroaromatic Compounds in Aquatic Sediments Under a Variety of Enrichment Conditions," *Appl. Environ. Microbiol.*, *55*, 1466–1471

Shelton, D. R., and J. M. Tiedje, 1984, "Isolation and Partial Characterization of Bacteria in an Anaerobic Consortium that Mineralizes 3-Chlorobenzoic Acid," *Appl. Environ. Microbiol.*, *48*, 840–848

Shiu, W. Y., W. Doucette, F. A. P. C. Gobas, A. Andren, and D. Mackay, 1988, "Physical–Chemical Properties of Chlorinated Dibenzo-*p*-dioxins," *Environ. Sci. Technol.*, *22*, 651–658

Sijm, D., 1992, "Influence of Biotransformation on Bioaccumulation and Toxicity of Chlorinated Aromatic Compounds in Fish," *Ph.D. Thesis*, Rijksuniversiteit te Utrecht, The Netherlands

Smith, L. M., T. R. Schwartz, K. Feltz, and T. J. Kubiak, 1990, "Determination and Occurrence of AHH-Active Polychlorinated Biphenyls, 2,3,7,8-Tetrachlorodibenzo-p-dioxin and 2,3,7,8-Tetrachlorodibenzofuran in Lake Michigan Sediment and Biota. The Question of Their Relative Toxicological Significance," *Chemosphere, 21*, 1063–1086

Speitel Jr, G. E., C.-J. Lu, M. Turakhia, and X.-J. Zhu, 1989, "Biodegradation of Trace Concentrations of Substituted Phenols in Granular Activated Carbon Columns," *Environ. Sci. Technol., 23*, 68–74

Squillace, P. J., and E. M. Thurman, 1992, "Herbicide Transport in Rivers: Importance of Hydrology and Geochemistry in Nonpoint-Source Contamination," *Environ. Sci. Technol., 26*, 538–545

Stanlake, G. J., and R. K. Finn, 1982, "Isolation and Characterization of a Pentachlorophenol-Degrading Bacterium," *Appl. Environ. Microbiol., 44*, 1421–1427

Struijs, J., and J. E. Rogers, 1989, "Reductive Dehalogenation of Dichloroanilines by Anaerobic Microorganisms in Fresh and Dichlorophenol-Acclimated Pond Sediment," *Appl. Environ. Microbiol., 55*, 2527–2531

Suflita, J. M., A. Horowitz, D. R. Shelton, and J. M. Tiedje, 1982, "Dehalogenation: A Novel Pathway for the Anaerobic Biodegradation of Haloaromatic Compounds," *Science, 218*, 1115–1117

Suntio, L. R., W. Y. Shiu, D. Mackay, J. N. Seiber, and D. Glotfelty, 1988a, "Critical Review of Henry's Law Constants for Pesticides," *Rev. Environ. Contam. Toxicol., 103*, 1–59

Suntio, L. R., W. Y. Shiu, and D. Makay, 1988b, "A Review of the Nature and Properties of Chemicals Present in Pulp Mill Effluents," *Chemosphere, 17*, 1249–1290

Ten Hulscher, Th. E. M., L. E. van der Velde, and W. A. Bruggeman, 1992, "Temperature Dependence of Henry's Law Constants for Selected Chlorobenzenes, Polychlorinated Biphenyls and Polycyclic Aromatic Hydrocarbons," *Environ. Toxicol. Chem., 11*, 1595–1603

Tse, G., H. Orbey, and S. I. Sandler, 1992, "Infinite Dilution Activity Coefficients and Henry's Law Coefficients of Some Priority Water Pollutants Determined by a Relative Gas Chromatographic Method," *Environ Sci. Technol., 26*, 2017–2022

Tsuchiya, T., and T. Yamaha, 1984, "Reductive Dechlorination of 1,2,4-Trichlorobenzene by *Staphylococcus epidermis* Isolated from Intestinal Contents of Rats," *Agric. Biol. Chem., 48*, 1545–1550

Vandenberg, P. A., R. H. Olsen, and J. E. Colaruotolo, 1981, "Isolation and Genetic Characterization of Bacteria that Degrade Chloroaromatic Compounds," *Appl. Environ. Microbiol., 42*, 737–739

Vannelli, T., M. Logan, D. M. Arciero, and A. B. Hooper, 1990, "Degradation of Halogenated Aliphatic Compounds by the Ammonia-Oxidizing Bacterium *Nitrosomonas europaea*," *Appl. Environ. Microbiol., 56*, 1169–1171

Vogel, T. M., and P. L. McCarthy, 1985, "Biotransformation of Tetrachloroethylene to Trichloroethylene, Dichloroethylene, Vinyl Chloride, and Carbon Dioxide under Methanogenic Conditions," *Appl. Environ. Microbiol., 49*, 1080–1083

Wackett, L. P., G. A. Brusseau, S. R. Householde, and R. S. Hanson, 1989, "Survey of Microbial Oxygenases: Trichloroethylene Degradation by Propane-Oxidizing Bacteria," *Appl. Environ. Microbiol., 55*, 2960–2964

Winter, B., A. Fiechter, and W. Zimmermann, 1991, "Degradation of Organochlorine Compounds in Spent Sulfite Bleach Plant Effluents by *Actinomycetes,*" *Appl. Environ. Microbiol., 57,* 2858–2863

Wolf, S. D., R. R. Lassiter, and S. E. Wooten, 1991, "Predicting Chemical Accumulation in Shoots of Aquatic Plants," *Environ. Toxicol. Chem., 10,* 665–680

Woods, S. L., J. F. Ferguson, and M. M. Benjamin, 1989, "Characterization of Chlorophenol and Chloromethoxybenzene Biodegradation during Anaerobic Treatment," *Environ. Sci. Technol., 23,* 62–68

Xie, T.-M., Abrahamsson, K. Fogelqvist, and B. Josefsson, 1986, "The Distribution of Chlorophenolics in a Marine Environment," *Environ. Sci. Technol., 20,* 457–463

Yin, C., and J. P. Hassett, 1989, "Fugacity and Phase Distribution of Mirex in Oswego River and Lake Ontario Waters," *Chemosphere, 19,* 1289–1296

You, I.-S., and R. Bartha, 1982, "Stimulation of 3,4-Dichloroanilne Mineralization by Aniline," *Appl. Environ. Microbiol., 44,* 678–681

Young, L. Y., and M. M. Häggblom, 1991, "Biodegradation of Toxic and Environmental Pollutants," *Current Opinion Biotechnol., 2,* 429–435

Yu-yun, T., W. Thumm, M. Jobelius-Korte, A. Attar, D. Freitag, and A. Kettrup, 1993, "Fate of Two Phenylbenzoylurea Insecticides in an Algae Culture System (*Scenedesmus Subspicatus*)," *Chemosphere, 26,* 955–962

Zitomer, D. H., and R. E. Speece, 1993, "Sequential Environments for Enhanced Biotransformation of Aqueous Contaminants," *Environ. Sci. Technol., 27,* 227–244

11

Environmental Fate of Chlorinated Organics

Heidelore Fiedler and Christoph Lau (*Bayreuth, Germany*)

Ecotoxicology, Edited by Gerrit Schüürmann and Bernd Markert.
ISBN 0-471-17644-3   © 1998 John Wiley & Sons, Inc. and Spektrum Akademischer Verlag.

11.1 SUMMARY

This chapter gives a short overview of classes of chlorinated organics and the major use pattern of chlorinated xenobiotics. In addition, environmental effects and ecotoxicological findings for some classes of chlorinated anthropogenic organics such as short-chain chlorinated hydrocarbons (Section 11.4), polychlorinated dibenzo-*p*-dioxins and polychlorinated dibenzofurans (Section 11.5), and polychlorinated biphenyls (Section 11.6) are summarized, and the relevance of long-range transport phenomena (Section 11.7) is discussed.

Sections 11.8 and 11.9 describe a screening process to identify potentially harmful chlorinated organics in the aquatic environment as an example for a preliminary risk assessment. This ecotoxicologically based model includes environmental parameters such as Henry's Law constant (H) to determine if a given chemical can be found in the aqueous phase or preferentially in the gaseous phase. Moreover, to pose a potential hazard, the chemical per se has to be toxic. Thus, as a second key parameter, the minimum lethal concentration for an aquatic organism was selected to rank organochlorines.

11.2 INTRODUCTION

Chlorinated organic compounds still attract considerable scientific attention and publicity as these substances once industrially or naturally produced can be found in many environmental matrices. "God created 91 elements, man a little more than a dozen and the devil one—chlorine!" Taken out of context, this remark of O. Hutzinger, University of Bayreuth, Germany (Hutzinger 1990), has often been cited by critics and opponents of the chlorine industry as evidence of how dangerous the world considers the element chlorine to be.

Undoubtedly every scientist knows that life without chlorine is unthinkable. However, the most outstanding property of chlorine, its reactivity, is both its blessing and its curse. Man, who tries to achieve almost everything that is within his intellectual and technical realm or possibilities, has capitalized on this property and developed a chlorine chemistry which has provided the basis for an impressive industry. Admittedly, man seems to have maneuvred himself into the position of Goethe's "sorcerer's apprentice: He can perform magic, but he hasn't found the right formula to keep the genie he has summoned under control". An example are the often cited CFC (chlorofluorocarbons). These have outstandingly useful properties and are toxicologically harmless. At the time of their launch around 1950, any potentially harmful effects for the world could not have been predicted.

There have been many serious discussions about chlorine chemistry in the past. Regrettably, the problems were made public too early, i.e., before all the scientific facts could be collected and evaluated. The solution to the problem does not lie in a crusade against chlorine or a total ban, but rather in a reasonable and scientific point of departure, as the following step-by-step course of action might represent:

1. Based on known physicochemical parameters, the influence of the structure on the environmental impact and toxicological potential should be investigated. It is shocking how little we know so far about the theoretical basis of the influence of chlorine on biodegradability, solubility in fat, and toxicological behavior. At least the first signs of progress in this field are in sight.

2. The literature on toxicity data should be carefully evaluated. For screening purposes, quantitative structure–activity relationships (QSARs) should be used to calculate toxicity and environmental fate data for every compound.

3. The organochlorine compounds found in the environment should be carefully recorded, and their concentrations in the different matrices should be more precisely investigated.

4. The significance of chlorine chemistry, including its economic aspects, of the various chlorine processes, and of the reactions in which chlorine plays a role should be summarized and evaluated.

5. The application of chlorinated products, their life cycle from production and application to their final disposal, and their possible impact on man and the environment should be outlined and evaluated. Here, too, first steps have been taken, as noted in the final report of the parliamentary hearing of the German Bundestag on the "Protection of Man and Environment"—albeit only for a few substances.

6. An objective risk–benefit analysis of chlorine products should be conducted.

7. In addition, what, where, and how chlorine-containing products can be replaced should be examined and then they should be replaced—provided that it is ecologically necessary and economically feasible. Should other substances show an obvious potential to damage man and the environment that is not controllable, their production must be stopped immediately, as in the past with polychlorinated substances such as polychlorinated biphenyls (PCB) and pentachlorophenol (PCP).

8. Above all, as soon as the scientific world has worked out and evaluated new data, the public should be informed objectively, openly, and without bias.

The term "chlorinated organics" comprises a wide range of chlorine-containing substances with highly diverse physical and chemical properties. Released into the environment from different sources, these compounds are transported via various mechanisms and having reached their final compartment, they can undergo transformations such as biodegradation, and physical, chemical, and photochemical degradation. Transformation processes can also take place during transport. Although CFC are of considerable concern due to their potential for ozone depletion and global warming, they will not be further discussed in this section.

11.3 CLASSIFICATION OF CHLORINATED ORGANICS

Chlorinated organic substances belong to a variety of structurally diverse chemicals with a wide range of physicochemical properties. They can be divided either by origin (natural or anthropogenic sources) or by their chemical backbone. This section gives a brief description of the class of chlorinated organic substances (see also Fiedler et al. 1994).

11.3.1 Naturally Occurring Chlorinated Organic Compounds

Biogenic Chlorinated Compounds

The volatilization of chlorine from sea-salt aerosol is believed to be the major source of inorganic chlorine gas in the troposphere. The release mechanism is likely to be connected with an acidification of deliquescent aerosol by sulfuric acid (H_2SO_4) and nitrous acid (HNO_2), followed by direct volatilization of hydrochloric acid (HCl). Another possibility is heterogeneous reactions involving the reaction of particulate chlorine with nonacidic nitrogen gases, ozone, free radicals, or moist semiconductors in the marine boundary layer (Keene et al. 1993).

Gribble (1992) summarized earlier reviews on halogenated natural products and presented a large number of naturally occurring organochlorine compounds that were discovered and investigated throughout the past decade. These compounds originate from either biosynthesis or incomplete combustion. According to Axegård et al. (1993), approximately 1500 halogenated compounds naturally exist in both terrestrial and marine environments, e.g., in humic substances, including halogenated phenols, ketones, fatty acids, terpenes, and a large variety of unsaturated compounds. Most compounds result from enzyme-mediated reactions by organisms such as bacteria, fungi, algae, lichens, and higher plants.

Chloromethane (CH_3Cl), produced by marine algae, is the only significant naturally occurring organochlorine compound. The natural production rate of CH_3Cl is estimated at 5 000 000 t/year, mostly produced by marine algae.

Natural production of chloroform thus dominates anthropogenic emissions which are estimated to be in the range of 26 000 t/year (Gribble 1992). The bulk of 2,4,6-trichlorophenol found throughout the world is thought to be from ocean and soil production, not from anthropogenic sources.

Lake sediments of the 13th century as well as several thousand years old groundwater contain many volatile halogenated organic compounds. The content of halogenated organic material in an industrially unaffected peat bog was found to be 300 times higher than the annual (wet and dry) deposition from the atmosphere. Also, a decrease in soil pH seems to correlate with an increase in the natural abundance of halogenated organic compounds; hence, acid rain may exacerbate this phenomenon (Axegård et al. 1993).

Unfortunately, there are almost no reliable estimates available for production rates of organohalogens from nonanthropogenic processes. Thus, apart from a few exceptions, it seems impossible to judge the relative contribution of these sources to the overall environmental load of chlorinated organics. Many chlorinated alkanes were found to be produced by various organisms. Usually, they are of low molecular weight, methylene chloride (chloroform, CH_3Cl) being the most prominent example. Likewise, ketones, alcohols, terpenes, alkaloids, peptides, and phenols are produced in considerable amounts in the natural environment (Gribble 1992).

Also terrestrial organisms including fungi, lichens, bacteria, higher plants, and animals are capable of synthesizing chlorinated compounds. Reviewing the formation of these compounds in plants, Engvild (1986) found that polyacetylenes, thiophenes, and sesquiterpenes from *Asteraceae* account for more than 50% of the total 130 chemicals identified. Soil organisms were found to be involved in the production of chlorinated humic substances (Asplund and Grimvall 1991). In an investigation of 15 higher plants such as peas and beans, 11 contained growth hormones with two chlorine atoms per benzene group (Vannerberg and Widén 1993).

Biochlorination reactions are catalyzed by chloroperoxidase enzymes. Hydrogen peroxide is utilized to oxidize the chloride anion to hypochlorite. This was confirmed by a laboratory experiment demonstrating that the chloroperoxidase enzyme facilitates the production of 2,4,6-trichlorophenol if hydrogen peroxide, chloride ions, and phenol are available (Hodin et al. 1991).

Hoekstra and Leer (1993a, 1993b) also propose a natural production of low molecular weight chlorinated organic compounds in soils, based on soil studies in four National Parks in the Netherlands. They investigated soil air concentrations of chloroform and several other chlorinated compounds which were believed to be of solely anthropogenic origin. Concentration in soil air exceeded the concentrations in ambient air by two to 45 times. Modelling the transport of organohalogens in the river Rhine, Germany, they also concluded that up to 50% of the halogenated organic substances (15 µg/L) in the river originate from diffusive sources other than direct industrial and atmospheric input (Hoekstra and Leer 1993b). They suggest soils bounding the river Rhine as a natural source for these substances.

Similarly, a number of chlorinated aromatic degradation products have been identified in industrially unaffected environments, including water and sediment samples of surface waters and groundwater, whose structure very much coincided with those found in chlorine bleaching kraft mill effluents (Dahlman et al. 1993a). However, relative concentrations of the different classes of compounds considerably diverge among industrially polluted and nonpolluted samples.

One difficulty in such analyses is that a clear distinction between natural and anthropogenic sources by, for example, atmospheric deposition, is not yet possible. This is in part due to the fact that natural sources for these compounds in soils and mechanisms for their, e.g., enzyme-mediated, production have not yet been properly identified (Svenson et al. 1989). Dahlmann et al. (1993a, 1993b) suggest oxidative degradation of fulvic acids from surface and groundwater as one potential natural source for aromatically bound halogens. There is an ongoing discussion with respect to natural vs industrial production of chlorinated organic aromatics found in industrially unpolluted environments.

Chlorinated Natural Combustion Products

Under conditions of high temperature and/or high pressure and in the presence of chlorine, incomplete combustion processes give rise to the production of numerous chlorinated organic compounds. Amongst these are polychlorinated benzenes, dibenzo-p-dioxins, dibenzofurans, biphenyls, and naphthalenes. Such naturally occurring thermal processes include volcanic eruptions and forest fires, which may release comparably large amounts. However, no quantification has been carried out so far.

11.3.2 Chlorinated Compounds of Anthropogenic Origin

Production and Use

The world reservoirs of stone salt are estimated to be 3.7×10^{12} t, and oceans contain approximately 50×10^{15} tons of sodium chloride. Presently, annual consumption rates of chlorine are 200×10^6 tons (Bayer 1995). In Germany, 3.17×10^6 tons of chlorine were produced in 1994.

About 50% of the chlorine output is used within the respective chemical factories to manufacture chlorine-free products. If the proper allowances are made for safety and emission control as established in industrial countries, the entire production should have no permanent negative effects on the environment. Approximately 50% of chlorine produced is used to manufacture products containing chlorine, whereby about 50% of this is allotted to polyvinyl chloride (PVC) production. The rest is used for the production of a large number of chemicals and pharmaceuticals (for more details, see below).

An overview of the world chlorine production figures for 1994 is given in Figure 11.1. The world chlorine demand by the consuming sector is shown in Table 11.1 (Kirk-Othmer 1991).

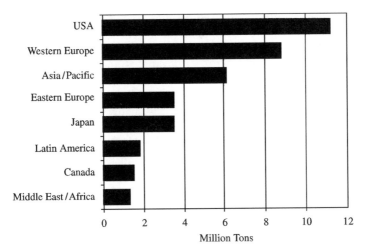

Figure 11.1 World chlorine production in 1994 (Bayer 1995).

Table 11.1 World chlorine demand by consuming sector for the year 1987

Chlorine usage	Quantity consumed 10^3 t	%
Vinyl chloride monomer	9012	26.1
Pulp and paper	4599	13.3
Propylene oxide	1990	5.8
Water treatment	1237	3.6
Carbon tetrachloride	934	2.7
Perchloroethylene	687	2.0
Hypochlorite	681	2.0
Epichlorohydrin	600	1.7
1,1,1-Trichloroethane	560	1.6
Methylene chloride	520	1.5
Chloroform	464	1.3
Methyl chloride	459	1.3
Ethylene dichloride (solvent)	401	1.2
Trichloroethylene	368	1.1
Chlorobenzene	355	1.0
Chloroprene	246	0.7
Bromine	129	0.4
Ethylene dichloride (trade)	50	0.1
Chlorinated polyethylene	16	<0.1
Miscellaneous organic	4988	14.5
Miscellaneous inorganic	6215	18.0
Total	34511	100

Source: Kirk-Othmer (1991).

A compilation of production figures and uses of economically important chlorinated organic compounds is given in Table 11.2. Chlorine is also vital in chemicals that do not contain chlorine in the final product, e.g., propylene, titanium oxide, some aluminum salts, isocyanates, and almost all silicones. For the production of these products the chlorinated compound is used as an adduct or an intermediate within the chemical process; therefore it can be assumed that the release of chlorines to the environment is negligible.

Because some of the chlorinated organic compounds are of great economic importance, they are produced at considerable rates. The degree to which these chemicals are released into the environment depends on their actual use patterns and the chemical nature of the substance. However, it has to be borne in mind that production figures for intermediate products of the chemical industry, for example, are not equivalent to the amounts released into the environment. This is different for other groups of organochlorines, such as pesticides, solvents, pharmaceuticals, and cleaning agents. These products are often introduced into the environment intentionally by spraying. A brief description of commercially important chloro-organics is given below.

Chlorinated Open-Chain hydrocarbons. This group can be subdivided into short-chain compounds and chlorinated paraffins. Short-chain hydrocarbons are chlorinated compounds with up to four carbon atoms. They are used as foaming agents, solvents, cleaning and degreasing agents, dyes, and in pharmaceuticals and pesticides (Table 11.2). The group of chlorinated paraffins comprises all of those compounds that are formed in chlorination reactions of C_{10}–C_{30} paraffins. Zitko (1980) pointed out that chlorination of one particular paraffin typically yields a mixture of compounds with various degrees of chlorination with an average chlorine content ranging from 10% to 70%. Polychlorinated paraffins have been used as plasticizers, lubricants, flame retardants, and in paints. They are available under several trade names such as Arubren, Cereclor, and FLX. Many applications of chlorinated paraffins are "open-ended", meaning that these compounds are eventually released into the environment. In 1986 the world production of paraffins amounted to over 300 kt (WHO/IARC 1990).

Polymers. A number of chlorinated organic chemicals are used for the production of polymers. Large amounts of vinyl chloride are necessary for the production of, for example, PVC (955 000 t in Germany in the year 1992; Bayer 1995). The amount of chlorine used in PVC production has decreased over the last few years. Thus, today, PVC production is increasing for long-living applications while the amount used for short-living purposes is declining (Nolte and Joas 1992). Epichlorohydrin (1-chloro-2,3-epoxypropane) also is a major industrial chemical used in the production of epoxide plastics and resins. The annual production in Germany in 1992 amounted to 855 000 t (Bayer 1995).

Table 11.2 Production figures and uses of economically important chlorinated short-chain hydrocarbons

Chemical	Annual production	Uses, technical application/intermediate in synthesis of	Miscellaneous
Chloromethane	163 kt (FRG 1987)[a]	Foaming agents,[a] dyes, pharmaceuticals	Most of atmospheric load is of natural origin
Dichloromethane	200 kt (FRG 1987)[a]	Decaffeination, metal, cleaning and degreasing[a]	
Chloroform	164 kt (USA 1982)[a]	Solvent, production of CFC 22[a]	By-product in chlorination of pulp[g] and drinking water
Carbon tetrachloride	270 kt (USA 1982)[b]	Solvent, cleaning agent, production of CFC[b]	
Chloroethane	69 kt (USA 1988)[c]	Foaming agent, alkylating agent	
Vinyl chloride	10 Mt (world 1987)[d]	Production of PVC	
1,2-Dichloroethane	5.4 Mt (USA 1983)[b]	Solvent, lead scavenger in fuels, vinyl chloride	
	1.8 Mt (FRG 1987)[a]		
1,1-Dichloroethene	100 kt (W. Eur. 1987)[d]	Precursor of trichloroethene, copolymer in PVC production	
1,1,1-Trichloroethane	140 kt (FRG 1982)[e]	Extracting agent	
1,1,2-Trichloroethane	186 kt (USA 1986)[c]	Production of vinylidene chloride	
Trichloroethylene	160 kt (FRG 1984)[e]	Solvent, degreaser	Most uses abolished in FRG in 1993[a]
Tetrachloroethane	40 kt (FRG 1986)	1,1,2,2-Tetrachloroethane is intermediate in production of trichloroethylene[i]	
Tetrachloroethene	250 kt (USA 1982)[b]	Metal degreasing, textile processing; production of CFC, dichlorvos	
	150 kt (FRG 1982)[a]		
3-Chloro-1-propene	95 kt (1987 FRG)[a]	Pesticide, production of epichlorohydrin	
	550 kt (world 1982)[d]		
1,2-Dichloropropane	20 kt (world 1985)[d]	Solvent	
1,3-Dichloropropene		Pesticide	
Hexachlorobutadiene	14 kt (EEC 1986)[f]	Heat conduction	By-product in chlorination of hydrocarbons[h]

Sources: [a]Nolte and Joas (1992); [b]Ware (1988); [c]IARC (1991); [d]Rippen (1990); [e]Dieter (1990); [f]Vogl (1987); [g]Kringstad and Lindström (1984); [h]Koch (1985); [i]Mross and Konietzko (1991).

W. Eur., Western Europe.

Chlorinated Aromatic Compounds. Chlorinated aromatic compounds can be classified into seven different chemical groups: chlorinated benzenes (PCBz), chlorinated phenols (PCPh), PCB, polychlorinated naphthalenes (PCN), polychlorinated terphenyls (PCT), chlorinated anilines, and polychlorinated dibenzo-*p*-dioxins and dibenzofurans (PCDD/PCDF). Whereas the compounds of the first six groups have been or are industrially produced, PCDD/PCDF have never been produced on a technical scale nor served any useful purpose. As PCDD/PCDF and PCB are considered to be of special environmental concern and their characteristics are quite well understood, both classes of compounds will be described in more detail (see Sections 11.5 and 11.6).

Lower PCBz predominantly serve as solvents or intermediates for further syntheses, whereas higher PCBz have often been used as pesticides. Mono- and dichlorinated benzenes serve as solvents and as intermediates in the synthesis of, for example, pesticides. Worldwide 300 000 t were produced in 1988, and 40 000 t were produced in Germany in 1986 (Rippen 1990). The perchlorinated compound, hexachlorobenzene (HCB), is produced as an effective fungicide (annual production in 1983 in Germany: 3500 t; Vogl 1987). In addition, HCB is a by-product of pesticide and tetrachloroethene synthesis. Moreover, HCB is a characteristic substance produced by almost all incomplete combustion processes, e.g., waste incineration and combustion of fossil fuels.

PCPh serve as pesticides due to their toxic effects on living organisms. By virtue of their fungicidal properties, many of these compounds serve primarily as wood, leather, and textile preservatives, and also as slimicides in saw mills. In addition, PCPh are by-products of chlorination reactions in chlorine bleaching of pulp (Kringstad and Lindström 1984) and in the disinfection of drinking water (Strobel and Dieter 1990; White 1986). Fungicides based on chlorophenols include 2,4,6-trichlorophenol (worldwide annual production rate of ca. 10 000 t; Vogl 1987), PCP (worldwide production of 34 000 t in 1984; Korte 1987, 1992). 2,4-di- and 2,4,5-trichlorophenol are used in the production of phenoxyacetic acid herbicides at worldwide rates of 70 000 t/year and 12 000 t/year, respectively (Vogl 1987). Chlorophenols can be contaminated with polychlorinated diphenyl ethers (PCDE), PCDD, and PCDF.

Chlorinated anilines mainly serve as intermediates in the production of pesticides, dyes, and pharmaceuticals and are produced at relatively low rates compared to other aromatic organochlorines (less than 10 000 t/year). Chlorinated anilines are found in the environment as degradation products of chlorinated pesticides such as the phenyl ureas (MacRae 1989).

PCN are a group of chemicals differing in degree and position of chlorination. They are still industrially produced at a rate of 5000 t/year and—similar to the PCB (see below)—mostly used as capacitor fluids and flame retardants (Haas 1992). Moreover, PCN are released into the environment as constituents of flue gas in waste incineration (Wienecke et al. 1992; Mariani et al. 1992) and can be found in fly ash and bottom ash from incinerators.

PCT have been used as adhesives and sealants in combination with PCB.

Pesticides. Many pesticides used in the past were chlorinated organic compounds. Pesticides are characterized by the fact that they are intentionally released into the environment. Usually, they are designed to be chemically stable to some degree in order to be effective for a prolonged period of time. Hence, nondegradable pesticides may partition in the environment and may be transported far away from the site originally targeted by their application.

Many types of pesticides have been used in considerable amounts. Today, many chlorinated pesticides, such as mirex, aldrin, toxaphene, and trichlorophenoxyacetic acid (2,4,5-T) are virtually banned in industrialized countries. However, there is a disparity between bans and production globally; for example, DDT is still produced and applied in the third world countries as is lindane. Chlorinated pesticides comprise the groups of the phenoxyacetic acid herbicides, e.g., 2-methyl-4-chlorophenoxyacetic acid (MCPA), 2,4-dichlorophenoxyacetic acid (2,4-D), and 2,4,5-T; the sulfonylurea herbicides, i.e., chlorsulfuron; hexachlorocyclohexane isomers, i.e., lindane (γ-HCH), chlorinated cyclodienes (i.e. aldrin), camphenes (toxaphenes), indenes (chlordane and heptachlor), DDT and its degradation products, as well as other chlorinated hydrocarbons such as dibromochloropropane, dichloropropene, mirex, and HCB. Moreover, many of the synthetic pyrethroid insecticides are chlorinated, i.e., bifenthrin and deltamethrin. Atrazine and simazine belong to the triazine herbicides and dichlorvos and chlorfenvinphos to the organophosphate insecticides. Table 11.3 gives a summary of the legal restrictions for production and use of organochlorine pesticides (VCI 1994). For more information on environmental risk assessment for pesticides, see Section 11.10.

Chlorinated Compounds in Effluents of the Pulp and Paper Industry. As pulp mill effluents are directly discharged into aquatic ecosystems, i.e., a treatment facility or water, their fate has been subject to various thorough investigations. To summarize, conventionally pulp is bleached in a sequence of different steps using chlorine or chlorine dioxide (Kringstad and Lindström 1984). In this process, chlorinated compounds are formed. By discharging the spent bleaching and extraction liquids, pulp and paper mills release significant amounts of chlorinated organic compounds into the environment (Reeve and Earl 1989).

However, modern bleaching technology helps to significantly reduce the formation of organically bound chlorine (OCl). This reduction is largely due to the fact that the use of chlorine gas has virtually been stopped. Moreover, low chlorine bleaching technology and operation of biological treatment plants have been implemented in most mills (Jokela et al. 1993). As a consequence, the adsorbable organic halogen (AOX) discharged from, e.g., Finnish pulp mills into receiving waters decreased from 2.7 kg per ton of pulp in 1989 to 1.7 kg per ton of pulp in 1991 (average for soft- and hardwoods). Sweden plans on further reductions to 0.5 kg/t in the year 2004 and 0.1 kg/t in 2010 (Reeve and Earl 1989), and the province of Ontario, Canada, recommended to cut the OCl production rate down to less than 1.5 kg/t by 1993. The US will basically

Table 11.3 History of German legislation for use of organochlorine pesticides

Aldrin	Since 1971 only for use on soils in vineyards Since 1974 only with permission by authorities 1979 Termination of the last permission 1980 Ban of application
Dieldrin	1971 Termination of last permission 1974 Ban of application
DDT	1972 Ban of production, import, export, marketing, and application (DDT-Ordinance)
Heptachlor	Since 1971 only for treatment of seeds of beets 1981 Termination of last permission 1981 Ban of application
Hexachlorobenzene	Since 1974 only for treatment of seeds for wheat and only with permission of authorities 1973 Termination of last permission 1977 Ban of application
HCH (technical grade)	Since 1974 only for abatement of bark-beetle in forestry. 1977 Ban of application
Lindane	Permission only for abatement of certain insects in beets, ants in gardens, and insects and scale insects in ornamental plants, under glass, and in gardens as well as on indoor plants

remove chlorine gas by 1998 resulting in estimated AOX emissions of under 25 000 t. In 1984 the world production of OCl was estimated at about 250 000 t/year, assuming that 5 kg chlorinated organics are formed per ton of pulp produced (Kringstad and Lindström 1984).

Chlorinated Compounds Formed by Chlorination of Drinking Water. Chlorination of water to provide permanent protection of drinking water distribution systems is an important process and, thus, will certainly be further applied in the future. Usually the chlorine species Cl_2 and ClO_2 serve as disinfectants in drinking water by oxidizing undesirable substances (White 1986; Strobel and Dieter 1990). Chlorine is added to raw water (prechlorination) as well as to drinking water at the distribution stage. Chlorine addition after treatment uses concentrations of about 0.5–2 mg/L (WHO/IARC 1991). In Germany, drinking water regulations require a minimum content of 0.1 mg/L of free chlorine after treatment (Nolte and Joas 1992). Sometimes effluent water is also chlorinated after treatment and before being discharged into the environment (Klein et al. 1991).

Addition of chlorine to water results in a number of by-products, such as chloramines, chlorophenolic compounds (Schlett et al. 1990), trihalomethanes (e.g., chloroform in the range of 10–30 µg/L; Strobel and Dieter 1990), chlorinated acetic acids (1–100 µg/L), chlorinated aldehydes and ketones, chlorinated acetonitriles (IARC 1991; Klein et al. 1991), and HCB (Hub et al. 1990). The typical range for total chlorinated organics obtained from chlorinated drinking water is 10–250 µg/L.

Reaction of OCl with humic substances results in the formation of a wide range of chlorinated organic compounds. Among these are haloforms, chlorinated acetones, butanones, acetic acids, acetaldehydes, phenols, acetonitriles, and furanone derivatives. Moreover, there is also evidence for the formation of chlorinated ethanes, ethenes, and aromatics (Strobel and Dieter 1990). Many of the above compounds have been found to be mutagenic in the Ames assay (Strobel and Dieter 1990) with 3-chloro-4-dichloromethyl-5-hydroxy-2(5H)-furanone (MX) contributing most to the mutagenicity of the chlorinated compounds formed (Langvik et al. 1991).

Chlorination of wastewater has been found to produce chloroform, dichlorobromomethane, dibromochloromethane, tri- and tetrachloroethane (Klein et al. 1991), as well as chlorinated octylphenol polyethoxylate residues (Ball et al. 1989). Combination with other raw water pretreatment techniques for the destruction of organic material (Strobel and Dieter 1990) seems a viable alternative. Chlorine preparations also contains impurities, such as carbon tetrachloride (WHO/IARC 1991).

11.4 SHORT-CHAIN CHLORINATED HYDROCARBONS

11.4.1 Chemical Identity and Environmental Levels

Short-chain aliphatic hydrocarbons (SCAH) are industrially produced at a total amount of about 15 million tons annually. They are mainly used in closed systems as monomers for polymer production and as intermediates in chemical syntheses; however, they are also used in emissive applications as solvents and propellants.

The technically most important and environmentally most abundant SCAH are (Frank 1993):

- The tropospherically and metabolically inert anthropogenic *CFC*

- The C_1 and C_2-*chlorocarbon* solvents dichloromethane, 1,1,1-trichloroethane, trichloroethene, and tetrachloroethene with long to intermediate environmental stability and intermediate metabolic reactivity

- The *polychloromethanes* tetrachloromethane and trichloromethane, emitted inadvertently or formed as secondary chlorination products

- The carcinogenic C_2-*halocarbons* 1,2-dichloroethane and vinylchloride which may occur locally in ambient air at levels up to several micrograms per cubic meter

- The natural methylhalides methylchloride, methylbromide, and methyliodide

- Since recently, various *partially halogenated chlorofluorocarbons* (*HCFC*) mostly C_2-halocarbons with atmospheric lifetimes in the range of a few years.

SCAH can be analytically monitored with high selectivity and high sensitivity. Most SCAH are of anthropogenic origin and, thus, are considered as "benchmark chemicals", i.e., model compounds for predicting the environmental fate of compounds in the environment. The ecotoxicological effects of these compounds are interesting because metabolic activation and abiotic transformation reactions of several SCAH generate reactive secondary pollutants which themselves exhibit considerable reactivity and potentially elevated toxicity. Chloroethenes can be metabolized oxidatively to epoxides and reductively to vinyl chloride, while atmospheric oxidation of SCAH entails numerous reactive carbonyl chlorides, acyl chlorides, and haloaliphatic acids.

When dealing with short- and medium-lived SCAH, it should be noted that the atmospheric concentration of these compounds fluctuate over two orders of magnitude. Thus, a large number of measurements are necessary to obtain representative data. Within the last two decades advances in automatization and miniaturization of chromatographic methods have been made, and therefore, e.g., the hypothesis of the ozone-depleting potential of CFC could be analytically verified. Despite all advances in analytical capabilities, however, there is still a need for a better monitoring of the environmental fate of SCAH, for identification of potentially toxic secondary products, and for assessing the ecotoxicological potential of these reaction products. Airborne haloacetic acids have only recently been recognized as ecotoxicologically relevant compounds, although they are amongst the most potent algal toxicants.

11.4.2 Chemical and Toxicological Properties

SCAH are often treated as a single class of compounds as:

- Most halocarbons are predominantly anthropogenic (for exceptions see Section 11.3.1);
- They have similar physical and chemical properties: relatively nonpolar, volatile, and lipophilic;
- They can be analytically detected with high selectivity and sensitivity.

Due to their chemical stability in organisms and in the environment and the peculiar toxicity of some of the SCAH, this class of compounds was commonly considered to be problematic. The chemical, toxicological, and environmental properties of SCAH vary considerably with their chemical structure and in particular with the respective halogen atom (F, C, Br, I). Physicochemical properties and the chemical and metabolic reactivities of short-chain organochlorines range from chemically stable to highly reactive, from metabolically inert to high highly toxic.

Some common features of all organohalogens are:

- The atomic volume of halogens are similar to the related noble gases;

- The outer p-electron shells are filled in the bonded state;
- Halogens have a high electron affinity;
- The carbon–halogen bond strengths are high, decreasing from F to I;
- The C–Hal bonds are polarized, increasing from F to I.

The first two characteristics allow weak interactions with water dipoles, high lipophilicity, and small Henry constants. The last three features influence the biotic and abiotic reactivities of the SCAH; e.g., the use of alkyl halides in chemical synthesis as electrophilic intermediates. These three features are also responsible for the increasing carcinogenicity: methyl fluoride is noncarcinogenic; methyl chloride is a weak carcinogen; for methyl bromide there exists strong evidence for carcinogenicity, whereas methyl iodide is a known carcinogen. Vicinal or geminal electronegative substituents, such as chlorine (halogen), oxygen, or sulfur, increase the electrophilicity of a compound and, hence, carcinogenicity. However, further chlorine atoms decrease oxidative reactivity towards biotic and abiotic oxidations. The latter effect results in inflammability and low toxicity and is the reason for using halocarbons in many technical applications. Polyhalogenated alkanes tend to participate in reductive metabolism and are converted to toxic radical intermediates. In a vinylic position, halogens first activate double bonds (vinylchloride); at higher substitution, e.g., tetrachloroethene, double bond reactivity decreases, especially when d-electron shell participation is possible (Cl, Br). Consequently, biotic and abiotic oxidation rates decrease with increasing halogen/chlorine substitution (Frank 1993).

11.4.3 Ecotoxicological Relevance

Some SCAH elicit ecotoxicological effects at concentrations several orders of magnitude lower than the thresholds considered safe for human exposure. The uptake and effects on C_1- and C_2-halocarbons upon plants have been studied in detail, as forest trees constitute more than 90% of the global biomass. In this context, not only SCAH but also haloacetic acids, probable atmospheric degradation products of SCAH, were taken into account. The relevance of the latter class of compounds is further demonstrated by the fact that some of the haloaliphatic acids, e.g., trichloro- and monochloroacetic acids and derivatives of these, have been or are still in use as herbicides. Monochloroacetic acid is one of the strongest algal toxicants known. It has been found in needle and leaf samples collected from various locations in central and northern Europe at levels between 10 and 100 ppb (the EC_{10} in algal multiplication tests is in a similar range).

Schröder (1993) studied the effect of tri- and tetrachloroethene, in particular, on plant uptake, their effects on xenobiotic-metabolizing enzymes, and the effects on photosynthetic pigment patterns. The uptake of SCAH into spruce needles was monitored in field experiments at ambient air concentrations of approximately 0.5 to 1 ppbv as well as in fumigation studies with considerably higher concentrations. Despite their physicochemical similarities, different uptake rates were observed for tri- and tetrachloroethene: whilst the perchloroethene uptake into the spruce needles proceeded at rates predicted from the stomatal conductance of water vapor, trichloroethene uptake was far below the predicted values. The results indicate that a significant amount of cuticular deposition is superimposed on the stomatal uptake of perchloroethene, while trichloroethene uptake seems to be limited due to some internal resistances. Injuries such as rapid changes in photosynthesis, respiration, and transpiration of the trees, as well as changes in the pigment pattern could be seen at levels of 25 ppbv. However, it should be remembered that atmospheric background concentrations in rural areas are usually 25- to 50-fold lower. This finding indicates that direct phototoxicity of chlorocarbons is probably less relevant than the impact of secondary pollutants such as haloacetates.

So far, no effects of SCAH, such as trichloromethane (chloroform), trichloroethene, tetrachloroethene, and trichloroacetate, on soil invertebrates were detected at levels typical for noncontaminated soil air when analyzing for population densities and diversity of invertebrate coenosis in forest soils (Roth 1993). However, model experiments using concentrations of approx. 30 mg/m^3 (1000-fold above background concentration) showed significant decreases in the population densities of nematodes and enchytraeids.

11.4.4 Biological Degradation

Several laboratory studies have shown that SCAH can be dehalogenated and even mineralized by bacteria at considerable rates. Short-chain aliphatic halocarbons with only one or two chlorine substituents are excellent growth and energy substrates for aerobic bacteria, such as certain species of *Pseudomonas*, *Hyphomicrobium*, *Arthrobacter*, and *Xanthobacter*. Some of the SCAH, e.g., dichloromethane, 1,2-dichloroethane, and vinyl chloride, can be utilized as the only carbon and energy source. Hydrolysis is the main dehalogenation pathway. The dehalogenation of 1,2-dichloroethane is catalyzed by two different halidohydrolases.

The second dehalogenation process is catalyzed by mono- and dioxygenases. These enzyme systems transform SCAH without being able to utilize them. All of the oxygenases form toxic epoxides, inhibiting the dechlorinating bacteria during the oxidation of the SCAH. Amongst others, this disadvantage might be responsible for the lack of technical applications of cometabolic dechlorination.

From thermodynamical considerations, highly chlorinated hydrocarbons are expected to be more easily dechlorinated under anaerobic conditions. Tetrachloroethene (perchloroethene), for example, is stable under aerobic conditions. Under strictly anaerobic conditions and in the presence of a suitable electron donor, however, this compound is reductively dechlorinated at high rates. Generally, low-chlorinated metabolites such as *cis*-1,2-dichloroethene and vinyl chloride accumulate during this process due to reduced reduction rates for the chlorinated intermediates. At least two anaerobic bacterial species are involved in the complete dehalogenation process of perchloroethene to ethene.

Presently, no biological processes are available to purify water contaminated with polychlorinated hydrocarbons. One of the most abundant contaminant in polluted groundwater is tetrachloroethene (perchloroethene, PCE); PCE is persistent under aerobic conditions. Due to the accumulation of toxic metabolites such as vinyl chloride during the aerobic dechlorination of PCE, a direct application of reductive dechlorination is not feasible. A two-step treatment based on anaerobic degradation of PCE followed by aerobic degradation of vinyl chloride and other low-chlorinated compounds can be a satisfactory procedure.

11.4.5 Photooxidation of Perchloroethene and Trichloroacetyl Chloride

The present knowledge of the degradation products of short-chain chlorinated organic compounds in the atmosphere is still limited (Behnke and Zetsch 1993). Most of the chlorinated solvents emitted are degraded in the troposhere by attack of OH, and only minor portions can reach the stratosphere. In smog chamber experiments, carbon tetrachloride (CCl_4) was formed as a product from photochemical degradation of perchloroethylene (PER) in air (Behnke and Zetsch 1993). The results suggest that CCl_4 is formed by heterogeneous photochemical reactions processes, e.g., at reactor walls or on the surfaces of aerosols. The experimental observations lead to the conclusion that the tropospheric yield of CCl_4 from PER will depend on the concentration level of atomic chlorine in the troposphere, and on the concentration, nature, and pH of the aerosol.

11.5 POLYCHLORINATED DIBENZO-*p*-DIOXINS AND POLYCHLORINATED DIBENZOFURANS (PCDD/PCDF)

11.5.1 Introduction

Polychlorinated dibenzo-*p*-dioxins (PCDD) and polychlorinated dibenzofurans (PCDF) are environmental contaminants which are detectable in almost all compartments of the global ecosystem in trace amounts. These compound classes in particular have caused major environmental concern. In contrast to PCB, PCN, and other polychlorinated contaminants, PCDD/PCDF never were produced intentionally. They are formed as by-products of numerous industrial activities and all combustion processes (Fiedler et al. 1990). Besides the anthropogenic sources, an enzyme-mediated formation of PCDD and PCDF from 2,4,5- and 3,4,5-trichlorophenol has been demonstrated in vitro (Öberg et al. 1990; Wagner et al. 1990).

First risk assessments only focused on the most toxic congener, the 2,3,7,8-tetrachlorodibenzo-*p*-dioxin (2,3,7,8-Cl_4DD = 2,3,7,8-TCDD). Soon it was recognized, though, that all PCDD/PCDF substituted at least in position 2, 3, 7, or 8 are highly toxic and, thus, major contributors to the overall toxicity of the dioxin mixture. In addition, despite the complex composition of many PCDD/PCDF-containing "sources", only congeners with substitutions in the lateral positions of the aromatic ring, namely the carbon atoms 2, 3, 7, and 8, persist in the environment and accumulate in food chains.

11.5.2 Toxicity of PCDD/PCDF

Many regulatory agencies developed so-called toxicity equivalency factors (TEF) for risk assessment of complex mixtures of PCDD/PCDF (Kutz et al. 1990). As a result of the NATO/CCMS Study on Chlorinated Dioxins and Related Compounds the so-called international toxicity equivalency factors (I-TEF) have been developed (Table 11.4). The TEF are based on acute toxicity values from in vivo and in vitro studies. This approach is based on evidence that there is a common, receptor-mediated mechanism of action for these compounds. However, the TEF approach has its limitations due to a number of simplifications. Although the scientific basis cannot be considered as solid, the TEF approach has been developed as an administrative tool and allows the conversion of quantitative analytical data for individual PCDD/PCDF congeners into a single toxic equivalent (TEQ). TEF particularly aid in expressing cumulative toxicities of complex PCDD/PCDF mixtures as one single TEQ value. Today, almost all literature data are reported in I-TEQ.

It should be noted that TEF are interim values and administrative tools. They are based on the present state of knowledge and should be revised as new data become available. Recently, the Science Advisory Board of the US Environmental Protection Agency (US-EPA) confirmed the use of the TEF approach for risk assessment within EPA's Dioxin Reassessment (EPA 1995).

Table 11.4 Toxicity equivalency factors (TEF) for PCDD/PCDF

Congener	I-TEF	Congener	I-TEF
$2,3,7,8\text{-}Cl_4DD$	1	$2,3,7,8\text{-}Cl_4DF$	0.1
$1,2,3,7,8\text{-}Cl_5DD$	0.5	$2,3,4,7,8\text{-}Cl_5DF$	0.5
		$1,2,3,7,8\text{-}Cl_5DF$	0.05
$1,2,3,4,7,8\text{-}Cl_6DD$	0.1	$1,2,3,4,7,8\text{-}Cl_6DF$	0.1
$1,2,3,7,8,9\text{-}Cl_6DD$	0.1	$1,2,3,7,8,9\text{-}Cl_6DF$	0.1
$1,2,3,6,7,8\text{-}Cl_6DD$	0.1	$1,2,3,6,7,8\text{-}Cl_6DF$	0.1
		$2,3,4,6,7,8\text{-}Cl_6DF$	0.1
$1,2,3,4,6,7,8\text{-}Cl_7DD$	0.01	$1,2,3,4,6,7,8\text{-}Cl_7DF$	0.01
		$1,2,3,4,7,8,9\text{-}Cl_7DF$	0.01
Cl_8DD	0.001	Cl_8DF	0.001

Source: Kutz et al. (1990).

11.5.3 Environmental Behavior

A literature survey on PCDD/PCDF in the aquatic environment by Fletcher and McKay (1993) as well as analyses of core profiles from Lake Huron by Czuczwa and Hites (1984) assume that PCDD/PCDF do not degrade once incorporated in the sediment. This is supported by an investigation by Knutzen and Oehme (1989) in Frierfjord, Norway. They found that there was little difference in PCDD/PCDF concentrations between deep anaerobic sediments and shallower aerobic sediments, thus suggesting consistency under varying redox conditions. Due to the association of PCDD/PCDF with organic matter it is likely that the degradation and the movement of organic carbon in the sediment determines the mobility of PCDD/PCDF.

Crustacea such as crabs and shrimp have proven to be useful biomonitors of PCDD/PCDF contamination in the aquatic environment. These species are able to take up all congeners from the water/sediment and to preserve them almost completely, whereas fish selectively accumulate the 2,3,7,8-substituted congeners (Oehme et al. 1990).

Atmospheric Transport

During the last few years there has been a growing recognition of the importance of aerial transport of PCDD/PCDF. Ambient air data from Germany, the United Kingdom, and Japan showed that PCDD/PCDF exhibit seasonal trends with higher concentrations in the winter months and lower levels during summer. Comparative measurements by Wallenhorst et al. (1995) have shown that in terms of I-TEQ approximately the same amount of PCDD/PCDF is found in the gas phase and bound to particles.

Biological Formation and Degradation

Biological formation of PCDD/PCDF from chlorinated precursors was discussed for compost and sewage sludge and questions on the possibility of a biogenic formation arose for sediments and soils (especially forest soils). Based on the results of Öberg et al. (1992) the rate of conversion of pentachlorophenol (PCP, the most suitable precursor) to PCDD is in the low ppm range (Table 11.5). Consequently, a chlorinated precursor present in an environmental matrix, such as soil or sediment, at ppm concentrations should be converted to not more than ppt levels of high-chlorinated PCDD (Cl_7DD and Cl_8DD). In other words, ppm concentrations of chlorophenols would generate ppt levels of Cl_7DD and Cl_8DD or ppq concentrations in TEQ. Thus, based on present knowledge, biological formation of PCDD from chlorinated phenols under environmental conditions are negligible.

Accumulation of PCDD/PCDF in aquatic and terrestrial biota and subsequent transfer to organisms of higher trophic levels occur as a result of their hydrophobicity and recalcitrant character. Thus, PCDD and PCDF are considered to be rather persistent in the environment. Biodegradation under

Table 11.5 ^{13}C-PCDD (total pg) found in sludge samples after spiking with ^{13}C-PCP and proportion of ^{13}C-PCP converted to ^{13}C-PCDD

Sample	^{13}C-1234679 Cl$_7$DD	^{13}C-1234678 Cl$_7$DD	^{13}C-Cl$_8$DD	Total ^{13}C-PCDD	^{13}C-PCDD ^{13}C-PCP
Activated sludge*	274	3726	2188	6187	3.5 ppm
Sedimented sludge*	310	1395	2597	4302	2.4 ppm
Sedimented sludge**	403	2494	3017	5913	3.4 ppm

Source: Öberg et al. (1992).

Incubation temperature was 22°C for * samples and 10°C for ** samples.

aerobic conditions appears to be restricted to congeners with four or fewer chlorines (Quensen and Matsumura 1983; Parsons and Storms 1989; Zeddel et al. 1994). Once adsorbed to the organic carbon of sediments and soils, it is assumed that PCDD and PCDF will persist also under anaerobic conditions. For polychlorinated biphenyls (PCB) and benzenes however, substantial changes in the concentrations in sediments caused by microbial processes was reported (Brown et al. 1987; Quensen et al. 1988). Beurskens et al. (1995) demonstrated that anaerobic microorganisms enriched from River Rhine sediments are able to remove chlorine substituents from PCDD. 1,2,3,4-tetra-chlorodibenzo-*p*-dioxin (1,2,3,4-Cl$_4$DD) was reduced to 1,2,3-trichlorodibenzo-*p*-dioxin (1,2,3-Cl$_3$DD) and 1,2,4-trichlorodibenzo-*p*-dioxin (1,2,4-Cl$_3$DD). These compounds were further degraded to 1,3-dichlorodibenzo-*p*-dioxin (1,3-Cl$_2$DD), 2,3-dichlorodibenzo-*p*-dioxin (2,3-Cl$_2$DD), and traces of 2-mono-chlorodibenzo-*p*-dioxin (2-ClDD). The rate for degradation was determined as $t_{1/2} = 15.5$ days. Biodegradation of 2,3,7,8-substituted PCDD or PCDF has not yet been reported.

Exposure Data

EPA's dioxin reassessment document revealed that for some media there is hardly any information available for the US (EPA 1994). For example, there are no deposition data and more data is needed for environmental levels in air, soil, sediments, and water. Although the daily intake of PCDD/PCDF is estimated to be about 90% via food ingestion, there is not enough food data available (e.g., intake via milk is based on two analyses).

Understanding the environmental fate of PCDD/PCDF is fundamental to evaluating human exposure. Empirical measurements of intermedia transfers, environmental degradation and clearance rates, as well as bioaccumulation factors are fundamental to apply mathematical models to simulate such events. Although the TEQ approach was developed and proven to be a helpful tool for risk assessment, input data for models and exposure assessment have to be congener specific. For further information on terrestrial models (air → grass → cow milk) see McLachlan et al. (1995).

Table 11.6 Congener-specific migration rates for transfer from grass fodder to cow milk (%)

	Fürst et al. (1992)	McLachlan (1992)	Blüthgen et al. (1994)
2,3,7,8-Cl$_4$DD	35	36	36
1,2,3,7,8-Cl$_5$DD	19	32	15
1,2,3,4,7,8-Cl$_6$DD	13	16	8
1,2,3,6,7,8-Cl$_6$DD	13	15	n.a.
1,2,3,7,8,9-Cl$_6$DD	4.5	15	n.a.
1,2,3,4,6,7,8-Cl$_7$DD	1.5	3	n.a.
Cl$_8$DD	0.8	4	n.a.
2,3,7,8-Cl$_4$DF	1.5	7	< 1
1,2,3,7,8-Cl$_5$DF	1	5	n.a.
2,3,4,7,8-Cl$_5$DF	15	33	2 7
1,2,3,4,7,8-Cl$_6$DF	6	15	10
1,2,3,6,7,8-Cl$_6$DF	6	15	n.a.
1,2,3,7,8,9-Cl$_6$DF	n.a.	n.a.	n.a.
2,3,4,6,7,8-Cl$_6$DF	7	14	n.a.
1,2,3,4,6,7,8-Cl$_7$DF	2	3	< 1
1,2,3,4,7,8,9-Cl$_7$DF	n.a.	8	n.a.
Cl$_8$DF	0.5	2	n.a.
I-TEQ	11.5	21	19.2

Source: Fiedler (1995).

n.a., No data available.

Most information is available on congener-specific transfer data determined for fodder → cow (milk) transfer in three studies (Fürst et al. 1992; McLachlan 1991; Blüthgen et al. 1994). The results are presented in Table 11.6.

Environmental Transformation Processes

Volatilization of gaseous PCDD/PCDF from contaminated soils was proven by plant uptake (Prinz et al. 1993). Contamination of leafy plants via a soil → plant transfer was studied in open-field, greenhouse, and chamber experiments. A transfer of PCDD/ PCDF from soil to plant was detected under the micro-climate conditions; however, due to open air turbulence an influence of the soil could not be detected in the plants (carrot leaves) in open-field experiments. As shown in greenhouse experiments, this pathway may be important for vegetation growing close to topsoil (Prinz et al. 1993; DECHEMA 1995).

The EPA Dioxin Reassessment could not give any information on degradation or formation processes of PCDD/PCDF under environmental conditions. The Exposure Panel at the Science Advisory Board Dioxin Reassessment Review Committee (Washington, DC, May 15–16, 1995) concluded that there are major gaps in our knowledge of environmental transformation processes:

- There are no photolysis data at all.
- There are gaps between emissions and sinks—nobody has any idea of what causes the changes in profiles and patterns of PCDD/PCDF.

More specifically, the discussions at the Science Advisory Board Dioxin Reassessment Review Committee and the results of a workshop in Germany (DECHEMA 1995) can be summarized as follows:

- Background samples show differences between the homologue and congener profiles and patterns between sources and environmental sinks. For example, the most abundant homologues emitted from combustion sources are Cl_4/Cl_5DF; however, in sinks such as soil, Cl_8DD is dominating. Moreover, the profiles in ambient air samples are different from those of emission samples. These findings are taken as indirect evidence that photolysis occurs in the vapor phase.
- There is hardly any information available on photolytic degradation of PCDD/PCDF in air and soil. Photolysis is either direct or via hydroxyl radicals. Particle-phase photolysis does not seem to occur (Koester and Hites 1992b); however, only one experiment has been done and no congener-specific data are available. In one poor paper it has been shown that the TCDD half-life is only a few seconds (suggesting direct photolysis). More important are OH radicals; however, no experiment has addressed this question. Atkinson (1987) determined a half-life of 1–40 days for Cl_4-Cl_8DD. Thus, reaction rates for OH-radical-induced photodegradation are much slower than for direct photolysis. Moreover, the photolytic degradation products of PCDD/PCDF are unknown.
- In contrast, Rappe (1992) and Vollmuth et al. (1994) reported on the photolytic formation of PCDD/PCDF.
- There is no information available on the microbial degradation of PCDD/PCDF under environmental conditions. There are indications for reductive dechlorination of higher chlorinated PCDD under methanogenic conditions in sediments and sewage sludge. So far, degradation under aerobic conditions was only proven for *mono-* and *di-* but not for higher chlorinated dibenzo-*p*-dioxins.
- It has been shown that peroxidases are capable of synthesizing PCDD/PCDF from precursors such as chlorophenols (Öberg et al. 1993; 1992; Öberg and Rappe 1992; Öberg et al. 1990; Wagner et al. 1992). The formation of especially Cl_7DD and Cl_8DD during the composting process has been proven in many experiments (for a summary, see Fiedler 1994). The overall effect found was that the I-TEQ increases by about 1–2 ppt during the composting process (when comparing the I-TEQ of the input material with that of the final compost and after correction for the loss of organic matter during composting). The increase seems to be independent of the PCDD/PCDF level in the input material; thus, first results reporting a three-to eightfold increase

in the I-TEQ during composting could not be verified when experiments were performed using higher contamination ranges.

- The decrease in 2,3,7,8-Cl_4DD in the topsoil of Seveso (DECHEMA 1995; Bertazzi et al. 1994), which was observed during approx. the first 6 months after the accident, was explained by photodegradation and volatilization. It was not possible to differentiate between these two degradation pathways. Over the subsequent 5 years, the disappearance of 2,3,7,8-Cl_4DD was much slower; thus, a half-life of 9.1 years was calculated for this compound. The decrease was attributed to translocation processes rather than to biodegradation.

11.5.4 National Inventories

Several nations calculated national emissions from known dioxin sources. Table 11.7 summarizes the national emission estimates. Although there are large differences between the PCDD/PCDF amounts released from all known sources for the various countries, it can be seen that the total masses roughly correlate with anthropogenic activities. Thus, in the less populated countries such as Sweden and the Netherlands, the dioxin emissions are smaller than in the USA where a population of 240 million Americans seem to produce higher amounts of PCDD/PCDF. The high degree of industrialization can be seen for the Netherlands, the United Kingdom, and Germany. Estimates for future emissions assume that technological improvements for purification of off-gases will have taken place and will result in a sharp decrease in PCDD/PCDF emissions to the air.

Mainly the same sources were considered in all mass balances. However, some differences can be observed:

- Lime burning for primary production was considered to be the major source of atmospheric emissions in Sweden. However, Sweden is the only country to include this type of industry.
- Although aware of the fact that crematories are known dioxin sources, such emissions were not included in the inventories of Germany and the USA.

Table 11.7 Summary of PCDD/PCDF national emission inventories (g TEQ/year)

Sweden		The Netherlands		USA	United Kingdom		Germany	
1990	1993	1991	2000	1987–1992 approx.	Present	Future	1985–1990	1993–1995
31.8–115	21.6–88	484	< 58	3774–34278	559–1099	112–345	1166–1646	452–656

Source: Fiedler (1995).

- The Netherlands is the only country to attribute significant emissions to chemical production processes (25 g I-TEQ/year).
- Only the USA estimate has a high number for diesel-truck-derived PCDD/PCDF emissions (27–270 g I-TEQ/year). However, this estimate has a low confidence rating.

There still seems to be large difficulties in estimating PCDD/PCDF emissions from residential homes. No sound information is available on the contribution from accidental and uncontrolled fires, such as burning homes (Germany: 1–10 g I-TEQ/year), forest fires (USA: 27–270 g I-TEQ/year), landfill fires (Sweden: 2.8–30 g Nordic-TEQ/year), burning of straw, volcanic eruptions, etc., and on diffuse losses from manufacturing processes.

11.5.5 National Mass Balances

In the USA, there are deposition data from just two places (Koester and Hites 1992). Total fluxes of wet and dry deposition given as the sum of PCDD and PCDF were determined for the cities of Indianapolis and Bloomington, Indiana, USA (Table 11.8). Due to the lack of congener-specific deposition data, no I-TEQ could be calculated. However, due to the difference in physicochemical properties, congener-specific information is necessary to correctly estimate such fluxes. The US data do not consider gaseous PCDD/PCDF. For Germany, there is a much better database on air concentrations of PCDD/PCDF available (e.g., see Fiedler 1995). A mass balance for PCDD/PCDF on a TEQ basis was performed by Wintermeyer and Rotard (1994). The authors used a tiered approach and divided the country into regions of low and higher deposition; the results are shown in Table 11.9. Inclusion of the gaseous part of the aerial PCDD/PCDF would result in a total atmospheric deposition of 2000–10000 g TEQ/year.

As for all other countries, e.g., Sweden (Rappe 1991) and the United Kingdom (Harrad and Jones 1992), the estimated deposition for the USA and Germany

Table 11.8 Deposition estimates for the USA

	Wet deposition $ng/(m^2 \cdot year)$	Dry deposition $ng/(m^2 \cdot year)$	Total $ng/(m^2 \cdot year)$
Indianapolis	220	320	540
Bloomington	210	160	370
	Remote areas $ng\ I\text{-}TEQ/(m^2 \cdot year)$	Populated areas $ng\ I\text{-}TEQ/(m^2 \cdot year)$	Total $g\ TEQ/year$
US-EPA	1	2–6	20000–50000

Source: Fiedler (1995); EPA (1995).

Table 11.9 Annual deposition in former West Germany

Region	Deposition pg TEQ/(m^2·day)	Deposition ng TEQ/(m^2·year)	Total area (km^2)	Input g TEQ/year
Agricultural and forestry use	5–20	1.8–7.3	218 000 (87.2%)	400–1600
Residential/ industrial			30 000 (12.2%)	
	20–100	7.3–36.5	3000 (10%)	200–1000
	100–500	36.5–183	540 (1.8%)	200–800
	500–3000	183–1100	120 (0.4%)	200–1100
Total				1000–4500

Source: Wintermeyer and Rotard (1994).
All concentrations refer to deposition measurements (dust) and do not include PCDD/PCDF in the gas phase.

exceed the estimated emissions by a factor of 10 or more. Thus, presently all mass balances, independent of the quality of data used to calculate emissions and deposition (gaseous deposition as the gas-phase PCDD/PCDF are included in all emission estimates!), face the same unsolved problem: annual deposition $_{PCDD/PCDF}$ ≫ annual emissions$_{PCDD/PCDF}$.

Several reasons have been put forward to explain this discrepancy (see e.g., Fiedler 1995):

- Emission data are not representative for the known sources (in general, the emissions are underestimated for several sources).
- Deposition data are not representative.
- There are additional unidentified sources (especially diffuse emissions).
- Additional deposition from emissions generated outside the country (long-range transport).
- Resuspension and deposition from reservoirs.
- Atmospheric transformation processes.

11.5.6 Reservoirs

So far, hardly any country has done a reservoir inventory for PCDD/PCDF. In other words, there is almost no knowledge about the total amounts of PCDD/PCDF present in sinks such as sediments of harbors, rivers, lakes, oceans, landfills (including residues of accidental fires, fly ash, residues and sludges from chemical manufacture), and contaminated soils from (chemical) production sites. It is thought that these reservoirs contain large amounts of PCDD/PCDF. Although these reservoirs may be highly contaminated with PCDD/PCDF, the physicochemical properties of these compounds imply that dioxins and furans will stay absorbed to organic carbon of soils or other particles. However, mobilization can occur in the presence of lipophilic solvents (which leads to leaching into deeper layers of soils and/or groundwater) or in cases of erosion or runoff by rain from topsoil (which results in translocation into the neighborhood). Although there is little information available, experience has shown that PCDD/PCDF transport due to soil erosion and runoff does not play a major role in environmental contamination and human exposure (Fiedler 1995).

A major translocation of PCDD/PCDF can occur during excavation of dioxin-contaminated landfills for remedial action or dredging of harbor and river sediments. Such action can lead to human exposure when the sediments are brought to areas of farming or horticultural use (of relevance for terrestrial food chain).

According to Hagenmaier and Krauß (1993) two-thirds of the PCDD/PCDF present in southwest Germany is present in forest soils, indicating that atmospheric deposition into forest ecosystems is a key factor in the environmental fate of these compounds. However, so far there is no consensus as to how this deposition occurs.

Other reservoirs include the former use of PCDD/PCDF-contaminated products such as 2,4,5-T, PCB, and pentachlorophenol/-phenate (PCP/PCP-Na). Although there are estimates of the total amount of these compounds produced for various purposes it seems to be impossible to deduce from these numbers a quantitative impact of PCDD/PCDF on the environment or humans (Fiedler 1995).

11.6 POLYCHLORINATED BIPHENYLS (PCB)

11.6.1 Introduction

Polychlorinated biphenyls (PCB) comprise a class of 209 individual compounds. The two main sources of PCB are

- Commercial production
- By-products in combustion processes.

PCB are produced by chlorination of biphenyls; their commercial production started about 60 years ago. The total amount produced worldwide is estimated at 1.5 million tons (Ivanov and Sandell 1992; Rantanen 1992).

Depending on the degree of chlorination of the PCB, their physicochemical properties, such as inflammability or electric conductivity, brought about a wide field of application. Thus, PCB have been used as electric fluids in transformers and capacitors, pesticide extenders, adhesives, dedusting agents, cutting oils, flame retardants, heat transfer fluids, hydraulic lubricants, sealants, paints, and in carbonless copy paper.

Some of their applications resulted in a direct or indirect release of PCB into the environment. Relatively large amounts were released due to inappropriate disposal practices, accidents, and leakages from industrial facilities. Hansen (1987) estimates that in 1987, about 400 000 t PCB were released into the environment. PCB were marketed with respect to percentage of their chlorine content (by weight) and were available under several trade names, e.g., Clophen (Bayer, Germany), Aroclor (Monsanto, USA), Kanechlor (Kanegafuchi, Japan), Santotherm (Mitsubishi, Japan), and Phenoclor and Pyralene (Prodolec, France).

11.6.2 Legal Situation

After the impact of PCB on the environment was recognized, the most prominent producers, Monsanto and Bayer, voluntarily stopped their PCB production in 1977 and 1983, respectively (DFG 1988). In Germany, production, marketing, and use of PCB, polychlorinated terphenyls (PCT), and vinyl chloride (VC) are prohibited by law since 1989. The law includes also mixtures containing more than 50 mg/kg PCB, e.g., recycled waste oils. The US Environmental Protection Agency took one further step: in 1979, a ban on commercial PCB production was implemented; today, the use of PCB is restricted to completely enclosed systems (Allen 1990). In Germany, capacitors and transformers with a PCB content of more than 1 L had to be replaced by 1993 and 1997, respectively (Wagner 1990). Because there is evidene for cancerogenic effects, PCB are an area of risk assessment that still requires further study (Safe 1994; Ahlborg et al. 1994).

11.6.3 PCB Levels in the Environment

PCB have been identified in almost every environmental compartment or matrix. Detected levels depend on the nature and location of the particular environmental sample. However, congener-specific analytical procedures for qualitative and quantitative detection of PCB have not been developed to the same extent as, for example, those for PCDD/PCDF.

When compared with other chemicals, PCB have very high K_{OW} values: log K_{OW} are in the range of 4.5 for monochlorobiphenyls to > 8 for higher chlorinated PCB. Consequently, PCB tend to adsorb to nonpolar surfaces and accumulate in lipophilic matrices along the aquatic and terrestrial food chain.

Commercial PCB, as well as environmental extracts, contain complex mixtures of congeners. PCB mixtures found in environmental matrices usually do not resemble commercial PCB mixtures. For example, a congener-specific analysis of PCB showed remarkable differences between a commercial Aroclor 1260 mixture and human breast milk (Safe 1990, 1994; Norén and Lundén 1991). These differences are due to the fact that the most abundant PCB in commercial mixtures are *ortho*-substituted congeners which are readily degradable.

However, smaller amounts of the so-called dioxin-like PCB, namely the coplanar (= non-*ortho*-substituted) and mono-*ortho*-substituted congeners, are present in commercial mixtures as well. The latter are very stable and resistant to biodegradation and metabolism. Moreover, it is well known that lower chlorinated PCB can volatilize and are, thus, more susceptible to atmospheric removal processes (Mackay et al. 1992).

Generally, the PCB levels found in environmental matrices are higher than the levels of PCDD/PCDF. This is due to the fact that besides thermal formation significant amounts of PCB were and still are released via diffuse emissions from industrial products. Input of PCB into soil occurs—as for other lipophilic chemicals—from spills, direct application of, e.g., sludges, or via dry and wet deposition. The organic carbon of soil is the natural sink for such nonpolar lipophilic substances. Due to the strong affinity for organic carbon, PCB are quite immobile in soils. As PCB are persistent compounds, soils possess a memory effect and remember inputs for a long time as well as long-term diffuse inputs.

Henry's Law constants are 0–1 for mono- and dichlorobiphenyls; thus, these substances are found preferentially in the gas phase. Due to the low water solubility they are not washed out with rainwater from the atmosphere. Higher chlorinated biphenyls are (completely) adsorbed to particulates and, thus, can be removed from the atmosphere by capture of aerosols in raindrops. These two effects result in a relative accumulation of lower chlorinated PCB in the atmosphere (Duinker and Bouchertall 1989). Air concentrations are in the picogram per cubic meter to nanogram per cubic meter range, with lower levels in remote and rural areas. Background air levels in the USA were constant in the range of 1 ng/m^3 over several years, with tri- and tetrachlorinated congeners dominating.

PCB in the Great Lakes display a more complex behavior. They were found to volatilize where a river discharges relatively high PCB loads into Green Bay, Wisconsin (Achman et al. 1993). Baker and Eisenreich (1990) calculated an average PCB volatilization rate from Lake Superior which approximately equals their atmospheric deposition. His findings support the conceptual model that these compounds permanently cycle between atmosphere and natural waters (Mackay et al. 1986). According to this model, PCB dissolved in raindrops or sorbed to particulates are washed out of the atmosphere by rain (for vapor–particle partitioning, see Bidleman 1988). This input of PCB into surface waters results in a fugacity gradient towards the atmosphere, which in turn drives volatilization.

Recent PCB data exist for sediments and suspended particles in German rivers (Breitung 1994). Along the river Saar it was found that close to locations with heavy industry (coal mining and steel industry) PCB and Ugilec (a commercial mixture of tetrachlorinated 2-methyl-diphenylmethanes) levels were higher than normal. Differentiation in depth showed that Ugilecs were only found in more recent sediments whereas PCB could be detected dow to 1.2 m, with higher concentrations in the older sediments.

11.6.4 Environmental Fate of PCB

Biodegradation

Biodegradation by microorganisms may occur via three different mechanisms:

- Aerobic respiration in the presence of oxygen, involving reaction with mono- and dioxygenases; in a final step H_2O is incorporated
- Anaerobic respiration under exclusion of oxygen; inorganic substances such as nitrate, sulfate, and carbon monoxide act as electron acceptors
- Fermentation under exclusion of oxygen; the organic compound to be degraded acts as an electron acceptor.

Aerobic Degradation. In general, bacteria cannot use chlorinated aromatic hydrocarbons as substrate. Present knowledge assumes that bacteria growing on nonchlorinated biphenyl are capable of causing chemical reactions on the chlorinated ring system as well (Kohler et al. 1992). However, some microorganisms are capable of using lower chlorinated PCB as a carbon source. Thus, *Acinetobacter* sp. P6, *Achromobacter* sp. B 218, and *Bacillus brevis* B 257 can grow on 4-chlorobiphenyl as the only carbon source. The main degradation product is 4-chlorobenzoic acid. In general, formation of chlorinated benzoic acids is the major degradation pathway for PCB (Rochkind et al. 1986). Other microorganisms capable of biodegrading PCB belong to the genera *Acetobacter*, *Alcaligenes*, and *Pseudomonas*.
Some general conclusions can be drawn (Rochkind et al. 1986):

- Increasing the number of chlorine substituents decreases biodegradation of PCB.
- Two chlorines in *ortho* position on the same or on different aromatic rings of the biphenyl molecule significantly inhibit biodegradation (exception: 2,4,6-trichlorobiphenyl, which is readily biodegradable by *Acinetobacter*).
- If there are chlorine substituents on both rings, the ring with less chlorine atoms will be hydroxylated first.
- Biphenyls that have chlorine substituents only on one ring system are metabolized more rapidly than a chlorobiphenyl with the same number of chlorine atoms but on both rings; thus, 3,4-dichlorbiphenyl will be better metabolized than 3',4-dichlorobiphenyl.
- PCB metabolism is facilitated when a carbon atom with a chlorine substituent is situated between two unsubstituted carbon atoms.
- Higher chlorinated congeners having a 2,3,4-trichlorophenyl group are resistant to biological degradation.
- Ring cleavage occurs preferentially in the unsubstituted ring (Furukawa et al. 1978).

- PCB containing chlorines at positions 2 and 3, e.g., 2,2′,3,3′-tetrachloro-biphenyl, 2,2′,3,5′- tetrachlorobiphenyl, and 2,2′,3′,4,5-pentachlorobiphenyl are more readily biodegraded than other tetra- and pentachlorinated biphenyls.

Anaerobic Degradation. Polychlorinated biphenyls are extremely resistant to conventional aerobic transformation, but they will undergo anaerobic reductive dechlorination. Studies of PCB contamination in Hudson River sediment demonstrate that anaerobic environments yield markedly lower levels of tri-, tetra-, and pentachlorobiphenyls and higher levels of mono- and dichlorobiphenyls than aerobic environments (Brown 1987). Many of the less chlorinated PCB then are aerobically biodegradable, and are therefore generally less toxic than highly chlorinated PCB.

Kong and Sayler (1983) reported that under laboratory conditions monochlorinated biphenyls can be degraded and totally mineralized by mixed bacterial cultures from river sediments. Mineralization rates of between 1 and 2 µg mL^{-1} day^{-1} have been observed at concentrations of 30 µg mL^{-1}. For PCB, the degradation rate is inversely related to the degree of chlorination (Young and Häggblom 1991); thus, highly chlorinated congeners are more readily dechlorinated than lower chlorinated congeners (Neilson 1990).

There is evidence that not only the number of chlorine substituents determines degradation rates, but also their position. Reductive dechlorination predominantly reduces chlorine in *meta* and *para* positions (Young and Häggblom 1991), resulting in accumulation of *ortho*-chlorinated congeners. Addition of organic substrates, such as methanol, glucose, or acetone stimulated dechlorination, whereas little stimulation was observed in cases with no organic additives (Nies and Vogel 1990; Ye et al. 1992).

Metabolism

PCB do not have reactive functional groups; thus, these lipophilic molecules have to be hydroxylated first to make them more polar and consequently subject to excretion. The rate-limiting step in the elimination of PCB is that of metabolism, which primarily occurs by the hepatic P-450-dependent monooxygenase system. Hydroxylated products are the major PCB metabolites, and, based on available studies, it can be concluded that hydroxylation mainly occurs at *para* or *meta* positions if these sites are unsubstituted. The chlorine content, the substitution pattern, and the presence of certain isoenzymes of the cytochrome-P-450 system are important factors in determining the transformation rate of PCB (Sipes and Schnellmann 1987). In general, metabolization of PCB decreases with increasing number of chlorine atoms present and with decreasing number of adjacent unsubstituted carbon atoms. Isoenzymes capable of metabolizing phenobarbital (PB) were found to metabolize the non-dioxin-like PCB, whereas coplanar, non-*ortho* or mono-*ortho*-substituted (dioxin-like) PCB can induce isoenzymes capable of metabolizing 3-methylcholanthrene (MC; Safe

1994). Commercial PCB, such as Aroclor 1254, induce both MC- and PB-inducible monooxygenases.

Besides hydroxylated metabolites with subsequent conjugation, sulfur-containing metabolites, e.g., methyl sulfones, and partially dechlorinated metabolites have also been identified. Methyl sulfones have been shown to selectively accumulate in the Clara cells of rat lung and in lung tissues of mice. Methyl sulfonyl metabolites of 2,4′,5-, 2,2′,4,5-′, and 2,2′,4,5,5-PCB have also been found in liver, adipose, and fetal tissues and have been identified in both environmental samples and in human milk (Ahlborg et al. 1992).

Due to the low transformation and excretion rates of PCB, certain congeners accumulate in organisms. Persistent congeners, such as 2,2′,4,4′,5,5′-hexachlorobiphenyl (PCB 153), were found to promote tumors in rats, whereas 2,2′,3,3′,6,6′-hexachlorobiphenyl (PCB 136) is easily degraded.

Bioconcentration

Substances of low biological degradability tend to accumulate throughout trophic levels of the food net. For example, concentrations of total PCB increases with the trophic level. Only concentrations of PCB in sediments are higher than levels in the subsequent trophic levels. It has been demonstrated that chlorinated dibenzo-p-dioxin and dibenzofuran congeners accumulate with considerable species differentiation. Contribution of the dioxin-like PCB congeners #77, 105, and 126 to the total TEQ is substantially greater than that of PCDD/PCDF (even in cases of known PCDD/PCDF contamination; Smith et al. 1990). In all cases an exchange between trophic levels (sediment → algae → plankton → planktivores → piscivorous fish → piscivorous birds) resulted in an increase in both total PCB concentration and dioxin-like TEQ (for TEF/TEQ of PCB, see Section 11.6.5; Jones et al. 1993).

11.6.5 Toxicity of PCB

Commercial mixtures, as well as the individual PCB congeners, elicit a broad spectrum of biochemical and toxic responses, some of which are similar to those caused by 2,3,7,8-tetrachlorodibenzo-p-dioxin. Due to the fact that dioxin-like compounds normally exist in environmental and biological samples as complex mixtures of congeners, the concept of TEQ has been developed to simplify risk assessment and regulatory control. In applying this concept, relative toxicities of dioxin-like compounds in relation to the reference compound (2,3,7,8-tetrachlorodibenzo-p-dioxin, 2,3,7,8-Cl$_4$DD) were determined on the basis of results obtained in in vivo and in vitro studies (Table 11.10). Three coplanar PCB, namely 3,3′,4,4′-tetrachlorobiphenyl, 3,3′,4,4′,5-pentachlorobiphenyl, and 3,3′,4,4′,5,5′-hexachlorobiphenyl, exhibit dioxin-like effects, such as Ah-receptor agonist activity (Ah = aryl hydrocarbon; Safe 1990, 1994; Ahlborg et al. 1992, 1994). PCB 170 and PCB 180 are included because they are active as inducers of

Table 11.10 Proposed TEF for coplanar and mono- and di-*ortho*-substituted PCB

Congener	Substitution	IUPAC no.	TEF
Non-*ortho* substituted PCB	3,3',4,4'-TetraCB	77	0.0005
	3,3',4,4',5-PentaCB	126	0.1
	3,3',4,4',5,5'-HexaCB	169	0.01
Mono-*ortho* substituted PCB	2,3,3',4,4'-PentaCB	105	0.0001
	2,3,4,4',5-PentaCB	114	0.0005
	2,3',4,4',5-PentaCB	118	0.0001
	2',3,4,4',5-PentaCB	123	0.0001
	2,3,3',4,4',5-HexaCB	156	0.0005
	2,3,3',4,4',5'-HexaCB	157	0.0005
	2,3',4,4',5,5'-HexaCB	167	0.00001
	2,3,3',4,4',5,5'-HeptaCB	189	0.0001
Di-*ortho* substituted PCB	2,2',3,3',4,4',5-HeptaCB	170	0.0001
	2,2',3,4,4',5,5'-HeptaCB	180	0.00001

Source: Ahlborg et al. (1994).
IUPAC, International Union of Pure and Applied Chemistry; CB, chlorinated benzene.

Table 11.11 Toxic equivalents (TEQ) calculated for fish, cow milk, and human milk samples using the interim WHO/ICPS TEF

TEQ	Human milk	Cow milk	Salmon
Sum of TEQ for non-*ortho* PCB	10.3	2.4	67.7
Sum of TEQ for mono-*ortho* PCB	10.1	0.4	46.8
Sum of TEQ for di-*ortho* PCB	0.6	0.04	8.3
Total TEQ for PCB	21.0	2.8	122.8
Total TEQ for PCDD/PCDF	20.6	5.6	56.0

Source: Ahlborg et al. (1994).

EROD (ethoxyresorufin-*o*-deethylase) activity and are present in significant amounts in environmental samples.

To illustrate the consequences of the recommended PCB TEF, the contribution of dioxin-like PCB and PCDD/PCDF to the total TEQ was calculated for some matrices (Table 11.11). As can be seen from Table 11.11, the contribution from the PCB to the total TEQ for PCDD/PCDF plus PCB is between 50% and 200%.

11.7 GLOBAL DISTRIBUTION—LONG-RANGE TRANSPORT

Today it is known that many chlorinated organics and other stable compounds are distributed on a global scale through atmospheric transport. A general tendency in these transport patterns is that different substances are evaporated and spread to the atmosphere at latitudes with warmer climates and then condense and fall out closer to the poles (global condensation). Consequently, areas close to the north and south poles receive a disproportionate share of this fall out. An indication of this phenomenon is that several chlorinated pesticides long banned in countries such as Sweden are found—although at relatively low levels—in environmental compartments of this country. Examples of these substances are chlordane, toxaphene, and hexachlorocyclohexane, i.e., compounds that are still being produced in other countries. For PCB, an annual fall out of 4 t was estimated for Sweden in 1994 (KEMI 1994).

The presence of persistent polychlorinated compounds in remote areas such as the Arctic and Antarctic has been reported. PCB, PCDD/PCDF, HCH, and HCB were found in marine organisms, e.g., in seal blubber and pinniped milk, as well as in lake and sea sediments. The occurrence of mainly man-made organochlorines in regions far away from industrialized and densely populated areas indicates that atmospheric transport is an important route for dispersing these compounds (Oehme et al. 1995). All three groups of compounds, PCB, PCDD/PCDF, and HCB, have the same source areas: densely populated and industrialized regions. From these source regions, organochlorines are transported via various mechanisms.

Norstrom (1994) summarized the present knowledge of the occurrence of persistent chlorinated organic compounds in the Arctic aquatic environment as follows. Chlorinated hydrocarbons (CHC), such as the pesticides chlordane and toxaphene (polychlorinated camphenes, PCC), hexachlorocyclohexanes (HCH), hexachlorobenzene (HCB), and DDT, and industrial chemicals such as polychlorinated biphenyls (PCB) have been identified in air, snow, ocean water, and biota in the marine ecosystem. Although most open uses of these chemicals were curtailed in many industrial countries, a considerable fraction of these compounds is still cycling in the ecosphere. Thus, it was estimated that 20% of the world production of PCB, 230 000 t, is present in the upper layers of the ocean and that 790 t are in the open ocean atmosphere. Large quantities of chlorinated pesticides continue to be used in less developed countries, especially in the southern hemisphere. Although there is little information on the production amounts and releases of organochlorines from Russia and China, these areas are undoubtedly major contributors to the environmental burden with CHC.

CHC can be remarkably resistant to degradation under environmental conditions and they tend to have long environmental half-lives. Chlorinated pesticides and PCB have sufficiently high vapor pressures that they readily volatilize when spread over a large surface area such as soil or water. Atmospheric residence time of PCB has been calculated to be in the order of a few months. Henry's Law constants of the above-mentioned compounds are in the range of 0.1–50 $Pa \cdot m^3/mol$ and thus will allow that these substances will evaporize and cycle back and forth between land or surface waters and air. These processes lead to a global distribution. The "cold finger effect" as discussed by Ottar (1981) will result in the fact that the Arctic and Antarctic regions will become the sinks for

Table 11.12 Average levels of organochlorines in the arctic marine–environment

Matrix, dimension	HCH	HCB	PCC	DDT	Chlordanes	PCB
Air, ng/m^3	0.58	0.19	0.044	<0.001	0.006	0.014
Snow, ng/L	1.72	<0.002	0.085	<0.01	0.06	0.086
Seawater (surf.), ng/L	4.3	0.028	0.36	<0.001	0.004	0.007
Seawater (deep), ng/L	0.51	0.01	0.11	<0.002	0.005	<0.014
Zooplankton, µg/g lipid	0.08	0.02	0.06	0.06	0.06	0.11
Amphipods, µg/g lipid	0.5	0.17	NA	<0.35	0.43	<0.44
Cod, µg/g lipid	0.58	0.2	1.84	0.26	0.19	0.23
Beluga, µg/g lipid	0.25	0.5	3.11	2.82	1.76	3.79
Ringed seal, µg/g lipid	0.23	0.03	0.32	0.5	0.4	0.55
Polar bear, µg/g lipid	0.51	0.27	ca. 0.4	0.4	3.7	5.4
Human milk, µg/g lipid	NA	0.14	NA	1.21	NA	1.05

Source: Norstrom (1994).
NA, Not available.

organochlorines due to the distillation of these compounds from warmer to colder regions.

Average concentrations of the major classes of CHC in the marine environment of the Arctic are given in Table 11.12. HCHs and HCB are the dominant CHC in air, followed by the PCC. DDT, chlordane, and PCB are one order of magnitude lower. Because of its higher Henry's Law constant, HCB is less dominating in snow and seawater. The high levels of HCH in the Arctic Ocean water support the distillation theory. Moreover, sources in Asia might have a contribution. The effect of the higher lipophilicity of compounds such as DDT, chlordanes, and PCB goes along with bioconcentration. The concentrations of these compounds increase along the aquatic food chain from plankton to beluga whales. The only exception are the HCH which do not tend to strongly bioconcentrate (log K_{OW}).

Generally, the concentrations in the Antarctic are lower than those in the Arctic. This finding is reasonable as more than 80% of the industrialized regions which act as sources for PCDD/PCDF and PCB are located in the northern hemisphere. The slow interhemispheric air exchange (about 1–2 years) reduces the atmospheric transport of semivolatile compounds from the northern to the southern hemisphere. Oehme et al. (1994) analyzed air samples from both regions and found some interesting differences: levels of HCH and HCB were lower in Antarctic air. Whereas in Arctic air, α-HCH was dominating, the γ-isomer dominated in the Antarctic. This indicates that preferentially the pure γ-HCH has been applied in the southern hemisphere whereas in the northern hemisphere more of the technical mixture (80%–85% α-HCH) was used. HCB is mostly of anthropogenic origin—from incomplete combustion processes and utilization as pesticide—and its atmospheric half-life of 1–2 years leads to an almost homogeneous distribution on both hemispheres. Four chlordane compounds were identified at similar concentrations in both hemispheres. The PCDD/PCDF pattern found in Antarctic fur seal blubber was significantly different from that found in Arctic ringed seals and harp seals. Reasons might be differences in the emission patterns in the northern and southern hemispheres, different food habits, or interspecies variations. Amongst the PCB patterns, PCB 77, PCB 126, and PCB 169 were similar in Antarctic fur seals when compared to the Arctic harp seal. However, the levels—in TEQ—were lower in the Antarctic seals by a factor of about 5. This is less than the 1–2 orders of magnitude reported in earlier studies (Bacon et al. 1992; Luckas et al. 1990). Amongst the 2,3,7,8-substituted PCDD/PCDF, 2,3,7,8-Cl_4DF was found to be the most abundant congener in many aquatic organism in the Arctic (Paasivirta et al. 1994; Oehme 1995).

Finally it should be mentioned that distribution of chlorinated organic compounds also occurs via economic pathways and international trade of goods. Examples are textiles, leather, wood, and packaging materials. Consequently, a ban of a given chloro-organic compound by one or a few countries— as, e.g., for pentachlorophenol in Germany and DDT and PCB in several industrialized countries—does not protect a country from "pollution" with the given compound or its contaminants (e.g., PCDD/PCDF). Moreover, it is difficult to control the concentrations of chemicals in these goods that pass the country's borders.

The same applies to transboundary atmospheric "imports" of unwanted pollutants: stringent emission limits set by authorities of a country do not prevent that country from importing atmospheric contaminants from neighboring countries. Thus, in many respects, the spread of persistent chloro-organic chemicals requires international solutions.

11.8 ORGANOCHLORINE SUBSTANCES IN THE MARINE ENVIRONMENT

11.8.1 Introduction

There is widespread concern that organochlorine compounds have a general potential for producing adverse effects in the environment. Consequently, screening procedures have to be developed to identify those organochlorine compounds that have to be considered as being the most potentially harmful for a given environment. Fiedler and Hutzinger (1990) developed a simple ecotoxicological model to identify organochlorine compounds potentially harmful for the aquatic environment. The main evaluation procedure and major findings are summarized in the following sections.

11.8.2 Criteria Important for an Environmental Impact Assessment

As organochlorine compounds cannot be reviewed as a single class of substances they have been grouped according to their chemical backbone as shown in Figure 11.2. The chemical structure is an imporant parameter and such information can help to predict environmental behavior. A chlorine substituent, in particular, has an influence on key parameters of a hydrocarbon backbone as

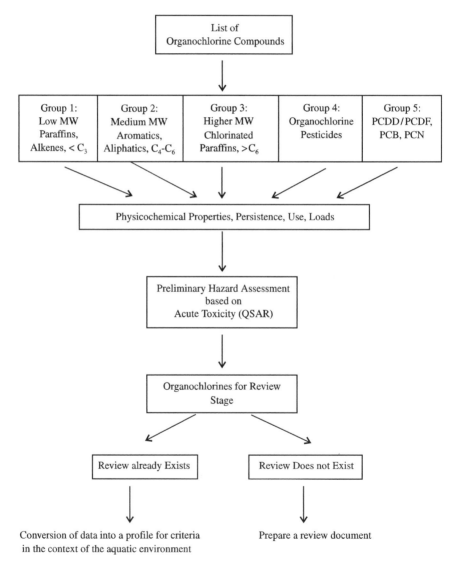

Figure 11.2 Recommended procedure for reviewing organochlorine compounds (Fiedler and Hutzinger 1990).

Table 11.13 Influence of chlorine substituents on the physicochemical properties of hydrocarbons

Compound	Number of chlorine atoms	S (mg/L)	$\log K_{OW}$
Benzene	0	1780	2.13
Hexachlorobenzene	6	0.006	6.18
Phenol	0	82 000	1.45
Pentachlorophenol	5	14	3.7
Biphenyl	0	5.9–7.5	3.89
PCB 209	10	0.000004	8.23
Dibenzo-p-dioxin	0	0.842	4.3
2,3,7,8-Cl$_4$DD	4	0.000008	7
Cl$_8$DD	8	0.0000004	8.2

S, Water solubility; K_{OW}, octanol/water partition coefficient.

shown in Table 11.13. As can be seen from the table, the water solubility dramatically decreases with increasing number of chlorine substituents in the molecule and the logarithm of the 1-octanol/water partition coefficient (K_{OW}) increases from the parent compound to the fully chlorinated compounds. Consequently, perchlorinated compounds are hardly found in water and tend to accumulate in organisms and adsorb to organic carbon.

In addition, chemical, physical, and biological data on a substance will allow the estimation of a compound's fate in the environment. Such basic data and information are:

$\log K_{OW}$:   The logarithm of the octanol/water partition coefficient is a key parameter of the environmental fate of organic chemicals. It has been proven that for chlorinated compounds, the $\log K_{OW}$ can be used for the estimation of water solubility, soil/sediment adsorption, and bioconcentration for aquatic life. In most risk assessments, substances with a $\log K_{OW} > 3$ are considered to have a high potential for causing harm. (According to McCarthy (1994), for regulatory purposes, a compound is defined as bioaccumulative when it accumulates to body levels greater than 1000 times the environmental level in water or if it has a $\log K_{OW}$ of greater than 4.0.)

LC_{50}:   The LC_{50} value is the concentration of a chemical in water which causes the death of 50% of the organisms tested over a specified time (e.g., 96 h); an EC_{50} records a sublethal effect such as immobility. For risk assessment, substances can be considered as (highly) toxic to aquatic organisms if they have a recorded $LC_{50} < 10$ mg/L.

Degradability: This parameter has to be taken into account when considering the loss of a chemical in the aquatic environment. Chemicals with half-lives longer than 1 week can be considered as persistent. (According to McCarthy (1994), for regulatory purposes, a compound was defined as persistent when its half-life of presence in the environment is greater than 8 weeks.)

Use-effluent: This is an important criterion for hazard assessment; however, it cannot be expected that appropriate data will be available for all chemicals. If the use of a chemical and the production tonnage is known, or if it is known that the substance is discharged from known manufacturing processes in a manner relevant to the aquatic environment, such information will be taken into account.

From various data sources, such as databanks and books (e.g., Verschueren 1987; Pesticides and Manual 1987; chemical catalogues), a total of 811 substances was assembled and grouped into five lists. It has to be noted that chemicals which are obviously used only in laboratories have been excluded.

Group 1—compounds C_1–C_3; 62 compounds
Group 2—compounds C_4–C_6; 232 compounds
Group 3—compounds $> C_6$; 301 compounds
Group 4—pesticides: 69 compounds
Group 5—209 PCB + 210 PCDD/PCDF

For compounds listed in Groups 4 and 5 it was concluded by the UNEP/IRPTC working group (see Fiedler and Hutizinger 1990) that these classes of compounds will need special attention and evaluation. For more information on these substances, see Sections 11.9 (Pesticides), 11.5 (PCDD/PCDF), and 11.6 (PCB).

11.8.3 Generation of Priority Compounds

From the compounds assigned to Groups 1 to 3, three lists have been generated according to their relevance for potenitally harmful effects to the aquatic environment. These three lists—LIST IN (potentially harmful compounds), LIST OUT (potentially harmless compounds), and LIST NODATA (compounds for which no experimental (analytical and/or toxicological) data were available)—were based on an ecotoxicological risk assessment and not on basic structural data (number of C atoms).

LIST IN (potentially harmful compounds). All substances meeting at least two of the criteria given below were considered as potentially harmful to the aquatic environment:

$\text{Log} K_{OW} > 3$
Toxicity $LC_{50} < 10$ mg/L
Persistence > 1 week
As a minor criterion, > 1000 t/year produced or generated in manufacturing processes.

A total of 79 substances belong to this category (Table 11.14); amongst these are tri- and tetrachlorothanes, carbon tetrachloride, chlorophenols, chlorobenzenes, chlorotoluenes, and chloronitrotoluenes.

Table 11.14 List of potentially harmful organochlorine substances selected on the basis of existing data

Group 1: C_1 to C_3 compounds

1,1,1,2-Tetrachloroethane	1,1,2-Trichloroethylene	Methylchloride
1,1,1-Trichloroethane	Chloroform	Pentachloroethane
1,1,2,2-Tetrachloroethane	Dichloromethane	Tetrachloromethane
1,1,2,2-Tetrachloroethylene	Epichlorohydrin	Trichloroethene
1,1,2-Trichloroethane	Hexachloroethane	Vinylchloride

Group 2: C_4 to C_6 compounds

1-Chloro-4-nitrobenzene	1-Chloro-4-nitrobenzene	3-Chlorophenol
1-Chlorobutane	1-Chlorobutane	4-Chloroaniline
2,3,4,5-Tetrachlorophenol	2,3,4,5-Tetrachlorophenol	4-Chlorophenol
2,3,4,6-Tetrachlorophenol	2,3,4,6-Tetrachlorophenol	Chlorobenzene
2,3,5,6-Tetrachlorophenol	2,3,5,6-Tetrachlorophenol	Hexachlorobenzene
2,4-Dichlorophenol	2,4-Dichlorophenol	Hexachlorocyclohexane
2,5-Dichlorophenol	2,5-Dichlorophenol	Hexachlorocyclopentadiene
2,3-Dichlorophenol	2,3-Dichlorophenol	Pentachlorobenzene
2-Chloroaniline	2-Chloroaniline	Pentachlorophenol
2-Chlorophenol	2-Chlorophenol	Pentachloropyridine
1-Chloro-2-nitrobenzene	3,4,6-Trichlorocatechol	Tetrachlorocatechol
1-Chloro-3-nitrobenzene	3-Chloroaniline	

Group 3: compounds with more than six carbons

DL-3-(α-Acetonyl-p-chloro-benzyl)-benzylchloride
1-(o-Chlorophenyl)-1-(p-chlorophenyl)-2,2-dichloroethane (p,p'-DDD)
1,1'(Dichloroethylidene)-bis[4-Chlorobenzene] (p,p'-DDE)
1-(o-Chlorophenyl)-1-(p-chlorophenyl)-2,2,2-trichloroethane (p,p'-DDT)

2,3-Dichlorotoluene	2-Chloro-4-nitrotoluene	4-Chlorotoluene
2,4,5-Trichlorophenoxyacetic acid	2-Chlorotoluene	$\alpha,\alpha,2,6$-Tetrachlorotoluene
2,4,5-Trichlorotoluene	3,4,5-Trichloroguaiacol	Benzotrichloride
2,4-Dichloroacetophenone	3,4,6-Trichloroguaiacol	m-Chlorobenzoic acid
2,4-Dichlorotoluene	3,4-Dichlorotoluene	o-Chlorobenzoic acid
2,5-Dichlorotoluene	3-Chlorotoluene	Octachlorostyrene
2,6-Dichlorobenzonitrile	4,5,6-Trichloroguaiacol	p-Chlorobenzoic acid
2,6-Dichlorotoluene	4-Chlorostyrene	Tetrachloroguaiacol

Source: Fiedler and Hutziner (1990).

LIST OUT (harmless compounds). This list includes all substances which (1) do not meet the above-mentioned criteria and (2) decompose in water, are highly volatile, or are highly water-soluble and, thus, do not tend to bioaccumulate, or (3) are used in closed industrial processes or in small amounts as additives. A total of 351 substances were assigned to this LIST OUT.

LIST NODATA or INSUFFICIENT DATA (no experimental data available). All substances—a total of 184 compounds—for which no sufficient experimental data could be found in the literature were included in the LIST NODATA. For a preliminary risk assessment, we classified these compounds by calculating the compound's K_{OW} as a key parameter for further hazard assessment. Additionally, compounds belonging to the same class or family of compound (e.g., trichloronitrobenzenes) have been included. Finally, 93 organochlorine substances were found to have a potential for being harmful to the aquatic environment ($K_{OW} > 3$; see Table 11.15) and 91 compounds were found to be less harmful ($K_{OW} < 3$).

In a final step, the existing literature has to be evaluated for the priority compounds. In cases of existing reports, these can be used for further risk assessment. Such reports are generated for example by the European Chemical Industry Ecology and Toxicology Centre (ECETOC), the International Agency for Research on Cancer (IRPTC) and International Programme on Chemical Safety (IPCS) in Switzerland, Beratergremium Umweltrelevante Altstoffe (BUA) in Germany, Environmental Protection Agency (EPA), Agency for Toxic Substances and Disease Registry (ATSDR), American Conference of Governmental Industrial Hygienists (ACGIH) in the United States, etc. If no detailed information on the compounds of interest is available such information has to be generated and evaluated.

11.9 ENVIRONMENTAL RISK ASSESSMENT FOR CHLORINATED PESTICIDES

11.9.1 General Aspects

Pesticides, including insecticides, herbicides, fungicides, and rodenticides, are widely used in many industrialized and developing countries. Generally, pesticides are end products, commercially marketed, and directly applied without further purification. It is well known that depending on their mode of application they generally enter the environment directly either as a solid (fumigation, dust) or as a solution (diluted in water, spray). Exposure may also occur during manufacture and formulation. Since pesticides are ubiquitous and those containing chlorine are still being used they may enter surface waters. Therefore, it is necessary to perform a risk assessment by ranking pesticides according to their potential to damage the marine environment. We have used the class of chlorinated pesticides as examples. Due to partially limited data we used quantitative

Table 11.15 Calculated log K_{OW} for compounds NODATA

Compounds IN (log K_{OW} > 3.0)

Octachloronaphthalene	7.96	4-Chloro-3-methylphenol	3.10
1,2,3,5,6,7-Hexachloronaphthalene	6.77	4-Chlorosalicylic acid	3.10
α,α′,2,3,5,6-Hexachloro-*p*-xylene	6.72	2,3,4,5-Tetrachloronitrobenzene	3.80
Hexachlorophene	6.47	1,2,3-Trichloro-4-nitrobenzene	3.80
Pentachloronaphthalene	6.17	4-Chloro-3,5-dimethylphenol	3.78
α,α,α,α′,α′,α′-Hexachloro-*p*-xylene	6.10	2-(2,4,5-Trichlorophenoxy)propionic acid	3.74
1,2,3,4-Tetrachloronaphthalene	5.58	1,1,4,4-Tetrachloro-1,3-butadiene	3.71
9-(Chloromethyl)anthracene	5.36	3,5-Dichloroanisole	3.70
4-Chloromethylbiphenyl	5.09	4,7-Dichloroquinoline	3.70
Trichloronaphthalene	4.98	1,1,2,3,4,4-Hexachlorobutane	3.69
2-Chloro-4-(1,1-dimethylethyl)phenol	4.96	1-Chloroanthraquinone	3.64
2,5-Dichloro-*p*-xylene	4.87	1-Chlorohexane	3.58
1,5-bis(Chloromethyl)-naphthalene	4.85	3,4,5-Trichloroaniline	3.58
2,4,6-Trimethylbenzyl chloride	4.84	2,4,5-Trichloroaniline	3.58
5,6-bis(Chloromethyl)-toluene	4.56	2,4,6-Trichloroaniline	3.58
Pentachloronitrobenzene	4.40	2-Chlorostyrene	3.58
Tetrachloro-*p*-benzoquinone	4.40	3-Chlorostyrene	3.58
N-(*p*-Chlorobenzhydryl)piperazine	4.38	4-Chlorostyrene	3.58
2,4-Dichloro-3,5-dimethylphenol	4.36	2,3,6-Trichlorophenol	3.57
2,3,4,5-Tetrachloroaniline	4.33	2,4,6-Trichlorophenol	3.57
1,5-Dichloroanthraquinone	4.30	1,1,3,4,4-Pentachlorobut-2-en	3.47
2,6-Dichlorostyrene	4.29	3,4-Dichlorobenzoic acid	3.46
3,5-Dichloromethylbenzene	4.22	2-Chloroanthraquinone	3.46
2,4-Dichloro-1-methylbenzene	4.22	3,5-Dichlorobenzoic acid	3.46
1,8-Dichloroanthraquinone	4.19	2,6-Dichloro-3-methylaniline	3.44
4-Chloro-*o*-xylene	4.15	2,3-Dichloroanisole	3.42
3,4,5-Trichlorophenol	4.13	a,*p*-Dichloroanisole	3.42
α,2,6-Trichlorotoluene	4.13	*p*-Chlorothiophenol	3.39
2,3,5,6-Tetrachloroaniline	4.10	*m*-Chlorothiophenol	3.39
1-Chloronaphthalene	4.03	3,4-Dichlorophenol	3.35
1-(Chloromethyl)naphthalene	3.88	3,5-Dichlorophenol	3.35
1,1,2,4,4-Pentachloro-1,3-butadiene	3.87	α-Chloro-*m*-xylene	3.35
α,α′-Dichloro-*o*-xylene	3.87	α-Chloro-*p*-xylene	3.35
α,α′-Dichloro-*p*-xylene	3.87	2,3,4-Trichloroaniline	3.33
2,3,5-Trichlorophenol	3.85	1,6-Dichlorohexane	3.29
2,3,4-Trichlorophenol	3.85	α,ȧ-Dichloro-*m*-xylene	3.27
2,4,5-Trichlorophenol	3.85	2,4-Dichloroguaiacol	3.23
2,3,5,6-Tetrachloronitrobenzene	3.80	Benzylidenechloride	3.22
Chlorocyclohexane	3.19	1,2,4-Trichloro-5-nitrobenzene	3.20
α-Chlorostyrene	3.17	Dichlorodimethoxybenzene	3.20
2,6-Dichloroanisole	3.14	Dichlorocyclohexane	3.19
4-Chloro-2-methylphenol	3.13	2,5-Dichloronitrobenzene	3.03
3,5-Dichlorobenzaldehyde	3.11	1,2,4,5-Tetrachloropentane	3.03
2,4-Dichlorobenzaldehyde	3.11	3,3′-Dichlorobenzidine	3.02
Pentachlorobutane	3.10	1,4-*cis*-Dichlorocyclohexane	3.02
5-Chlorosalicylic acid	3.10	3,5-Dichlorobenzonitrile	3.00

Table 11.15 (*Continued*)

Compounds *OUT* (log K_{OW} < 3.00)

5-Chloro-1,3-dimethoxybenzene	2.99	2-Chloroquinoline	2.52
2,6-Dichloro-4-nitroaniline	2.96	3,4-Dichlorophenylisocyanate	2.50
2,4,5-Trichlorophenoxy acetic acid	2.95	2,3-Dichloro-1,3-butadiene	2.49
3,4,6-Trichloro-2-nitrophenol	2.93	*o*-Nitrobenzyl chloride	2.45
2,4-Dichlorobenzoic acid	2.90	4-Chloro-6-nitro-*m*-cresol	2.41
2,5-Dichlorobenzoic acid	2.90	*bis*-4-Chlorobutylether	2.41
2-Chloro-5-methylphenol	2.85	2,6-Dichlorobenzoic acid	2.34
2,3-Dichlorophenoxyacetic acid	2.84	2,4-Dichloro-6-nitrophenol	2.33
3,4-Dichlorophenoxyacetic acid	2.84	2,5-Dichloro-4-nitrophenol	2.33
4-Chloroquinoline	2.82	Benzoylchloride	2.29
2,4-Dichloroaniline	2.80	1,4-Dichlorobutane	2.24
3,5-Dichloroaniline	2.79	*p*-Chlorophenoxyacetic acid	2.18
2,5-Dichloroaniline	2.79	2-Chloro-4-nitroaniline	2.17
2,6-Dichloroaniline	2.79	*o*-Chlorobenzylidene malononitrile	2.13
2,6-Dichlorophenol	2.79	DL-3-(α-acetonyl-*p*-chloro-benzyl)-	
1,1,3,5-Tetrachloropentane	2.79	4-hydroxy-coumarin (Coumachlor)	2.10
3,4-Dichloroaniline	2.79;	*m*-Chlorophenylisocyanate	1.91
	2.69	*p*-Chlorophenylisocyanate	1.91
Tetrachloro-*o*-benzoquinone	2.76	4-Chloro-3-nitroanisole	1.90
6,9-Dichloro-2-methoxyacridine	2.74	2,4-Dichloro-6-nitroaniline	1.88
2,3-Dichloroaniline	2.70	4,5-Dichloro-2-nitroaniline	1.88
Benzylchloride	2.70	5-Chloro-2-hydroxyaniline	1.81
Benzylidenechloride	2.70	4-Chloro-1-naphthol	1.81
2-Chloro-6-nitrotoluene	2.69	5-Chloro-2-hydroxyaniline	1.81
4-Chloro-3-nitrotoluene	2.69	1,1-Dichloroethane	1.79;
5-Chloro-2-nitrotoluene	2.69		1.43
Chloromethylnitrobenzene	2.69	5-Chloro-2-nitrophenol	1.73
2,4-Dichlorophenoxyacetic acid	2.68	2-Chloro-4-nitrophenol	1.73
4-Chloro-*N*-methylaniline	2.66	Chloro-4-nitrophenol	1.73
4-Chloro-2-nitroaniline	2.64	Chloro-4-nitrophenol	1.73
Chlorocyclopentane	2.63	Trans-1,4-dichlorobutene-2	1.69
3,4-Dichlorocatechol	2.62	1,4-dichlorobutene-2	1.69
2,3-Dichloronitrobenzene	2.61	1-Chloro-3,4-dinitrobenzene	1.53
3,4-Dichloronitrobenzene	2.61	2-Chlorocyclopentanone	1.51
2,4-Dichloronitrobenzene	2.61	1,1-Dichloroethylene	1.48
1,3-Dichloro-5,5-dimethylhydantoin	2.60	1,2-Dichloroethane	1.48
5-Chloro-2-methoxyaniline	2.58	1,2-Dichloroethylene	1.48
3-Chloro-4-methoxyaniline	2.58	2-Chloropyridine	1.44
2-Chloro-5-methylaniline	2.58	3-Chloropyridine-	1.44
2-Chloro-4-methylaniline	2.58	Chloronitrocyclohexane	1.44
3-Chloro-2-methylaniline	2.58	2,3-Dichloroquinoxaline	1.32
3-Chloro-4-methylaniline	2.58	2-Chloro-5-nitroaniline	1.28
4-Chloro-2-methylaniline	2.58	4-Chloro-3-nitroaniline	1.12
5-Chloro-2-methylaniline	2.58	2-Chlorocyclohexanone	0.81
2-Chloro-6-methylaniline	2.58	Chlorofluoromethane	0.81
2-Chloro-6-methylaniline	2.58	2-Chloro-4,6-diamino-1,3,5-triazine	0.00
3-Chloro-5-methoxyphenol	2.56	D-Chloramphenicol	0.13

Source: Fiedler and Hutziner (1990).

structure–activity relationships (QSARs) to make predictions about the potential threat of chlorinated pesticides towards aquatic organisms and to rank these substances.

11.9.2 Toxicity Ranking System for Chlorinated Pesticides

On the basis of physicochemical data, such as water solubility and vapor pressure, as well as the acute toxicity (LC_{50}), an ecotoxicological model was developed for preliminary hazard assessment (Fiedler et al. 1989; Fiedler and Schramm 1990). By use of the reciprocal product from the decadic logarithm of Henry's Law constant (log H) and the lethal concentration (LC_{50}) a suitable ranking system was applied to predict potential damage to aquatic organisms via application of pesticides.

As far as possible basic data were collected from the literature (Verschueren 1987; Worthing and Walker 1987; Rippen 1989; Witte et al. 1988; IARC 1983; AQUIRE 1989). In cases where no reported data existed we calculated the chemical and physical properties and the environmental partitioning by use of a QSAR system (Hunter et al. 1987). The QSAR system estimated chemical properties, environmental fate, and toxicity of organic compounds using models based upon chemical structure.

For a preliminary hazard assessment the values for Henry's Law constant H and the LC_{50} values for fish will indicate whether a pesticide has to be considered as potentially toxic to aquatic organisms. In addition, the LC_{50} value is a direct measure of the toxicity of a substance. Compounds with a high Henry's Law constant will not be found in water and vice versa. The Henry's Law constant can be calculated as follows:

$$H = \frac{V_p \ (atm)}{S \ (mol/m^3)} \tag{11.1}$$

To get a suitable equation we formed the reciprocal product of H (atm $\cdot$ m^3 $\cdot$ mol) and LC_{50} (mg $\cdot$ L):

$$F = \frac{1}{H \cdot LC_{50}} \tag{11.2}$$

Thus, a high F value indicates a high toxic potential of this compound in a water environment. Whereas a small F value indicates little toxic potential.

11.9.3 Results

For 13 of the 70 pesticides (chloralose, chlormephos, chlorthiophos, coumachlor, crimidine, 2,4-DES-sodium, dialifos, endosulfan, heptachlor, leptophos, profenfos, triallat, and trichloronate), no F factors could be obtained due to either lack of Henry's Law constant (water solubility and/or vapor pressure) or missing data for aquatic toxicity. The F factors for 57 pesticides were ranged from 2 200 000 000 000 (endrin) to 3 (allylchloride; Table 11.16).

In a later hazard assessment, further organochlorine compounds were included. It could be shown that 2,4,5-trichlorophenoxyacetic acid (2,4,5-T) was the most toxic compound, whereas 2,3,7,8-tetrachlorodibenzo-p-dioxin (2,3,7,8-Cl_4DD) was found to be less dangerous. The result is based on the fact that (a) 2,4,5-T exhibits a much greater water solubility than 2,3,7,8-Cl_4DD and (b) 2,4,5-T is quite toxic per se.

Table 11.16 Toxicity ranking system based on log H and LC_{50}

	LC_{50} (mg/L)	log H (atm m^3/L)	Toxicity ranking factor (F)
Endrin	0.00006	−8.12	2 200 000 000 000
Captafol	0.021	−9.74	262 000 000 000
Carbophenothion	0.0044	−8.3	49 900 000 000
Lindane	0.001	−7.29	19 400 000 000
Dieldrin	0.008	−8.12	16 500 000 000
Chloroxuron	0.43	−9.72	12 200 000 000
Fenvalerate	0.00069	−6.72	7 610 000 000
Permethrin	0.0002	−5.76	2 870 000 000
DDT	0.08	−7.08	1 505 000 000
Chlorfenvinphos	0.045	−7.61	905 000 000
Iodofenphos	0.016	−6.44	172 000 000
Fenarimol	0.91	−8.18	166 000 000
Chlordane	0.0014	−5.03	107 000 000
Chlorpyrifos	0.0024	−5.37	97 000 000
Chloropropylate	1.2	−7.99	81 400 000
Chlorothalonil	0.11	−6.93	77 400 000
Captan	0.0034	−5.88	22 300 000
Metolachlor	2	−7.57	18 600 000
Iprodione	6.7	−8.04	16 700 000
Pentachlorophenol	0.17	−6.18	8 950 000
Diuron	4.0	−7.59	7 730 000
Propanil	0.35	−5.88	2 170 000
Bromophos	0.05	−5	2 000 000
MCPB (4-(4-Chloro-o-tolyl)-butyric acid)	0.070	−4.85	1 010 000
Chlordimeform	7.1	−6.77	829 000
Chlorobenzilate	0.299	−5.35	749 000
Fenoprop	0.35	−5.40	718 000
Heptenophos	13.1	−6.95	682 000
2,4,6-Trichlorophenol	0.1	−4.81	646 000
Methoxychlor	0.052	−4.44	530 000
Dicofol	0.053	−4.38	453 000

Table 11.16 (*Continued*)

	LC_{50} (mg/L)	Log H (atm m^3/L)	Toxicity ranking factor (F)
Propachlor	1.3	−5.44	212 000
Flamprop-isopropyl	2.5	−5.24	69 500
Aldrin	0.018	−2.88	41 800
Hexachlorobenzene	0.05	−3.29	38 700
2,4-DB (2,4-Dichlorophenoxy-acetic acid)	2	−4.87	37 100
Atrazine	10	−5.54	35 400
Diallate	8.2	−5.41	31 300
Flampropmethyl	4.7	−5.12	28 000
Thiobencarb	1.2	−3.97	27 800
Neburon	0.6	−4.19	25 800
Quinonamid	5	−4.98	19 100
2,4,5-Trichlorophenol	3.1	−4.77	16 200
Clorfenethol	1.4	−4.25	12 700
Dichlorvos	0.27	−3.50	11 700
Trietazine	0.85	−3.65	5260
Chloropicrin	1	−3.49	3090
2,3-Dichloro-*N*-(4-fluorophenyl) maleimide	5.6	−4.00	1790
2,4-Dichlorophenol	7.4	−3.96	1230
2,4-D	1	−2.85	708
Methazole	3	−2.62	139
MCPA (4-(4-Chloro-o-tolyl)acetic acid)	232	−3.52	133
Terbutylazine	4.6	−2.62	91
2,3,4,6-Tetrachlorophenol	0.25	−1.17	59
Allylchloride	10	−2.49	3

Source: Fiedler and Schramm (1990).
LC_{50}, Lethal concentration, i.e., concentration of a chemical that causes the death of 50% of the test population at 96 h exposure time.
H, Henry's Law constant; ratio of the concentration of a chemical in air to that in water.

11.9.4 Discussion

The preliminary hazard assessment as presented here basically consists of an evaluation of potential exposure and the potential effects of that exposure. With the help of physicochemical properties, such as water solubility and vapor pressure, environmental data such as Henry's Law constant can be calculated. Toxicity data from acute toxicity tests (e.g., lethal concentration for aquatic organisms, LC_{50}) can give basic information on the activity of a molecule. With the help of these data models can be set up that allow evaluation of the potential hazard in water for these substances. Combining these results with information on uses and loads likely to be introduced into the aquatic environment will provide a helpful tool for risk assessment. When evaluative models are used it should be remembered that predictions are not always correct. However, experimental data can also be associated with large errors, and moreover costs are likely to be much higher. Models and ranking systems are helpful tools when compounds with potential risks have to be identified from large numbers of substances. The results of such evaluation will provide a helpful tool in minimizing the impact of potentially dangerous substances on the environment.

11.10 REFERENCES

Achman, D. R., K. C. Hornbuckle, and S. J. Eisenreich, 1993, "Volatilization of Polychlorinated Biphenyls from Green Bay, Lake Michigan". *Environ. Sci. Technol.*, *27*, 75–86

Ahlborg, U. G., C. G. Becking, L. S. Birnbaum, A. Brouwer, H. J. G. M. Derks, M. Feeley, G. Golor, H. Hanberg, J. C. Larsen, A. K. D. Liem, S. H. Safe, C. Schlatter, F. Waern, M. Younes, and E. Yrjänheikki, 1994, "Toxic Equivalency Factors for Dioxin-like PCBs – Report on a WHO-ECEH and IPCS Consultation, December 1993," *Chemosphere*, *28*, 1049–1067

Ahlborg, U. G., A. Hanberg, and K. Kenne, 1992, "Risk Assessment of Polychlorinated Biphenyls (PCB)," Institute of Environmental Medicine, Karolinska Institutet Stockholm, Sweden, Nord, *26*

Allen, T. K., 1990, "Developments in the Federal Regulation of PCBs since October 1987," in *Proc. 1989 EPRI PCB Seminar* (G. Addis, Ed.), Electric Power Research Institute, Palo Alto, (CA), USA

Aquire Database – Aquatic Information Retrieval Toxicity Data Base, 1989, US Environmental Protection Agency, Duluth, USA

Asplund, G., and A. Grimvall, 1991, "Organohalogens in Nature," *Environ. Sci. Technol.*, *25*, 1346–1350

Atkinson, R., 1987, "Estimation of OH Radical Reaction Rate Constants and Atmospheric Lifetimes for Polychlorobiphenyls, Dibenzo-*p*-dioxins and Dibenzofurans," *Environ. Sci. Technol.*, *21*, 305–307

Axegård, P., O. Dahlman, I. Haglind, B. Jacobson, R. Mörck, and L. Strömberg, 1993, "Pulp Bleaching and the Environment – the Present Situation," STFI-Meddlelande, A 994, Sweden

Bacon, C. E., W. M. Jarman, and D. P. Costa, 1992, "Organochlorine and Polychlorinated Biphenyl Levels in Pinniped Milk from the Arctic, the Antarctic, California and Australia," *Chemosphere*, *24*, 779–791

Baker, J. E., and S.J. Eisenreich, 1990, "Concentrations and Fluxes of Polycyclic Aromatic Hydrocarbons and Polychlorinated Biphenyls across the Air–Water Interface of Lake Superior," *Environ. Sci. Technol.*, *24*, 342–352

Ball, H. A., M. Reinhard, and P. L. McCarty, 1989, "Biotransformation of Halogenated and Nonhalogenated Octylphenol Polyethoxylate under Aerobic and Anaerobic Conditions," *Environ. Sci. Technol.*, *23*, 951–961

Bayer, AG (Ed.), 1995, "Chemie mit Chlor," in *Chancen-Risiken-Perspektiven*, Leverkusen, Germany

Behnke, W., and C. Zetsch, 1993, "Tropospheric Photooxidation of Perchloroethene and Trichloroacetyl Halocarbons in Conifers," *Organohalogen Compd.*, *14*, 273–276, Federal Environmental Agency, Vienna, Austria

Bertazzi, P. A., and A. di Domenico, 1994, "Chemical, Environmental and Health Aspects of the Seveso, Italy, Accident," in: *Dioxins and Health*, (A. Schecter, Ed.), Plenum Press, New York and London, pp. 587–632

Beurskens, J. E. M., M. Toussaint, J. de Wolf, J. M. D. van der Steen, P. C. Slot, L. C. M. Commandeur, and J. R. Parsons, 1995, "Dehalogenation of Chlorinated Dioxins by an Anaerobic Microbial Consortium from Sediment," *Environ. Toxicol. Chem.*, *14*, 393–343

Bidleman, T. F., 1988, "Atmospheric Process," *Environ. Sci. Technol.*, *22*, 361–367

Blüthgen, A., W. Heeschen, and U. Ruoff, 1994, "Zum Carry-Over toxikologisch relevanter Polychlordibenzodioxin- und -furankongenere in die Milch laktierender Kühe," *Kieler Milchwirtschaftl. Forschungsberichte*, *46*, 130–150

Breitung, V., 1994, "Belastung des Saar mit Ugilec 141 und PCB," *Vom Wasser 79*, 39–47

Brown, J. F., D. L. Bedard, M. J. Brennan, J. C. Carnahan, H. Feng, and R. E. Wagner, 1987, "Polychlorinated Biphenyl Dechlorination in Aquatic Sediments," *Science, 236*, 709–712

Czuczwa, J. M., and R. A. Hites, 1984; "Environmental Fate of Combustion-Generated Polychlorinated Dioxins and Furans," *Environ. Sci. Technol.*, *18*, 444–450

Dahlman, O., A. Reimann, P. Ljungquist, R. Mörck, C. Johansson, H. Borén, and A. Grimvall, 1993a, "Characterization of Chlorinated Aromatic Structures in High Molecular Weight BKME-Materials and in Fulvic Acids from Industrially Unpolluted Waters," *Paper presented at the 4th IAWQ Symp.*, June 1993

Dahlman, O., R. Mörck, P. Ljungquist, A. Reichmann, C. Johansson, H. Borén, and A. Grimvall, 1993b, "Chlorinated Structural Elements in High Molecular Weight Organic Matter from Unpolluted Waters and Bleached-Kraft Mill Effluents," *Environ. Sci. Technol.*, *27*, 1616–1620

DECHEMA, 1995, "Dioxins and Phthalates in Soil – A Critical and Comparative Assessment," compiled by W. Klein, W. Kördel, G. H. M. Krause, and J. Wiesner, Frankfurt, Germany, March 1995

DFG, 1988, "Polychlorierte Biphenyle, Bestandsaufnahme über Analytik, Vorkommen, Kinetik und Toxikologie," Deutsche Forschungsgemeinschaft, Verlag Chemie, Weinheim, Germany

Duinker, J. C., and F. Bouchertall, 1989, "On the Distribution of Atmospheric Polychlorinated Biphenyl Congeners between Vapor Phase, Aerosols, and Rain," *Environ. Sci. Technol.*, *23*, 57–62

EPA, 1995, Science Advisory Board, May 1995, Herndon, VA, USA

EPA, 1994a, "Estimating Exposure to Dioxin and Related Compounds," Vols 1–3, United States Environmental Protection Agency, Washington, DC, USA, EPA/600/6-88/005

Engvild, K. C., 1986, "Chlorine-Containing Natural Compounds in Higher Plants," *Phytochemistry*, *25*, 781–791

Fiedler, H., 1995, "EPA DIOXIN-Reassessment: Implications for Germany," *Organohalogen Compd.*, *22*, 209–228, ECO-INFORMA Press, Bayreuth, Germany

Fiedler, H., 1994, "Dioxine im Biokompost," *Organohalogen Compd.*, *18* (124 pp.), ECO-INFORMA Press, Bayreuth, Germany

Fiedler, H., 1993, "Formation and Sources of PCDD/PCDF," *Organohalogen Compd.*, *11*, 221–228, Federal Environmental Agency, Vienna, Austria

Fiedler H., and O. Hutzinger, 1990, "Organochlorine Compounds in the Marine Environment," Study for UNEP/IRPTC, Contract No. G/CON/89-01

Fiedler, H., H. Hoff, J. Tolls, C. Mertens, A. Gruber, and O. Hutzinger, 1994a, "Environmental Fate of Organochlorines in the Aquatic Environment," *Organohalogen Compd.*, *15*, ECO-INFORMA Press, Bayreuth, Germany (199 pages)

Fiedler, H., K. Fricke, and H. Vogtmann, 1994b, "Bedeutung polychlorierter Dibenzo-*p*-Dioxine und polychlorierter Dibenzofurane (PCDD/PCDF) in der Abfallwirtschaft," *Organohalogen Compd.*, *17*, 156 pp

Fiedler, H., G. Herrmann, K.-W. Schramm, and O. Hutzinger, 1989, "Application of QSARs to Predict Potential Aquatic Toxicities of Organochlorine Pesticides," *Toxicol. Environ. Chem.*, *26*, 157–160

Fiedler, H., and K.-W. Schramm, 1990, "QSAR Generated Ranking System for Organochlorine Compounds in the Marine Environment," *Organohalogen Compd.*, *4*, 391–394, ECO-INFORMA Press, Bayreuth, Germany

Fiedler, H., C. Lau, S. Schulz, C. Wagner, O. Hutzinger, and K. T. von der Trenck, 1995, "Stoffbericht Polychlorierte Biphenyle (PCB)," *Handbuch Altlasten und Grundwasserschadensfälle*, Berichtsnummer 16/95. Landesanstalt für Umweltschutz Baden-Württemberg, Germany

Fiedler, H., M. Hub, and O. Hutzinger, 1993, "Stoffbericht Hexachlorcyclohexan (HCH)," in *Handbuch Altlasten – Texte und Berichte zur Altlastenbearbeitung 9/93* (254 pp.)

Fiedler, H., O. Hutzinger, and C. Timms, 1990, "Dioxins: Sources of Environmental Load and Human Exposure," *Toxicol. Environ. Chem.*, *29*, 157–234

Fletcher, C. L., and W. A. McKay, 1993, "Polychlorinated Dibenzo-*p*-dioxins (PCDDs) and Dibenzofurans (PCDFs) in the Aquatic Environment – A Literature Review," *Chemosphere*, *26*, 1041–1069

Frank, H., 1993, "Short-chain Aliphatic Halocarbons: Environmental Levels and Ecotoxicological Properties," *Organohalogen Compd.*, *14*, 267–270, Federal Environmental Agency, Vienna, Austria

Fürst, P., G. H. M. Krause, D. Hein, and T. Delschen, 1992, "Influence of PCDD/PCDF Levels in Grass and Soil on the Contamination of Cow's Milk," *Organohalogen Compd.*, *8*, 333–336, Finnish Institute of Occupational Health (Ed.), Helsinki, Finland

Furukawa, K., K. Tonumura, and A. Kamibashi, 1978, "Effect of Chlorine Substitution on the Biodegradability of Polychlorinated Biphenyls," *Appl. Environ. Microbiol.*, *35*, 223–227

Gribble, G. W., 1992, "Naturally Occurring Organohalogen Compounds – A Survey," *J. Nat. Prod.*, *55*, 1353–1395

Haas, R., 1992, "Polychlorierte Naphthaline bei Rüstungsaltlasten," *UWSF-Z. Umweltchem. Ökotox.*, *4*, 350–351

Hagenmaier, H., and P. Krauß, 1993, "Attempts to Balance Transport and Fate of PCDD/PCDF for Baden-Württemberg," *Organohalogen Compd.*, *12*, 81–84, Federal Environmental Agency, Vienna, Austria

Hansen, L. G., 1987, "Environmental Toxicology of Polychlorinated Biphenyls," in *Environmental Toxin Series 1*, (S. Safe and O. Hutzinger, Eds.), pp. 15–48, Springer, Berlin, Heidelberg, New York, London, Paris, Tokyo

Harrad, S., and K. C. Jones, 1992, "A Source Inventory and Budget for Chlorinated Dioxins and Furans in the United Kingdom Environment," *Sci. Tot. Environ.*, *126*, 89–107

Hodin, F., H. Boren, A. Grimvall, and S. Karlsson, 1991, "Formation of Chlorophenols and Related Compounds in Natural and Technical Chlorination Processes," *Water Sci. Technol.*, *24*, 403–410

Hoekstra, E. J., and E. W. B. de Leer, 1993a, "AOX-Levels in the River Rhine: 50 Percent of Natural Origin," in *Integrated Soil and Sediment Research: A Basis for Proper Protection*, H. J. P. Eijsackers, and T. Hamers (Eds.), Kluwer Academic Publishers, The Netherlands, pp. 93–95

Hoekstra, E. J., and E. W. B. de Leer, 1993b, "Natural Production of Chlorinated Organic Compounds in Soil," in *Integrated Soil and Sediment Research: A Basis for Proper Protection*, H. J. P. Eijsackers, and T. Hamers (Eds.), Kluwer Academic Publishers, The Netherlands, pp. 96–98

Hub, M., H. Fiedler, and O. Hutzinger, 1990, "Stoffverhalten von Hexachlorbenzol unter besonderer Berücksichtigung der Altlastenproblematik," Report prepared for Environmental Protection Agency Baden-Württemberg, Karlsruhe, Germany

Hunter, R. S., F. D. Culver, J. R. Hill, and A. Fitzgerald, 1987, "QSAR System Manual," Institute for Biological and Chemical Process Analysis, Montana State University, Bozman, Montana, USA

Hutzinger, O., 1990, Editorial. *UWSF-Z. Umweltchem. Ökotox.*, *2*, 61

IARC, 1983, "Monographs on the Evaluation of the Carcinogenic Risk of Chemicals to Humans, Miscellaneous Pesticides," Vol 30, World Health Organization, International Agency for Research on Cancer, Lyon, France

Ivanov, V., and E. Sandell, 1992, "Characterization of Polychlorinated Biphenyl Isomers in Sovol and Trichlorodiphenyl Formulations by High-Resolution Gas Chromatography with Electron Capture Detection and High-Resolution Gas Chromatography-Mass Spectrometry Techniques," *Environ. Sci. Technol.*, *26*, 2012–2017

Jokela, J. K., M. Laine, M. Ek, and M. Salkinoja-Salonen, 1993, "Effect of Biological Treatment on Halogenated Organics in Bleached Kraft Pulp Mill Effluents Studied by Molecular Weight Distribution Analysis," *Environ. Sci. Technol.*, *27*, 547–557

Jones, P. D., G. T. Ankley, D. A. Best, R. Crawford, N. De Galan, J. P. Giesy, T. J. Kubiak, J. P. Ludwig, J. L. Newsted, D. E. Tillit, and D. A. Verbrugge, 1993, "Biomagnification of Bioassay Derived 2,3,7,8-Tetrachlorodibenzo-*p*-dioxin Equivalents," *Chemosphere*, *26*, 1203–1212

Keene, W. C., J. R. Maben, A. A. P. Pszenny, and J. N. Galloway, 1993, "Measurement Technique for Inorganic Chlorine Gases in the Marine Boundary Layer," *Environ. Sci. Technol.*, *27*, 866–874

KEMI, 1994, "Chlorine and Chlorinated Compounds. Use and Risks – Need for Precautionary Measures," Report Summary, National Chemicals Inspectorate, Stockholm, Sweden

Kirk-Othmer, 1991, *Encyclopedia of Chemical Technology*, 4th Edn. Wiley, New York

Klein, S., E. Stottmeister, H. Hermenau, and P. Hendel, 1991, "Leichtflüchtige Halogen-kohlenwasserstoffe (LHKW) nach Chlorungen in einer Großkläranlage," *UWSF-Z. Umweltchem. Ökotox.*, *3*, 137–138

Knutzen, J., and M. Oehme, 1989, "Polychlorinated Dibenzofuran (PCDF) and Dibenzo-*p*-dioxin (PCDD) Levels in Organisms and Sediments from the Frierfjord, Southern Norway," *Chemosphere*, *19*, 1897–1909

Koch, R. 1985, *Umweltchemikalien*, Verlag Chemie, Weinheim, Germany

Koester, C. J., and R. A. Hites, 1992a, "Wet and Dry Deposition of Chlorinated Dioxins and Furans," *Environ. Sci. Technol.*, *26*, 1374–1382

Koester, C. J., and R. A. Hites, 1992b, "Photodegradation of Polychlorinated Dioxins and Dibenzofurans Adsorbed on Fly Ash," *Environ. Sci. Technol.*, *26*, 502–507

Kohler, H.-P. E., D. Kohler-Staub, and A. C. Alder, 1992, "Mikrobielle Umwandlungen polychlorierte Biphenyle (PCB)," *GAIA 1*, 153–165

Kong, H.-L., and G. S. Sayler, 1983, "Degradation and Total Mineralization of Mono-halogenated Biphenyls in Natural Sediment and Mixed Bacterial Culture," *Appl. Environ. Microbiol.*, *49*, 666–672

Korte, F., 1987 and 1992, *Lehrbuch der ökologischen Chemie, 2nd and 3rd Edn.*, Thieme Verlag, Stuttgart, Germany

Kringstad, K. P., and K. Lindström, 1984, "Spent Liquors from Pulp Bleaching," *Environ. Sci. Technol.*, *18*, 236A–248A

Kutz, F. W., D. G. Barnes, E. W. Bretthauer, D. P. Bottimore, and H. Greim, 1990, "The International Toxicity Equivalency Factor (I-TEF) Method for Estimating Risks Associated with Exposures to Complex Mixtures of Dioxins and Related Compounds," *Toxicol. Environ. Chem.*, *26*, 99–110

Langvik, V.-A., O. Hormi, L. Tikkanen, and B. Holmbom, 1991, "Formation of the Mutagen 3-Chloro-4-(Dichlormethyl)-5-Hydroxy-2(5H)-Furanone and Related Compounds by Chlorination of Phenolic Compounds," *Chemosphere*, *22*, 547–555

Luckas, B., W. Vetter, P. Fischer, G. Heidemann, and J. Plötz, 1990, "Characteristic Chlorinated Hydrocarbon Patterns in the Blubber of Seals from Different Marine Regions," *Chemosphere*, *21*, 13–19

Mackay, D., W. Y. Shiu, and K. C. Ma, 1992, *Illustrated Handbook of Physical-Chemical Properties and Environmental Fate for Organic Chemicals*, Vols I & II, Lewis Publishers Inc., Boca Raton, FL, USA

Mackay, D., S. Paterson, and W. H. Schroeder, 1986, "Model Describing the Rates of Transfer Processes of Organic Chemicals between Atmosphere and Water," *Environ. Sci. Technol.*, *20*, 810–816

MacRae, I. C., 1989, "Microbial Metabolism of Pesticides and Structurally Related Compounds," in *Rev. Environ. Contam. Toxicol.*, (G. W. Ware, Ed.), Vol 109, pp. 1–88, Springer, Berlin, Germany

Mariani, G., E. Benfenati, R. Fanelli, A. Nicoli, E. Bonfitto, and S. Jacopone, 1992, "Incineration of Agro-Industrial Wastes and Macro- and Micropollutants Emission," *Chemosphere*, *24*, 1545–1551

McCarthy, L. S., 1994, "Chlorine and Organochlorines in the Environment: A Perspective," *Can. Chem. News*, *23*, 22–25

McLachlan, M. S., 1991, "Die Anreicherung von PCDD/F in Nahrungsketten," *Organohalogen Compd.*, *6*, 183–211, ECO-INFORMA Press, Bayreuth, Germany

McLachlan, M. S., K. Welsch-Pausch, J. Tolls, and G. Umlauf, 1995, "Pathways and Mechanisms of PCDD/F-Input into Agricultural Food Chains," *Organohalogen Compd.*, *6*, 81–104, ECO-INFORMA Press, Bayreuth, Germany

Mross, K. G., and J. K. Konietzko, 1991, "Chlorinated Ethanes," in *Environmental Chemistry 3 G*, (O. Hutzinger, Ed.), Springer, Heidelberg, Germany, pp. 142–193

Neilson, A. H., 1990, A Review: The Biodegradation of Halogenated Organic Compounds, *J. Appl. Bacteriol.*, *69*, 445-470

Nies, L., and T. M. Vogel, 1990, "Effects of Organic Substrates on Dechlorination of Aroclor 1242 in Anaerobic Sediments," *Appl. Environ. Microbiol.*, *56*, 2612–2617

Nolte, R. F. and R. Joas, 1995, Handbuch Chlorchemie I-1992 – Produkte, Unternehmen, Stoffflüsse. UBA-Texte 25/95. Erich Schmidt, Berlin, Germany, April 1995

Norén, K., and Å. Lunden, 1991, "Trend Studies of Polychlorinated Biphenyls, Dibenzo-*p*-dioxins and Dibenzofurans in Human Milk," *Chemosphere*, *23*, 1895–1901

Norstrom, R. J., 1994, "Chlorinated Hydrocarbon Contaminants in the Arctic Marine Environment," *Organohalogen Compd.*, *20*, 541–544

Öberg, L., N. Wagman, R. Andersson, and C. Rappe, 1993, "*De novo* Formation of PCDD/Fs in Compost and Sewage Sludge – a Status Report," *Organohalogen Compd.*, *11*, 297–302, (Federal Environmental Agency, Eds.), Vienna, Austria

Öberg, L. G., R. Andersson, and C. Rappe, 1992, "*De novo* Formation of Hepta- and Octachlorodibenzo-*p*-dioxins from Pentachlorophenol in Municipal Sewage Sludge," *Organohalogen Compd.*, *9*, 351–354, Finnish Institute of Occupational Health, Helsinki, Finland

Öberg, L., C. Rappe, 1992, "Biochemical Formation of PCDD/F from Chlorophenols," *Chemosphere*, *25*, 49–52

Öberg, L. G., B. Glas, S. E. Swanson, C. Rappe, K. G. Paul, 1990, "Peroxidase-Catalyzed Oxidation of Chlorophenols to Polychlorinated Dibenzo-*p*-dioxins and Dibenzofurans," *Arch. Environ. Contam. Toxicol.*, *19*, 930–938

Oehme, M., 1995, "Globaler atmosphärischer Langtransport von Dioxinen und verwandten Verbindungen," Tagungsband zum Dübendorfer Dioxintag am 9.3.1995, Eidgenössische Materialprüfungs- und Forschungsanstalt (EMPA), Dübendorf, Switzerland, pp. 157–185

Oehme, M., M. Schlabach, and I. Boyd, 1995, "Polychlorinated Dibenzo-*p*-dioxins and Dibenzofurans and Coplanar Biphenyls in Antarctic Fur Seal Blubber," *Ambio*, *24*, 41–46

Oehme, M., J.-E. Haugen, R. Kallenborn, and M. Schlabach, 1994, "Polychlorinated Compounds in Antarctic Air and Biota: Similarities and Differences Compared to the Arctic," *Organohalogen Compd.*, *20*, 523–528

Ottar, B., 1981, "The Transfer of Airborne Pollutants to the Arctic Region," *Atmos. Environ.*, *15*, 1439–1445

Paasivirta, J., J. Koistinen, T. Rantio, and P. J. Vuortinen, 1994, "Persistent Organochlorine Compounds in Arctic and Baltic Fish," *Organohalogen Compd.*, *20*, 529–532

Parsons, J. R., M. C. M. Storms, 1989, "Biodegradation of Chlorinated Dibenzo-*p*-dioxins in Batch and Continuous Cultures of Strain JB1," *Chemosphere*, *19*, 1297–1308

Pesticide Manual: *The Pesticide Manual, A World Compendium*, Glasshouse Crops Research Institute, (C. R. Worthing, Ed.), Boots Company Limited, Nottingham, UK

Prinz, B., G. H. M. Krause, and L. Radermacher, 1993, "Standards and Guidelines for PCDD/PCDF – An Integrated Approach with Special Respect to the Control of Ambient Air Pollution," *Chemosphere*, 27, 491–500

Quensen, J. F., and F. Matsumura, 1983, "Oxidative Degradation iof 2,3,7,8-Tetra-chlorodibenzo-*p*-dioxin by Microorganisms," *Environ. Toxicol. Chem. 2*, 261–268

Quensen, J. F., J. M. Tiedje, and S. A. Boyd, 1988, "Reductive Dechlorination of Polychlorinated Biphenyls by Anaerobic Microorganisms from Sediments," *Science*, 242, 752–754

Rantanen, J., 1992, "Industrial and Environmental Emergencies; Lessons Learned," *Organohalogen Compd.*, 10, 291–294. Finnish Institute of Occupational Health, Helsinki, Finland

Rappe, C., 1992, "Sources of Exposure, Environmental Levels and Exposure Assessment of PCDDs and PCDFs," *Organohalogen Compd.*, 9, 5–8, Finnish Institute of Occupational Health (Eds.), Helsinki, Finland

Rappe, C., 1991, "Sources of Human Exposure to PCDDs and PCDFs," In *Biological Basis for Risk Asssessment of Dioxins and Related Compounds*," (M. Gallo, R. Scheuplein, K. van der Heijden, Eds.) Banbury Report 35, Cold Spring Harbor Laboratory Press, Plainview, NY, USA

Reeve, D. W., P. F. Earl, 1989, "Chlorinated Organic Matter in Bleached Chemical Pulp Production: Part 1," *Pulp Pap. Can.*, 90, 128–132

Rippen, G., 1990, *Handbuch Umweltchemikalien*, Vols 3–5, Loseblatt-Datensammlung, ecomed Verlag, Landsberg, Germany

Rochkind, M. L., J. W. Blackburn, and G. S. Sayler, 1986, *Microbial Decomposition of Chlorinated Aromatic Compounds*, EPA/600/2-86/090, 1986. US Environmental Protection Agency, Cincinnati, Ohio, USA

Roth, M., 1993, "Effects of Short Chain Halocarbons on Invertebrates of Terrestrial and Aquatic Ecosystems," *Organohalogen Compd.*, 14, 285–286

Safe, S., 1994, "Polychlorinated Biphenyls (PCBs): Environmental Impact, Biochemical and Toxic Responses and Implications for Risk Assessment," *CRC Crit. Rev. Toxicol.*, 24, 87–149

Safe, S., 1990, "Polychlorinated Biphenyls (PCBs), Dibenzo-*p*-dioxins (PCDDs), Dibenzofurans (PCDFs), and Related Compounds: Environmental and Mechanistic Considerations which Support the Development of Toxic Equivalency Factors (TEFs)," *CRC Crit. Rev. Toxicol.*, 21, 51–88

Schlett, S., H. Fiedler, and O. Hutzinger, 1990, "Toxic and Persistent Organic Compounds: Distribution and Health Effects in Humans: Pentachlorophenol," Report for Commission of the European Communities, Joint Research Centre Ispra, Contract No. 3521-88-10-ED ISPD, October 1990

Schröder, P., 1993, "On the Uptake and Possible Detoxification of Short-chain Aliphatic Halocarbons in Conifers," *Organohalogen Compd.*, 14, 277–279

Sipes, J. G., and R. G. Schellmann, 1987, "Biotransformation of PCBs: Metabolic Pathways and Mechanisms," in *Polychlorinated Biphenyls.* (O. Hutzinger, and S. Safe, (Eds.), *Environmental Toxin Series, Vol. 1*, Springer, Berlin, Germany, pp. 133–145

Smith, L. M., T. R. Schwartz, K. Feltz, and T. J. Kubiak, 1990, "Determination and Occurrence of AHH-active Polychlorinated Biphenyls, 2,3,7,8-Tetrachloro-*p*-dioxin and 2,3,7,8-Tetrachlorodibenzofuran in Lake Michigan Sediment and Biota. The Question of Their Relative Toxicological Significance," *Chemosphere*, *21*, 1063–1086

Strobel, K., H. H. Dieter, 1990, "Toxicological Risk/Benefit-Aspects of Drinking Water Chlorination and of Alternative Disinfection Procedures," *Wasser-Abwasser-Forsch.*, *23*, 152–162

Svenson, A., L.-O. Keller, and C. Rappe, 1989, "Enzyme Mediated Formation of 2,3,7,8-Tetrasubstituted Chlorinated Dibenzodioxins and Dibenzofurans," *Environ. Sci. Technol.*, *23*, 900–902

Vannerberg, N.-G., and E. Widén, 1993, *Nordic Scientists Hearing on Organochlorines*, Stockholm, 2 June 1993, Notes, Eka Nobel, Nobel Industries Sweden

VCI, 1944, Association of Chemical Manufacturers, Frankfurt, Germany, (Personal Communication)

Verschueren, K., 1987, *Handbook of Environmental Data on Organic Chemicals*, Van Nostrand Reinhold Company, New York, USA

Vogl, J., 1987, *Handbuch des Umweltschutzes*, 2nd Edn., ecomed Verlag, Landsberg, Germany

Vollmuth, S., A. Zajc, and R. Niessner, 1994, "Formation of Polychlorinated Dibenzo-*p*-dioxins and Polychlorinated Dibenzofurans during Photolysis of Pentachlorophenol-Containing Water," *Sci. Environ. Technol.*, *28*, 1145–1149

Wagner, U. K., 1990, "PCB-Situation in Western Europe," in *Proc. 1989 EPRI PCB Seminar* (G. Addis, Ed.), Electric Power Research Institute, Palo Alto, CA, USA

Wagner, H. C., K.-W. Schramm, and O. Hutzinger, 1990, "Biogenes polychloriertes Dioxin aus Trichlorphenol," *UWSF – Z. Umweltchem. Ökotox.*, *2*, 63–65

Wallenhorst, T., P. Krauss, H. Hagenmaier, 1995, "PCDD/F in Ambient Air and Deposition in Baden-Württemberg, Germany," *Organohalogen Compd.*, *24*, 157–161

Ware, G. W. (ed.), 1988, *Rev. Environ. Contam. Toxicol.*, *106*, 1–233

White, G. C. (Ed.), 1986, *The Handbook of Chlorination*, 2nd Edn., Van Nostrand Reinhold, New York, USA

WHO/IARC, 1991, "Monograph on the Evaluation of Carcinogenic Risks to Humans, Vol 52, Chlorinated Drinking Water, Chlorination By-Products, Some Other Halogenated Compounds," International Agency for Research on Cancer, Lyon, France

WHO/IARC, 1990, "IARC Monographs on the Evaluation of Carcinogenic Risks to Humans," *Chlorinated Paraffins*, *48*, 55–72

Wienecke, J., H. Kruse, and O. Wassermann, 1992, "Organic Compounds in the Waste Gasification and Combustion Process," *Chemosphere*, *25*, 437–447

Witte, I., R. Jähne, R. Weinert, K. Köbrich, and H. Jacobi, 1988, *Gefährdungen der Gesundheit durch Pestizide*, Fischer Taschenbuch Verlag, Frankfurt am Main, Germany

Worthing, C. R., and S. B. Walker, (Eds.), 1987, *The Pesticide Manual, A World Compendium*, 8th Edn., Glasshouse Crops Research Institute, Boots Company Limited, Nottingham, UK

Ye, D., J. F. Quensen, J. M. Tiedje, and S. A. Boyd, 1992, "Anaerobic Dechlorination of Polychlorobiphenyls (Aroclor 1242) by Pasteurized and Ethanol-Treated Microorganisms from Sediments," *Appl. Environ. Microbiol.*, *58*, 1110–1114

Young L. Y., and M. M. Häggblom, 1991, "Biodegradation of Toxic and Environmental Pollutants," *Cur. Opin. Biotechnol.*, *2*, 429–435

Zeddel, A., A. Majcherczyk, A. Hüttermann, K. Ballschmiter, H. Fiedler, and M. McLachlan, 1994, "Die Rolle von Weißfäulepilzen bei Abbau und Neubildung von chlororganischen Verbindungen in Festphasensystemen," *Organohalogen Compd.*, *18*, 89–100, ECO-INFORMA Press, Bayreuth, Germany

Zitko, V., 1980, "Chlorinated Paraffins," in *Environmental Chemistry 3 A*, (O. Hutzinger, Ed.), Springer, Heidelberg, Germany, pp. 149–156

12

Ecochemistry of Toxaphene

Mehmet Coelhan (*Kassel, Germany*)

and

Harun Parlar (*Freising-Weihenstephan, Germany*)

12.1 SUMMARY

Toxaphene, a halogenated hydrocarbon insecticide produced by chlorination of camphene, consists of a poorly defined mixture of at least 180 substances. The formulas for

Ecotoxicology, Edited by Gerrit Schüürmann and Bernd Markert.
ISBN 0-471-17644-3   © 1998 John Wiley & Sons, Inc. and Spektrum Akademischer Verlag.

most of these are $C_{10}H_{18-n}Cl_n$ or $C_{10}H_{16-n}Cl_n$, where n is 6–10. Although the chromatographic behavior of the toxaphene components is extremely similar, more than 60 components have been isolated up to now, mostly hepta- to decachloro derivatives of bornane. Experiments have shown that, depending on their structure, some toxaphene components are rapidly degraded in various environmental systems, while others are highly persistent.

12.2 INTRODUCTION

Toxaphene was developed in 1945 by Hercules Inc., USA, and has been one of the most widely used insecticides in the United States since that time. It played an increasing role in pest control in agriculture and forestry during the years following the limitation of DDT and the cyclodiene insecticides. The range of application covered nearly all areas of agriculture. Concentrations ranging from 0.5 to 10 kg per ha were recommended. During the years following its introduction in 1946, cumulative world use up to 1974 was 450 000 tons, and cumulative use between 1950 and 1993 has been estimated at 1.33 million tons (Voldner and Lie 1993). Although toxaphene has been banned in the USA and the EC since 1982, it still belongs to the most important insecticides especially in cotton-growing countries. Two-thirds of its production are used for cotton pest control, while other uses include application on vegetables, especially soybeans. Because of its persistence in the environment—its half-life in soil being about 10 years—and its mobility, toxaphene nowadays can be considered as being dispersed throughout the world.

On comparing its accumulation in different environmental compartments, it becomes apparent that soil is most heavily contaminated at application areas. The easy sorption of toxaphene to organic soil constituents results in rather slow microbial degradation. Although toxaphene is only slightly volatile, long-range transport has been found to occur. At high temperatures, transport in gaseous form prevails while at moderate temperatures toxaphene is transported to a great extent bound to aerosols. By wet deposition as well as photo- and biodegradation, the composition of the toxaphene mixture can be profoundly changed; this led to difficulties in quantification when the technical mixture was used as a standard. Only recently have enough single components been isolated and identified to allow exact quantitative analysis and comparison of toxicological and environmental behavior.

12.3 COMPOSITION OF TOXAPHENE

Toxaphene is produced by passing chlorine gas through a solution of technical grade camphene in carbon tetrachloride under UV irradiation. The product obtained is marketed without further purification. It contains 67%–69% chlorine, corresponding to the empirical formula $C_{10}H_{10}Cl_8$. Due to the lack of selectivity of chlorination, the number of isomers increases with a nonuniform degree of substitution. An example is the addition of chlorine to the double bond of camphene, which follows an unspecific pathway. This reaction has been thoroughly investigated (Tishchenko and Uvarov 1953; Jennings and Herschbach 1965; Ghiurdoglu et al. 1957; Parlar et al. 1977). Despite contradictory evidence, it has long been accepted that the addition of a halogen to the double bond leads to 2-*exo*,10-dichlorobornane. From this reaction, it is assumed that toxaphene consists mostly of higher substituted chlorobornanes (Figure 12.1a). This assumption is further substantiated by the fact that an equivalent amount of HCl is evolved during the preparation. In this case, 2-*exo*- as well as 2-*endo*-chlorobornane are formed in a Wagner-Meerwein rearrangement.

Contrary to former suggestions concerning the occurrence of chlorobornenes in technical toxaphene, the recent results allow the postulation that the unsaturated components formed during the production and the photochemical transformation of toxaphene are mainly chlorocamphenes (Figure 12.1b). A comparison of models of chlorobornene and chlorocamphene reveals that the 6-membered ring of a bornene is very strongly stressed on account of the two sp^2-hybridized C atoms and that the camphene structure must be favored energetically. As a consequence, the bornadiene derivatives mentioned in the literature (Saleh 1983) are possibly camphedienes formed by elimination of hydrogen chloride between C_2 and C_3. The existence of extremely unfavorable compounds such as bornadienes in radical equilibrium reactions cannot be expected.

The camphenes could be formed by two different routes. On the one hand, during the first step in the preparation of toxaphene, the major product 2-*exo*,10-dichlorobornane is accompanied by *cis*- and *trans*-10-chlorocamphene,

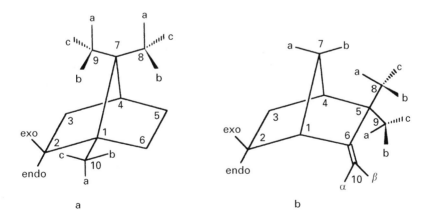

Figure 12.1 Numbering of the bornane (a) and camphene (b) skeleton without Cl-substitution. Since the groups C_8, C_9, and C_{10} are not able to rotate, the a, b, c nomenclature has been introduced to distinguish possible stereoisomers.

which normally reacts further to 2-*exo*,10,10-trichlorobornane and 10-chloro-tricyclen, which partly isomerizes to 2-*exo*-chlorocamphene (Parlar et al. 1978; Casida et al. 1974; Burhenne et al. 1993). Perhaps a small percentage is further chlorinated without the double bond being attacked. With excessive chlorine, 2-*exo*,10,10-trichlorocamphene is formed. However, it is more likely that there is some sort of photochemically induced radical Wagner-Meerwein rearrangement of more highly chlorinated bornanes, leading to camphenes by the elimination of HCl or Cl_2. The condition for this is the existence of a CHCl group in the 2- or 6-position and, in the case of a Cl_2 elimination, at least one chlorine atom in the 10-position. The resulting chlorocamphenes can then be chlorinated to a higher extent without destroying the double bond because of the shielding effect of the already existing chlorines in the molecule.

Some general postulates can be formulated on the basis of the numerous bornane structures that have been identified (Hainzl et al., 1995). The maximum number of chlorine atoms is two on each of the tentacles 8, 9, and 10, whereas the total maximum number at the positions 8 and 9 together is three. Two CCl_2 groups are never found next to each other or directly opposite each other. It has always been assumed that the bridge head at position 4 is never substituted with a chlorine atom (Wolinsky 1961). These findings mean that the total number of possible chlorinated bornane derivatives is perhaps significantly lower than expected.

12.4 ISOLATION AND IDENTIFICATION OF ENVIRONMENTALLY RELEVANT TOXAPHENE COMPONENTS

Former attempts to separate all of the constituents in the commercial product by several chromatographical methods, e.g. thin-layer and column chromatography, high-pressure liquid chromatography (HPLC), or preparative gas chromatography have led to unsatisfactory results. Only few components could be isolated using these methods (Turner et al. 1975; Matsumura et al. 1975; Landrum et al. 1976; Holmstead et al. 1974). UV-irradiation of a solution of technical toxaphene in oxygen-free *n*-hexane at 230 nm for 2 h led to a partly dechlorinated mixture, which has been used for the quantification of fish samples (CB standard) (Lach and Parlar 1990). Figure 12.2 shows the electron capture detector (ECD) chromatogram of this photochemically modified mixture (Figure 12.2b) compared with that of technical toxaphene (Figure 12.2a). The peaks that were unequivocally identified on the basis of high resolution gas chromatography mass spectrometry (HRGC-MS) experiments are numbered accordingly. Because this standard (Figure 12.2b) very much resembles the toxaphene residues obtained from different fish and fish products, it is apparent that many of the components of this standard play an important role in the marine environment.

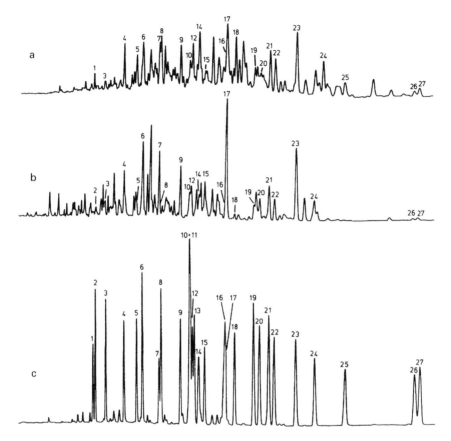

Figure 12.2 Comparison of the GC-ECD chromatograms of technical toxaphene (a), the CB standard (b), and the mixture of the isolated congeners (c). Conditions: Varian 3400 gas chromatograph (Injection: on-column, 120° to 230°C with 200°C/min; column: 60 m DB-5, fused silica, i.d. 0.25 mm, film thickness 0.32 μm; EC detector 320°C; carrier gas: 1.5 ml/min; temperature program (column): 120°C (0 min)—10°C/min—180°C (1 min)—5°C/min—250°C) (Hainzl et al. 1995).

With the help of liquid chromatography followed by a HPLC step on a C18 stationary phase the isolation of 25 chlorinated C_{10}-terpenes (the compounds 1–16, 18–24, 26, 27 in Figure 12.2c) from this mixture has recently been achieved (Hainzl 1994; Hainzl et al. 1995). The chlorobornanes 17 and 25 were prepared by isolation after specific chlorination of 2-*exo*,10-dichlorobornane (Burhenne 1993) and their spectroscopic behavior was investigated. In Figure 12.2c, a typical GC-ECD chromatogram is shown of a mixture of all 27 isolated compounds part of which represents the major components of the extracts obtained from environmental samples. For example, the components 4, 6, 7, 9, 17, and 23 are present in nearly all extracts of the fish samples originating from the North Atlantic (Burhenne et al. 1993; Lach et al. 1991; Seiber et al. 1975).

Compound 6 (2-endo,3-exo,5-endo,6-exo,8b,8c,10a,10b-octachlorobornane) and compound 17 (2-endo,3-exo,5-endo,6-exo,8b,8c,9c,10a,10b-nonachlorobornane, Tox Ac) are the most frequently detected toxaphene components. Their concentrations in fish samples from the North Sea are similar to or even higher than those of major polychlorinated biphenyl (PCB) congeners and chlordane components (Muir et al. 1992; Vetter et al. 1992; Bidleman et al. 1993). Especially in cod liver oil and salmon oil, high concentrations of these components have been detected (Lach et al. 1991). In beluga whale fat they could even be isolated in pure form (Stern et al. 1992). Assuming a bioconcentration factor (BCF) of 10^5–10^6 (Schnitzerling 1992) and an average level of about 50 µg/kg of the compound 17 in fish samples a concentration in North Atlantic surface water of about 0.01 ng/L seems to be possible.

The compounds 8, 10, 14, 16, 22, and 25 have already been isolated previously from technical toxaphene or mixtures similar to toxaphene (Anagnostopoulos et al. 1974; Khalifa et al. 1974; Landrum et al. 1976; Chandurkar et al. 1978). The most important of these components are compound 8 (Tox B) and compound 14 (Tox A), a 1:1 mixture of two octachlorobornanes (Saleh 1991). Although many research groups have investigated the contamination with toxaphene of various samples, such as fish, liver oil, fish oil, animal fat, human tissues, food stuff, rain water, and soil (Lach et al. 1991; Vetter et al. 1992; Heinisch et al. 1991; Lach and Parlar 1991; Hainzl et al. 1993), these compounds could not be detected in significant amounts. Their concentrations in fish samples from the North Atlantic only lie between not detectable and 3 µg/kg (Xu 1994), indicating that these main components of the technical mixture are probably not stable in the environment and react rapidly to form conversion products (Korte et al. 1979).

The percentage of the substances isolated in technical toxaphene and photochemically modified toxaphene cannot be given precisely, because even with the knowledge of their response factors the possibility of their coelution makes it impossible to quantify them correctly by peak area integration (Figure 12.2, chromatograms a and b). However, neglecting this restriction, the isolated congeners make up nearly 50% of the ECD peak area in technical toxaphene and about 57% in the photochemically modified mixture. In Table 12.1, known chlorobornanes and camphenes and their structures are given.

12.5 OCCURRENCE OF TOXAPHENE IN THE ENVIRONMENT

12.5.1 Plants

Numerous residue data on toxaphene-treated plants have shown that toxaphene may persist on plants. The half-life on leafy crops is reported to be 5–13 days, depending on species, growth rate, formulation, and weather (FAO/WHO 1969). After application at recommended rates, residues of about 2 ppm may be expected after 30 days (Maier-Bode 1965).

Residue decline studies have been done with alfalfa, range grass, and winter wheat treated with toxaphene emulsion spray (Carlin et al. 1974). Toxaphene residues were measured by electron capture gas chromatography after partial dehydration by sodium in liquid ammonia. All residues were identified as toxaphene components by GC-ECD. ^{36}Cl-toxaphene has been applied to cotton plants which then were maintained in a closed all-glass system for 7 days. Samples from various sites throughout the system were analyzed by GC-ECD and liquid scintillation counting procedures. Although ^{36}Cl-toxaphene components were distributed throughout the apparatus, the majority of the residues were detected on the plants. No evidence of metabolism could be found. The mechanism of loss was suggested to be volatilization.

To study the volatilization of toxaphene from treated plants into the atmosphere, Nash et al. (1977) applied it to cotton plants at weekly intervals for 6 weeks in a closed chamber (agro-ecosystem) and monitored the evaporation for 90 days. During the first week, the volatilization seemed to follow log concentration with log time, and afterwards, log concentration with linear time. From a first-order equation the calculated half-life for volatilization of toxaphene was 15.1 days. Willis et al. (1980, 1983, 1985) examined evaporation from cotton fields after application by aircraft and found a loss of 26% after just 5 days. Maximum loss rate was observed in the afternoon, after slight rain, or after drying of dew from the leaves. Generally, during summer the loss of applied toxaphene amounts to 50%, while during spring only ca. 10% is evaporated (Voldner and Schroeder 1990).

Bioaccumulation by plants have been investigated only for lake vegetations (Johnson et al. 1966; Terriere et al. 1966). BCFs found vary between 500 and 3000.

12.5.2 Animals

During the last few years, toxaphene residues were detected in many fish species, especially from the North Sea and the North Atlantic (Lach et al. 1991; Xu 1994; Zell and Ballschmiter 1980; Widequist et al. 1984; Gooch and Matsumura 1985; Onuska and Terry 1989; Lach and Parlar 1990; Xu et al. 1992). Average concentrations in different fish species are given in Table 12.2 and 12.3. Because of the rather high octanol–water partition coefficient of 5.5 (Isnard and Lambert 1989), bioaccumulation can be expected. In aquatic animals, bioconcentration factors of 1000–2000 for invertebratae and 10 000–76 000 for trouts have been found (Terriere et al. 1966; Verschueren 1983; Table 12.4). Accumulation through the food chain has been reported. From plankton to snails and water fleas to fish and birds the increase in toxaphene concentration was from 0.2 to 39 ppm (Matthias 1992). Selective accumulation of single persistent toxaphene components also occurred (Lach et al. 1991; Xu 1994; Matthias 1992), especially in fish. In *Lota lota* and *Monodon monoceros* only two of the chlorobornanes were found; the concentration in liver fat of *Lota* was between 0.8 and 2.3 ppm and in blubber of *Monodon* between 1.9 and 13.2 ppm, respectively (Bidleman et al. 1993).

Table 12.1 Selection of already identified chlorobornanes and camphenes

No. of compound	Name	Structure	References
1	(±)-2-*exo*,3-*endo*,6-*exo*,8,9,10-hexachlorobornane		Burhenne (1993) Black (1974) Saleh et al. (1977) Fingerling (1996)
	(1S,2R,3R,4R,6R), 2-*exo*,3-*endo*,6-*exo*,8,9,10-hexachlorobornane		
2	(+)-2-*exo*,3-*endo*,6-*endo*,8,9,10-hexachlorobornane		Burhenne (1993) Saleh et al. (1977)
3	(±)-2,2,5-*endo*,6-*exo*,8c,9b,10a-heptachlorobornane (toxicant B) (Parlar No. 32)		Jennings and Herrschbach (1965) Hainzl (1994) Burhenne (1993) Turner et al. (1977)

Table 12.1 (*Continued*)

No. of compound	Name	Structure	References
4	(±)-2-*endo*,3-*exo*,5-*exo*,6-*exo*,9,10,10-heptachlorobornane		Parlar (1988)
5	(±)-2-*endo*,3-*endo*,5-*exo*,6-*exo*,9,10,10-heptachlorobornane		Parlar (1988)
6	(±)-2,2,5,5,9c,10a,10b-heptachlorobornane (Parlar No. 21)		Hainzl (1994)

Table 12.1 (*Continued*)

No. of compound	Name	Structure	References
7	(±)-2-*exo*,3-*exo*,5,5,8,9,10,10-octachlorobornane		Black (1974) Chandurkar and Matsumura (1979a,b)
8	(±)-2-*endo*,3,3,5-*exo*,6-*exo*,9,10,10-octachlorobornane		Anagnostopoulos et al. (1974)
9	(±)-2,2,5-*endo*,6-*exo*,8b,8c,9c,10a-octachlorobornane (toxicant A₁) (Parlar No. 42)		Hainzl (1994) Saleh et al. (1977)

Table 12.1 (*Continued*)

No. of compound	Name	Structure	References
10	(±)-2,2,5-*endo*,6-*exo*,8c,9b,9c,10a-octachlorobornane (toxicant A₂) (Parlar No. 42)		Hainzl (1994) Saleh et al. (1977)
11	(±)-2-*endo*,3-*exo*,5-*exo*,6-*exo*,8,9,10,10-octachlorobornane		Parlar (1988) Parlar et al. (1976) Parlar and Korte (1983)
12	(±)-2-*endo*,3-*endo*,5-*exo*,6-*exo*,8,9,10,10-octachlorobornane		Parlar (1988) Parlar et al. (1976) Parlar and Korte (1983)

Table 12.1 (*Continued*)

No. of compound	Name	Structure	References
13	(±)-2,2,5,5,6-*exo*,8,9,10-octachlorobornane		Turner et al. (1977)
14	(±)-2,2,3-*exo*,5-*endo*,6-*exo*,8c,9b,10a-octachlorobornane (Parlar No. 39)		Hainzl (1994) Turner et al. (1977)
15	(±)-2,2,5-*endo*,6-*exo*,8c,9b,10a,10b-octachlorobornane		Hainzl (1994) Turner et al. (1977)

Table 12.1 (*Continued*)

No. of compound	Name	Structure	References
16	(±)-*2-endo,3-exo,5-endo,6-exo,8b,8c,10a,10c*-octachlorobornane (Parlar No. 26)		Hainzl (1994) Stern et al. (1992) Vetter et al. (1994)
17	(±)-*2,2,5,5,9b,9c,10a,10b*-octachlorobornane (Parlar No. 38)		Hainzl (1994)
18	(±)-*2-endo,3-exo,5-endo,6-exo,8b,9c,10a,10c*-octachlorobornane (Parlar No. 40)		Hainzl (1994)

Table 12.1 (*Continued*)

No. of compound	Name	Structure	References
19	(±)-2-*exo*,3-*endo*,5-*exo*,8c,9b,9c,10a,10b-octachlorobornane (Parlar No. 41)		Hainzl (1994)
20	(±)-2-*exo*,5,5,8c,9b,9c,10a,10b-octachlorobornane (Parlar No. 44)		Hainzl (1994)
21	(±)-2,2,5,5,8c,9b,10a,10b-octachlorobornane (Parlar No. 51)		Hainzl (1994)

Table 12.1 (*Continued*)

No. of compound	Name	Structure	References
22	(±)-2-endo,3,3,5-exo,6-exo,8,9,10,10-nonachlorobornane		Anagnostopoulos et al. (1974)
23	(±)-2,2,5-endo,6-exo,8b,8c,9c,10a,10c-nonachlorobornane (Parlar No. 56)		Hainzl (1994) Turner et al. (1977)
24	(±)-2,2,3-exo,5-endo,6-exo,8,9,10,10-nonachlorobornane		Turner et al. (1977)

Table 12.1 (*Continued*)

No. of compound	Name	Structure	References
25	(±)-2-*endo*,3-*exo*,5-*endo*,6-*exo*,8b,8c,9c,10a,10c-nonachlorobornane (toxicant Ac) (Parlar No. 50)		Burhenne (1993) Stern et al. (1992) Vetter et al. (1994)
26	(±)-2,2,5,5,8c,9b,9c,10a,10b-nonachlorobornane (Parlar No. 62)		Hainzl (1994) Burhenne (1993)
27	(±)-2-*endo*,3-*exo*,5-*endo*,6-*exo*,8,9,9,10,10-nonachlorobornane		Burhenne (1993)

Table 12.1 (*Continued*)

No. of compound	Name	Structure	References
28	(±)-2-exo,3-endo,5-exo,6-exo,8b,8c,9c,10a,10a-nonachlorobornane (Parlar No. 63)		Burhenne (1993)
29	(±)-2,2,3-exo,5-endo,6-exo,8c,9b,9c,10a-nonachlorobornane		Hainzl (1994)
30	(±)-2,2,3-exo,5,5,8c,9b,10a,10b-nonachlorobornane (Parlar No. 58)		Hainzl (1994)

Table 12.1 (*Continued*)

No. of compound	Name	Structure	References
31	(±)-2,2,5-*endo*,6-*exo*,8c,9b,9c,10a,10b-nonachlorobornane (Parlar No. 59)		Hainzl (1994)
32	(±)-2,2,5,5,6-*exo*,8c,9b,9c,10a-nonachlorobornane		Hainzl (1994)
33	(±)-2-*endo*,3,3,5-*endo*,6-*endo*,8,9,9,10,10-decachlorobornane		Anagnostopoulos et al. (1974)

Table 12.1 (*Continued*)

No. of compound	Name	Structure	References
34	(±)-2,2,3-*exo*,5-*endo*,6-*exo*,8c,9b,9c,10a,10b-decachlorobornane		Burhenne (1993)
35	(±)-2,2,5,5,6-*exo*,8c,9b,9c,10a,10b-decachlorobornane (Parlar No. 69)		Hainzl (1994) Burhenne (1993)
36	(±)-2,2,3-*exo*,5,5,8c,9b,9c,10a,10b-decachlorobornane		Hainzl (1994) Burhenne (1993)

Table 12.1 (*Continued*)

No. of compound	Name	Structure	References
37	(±)-2,2,3-*exo*-trichloro,5,5,-bis(chloromethyl), 6-(E)-chloromethylen,8,9,10-trinorbornane (Parlar No. 11)	37	Hainzl (1994)
38	(±)-5-*exo*,6-*endo*-dichloro,2-*endo*-chloromethyl, 3-(E)-chloromethylen-2-dichloromethyl, 8,9,10-trinorbornane (Parlar No. 12)	38	Hainzl (1994)
39	(±)-5-*exo*,6-*endo*,7-anti-trichloro,2,2-bis(chloromethyl,3-(E)- chloromethylen,8,9,10-trinorbornane (Parlar No. 15)	39	Hainzl (1994)

Table 12.1 (Continued)

No. of compound	Name	Structure	References
40	(±)-2,2,3-exo-trichloro,5-endo-chloromethyl, 6-(E)-chloromethylen,5-dichloromethyl,8,9,10-trinorbornane (Parlar No. 25)	40	Hainzl (1994)
41	(±)-2,2,3-exo-trichloro,6-(E)-chloromethylen, 5,5,-bis(dichloromethyl),8,9,10-trinorbornane (Parlar No. 31)	41	
42	(±)-2-exo,5-exo,6-exo-trichloro,3,3-bis(chloromethyl), 2-dichloro-methyl,8,9,10-trinorbornane	42	Landrum et al. (1976)

Source: Coelhan and Parlar (1996).

Table 12.2 Toxaphene residues in fish and fish products (ng/g)

Sample (origin)	Wet weight basis	Fat basis
Halibut filet (Norway)	89	1780
Redfish filet (Greenland)	103	3433
Trout filet (Germany)	30	1500
Smoked dogfish (North Sea)	145	897
Smoked dogfish (Spain)	201	917
Lumpfish spawn (Germany)	16	479
Lumpfish spawn (Iceland)	23	611
Cod liver oil (USA)	NA	1622
Cod liver oil (Iceland)	NA	2752
Fish oil (Germany)	NA	129

Source: Xu et al. (1992).
NA, Not applicable.

Table 12.3 Concentration of three single toxaphene components in fish and fish products

Sample (origin)	Concentration (μg/kg)				
	No. 26	No. 50	No. 62	$\sum$ 26,50,62	Tox_{tech}[a]
Caviar substitute (Iceland)	58 ± 21	64 ± 28	28 ± 12	150	1490 ± 590
Caviar substitute (Germany)	33 ± 14	36 ± 15	14 ± 4	83	909 ± 310
Fish oil 1 (Germany)	10 ± 4	11 ± 4	2 ± 1	23	255 ± 85
Fish oil 2 (Japan)	4 ± 2	6 ± 2	2 ± 1	12	130 ± 40
Salmon oil 1	33 ± 13	60 ± 28	9 ± 3	102	1100 ± 330
Salmon oil 2 (Norway)	11 ± 4	40 ± 16	1 ± 1	52	540 ± 280
Cod liver oil 1 (Germany)	220 ± 75	343 ± 98	171 ± 63	734	2450 ± 720
Cod liver oil 2 (Germany)	210 ± 70	405 ± 240	80 ± 30	695	2730 ± 925
Cod fish (Greenland)	50 ± 24	82 ± 33	50 ± 24	182	1920 ± 620
Red fish (Greenland)	7 ± 2	15 ± 8	8 ± 4	30	310 ± 120
Halibut (Germany)	4 ± 2	12 ± 5	3 ± 1	19	200 ± 75

Source: Xu (1994).
[a] Quantification with irradiated toxaphene as the standard.

Table 12.4 Bioconcentration of toxaphene in different species

Species		Exposure time (days)	BCF
Molluscs:			
Crassostrea virginica		4	9000–15200
Crustacean:			
Penaeus duorarum		4	400–800
Palaemonetes pugio		4	800–1200
Fish:			
Cyprinodon variegatus		4	3100–20600
Fundulus similis	juveniles	28	23700–60000
	adult	14	4200–6800
	fry	28	13300–33300
Lagodon rhomboides		4	3800–3900
Brook trout (fry)		150	15000–20000
		15	76000

Source: Verschueren (1983).

Of all toxaphene components, Tox Ac shows the highest accumulation potential (van der Falk and Wester 1991). Uptake by aquatic organisms proceeds via food as well as directly via body surface (Hill and Wright 1978).

12.5.3 Soil

Residues of toxaphene in soil may persist for years; in this case also, volatilization is regarded to be the major mechanism of loss, though microbial degradation has also been confirmed (Parr and Smith 1974; Fingerling 1995). In cropland under regular use, the recovery rate 1–3 years after the last application is in the order of 10%–30%. Evaporation from the soil surface is rapid provided the surface is not dry. The process of volatilization is nearly stopped if the soil is cultivated or if the toxaphene is mixed with the soil (Stevens et al. 1970). In temperate climates, the half-life of toxaphene is about 10 years (Menzie 1972).

12.5.4 Air

The occurrence of atmospheric residues of toxaphene was first demonstrated at various locations in the United States (Stanley et al. 1971). The correlation to application patterns has been studied for 3 consecutive years (Manigold and Schulze 1969). Over the most intensive growing cotton areas of Mississippi, residue levels were highest in August, directly indicating the spray activities in this area. However, as toxaphene may be transported over long distances, it is found not exclusively in the air near the application sites. Long-range transport has been proved by detection of toxaphene in air samples collected over the western North Atlantic with mean concentrations of 0.63 ng/m^3, while those of *p,p'*-DDT were only 0.024 ng/m^3 (Bidleman and Olney, 1975). The ratio of these concentrations is close to that of the outdoor evaporation rates of the two pesticides.

12.5.5 Water and Sediments

In addition to transport through air, movement from soil to water may also be regarded as a way of toxaphene dispersion. The primary route of entry into water is through surface runoff. Nearly all toxaphene displacement takes place while bound to sediment, so that it is possible to measure toxaphene runoff and reduction to biologically inactive levels in water by sediment analysis. In surface waters of streams in the western USA, toxaphene was not found within 2 years of monitoring (1966–1968) among samples collected at 20 stations. In most cases, the time for degradation in water varies from a few days to a few months, depending on the size of the surface area of the water body, the organic matter content, the sediment load, and the toxaphene concentration (Stevens et al. 1970).

Under normal conditions, transfer from soil to groundwater is not likely. Numerous studies confirmed a significant reduction in toxaphene concentrations with increasing depth of soil; even years after treatment 95% of the amount applied was found nearly exclusively in the upper 20 cm (Nash and Woolson 1968; Swoboda et al. 1971; Gallagher et al. 1979; Jaquess et al. 1989). Nevertheless, due to its high persistence, movement to lower soil horizons may occur under unfavorable conditions, and one case of groundwater contamination by toxaphene has been demonstrated (LaFleur et al. 1973). The extent of vertical displacement of toxaphene depends on climate conditions and soil properties, such as humus content, grain size, pH, and water content. According to Stanley et al. toxaphene loss from a soil surface treated with 100 kg/ha seems to occur in two steps, of which the second, major one is about linear on a log residue vs log time plot (Stanley et al. 1971). Toxaphene was found in underlying groundwater within 2 months after application to the surface during 1 year of monitoring. However, it should be emphasized that such an extremely high level is unlikely to occur with normal use, except by accidents or in areas near manufacturing plants. In drinking water, toxaphene was found in 27 of 680 samples in the USA with only two samples being in the concentration range above 0.05 ppm. In more than 500 samples taken along the Mississippi and Missouri rivers toxaphene could not be detected (Manigold and Schulze 1969).

Accumulation studies in lake sediments gave a ratio of 700 (Terriere et al. 1966). However, when Mississippi river sediment was analyzed, toxaphene was not detected (Bidleman and Olney, 1975), not even in the delta area where the highest amounts of other pesticides had been reported by another author (Edwards 1970). Only two samples proved point source contamination resulting from manufacturing activities. It seems that point source contamination does not noticeably spread, as can be seen from the absence of toxaphene at down-stream localities and from other studies on toxaphene-contaminated factory outlets (Durant and Reimold 1972).

As in the case of residues on plants and in soil, disappearence of toxaphene from water is mostly due to volatilization. The rate of movement to the water surface is controlled by several factors including rates of diffusion and desorption. However, it has been observed that very small amounts of tox-aphene remained in lakes for more than 2 decades (Stevens et al. 1970).

12.6 PHOTODEGRADATION OF TOXAPHENE COMPONENTS

Irradiation with wavelengths above 290 nm only leads to a slight degradation of toxaphene components, which are, presumably, unchlorinated, as the GC-ECD peak pattern remains unchanged; these components amount to ca. 15% of technical toxaphene (Becker 1987). Below 290 nm, dechlorination and dehy-drochlorination especially of higher chlorinated components takes place, while dichlorobornanes are inert under these conditions (Parlar et al. 1983). Degrada-tion by irradiation proceeds via photolytical loss of one chlorine atom, preferen-tially in C_2-position, followed by abstraction of hydrogen from the solvent. Irradiation of single toxaphene components also gave only dechlorination products, with reaction rates depending on the structure (Burhenne 1993; Fingerling 1995; Parlar et al. 1983; Saleh and Casida 1978; Parlar 1988; Figure 12.3). Generally, during irradiation in solvents the bornane structure is preserved, and dechlorination takes place only at the ring, while the degradation

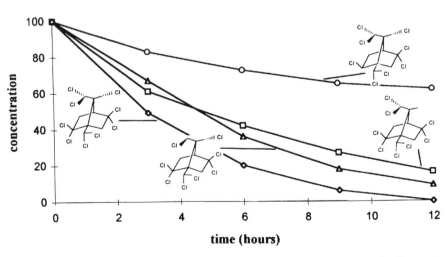

Figure 12.3 Photodegradation rate of four single toxaphene components (irradiation in *n*-hexane at 254 nm; Hg low-pressure lamp Vycor 250 mA/500 V) (Fingerling 1996).

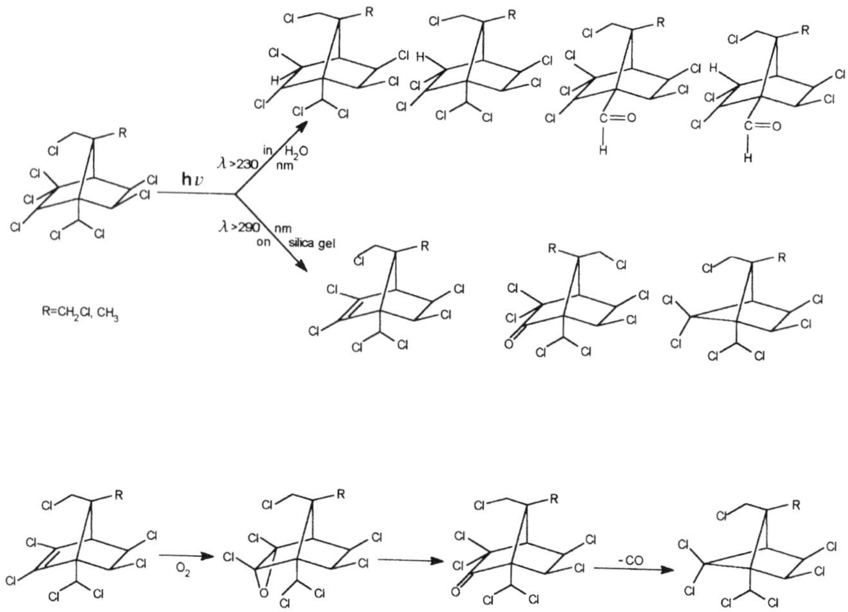

Figure 12.4 Photodegradation pathways of 2-*endo*,3,3,5-*exo*,6-*exo*,8,9,10,10-nonachlorobornane and 2-*endo*,3,3,5-*exo*,6-*exo*,8,10,10-octachlorobornane in water or adsorbed to silica gel at $\lambda > 290$ nm (Parlar 1988).

rate shows a positive correlation to the chlorine substitution at tentacles and bridgehead. Photolability seems to depend on the presence of a geminal dichloro group in C_2-position. The dechlorination rate is enhanced by an additional chlorine atom in C_3-position, but not by a dichloro group in C_5-position. Substances with only single chlorine atoms at all secondary ring atoms in alternating orientation have been found to be extremely photostable (Fingerling 1995). Contrary to this, irradiation of toxaphene adsorbed to silica gel results in mineralization to CO_2 and HCl (Durant and Reimold 1972; Becker 1987). Reductive dechlorination, dehydrodechlorination, and oxidation, presumably by reaction of chlorobornane radicals with hydroxyl or hydrogen radicals, has been observed. Adsorbed to silica gel, chlorobornenes are formed which can be easily excited and then react with oxygen, water, or peroxy radicals (Parlar 1988). Even lower chlorinated bornanes are degraded under these conditions (Korte et al. 1979; Figure 12.4).

12.7 DEGRADATION OF TOXAPHENE COMPONENTS IN SOIL

Under aerobic conditions, the degradation of toxaphene proceeds rather slowly, as residue analysis of aerated soil samples from a former toxaphene manufacturing plant gave the same GC-ECD peak pattern as technical tox-aphene (Fingerling 1995). Under anaerobic conditions, however, toxaphene is more easily degraded by microorganisms (Parr and Smith 1974; Saleh and Casida 1978; Clark and Matsumura 1979; Murthy et al. 1984; Mirsatari et al. 1987). GC-ECD analysis has shown that the highest chlorinated components, such as nona- and decachlorobornanes, are nearly completely degraded to lower chlorinated bornanes. The same applies to part of the octachlorobornanes, while hexa- and heptachlorobornanes are accumulated (Murthy et al. 1984; Mirsatari et al. 1987). In laboratory experiments with technical toxaphene in a loamy soil under anaerobic conditions there was a significant shift of the GC-ECD peak pattern of toxaphene towards lower retention times even after 1 week (Finger-ling 1996). After 3 months, an additional decrease in detectable components could be observed, the main components being two hepta- and one hexa-chlorobornane. During the subsequent period the peak pattern became even more simple, with only 12 components left after 5 months. The main component was 2-*exo*,5-*endo*,6-*exo*,8,9,10-hexachlorobornane, but hexachlorobornenes and a tetrachlorocamphenone were also found.

Studies on the degradation of single toxaphene components in loamy soil under anaerobic conditions confirmed the dependence of microbial dechlorina-tion rate on the chlorine substitution at the six-membered ring (Fingerling 1995; Figures 12.5, 12.6). Components with only one chlorine atom at each C atom in alternating orientation were highly persistent, while components with geminal dichloro groups at the ring were rather labile, especially when the dichloro group was localized at the C_2 atom. Like photoreactions, microbial degradation was not enhanced by the presence of a second geminal dichloro group. Dehy-drochlorination was little, but oxidation occurred under special conditions and reductive dechlorination took place not only at the ring, but also at the bridgehead and tentacles. Product ratios for the degradation of several chloro-bornanes under sulfate-reducing conditions are given in Table 12.5. The mecha-nism of reductive dechlorination has not yet been elucidated, but it may proceed by cometabolism. A dead-end metabolite is (1S,2R,3R,4R,6R),2-*exo*,3-*endo*,6-*exo*,8,9,10-hexachlorobornane, which has been found by residue analysis of different samples in rather high concentrations. Whether this is due to accumu-lation of this component or to a high proportion of components in technical toxaphene with a chlorine substitution leading to this product is not yet known.

12.8 DEGRADATION OF TOXAPHENE BY ANIMALS

As most of the single toxaphene components have only become available during recent years, most studies of excretion and metabolism have been made with normal or radiolabelled technical toxaphene. Experiments with [36]Cl-toxaphene showed that rats excreted ca. 53% of a single oral dose of 20 mg/kg during the first 9 days (Crowder and Dindal 1974). The greater part of the radiolabelled

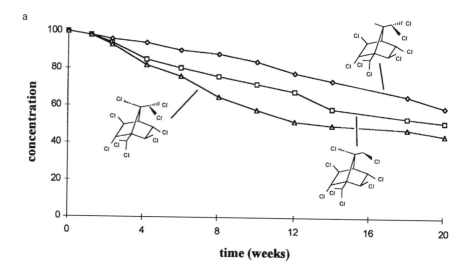

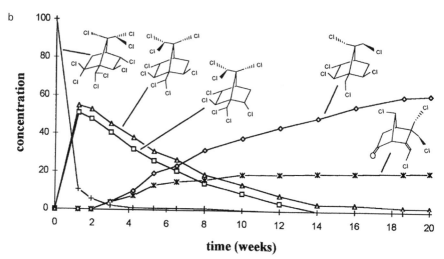

Figure 12.5a–d Degradation rate and products of single toxaphene components in soil under anaerobic conditions (Fingerling 1995).

metabolites was of ionic nature, as it was found in the hydrous part of the excrements. Other investigations confirmed the ^{36}Cl ion as main metabolite (Ohsawa et al. 1975). Excretion rates were 76% for ^{36}Cl-toxaphene and 57% for ^{14}C-toxaphene after 14 days. As 2% of activity of the ^{14}C-toxaphene applied was found in exhaled CO_2, total degradation of at least part of the toxaphene components is possible. Three percent was excreted unchanged, while 5%–10% of the amount excreted was completely and 27% partly dechlorinated.

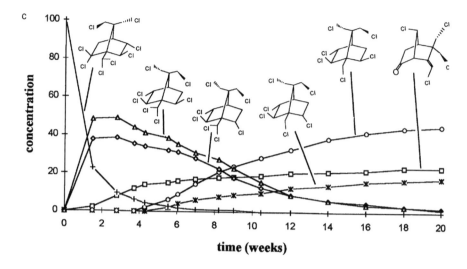

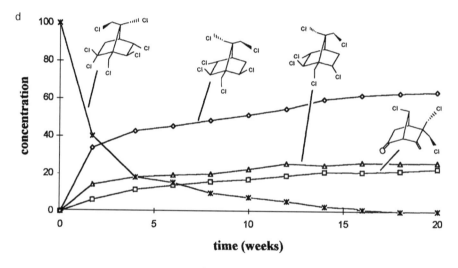

Figure 12.5c,d

Oxidative metabolism seems to be important for total biotic degradation of toxaphene. NADPH-dependent mixed-function oxidases from rat liver microsomes were able to quickly transform ^{14}C-toxaphene to more polar metabolites (Chandurkar and Matsumura 1979a,b). The introduction of hydroxyl groups was confirmed by the finding of glucuronides, the conjugation products of hydroxyl-containing xenobiotics. In contrast, experiments with isolated rat liver treated with Tox B and NADPH yielded no hydroxyl-containing products, while Tox C (2-*endo*,3,3,5-*exo*,6-*exo*,8,9,10,10-nonachlorobornane) was transformed to oxidized as well as reduced products. Only

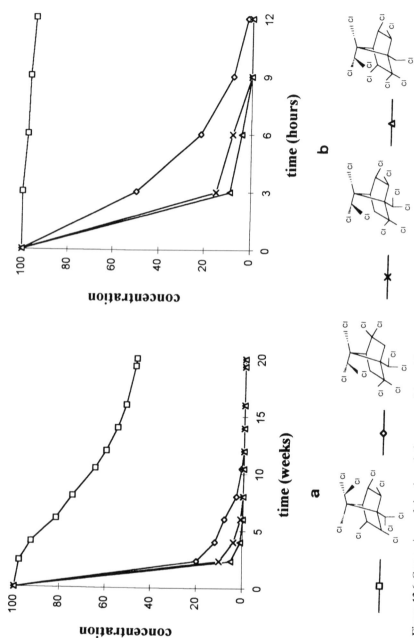

Figure 12.6 Comparison of the degradation rates of four chlorobornanes under anaerobic conditions in soil (a) and under irradiation ($\lambda = 254$ n) (b) (Fingerling 1995).

Table 12.5 Degradation rate and products of single toxaphene components under sulfate-reducing conditions (E_0 ca. 0.2 V) in soil

Compound	Products (%)			
	Reductive dechlorination	Dehydro-dechlorination	Vicinal didechloration	Oxygen containing products
2,2,5-endo,6-exo,8,9,10-Heptachlorobornane (Tox B)	72	3	0	25
2,2,5-endo,6-exo,8,8,9,10-and 2,2,5-endo,6-exo,8,9,9,10-Octachlorobornane (Tox A)	85	3	0	12
2,2,5-endo,6-exo,8,9,10,10-Octachlorobornane	80	5	0	15
2,2,5-endo,6-exo,8,8,9,10,10-Nonachlorobornane	88	2	0	10
2,2,5-endo,6-exo,8,9,9,10,10-Nonachlorobornane	89	2	0	9
2-exo,5,5,8,9,9,10,10-Octachlorobornane	92	0	0	1
2,2,5,5,8,9,10,10-Octachlorobornane	88	5	0	2
2,2,5,5,8,9,9,10,10-Nonachlorobornane	90	6	0	4
2,2,3-exo,5-endo,6-exo,8,9,10-Octachlorobornane	87	7	0	4
2,2,3-exo,5-endo,6-exo,8,9,9,10-Nonachlorobornane	10	0	90	0
2,2,3-exo,5-endo,6-exo,8,9,9,10,10-Decachlorobornane	7	0	93	0
2,2,3-exo,5,5,8,9,10,10-Nonachlorobornane	8	0	92	0
2-endo,3-exo,5-endo,6-exo,8,8,10,10-Octachlorobornane	12	0	88	0
2-endo,3-exo,5-endo,6-exo,8,9,10,10-Octachlorobornane	32	0	0	0
2-endo,3-exo,5-endo,6-exo,8,9,10,10-Octachlorobornane	42	0	0	0
2-endo,3-exo,5-endo,6-exo,8,9,9,10,10-Nonachlorobornane (Tox Ac)	48	0	0	0

Source: Fingerling (1995).

one of the six metabolites could be identified as 2-*exo*,3-*exo*,5,5,8,9,10,10-octachlorobornane. Derivatization experiments indicated the additional involvement of glutathione S-transferase.

Comparisons of the metabolism of toxaphene and Tox B by six different species of mammals gave reductive dechlorinations and dehydrochlorinations as dominant reactions, the transformation rates depending on the species (Saleh et al. 1979). Main products of the metabolism of Tox B were 2-*exo*,3-*endo*, 6-*exo*,8,9,10-hexachlorobornane and 2-*exo*,3-*endo*,6-*endo*,8,9,10-hexachlorobornane; 2,5-*endo*,6-*exo*,8,9,10-hexachloroborn-2-ene was found in smaller amounts. All three metabolites were also found to be excreted by animals treated with technical toxaphene. With rats and primates, they even were the main metabolites. As the amount of Tox B in technical toxaphene is rather small, other toxaphene components seem to give the same products.

Fish and aquatic mammals show a very selective metabolism. GC-ECD peak patterns of toxaphene residues in fish liver or oils resemble those of photodechlorinated toxaphene (Lach and Parlar 1990; Lach et al. 1991), which may indicate dechlorination and dehydrodechlorination of higher chlorinated bornanes though unselective accumulation of abiotically degraded toxaphene is also possible. In contrast, in narwhale (*Monodon monoceros*) or Weddell seal (*Leptonychotes weddellii*) only two toxaphene components were detected, which seem to be the only resistant ones (Muir et al. 1988; 1992; Vetter 1993).

12.9 DEGRADATION OF TOXAPHENE BY PLANTS

Uptake of toxaphene by plants, such as soybeans or Helianthus cultivated in water, is rather limited (Fingerling 1995). Like other chlorocarbon insecticides, toxaphene strongly adsorbs to the outer membranes of the root (Deutsche Forschungsgemeinschaft 1994). This results in a concentration gradient of ca. 80:1 between root and axis (Fingerling 1995). Translocation proceeds with the transpiration stream along the xylem. While the GC-ECD peak pattern of residues extracted from the roots was nearly identical to that of technical toxaphene, indicating nothing but adsorption, the pattern of extracts from leaves or the axis showed selective accumulation of two toxaphene components, Tox B and Tox Ac. Tox B is rather polar and, normally, easily metabolized, though transformation could not be observed in plants during the first 3 weeks after application; Tox Ac is unpolar and known for its persistence, which seems to be true also for plants. Whether Tox Ac is translocated in spite of its lipophilicity or whether its structure allows an ambivalent behavior is not clear.

Selective metabolism has been confirmed by studies on pelargonium (Fingerling 1995). Uptake in a closed chamber experiment was from the gas phase through the stomata. Residue analysis of extracts from the leaves showed GC-ECD peak pattern changes indicating dechlorination and dehydrochlorination as in other organisms. The dominant peaks of those remaining were a hexa-, a hepta-, and an octachlorobornane. The structure of the heptachloro component could not be identified. The hexachlorobornane was identical to 2-*exo*,3-*endo*,6-*exo*,8,9,10-hexachlorobornane, which is the major product of microbial degradation in soil under anaerobic conditions; the octachloro compound was identical to 2-*endo*,3-*exo*,5-*endo*,6-*exo*,8,8,10,10-octachlorobornane, which is very stable against photochemical and microbial degradation.

12.10 TOXICITY, MUTAGENICITY, AND CANCEROGENICITY OF TOXAPHENE

Toxaphene is moderately toxic for mammals, with LD_{50} values varying between about 20 and 200 mg/kg depending on the species (Table 12.6). Aquatic animals are much more sensitive to toxaphene, which for a short time was used as a piscicide.The highest acute toxicity has been found towards saltwater fish, with lethal concentrations in the range of 0.5–8.2 µg/L. Freshwater fish and molluscs are more tolerant, but the LC_{50} values for crustaceans are generally as low as

Table 12.6 Acute toxicity of toxaphene in different species

Species	LC_{50} (96 h, mg/L)	LD_{50} (oral, mg/kg)
Molluscs:		
Crassostrea virginica	16	
Crustaceans:		
Gammarus fasciatus	26	
Penaeus duorarum	1.4	
Palaemonetes pugio	4.4	
Fish:		
Salmo gairdneri	10.6	
Cyprinodon variegatus	1.1	
Lagodon rhomboides	0.5	
Birds:		
Quail		80–100
Pigeon		200–250
Mammals:		
Guinea pig		69–375
Rat (♀)		80
Rat (♂)		90–125
Dog		15–40
Sheep		100
Human (estimated)		60

Source: Maier-Bode (1965), Verschueren (1983).

those values for saltwater fish. Sublethal concentrations give rise to abnormal growth, reduced productivity, spine anomalies, and degeneration of inner organs (US EPA 1980).

Like all chlorocarbon insecticides, toxaphene is a neurotoxin, presumably excerting its convulsant action by interfering with the gamma-aminobutyric acid (GABA)-receptor-associated chloride ionophore (Brooks 1992), though the connection between binding to the target and lethal action in the insect is not always clear (von Keyserlingk and Willis 1992). Studies with single components are rare, but the influence of the composition of the toxaphene mixture on its toxicity has long been known. Two fractions of toxaphene obtained by recrystallization from methanol were significantly different in toxicity, indicating a synergistic effect (Parlar et al. 1983). Tox A and Tox B are much more toxic for mice than technical toxaphene (von Keyserlingk and Willis 1992). The same is true for Tox Ac. Possibly, the dominant feature determining the activity is the chlorine substitution at C-8 and C-9. Some results of investigation of structure–activity relationships are shown in Table 12.7.

Technical toxaphene behaves as a mutagen in the AMES test with the strain TA-100 of *Salmonella typhimurium*, and it is considered to be carcinogenic. Induction of microsomal oxygenases of the liver has been ascertained as well as the increase in malignant liver and glandular tumors in rodents in long-term exposure experiments. In humans, the estimated cancer threshold value after ingestion of toxaphene is 1.13 mg kg^{-1} day^{-1} (US EPA 1987).

Table 12.7 Acute toxicity of different toxaphene components in relation to Tox B

Component	Substitution[a]	Dehydrochl./ Dechlor.[a]	Relative toxicity		
			Mouse	House-fly	Goldfish
Tox B			100	100	100
Octachlorobornane	3-*exo*		< 75	62	7
	5-*exo*		ca. 312	44	22
	8		2272	209	264
	9		> 3000	< 371	> 527
	10		< 75	14	8
Nonachlorobornane	3-*exo*,10		n.a.	12	3
	8, 10		n.a.	19	8
Hexachlorobornane		2-*endo*	n.a.	32	60
		2-*exo*	n.a.	5	3
Hexachlorobornane		2, 3		32	11
		5, 6	ca. 115	5	< 3

Source: Saleh et al. (1977).

[a]Chlorine substitution or dehydrochlorination/dechlorination in comparion with Tox B.
n.a., Data not available.

12.11 STRUCTURE–DEGRADABILITY RELATIONSHIP OF TOXAPHENE COMPONENTS

When comparing the structures of isolated chlorobornanes, various structural relationships can be detected (Hainzl et al. 1995). If the retention volume in silica gel chromatography is taken as a measure of the polarity of chloroterpenes and, correspondingly, their retention time on HPLC as a measure of their lipophilicity, compound 8 (Tox B) is by far the most polar bornane derivative, followed by the octachlorobornane 12, while compounds 26 and 21 possess the greatest hydrophobicity. Here again the behavior of the three compounds with methyl groups in the 9-position is noteworthy. All three possess a degree of hydrophobicity unusual for hepta- or octachlorobornanes, being comparable to that of decachlorobornanes. Almost the same behavior can be observed with the octachlorocamphene 7, which is another reason for the enrichment of these compounds in lipophilic environmental samples (Holmstead et al. 1974; Lach and Parlar 1990). Comparing the compounds 6 and 12, it becomes evident how a small structural difference can lead to an enormous change in polarity. Nevertheless, it has not yet been possible to describe relationships between polarity and particular structural elements of the chloroterpenes without restrictions.

The photochemical stability of compounds with a CH_3 group in the 8-9-position is noteworthy. All three compounds (4, 6, and 9) are present in the photochemically modified toxaphene standard; hence, they are very slowly decomposed by irradiation with UV light. This may be one of the reasons why these substances play such an important role in toxaphene residues. Substance 6, in particular, is found together with Tox Ac (17) as a major component in nearly all environmental samples, especially in fish oils (Muir et al. 1992). The chlorinated camphene derivative 7 also appears to be photostable, while all decachlorobornanes (25–27) are unstable.

Six of the compounds isolated from the photochemically modified CB standard are camphene derivatives. As far as is known, none of these compounds seems to play an important role in environmental systems, but the situation would perhaps be totally different if toxicological data were available. Furthermore, taking the relatively low detector response factors into account, some of these compounds, such as chlorocamphene 7, should not be underestimated with respect to their environmental effect.

12.12 REFERENCES

Anagnostopoulos, M. L., H. Parlar, and F. Korte, 1974, "Ecological Chemistry LXXI. Isolation and Toxicology of some Toxaphene Components," *Chemosphere*, 3, 65–70

Becker, F., 1987, "Entwicklung einer Arbeitsmethode zur Bestimmung von Toxaphen-Rückständen in Umweltproben," *Thesis*, Technical University Munich, Germany

Bidleman, T. F., and C. E. Olney, 1975, "Long Range Transport of Toxaphene Insecticide in the Atmosphere of the West North Atlantic," *Nature*, *257*, 475–477

Bidleman, T. F., M. D. Walla, D. C. G. Muir, G. A. Stern, 1993, "Selective Accumulation of Polychlorocamphenes in Aquatic Biota from the Canadian Arctic," *Environ. Toxicol. Chem.*, *12*, 701–709

Black, D. K., 1974, 168th National American Society Meeting, Atlantic City, New Jersey, USA

Brooks, G. T., 1992, "Progress in Structure-Activity on Cage Convulsants and Related GABA Receptor-Chloride Ionophore Antagonists," in *Insecticides: Mechanism of Action and Resistance* (D. Otto, and B. Weber, Eds), Intercept, Andover, UK, pp. 237–242

Burhenne, J., 1993, "Darstellung ökotoxikologisch relevanter Chlorbornane als Referenzsubstanzen für die rückstandsanalytische Bestimmung des Insektizids Toxaphen," *Thesis*, University of Kassel, Germany

Burhenne, J., D. Hainzl, L. Xu, B. Vieth, L. Alder, and H. Parlar, 1993, "Preparation and Structure of High-Chlorinated Bornane Derivatives for the Quantification of Toxaphene Residues in Environmental Samples," *Fresenius J. Anal. Chem.*, *346*, 779–785

Carlin, F. J., J. J. Ford, and R. G. Kangas, 1974, 168th ACS National Meeting Atlantic City, New Jersey, USA, September 9–13

Casida, J. E., R. L. Holmstead, S. Khalifa, J. R. Knox, T. Ohsawa, K. J. Palmer, and R. Y. Wong, 1974, "Toxaphene Insecticide, Complex Biodegradable Mixture," *Science*, *183*, 520–521

Chandurkar, P. S., and F. Matsumura, 1979a, "Metabolism of Toxaphene in Rats," *Arch. Environ. Contam. Toxicol.*, *8*, 1–24

Chandurkar, P. S., and F. Matsumura, 1979b, "Metabolism of Toxicant B and Toxicant C of Toxaphene in Rats," *Bull. Environ. Contam. Toxicol.*, *21*, 539–547

Chandurkar, P. S., F. Matsumura, and T. Ikeda, 1978, "Identification and Toxicity of Toxicant Ac, a Toxic Compound of Toxaphene," *Chemosphere*, *6*, 123–130

Chiurdoglu, G., C. Goldenberg, and J. Geeraerts, 1957, "Bredt's Rule. About Tichtchenko's 2-Chloromethylcamphenile," *Bull. Soc. Chim. Belges*, *66*, 200–208

Clark, J. M., and F. Matsumura, 1979, "Metabolism of Toxaphene by Aquatic Sediment and a Camphor Degrading Pseudomonad," *Arch. Environ. Toxicol.*, *8*, 285–298

Coelhan, M., and H. Parlar, 1996, "The Nomenclature of Chlorinated Bornanes and Camphenes Relevant to Toxaphene," *Chemosphere*, *32*, 217–228

Crowder, L. A., and E. F. Dindal, 1974, "Fate of ^{36}Cl-toxaphene in Rats," *Bull. Environ Contam. Toxicol.*, *12*, 320–327

Deutsche Forschungsgemeinschaft, 1994, *Ökotoxikologie von Pflanzenschutzmitteln. Sachstandsbericht*, VCH, Weinheim, Germany

Durant, C. J., and R. J. Reimold, 1972, "Effects of Estuarine Dredging of Toxaphene-Contaminated Sediments in Terry Creek, Brunswick, Ga-1971," *Pestic. Monit. J.*, *6*, 94–96

Edwards, C. A., 1970, "Persistent Pesticides in the Environment," *Critical Reviews in Environmental Control*, *1*, CRC, Cleveland, Ohio, USA, pp. 7–67

FAO/WHO, 1969, 1968, "Evaluations of Some Pesticide Residues in Food," pp. 267–283

Fingerling, G., 1995, "Umwandlung von isolierten Toxaphenkomponenten unter abiotis-chen und biotischen Bedingungen," *Thesis*, University of Kassel, Germany

Gallagher, J. L., S. E. Robinson, W. J. Pfeiffer, and D. M. Seliskar, 1979, "Distribution and Movement of Toxaphene in Anaerobic Saline Marsh Soils," *Hydrobiologia*, *63*, 3–9

Gooch, J. W., and F. Matsumura, 1985, "Evaluation of the Toxic Components of Toxaphene in Lake Michigan Lake Trout," *J. Agr. Food Chem.*, *33*, 844–848

Hainzl, D., 1994, "Isolierung und Identifizierung von C10-Chlorterpenen aus dem Insek-tizid Toxaphen," *Thesis*, University of Kassel, Germany

Hainzl, D., J. Burhenne, H. Barlas, and H. Parlar, 1995, "Spectroscopical Character-ization of Environmentally Relevant C_{10}-Chloroterpenes from a Photochemically Modified Toxaphene Standard," *Fresenius J. Anal. Chem.*, *351*, 271–285

Hainzl, D., J. Burhenne, and H. Parlar, 1993, "Isolation and Characterization of Environ-mental Relevant Single Toxaphene Components," *Chemosphere*, *27*, 1857–1863

Heinisch, E., A. Kettrup, A. Jumar, S., Wenzel-Klein, J. Stechert, P. Harmann, and P. Schaffer, 1991, "Zu einigen Folgen der Anwendung von Toxaphen im chemischen Pflanzenschutz der ehemaligen DDR," Institut für Ökosystemforschung, Magdeburg und Institut für Ökologische Chemie der GSF Neuherberg; in *Ökologisch-chemische und ökotoxikologische Fallstudien über organische Spurenstoffe und Schwermetalle in Ost-Mitteleuropa* (Schadstoffatlas Osteuropa, E. Heinisch, A. Kettrup, S. Wenzel-Klein, Eds.), Ecomed, Landsberg, Germany, 1994

Hill, I. R., and S. J. L. Wright, 1978, "The Behaviour and Fate of Pesticides in Microbial Environments," in *Pesticide Microbiology* (I. R. Hill, and S. J. L. Wright, Eds.), Academic Press, New York, USA

Holmstead, R. L., S. Khalifa, and J. E. Casida, 1974, "Toxaphene Composition Analysed by Combined Gaschromatography-Chemical Ionisation Mass Spectrometry," *J. Agr. Food Chem.*, *22*, 939–944

Isnard, P., and S. Lambert, 1989, "Aqueous Solubility and *n*-Octanol/H_2O-Partition Coefficient Correlations," *Chemosphere*, *18*, 1837–1857

Jaquess, A. B., W. Winterlin, and D. Peterson, 1989, "Feasibility of Toxaphene Transport Through Sandy Soil," *Bull. Environ. Contam. Toxicol.*, *42*, 417–423

Jennings, B. H., and G. B. Herschbach, 1965, "The Chlorination of Camphene," *J. Org. Chem.*, *30*, 3902

Johnson, W. D., G. F. Lee, and D. Spyridakis, 1966, "Persistence of Toxaphene in Treated Lakes," *Air Wat. Pollut. Int. J.*, *10*, 555–560

Keyserlingk, H. C. von, and R. J. Willis, 1992, "The GABA-Activated Cl^- Channel in Insects as Target for Insecticide Action – a Physiological Study," in *Insecticides: Mechanism of Action and Resistance* (D. Otto, and B. Weber, Eds.), Intercept, Andover, UK, 205–236

Khalifa, S., R. L. Holmstead, and J. E. Casida, 1976, "Toxaphene Degradation by Iron(II) Protoporphyrin Systems," *J. Agr. Food Chem.*, *24*, 277–282

Khalifa, S., T. R. Mon, J. L. Engel, and J. E. Casida, 1974, "Isolation of 2,2,3-endo,6-exo,8,9,10-Heptachlorobornane and an Octachloro Toxicant from Technical Toxa-phene," *J. Agr. Food Chem.*, *22*, 653–657

Korte, F., I. Scheunert, and H. Parlar, 1979, "Toxaphene (Camphechlor). A Special Report," *Pure Appl. Chem.*, *51*, 1583–1601

Lach, G., and H. Parlar, 1990, "Quantification of Toxaphene Residues in Fish and Fish Products Using a New Analytical Standard," *Chemosphere*, *21*, 29–34

Lach, G., and H. Parlar, 1991, "Comparison of Several Detection Methods for Toxaphene Residue Analysis," *Toxicol Environ. Chem.*, *31*, 209–219

Lach, G., U. Ständecke, B. Pletsch, L. Xu, and H. Parlar, 1991, "Ein Beitrag zur Quantifizierung von Toxaphenrückständen in Fischölen," *Z. Lebensm. Unters. Forsch.*, *192*, 440–444

LaFleur, K. S., G. A. Wojeck, and W. R. McCaskill, 1973, "Movement of Toxaphene and Fluometuron Through Dunbar Soil to Underlying Ground Water," *J. Environ. Quality*, *2*, 515–518

Landrum, P. F., G. A. Pollock, J. N. Seiber, H. Hope, and K. L. Swanson, 1976, "Toxaphene Insecticide: Identification and Toxicity of a Dihydrocamphene Component," *Chemosphere*, *5*, 63–69

Maier-Bode H., 1965, *Pflanzenschutzmittel-Rückstände*, Ulmer, Stuttgart, Germany

Manigold, D. B., and J. A. Schulze, 1969, "Pesticides in Water. Pesticides in Selected Western Streams – a Progress Report," *Pestic. Monit. J.*, *3*, 124–135

Matsumura, F., R. W. Howard, and J. O. Nelson, 1975, "Structure of the Toxic Fraction A of Toxaphene," *Chemosphere*, *4*, 271–276

Matthias, A., 1992, "Probleme bei der Charakterisierung von Toxaphenrückständen in Muttermilch," *Diploma Thesis*, University of Kassel, Germany

Menzie, C. M., 1972, "Fate of Pescticides in the Environment," *Annu. Rev. Entomol.*, *17*, 199

Mirsatari, S., M. McChesney, A. Craigmill, W. Winterlin, and S. Seiber, 1987, "Anaerobic Microbial Dechlorination: An Approach to On-Site Treatment of Toxaphene-Contaminated Soil," *J. Environ. Sci. Health B*, *22*, 663–690

Muir, D. C. G., C. A. Ford, N. P. Grift, R. E. A. Stewart, and T. F. Bidleman, 1992, "Organochlorine Contaminants in Narwhal (*Monodon monoceros*) from the Canadian Arctic," *Environ. Pollut.*, *75*, 307–316

Muir, D. C. G., R. J. Norstrom, and M. Simon, 1988, "Organochlorine Contaminants in Arctic Marine Food Chains: Accumulation of Specific Polychlorinated Biphenyls and Chlordane-Related Compounds," *Environ. Sci. Technol.*, *22*, 1071–1079

Murthy, N., W. Lusby, J. Oliver, and P. Kearney, 1984, "Degradation of Toxaphene Fractions in Anaerobic Soil," *J. Nucl. Agr. Biol.*, *13*, 16–17

Nash, R. G., M. C. Beall, and W. G. Harris, 1977, "Toxaphene and 1,1,1-Trichloro-2,2-bis (*p*-Chloro-Phenyl)Ethane (DDT) Losses from Cotton in an Agroecosystem Chamber," *J. Agr. Food Chem.*, *25*, 336–341

Nash, R. G., and E. A. Woolson, 1968, "Distribution of Chlorinated Insecticides in Cultivated Soil," *Soil Sci. Soc. Amer. Proc.*, *32*, 525–527

Ohsawa, T., J. R. Knox, S. Khalifa, and J. E. Casida, 1975, "Metabolic Dechlorination of Toxaphene in Rats," *J. Agr. Food Chem.*, *23*, 98–106

Onuska, F. I., and K. A. Terry, 1989, "Quantitative High-Resolution Gas Chromatography and Mass Spectrometry of Toxaphene Residues in Fish Samples," *J. Chromatogr.*, *471*, 161–171

Parlar, H., 1988, "Photoinduced Reactions of Two Toxaphene Compounds in Aqueous Medium and Adsorbed on Silica Gel," *Chemosphere*, *17*, 2141–2150

Parlar, H., S. Gäb, S. Nitz, and F. Korte, 1976, "Zur Photochemie des Toxaphens: Reaktionen von chlorierten Bornanderivaten in Lösung und adsorbiert an Kieselgel," *Chemosphere*, 5, 333–338

Parlar, H., and F. Korte, 1983, "Ökochemische Bewertung des Insektizids Toxaphen, Teil II. Verhalten unter biotischen und abiotischen Bedingungen," *Chemosphere*, 12, 927–934

Parlar, H., D. Kotzias, and F. Korte, 1983, "Ökochemische Bewertung des Insektizids Toxaphen, Teil IV. Vorkommen und Analytik der Toxaphenkomponenten," *Chemosphere*, 12, 1453–1458

Parlar, H., S. Nitz, S. Gäb, and F. Korte, 1977, "A Contribution to the Structure of the Toxaphene Components. Spectroscopic Studies on Chlorinated Bornane Derivatives," *J. Agr. Food Chem.*, 25, 68–72

Parlar, H., S. Gäb, S. Nitz, A. Michna, and F. Korte, 1978, "Ein Beitrag zur Zusammensetzung des Insektizides "Toxaphen". Gaschromatographisch-Massenspektroskopische Charakterisierung eines öligen und eines kristallinen Anteils des technischen Toxaphens und dessen insektizide Wirksamkeit," *Z. Naturforsch.*, 33b, 915–923

Parr, J. F., and S. Smith, 1976, "Degradation of Toxaphene in Selected Anaerobic Soil Environments," *Soil Sci.*, 121, 52–57

Pollock, G. A., and W. W. Kilgore, 1978, "Toxaphene," *Res. Rev.*, 50, 87–140

Saleh, M. A., 1983, "Capillary Gas Chromatography-Electron Impact and Chemical Ionisation Mass Spectrometry of Toxaphene," *J. Agr. Food Chem.*, 31, 748–751

Saleh, M. A., 1991, "Toxaphene, Chemistry, Biochemistry, Toxicity and Environmental Fate," *Rev. Environ. Contam. Toxicol.*, 118, 1–85

Saleh, M. A., and J. E. Casida, 1978, "Reductive Dechlorination of the Toxaphene Component 2,2,5-endo,6-exo,8,9,10-Heptachlorobornane in Various Chemical, Photochemical and Metabolic Systems," *J. Agr. Food Chem.*, 26, 583–590

Saleh, M. A., R. F. Skinner, and J. E. Casida, 1979, "Comparative Metabolism of 2,2,5-endo,6-exo,8,9,10-Heptachlorobornane and Toxaphene in Six Mammalian Species and Chickens," *J. Agr. Food Chem.*, 27, 731–737

Saleh, M. A., W. V. Turner, and J. E. Casida, 1977, "Polychlorobornane Components of Toxaphene, Structure Toxicity Relations and Metabolic Reductive Dechlorination," *Science*, 198, 1256–1258

Schnitzerling, S., 1992, "Betrachtungen des Akkumulationsverhaltens von Toxaphen bei aquatischen Organismen am Beispiel des Zebrabärblings," *Diploma Thesis*, University of Kassel, Germany

Seiber, J. N., P. F. Landrum, S. C. Madden, K. D. Nugent, and W. L. Winterlin, 1975, "Isolation and Gas Chromatographic Characterization of some Toxaphene Components," *J. Chromatogr.*, 114, 361–368

Stanley, C. W., J. F. I. Barney, M. R. Helton, and A. R. Yobs, 1971, "Measurement of Atmospheric Levels of Pesticides," *Environ. Sci. Technol.*, 5, 430–435

Stern, G. A., D. C. G. Muir, C. A. Ford, N. P. Grift, E. Dewailly, T. F. Bidleman, and M. D. Walla, 1992, "Isolation and Identification of Two Major Recalcitrant Toxaphene Congeners in Aquatic Biota," *Environ. Sci. Techn.*, 26, 1838–1840

Stevens, J. L., C. W. Collier, and D. W. Woodham, 1970, "Pesticides in Soil. Monitoring Pesticides in Soils from Areas of Regular, Limited, and no Pesticide Use," *Pestic Monit. J.*, 4, 145–166

Swoboda, A. R., G. W. Thomas, F. B. Cady, R. W. Baird, and G. W. Knisel, 1971, "Distribution of DDT and Toxaphene in Housten Black Clay on Three Watersheds," *Environ Sci. Technol.*, *5*, 141–145

Terriere, L. C., U. Kiigemagi, A. R. Gerlach, and R. L. Borovicka, 1966, "The Persistence of Toxaphene in Lake Water and its Uptake by Aquatic Plants and Animals," *J. Agr. Food Chem.*, *14*, 66–69

Tishchenko, D., and I. Uvarov, 1953, "Structure and some Transformations of Camphene Dichloride," *J. Gen. Chem. USSR*, *23*, 1407–1414

Turner, W. V., J. L. Engel, and J. E. Casida, 1977, "Toxaphene Components and Related Compounds, Preparation and Toxicity of some Hepta-, Octa- and Nonachlorobornanes, Hexa- and Heptachlorobornenes and a Hexachlorobornadiene," *J. Agr. Food Chem.*, *25*, 1394–1401

Turner, W. V., S. Khalifa, and J. E. Saleh, 1975, "Toxaphene Toxicant A. Mixture of 2,2,5-endo,6-exo,8,8,9,10-Octachlorobornane and 2,2,5-endo,6-exo,8,9,9,10-Octachlorobornane," *J. Agr. Food Chem.*, *23*, 991–994

US EPA, 1980, "Ambient Water Quality Criteria for Toxaphene," US Environmental Protection Agency, Washington DC, USA, Rept. 440/5-80-076

US EPA, 1987, "Health Effects Assessment for Toxaphene," EPA 600/8-88/06S, US Environmental Protection Agency, Washington DC, USA, Govt. Rept. Announce Index (US) 88(13)

Valk, F. van der, and P. G. Wester, 1991, "Determination of Toxaphene in Fish from Northern Europe," *Chemosphere*, *22*, 57–66

Verschueren, K., 1983, *Handbook of Environmental Data on Organic Chemicals*, Van Nostrand Reinhold, New York, USA

Vetter, W., 1993, "Zur Isolierung und Charakterisierung von Einzelstandards für die Bestimmung von Rückständen polychlorierter Multikomponentengemische," *Thesis*, University of Hohenheim, Germany

Vetter, W., B. Luckas, and M. Oehme, 1992, "Isolation and Purification of Two Main Toxaphene Congeners in Marine Organisms," *Chemosphere*, *25*, 1643–1652

Vetter, W., G. Scherer, M. Schlabach, B. Lukas, and M. Oehme, 1994, "An Unequivalent ^{1}H NMR Structural Assignment of TOX8 and TOX9, the Two Most Abundant Toxaphene Congeners in Marine Mammals," *Fresenius J. Anal. Chem.*, *349*, 552–558

Voldner, E. C., and W. H. Schroeder, 1990, "Long Range Atmospheric Transport and Deposition of Toxaphene," in *Long Range Transport of Pesticides* (D. A. Kurtz, Ed.), Lewis Publishers, Chelsea, Michigan, USA

Voldner, E. C., and Y. F. Lie, 1993, "Global Usage of Toxaphene," *Chemosphere*, *27*, 2073–2078

Widequist, U., B. Jansson, L. Reutergårdh, and G. Sundström, 1984, "The Evaluation of an Analytical Method for Polychlorinated Terpenes (PCC) in Biological Samples Using an Internal Standard," *Chemosphere*, *13*, 367–379

Willis, G. H., L. L. McDowell, L. A. Harper, L. M. Southwick, and S. Smith, 1983, "Seasonal Disappearence and Volatilization of Toxaphene and DDT from a Cotton Field," *J. Environ. Qual.*, *12*, 80–85

Willis, G. H., L. L. McDowell, S. Smith, L. M. Southwick, and E. R. Lemon, 1980, "Toxaphene Volatilization from a Mature Cotton Canopy," *Agron. J.*, *72*, 627–631

Willis, G. H., L. L. McDowell, L. M. Southwick, and S. Smith, 1985, "Toxaphene, Methyl-Parathion, and Fenvalerate Disappearence from Cotton Foliage in the Mid-South," *J. Environ. Qual.*, *14*, 446–449

Wolinsky, J., 1961, "Ring Enlargement Produced by Alkaline Fusion of ω-Bromocamphene," *J. Org. Chem.*, *26*, 704–711

Xu, L., 1994, "Quantifizierung von Toxaphenrückständen mit Hilfe eines neuen Standards und umweltrelevanten Toxaphenkomponenten in Fisch und Fischprodukten," *Thesis*, University of Kassel, Germany

Xu, L., U. Ramus, B. Pletsch, and H. Parlar, 1992, "Quantification of Toxaphene Residues in Fish and Fish Products," *Fresenius Envir. Bull.*, *1*, 58–63

Zell, M., and K. Ballschmiter, 1980, "Global Occurrence of Hexachlorobenzene (HCB) and Polychlorocamphenes (PCC) in Biological Samples," *Fresenius J. Anal. Chem.*, *300*, 387–402

13

Specimen Banking as an Environmental Surveillance Tool

Antonius Kettrup and Petra Marth
(Oberschleißheim, Germany)

13.1 SUMMARY

Thousands of chemicals are traded on the market, but only in a few cases is information available on their distribution and effects on man and the environment. In addition to monitoring actual concentrations of chemical, it is, thus, necessary to establish an environmental specimen bank (ESB) for the retrospective monitoring of chemicals in the future. The bases of an ESB are representative bioindicators of systematically collected biological systems and environmental samples. As a result of the extremely low temperature ($T < -150°C$) it is guaranteed that the samples are not subject to chemical changes during long-term storage. The German project is described here, including selected results related to environmental trend monitoring.

13.2 INTRODUCTION

According to the "European Inventory of Existing Commercial Substances" (EINECS; GDCh/BUA 1987), more than 100 000 different chemical substances

Ecotoxicology, Edited by Gerrit Schüürmann and Bernd Markert.
ISBN 0-471-17644-3   © 1998 John Wiley & Sons, Inc. and Spektrum Akademischer Verlag.

are produced worldwide. Every year 1000–2000 new chemicals enter the market. For most of them we lack sufficient information about their effects on man, animals, and plants and about their further reaction and fate in the environment (SRU 1987). New technologies *always* produce unintended and unpredicted wastes and impacts. In most cases the introduction of chemicals into the environment represents an irreversible step. A considerable number of chemicals reaching the environment do not degrade at all or only very slowly. They accumulate in the environment and, having distributed, certain pollutants become ubiquitous, e.g., polychlorinated biphenyls (PCBs) and chlorinated insecticides. Persistent biologically active chemicals, sometimes at concentrations below our ability to analyze or detect, can pose serious pervasive and possibly irreversible threats to human health and the integrity of the biosphere (Lewis 1988).

Numerous industrial countries have passed laws in order to assess the hazards of chemicals for man and the environment. Investigations on forecasting the impacts of chemicals are, however, afflicted by special costs and problems (Figure 13.1; Wagner 1994a).

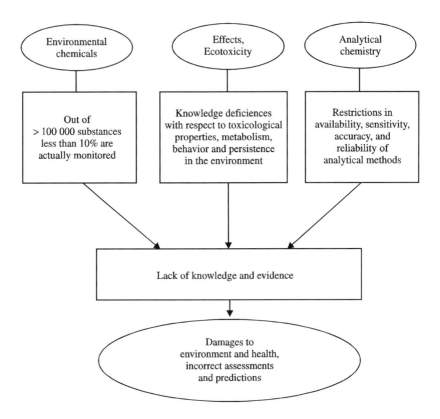

Figure 13.1 Deficits in assessment and control of environmental chemicals (Wagner 1994a).

Living organisms and their environments are enormously heterogeneous, complex, and open systems characterized by poorly defined boundaries. Under the best circumstances we can evaluate most quantitative features of biological systems only for certain properties and only by using certain samples or subsets of the class. Results are always statistical in nature and extrapolation from the sample to the class is necessary. Many environmental factors that cannot be reproduced or represented in the laboratory or by models can modify the environmental behavior and effects of chemicals. The error in forecasts based upon trend extrapolations without knowing the relationships of environmental variables may become enormously enlarged with time. Recognizing the actual form of a trend among reasonable alternatives is difficult and often subjective. Thus, the level of uncertainty of most forecasts and assessment of chemical impacts upon man and the environment is often quite high. Even today we repeatedly confront serious and unexpected consequences of our technologies, products, and wastes. Effective damage protection and contaminant risk strategies can be developed by politicians only if they have reliable basic information at their disposal.

13.3 BIOINDICATORS

The idea that organisms can provide an indication of the quality of their environment is widespread and at least as old as agriculture (Thalius 1588). It is not possible to establish any clear definition for the term "bioindicator" from the large amount of literature published in the last decades. The following definitions are suggested in agreement with many European authors (Wittig 1993):

- Bioindication is the use of an organism (a part of an organism or a society of organisms) to obtain information on the quality (of a part) of its environment. Organisms which are able to give information on the quality (of a part) of its environment are bioindicators.
- Biomonitoring is the continuous observation of an area with the help of bioindicators, which may also be called biomonitors. With the aid of organisms trends in time and space concerning the distribution and ecological effects of environmental chemicals can be observed by a semiquantitative evaluation of the results.

Biological samples from the environment are mainly used and analyzed as representatives for larger entities or similar or related environmental compartments. This requires the selection of standardized (bio)indicator systems, which react with known specificity and sensitivity to environmental chemicals and have the capability of spatial and/or temporal integration. Such indicator systems can be efficiently and reproducibly analyzed and evaluated. The results can then be extrapolated to other sensitive targets in the environment, which are often extremely variable with respect to the space, time and physiology. Bioindicator systems are preferred in such cases where potential integral effects of complex or unknown immission types have to be detected and quantified. Such effects may occur on different levels, from specific organs of single organisms to whole ecosystems. Bioindicators are also preferred in such cases where they offer advantages due to their high sensitivity towards a broad spectrum of substances or because of their ability to accumulate a substance over an extended period of time or to integrate its influence in an area of known and relevant size. This is the case if the sensitivity of available analytical methods for dangerous substances is too low to find them in environmental compartments such as air, water and soils.

In addition to analyzing the concentrations of toxic substances and their metabolites, biological specimens can also be analyzed for essential components and a broad spectrum of possible biochemical, physiological, morphological and/or genetical effects. Organisms and biological communities normally do not react to single components or substances in their environment. Rather, they show the combined effects of all acting substances and environmental factors. Decisive for the use of biological specimens is their ecotoxicological relevance, i.e., the relevance or indicator function of the effects for other living organisms and communities including man.

13.4 IDEA OF ENVIRONMENTAL SPECIMEN BANKING

With respect to effects of pollutants, their quantities, and distribution under natural conditions the aquisition of reliable information requires a systematic program of environmental monitoring in which concentrations of hazardous chemical substances are measured in suitable environmental specimens of various trophic stages and food chains. However, actual monitoring of the environment can only be as good as our present knowledge and as analytical possibilities allow. Among the multitude of substances found in the environment only those can be monitored which have already been recognized to be hazardous. The present assessment of the measured environmental concentrations of hazardous substances—and thus the quality of regulatory decisions—suffers from the fact that no results are available on pollutant burdens of former times or that the data which are available are ambiguous (Kayser et al. 1982).

At the beginning of the 1970s the idea of using biological samples as reference material to furnish proof of environmental pollution was put forward by Frederick Coulsten of Albany Medical College, Albany, New York and Friedhelm Korte of the Institute of Ecological Chemistry, GSF-Research Centre, Munich-Neuherberg. In an environmental specimen bank (ESB) carefully selected, relevant environmental samples are stored systematically at temperatures below $-150°C$ immediately after collection. In this way no or only small chemical changes occur over a long period of time. Baseline levels of contaminants in the environment can be established by taking samples at the present time for future use in ecological chemistry research. Long-term storage of samples with indicator functions represents a necessary complement to the actual monitoring of the environment and a safety net in the assessment of chemical risk.

A systematically established archive of frequently collected representative environmental specimen samples fulfill the following important functions (Lewis 1988; Kayser et al. 1982; Wise and Zeisler 1984; Keune 1993):

- They may be used for the determination of the environmental concentrations of those substances, which, at the time of storage, were not recognized to be hazardous or which at present cannot be analyzed with adequate accuracy (retrospective monitoring).

- They may serve as reference samples for the documentation of the improvement of analytical efficiency and for the verification of previously obtained monitoring results.

- Early detection of environmental increases in hazardous chemicals thought to be under control is possible. Also, the effectiveness of restrictions, regulations, or management practices that have been applied to the community, environment, or to the manufacture, distribution, disposal, or use of toxic chemicals can be assessed.

- Depending upon the analysis and evaluation of stored materials an ESB can save considerable time and money when unexpected impacts are observed.

- Sources of chemicals may be identified. Often, by the time a chemical is recognized as a health or environmental problem it is sufficiently widespread to defy identification of the principal sources or pathways.

- An ESB is a useful tool for providing reliable data on pollutant burdens of earlier times, because assessments and regulatory decisions are usually limited by inconsistencies and ambiguities in available data.

In Germany the Federal Ministry for Research and Technology supported a comprehensive pre- and pilot phase of an ESB between 1976 and 1984. During this period the technical feasibility regarding the sampling of different species, handling and shipping of samples, deep freezing, homogenization, ultra trace analysis, packing materials, logistics, storage temperature, and documentation was confirmed (Boehringer 1988). The results were so encouraging that in 1985 the German government decided to set up a permanent ESB under the responsibility of the Federal Ministry for the Environment, Nature Conservation and Reactor Safety (BMU), coordinated by the Federal Environmental Agency (Umweltbundesamt). Two specimen banks are subsumed under the general heading of the German ESB:

- the Specimen Bank for Environmental Specimens at the Institute of Applied Physical Chemistry of the Research Center Jülich (KFA)

Table 13.1 Participating institutions of the German ESB

Institution	Task
KFA Research Center Jülich, Institute of Applied Physical Chemistry	Specimen bank for environmental specimens Central banking facilities Logistics Element analysis Sampling of marine specimens
University of Münster, Institute of Pharmacology and Toxicology	Specimen bank for human tissues Sampling, characterization, storage and analysis of human samples Data banking system
GSF Research Center Neuherberg Institute of Ecological Chemistry	Analysis of chlorinated hydrocarbons
Biochemical Institute for Environmental Carcinogens, Grosshansdorf	Analysis of polycyclic aromatic hydrocarbons
University of Saarland, Saarbrücken Institute of Biogeography	Selection and characterization of areas and specimen types Sampling of terrestrial and limnic specimens Deal with ecological questions

• the Specimen Bank for Human Organ Specimens at the Institute of Pharmacology and Toxicology of the University of Münster.

The work is distributed among five institutions depending on their special scientific capabilities. In Table 13.1 the participating institutions of the German ESB and their responsibilities are summarized.

In addition, an international cooperation of ESBs in the Federal Republic of Germany, USA, Canada, Japan, Finland, Sweden, Norway, and Denmark has been established (Wise et al. 1988; Stoeppler and Zeisler 1993).

13.5 REALIZATION

13.5.1 Sampling

Selection of Sampling Areas

Sampling areas have been chosen so as to form a national network of ecological assessment parks coordinating environmental specimen banking with long-term ecological research and environmental monitoring (Lewis 1985, 1987; Lewis et al. 1989). An overall concept has been developed by a committee of experts under the auspices of the BMU, taking into consideration different types of ecosystems with corresponding representative sampling areas according to the following criteria:

- stability of utilization
- assured long-term use
- sufficient minimal size
- availability of suitable samples
- practicability, e.g., accessibility, public ownership (National Park), no conflict with the protection of biotopes and species, high level of information, nearby suitable institutions for research.

The list of sampling areas (Figure 13.2) comprises the major ecosystems and habitat types that occur within the Federal Republic of Germany including

- freshwater and marine ecosystems
- urban industrial ecosystems
- forest and agricultural ecosystems
- near nature ecosystems.

The continuous sampling is carried out every year in order to utilize the analytical, biometric, and meteorological data for real-time monitoring.

Selection of Specimen Types

The selection and assignment of representative specimen species of the terrestrial, limnic, and marine ecosystems for the ESB was undertaken by a committee of experts in consideration of the above-mentioned indicator functions so that a broad spectrum of different types of matrices (all trophic levels) and media (air, sediment, soil) with environmentally relevant concentrations of xenobiotics is available (Figure 13.2; Lewis 1985; Lewis et al. 1989). The following requirements must be fulfilled by a matrix for use as a bioindicator:

- The chemicals must be accumulated to levels comparable to those occurring in the environment.
- Contamination trends in the environment must correspond to those in the matrix.
- The matrix should have a widespread distribution and must be available in time and place to a sufficient extent.

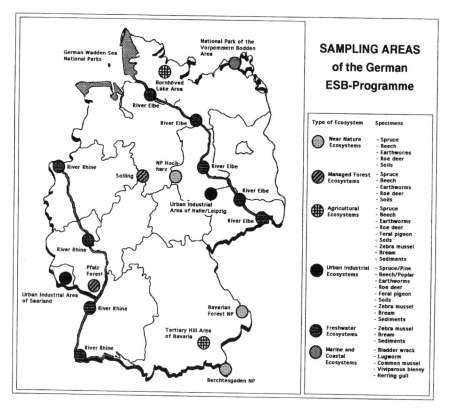

Figure 13.2 Sampling areas and specimens of the German ESB program (Klein et al. 1994).

- The organism should be sedentary and easy to identify.
- The species should accumulate the pollutant without being killed or rendered incapable of reproduction.

Standard Operation Procedures

Standardized sampling guidelines in the sense of standard operating procedures are the basis for the comparability, reliability, and repeatability of the banked samples. They contain detailed instructions for:

- selection of sampling sites and specimens
- sampling
- providing cover for repeatability of sampling
- area and sample characterization
- sample treatment and long-term storage
- documentation of sampling and storage conditions

- chemical analysis
- data processing and evaluation
- quality assurance (Wagner 1994a).

Nevertheless, sampling of biological and other environmental specimens is always influenced by factors which may modify the exposure as well as the accumulation behavior of the specimen types in relation to xenobiotics, e.g., climatic factors, weather conditions, and changes in the population sampled or in the structure of the whole ecosystem (Wagner 1994b, 1995; Klein and Paulus 1995).

Ecological and biometrical sample characterization provides basic information about changes in the quality of the sampled material and its comparability with previous and subsequent samples from the same area or the same specimen type sampled in different areas. Biological sample characterization can also give information about ecological and ecotoxicological effects on the population sampled. As an example, Figure 13.3 demonstrates changes in the condition index, i.e., the relation between body weight and length of bream caught in different sites of the river Elbe from 1991 to 1995. While the values from 1991 show high variability, standardization of the sampling to a defined age class yielded more reproducible values, which demonstrate a clear increase in the condition index downstream of the Elbe river and also increasing values in some of the sampling sites (Wagner et al. 1996).

Another key for the understanding and evaluation of analytical results is the land use structure of the sampling area and its annual changes. This means that

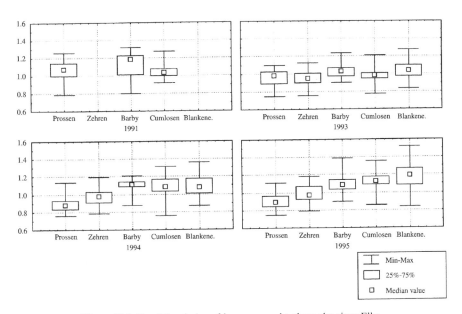

Figure 13.3 Condition index of breams caught along the river Elbe.

not only the potential sources of chemical pollution have to be detected, characterized, and monitored but also sources of nutritional and physical disturbance by man have to be investigated. Figure 13.4 shows a computer map of the urban industrial agglomeration of the Saar region in western Germany as an example of land use mapping in extended sampling areas as a basis for the detection of temporal changes and the interpretation of analytical results. However, in agricultural areas the land use pattern has to be mapped in much more detail and actualized each year in order to recognize the potential and actual emissions and effects emerging from each unit of used area (Müller et al. 1996).

In environmental specimen banking samples of different specimen types and ecosystems are frequently sampled, characterized, processed, and stored with considerable effort in order to take the precautions necessary for deferred analysis of initially unknown substances or parameters. Quality assurance is therefore an absolute necessity and an innovative challenge in environmental specimen banking. Errors made during the sampling in the field, transportation, and sample pretreatment can seldom be recognized and never corrected afterwards during the subsequent analytical measurements. Thus, the quality assurance system for an ESB includes the whole process from planning, sampling, ecological and biometrical characterization, packing, transportation, storage, homogenization, and subsampling up to the analytical procedures and evaluation of the results (Klein et al. 1994; Paulus et al. 1995). A flow chart of the entire preparation procedure for the final long-term storage of an environmental specimen is shown in Figure 13.5. On average 2.5 kg of material per specimen per sampling site was collected, producing nearly 250 standardized subsamples of approximately 10 g.

13.5.2 Analytical Sample Characterization

The choice of pollutants or classes of pollutants for the analytical sample characterization was made according to ecotoxicological importance.

Inorganic Analysis

The analytical procedures for inorganic analysis of ESB samples were selected with emphasis on trace analysis capability and applicability to very complex biological matrices. During the pilot phase sample preparation was optimized to suit the matrix to be analyzed and the detection technique to be used (Emons 1994; Standard Operating Procedures of the German Environmental Specimen Bank, in press). The following four groups of analytical methods are applied for trace analysis:

- atomic spectrometry: electrothermal graphite furnance atomic absorption spectrometry (GF-AAS), cold vapor atomic absorption spectrometry (CV-AAS), hydride AAS, inductive coupled plasma atomic emission spectrometry (ICP-AES)

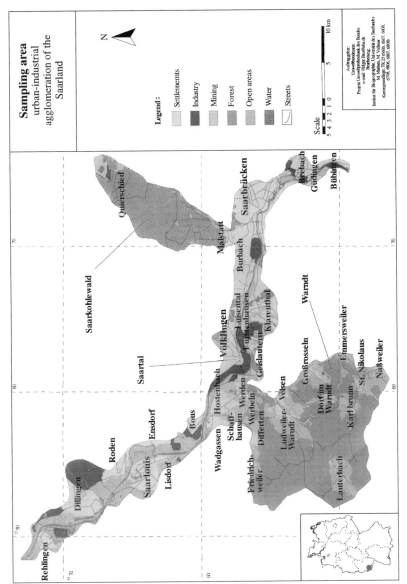

Figure 13.4 Computer map of the industrial agglomeration of the Saar region.

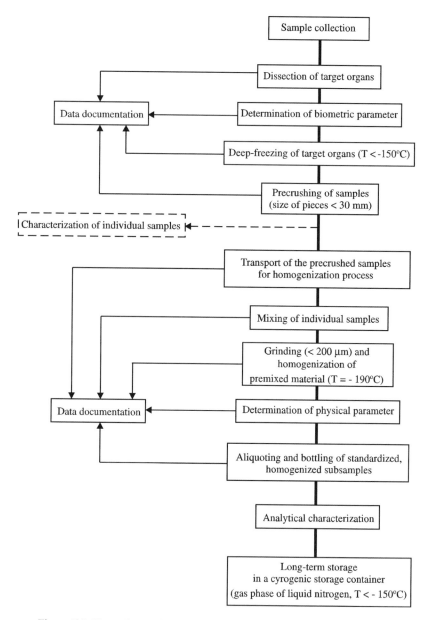

Figure 13.5 Flow scheme of the sample preparation steps of the German ESB.

- mass spectrometry: isotope dilution mass spectrometry (IDMS), inductive coupled plasma mass spectrometry (ICP-MS)
- electrochemical methods: potentiometric stripping analysis (PSA), voltametry
- radiochemical methods: instrumental neuron activation

- analysis (INAA), prompt gamma cold neutron activation analysis (PGCNAA)

An important aspect of the ESB consists in the long-term monitoring of heavy metals in biological samples in different regions in Germany. For example, a clear decline in mercury pollution in the estuary region of the Elbe in the past few years can be documented on the basis of samples of herring gull eggs (*Larus argentatus*) collected in the Trischen bird sanctuary (Figure 13.6; Schwuger 1994; UPB 1996). More than 90% of the mercury was present in the form of the highly toxic methyl mercury. Before the unification of the two German states (1988/89) the mercury concentrations in herring gull eggs from the island Trischen was twice as high as for the subsequent years (1991–1995). The decreasing temporal trend is probably associated with the closure of industrial plants in the upper regions of the Elbe and its tributaries. As shown in Figure 13.6, it is obvious that birds living in the estuary of the river Elbe (Trischen) exhibit higher uptakes of Hg than species living in the estuary of the River Weser (Mellum). This demonstrates the influence of mercury input from the river Elbe into the North Sea.

Polycyclic Aromatic Hydrocarbons (PAHs)

The determination of PAHs is of very high importance because they are considered to be the most relevant class of environmental carcinogens (Grimmer

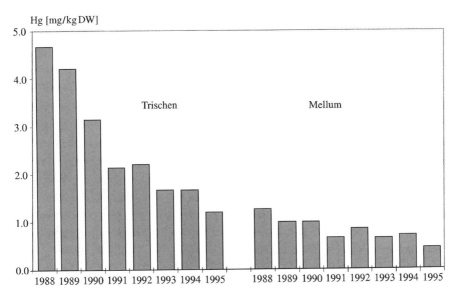

Figure 13.6 Time trend of mercury concentration in herring gull eggs from the islands Trischen and Mellum (UPB 1996).

1993). PAH emission sources result from incomplete fossile combustion, wood burning, and waste gases of industrial and private combustion processes. The continuous measurement of PAH concentration in various selected environmental matrices representing the terrestrial, aquatic, and atmospheric ecosystem provides an opportunity to recognize trends in environmental pollution.

For PAH detection sample homogenates are extracted with toluene or cyclohexane by Soxhlet. Interferences are removed by liquid/liquid distribution and chromatography on silica and Sephadex LH20. The detection of PAHs is carried out by gas chromatography equipped with FID (flame-ionization detector) or MS/SIM (mass spectrometry/single ion monitoring) detectors (Grimmer et al. 1996).

Spruce and pine sprouts are passive samplers, reflecting atmospheric pollution by PAHs (Jacob et al. 1997). For example, the benzo[a]pyrene (B[a]P)-concentration of spruce sprouts from the industrialized area of Saarland decreased by a factor of 3 within 10 years (Figure 13.7). The same trend in pine sprouts has been observed for a sampling area in eastern Germany (Dübener Heide) during the period 1991–1995 with B[a]P declining from 3.5 μg/kg to 1.5 μg/kg. Similar results were obtained with poplar leaves from Halle, whereas samples from Leipzig showed no consistent temporal trend (Figure 13.8). These findings show that the reduction of pollution by technical improvements such as modern vehicle conceptions as well as improved domestic and industrial combustion devices in the past were successful.

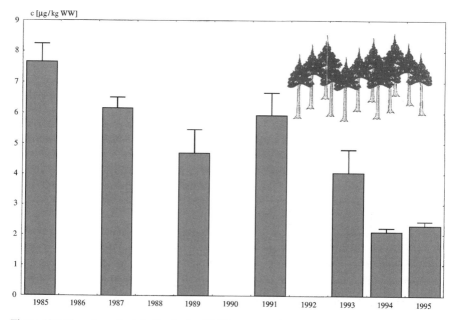

Figure 13.7 Time-dependent decline in the B[a]P concentration of spruce sprouts from Saarland-Warndt during 1985–1995.

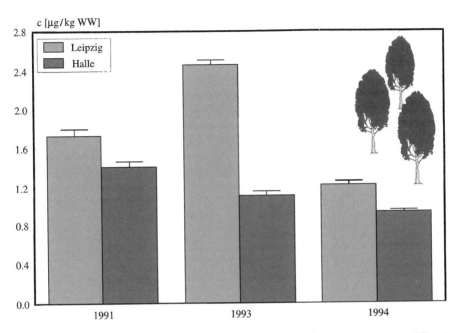

Figure 13.8 Time course of the B[*a*]P concentration of poplar leaf homogenate from two different sampling locations of Dübener Heide (former East Germany) during 1991–1995.

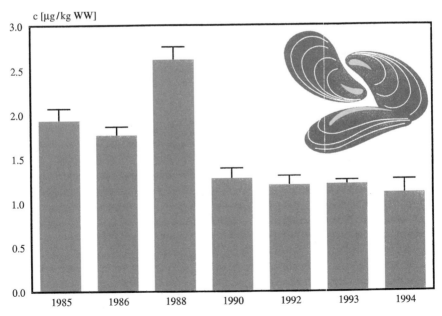

Figure 13.9 Time course of the B[*a*]P concentration of mussel homogenate from Eckwarderhörne (North Sea) from 1985 to 1995.

Mussels (*Mytilus edulis*), which are used as a bioindicator of the marine environment for the North Sea (Eckwarderhörne), likewise exhibited a decline in the PAH concentration, although to a lesser extent (Jacob et al. 1997). With the exception of 1988 when higher concentrations were measured, B[*a*]P concentration decreased from 1.9 µg/kg in 1985 to 1.2 µg/kg in 1990 and remained practically constant since then as presented in Figure 13.9. Long-distance transfer by atmospheric pollutants from other countries and the dilution effect play an important role for the contamination of the North Sea.

Chlorinated Hydrocarbons (CHC)

Numerous chlorinated insecticides and industrial chlorinated hydrocarbons (e.g., polychlorinated biphenyls (PCBs) and polychlorinated dibenzo-*p*-dioxins and dibenzofurans (PCDD/F)) are extremely resistant to degradation in the environment. Residues of these xenobiotics are still detected throughout the world, although most of them have been banned since the 1970s. Because of their toxicological properties and accumulation effects, long-term studies on their residue levels are essential to estimate the environmental contamination in the past and to predict future trends.

For CHC determination the samples are mixed with anhydrous sodium sulfate/seasand to form a free-flowing product, which is extracted with *n*-hexane/acetone (2 : 1 v/v) in an extraction column. Clean-up procedures are performed with gel permeation and high-performance liquid chromatography. Quantification is carried out by high-resolution gas chromatography (HRGC), with electron capture detection using two columns of different polarity. More details of the applied method are given by Oxynos et al. (1992).

Herring gull eggs are a suitable bioindicator for lipophilic xenobiotics since they reflect the local and temporal environmental conditions (Oxynos et al. 1993). The time series of herring gull eggs at the sampling locations on the islands of Mellum and Trischen are shown in Figure 13.10. The location influenced by the river Elbe (Trischen) exhibited higher concentrations than the Mellum location, which is influenced by fluxes of pollutants in the river Weser. The significantly higher DDE concentrations in eggs from Trischen in 1989 can be explained by DDT applications in the former GDR in the late 1980s.

Figure 13.11 summarizes the PCDD/F levels of herring gull eggs from 1988 to 1993 at Trischen and Mellum (Schramm et al. 1996). A decreasing trend of the contaminants is obvious. This development shows the result of legislative actions to minimize dioxin emissions, e.g., the banning of sea burning of hazardous waste.

Investigations of sediments and fish (bream) along the river Elbe (Figure 13.12) in 1991 have shown that the eastern sampling sites (Prossen, Dresden) have been heavily contaminated by hexachlorobenzene (HCB), octachlorostyrene (OCS), DDT metabolites, and PCBs (Figure 13.13; Oxynos et al. 1995). This

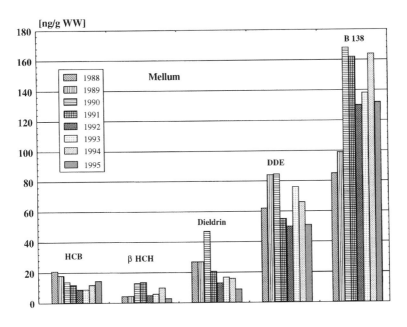

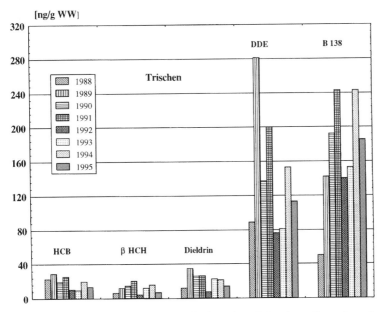

Figure 13.10 Time trend of selected chlorinated hydrocarbons in herring gull eggs from the islands Trischen and Mellum.

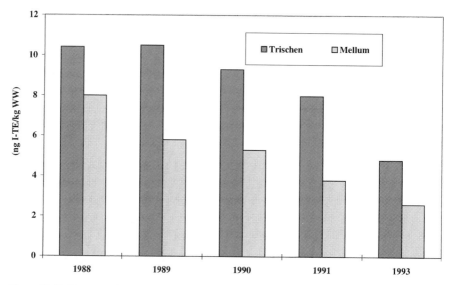

Figure 13.11 Time trend of PCDD/F I-ITE in herring gull eggs from the islands Trischen and Mellum.

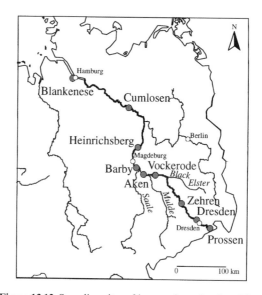

Figure 13.12 Sampling sites of bream along the river Elbe.

observation is probably a result of the considerable pollution of the river Elbe by the industrialized areas (e.g., Pardubice, Neratovice, Usti) of the former CSFR (Nesmerak 1993). Furthermore, sediments at the Dresden station are highly contaminated because they were strongly subjected to effluents of a pulp

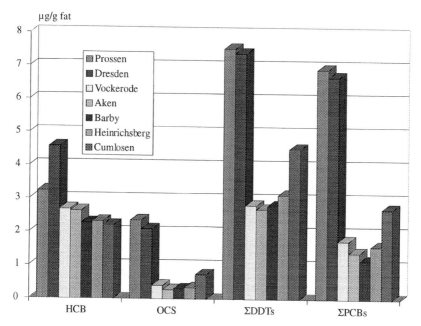

Figure 13.13 Levels of major contaminants related to fat content in bream muscle tissue from the river Elbe in 1991.

mill and chemical industries up to 1991 (Internationale Kommission zum Schutz der Elbe 1992). A 30% decrease in the CHC concentrations in bream muscle tissue was found between Dresden and Vockerode stations; further downstream, the total CHC burden remained nearly constant. Elevated levels of hexachlorocyclohexane isomers (HCHs) in bream muscle and livers were observed downstream between Aken and Heinrichsberg due to influxes of the rivers Mulde and Saale as well as discharges from pesticide plants located in Magdeburg (Marth et al. 1997).

A comparison of CHC patterns (Figure 13.14) in bream livers reveals significant differences between different ecosystems. DDT metabolites and PCBs make up to one-third of the total CHC burden of bream livers from the limnic ecosystem Elbe. PCBs were the major organochlorine contaminants (70%) in breams of a typical industrialized area (Saarland) in contrast to agricultural areas of eastern Germany, where DDT metabolites were the dominant pollutants (80%). This was caused by the continued application of DDT in the former GDR after DDT was banned in western Europe. Similarly, significantly higher concentrations of DDT metabolites were found in pigeon eggs from Leipzig (former East Germany) than in pigeon eggs from western German locations. Fortunately, a decreasing trend can be observed (Figure 13.15) for these pollutants.

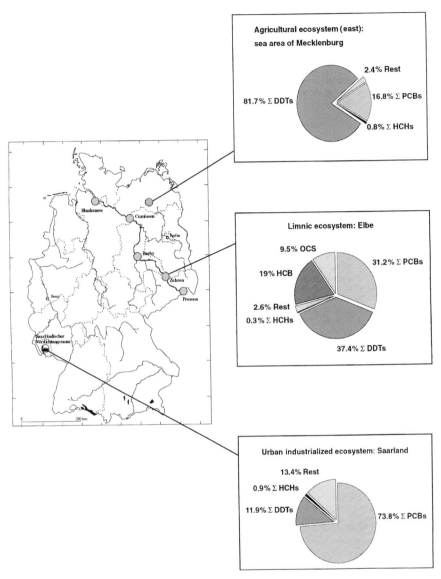

Figure 13.14 Comparison of different ecosystems with bream (*Abramis brama*) as bioindicator.

13.6 CONCLUSION AND FUTURE PERSPECTIVES

Twenty years of practical experience in environmental specimen banking have demonstrated that the concept of long-term storage of biological specimens for retrospective analysis complements traditional environmental pollution monitoring. The ESB can serve as a valuable resource for the assessment of

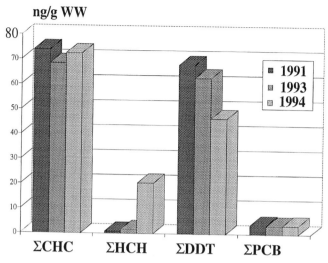

Figure 13.15 Temporal comparison of chlorinated hydrocarbons in pigeon eggs from Leipzig.

long-term trends of pollutants affecting human and environmental health, in particular for those pollutants that were previously unnoticed.

The effects of political decisions about the emission or other regulations on pollutants (e.g., introduction of unleaded fuel, ban on sea burning of hazardous waste) are monitored by the ESB. In addition, illegal applications of chemicals have been identified, e.g., DDT applications in the former GDR.

The generated standard operating procedures for analytical and sampling procedure are well documented, similar to the demands of good laboratory practice or the European standards for analytical work. Thus, in this field it is possible to identify and avoid possible errors in the future. Due to the high quality standard of sampling and analysis the collection is of particular value for future analytical work on environmental contaminants of which we presently may know very little.

The increasing amount of generated data allows the verification with very high accuracy of many relationships discovered in the past in environmental research, i.e., bioaccumulation, biomagnification, distribution, transport, and degradation of chemicals. The conceptional development of the ESB has resulted in a pool of knowledge which can be used for future decisions and recommendations for researchers and politicians. As an instrument for science, administration, and management, an ESB can support analytical and environmental research and monitoring generally in many ways and make them more effective and reliable, e.g., by suppling reference materials for environmental analysis; by preserving authentic records as an archive for long-term comparisons of environmental change; and by securing and perpetuating evidence in the fields of biology, medicine, forensic medicine, biotechnology, deposition or conversion of problematic wastes, and environmental planning and risk assessment.

For future scientific activities the following goals are recommended:

- extension of the set of chemicals which are analyzed
- selection of the most important bioindicators to minimize costs and labor
- creation of a pattern library for sources and bioindicators by normalizing and archiving chromatographic raw data and applying multivariate statistics.

One of the major tasks for the ESB in the future will be to make ties with ESBs of other nations and in cooperation with these establish an international forum for exchange of information and samples.

13.7 REFERENCES

Boehringer, U. R., 1988, *Umweltprobenbank: Bericht und Bewertung der Pilotphase.* Bundesministerium für Forschung und Technologie, Umweltbundesamt Berlin, Springer, Berlin, Germany

Emons, H., 1994, "Inorganic Analysis Within the German Environmental Specimen Bank," in *German–Egyptian Seminar on Environmental Research* (H. F. Aly, and D. Nentwich, Eds.): Cairo, Aegypten, 21–23 March 1994, Bilateral Seminars of the International Bureau, Forschungszentrum Jülich GmbH, Germany, Vol. 19, pp. 195–204

GDCh/BUA, 1987, Altstoffbeurteilung, p. 32

Grimmer, G., 1993, "Relevance of Polycyclic Aromatic Hydrocarbons as Environmental Carcinogens," in *Polycyclic Aromatic Compounds 13th Int. Symp. PAH* (P. Garrigues, and M. Lamote, Eds.), 1–4 October 1991, Bordeaux, Gordon and Breach, London, UK, pp. 31–41

Grimmer, G., J. Hildebrandt, J. Jacob, and K.-W. Naujack, 1996, "Standard Operating Procedure for the Analysis of Polycyclic Aromatic Hydrocarbons (PAH) in Various Matrices," in *Federal Environmental Specimen Bank: Standard Operating Procedures for Sampling, Transport, Storing, and Chemical Characterization of Environmental Specimens and Human Organic Specimens* (in German), (Umweltbundesamt, Eds.), Erich Schmitt Verlag, Berlin, Germany, Chapter 12, p. 18

Internationale Kommission zum Schutz der Elbe (IKSE), 1992, Erste Auswertung der Abwasserlasten von industriellen Direkteinleitern aus drei ausgewählten Industriezweigen im Einzugsgebiet der Elbe im Jahre 1991 gegenüber 1989, Magdeburg, Germany

Jacob, J., G. Grimmer, and J. Hildebrandt, 1997, "Long-term Decline of Atmospheric and Marine Pollution by Polycyclic Aromatic Hydrocarbons (PAH) in Germany," *Chemosphere, 34,* 2099–2108

Kayser, D., R. U. Boehringer, and F. Schmidt-Bleek, 1982, "The Environmental Specimen Banking Project of the Federal Republic of Germany (pilot phase)," *Environ. Monitoring Assessment, 1,* pp. 241–255

Keune, H., 1993, "Environmental Specimen Banking (ESB): An Essential Part of Integrated Ecological Monitoring on a Global Scale," *Sci. Tot. Environ., 139/140,* 537–544

Klein, R., and M. Paulus, (Eds.), 1995, *Umweltproben für die Schadstoffanalytik im Biomonitoring-Standards zur Qualitätssicherung bis zum Laboreingang*, Gustav Fischer Verlag, Jena, Germany

Klein, R., M. Paulus, G. Wagner, and P. Müller, 1994, "Das ökologische Rahmenkonzept zur Qualitätssicherung in der Umweltprobenbank des Bundes," in *Biomonitoring und Umweltprobenbank, Teil I* (M. Paulus, R. Klein, G. Wagner, and P. Müller, Eds.), Beitragsserie in der UWSF – Z. *Umweltchem. Ökotox., 6,* 223–221

Lewis, R. A., 1985, *Richtlinien für den Einsatz einer Umweltprobenbank in die Praxis. Umweltforschungsplan des Bundesministers des Innern*, Chr. Eschl.-Verlag, Saarbrücken, Germany

Lewis, R. A., 1987, "Guidelines for Environmental Specimen Banking with Special Reference to the Federal Republic of Germany: Ecological and Managerial Aspects," U.S. Department of the Interior, National Park Service, *U.S. MAB Report, 12,* 182 pp

Lewis, R. A., 1988, "Remarks on the Status of Environmental Specimen Banking in Relation to Health and Environmental Assessment," *11th U.S.–German Seminar of State and Planning on Environmental Specimen Banking*, Bayreuth, Bavaria, Germany, May 1–3

Lewis, R. A., M. Paulus, C. Horras, B. Klein, 1989, Auswahl und Empfehlung von ökologischen Umweltbeobachtungsgebieten in der Bundesrepublik Deutschland, *MaB-Mitt. 29*

Marth, P., K. Oxynos, J. Schmitzer, K.-W. Schramm, and A. Kettrup, 1997, "Levels of Chlorinated Hydrocarbons (CHC) in Breams (*Abramis brama*) from the River Elbe (a Contribution to the Federal Environmental Specimen Bank)," *Chemosphere, 34,* 2183–2192

Müller, P., G. Wagner, M. Paulus, and R. Klein, 1996, "Biological Environmental Specimen Banking as Precondition for Intelligent Environmental Monitoring," BESBE-2, Stockholm, Sweden

Nesmerak, I., 1993, "Kontamination der Elbe aus dem Gebiet der Tschechischen Republik und der Moldau mit organischen Schadstoffen," in *Schadstoffatlas Osteuropa* (E. Heinisch, A. Kettrup, and S. Wenzel-Klein, Eds.), Ecomed-Verlag, Landsberg, Germany, pp. 167–170

Oxynos, K., J. Schmitzer, H. W. Dürbeck, and A. Kettrup, 1992, "Analysis of Chlorinated Hydrocarbons (CHC) in Environmental Samples," in *Specimen Banking*, (M. Rossbach, J. D. Schladot, and P. Ostapczuk, Eds.), Springer, Berlin, Germany, p. 127

Oxynos, K., J. Schmitzer, and A. Kettrup, 1993, "Herring Gull Eggs as Bioindicator for Chlorinated Hydrocarbons," *Sci. Tot. Environ., 139/140,* 387–398

Oxynos, K., K.-W. Schramm, P. Marth, J. Schmitzer, and A. Kettrup, 1995, "Chlorinated Hydrocarbons- (CHC) and PCDD/F-levels in Sediments and Breams (*Abramis brama*) from the River Elbe (Contribution to the German Environmental Specimen Bank)," *Fresenius J. Anal. Chem., 353,* 98–100

Paulus, M., C. Horras, B. Klein, and R. A. Lewis, 1990, "Vertiefte Auswahl von Probenahmeregionen für die Umweltprobenbank und ökologische Beratung zu ihrem Betrieb," Umweltforschungsplan des Bundesministers für Umwelt, Naturschutz und Reaktorsicherheit. Anschlußbericht zum BMU-Forschungsvorhaben 10808001, Saarbrücken, Germany

Paulus, M., R. Klein, M. Zimmer, J. Jacob, and M. Rossbach, 1995, "Die Rolle der biometrischen Probencharakterisierung in der Umweltanalytik am Beispiel der Fichte (*Picea abies*)," in *Biomonitoring und Umweltprobenbank*, Teil VI (M. Paulus, R. Klein, G. Wagner, and P. Müller, Eds.), Beitragsserie in der UWSF – Z. *Umweltchem. Ökotox.*, 7, 236–244

Schwuger, M. J., 1994, "Environmental Specimen Bank of the Federal Republic of Germany – Significance of Surfactants," in *German–Egyptian Seminar on Environmental Research* (H. F. Aly, and D. Nentwich, Eds.), Cairo, Egypt, 21–23 March 1994, Bilateral Seminars of the International Bureau, Forschungszentrum Jülich GmbH, Germany, Vol. 19, 159–194

Schramm, K.-W., A. Kettrup, J. Schmitzer, P. Marth, and K. Oxynos, 1996, "Environmental Specimen Bank – A Useful Tool for Prospective and Retrospective Environmental Monitoring," *TEN*, 3, 43–49

SRU (Rat der Sachverständigen für Umweltfragen), 1987, "Umweltgutachten 1987," Deutscher Bundestag, Drucksache 11/1569 und Verlag Kohlhammer, Stuttgart/ Mainz, Germany

Stoeppler, M., and R. Zeisler, (Eds.), 1993, "Biological Environmental Specimen Banking," *Sci. Tot. Environ.*, BESB special issue, *139/140*

Thalius, J., 1588, *Sylvia hercynia, sive catalogus plantarum sponte nascentium in montibus*, Frankfurt/M, Germany

Umweltbundesamt, (Eds.), 1996, *Federal Environmental Specimen Bank: Standard Operating Procedures for Sampling, Transport, Storing, and Chemical Characterization of Environmental Specimens and Human Organic Specimens* (in German), Erich Schmitt Verlag, Berlin, Germany

UPB, 1996, "Jahresbericht der Bank der Umweltproben 1995," Jülich, Germany

Wagner, G., 1994a, "Environmental Specimen Banking (ESB) in the Federal Republic of Germany – An Instrument for Long-Term Environmental Monitoring, Assessment, and Research," in *Ecoinforma '94* (K. Alef, W. Blum, S. Schwarz, A. Riss, H. Fiedler, and O. Hutzinger, Eds.), 5, Bayreuth, Germany, pp. 457–462

Wagner, G., 1994b, "Biologische Umweltproben," in Stoeppler, M. (eds.), *Probenahme und Aufschluß*, Springer Labormanual, Berlin, Heidelberg, Germany

Wagner, G., 1995, "Basic Approaches and Methods for Quality Assurance and Quality Control in Sample Collection and Storage for Environmental Monitoring," *Sci. Tot. Environ.*, 176, 63–71

Wagner, G., R. Klein, K. Nentwich, M. Paulus, J. Sprengart, R. Wüst and P. Müller, 1996, "Umweltprobenbank des Bundes: Beiträge zur Probenahme und Probenbeschreibung," Jahresbericht 1995, Saarbrücken 1996, Germany

Wise, S. A., and R. Zeisler, 1984, "The Pilot Environmental Specimen Bank Program," *Environ. Sci. Technol.*, 18, 302A–307A

Wise, S. A. R. Zeisler, and G. M. Goldstein, (Eds.), 1988, "Progress in Environmental Specimen Banking," NBS Special Publication 740, U.S. Dep. of commerce, U.S. Government Printing Office, Washington, USA

Wittig, R., 1993, "General Aspects of Biomonitoring Heavy Metals by Plants," in *Plants as Biomonitors – Indicators for Heavy Metals in the Terrestrial Environment*, (B. Markert, Ed.), VCH, Weinheim, Germany, pp. 3–27

PART 3

Bioaccumulation and Biological Effects of Chemicals

14

Bioaccumulation of Chemicals by Aquatic Organisms

Des W. Connell (*Nathan, Australia*)

14.1 SUMMARY

The bioaccumulation of persistent organic chemicals by aquatic organisms can occur from food or the ambient water, but the latter source is generally dominant in most situations. Bioaccumulation of chemicals from water, described as bioconcentration, occurs by passive diffusion from the ambient water across the gills into the circulatory fluid to be deposited in lipid tissues. The chemicals most susceptible to bioaccumulation are the chlorohydrocarbons and polyaromatic hydrocarbons with $\log K_{ow}$ values between 2 and 6.5. Bioconcentration can be characterized by the bioconcentration factor (concentration in biota/concentration in water at equilibrium). Octanol provides a reasonable surrogate for biota lipid in most situations; thus, bioconcentration behavior with aquatic organisms can generally be predicted by relationships with the octanol/water partition coefficient. This can be extended to some aquatic infauna, particularly aquatic worms, utilizing a three-phase model of bioconcentration. Bioaccumulation in the

Ecotoxicology, Edited by Gerrit Schüürmann and Bernd Markert.
ISBN 0-471-17644-3   © 1998 John Wiley & Sons, Inc. and Spektrum Akademischer Verlag.

natural environment is influenced by a variety of factors such as patterns of migration and other biological factors, but the laboratory-based models provide a valuable basis for interpretation of observed bioaccumulation behavior.

14.2 INTRODUCTION

Early in the history of investigation of hazardous chemicals in the environment it was recognized that some chemicals could be taken up by organisms, particularly aquatic organisms, and retained, leading to much higher biotic than abiotic concentrations (Connell 1990). Since that time a good understanding of many of the factors governing bioaccumulation has been developed and methods to predict bioaccumulation in aquatic systems devised.

A wide range of different classes of chemicals can be discharged to the aquatic environment in industrial discharges, sewage, and stormwater run-off. When these substances enter the aquatic environment a variety of processes occur which result in their distribution to different environmental phases and the exposure of biota (Figure 14.1). It is important to note that a set of abiotic partitioning processes are developed which are shown in Figure 14.1: the water/sediment process, the air/water process, and the air/soil process. The water/sediment process can be characterized by the partition coefficient, K_D,

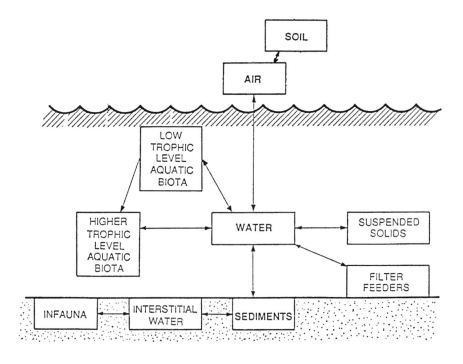

Figure 14.1 Diagrammatic representation of the distribution of a chemical in an aquatic system.

and in terms of organic carbon, K_{OC} (Karickhoff 1985). The air/water process can be characterized by the Henry's Law Constant (H) which is the air/water partition coefficient. The air/soil process involves a terrestrial phase and can be interpreted in terms of partition mechanisms.

Current evidence suggests that some aquatic biota bioaccumulate compounds as a result of direct partitioning with ambient water, which is usually described by the term *bioconcentration*. For example, in Figure 14.1 the aquatic system includes biota/water partitioning, filter feeders/water partitioning, and infauna/interstitial water partitioning which can usually be considered to be a direct partition process involving three phases. These systems would then be expected to be characterized by partition coefficients which could be used to interpret the bioconcentration process. On the other hand the transfer of chemicals from low trophic level aquatic biota to higher trophic level biota cannot be considered to be a simple direct partitioning process, although partitioning would most likely be involved. This transfer occurs during the consumption of low trophic level aquatic biota as food by higher trophic level aquatic biota and is usually described by the term *biomagnification*. An example of this would be the consumption of phytoplankton by some species of fish. In many situations the means whereby an organism acquires a chemical is not known and the term *bioaccumulation* is used.

This chapter briefly reviews the mechanisms and processes affecting the bio-accumulation of chemicals by aquatic organisms.

14.3 SIGNIFICANCE OF BIOMAGNIFICATION IN AQUATIC SYSTEMS

Both bioconcentration and biomagnification must operate with aquatic organisms (Connell 1990). Air-breathing aquatic mammals, such as whales, dolphins, etc. lack an organism/water exchange interface, since exchange cannot occur through the skin; thus, biomagnification is the only mechanism involved. In contrast, autotrophic organisms, such as phytoplankton and some bacteria, draw their food, as well as other chemical components, directly from dissolved substances in the ambient water. With these organisms bioconcentration is the only possible mechanism. With other aquatic organisms current evidence suggests that bioconcentration is the dominant mechanism of bioaccumulation. Thus, in the remainder of this chapter attention is focussed principally on bioconcentration.

14.4 INFLUENCE OF TYPE OF COMPOUND ON BIOCONCENTRATION

A compound should persist for a significant period to allow bioconcentration to occur. The level of persistence must allow the concentration of the chemical to increase above the level in the ambient environment. Generally persistence has been evaluated by experiments in soil, and the resulting data suggest that soil persistence in the order of years is required with a compound. If a compound is subject to biodegradation then the bioconcentration would be expected to be reduced in accord with the amount of loss of compound which results. Thus, for example, alkanes exhibit little bioaccumulation despite having suitable physicochemical properties because they lack persistence in biota. The group with the most suitable persistence is the chlorohydrocarbon group and to a lesser extent the group of polyaromatic hydrocarbons.

Bioconcentration is related to the octanol/water partition coefficient (K_{OW}) with compounds where the properties of n-octanol resemble those of biota lipid. Generally n-octanol is a reasonable match for lipid for compounds having $\log K_{OW}$ values, from 2 to about 6.5, and these compounds are referred to as lipophilic compounds. However, the different chemical nature of octanol and biota lipid may result in differences in bioconcentration behavior, even with compounds having $\log K_{OW}$ between 2 and 6.5. This could be due to the presence of specific active chemical groups in the test compounds which interact with octanol and lipid in different ways. Compounds which are ionized exhibit low bioaccumulation in the ionic form. Molecular dimensions may also influence the bioaccumulation process. In general the chlorohydrocarbons and polyaromatic hydrocarbons are relatively neutral compounds and exhibit the most consistent behavior.

14.5 THE MECHANISM OF BIOCONCENTRATION AND BIOMAGNIFICATION

Baughman and Paris (1981) have reviewed the available information on bioconcentration and concluded that bioconcentration of lipophilic compounds by microorganisms occurs as a result of partitioning between water and the microorganism. Various authors, e.g., Kerr and Vass (1973) and Sondergren (1968), have shown that lipophilic compounds adsorb onto the outer surface and then diffuse internally within the cell. Mortimer and Connell (1993) investigated bioconcentration by juvenile crabs and suggested that passive diffusion from water is the process involved with these organisms. With fish, a number of authors have shown that gills are the site of uptake and partitioning of lipophilic compounds, as reviewed by Connell (1988). The evidence indicates that the route of uptake and loss of lipophilic compounds with aquatic organisms is generally through the oxygen uptake route. The gills are the primary site; then partitioning with the circulatory fluid occurs, resulting in deposition of the lipophilic compound in biota lipid. Subsequently, the compound can be metabolized, generally to more oxygenated and water-soluble forms, and excreted as indicated in Figure 14.2. The common mechanism of uptake and loss through the oxygen pathway suggests that all aquatic organisms can be treated similarly as

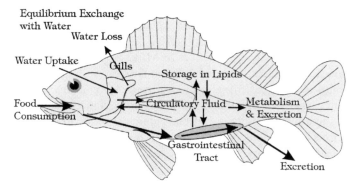

Figure 14.2 Diagrammatic representation of the routes of uptake and clearance of a lipophilic chemical by fish.

regards the water/organism partition process. This indicates that biota fat/water partitioning is effectively the dominant factor in the physical process resulting in the uptake and accumulation of lipophilic compounds from water by aquatic organisms (Mackay 1982).

Biomagnification is a more complex process but can be diagrammatically represented in a simplified form as shown in Figure 14.2. In this process lipophilic contaminants in food can be seen as partitioning with circulatory fluid and being deposited in lipid tissues by routes similar to the bioconcentration process.

14.6 INFLUENCE OF SOME BIOLOGICAL CHARACTERISTICS ON BIOCONCENTRATION

The biotic lipid phase is the dominant concentrating phase for lipophilic compounds and the aqueous and other phases are of little significance with aquatic organisms (Mackay 1982). Thus the lipid content of an organism is a major factor influencing the amount of chemical bioconcentrated from water. The concentration of bioconcentrated chemical would be expected to be in direct proportion to the lipid content. Organisms and groups of organisms exhibiting low lipid content exhibit correspondingly low bioconcentration capacity.

Apart from the physical process outlined above, the persistence of a chemical in an organism is a major factor influencing bioconcentration as previously discussed. The uptake of lipophilic compounds by biota usually results in the induction of mixed function oxidase (MFO) enzyme systems in the exposed biota. The MFO systems stimulate oxidation of lipophilic compounds, leading to their removal from the organism. Not all organisms have the same capacity to respond to exposure to lipophilic compounds in this way. In fact aquatic organisms can show considerable variation in MFO activity between species (e.g., Connell and Miller 1981). This may relate to particular metabolic characteristics of the species or to factors such as previous exposure patterns to lipophilic compounds. Such differences in MFO activity may lead to different bioaccumulation characteristics for different species.

The habitat of an aquatic organism, or group of organisms, has an effect on bioaccumulation characteristics. Of particular importance are aquatic infauna, which are fauna which reside in bottom sediments in aquatic areas, including aquatic worms and various other organisms. Bioaccumulation with this group can be best understood by a three-phase model as outlined in Figure 14.1 for sedimentary infauna. This indicates that there are two partition processes involved in bioaccumulation within the sedimentary system. Compounds are released from the sediment by the sediment/interstitial water partitioning system and then there is uptake and concentration by the interstitial water/infauna partition process. The interstitial water/infauna process is bioconcentration, which parallels the water/fish process and would be expected to exhibit similar characteristics.

14.7 QUANTITATIVE STRUCTURE–ACTIVITY RELATIONSHIPS FOR BIOCONCENTRATION

Bioconcentration is characterized by the bioconcentration factor (K_B), which is the ratio of the concentration of the compound in biota (C_B) to the concentration in water (C_W). The process is complex but largely governed by the biota lipid/water partition process, and as a result it has been found that the most useful characteristic for the prediction of the bioconcentration factor is the n-octanol/water partition coefficient (K_{OW}). Despite the simplicity of the octanol/water system as compared with the bioconcentration process, the K_{OW} value provides a reasonably good estimate of the bioconcentration capacity of lipophilic compounds within certain limits. The success of the system is largely dependent on the similarity of octanol to biota lipid since water is common to both systems. Other parameters, including aqueous solubility and Randic Indices, have also been used to develop bioconcentration quantitative structure–activity relationships (QSARs), but usually these are less successful.

Compounds which have $\log K_{OW}$ values less than 2 usually bioconcentrate more than would be expected from the K_{OW} values. This is a result of nonlipoid tissue becoming increasingly important due to the decreasing lipophilicity and increasing water solubility of these compounds. On the other hand compounds with $\log K_{OW}$ greater than 6.5 bioconcentrate to a lesser extent than would be expected from the K_{OW} values. With these compounds the biota lipid and octanol differ in their solubility properties, resulting in a reduction in the solubility of these compounds in lipid and a reduced bioconcentration factor. Other properties, such as the size of the molecule, may also be factors causing reduced bioconcentration.

A QSAR for bioconcentration can be derived as follows. If octanol is a perfect surrogate for biota lipid, then

$$K_B = K_{OW} y_L \tag{14.1}$$

and

$$\log K_{OW} = 1 \log K_{OW} + \log y_L \tag{14.2}$$

where y_L is the fraction of lipid present in the biota.

The empirical results obtained from bioconcentration experiments are usually expressed in the following general form:

$$\log K_B = a \log K_{OW} + b \tag{14.3}$$

and

$$K_B = 10^b K_{OW}^a \tag{14.4}$$

where a and b are empirical constants.

Based on the outline above, constant a is an empirical constant expressing the nonlinearity of the relationship and indicates how well octanol represents the biota lipid, and $b = \log y_L$. This indicates that under perfect conditions y_L is always less than unity, which means that constant b should always be negative. Some of the empirical relationships, generally derived from experiments done in aquaria, are shown in Table 14.1.

There is a large volume of data available on fish which is reflected by the data reported in Table 14.1. Mackay (1982), Davies and Dobbs (1984), Connell and Hawker (1988), Connell and Schuurmann (1988), and Schuurmann and Klein (1988) collected and collated sets of data and in some cases evaluated the accuracy and application of this material to the bioconcentration relationships. These relationships in most cases were derived from the use of chlorohydrocarbons and polyaromatic hydrocarbons. Kenaga and Goering (1980) reported that the lipid content of fish ranges from 1% to about 16% depending on a variety of factors, which is in accord with the lipid equivalent to constant b shown in Table 14.1. The results suggest that with effectively nonbiodegradable compounds, principally the chlorohydrocarbons and polyaromatic hydrocarbons, octanol provides a reasonable representation of biota lipid for compounds with $\log K_{OW}$ values between 2 and 6.5.

14.8 KINETICS OF BIOCONCENTRATION

A theoretical treatment of the kinetics of uptake and clearance of environmental contaminants was developed by Moriarty (1975) based on models developed in

Table 14.1 Characteristics of some relationships between $\log K_B$ and $\log K_{OW}$ for various biota

Biota	Constant a	b	Lipid equivalent to constant b (% wet weight)	Actual lipid (% wet weight)	Range of $\log K_{OW}$
Microorganisms	0.91	−0.36	44	n.a.	3–7
Daphnids	0.90	−1.32	4.8	n.a.	2–8
Polychaetes and oligochaetes	0.99	−0.60	25	n.a.	4–8
Fish	0.94	−1.00	10	1–16	3–6
Fish	0.98	−1.36	4.3	1–16	1.5–6.5
Fish	1.00	−1.32	4.8	1–16	0.5–6.0
Fish	0.95	−1.06	8.8	1–16	2–6
Molluscs	0.84	−1.23	5.9	1.2–1.8	3.5–8

Source: Connell (1990).
Note: $\log K_B = a \log K_{OW} + b$.
n.a., Data not available.

pharmacology. The most useful model for the kinetics of bioconcentration is based on the single-compartment approach utilizing a single compartment to represent an organism. Moriarty (1975) has also used a two-compartment system, with one compartment to represent a peripheral system and the other, a central system within an organism. Applied to available data he found that this gave a better explanation for the kinetics of bioconcentration than the single-compartment approach, but the increase in complexity makes this approach more difficult to apply effectively.

If bioconcentration is dominated by exchange through the gills, then the food route can be disregarded. Thus, bioconcentration can be seen as the balance between water uptake and water loss as shown in Figure 14.2. While more complex processes follow exchange through the gills the system is represented by the water/lipid process. This process proceeds by first-order kinetics characterized by the rate constants k_1 and k_2. Clearance is a physical process due to the reverse movement of molecules as a result of the concentration of compound in the organism. If the compound involved is lipophilic and nondegradable then metabolism and excretion can be regarded as negligible and the rate of increase in concentration in biota is expressed by rate of change in biotic concentration = rate of uptake − rate of clearance. Thus

$$\frac{dC_B}{dt} = k_1 C_W - k_2 C_B \tag{14.5}$$

where C_B is concentration in the biota and C_W, the concentration in water.

Since the amount of compound in the water represents a large reservoir compared with the relatively low amount that can be taken up by biota, C_W can

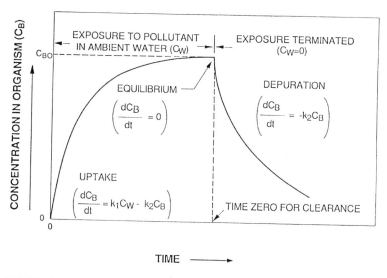

Figure 14.3 Uptake and clearance patterns of a lipophilic compound over time in the bioconcentration process according to first-order kinetics.

be regarded as constant in any particular situation. By integration and rearrangement of the Equation 14.5

$$C_B = k_1/k_2 C_W (1 - e^{-k_2 t}) \tag{14.6}$$

This predicts that C_B will exhibit an increase in concentration with time but with a declining rate of increase as shown in Figure 14.3. Thus, t continues to increase until $e^{-k_2 t}$ is effectively zero and the C_B curve is effectively parallel to the time axis. At this time

$$C_B = (k_1/k_2) C_W \tag{14.7}$$

and

$$\frac{C_B}{C_W} = \frac{k_1}{k_2} = K_B$$

If exposure to the compound is terminated, for example by transfer to uncontaminated water, then $C_W = 0$ and so $k_1 C_W = 0$ and

$$\frac{dC_B}{dt} = -k_2 C_B \tag{14.8}$$

Thus, while before both uptake and clearance were operating, now uptake does not occur and only clearance is in operation. By integration and rearrangement

$$C_B = C_{BO}e^{-k_2 t} \quad \text{and} \quad \ln C_B = \ln C_{BO} - k_2 t \qquad (14.9)$$

Thus, applying first-order kinetics, the persistence of a compound in an organism can be characterized as a half-life which is due to the physical loss of compound and not by biodegradation and excretion. If these processes are significant then a different half-life, most likely considerably shorter, would be in effect.

In the previous section the relationship between $\log K_B$ and $\log K_{OW}$ for aquatic organisms was described and takes the general form of Equation 14.3. Since $k_1/k_2 = K_B$ from Equation 14.7 then

$$\log(k_1/k_2) = a \log K_{OW} + b \qquad (14.10)$$

This means that the ratio of the uptake and clearance rate constants is proportional to the $\log K_{OW}$ value and suggests that there may be a possible relationship between the rate constants individually and $\log K_{OW}$. In fact, Hawker and Connell (1986) have investigated the data on fish, molluscs, and daphnids and found relationships of the following general form

$$\log(1/k_2) = x \log K_{OW} + y \qquad (14.11)$$

and it can be shown that

$$\log k_1 = (a - x) \log K_{OW} + (b - y) \qquad (14.12)$$

However, it is important to note that these relationships have been established over the range of $\log K_{OW}$ values from about 2 to about 6.5 and may not be applicable outside this.

Since the $\log K_B$ to $\log K_{OW}$ relationship requires that equilibrium be established it is important to evaluate the time period needed to reach this stage. The theoretical time period to reach equilibrium occurs when $e^{-k_2 t} = 0$ and thus when t is infinity. However, effective equilibrium is reached at t_{eq}, when C_B is 0.99 of the C_B at infinity. Thus from Equation 14.6 it can be shown that

$$t_{eq} = 4.605/k_2 \qquad (14.13)$$

Using the appropriate values for k_2 the time to establish effective equilibrium can be calculated.

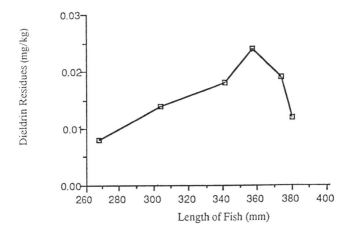

Figure 14.4 Plot of the mean concentration of dieldrin in Clarence River (Australia) mullet against fish length.

14.9 BIOACCUMULATION IN THE NATURAL ENVIRONMENT

The QSARs and other relationships have been substantially based on results obtained from controlled experiments conducted in the laboratory. In the natural environment a range of factors can operate which may lead to variations from the laboratory situation. A factor of particular importance is the movement pattern of fish. Some fish have seasonal migrations from rivers, where exposure to contaminants occurs, to the open sea where there is clearance of bioconcentrated compounds. Examples of this effect have been reported by Cullen and Connell (1992), who found that the concentrations of chlorohydrocarbon pesticides in the sea mullet (*Mugil cephalus*) and the whiting (*Sillago ciliata*) increased with age until maturity when migration to the open sea occurred with a resultant fall in contaminant concentration (Figure 14.4). A range of other biological and ecological factors may need to be taken into account to provide a reliable evaluation of bioconcentration in the natural environment.

14.10 CONCLUSIONS

Bioaccumulation occurs with persistent lipophilic organic compounds and can result from the uptake and retention of contaminants in food and water. However, the pathway from water seems to be dominant in most situations. It is likely that the QSARs established with fish can be extended to other aquatic organisms. However, caution must be exercised in applying these relationships since they are most applicable with the chlorohydrocarbons and deviations may occur with other substances. Also, these relationships depend on a lack of biodegradation within the organism during the bioaccumulation process. The least biodegradable compounds are the chlorohydrocarbons, but biodegradation may occur to different extents with different organisms since these may have differing capacities to carry out this process. As a result, different organisms may possibly exhibit different capacities to bioaccumulate various compounds.

14.11 REFERENCES

Baughman, G. L., and D. F. Paris, 1981, "Microbial Bioconcentration of Organic Pollutants from Aquatic Systems – a Critical Review," *CRC Crit. Rev. Microbiol.*, 205–227

Connell, D. W., 1988, "Bioaccumulation Behaviour of Persistent Organic Chemicals with Aquatic Organisms," *Rev. Environ. Contam. Toxicol.*, *101*, 117–154

Connell, D. W., 1990, *Bioaccumulation of Xenbiotic Compounds*, CRC Press Inc., Boca Raton, FL, USA

Connell, D. W., and D. W. Hawker, 1988, "Use of Polynomial Expressions to Describe the Bioconcentration of Hydrophobic Chemicals by Fish," *Ecotoxicol. Environ. Safety*, *16*, 242–257

Connell, D. W., and G. J. Miller, 1981, "Petroleum Hydrocarbons in Aquatic Ecosystems – Behaviour and Effects of Sub-lethal Concentrations," *CRC Crit. Rev. Environ. Control*, *11*, 37–104

Connell, D. W., and G. Schüürmann, 1988, "Evaluation of Various Molecular Parameters as Predictors of Bioconcentration in Fish," *Ecotoxicol. Environ. Safety*, *15*, 324–335

Cullen, M. C., and D. W. Connell, D. W. 1992, "Bioaccumulation of Chlorohydrocarbon Pesticides by Fish in the Natural Environment," *Chemosphere*, *25*, 1579–1587

Davies, R. P., and A. J. Dobbs, 1984, "The Prediction of Bioconcentration in Fish," *Water Res.*, *18*, 1253–1262

Harding, G. C. H., and W. P. Vass, 1978, "Uptake from Seawater of DDT by Marine Planktonic Crustacea," *J. Fish. Res. Bd. Can.*, *36*, 247–254

Hawker, D. W., and D. W. Connell, 1986, "Bioconcentration of Lipophilic Compounds by Aquatic Organisms," *Ecotoxicol. Environ. Safety*, *11*, 184–197

Kerr, S. R., and W. P. Vass, 1973, "Pesticide Residues in Aquatic Invertebrates," in *Environmental Pollution by Pesticides*, (C. A. Edwards, Ed.), Plenum Press, London, UK, pp. 134–180

Mackay, D., 1982, "Correlation of Bioconcentration Factors," *Environ. Sci. Technol.*, *16*, 274–278

Moriarty, F. 1975, "Exposure and Residues," in *Organochlorine Insecticides: Persistent Organic Pollutants* (F. Moriarty, Ed.), Academic Press, London, UK, pp. 29–72

Mortimer, M. R., and D. W. Connell, 1993, "Bioconcentration Factors and Kinetics of Chlorobenzenes in a juvenile crab (*Portunus pelagicus*(L))," *Aust. J. of Mar. Freshwater Res.*, *44*, 565–576

Schüürmann, G., and W. Klein, 1988, "Advances in Bioconcentration Prediction," *Chemosphere*, *17*, 1551–1559

Sondegren, A., 1968, "Uptake and Accumulation of DDT by *Chlorella* sp.," *Oikos.*, *19*, 126–134

15

Metal Bioaccumulation in Freshwater Systems: Experimental Study of the Actions and Interactions Between Abiotic and Contamination Factors

Alain Boudou, Béatrice Inza, Sylviane Lemaire-Gony, Régine Maury-Brachet, Muriel Odin, and Francis Ribeyre (Talence, France)

15.1 SUMMARY

In natural conditions, abiotic factors and their quasi-permanent variations, both in time and space, strongly influence the chemical fate of metals and their bioavailability, which control accessibility to the biological barriers at the interface between organisms and the external medium and also uptake across these structures (epithelia, cytoplasmic membranes). This chapter gives several examples of results from the contamination of indoor microcosms with cadmium and two mercury compounds (inorganic Hg and methyl-

Ecotoxicology, Edited by Gerrit Schüürmann and Bernd Markert.
ISBN 0-471-17644-3   © 1998 John Wiley & Sons, Inc. and Spektrum Akademischer Verlag.

mercury). The ecotoxicological models are based on a mixed biotope: water column and natural sediment. Two species were selected to illustrate both the methodological bases and the principal types of results obtained: the nymphs of the mayfly *Hexagenia rigida*, a burrowing species, and a bivalve mollusc, the Asiatic clam *Corbicula fluminea*. The experimental approach is based on two principal steps: an exploratory phase, in order to reveal and quantify the actions of and interactions between a large number of ecotoxicological factors using experimental factorial designs; and an in-depth or mechanistic phase, based on a limited number of conditions, selected according to their relevance for the research objectives, where analysis levels and criteria are diversified.

15.2 INTRODUCTION

Within aquatic systems, the bioaccumulation of trace metals results from the actions of and interactions between three sets of factors: contamination factors, relating to the exposure conditions from the direct and/or trophic route; abiotic factors, corresponding to the physicochemical characteristics of the biotopes (water column and sediments); biotic factors, representing the structural and functional specificities of the living organisms (Boudou and Ribeyre 1989). Analyzed at the uni- or multicellular organism level, the general scheme for metal bioaccumulation is based on several basic components: uptake mechanisms, involving adsorption and absorption processes at the biological barrier level (cell membranes, epithelial structures, i.e., gills, gut wall, integument); distribution to internal compartments (cell organelles; organs and tissues) as a result of metal transport within the cell or the organism, via the circulatory system, in combination with storage or sequestration in these compartments (metal binding proteins, intracellular granules, etc.); metal release, via a large variety of more or less specialized mechanisms, such as renal and digestive pathways or molting processes.

From an ecotoxicological point of view, metal bioavailability takes on a fundamental importance with respect to bioaccumulation mechanisms. Closely related to metal partitioning in the surrounding environment, bioavailability depends on the biogeochemical processes which control metal accessibility to biological barriers at the interface between living organisms and the external medium and also on metal uptake across these biological structures (Figure 15.1).

Abiotic factors and their quasi-permanent variations both in time and space strongly influence the chemical fate of contaminants in natural conditions. Thus, for example, the complexation reactions between trace metals and inorganic and organic ligands within the dissolved and particulate phases in the water column and sediment are controlled by the physicochemical characteristics of the medium: temperature, pH, pCl, turbidity, water hardness, dissolved organics, etc. Suspended particles are characterized by a very high fixation capacity; they therefore contribute to metal transport in running waters and/or to metal accumulation in sediments, with extremely long residence times. Several mass

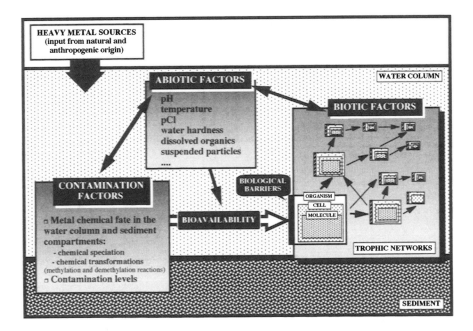

Figure 15.1 Ecotoxicological approach to investigate metal bioaccumulation mechanisms at the ecosystem level—actions of and interactions between the three fundamental sets of ecotoxicological factors: contamination factors, abiotic factors, and biotic factors.

balance studies which take into account the different abiotic and biotic compartments of freshwater ecosystems have shown that more than 90% of the metal burden, indeed 95%, is localized in the superficial sediment (Campbell et al. 1988; Kudo 1989; Watras et al. 1994). For many metals, chemical speciation reactions within the dissolved phase are able to generate a very high diversity of chemical species with a range of different properties, such as their capacity to bind to membrane sites or their absorption ability via the different transport routes through the biological barriers (Bernhard et al. 1986). This is the case with, for example, inorganic mercury (HgII) and methylmercury (MeHg), which are able to produce a large set of anionic, cationic, and neutral chemical species, depending on pH and pCl conditions (Figure 15.2). For MeHg, the two species CH_3HgCl and CH_3HgOH are characterized by different octanol/water distribution coefficients, 1.7 and 0.2, respectively, leading to marked differences between their diffusion fluxes through the phospholipidic bilayers of the cell membranes (Bienvenue et al. 1984; Faust 1992). The presence of these chemical species and their relative abundance can be greatly modified by small variations in the physicochemical characteristics of the medium via chemical equilibrium shifts. From an ecotoxicological point of view, the bioavailability of trace metals and their bioaccumulation capacities depend on the microenvironment characteristics at the sites of uptake, where the nature and properties of the ligands and

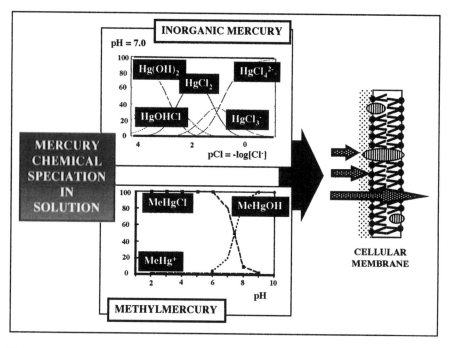

Figure 15.2 Chemical speciation of inorganic mercury and methylmercury in solution, according to pH and pCl variations, and fundamental links with metal accessibility to the cellular membrane interface, binding to proteins and phospholipid polar heads, and transport through the hydrophobic barrier.

the reaction conditions can differ greatly from those in the surrounding water, even when this is at a distance of only a few millimeters, or even a few micrometers. This is true, for example, in the case of the increase in pH in the branchial microenvironment in rainbow trout, attributable to the expired CO_2 and ammonia, which contrasts with the acidic condition of the medium. These local changes on the gill lamella surface are able to induce formation of metal precipitates, leading to marked respiratory and ionoregulatory impairment (Roesijadi and Robinson 1994). Similar processes occur within the sediment compartment, in relation to the biogeochemical gradients in the oxic and anoxic layers, and to the heterogeneity of the components of the particulate phase. Considering the large quantities of metal stored in the superficial sediment layers, chemical speciation processes play a fundamental role in this compartment with regard to the transfer potentialities towards the benthic organisms and also to the water column via release processes (diffusion, bioturbation, etc.; Tessier and Campbell 1988).

Abiotic factors act simultaneously on living organisms. Depending on the adaptative capacities of the individuals, they can induce more or less severe structural and functional disturbances. Among these factors, temperature is significant, given the heterothermal natural of the individuals. Thus, temperature changes in the external medium lead very rapidly to a modification of the internal temperature, which can, according to the degree of thermal stress, induce adaptive responses on a global scale (e.g., respiratory and circulatory functions) and at cellular and molecular levels (enzymatic activities, membrane fluidity, etc.). Several other physicochemical factors, within their variation ranges of natural or anthropogenic origin, can act directly or indirectly on living organisms, e.g., dissolved oxygen, pH, and water hardness.

Lastly, biotic factors contribute to the heterogeneity of ecotoxicological responses to a homogeneous contamination source. Interspecies differences within aquatic biocenoses sometimes lead to very wide differences in bioaccumulation capacities, from the direct uptake route via the surrounding medium, or trophic transfers, via ingested prey. These differences are due to specific features, e.g., anatomical, physiological, biochemical, and behavioral; to the position of the species within the trophic networks (primary producers, herbivores, carnivores, detritivores, etc.); to the macro- and microhabitats (pelagic or benthic species); etc. The fundamental biology and ecology of most aquatic species are poorly understood: this can considerably limit the extent of interpretative analysis possible of the results of *in situ* studies or experimental approaches. Similarly, at the inter-individual level, the genetic heterogeneity within populations and differences due to other factors, such as age or development stage, can lead to marked differences both in bioaccumulation capacities and in toxicological effects.

Thus, whatever the biological integration level studied, bioaccumulation mechanisms result from the actions of and interactions between a very large set of ecotoxicological factors, which necessarily lead to a very considerable qualitative and quantitative variability. *In situ* studies are confronted with the very great complexity of these phenomena. Data obtained at the field level provide a more or less detailed picture of the contamination levels of abiotic and biotic compartments and of their evolution in the short, medium or long term. These data are a primordial representation of "field reality", but often they cannot be used to analyze and quantify the direct and indirect effects of ecotoxicological factors, owing to their extreme diversity, their variations, and interactions. They must be complemented by experimental studies, set up under controlled conditions. The common objective between these different approaches is to increase knowledge in order to advance and assist interpretation.

In this chapter, we present our experimental study of mercury compound—inorganic Hg (HgII) and methylmercury (MeHg)—and cadmium (Cd) bioaccumulation and transfer within freshwater systems. The methodological approach is based on indoor microcosms, consisting of a mixed biotope—water column and natural sediment—and one or several aquatic species occupying key positions in relation to the metal transfer routes from the two initial contamination sources: water column or sediment. Two biological models were selected to illustrate both the methodological bases and the principal types of results obtained: nymphs of the mayfly *Hexagenia rigida*, a burrowing species living in the superficial sediment layers and ingesting large quantities of substrate; and a bivalve mollusc, the Asiatic clam *Corbicula fluminea*, which colonizes the first centimeters of sediment in lotic or lentic conditions and filters large volumes of water for respiratory and nutritive purposes.

In this experimental approach priority is given to the study of ecotoxicological factors, which are considered separately and in interaction. Thus, with adapted and automatized equipment, the protocols enable us to take into account simultaneously a large number of contamination factors—metal chemical forms and species, initial contamination sources, contamination levels, exposure duration, etc.—and abiotic factors, i.e., temperature, pH, photoperiod, sediment characteristics, etc. The comparative analysis of several, even hundreds, of ecotoxicological conditions enables us during the exploratory phases to identify and quantify the actions of and interactions between the ecotoxicological factors. A second experimental phase, using a limited number of conditions selected according to their relevance with regard to the research objectives, consists of a more in-depth analysis of the mechanisms involved—a vertical approach—by diversifying the analysis levels and criteria.

15.3 METHODOLOGICAL BASES

The indoor microcosms consist of glass tanks ($12 \times 12 \times 30$ cm) lined with a plastic bag (Plastiluz, alimentary standard) and containing a 5-cm-deep sediment compartment, 2.9 L of dechlorinated tap water, and four molluscs or four mayfly nymphs (Figure 15.3).

The sediment was collected from the banks of the Garonne river, upstream from Bordeaux, France. It is a homogeneous silt, rich in clays (75%–80%), with an average 2% organic carbon content. Wet weight (ww)/dry weight (48 h desiccation at 60°C) ratio was 2.1 ± 0.2. Background metal concentrations in the sediment were 140 ± 15 µg Hg/g and 480 ± 32 µg Cd/g. This substrate was directly used in the experimental units (EUs) containing the *H. rigida* nymphs; for *C. fluminea*, this sediment was blended with pure sand (98% silica; granulometry: 0.8–1.4 mm; SILAQ, France) in a 50/50 (ww basis) mixture and homogenized by mechanical mixing in order to obtain a substrate well adapted to the burrowing of molluscs.

The general chemistry of the tap water is: pH = 7.5; resistivity = 2470 ohm/cm; $HCO_3 = 232$ mg/L; $Cl = 16.0$ mg/L; $SO_4 = 37.5$ mg/L; $Ca = 53.6$ mg/L; $Mg = 12.2$ mg/L; $NH_4 < 0.01$ mg/L; $NO_2 = 0.09$ mg/L; $NO_3 = 1.8$ mg/L; $PO_4 < 0.05$ mg/L.

The EUs were contaminated from the water column or the sediment source. For the water source, the procedure selected is based on daily additions of mercury or cadmium to the water column from concentrated aqueous solutions, in order to compensate for the decrease in metal concentrations in this compartment due to adsorption on the tank walls, transfers to the sediment superficial

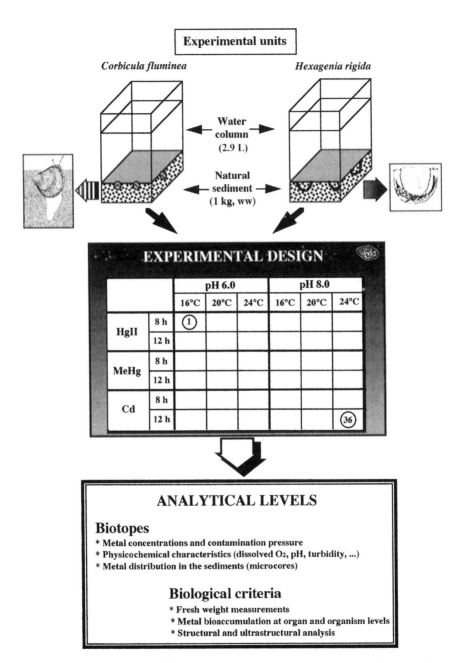

Figure 15.3 Methodological bases of the experimental approach to investigate metal bioaccumulation and transfers between abiotic and biotic compartments of indoor microcosms—quantification of the actions of and interactions between the ecotoxicological factors.

layers, bioaccumulation in the organisms, and volatilization. Two strategies were defined, according to the objectives of the experiment and to the material constraints imposed by the number of EUs set up simultaneously (Ribeyre 1991):

(1) Daily additions of identical amounts of metal to the EUs during the whole experiment, whatever the abiotic conditions studied. These additions are generally similar to the initial additions corresponding to the nominal concentrations selected. In this case, all of the EUs corresponding to the same contamination condition received the same amount of metal at the end of the experiment. However, due to the direct and indirect effects of the abiotic or biotic factors studied, the contamination pressures can differ widely; there are differences, sometimes very marked, in the evolution of the metal concentrations in the water column (Figure 15.10). In order to take into account these differences during the analysis of the bioaccumulation results, metal determinations were carried out periodically on water samples collected from the EUs. From these a global index was calculated for each experimental condition, which is representative of the contamination pressure and called "concentration-days equivalent" (CDE; Ribeyre 1993):

$$\text{CDE} \ (\mu g \, L^{-1} \, day) = [(C_1 + C_0)/2](t_1 - t_0) + [(C_2 + C_1)/2](t_2 - t_1)$$
$$+ \cdots + [(C_j + C_i)/2](t_j - t_i)$$

Using this index, theoretical bioaccumulation values are calculated by dividing the mean metal concentrations measured in the organisms by the corresponding CDEs. These correlated values enable us to compare bioaccumulation capacities for similar exposure conditions; this estimate presupposes a strict proportionality between the metal concentrations in the medium and those measured in the organisms.

(2) Metal additions in the water column adjusted according to the decrease in the concentrations between the daily additions. The decrease in metal concentration was determined in water samples collected from each EU at the end of 24-h cycles. This procedure enabled a relatively constant contamination pressure to be maintained during the exposure period; nevertheless, the amounts of metal added sometimes strongly differed between the EUs, according to the direct and indirect effects of the abiotic conditions on the metal decrease between the cyclical additions and/or of the structure of the experimental systems.

   The contamination of the sediment source was achieved by metal additions from concentrated aqueous solutions. Homogeneity of the metal distribution was obtained by mechanical mixing; samples were taken from each sediment batch to check contamination levels and metal distribution throughout the substratum.

The organisms were introduced into the EUs 10 days after the water and sediment had been added. This amount of time is needed to allow the physicochemical conditions to stabilize.

C. fluminea specimens were collected from a sampling station on the banks of the Lac de Sanguinet (Gironde, France). This bivalve currently exhibits very strong invasive dynamics in rivers and lakes in the southwest of France; its presence in Europe was first noted in 1981 in the Dordogne river, near the Gironde estuary (France) and in Portugal (Araujo et al. 1993). Note that this species, which originates from Asia, was introduced into North America some-time prior to 1938, it has subsequently spread to every major river basin south of 40°N latitude and has become a major component of the benthic communities and a pest to industrial and domestic water supply systems (Counts 1986; Britton and Morton 1982). Large batches of molluscs are collected *in situ* and stored in the laboratory for several weeks; they are fed with periodic additions of phytoplankton algae (Foe and Knight 1986). Screening based on shell measure-ments produces homogeneous populations for the experiments (average length of shell: 1.5–2 cm). Like many filter-feeding bivalves, *C. fluminea* lives buried in the superficial sediment layers, the burrowing being due on foot and shell valve movements (Figure 15.4). Permanent exchanges with the water column, for respiratory and nutritive purposes, are due to ciliary currents on the surface of the ctenidia, labial palps, and mantle; the inhalant and exhalant siphons control water flow into and out of the internal mantle cavity (Britton and Morton 1982; Reid et al. 1993). In response to mechanical or chemical perturbations, behav-ioral modifications are observed, valve closure, limitation of exchange with the surrounding environment, etc. (Doherty et al. 1987; Ham and Peterson 1994). In many ways, *C. fluminea* is an excellent biological model for ecotoxicological experiments, owing to the ease of collection of large samples in the field and of storage in the laboratory, and to its intense filtration capacity at the water/sediment interface (Britton and Morton 1982; Graney et al. 1984; Way et al. 1990).

H. rigida nymphs were obtained from a mass culture initiated in the laborat-ory from eggs collected during the summer on the banks of Lake Winnipeg (Freshwater Institute, Winnipeg, Canada) and stored at 4°C. This burrowing species, which is found abundantly throughout the North American continent, prefers to colonize lentic biotopes. Average estimated densities are between 60 and 300 individuals per m^2, the maximal densities reaching 2000 nymphs per m^2 (Flannagan 1979; McCafferty 1994). Hatching is carried out in dechlorinated tap water by raising the temperature from 4° to 20°C in 4°C stages, every 48 h (Friesen 1982; Saouter et al. 1991b). Newly hatched nymphs are transferred to large tanks (25 × 25 × 30 cm) containing a 5-cm sediment layer, identical to that used in the experiments. The nymphs are reared on organic matter in the sediment; additional powdered food is given once a week (TetraMin baby fish food, TetraWerke, Germany). The sediment is renewed every 2 months, which enables us to monitor nymph density and growth. Extensive growth heterogen-eity is observed in the culture tanks; in order to obtain a similar biomass in each

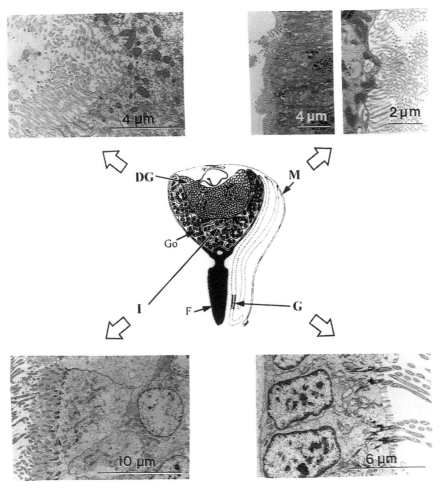

Figure 15.4 Schematic representation of a transversal section of *Corbicula fluminea* showing the various epithelia likely to be involved in contaminant uptake. The mantle (M) covers the soft parts of the organism and consists of two epithelium types: the inner epithelium (left) facing the body and the outer epithelium (right) facing the shell, both displaying short microvilli. The gill ctenidia (G) are located inside the mantle cavity and their epithelium is composed of different types of ciliated cells bearing short microvilli. The visceral mass is represented mainly by the association of the gonad (Go), the digestive gland (DG), and the intestine (I), the intestinal wall having absorbing surfaces. Other tissues studied for bioaccumulation are the musculous foot (F) and the kidney (not represented because of its posterior position; from Lemaire-Gony and Boudou, 1997).

EU, the nymphs were grouped into four weight classes: 25–35, 35–45, 45–55, 55–65 mg (ww). Next, one nymph per class was introduced into each EU (random assignment). This species has several advantages for experimental research in aquatic ecotoxicology: long larval life of between 6 months and 3 years depending on ecological conditions; large size, close to 3 cm in length when transformed into imagos, allowing organ and tissue sampling after micro-dissection; high tolerance to variations in the physicochemical characteristics of the medium; large bioaccumulation capacities of metals present in the super-ficial sediment layers, via trophic transfers from ingested sediment, and direct transfers from the interstitial water and water column, via the currents in the burrows (Figure 15.5).

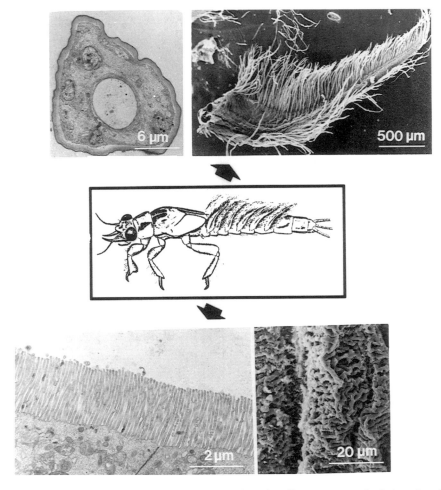

Figure 15.5 Biological barriers in *H. rigida* nymphs. The gills are composed of six pairs of filamentous lamellae (upper right) containing trachea and tracheoles (upper left). They are involved in the creation of the permanent water currents in the burrows. The macroscopic observation of the digestive tract shows macrovilli and grooves (lower right: proctodeum part), but the larger uptake area is represented by the mesenteron epithelial cells covered with macrovilli (lower left; from Saouter et al. 1991a, modified).

At the end of the exposure periods, the organisms were collected from the EUs and individually weighed (ww). The principal organs were collected from the two species using microdissection: gut, gills, and the rest of the body for the mayfly nymphs; mantle, gills, foot, visceral mass, and kidney for the molluscs. It is important to stress that the burrowing nymphs do not have their gut cleared after they are taken out of the contaminated EUs. Thus, the sediment contained in the digestive tract may represent a significant proportion of the bioaccumulated metal burden. Working on an estimation of the weight of the intragut sediment, which represents on average 12% of the nymph biomass (dw; Hare et al. 1989; Odin 1995a), it is possible to take into consideration the quantities of Hg or Cd contained in this compartment if we assume that the metal concentrations in the sediment within the digestive tract are identical to those in the surrounding sediment. After dissection, the biological samples were weighed and frozen in glass tubes before metal determinations. They were first digested by a nitric acid attack in a pressurized medium at 95°C for 3 h. Cd was analyzed with a Varian AA 20 spectrophotometer equipped with a GTA 96 graphite tube atomizer and autosampler. Total Hg determination was carried out by flameless atomic absorption spectrometry (Varian AA 475). A bromine salt treatment was applied to water samples and to the diluted digests before the addition of stannous chloride. The detection limit was 0.1 μg Hg/L. The accuracy of the two analytical procedures was monitored by periodic analyses of standard reference materials from BCR (Brussels, Belgium), KFA (Jülich, Germany), or IEAE (Monaco) in combination with the biological samples series.

Background metal concentrations in *H. rigida* nymphs after elimination of the intragut sediment were 120 ± 28 ng Cd/g and 118 ± 10 ng Hg/g (ww); for the molluscs, they were 85 ± 16 ng/g and 24 ± 2 ng/g, respectively.

The quantification of the actions and interactions of the various controlled abiotic factors—temperature, pH, photoperiod, ranges of metal concentrations, etc.—with regard to the bioaccumulation of the metals was based on the creation and implementation of complete factorial designs. In these designs two or three modalities are considered for each factor, with two replicates for each condition studied and different exposure periods.

The monitoring and regulation of the abiotic factors was achieved using automated systems. The EUs were placed in large water tanks (140 × 65 × 30 cm), which were themselves in enclosed containers. Each tank had thermoregulation equipment (heating and cooling systems), which is very efficient due to the large volume of water constantly stirred by submerged pumps ($\pm 0.2°C$). Light was artificially produced by two neon tubes (Sylvania F36W/GRO) positioned 45 cm above the surface of the EUs and operated by timer switches. Average light intensity at the water/air interface was $35 \pm 3\mu E/cm^2$ per second. The pH of the water column was regulated in each EU by automated equipment, with a central control point (AOIP-SAM 60, Paris) connected to electrodes placed in each acidified unit and to injection systems with which dilute acid solutions (H_2SO_4, 0.1%) were added to the units via electrically operated shutters. No aeration was provided in the EUs containing the *H. rigida* nymphs:

oxygen saturation in the water column varied within a range of 70% to 95%, corresponding to minimal concentrations greater than 6 mg O_2/L. For the molluscs, permanent aeration was provided in the upper layers of the water column. Several other parameters were periodically checked, notably the turbidity in the water column (turbidimeter ESD 800). The concentrations of the suspended particles were directly related to the bioturbation activity of the nymphs, which is in turn strongly influenced by abiotic factors (temperature, pH). Metal concentrations in the dissolved and particulate phases after filtration of the water samples at 0.45 μm were periodically measured in order to estimate metal partitioning and, indirectly, their bioavailability.

Data treatment was based on gradual stages, from an exhaustive graphic representation of the data, supplemented by additional elementary statistical parameters (average, SD, etc.), to the quantification of the actions and interactions of the different ecotoxicological factors taken into account, using multiple linear regression. The choice of complete experimental designs and orthogonal polynomials simplifies the interpretation of the effects of each regressor because of the independence of the regression coefficients. The regressor coding was defined according to the number of modalities (Snedecor and Cochran 1971). Depending on the variance/average ratios for each set of data, different types of transformation of the explained variables were applied (log Y, $1/Y$, $\sqrt{Y}$, etc.). F values were calculated with reference to the interreplicate variance. An α risk equal to 0.01 was adopted for the statistical significance of the effects observed.

15.4 EXAMPLES OF RESULTS

Bioaccumulation of mercury and cadmium was studied in two freshwater species—*Corbicula fluminea* and *Hexagenia rigida*—after exposure to the water column or sediment contamination source. From the numerous data obtained from the comparative study of the two chemical forms of mercury and cadmium, we have illustrated in Figure 15.6 the relations between the metal concentrations measured in the two biological models—*C. fluminea* and *H. rigida*—at the whole organism level, with reference to the two initial contamination sources of the experimental microcosms—water column and sediment—after 14 days of exposure. These results were obtained under similar experimental conditions with respect to both the structure of the EUs and the physicochemical characteristics of the biotopes, notably the temperature ($21°C \pm 0.2°C$) and pH (7.5 ± 0.3).

When the EUs were contaminated via the water source, the metal concentrations measured in the soft body of the molluscs (Figure 15.6A) reveal a significant bioaccumulation of MeHg, which is 7.5 times higher than that observed after exposure to inorganic Hg and about 15 times higher than the observed Cd bioaccumulation, for a similar level of water contamination (1.5 μg/L). The relations between the metal concentrations in the organisms and in the medium

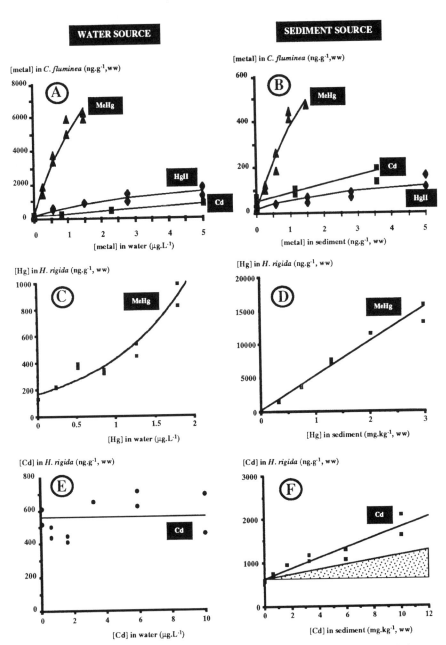

Figure 15.6 Comparative analysis of the bioaccumulation of inorganic Hg (HgII), methylmercury (MeHg), and cadmium (Cd) in *C. flumine* (A, B) and *H. rigida* (C–F) after 15 days of exposure as a function of the metal concentrations in the water column or sediment as initial contaminaton sources. The dotted area in F corresponds to the estimated [Cd] in the gut of the nymphs (data from Odin et al. 1995b; Inza et al. 1995, modified).

show a quasi-proportionality between these two criteria. After contamination of the EUs by the sediment source, the bioaccumulation levels are small relative to the metal concentrations in the sediment (Figure 15.6B). Thus, for MeHg, which exhibits the highest bioaccumulation capacities, the average concentration in the organisms is lower than 500 ng/g after exposure to the highest contamination level of the sediment (1.5 µg/g). The differences between the bioaccumulation of the two Hg chemical forms are close to a factor of 9; Cd concentrations in the bivalves, on the other hand, are twice as high as those obtained after contamination by inorganic Hg. Frequent measurements of the metal concentrations in the water column during the 2 weeks exposure to the sediment source show that the Cd fluxes between the sediment and the water column are higher than those observed after contamination with inorganic Hg; the indirect route of metal uptake, via the release processes, would appear to play a preponderant role in the contamination of the molluscs. As for the water source, the relations between the metal concentrations in the organisms and in the sediment reveal a quasi-proportionality between these two criteria.

The anatomical and physiological particularities of C. fluminea, notably the intense filtration activity at the water/sediment interface, lead us to favor the direct route of uptake and metal transfers from the aqueous phase. Thus, metal burden accumulated from the water source is much greater, giving [Hg] organisms/[Hg] contamination source ratios close to 4000 for MeHg, whereas they are close to only 0.3 after exposure to the sediment source.

Experimental study of the metal fate at the whole organism and organ levels when the exogenous contamination sources are abolished shows that the decrease in Cd concentrations is extremely low: for example, after 30 days of transfer in uncontaminated EUs, following the contamination phase of the molluscs from the water source enriched with Cd, no significant differences are observed in comparison with the average concentrations measured at the end of the contamination period. After exposure to MeHg, decontamination is also very low; the decrease observed at the whole organism level is less than 5% after 30 days. For the inorganic form of mercury, the decrease is close to 40%, the main part of the decontamination occurring during the first week (Inza et al. 1995). In the processes which contribute to ensuring a high level of persistence of cadmium in the molluscs, the metal binding proteins or metallothioneins (MTs) play an important role. They are induced in response to the metal's accessibility to the cytosol, especially within the visceral mass cells (Doherty et al. 1988; Baudrimont 1994). However, the quantitative determination of MTs after exposure of C. fluminea to inorganic Hg and MeHg by the saturation method with cold HgII revealed no significant increase in the concentrations of these hydrosoluble proteins (Baudrimont et al. 1997). Several other sequestration processes have been described in freshwater and marine bivalves, such as the presence of intracellular deposits (granules, concretions) within the cells of digestive and excretory tissues (Roesijadi and Robinson 1994).

Metal transfers between the two contamination sources and the burrowing mayfly nymphs lead to marked differences both at the level of bioaccumulation capacities and in respect of relationships between concentrations measured in the organisms and those which characterize the contamination pressures (Figures 15.6C–F). For the "water/MeHg" source, for example (Figure 15.6C), the relationship between the two concentrations is exponential, the estimated bioconcentration factors reaching a maximal value close to 500; for similar exposure conditions, the mean concentrations measured in the molluscs were about ten times higher. When the EUs are contaminated with Cd, no significant bioaccumulation is observed in comparison with the metal concentrations measured in the control individuals, despite the wide range of Cd concentrations in the water column, from 0 to 10 µg/L (Figure 15.6E).

In contrast, for the sediment source, MeHg and Cd transfers are greater, with a linear relation between concentrations measured in the organisms and in the sediment. For a similar initial level of contamination by the two metals, e.g., 3 mg/kg, Hg concentrations in the nymphs after deduction of the background levels are 45 times higher than Cd concentrations. As a comparison, note that the differences between the bioaccumulation of the two Hg chemical forms are between 20 and 40 times, in favor of MeHg (data not shown; Odin at al. 1994). It is important to underline the fact that metal determinations were carried out on nymphs whose gut had not been cleared after being taken from the EUs. Estimates of the relative Cd burden in the intragut give an average value of between 25% and 30%. If, on the other hand, the Cd concentrations measured in the control nymphs are deducted (Cd in the tissues +Cd in the intragut sediment), the relative average burden in the gut at the end of the experiment is then greater, close to 50%. In this case, the Cd concentrations that are really bioaccumulated from the sediment source are about 550 ng/g when the metal concentration in the sediment is 10 µg/g (Figure 15.6F). After exposure to the sediment source enriched with MeHg, average metal contents in the gut represent only 1.3%–1.8% of the total amount accumulated in the nymphs. As a comparison, when the sediment has been enriched with inorganic Hg, the relative metal burden in the gut is close to 25% (Odin et al. 1994). After contamination of the nymphs via the sediment source, a large proportion of the Cd is eliminated rapidly. The clearance of the digestive tract, which is carried out during the first few hours after the nymphs are transferred to new EUs, is followed by a significant metal loss from the other tissular compartments: the estimated half-life is close to 12 hours (Figure 15.7). For MeHg, however, the decrease in concentrations during the decontamination phase reveals a slow rate of decontamination, the estimated half-life being close to 24 days (Odin et al., 1997).

The differences observed between the bioaccumulation capacities of mercury compounds and cadmium can be related to their physicochemical properties and to the transport mechanisms which ensure the crossing of the biological barriers at the interface with the surrounding medium.

The anatomical structures of C. fluminea are typical of filter-feeding bivalves (Figure 15.4). The ctenidium on each side of the body comprises an outer and inner demibranch. It separates the mantle cavity into infra- and suprabranchial chambers, the former receiving incoming water via the inhalant siphon. The water flow is effected by the beating of the cilia on the ctenidia filaments and by the regular pumping movements of the two valves. The ctenidia have several functions: respiratory exchanges; water filtering, the sieved material being directed toward the mouth by the ciliated area on the frontal side; and incubation of the fertilized eggs within the inner demibranch (Britton and Morton 1982). The respiratory part of the gill filaments corresponds to the unciliated abfrontal area, based on a monolayer of thin pavement epithelial cells; the interface with the external medium displays small microridges and a well-developed glycocalyx (Figure 15.4). The left and right mantle lobes are fused ventral to the inhalant siphon; the mantle cavity lies outside the body of *Corbicula* and is contiguous with the external environment. Most of this organ is made of a thin lamella consisting of two epithelia: the outer layer facing the shell and the inner layer in contact with the mantle cavity. They are separated by hemolymph sinuses and connective tissue. The general inner surface of the mantle lobes is densely covered by cilia; they transport material rejected from the ctenidia and/or first rejected by the visceral mass (pseudofeces). The digestive barrier is made up of different parts: the mouth, the esophagus, a large capacious stomach, and the gut including the crystalline style diverticulum, the intestine, and the rectum. The intestinal wall is made of columnar ciliated cells, with abundant microvilli between the cilia and pinocytotic vesicles in the apical part of the cells (Lemaire-Gony and Boudou, in press).

Structural and ultrastructural analyses of the biological barriers in *H. rigida* nymphs reveals very marked specificities (Saouter et al. 1991a). The gills are made up of six pairs of external filamentous lamellae, containing tracheae and tracheoles, carried on the first six segments of the abdomen, with several hundred secondary ramifications on either side of the principal axis (Figure 15.5). They are in direct contact with the surrounding water, and exchanges are very important considering the wide area of this epithelium and the permanent water currents in the burrows due to the constant pulsing of the gills and the undulations of the abdomen. The nymphs' cutaneous coating is generally considered to be impermeable to many chemicals. Trace metals can be bound directly to the chitinous body surface and/or indirectly by complexation with sediment component deposits on the exoskeleton (Hare 1992). The nonsignificant bioaccumulation of Cd after nymph exposure via the direct route can be explained by the presence of the cuticular coating at cutaneous and gill lamella levels, which appears to present an impermeable barrier to metal uptake. It is also possible that decontamination processes predominate; these would contribute to the elimination of the metal adsorbed or absorbed through the biological barriers. This second hypothesis could explain the fact that a significant bioaccumulation of Cd by nymphs exposed via the water source is observed when the temperature of the medium is fairly low (data not shown; Odin et al. 1995b). MeHg, on the other hand, and to a lesser extent inorganic Hg, can cross these structures easily. The structural and ultrastructural properties of the gut barrier of *H. rigida* nymphs also has an important role to play in relation to bioaccumulation of the metals via the ingestion of contaminated sediments. The digestive tract consists of three fundamental parts: the foregut or stomodeum, which is extremely short and localized in the anterior part of the cephalic canal; the midgut or mesenteron; and the hindgut or proctodeum, which represents about one-quarter of the length of the midgut. Microvilli on the apical face of the mesenteron epithelial cells provide a very large surface area for nutrient absorption and potentially for metal binding and uptake. The proctodeum consists of elongated and very ornamented macrovilli, the apical face of the cells being covered with a 0.5-μm layer of chitin (Figure 15.5). This gut barrier plays an important role in discriminating between the two chemical forms of mercury. As demonstrated in several aquatic species (zooplankton crustacea, fish, etc.) and terrestrial mammals, it is relatively impermeable to inorganic Hg, but it does have a high capacity to bind this compound, especially at the lumen/microvilli interface where the cell-coat is abundant (Boudou et al. 1991). In contrast, absorption of MeHg across the digestive barrier is extremely efficient, with estimated transfer rates being as high as 95% or even 100% (Boudou and Ribeyre 1985). For Cd, the intestine, or to be more precise the mesenteron, represents the predominant route of entry into aquatic insects (Hare 1992). Our results show, however, that the metal bioaccumulated after contamination of the nymphs via the trophic route is very rapidly eliminated during the decontamination phase (Figure 15.7), reflecting the predominance of Cd adsorbed at the digestive barrier and easily eliminated when noncontaminated sediment is ingested, and/or the rapid turnover of the epithelial structures in relation to the frequency of the moults, for example.

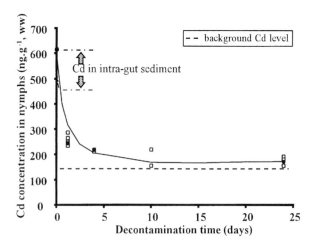

Figure 15.7 Evolution of Cd concentrations in *H. rigida* nymphs (whole organism level) during the decontamination phase, after exposure of the burrowing organisms via the sediment source during 14 days. Symbols correspond to the average values per EU (four nymphs/EU; two replicates/condition; data from Odin et al. 1997).

Whatever the level of complexity of the biological barriers, the cytoplasmic membrane represents the unit structure at the cellular level which controls the uptake of toxic products. Mercury crosses cell membranes mainly by diffusion through phospholipid bilayers; other transport processes have been observed, for example, the crossing of MeHg through the cell membranes of the blood–brain barrier by forming a MeHg–cysteine complex that structurally mimics the amino acid methionine (Clarkson 1994). The hydro- and liposolubility of inorganic Hg and MeHg chemical species play a fundamental role in their ability to reach, and in particular to cross, cytoplasmic membranes. The significance of the liposolubility of MeHg, very often put forward to explain its high bioaccumulation capacity, needs to be reevaluated as the octanol/water partition coefficients are fairly small (Major et al. 1991). Measurements of Hg diffusion through artificial phospholipid bilayers (BLM (bimolecular lipid membranes) models) and analysis of metal binding to membrane ligands using ^{199}Hg NMR clearly demonstrate the importance of chemical speciation reactions in solution: for example, the transmembrane fluxes were very similar for inorganic Hg and MeHg when the $HgCl_2$ and CH_3HgOH species predominated (Bienvenue et al. 1984; Delnomdedieu et al. 1992; Girault et al. 1995; Gutknecht 1981). The penetration of Cd across biological membranes is also dependent on chemical speciation phenomena, especially the preponderance of species Cd^{2+} and $CdCl_2$. In addition to diffusion processes, different transport mechanisms play an important role, e.g., calcium channels or specific SH-containing pathways for zinc or manganese (Foulkes 1991; Hinkle et al. 1994; Roesijadi and Robinson 1994).

The study of the distribution of metals in the main tissue compartments provides information on the predominance of various uptake routes, according to the different exposure conditions, and also on the processes of storage and interorgan transfers during the contamination and decontamination phases. In C. fluminea, analysis of the organotropism of mercury and cadmium in five tissue compartments—gills, mantle, foot, visceral mass (gut, digestive gland, gonad, heart), and kidney—after contamination from the water source revealed some marked specificities of the two mercury compounds and the Cd (Figure 15.8). There was a very high relative content of inorganic Hg in the visceral mass; for MeHg, metal distribution was homogeneous in relation to the relative weights of the organs, showing once again this compound's ability to cross the biological barriers which separate individuals from their surrounding environment. For Cd, the distribution was rather homogeneous, but the kidney appears to be a target organ for bioaccumulation, with concentrations there being particularly high.

After contamination of the molluscs by the sediment source, the organotropism of the metals was similar to that which characterized contamination by the water source (data not shown; Inza et al. 1997). These results confirm that the direct contamination route, via transfers from the water column, predominates in both exposure conditions studied. The metals contained in the sediment compartment are not very bioavailable to C. fluminea, unlike the case of the burrowing mayfly larvae, which ingest large quantities of

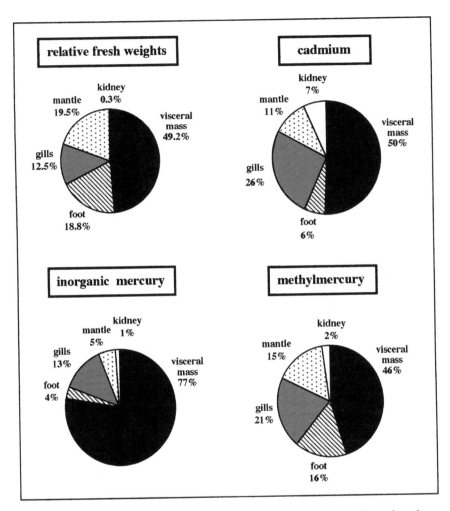

Figure 15.8 Relative fresh weights of five organs of *C. fluminea*—gills, mantle, kidney, visceral mass, foot—and relative contents of mercury and cadmium after 15 days exposure to the water column source enriched with inorganic Hg (HgII), methylmercury (MeHg), or cadmium (Cd) (data from Inza et al. 1995).

sediment for nutritional purposes. However, all of the processes which contribute to metal transfers between the superficial sediment layers and the water column—diffusion from the interstitial water, mechanical mixing, bioturbation, etc.—may contribute to an increase in the quantities of metal bioaccumulated from the sediment source. Thus, comparative studies based on different microcosms containing either simultaneously or separately the two species *C. fluminea* and *H. rigida* show that Cd concentrations in the molluscs are significantly greater when the burrowing activity of the nymphs leads to the sediment particles being returned to suspension and increases indirect transfers of the metal, via the water column (Andrès et al., manuscript in preparation).

In *H. rigida*, three compartments were studied: the gut after elimination of the intestinal sediment, the gills, and the rest of the body. After contamination of the nymphs by the water and sediment sources enriched in inorganic Hg or MeHg, the relative burden of metal in these three compartments reveal a very different distribution when considered in with respect to the properties of the biological barriers (Figure 15.9). For HgII, the gills, after exposure via the water source, and the gut, after exposure via the sediment source, contain 50% and 43%, respectively, of the metal burden bioaccumulated in the whole organism, whereas their relative biomasses are 6.3% and 6.7%, respectively (Saouter et al. 1991c, 1993). These results show the high accumulation capacity of this inorganic form of the metal in biological barriers and, at the same time, the limited amount of transfer to the other internal organs and tissues via absorption through epithelial structures and transport via the hemolymph. For MeHg, metal distribution in the nymphs reveals a high permeability of the branchial and intestinal barriers, with a very high relative burden in the rest of the body after exposure to the two contamination sources.

15.5 EFFECTS OF ABIOTIC FACTORS AND INTERACTIONS BETWEEN THESE FACTORS ON MERCURY BIOACCUMULATION IN BURROWING MAYFLY NYMPHS

In order to illustrate our experimental approach to determine the effects of the actions of and interactions between several physicochemical characteristics of the biotopes on metal bioaccumulation, results from the contamination of *H. rigida* nymphs via the water column source have been selected (Odin et al. 1995a). Three abiotic factors were taken into account: temperature, pH, and photoperiod. A complete factorial design, including combinations of the different levels for each factor—10°C, 18°C, and 26°C for the temperature; 5.0 and 7.5 for the pH; 6, 12, and 18 h of light per day for the photoperiod—represented 18 different ecotoxicological conditions studied simultaneously. The duration of the experiment was 15 days. During the first 5 days, the levels of the three factors were progressively adapted from the culture conditions in order to reach the levels selected. The procedure for the contamination of the water column was based on twice-daily additions of identical amounts of inorganic Hg to the EUs (8 µg Hg/day); thus, for the different experimental conditions the same amount of metal had been added by the end of the experiment. However, due to the direct and indirect effects of the three abiotic factors on the metal chemical fate in the water column, marked differences appeared between the contamination pressures, which could lead to a greater or lesser impact on Hg bioaccumulation in the nymphs. The evolution of Hg concentrations in the unfiltered water samples is illustrated in Figure 15.10. The evolution trends over the

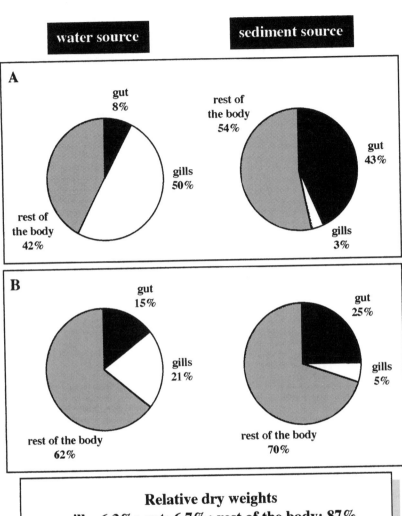

Figure 15.9 Relative mercury burden in the gills, gut, and rest of the body of *H. rigida* nymphs after contamination of the water column or sediment source with (A) inorganic mercury (HgII) and (B) methylmercury (MeHg) (data from Saouter et al. 1991c).

2 weeks showed marked differences according to temperature. The pH effect is very pronounced at 10°C: the average Hg concentrations at the end of the exposure period are close to 20 and 10 µg/L at pH 5.0 and 7.5, respectively. At 18°C and 24°C, the pH effect is no longer significant, with average Hg concentrations in the water column close to 8 and 4 µg/L, respectively. Photoperiod has very little effect on the contamination pressure. Hence, although all EUs

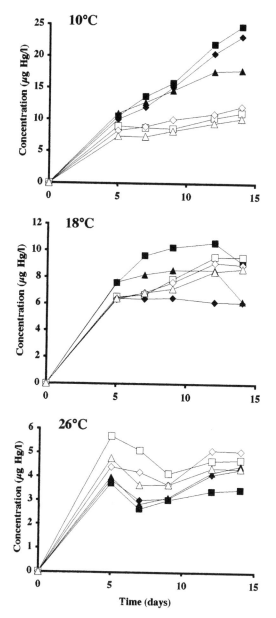

Figure 15.10 Evolution of Hg concentrations in the water column of the experimental units throughout the 15 days exposure period as a function of temperature, pH, and photoperiod ■ pH 5.0–6 h/24 h, □ pH 7.5–6 h/24h, ◆ pH 5.0–12 h/24 h, ◇ pH 7.5–12 h/24 h, ▲ pH 5.0–18 h/24 h, △ pH 7.5–18 h/24 h (from Odin et al. 1995a).

received identical twice-daily additions of metal, the combined actions of the temperature and pH gave rise to very marked differences between the metal concentrations in the water column and their evolution over the 2 weeks of exposure: the maximal difference is close to a factor of 5.

Hg bioaccumulation in the organisms, which is a function of the 18 abiotic conditions studied, is shown in Figure 15.11A. A multiple regression analysis, based on the average metal concentrations measured in four nymphs per EU (two replicates per condition), shows that the three factors exerted a significant influence, both alone and in combination, leading to complex phenomena. Acidification of the water column caused an increase in Hg concentrations in the nymphs, with the exception of the "10°C–12 h/24 h" condition: the greatest difference between the two pH conditions is close to twofold. As can be seen in the 3D plot, Hg concentrations in the nymphs are greatest between 10°C and 18°C. The effects of photoperiod are complex, with numerous interactions with the other two factors and square terms in the regression model: this factor has little effect on Hg bioaccumulation at 26°C for both pH modalities; at 10°C and 18°C, its effects are more pronounced.

In order to take into account the wide differences between the contamination pressures, the "concentration-days" equivalent index (CDE) has been calculated for each experimental condition after 15 days (data not shown). This index enables us to correct the measured concentrations in the nymphs and to compare the bioaccumulation data for theoretical and similar exposure conditions. Analysis of the "corrected" Hg concentrations in the nymphs (measured [Hg] organism/CDE) reveals very important modifications in the effects of the three abiotic factors on Hg bioaccumulation (Figure 15.11B): at pH 5.0, metal concentrations in *H. rigida* nymphs reach a maximum level of between 18°C and 26°C and are greatly affected by the three modalities of the photoperiod; at pH 7.5, the temperature effect is less pronounced, with maximal bioaccumulation values at the lower temperatures, i.e., between 10°C and 18°C.

The analysis of these results inevitably comes up against the complexity of the processes. Very little ecotoxicological data are currently available on interaction mechanisms from the biological point of view—combined effects of environmental factors on the physiology of the aquatic species, on the structural and functional properties of the biological barriers, or on exchanges with the surrounding environment—or on chemical speciation of the metals within the biotopes and their bioavailability. With our experimental conditions, information from measurements on samples collected in the water column and in the sediment compartments suggest several hypotheses.

For example, Hg partitioning in the water column is strongly influenced by the concentrations of suspended sediment particles, resulting from the bioturbation activity of the nymphs. Turbidity, measured after 5, 10, and 15 days, exhibits a marked influence of the combined effects of pH and temperature: the ratio between the two extreme NTU (nephelometric units) values is close to 10 (Figure 15.12A). Measurements of Hg concentrations in the dissolved phase, after filtration of the water samples at 0.45 μm, indicate that the proportion of the

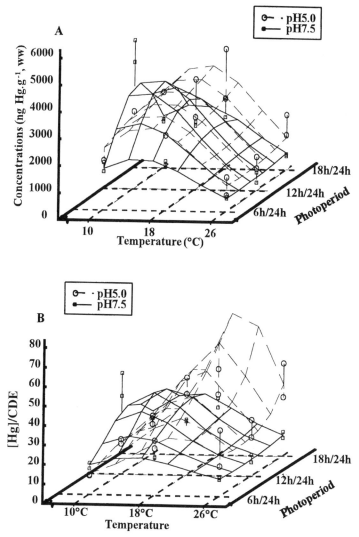

Figure 15.11 (A) Total Hg concentrations measured in *H. rigida* nymphs (whole organism level) after 15 days exposure to inorganic Hg via the water column source as a function of temperature, pH, and photoperiod. (B) Corrected Hg concentrations (measured [Hg]/CDEs) in *H. rigida* nymphs (whole organism level) after 15 days exposure to inorganic Hg via the water column source as a function of temperature, pH, and photoperiod. CDE, concentration-days equivalent index. The symbols on the 3D plots correspond to the average measured values (two replicates) and the plans to the multiple regression model ($\alpha = 0.01$; (A): contribution 91%; (B): contribution 91%) (from Odin et al. 1995a).

metal in this phase decreases very rapidly when the temperature increases; acidification of the water column induces an increase in this criterion, with marked interactions with the temperature (Figure 15.12B). Several experimental

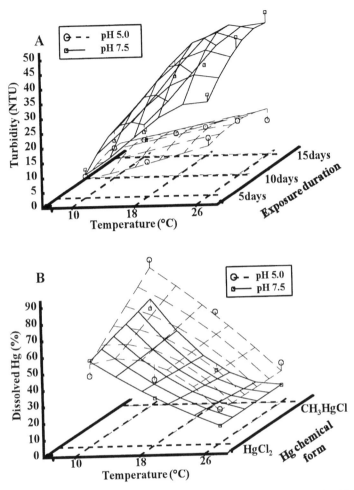

Figure 15.12 (A) Water column turbidity (nephelometric units), in relation to the bioturbation activity of *H. rigida* nymphs, as a function of temperature, pH and photoperiod (from Odin et al. 1995a). (B) Dissolved Hg fraction (%) in the water column after 15 days exposure as a function of Hg chemical form, temperature, and pH (from Odin et al. 1995a). The symbols on the 3D plots correspond to the average measured values (two replicates) and the two plans to the multiple regression model ($\alpha = 0.01$; (A): contribution 96%; (B): contribution 94%).

and field studies have shown that dissolved metals are inherently more available to aquatic species compared with metals bound to inorganic and organic suspended matter (Nelson and Campbell 1991; Roesijadi and Robinson 1994).

In addition to the part played by the partitioning of the metals between the dissolved and particulate phases, chemical speciation in solution, which is directly controlled by the physicochemical conditions of the biotopes, exerts a strong influence on bioavailability and bioaccumulation capacities. In the case of inorganic Hg, for example, in freshwater conditions, acidification favors the formation of the $HgCl_2$ species, which is better able to adsorb onto membrane sites, notably the primary amine groups on the phospholipid polar heads, and to cross the hydrophobic bilayers (Bienvenue et al. 1984; Delnomdedieu et al. 1989, 1992). The combined effects of temperature and pH on inorganic Hg bioaccumulation in the nymphs can also be exerted via the chemical transformations of this compound in the water column and sediment and also within the organisms, notably in the intragut compartment. Inorganic Hg may be methylated, with the production of monomethylmercury and/or dimethylmercury (volatile form), due to abiotic reactions (humic acids, for example) and in particular to the action of bacteria, notably the sulfate-reducing bacteria (SRB; Baldi 1993; Gilmour 1992; Matilianen 1995; Rudd 1995; Weber 1993). Inorganic Hg may also be transformed into elemental Hg (volatile form); this reduction is due to enzyme processes (Hg reductase) in several groups of aquatic bacteria (Barkay 1987; Moore et al. 1990). Under our experimental conditions, the procedure selected for the contamination of the water source, i.e., daily additions during 2 weeks of exposure, probably tends to minimize the significance of these tranformations. However, the two factors temperature and pH influence these processes, especially those which depend on bacteria. For example, recent results from *in situ* studies show that MeHg production in hypolimnion and sediment compartments is stimulated in acidic conditions, especially as a result of sulfuric acid deposition and effects on SRB (Gilmour 1992; Zhang and Planas 1994). The relative MeHg burden measured in zooplankton samples collected in the acidified part of the Little Rock Lake (Wisconsin, USA) show a marked increase in comparison with the reference basin: 90% and 30%, respectively (Watras and Bloom 1992). Little is known about the mechanisms involved in mercury transformations. Hg methylation and demethylation result from very complex processes, strongly dependent on ecological conditions and on the contamination levels of the biotopes. Research is currently being done in our laboratory to analyze these fundamental processes within the different compartments of the experimental microcosms and to quantify their effects on Hg bioaccumulation, using analytical methods based on the coupling of gas chromatography and cold vapor atomic fluorescence spectrometry after an aqueous ethylation step (Bloom 1989; Saouter and Blattmann 1994).

The example of *H. rigida* nymphs clearly demonstrates the high complexity of the processes and the importance of the direct and indirect effects of the ecotoxicological factors on bioaccumulation mechanisms. Similar experimental studies after exposure to the water column source enriched with MeHg or Cd, or to the sediment source, lead to similar conclusions: all of the factors considered produce effects of varying degrees of intensity, with the interactive phenomena producing sometimes considerable differences in the amounts of metal bioaccumulated (Odin 1995; Odin et al. 1995a,b; Inza et al. 1993, 1995). Previous experiments based on several other biological models, such as freshwater rooted macrophytes, have also revealed the importance of direct and indirect effects of the abiotic factors on mercury bioaccumulation (Maury-Brachet et al. 1990; Ribeyre and Boudou 1990).

15.6 CONCLUSION

Metal bioaccumulation and trophic transfers along aquatic food webs result from very complex mechanisms, both at the field and laboratory levels. Several major research goals are emerging now in experimental ecotoxicology, from fundamental and applied perspectives:

- intensify the development of new methodologies, giving priority to the specifities of the ecotoxicological approach, in particular the complexity of the biological structure of the indoor models and the variations in the abiotic and contamination factors;
- promote more fundamental research in order to understand the mechanisms involved, from the cellular and molecular levels up to the higher biological levels of integration;
- give priority to programmes which combine indoor and outdoor approaches, in order to increase the relevance and representativeness of the methodologies and to better define and select analysis levels and criteria;
- make studies more multidisciplinary, especially at the interface between environmental chemistry and the biological and ecological sciences.

15.7 REFERENCES

Andres, S., A. Boudou, and F. Ribeyre, 1997, "Experimental Study of the Effect of Bioturbation on Cadmium Bioaccumulation in the Asiatic Clam *Corbicula fluminea*," in press

Araujo, R., D. Moreno, and M. A. Ramos, 1993, "The Asiatic Clam *Corbicula Fluminea* (Müller, 1774) (Bivalvia: Corbiculidae) in Europe," *Am. Malacol. Bull., 10* (1), 39–49

Baldi, F., 1993, "Biological Removal of Inorganic Hg(II) as Gaseous Element Hg(0) by Continuous Culture of a Hg-Resistant *Pseudomas putida* Strain FB," *J. Microbiol. Biotechnol., 9*, 275–279

Barkay, T., 1987, "Adaptation of Aquatic Microbial Communities of Hg^{2+} Stress," *Appl. Environ. Microbiol., 53*, 2725–2732

Baudrimont, M., 1994, "Etude Expérimentale du Rôle des Métallothionéines dans les Mécanismes de Bioaccumulation du Cadmium et du Mercure par un Mollusque Bivalve d'eau Douce: *Corbicula fluminea*," DEA National de Toxicologie, Université Bordeaux I, France

Baudrimont, M., J. Métivaud, F. Ribeyre, and A. Boudou, 1997, "Metallothioneins or MTs-like response in the Asiatic clam *Corbicula fluminea* after experimental exposure to cadmium and inorganic mercury," *Environ. Toxicol. Chem., 16* (in press)

Bernhard, M., F. E. Brinckman, and P. J. Sadler, 1986, *The Importance of Chemical "Speciation," in Environmental Processes*, Springer, Berlin, Germany

Bienvenue, E., A. Boudou, J. P. Desmazès, C. Gavach, D. Georgescauld, J. Sandeaux, and P. Seta, 1984, "Transport of Mercury across Biomolecular Lipid Membranes: Effect of Lipid Composition, pH and Chlorie Concentrations," *Chem. Biol. Interact.*, 48, 91–101

Bloom, N. S., 1989, "Determination of Picogram Levels of Methylmercury by Aqueous Phase Ethylation, followed by Cryogenic Gas Chromatography with Cold Vapor Atomic Fluorescence Detection," *Can. J. Fish. Aquat. Sci.*, 46, 1131–1140

Boudou, A., M. Delnomdedieu, D. Georgescauld, F. Ribeyre, and E. Saouter, 1991, "Fundamental Roles of Biological Barriers in Mercury Accumulation and Transfer in Freshwater Ecosystems (Analysis at Organism, Organ, Cell and Molecular Levels)," *Water, Air, Soil Pollut.*, 56, 807–821

Boudou, A., and F. Ribeyre, 1985, "Experimental Study of Trophic Contamination of *Salmo gairdneri* by Two Mercury Compounds: Analysis at the Organism and Organ Levels," *Water, Air, Soil Pollut.*, 26, 137–148

Boudou, A. and F. Ribeyre, 1989, "Fundamental Concepts in Aquatic Ecotoxicology," in *Aquatic Ecotoxicology: Fundamental Concepts and Methodologies* (A. Boudou, and F. Ribeyre, Eds.), CRC Press, Boca Raton, USA, pp. 35–75

Britton, J. C., and B. Morton, 1982, "A Dissection Guide, Field and Laboratory Manual for the Introduced Bivalve *Corbicula fluminea*," *Malacol. Rev.*, pp. 1–82

Campbell, P. G. C., A. G. Lewis, P. M. Chapman, A. A. Crowder, W. K. Fletcher, B. Imber, S. N. Luoma, P. M. Stokes, and M. Winfrey, 1988, *Biologically Available Metals in Sediments*, National Research Council Canada, Ottawa, Canada

Clarkson, T. W., 1994, "The Toxicology of Mercury and its Compounds," in *Mercury Pollution: Integration and Synthesis* (C. J. Watras, and J. W. Huckabee, Eds.), Lewis Publishers, Chelsea, USA, pp. 631–642

Counts, C. L., 1986, "The Zoogeography and History of the Invasion of the United States by *Corbicula fluminea* (Bivalvia: Corbiculidae)," *Am. Malacol. Bull.*, 7–39

Delnomdedieu, M., A. Boudou, J. P. Desmazès, and D. Georgescauld, 1989, "Interaction of Mercury Chloride with the Primary Amine Goups of Model Membranes Containing Phosphatidylserine and Phosphatidylethanolamine," *Biochim. Biophys. Acta*, 986, 191–199

Delnomdedieu, M., A. Boudou, D. Georgescauld, and E. J. Dufourc, 1992, "Specific Interactions of Mercury Chloride with Membranes and other Ligands as Revealed by Mercury-NMR," *Chem. Biol. Interact.*, 81, 243–269

Doherty, F. G., D. S. Cherry, and J. Cairns Jr. 1987, "Valve Closure Responses of the Asiatic Clam *Corbicula fluminea* Exposed to Cadmium and Zinc," *Hydrobiologia, 153*, 159–167

Doherty, F. G., M. L. Failla, and D. S. Cherry, 1987, "Identification of a Metallothionein-Like, Heavy Metal Binding Protein in the Freshwater Bivalve, *Corbicula fluminea*," *Comp. Biochem. Physiol. C*, 87, 113–120

Doherty, F. G., M. L. Failla, and D. S. Cherry, 1988, "Metallothionein-Like Heavy Metal Binding Protein Levels in Asiatic Clams are Dependent on the Duration and Mode of Exposure of Cadmiun," *Water. Res.*, 22 (7), 927–932

Faust, B. C., 1992 "The Octanol/Water Distribution Coefficients of Methylmercuric Species: the Role of Aqueous-Phase Chemical Speciation," *Environ. Toxicol. Chem., 11*, 1373–1376

Flannagan, J. F., 1979, "The Burrowing Mayflies of Lake Winnipeg, Manitoba, Canada," *Proc. 2nd Int. Conf. on Ephemeroptera*), Winnipeg, Canada, August 1975 (K. Pasternak, and P. P. Sowa, Eds.), pp. 103–113

Foe, C., and A. Knight, 1986, "Growth of *Corbicula fluminea* (Bivalvia) Fed Artifical and Algal Dits," *Hydrobiologia, 133,* 155–164

Foulkes, E. C., 1991, "Further Findings on the Mechanism of Cadmium Uptake by Intestinal Mucosal Cells," *Toxicology, 70,* 261–270

Friesen, M. K., 1982, "*Hexagenia Rigida.*," in *Manuel d'élevage d'invertébrés Dulçaquicoles Chosis* (S. G. Laurence, Ed.), Publication Spéciale Canadienne des Sciences Halieutiques et Aquatiques, Ottawa, Canada, pp. 127–141

Gilmour, C. C., 1992, "Sulfate Stimulation of Mercury Methylation in Freshwater Sediments," *Environ. Sci. Technol., 26* (11), 2281–2287

Girault, L., P. Lemaire, A. Boudou, and E. J. Dufourc, 1995, "Inorganic Mercury Interactions with Lipid Components of Biological Membranes: ^{31}P-NMR Study of Hg(II) Binding to Headgroups of Micellar Phospholipids," *Water, Air, Soil Pollut., 80,* 95–98

Graney Jr. R. L., D. S. Cherry, and J. Cairns Jr., 1984, "The Influence of Substrate, pH, Diet and Temperature Upon Cadmium Accumulation in the Asiatic Clam (*Corbicula fluminea*) in Laboratory Artifical Streams," *Water Res., 18* (7), 833–842

Gutknecht, J., 1981, "Inorganic Mercury (Hg^{2+}) Transport Through Lipid Bilayer Membranes," *J. Membr. Biol., 61,* 61–66

Ham, K. D., and M. J. Peterson, 1994, "Effect of Fluctuating Low-Level Chlorine Concentrations on Valve-Movement Behavior of the Asiatic Clam (*Corbicula Fluminea*)," *Environ. Toxicol. Chem., 13* (3), 493–498

Hare, L., 1992, "Aquatic Insects and Trace Metals: Bioavailability, Bioaccumulation and Toxicity," *Crit. Rev. Toxicol., 22* (5/6), 327–369

Hare, L., P. G. C. Campbell, A. Tessier, and N. Belzile, 1989, "Gut Sediments in a Burrowing Mayfly (Ephemeroptera, *Hexagenia limbata*): Their Contribution to Animal Trace Element Burdens, Their Removal, and the Efficacy of a Correction for Their Presence," *Can. J. Fish. Aquat. Sci., 46,* 451–456

Hinkle, P. M., and M. E. Osborne, 1994, "Cadmium Toxicity in Rat Pheochromocytoma Cells: Studies on the Mechanism of Uptake," *Toxicol. Appl. Pharmacol., 124,* 91–98

Inza, B., 1993, *Etude en Ecotoxicologie Expérimentale des Processus de Bioaccumulation du Cadmium par un Mollusque Filtreur d'eau Douce: Corbicula fluminea.*, DEA National de Toxicologie, Université Bordeaux I, France

Inza, B., R. Maury-Brachet, J. M. Laporte, A. Boudou, and F. Ribeyre, 1995, "Experimental Study of Cadmium, Inorganic Mercury and Methylmercury Contamination of the Freshwater Mollusc *Corbicula fluminea*," 5th SETAC-Europe Congress, Copenhagen, June 1995, Abstracts, SETAC Europe, Brussels

Inza, B., F. Ribeyre, R. Maury-Brachet, and A. Boudou, "Experimental Study of Cadmium and Mercury Compounds Bioaccumulation in the Asiatic Clam *Corbicula fluminea*: Effects of the Contamination Levels of the Water Column and Sediment Sources," *Chemosphere* (in press)

Kudo, A., 1989, "Mercury in the Ottawa River (Canada)," in *Aquatic Ecotoxicology: Fundamental Concepts and Methodologies* (A. Boudou, and F. Ribeyre, Eds.), CRC Press, Boca Raton, USA, pp. 201–217

Lemaire-Gony, S., and A. Boudou, "Mantle and Gill Fine Structure in the Freshwater Asiatic Clam *Corbicula fluminea*," *Ann. Limnol.*, in press

Major, M. A., D. H. Rosenblatt, and Bostian, K. A., 1991, "The Octanol/Water Partition Coefficient of Methylmercuric Chloride and Methylmercuric Hydroxide in Pure Water and Salt Solutions," *Environ. Toxicol. Chem.*, *10*, 5–8

Matilianen, T., 1995, "Involvement of Bacteria in Methylmercury Formation in Anaerobic Lake Waters," *Water, Air, Soil Pollut.*, *80*, 757–764

Maury-Brachet, R., F. Ribeyre, and A. Boudou, 1990, "Actions and Interactions of Temperature and Photoperiod on Mercury Accumulation by *Elodea Densa* from Sediment Source," *Ecotoxicol. Environ. Safety*, *10*, 141–151

McCafferty, W. P., 1994 "Distributional and Classificatory Supplement to the Burrowing Mayflies (Ephemeroptera: Ephemeridae) of the United States," *Entomol. News.*, *105*, 1–13

Moore, M. J., M. D. Distefano, L. D. Zydowsky, R. T. Cummings, and C. T. Walsh, 1990, "Organomercurial Lyase and Mercuric Ion Reductase: Nature's Mercury Detoxification Catalysts," *Acc. Chem. Res.*, *23*, 301–308

Nelson, W. O., and P. G. C. Campbell, 1991, "The Effects of Acidification on the Geochemistry of Al, Cd, Pb and Hg in Freshwater Environments: A Literature Review," *Environ. Pollut.*, *71*, 91–130

Odin, M., A. Feurtet-Mazel, F. Ribeyre, and A. Boudou, 1994, "Actions and Interactions of Temperature, pH and Photoperiod on Mercury Bioaccumulation by Nymphs of the Burrowing Mayfly *Hexagenia rigida*, from the Sediment Contamination Source," *Environ. Toxicol. Chem.*, *13* (8), 1291–1302

Odin, M., 1995, "Transferts des Dérivés du Mercure et du Cadmium entre les Sédiments ou la Colonne d'eau et les Larves d'*Hexagenia rigida* (Ephéméroptères), en fonction des Conditions Expérimentales (Température, Photopériode, pH et Nature du Sédiment)," *Thèse* No. 1358, Université Bordeaux I, France

Odin, M., A. Feurter-Mazel, F. Ribeyre, and Boudou, A., 1995a, "Inorganic Mercury and Methylmercury Bioacumulation by Nymphs of the Burrowing Mayfly *Hexagenia rigida* from the Water Column as the Initial Contamination Source – Quantification of the Actions and Interactions with Temperature, pH and Photoperiod," *Toxicol. Environ. Chem.*, *48*, 213–244

Odin, M., F. Ribeyre, and A. Boudou, 1995b, "Cadmium and Methylmercury Bioaccumulation by Nymphs of the Burrowing Mayfly *Hexagenia rigida* from the Water Column and Sediment," *Environ. Sci. Pollut. Res.*, *2*, 145–152

Odin, M., F. Ribeyre, and A. Boudou, 1997, "Depuration Processes after Exposure of Burrowing Mayfly Nymphs (*Hexagenia rigida*) to Methylmercury and Cadmium from Water Column or Sediment – Effects of Temperature and pH," *Aquat. Toxicol.*, *37*, 125–137

Reid, R. G., R. F. McMahon, D. O. Foighil, and R. Finnigan, 1993, "Anterior Inhalant Currents and Pedal Feeding in Bivalves," *Veliger*, *35* (2), 93–104

Ribeyre, F., and A. Boudou, 1990, "Bioaccumulation of Two Mercury Compounds in Two Aquatic Plants (*Elodea Densa and Ludwigia Natans*): Actions and Interactions of Four Abiotic Factors," in *Use of Plants for Toxicity Assessment*, STP 1019, ASTM, Philadelphia, USA, pp. 97–113

Ribeyre, F., 1993, "Evolution of Mercury Distribution Within an Experimental System "Water-Sediment-Macrophytes (*Elodea Densa*)," *Environ. Technol.*, *14*, 201–214

Ribeyre, F., 1991, "Experimental Ecosystems: Comparative Study of Two Methods of Contamination of the Water Column by Mercury Compounds in Relation to Bioaccumulation of the Metal by Rooted Macrophytes (*Ludwigia nantans*)," *Environ. Technol.*, *12*, 503–518

Roesijadi, G., and W. E. Robinson, 1994, "Metal Regulation in Aquatic Animals: Mechanisms of Uptake, Accumulation and Release," in *Aquatic Toxicology: Molecular, Biochemical and Cellular Perspectives* (D. C. Malins, G. K. Ostrander, Eds.), CRC Press, Boca Raton, USA, pp. 387–420.

Rudd, J. W. M., 1995, "Sources of Methylmercury to Freshwater Ecosystems: A Review," in *Mercury as a Global Pollutant* (D. B. Porcella, J. W. Huckabee, B. Wheatley, Eds.), Kluwer, Dordrecht, The Netherlands, pp. 697–713

Saouter, E., and B. Blattmann, 1994, "Rapid and Sensitive Analyses of Organic and Inorganic Mercury by Atomic Fluorescence Spectrometry Using a Semi-Automatic Analytical System," *Anal. Chem.*, *66* (13), 36–45

Saouter, E., L. Hare, A. Boudou, F. Ribeyre, and P. G. C. Campbell, 1993, "Mercury Accumulation in the Burrowing Mayfly *Hexagenia rigida* (Ephemeroptera) Exposed to CH_3HgCl or $HgCl_2$ in Water and Sediment," *Water Res.*, *27* (6), 1041–1048

Saouter, E., R. Le Menn, A. Boudou, and F. Ribeyre, 1991a, "Structural and Ultrastructural Analysis of Gills and Gut of *Hexagenia rigida* nymphs in Relation of Contamination Mechanisms," *Tissue Cell*, *23*(6), 929–938

Saouter, E., F. Ribeyre, and A. Boudou, 1991c, "Synthesis of Mercury Contamination Mechanisms of a Burrowing Mayfly *Hexagenia rigida*: Methdological Bases and Principal Results," in *Heavy Metals in the Environment* (J. P. Vernet, Ed.), Elsevier, Amsterdam, The Netherlands, pp. 175–184

Saouter, E., F. Ribeyre, A. Boudou, and R. Maury-Brachet, 1991b, "*Hexagenia Rigida* (Ephemeroptera) as a Biological Model in Aquatic Ecotoxicology: Experimental Studies on Hg Transfers from Sediment," *Environ. Pollut.*, *69*, 51–67

Snedecor, G. W., and W. G. Cochran, 1971, *Méthodes Statistiques*, Association de la Coordination Technique Agricole, Paris, France

Tessier, A., and P. G. C. Campbell, 1988, "Partitioning of Trace Metals in Sediments," in *Metal Speciation: Theory, Analysis and Application* (J. R. Kramer, and H. E. Allen, Eds.), Lewis Publishers, Chelsea, USA, pp. 183–199

Watras, C. J., N. Bloom, R. J. M. Hudson, S. Gherini, R. Munson, S. A. Claas, K. A. Morrison, J. Hurley, J. G. Wiener, W. F. Fitzgerald, R. Mason, G. Vandal, D. Powell, R. Rada, L. Rislov, M. Winfrey, J. Elder, D. Krabbenhoft, A. W. Andren, C. Babiarz, D B. Porcella, and J. W. Huckabee, 1994, "Sources and Fates of Mercury and Methylmercury in Wisconsin Lakes," in *Mercury Pollution: Integration and Synthesis* (C. J. Watras, and J. W. Huckabee, Eds.), Lewis Publishers, Chelsea, USA, pp. 153–180

Watras, C. J., and N. S. Bloom, 1992, "Mercury and Methylmercury in Individual Zooplancton. Implications for Bioaccumulation," *Limnol. Oceanogr. 37*, 1313–1318

Way, C. M., D. J. Hornbach, C. A. Miller-Way, B. S. Payne, and A. C. Miller, 1990, "Dynamics of Filter Feeding in *Corbicula fluminea* (Bivalvia: Corbiculidae), *Can. J. Zool.*, *68*, 115–120

Weber, J. H., 1993, "Review of Possible Paths for Abiotic Methylation of Mercury(II) in the Aquatic Environment," *Chemistry*, *26* (11), 2063–2077

Zhang, L., and D. Planas, 1994, "Biotic and Abiotic Mercury Methylation and Demethylation in Sediments," *Bull. Environ. Contam. Toxicol.*, *52*, 691–698

16

Process-Oriented Descriptions of Toxic Effects

Sebastiaan A. L. M. Kooijman (Amsterdam, Netherlands)

Ecotoxicology, Edited by Gerrit Schüürmann and Bernd Markert.
ISBN 0-471-17644-3   © 1998 John Wiley & Sons, Inc. and Spektrum Akademischer Verlag.

16.1 SUMMARY

In this chapter I review the problems that are inherent to the traditional empirical descriptions of toxic effects of chemicals on individual organisms, such as NOEC and LC_{50}/EC_{50} values, and show how process-based descriptions can solve these problems. Such descriptions have three components: a toxicokinetics component, which relates external to internal concentrations; an effect component, which relates internal concentrations to changes in the value of a physiological target parameter; and an output component, which relates the value of the target parameter to the variable that is measured, such as body size, cumulative number of offspring, etc. The theory has been worked out for effects on survival, body growth, reproduction, and population growth. I show that the resulting data analysis is simple in terms of the parameters that have to be estimated if the dynamic energy budget theory is used for the identification of possible physiological target parameters and for the output component of effect models. The effect parameters are the no-effect concentration (NEC), the tolerance concentration (or killing rate in the case of effects on survival), and the elimination rate (considers the rate at which effects build up during exposure). All three parameters do not depend on exposure time. Patterns in the toxicity of chemicals, such as the effect of ionization (pH), body size, and octanol–water partition coefficient, can be derived on the basis of first principles.

16.2 INTRODUCTION

16.2.1 Ecotoxicology

Ecotoxicology developed from human toxicology and pharmacology; these roots are still clearly visible today. Animals originally served just as models for humans; only much later did the health of animals themselves gain some importance. The extension to algae, plants, and ecosystems is rather recent. This is partly due to the poor structure of ecology as a science to predict population and ecosystem dynamics. The scanty knowledge in this area is mostly based on what is called unstructured population dynamics, where individuals are treated as identical copies. Since pollutants affect individuals, not populations directly, theories on structured population dynamics had to be developed, where individuals differ in one or more respects from each other (age, size, energy reserves, toxic burden, etc). These theories are also essential to give population dynamics a firm physiological rooting. Processes that define the physiology of ecosystems, such as nutrient recycling, carbon budgets, etc., can only be understood as further consequences of population dynamics. Theories about structured population dynamics are still in their second decade of existence, and they are rather complex, which explains why they still play a minor role in ecological thinking. Fortunately, this is now rapidly changing.

Most interest in ecotoxicology derives from a costs–risk analysis where the aim is to minimize production costs of goods as well as effects of human-induced chemical pollution that results from this production. The scientific problem of assessing effects of toxicants is wider, however, and also of fundamental interest. Organisms have been struggling for survival in chemically unfriendly environments from their first existence on. The appearance of oxygen as a byproduct of photosynthesis in the atmosphere has probably been fatal for most Precambrian organisms; the production of cyanides, alkaloids, and other secondary products by plants obviously function to deter herbivores; the tannins of acorns effectively blocks digestion by the European red squirrel, for instance, but the American grey squirrel found a way to deal with this defence of the oak and so managed to replace the red squirrel in parts of Europe (MacDonald 1995); bacteria that quickly transform sugar into acetate for later consumption manage to suppress growth of competitors. Natural growth-suppressing compounds, such as penicillin, are intensively applied in medicine. The ability of the parasitic bacterium *Wolbachia* to induce parthenogenesis in normally sexually reproducing species (doubtlessly via chemical interference) has recently attracted a lot of attention (Moran and Baumann 1994). The use of poisons by snakes, wasps, millipedes, and many other organisms to kill is well known. Botulin, which is produced by the bacterium *Clostridium botulinum*, causes frequent casualties among fish and birds. Biology is full of examples of chemical warfare with sometimes striking responses and defense systems (Agosta 1995). These different fields of interest have to my knowledge not yet been brought together. The reason is probably that the pressure to quickly "solve" practical problems in risk assessment has always been intense and has dominated ecotoxicological methods. Science, however, would greatly benefit from a more fundamental approach to the problem of effects of toxicants (human induced *and others*) on organisms.

The development of ecotoxicology began with aquatic environments because of early human health problems with drinking water, and also because of the relative simplicity of this environment in being well mixed and relatively easy to standardize. There is no such a thing as a "standard soil" with a "standard humidity". Although the basic principles are the same in aquatic and soil environments, the problems of bioavailability, transport, and transformation are very difficult to disentangle from toxicity in soil environments. Although standardization benefits the comparability of toxicity results, routine toxicity testing provides good examples to illustrate that standardization of experiments that lack a firm scientific basis actually hampers the development of such a basis. Up to now applied aspects (i.e., risk assessment, cost control) rather than the scientific problem have controlled experimental design and the methods of ecotoxicology, a most undesirable situation. The slogan "no change, no progress" certainly applies to the present situation of toxicity testing.

16.2.2 Risk Assessment

Risk analysis is based on expected effects of chemical compounds in the environment. Spatial and temporal scales are very important. There is a great difference between point and diffuse emissions and between emissions that occur only at one given moment or that continue in time. This is due to the differences in the relative importance of aspects such as transport, chemical transformation, adaptation, and local effects of the chemicals which may eventually be repaired by recovery due to migration from distant areas. The most important chemical transformation to be considered in the translation from laboratory results into expected effects in the field relates to the phenomenon of bioavailability. Only a small fraction of the compound is directly available in the field due to binding to ligands (mainly dispersed organic matter), but the fraction that is not available can become available later. Such delays can be difficult to judge. The weakest topic in risk assessment, however, is the set of biological effects itself, because of the problem that we already have to predict situations without toxicants. The numerical behavior of real-world populations are frequently erratic and difficult to understand. Hence, deviations induced by a toxic compound can be difficult to recognize, especially when the effects are small. This does not imply that small deviations are not important in the long run. The unpredictable behavior is one reason why experiments with mesocosms do not always demonstrate effects that can be expected on the basis of single-species toxicity tests under well-controlled conditions (Kooijman 1988). These remarks point out that many factors contribute to expected effects in the field, but it is through the effects on biota that human-induced chemical pollution derives its interest.

Central to the problem of risk assessment is the fact that it is easy to demonstrate effects on individuals under controlled conditions, but the environmental problem occurs at the ecosystem level. We therefore need a reasoning how these levels of biological organization are interrelated. Energy budgets are central in this translation because they define the processes of competition, predation, and propagation, which are the core of population dynamics and so of ecosystem dynamics. We thus have to know how energy budgets are affected and how this translates to a deviation of an ecosystem behavior without toxicants. The latter is obviously far from straightforward, in view of the problems we already have to create models where a rich diversity of species is persistent. We are just beginning to understand how population dynamics in simple (homogeneous) environments relate to properties of energy budgets and how they change during a life span.

The significance of toxicant-induced mutations and teratogenic effects for environmental risk assessment is extremely poorly understood. Some workers even deny the environmental significance of mutagenic compounds, since mutations in somatic cells hardly effect the health of the organism (except humans) and mutations in gametes are not relevant because of the abundance of unaffected gametes. Below, I discuss the rationale behind the idea that mutagenic compounds reduce the life span of organisms via interaction with the process of aging.

If the translation of laboratory results to field predictions is feasible, it will only be so if the description of effects is based on mechanistic insight. Existing methods to describe effects of toxicants are purely descriptive, however. The most important parameters are concentrations for which the survival probability is half the value of the control (LC_{50}), or of concentrations for which some quantity of interest such as number of offspring is half the value of the control (EC_{50}). In Section 16.3 the scientific problems with such a description, which also affect the estimation of concentrations that have no effect at all, are discussed.

The most popular method to obtain an estimate for such a concentration is to identify the highest tested concentration, the effect of which does not differ significantly from the control (the no-observed effect concentration, NOEC). See Yanagawa et al. (1994) and Hothorn (1994) for recent statistical reviews. Apart from the minor problem that the possible values are restricted to the limited set of tested concentrations, the fact that the estimate depends on the unknown power of the statistical test that is used to spot significant deviations from the control can lead to very strange results (Skalski 1981; Kooijman 1981a,b; Hoekstra and van Ewijk 1983; Pack 1983; Stephan and Rogers 1985). It is quite possible that the NOEC is larger than the EC_{50} for instance! Because legislation aims to prevent effects by reducing the amount of emitted chemicals, NOECs still play a major role in applied ecotoxicology. Better alternatives that eliminate the statistical problems of NOECs are also discussed here.

After a brief description of standard methods to describe effects of toxicants, the analysis of the most relevant effects of compounds is discussed: survival, mutation, reproduction. These effects are the most important in relation to the quality and quantity of life. Population dynamics depends directly on the processes of reproduction and survival. Reproduction can, however, be affected directly or indirectly via feeding (assimilation), growth, and maintenance. Why and how these physiological processes are related is the subject of the dynamic energy budget (DEB) theory, which is also discussed briefly. Although the mechanisms and analyses apply to aquatic toxicity, terrestrial toxicity, and human toxicity, the discussion focuses on aquatic ecotoxicity.

As I proposed earlier (Kooijman 1981), the effect size is related to the concentration of toxicant in the organism, and consequently the kinetics (i.e., uptake and elimination behavior) of the compound is relevant for understanding the effects. Since kinetics of lipophilic compounds depend on the lipid content of the organism, and therefore on the feeding status and the energetics, we have a second reason why the DEB theory is basic for the understanding of effect sizes. Hence, after describing the standard methods, I first discuss elements of the DEB theory, then toxicokinetics, and subsequently the effects and consequences of these effects on population dynamics.

16.3 STANDARD DESCRIPTIONS

16.3.1 LC_{50} and Gradient

The results of standard tests on the lethality of toxicants usually give the number of surviving animals as a function of the concentration of toxicant, which has been constant (hopefully) during a standardized exposure. The control survival probability is typically larger than 90% and a sigmoid curve is fitted to the number of survivors as a function of the concentration of toxicant (in the water). This curve is usually the log-logistic or log-probit curve, which are both characterized by a 50% point (the LC_{50}) and a gradient parameter, which represents the maximum slope of the number of survivors as a function of the logarithm of the concentration in the water.

When counts have been made at different observation times, the LC_{50} generally decreases as a function of exposure time. This phenomenon can be described well on the assumption that the survival probability depends on the concentration in the organism and that the toxicant follows some simple kinetics (Kooijman 1981). The relationship between uptake and effects are well established (Crommentuijn et al. 1994). Death is certain as soon as the toxicant in the organism exceeds a certain individual-specific threshold value. Individuals vary in physiological condition, and therefore in threshold values. The threshold value of a particular individual is assumed to be a (random) trial from a bell-shaped frequency distribution, which leads to a sigmoid concentration–response curve for the number of survivors. The effect, then, is described deterministically at the level of the individual and stochastically at the level of the cohort of tested organisms.

16.3.2 Problems

The basic assumptions in the standard model reveal several problems that are inherent to this method of description.

- Extreme standardization of culture conditions practically eliminates the physiological differences between individuals, but experimental practice shows that the gradient parameter cannot be increased beyond a given maximum. There is an upper limit for the maximum slope of the concentration response curve. In other words, there is a rather substantial variation of threshold values between individuals. It appears that the effect is *stochastic* at the level of the individual, not deterministic.

- The distribution that describes the variation in threshold values (log-logistic, log-probit) represents a rather arbitrary choice from the large set of possible distributions. This is of little relevance to the estimation of the LC_{50} value itself, as long as the selected curve fits the data. The particular choice of response function is, however, of great importance when we wish to obtain "small" effect concentrations, such as the LC_1 or LC_5, from the estimated parameters (LC_{50} and the gradient). The smaller the effect level, the larger the confidence interval, and more important, the more the result depends on the specific choice of the model.

- At high concentrations, the standard LC_{50}/EC_{50} model makes the very unrealistic prediction that there are unaffected individuals at very high concentrations of a very toxic compound. This property is due to the infinitely long upper tail of the distribution of threshold values. This is inconsistent with physical chemistry, and, from long experience, we simply know that no individual survives prolonged exposure to very toxic compounds.

- Since the gradient parameter reflects the variation of the (logarithms of the) threshold values, it is independent of the exposure time. In practice, however, the gradient tends to increase with the exposure time. We cannot simply parameterize this phenomenon by choosing a time-dependent function for the gradient parameter, because we then run into the problem that for certain (low or high) concentrations, the survival probability will increase in time, which is obviously not possible. The only way to incorporate such phenomena is to go back to a process-oriented description for survival.

- As mentioned before, LC_{50}/EC_{50} values themselves depend on exposure time. The problem is not completely solved by standardizing toxicity tests to a fixed exposure period. (The standardized toxicity test with *Daphnia magna* lasts 48 h, independent of the type of compound that is tested.) Surfactants react quickly; if no effect shows up after a few hours of exposure, it is unlikely that any effect will show up at that concentration. The situation is totally different for toxicants such as cadmium. The LC_{50} for an animal as small as *Daphnia* still decreases after 3 weeks of exposure. The LC_{50}–time behavior depends on properties of the chemical as well as those of the organism (especially body size). This mixture of properties is most unfortunate in application of LC_{50}/EC_{50} in risk assessment and reduces the comparability of results of standardized tests.

16.3.3 EC_{50}

There is not much to say about models that are used to describe sublethal effects. The standard approach is to relate the quantity of interest, such as the cumulated number of offspring during a standardized test period, to the logarithm of the concentration of toxicant in a logistic way. Apart from the NOEC, this gives three parameters: the control value, the EC_{50}, and a gradient parameter. This model just serves the purpose of describing a very limited data set, without bothering about the foundations. The aim seems to be to obtain an EC_{50} value for the quantity of interest and compare it with other EC_{50} values of other compounds and/or effects. A good example of a frequently applied nonsense parameter is the EC_{50} for biomass of algae in a 3-day test of growth inhibition. The value depends on the arbitrarily chosen test period and the (control) population growth rate (and therefore the medium composition, the temperature, light conditions, turbulence, and alga species). Although these limitations are known (Nyholm 1985), this is apparently no reason to abandon such measures. Many people seem to think that standardization solves all problems and leave to the poor administrator the problem of risk assessment which requires an integration of different information.

16.3.4 Conclusion

The standard model is based on assumptions that are not realistic. The fact that LC_{50} or EC_{50} values as well as NOECs depend on exposure time in a way that depends on both compound and organism characteristics hampers their application. The use of repeated observations to detect deviations from the control is problematic without an adequate model for the appearance of effects. In general, such observations are statistically dependent. Moreover, the standard model is inconsistent with the concept of the NOEC, because the log-logistic as well as the log-probit concentration–response curve approaches the control response for decreasing concentrations only asymptotically. The standard model can be extended to include a no-effect concentration (NEC) as a parameter (Kooijman 1981). Such a model solves the problem of statistical dependence and of the unknown power of the test to spot deviations from the control response. This is because the null hypothesis states that the NEC equals zero, while the alternative hypothesis asserts that it is positive. Twenty years of routine application indicates that point estimates for NEC are positive in about 50% of the cases and, in less than 10% of the cases, the NEC differs significantly from zero. This poor performance is due to the gradient parameter, whose unknown value reduces the information content of a concentration that shows an obvious effect to "the NEC is smaller than that concentration." Response curves with a positive NEC and with a NEC of zero are too similar. I consider my previous attempts to improve the standard model a failure and advocate a radical rejection of standard model in favor of the DEB-based models that are discussed below.

Small effect concentrations have been proposed to replace the NOEC (see Bruce and Versteeg 1992; Pack 1993). Apart from the problem of defining "small," such parameters hardly solve the problem; due to the arbitrariness of the choice of response function. The problems become less pressing for moderate effect concentrations (e.g., LC_{10} of LC_{25}), but who wants to allow such effects to occur? The larger the effect, the more important it is to have a reliable translation of the effect into consequences to the ecosystem. Such reliable translations do not exist and it is very unlikely that they will exist in the near future.

In summary, I have to conclude that the LC_{50} and EC_{50} are parameters that have desirable statistical properties (Hoekstra 1993), but nobody should attach much ecological importance to their values. They are hardly relevant for risk assessment and they are based on a model with a shaky basis. Useful descriptions should be process oriented. The existence of extensive data bases for LC_{50}/EC_{50} values should not be a reason to continue the application of the standard model.

When considering alternative methods to characterize toxic effects, it might help to have a closer look at mutagenic effects first. Traditionally, these effects are considered as a completely different category, but I will show that it is possible to construct a framework for the characterization of toxic effects, in which effects on survival, growth, reproduction, *and* mutation fit naturally.

Several other effects, such as those on respiration and mineralization, can be understood in terms of the mentioned effects. The theoretical basis of this has been described by Kooijman (1995).

The most frequently used bioassay for testing the mutagenic effects of compounds is the Ames test. A series of petri dishes with rich medium, but a very small amount of histidine, is inoculated with a mutant of the bacterium *Salmonella typhimurium* that is not able to synthesize histidine itself. Different amounts of test compounds are added to the dishes and the number of revertant colonies is counted. The number of revertant colonies is usually a linear function of the concentration. It appears that back mutations to the wild type, which is able to synthesize histidine, only occurs if the mutation takes place during the short period of growth when histidine is available (van der Hoeven et al. 1990). The relevance of this bioassay for mammals (and in particular for humans) is increased by adding liver homogenate of rats to mimic metabolic transformations that might occur in mammalian cells. NOECs are usually not obtained for this test; the only interest is in the slope of the concentration–response line.

False positives can occur in the Ames test, if background mutations exist (e.g., as a result of autoclaving the medium) and if the compound affects the growth rate. In that case, the compound increases the exposure period to the mutagenic compounds in the medium because the growth process lasts longer at high concentrations.

Each molecule is assumed to have an equal probability to be mutated, which results in a linear concentration–response relationship. Below I show that this idea of independent effects of molecules extends to all kinds of toxic effects *if* we focus on the right physiological target process. If we do that, we can obtain the familiar sigmoid concentration–response relationships for the number of survivors, the number of offspring, the size of organisms, etc. The DEB theory will be used to identify the target processes and how they are interrelated. I also show how mutagenic effects relate to the process of aging within the context of this theory.

16.4 DYNAMIC ENERGY BUDGETS

Energy budgets have proven to provide a very useful vehicle for understanding the connections between food uptake, digestion, maintenance, growth, reproduction, and aging. Since these relationships change in a systematic manner during a life span, such budgets should be regarded as being dynamic. Generally three life stages should be distinguished: embryo (which does not feed or allocate material to reproduction), juvenile (which feeds, but still does not allocate material to reproduction), and adult (male/female or both). The DEB theory gives detailed quantitative descriptions for the basic processes during life, which are supposed to apply to all heterotrophs, from bacteria to humans to whales.

Unicellular organisms are treated as juveniles because they take up resources from the environment and do not allocate energy or material to reproduction. It is beyond the scope of this section to discuss the details of the theory; only the most basic axiomata are mentioned. It can be shown (Kooijman 1995) that a tight coupling exists between mass and energy fluxes in heterotrophs, which rests on the concept of homeostasis: the ability of organisms to keep the composition of the body constant, despite variations in the chemical environment. All mass fluxes must be weighted sums of three categories of energy fluxes: assimilation, growth, and dissipating energy fluxes such as maintenance, heating, etc. Autotrophs are more complex in this respect, but follow the same basic rules. Consequently, the processes of mineral recycling and other ecosystem physiological processes can all be conceived as consequences of dynamic energy budgets.

Food uptake by juveniles and adults is scaled as proportional to the surface area and depends hyperbolically on food density. The surface area is important because it is linked to encounter rates with food particles at low food density (filtering rates, searching rates) and to food processing rates at high food density (surface area of gut, which is about proportional to surface area of the whole animal). A hyperbolic relationship between food uptake and food density results if each food particle blocks the uptake of other food particles for some time, leading to a rejection of particles when the animal is "busy." This handling time may relate to mechanical handling of food particles, but also to the digestion process and/or some process deeper down the line of food processing. The maximum ingestion rate is realized when the animal spends all its time handling food particles, and no waiting time between subsequent particles is left.

Digestion efficiency is constant and so is independent of food density and the size of the organism. Material and energy that is derived from food is added to the reserves. The reserves in all three life stages are used at a rate that depends on the amount of reserves and the size of the organism. A fixed fraction of this mobilized material is used for maintenance plus growth; the rest is used for development and reproduction. (Reproduction only applies to adults.) The amount of energy that is required for maintenance is proportional to the volume of structural biomass that is synthesized as a result of resource allocation to growth. Thus, growth stops if the energy allocated to growth plus maintenance is fully consumed by the process of maintenance; reproduction is a parallel process.

The organism switches from embryo to juvenile and from juvenile to adult when it has invested a certain cumulative amount of energy into the increase in its state of maturity. The material allocated to reproduction is cumulated into a buffer and at the moment of reproduction this material is converted into one or more eggs. (Some organisms, such as most mammals, do not reproduce via eggs but via fetuses, which receive nutrition from the mother during development.) The structural biomass of eggs is negligibly small initially; the embryo develops at the expense of its reserves.

Unicellular organisms divide when they have invested a certain amount of energy in development. Due to the partitioning rule for energy that is mobilized from the reserves, this occurs at a certain structural biomass. More precisely, the cell starts to duplicate its DNA when it exceeds a certain structural biomass. This duplication process lasts a fixed period during which the cell continues to grow. Since the rate of growth depends on the amount of reserves, the cell size at division depends on the feeding conditions.

As a result of respiration (the use of oxygen to free energy from reserve materials), DNA is affected via free oxygen radicals with an efficiency that depends on the type and amount of antioxidants in the tissue. The exact link between respiration and energy fluxes on the basis of the above-mentioned rules is somewhat technical, but fully specified by the set of rules for energy uptake and use. When DNA is affected, there are two consequences. The original protein (usually enzyme) is no longer produced by the affected gene and that gene produces disfunctional proteins. Affected cells do not divide. For unicellular organisms this means a kind of death. For metazoans this means that the damage, measured as the cumulated amount of disfunctional proteins, can be diluted by growth, i.e., the increase of unaffected structural biomass. It appears that a very good agreement with experimental survival patterns results if we simply take the hazard rate to be proportional to the *cumulated* damage. The hazard rate at a certain age is defined as the instantaneous death rate, given survival up to that age. The survival probability at that age equals the exponent of minus the cumulated hazard rate up to that age. Notice, however, that aging is just one of the components that affect survival. The process of aging is described by a single parameter, called the aging acceleration. It is an acceleration because the hazard rate is proportional to cumulated damage; the cumulation process transforms a rate (dimension time^{-1}) into an acceleration (dimension time^{-2}).

This model for aging gives the correct predictions of the interrelationships between life span and feeding level for ectotherms, i.e., animals that do not allocate energy to heat their body to some preset temperature. Endotherms seem to decrease the efficiency of the defense system to catch free radicals during life. This can be understood in view of the fact that organisms use free radicals to build up genetic variability in the gametes, which are treated in the same way as somatic cells with respect to mutation frequencies. There is a trade off between the expected life span of individuals and genetic variability.

In summary, the DEB theory selects reserves, structural biomass, and cumulated damage as the state variables to describe the various processes quantitatively. A set of 11 parameters that determine all rates in the three life stages, given food density and temperature, is assumed to be fixed for a certain individual during its life span. Arrhenius relationships describe how rate parameters depend on temperature. It is beyond the scope of this chapter to explain why and how these 11 parameters tend to covary when we compare individuals of different species. Body size scaling relationships, which describe how physiological life history parameters change with adult body size, can be derived correctly on the basis of the structure of the theory, without using any empirical argument. Indeed, it explains why the respiration rate is roughly proportional to biomass raised to the power of 0.75; a well-known physiological enigma seems to be eliminated.

The DEB theory is at the heart of population dynamics. Populations are conceived as a group of individuals that obey the rules implemented in the DEB theory and, as a first approximation, only interact via feeding on the same resource. Population dynamical theories that account for differences between organisms in terms of age, size, etc. are called structured population dynamics (see Metz and Diekmann 1986 for an introduction). The laws of energy and mass conservation then determine how population size and structure (in terms of frequency distributions of body size and reserves) change in time, given a specification of the environment in which a population lives (Kooi and Kooijman 1994, 1994a, 1995). This is obviously rather complex for organisms that do not change in shape during growth, which makes their surface area proportional to their volume raised to the power of 2/3. We use parallel computers to analyze the dynamics of such populations. If the organisms do change in shape during growth, such that their surface area is proportional to their volume, the rules for population dynamics simplify to what is known as unstructured population models, i.e., models in which the individuals are combined into a total population biomass. Unicellular organisms have a very interesting intermediate position, because details about their growth from baby cell to mother cell, the size difference between which amounts to a factor of 2 only, do not greatly affect the population dynamics. We use these paradigms to find simple approximations for population dynamics and conditions under which these approximations hold. Dynamic transformation efficiencies for the conversion of food into biomass play a central role in simplified descriptions of structured population dynamics.

Populations can be tied into food chains and food webs, which can easily show a very complex dynamic behavior in simple environments. A theoretical microbial food chain of length 4 (i.e., a nonreproducing substrate, prey, predator, and top predator) can behave chaotically in a simple chemostat environment with special combinations of throughput rate and concentration of substrate in the feed. At this moment we are working on simplified approximate descriptions of summary statistics, such as total biomass of the food web, which might contribute to ecological insight. It seems that maintenance requirements set a natural maximum to the length of a food chain, for instance, but we still cannot derive this maximum as a function of the parameter sets of the species in the chain.

Although we are heading towards an understanding of the behavior of simplified ecosystems in terms of mass (nutrient) and energy fluxes in the compartments of primary producers (plant/algae), consumers (animals), and decomposers (bacteria/fungi), progress is slow and painstaking. It is only fair to mention that we are still far from having a scientifically sound understanding of ecosystem dynamics. A major obstacle is spatial heterogeneity. It seems that this component is very important for realistic population dynamics but very hard to incorporate and analyze in models.

16.5 TOXICOKINETICS

The simplest, and in many cases, most realistic model that describes the uptake and elimination process is the one-compartment model. It assumes that uptake rate is proportional to the concentration in the environment and elimination rate is proportional to the concentration in the tissue. Notice that the uptake rate of toxicants depends on the environmental concentration in the same way as food, provided that the "handling" time is negligibly short with respect to the period between subsequent arrivals of molecules. This seems realistic at low concentrations of toxicant, where the handling in this case refers to residence time in the exposed membranes.

At a constant concentration in the environment, this gives an exponentially satiating curve when the tissue concentration is plotted against exposure time. The ratio of the ultimate tissue concentration to the environmental concentration equals the ratio of the uptake to the elimination rate; this is known as the bioconcentration coefficient.

Uptake can be direct from the environment, for instance, via the respiratory system (gills or lung), the skin, or food. Food uptake is proportional to surface area, according to the DEB theory, and so the total uptake rate of the toxicant is proportional to surface area. This also holds for the elimination rate, which implies that the waiting time to reach a certain tissue concentration is proportional to the ratio of the volume to the surface area: the volumetric length of an organism. (The volumetric length is defined as the cubic root of the volume.) Effects of toxicants relate to tissue concentrations. The test on survival of 1-mm-long water fleas (*Daphnia*), which weigh some 0.2 mg (volume $\simeq 0.2$ mm^3), is standardized at 48 h, and of 40-mm-long young *Salmo*, which weigh some 85 mg (volume $\simeq 85$ mm^3), at 96 h. If the species have the same sensitivity and surface-area-specific kinetics, there must exist a factor of $\sqrt[3]{85/0.2} = 7.5$ between the lengths of the tests, rather than $96/48 = 2$. Little wonder that *Daphnia* is generally considered to be a sensitive test organism.

If an animal grows during exposure the kinetics become a bit more complicated due to the changing surface area and the dilution by growth (Figure 16.1). A *Daphnia* reproduction test is standardized at 21 days and starts with 0.8-mm-long neonates. After 21 days the daphnids are 4 mm (in the control), which means an increase in weight by a factor of $5^3 = 125$. With use of the DEB theory, correction for this substantial growth during exposure is straightforward.

Another cause of deviation from a one-compartment kinetics relates to changing lipid content of the animal due to changing feeding conditions, for instance. Animals such as the blue mussel *Mytilus edulis* accumulate lipids during the year; at spawning, they release most of it in gametes and drop in dry weight by more than a factor two. Lipophilic compounds accumulate in lipid-rich organisms rapidly. For this reason, eels (*Anguilla*) tend to have much higher loadings of lipophilic compounds than other fish. The DEB theory can be used

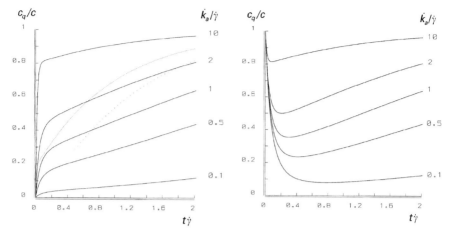

Figure 16.1 Uptake and elimination during growth. The left curves for the scaled tissue concentration start from $c_q(0) = 0$, the right curves start from $c_q(0) = c$, where c stands for the environmental concentration. The different curves correspond with different choices for the value of the elimination rate $\dot{k}_a$, relative to the von Bertalanffy growth rate $\dot{\gamma}$. The finely stippled curve represents (scaled) body length and the other stippled curve the (scaled) reproduction rate. The (scaled) lengths at the start of exposure and reproduction are realistic for the water flea *Daphnia magna* and the value $t\dot{\gamma} = 2$ corresponds to 21 d for *D. magna* at 20°C. All curves in both graphs have an asymptote at the value 1. If the product of the von Bertalanffy growth rate and the exposure time $t\dot{\gamma} > 0.4$, the curves in the left and right panels are almost identical; i.e., independent of the initial tissue concentration. The deviations from $c_q = c$ can therefore be attributed to "dilution by growth." For f denoting the scaled functional response (i.e., feeding rate), the change in scaled tissue concentration c_q and scaled length l is given by

$$\frac{d}{dt}c_q = c\dot{k}_a f/l - c_q\left(\dot{k}_a f/l + \frac{d}{dt}\ln l^3\right) \quad \text{and} \quad \frac{d}{dt}l = \dot{\gamma}(f - l)$$

to account for these changes on the basis of the assumption that partitioning of the compound over the various body fractions is fast with respect to the uptake and elimination rates. If partitioning is not fast, we have to turn to multicompartment models, but experience shows that experimental data hardly justify such a move. The problem is that the number of parameters is already large enough to obtain the flexibility to fit any observed data well, irrespective of the realism of the model. Application of multicompartment models is only justified if the compartments are identified and their burden measured.

The third class of deviations from a one-compartment model relate to metabolic transformations of the toxicant, which usually occurs in the liver. These transformations usually increase the hydrophilicity and therefore the elimination rate. They represent a mechanism by which the organism detoxifies the compound. Both the elimination and effects relate to the aqueous fraction, not to the fraction in the lipid storage of organisms, because the latter are metabolically inert. This is best illustrated by freshly laid birds eggs, which are almost pure reserves, the structural biomass of the embryo being negligibly small; such eggs hardly respire. During ontogeny (incubation), the respiration rate increases as the structural biomass of the embryo increases. This also means that effects are likely to show up during starvation when the reserves are used. Models for metabolic transformations are rather organism and compound specific.

If the transformation rate is proportional to the concentration in the aqueous fraction and partitioning is fast, we still are in the class of one-compartment models (with varying coefficients). This class of model is rather easy to implement in effect studies. The use of the extremely parameter-rich multicompartment models originate from pharmacology but usually have too much detail to apply to ecotoxicology.

16.6 EFFECTS

As I already mentioned in the introduction, organisms have evolved in a chemically varying environment; consequently they can cope with varying concentrations of any particular compound, as long as the variations are within a certain range. The upper boundary for this range, i.e., the internal no-effect concentration, might be zero for particular compounds. Each molecule of such compounds induces effects with a certain probability, but for most compounds, the upper boundary is a positive value. The lower boundary is zero for most compounds because they are not necessary for life. Elements such as copper are required, so that the lower boundary for copper is a positive value. Effects of a shortage of a compound resemble effects of an overdose in their kinetics. The founder of ecotoxicology, Sprague (1969), studied the effects of toxicants in bioassays using oxygen shortage as an example. Although many interrelationships exist between nutrition and toxic effects, the upper boundary of the tolerance range attracts most attention in ecotoxicology, due to its application in risk assessment studies, while ecology focuses on the lower boundary (see White 1993).

Discussions reveal that not every ecologist feels comfortable with the assumption of a tolerance range for chemicals. For temperature it is widely accepted, however. The fact that ecological observations are usually made without a detailed specification of the chemical environment implies that most ecologists do accept the existence of a tolerance range for "natural" compounds (whatever they might be). I hardly see the assumption of a tolerance range for toxicants as a problem, because we always can (and should) test the hypothesis that the upper boundary of the tolerance range (i.e., the internal no-effect concentration and therefore also the (external) no-effect concentration, NEC) equals zero. If we cannot reject this hypothesis, we have to accept the possibility that each molecule of that compound can have an effect. If this compound still has to be released into the environment, this might be a good reason to give priority to research on the magnitude of the ecological effect of such an emission. The primary purpose of routine toxicity testing is to set priorities for further research, not to predict ecological effect sizes.

Each physiological process has its own tolerance range for any compound. The upper boundaries can be ordered, which means that at the lowest tissue concentration range that produce effects, only one physiological process is affected, while at high tissue concentrations many processes are affected. As long as the partitioning of the compound over the various body fractions is fast with respect to the uptake/elimination kinetics for the whole animal, it is not essential to specify the tissue or organ in which the most sensitive physiological process is affected. This only becomes essential if the partitioning is slow. This makes multicompartment models as a basis for effect studies much more complex to apply: we have to know a great deal more. Notice that one-compartment models can handle different concentrations in different organs as long as partitioning is fast. Observed deviations from one-compartment kinetics with constant coefficients frequently relate to the variations in the coefficients, not necessarily to the presence of more compartments.

Basic to the description of small effects of toxicants is the notion that each molecule that exceeds the tolerance range contributes to the same extent to the effect. Interactions between the molecules only occur at higher tissue concentrations. This means that the effect size is, as a first approximation, a linear function of the tissue concentration. This point of view relates to the Taylor approximation describing how effect size relates to tissue concentrations. This function might be nonlinear, but we use only the first term of the Taylor approximation at the upper boundary of the tolerance range. The theorem by Taylor states that we can describe any nonlinear function in a given interval arbitrarily well with an appropriate polynomial function if we include enough higher order terms. Hence, when we want to improve the description of effects, if they happen to deviate from a linear relationship with tissue concentrations, we simply include the squared term, the cubed term, etc. Such improvements will rapidly become counterproductive because we increase the number of parameters that must be estimated and because higher tissue concentrations will affect more physiological processes. So we are increasing precision at the wrong points. Practice teaches that very good descriptions can be obtained by just making the relationship between effect size and the tissue concentration linear, even at rather high effect sizes, provided that we focus on the correct physiological process.

I will now discuss a selection of frequently occurring effects of toxicants that relate to the energy budgets of organisms, all based on the above-mentioned principles. Effects such as those on the immune system or resistance against parasites are not discussed here.

16.6.1 Effects on Survival

When accidental mortality is independent of age, the hazard rate due to this cause of mortality is constant and the corresponding survival probability equals the exponent of (minus) a constant times time.

The model for aging has proved to be successful when the hazard rate is taken to be proportional to cumulative damage. The resulting survival probability then depends on age in constant environments as the exponent of (minus) a constant times age cubed, apart from the problem of dilution by growth.

These mortality models suggest that effects of toxicants on survival can be modelled by making the relationship between the hazard rate and the tissue concentration linear, that is proportional to the tissue concentration that exceeds the threshold value (i.e., the upper boundary of the tolerance range; see Kooijman and Bedaux 1996a,b). The proportionality factor, called the killing rate, is a measure of the toxicity of the compound that is independent from the exposure time. In (acute) toxicity tests that are started with animals from a control culture, this cause of mortality results in a survival probability that equals the exponent of (minus) a constant times time squared if the exposure time is short with respect to the inverse elimination rate. When the animals reach steady state, so that the tissue concentration does not increase any longer, the survival probability equals the exponent of (minus) a constant times time. Since the hazard rate starts to increase from the control value at the moment that the NEC value is exceeded, we have to wait longer for effects at lower concentrations.

An important difference between the hazard and the standard model is that survival of a particular individual is *stochastic* in the hazard model and *deterministic* in the standard model. The choice for a stochastic formulation in this case can best be illustrated with two alternative models for the outcome of a tossing experiment with a dice. A very simple stochastic model just states that each of the six possible outcomes will occur with the same probability. A very complex deterministic model describes in detail how the dice is tossed, how it bounces on the table, and how it will eventually come to rest. With this model it is possible to predict the outcome deterministically. The reason why this complex deterministic model fails to be practical is that we cannot control the tossing movements in sufficient detail and because of imperfections in the microscopic detail of the surface of the table and in the elasticity of the dice. Similarly, the death of an individual has too many molecular roots for a deterministic description to be practical. The different individuals are thus treated as identical (stochastic) copies, rather than different (deterministic) copies.

It is of course possible to account for differences between individuals in the hazard-based model, which then appear as differences in parameter values for each individual. This makes sense for animals that are collected from the field, where differences in health, age, size, feeding conditions, and sex, all contribute to differences in sensitivity. The way to proceed is to describe this variation in the set of four parameters by some (multivariate) scatter distribution and obtain what is called a "mixture" in applied probability theory. The number of parameters in this scatter distribution is obviously larger than four and the resulting survival model can easily become complex. One must be prepared to estimate these extra parameters by increasing the number of observation times and tested concentrations. When the individual sizes are measured and the mortality observations specify the individuals, it is possible to correct for these differences in size in a rather straightforward way which does not increase the number of parameters.

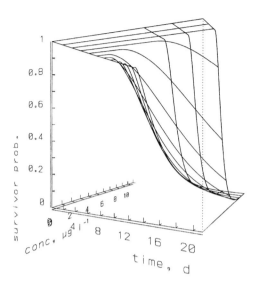

Figure 16.2 Toxic effects on survival. The elimination rate is taken here to be equal to the von Bertalanffy growth rate (0.1 day^{-1}) and the no-effect concentration is 1.5 μg L^{-1}.

The mortality is thus taken to be a function of the environmental concentration and the exposure time, which is described by the NEC, the killing rate, the elimination rate, and the control mortality rate. Figure 16.2 illustrates the response surface when growth during exposure is of importance, such as for the chronic (3 weeks) toxicity test with *Daphnia* (Figure 16.1). Even if the number of survivors are counted just once, we have to estimate all four parameters. If the exposure time is short with respect to the inverse of the elimination rate, we can only estimate the product of the elimination and the killing rate and are unable to translate the NEC for that test to an ultimate NEC. If the exposure time is large with respect to the inverse of the elimination rate, we end up with an exponential survival model that has the (ultimate) NEC and the killing and control mortality rates as parameters. So we can sandwich the four-parameter model between two three-parameter models. If the cultures are in good condition and the test has been done carefully, we can avoid control mortality, which further reduces the number of parameters. This low number of parameters is essential for routine applications because the data do not allow the estimation of a larger number of parameters. Although the model has better mechanistic underpinnings than the standard model, the number of parameters is lower because the gradient parameter and the LC$_{50}$ are both replaced by the killing rate.

If the environmental concentration is not constant, it is rather straightforward to correct for changes in these concentrations if they are measured functions of time. This is one of the advantages of a mechanistic model. If these changes can be calculated, for instance, if the compound disappears from the medium due to accumulation in the animal(s), it is in principle not necessary to measure the

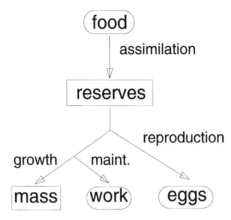

Figure 16.3 The powers as specified quantitatively by the DEB model for an ectotherm with body size and reserve density as state variables. Toxic compounds that affect reproduction can do so directly, or indirectly via assimilation, growth, and maintenance. The rounded boxes indicate sources or sinks.

decline as a function of time. One measurement at the end of the experiment will suffice.

The advantage of hazard-based modelling is that other causes of mortality can easily be built in, so that the significance of toxicant-induced mortality can be evaluated. This fits into the theory of competing risks (David and Moeschberger 1978), which is based upon the fact that an animal can die only once, despite many possible causes of death. A direct application in ecotoxicology is in the toxicity of mixtures of compounds that do not interact. The contributions of different compounds to the hazard rate just have to be added, i.e., the corresponding survival probabilities have to be multiplied.

Mutagenic compounds decrease the life span of animals very similar to the aging process. This can be illustrated with the experiments by Robertson and Salt (1981), who studied feeding, growth, and life span of the rotifer *Asplanchna* using the ciliate *Paramecium* as food. These ciliates had been cultured on lettuce, which is extremely rich in nitrate. Nitrate can be transformed into nitrite, which is mutagenic. It appears that aging acceleration has a perfect linear relationship with the feeding level, which strongly suggests that the effect of nitrite is very similar to that of free (oxygen) radicals and therefore affects the parameter "aging acceleration" linearly. This is consistent with the model for mutagenic effects in the Ames test for bacterial populations. I show in the next section that all sublethal effects of toxicants can be described by linear effects on parameter values.

16.6.2 Effects on Growth and Reproduction

The DEB model describes how resources (i.e., products derived from food) are allocated to reproduction. Toxic effects of chemicals change the allocation via the parameter values. Since the processes of assimilation (i.e., the combination of feeding and digestion), growth, maintenance, and reproduction are intimately interlinked, changes in any of these processes will result in changes in reproduction (Kooijman and Bedaux 1996b,c). Two classes for the mode of action of compounds are distinguished: direct and indirect effects on reproduction.

When reproduction is affected directly, assimilation, growth, and maintenance are not affected. There are two closely related routes within the DEB framework which affect reproduction directly. One is via survival of each ovum, and the other is via the energy costs of each egg.

The survival probability of each ovum is affected as has been discussed in the previous section in relation to effects on survival, except that the sensitive period is taken to be relatively short and fixed. (The case in which the sensitive period is long has already been discussed in the previous section.) The combination of an effect on the hazard rate of the ovum and a fixed sensitive period results in a survival probability that depends on the local environment of the ovum. This leads to another important difference to the previous section: the local environment of the ovum is the tissue of the mother, rather than the environmental concentration. Therefore the relevant concentration changes with time. The toxicity parameters that are relevant for the survival probability of an ovum are the NEC, as before, and the tolerance concentration, which is inversely proportional to the product of the killing rate and the length of the sensitive period. The elimination rate defines how the effect builds up during exposure.

The reproduction rate, in terms of number of eggs per time, equals the ratio of the energy allocated to reproduction to the energy costs of an egg. If the compound affects the latter, it can be modelled by making the energy costs a linear function of the tissue concentration. The model is mathematically different from the hazard model, but behaves quantitatively rather similar. It has the same three toxicity parameters: the NEC, a tolerance concentration, and elimination rate. One possible interpretation of the tolerance concentration is the EC_{50} for the energy costs per egg minus the NEC.

Figure 16.4 illustrates the response surface for the various effects on reproduction as functions of the exposure time and the environmental concentration.

Allocation to reproduction is initiated as soon as the cumulative investment into the increase of the state of maturity exceeds some threshold values. Since direct effects on reproduction only affect the translation from energy allocated to reproduction into number of offspring, these modes of action do not affect the time of onset of reproduction. Indirect effects on reproduction via assimilation, maintenance, and growth do delay the onset of reproduction. The occurrence of such delays is the best criterion to distinguish direct from indirect effects.

Figure 16.4 Direct (left panel) and indirect (right panel) effects on reproduction. The various tolerance concentrations are chosen such that the effect size is similar. The elimination rate is set equal to the von Bertalanffy growth rate, $\dot{k}_a = \dot{\gamma} = 0.1\,\mathrm{day}^{-1}$, and the no-effect concentration is $c_0 = 1.5\,\mu\mathrm{g\,L}^{-1}$.

hazard, $c_{\mathrm{H}} = 1.62\ \mu\mathrm{g}\ \mathrm{L}^{-1}$

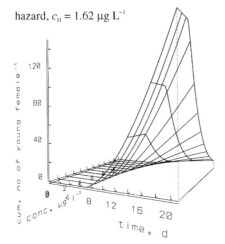

maintenance, $c_{\mathrm{M}} = 6.02\ \mu\mathrm{g}\ \mathrm{L}^{-1}$

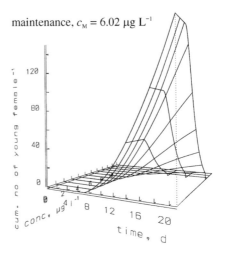

costs, $c_{\mathrm{R}} = 1.35\ \mu\mathrm{g}\ \mathrm{L}^{-1}$

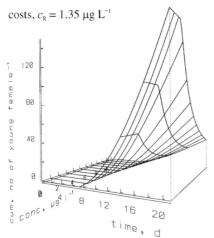

growth, $c_{\mathrm{G}} = 0.61\ \mu\mathrm{g}\ \mathrm{L}^{-1}$

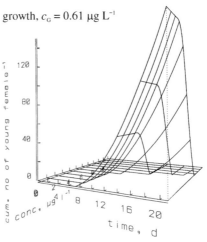

assimilation, $c_{\mathrm{A}} = 9.57\ \mu\mathrm{g}\ \mathrm{L}^{-1}$

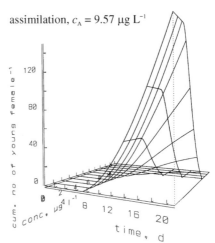

Indirect effects on reproduction all follow the same basic rules: the relevan parameter (surface-specific assimilation rate, volume-specific maintenance costs or volume-specific costs of growth) is taken to be a linear function of the tissue concentration. Since the assimilation rate represents a source of income, rathe than costs, it is assumed to *decrease* linearly with the tissue concentration, rathe than increase. The effects on the reproduction rate as a function of environ mental concentration and exposure time all work out rather similar and all have the same three toxicity parameters: NEC, tolerance concentration, and elimina tion rate. If growth has been measured during exposure, or if the sizes of the animals at the end of the exposure period have been measured, it is possible to identify the mode of action. Figure 16.6 illustrates the response surface for direc effects on growth as a function of the exposure time and the environmenta concentration, as well as for indirect effects on growth via assimilation and maintenance. The differences in effects on reproduction are too small to do this on the basis of effects on reproduction. Figure 16.7 compares the three indirec effect models fitted to the same data. It shows that the models differ little in terms of quality of fit.

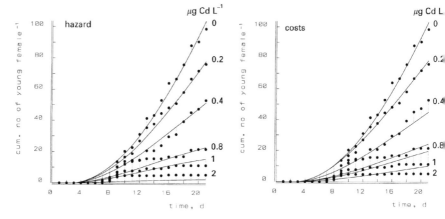

Figure 16.5 Direct effects on *Daphnia* reproduction: Cadmium. The mean cumulated number of young per female daphnid as a function of the exposure time to several concentrations of cadmium The fitted curves represent least squares fits of the hazard (left) and the cost (right) model for effects on reproduction. Given an elimination rate of $\dot{k}_a = 0.05$ day^{-1}, the estimated parameters are

	$c_0, \mu g\, L^{-1}$	$c_*, \mu g\, L^{-1}$	R_m, day^{-1}
Hazard	0.023	0.166	13.1
Cost	0.047	0.069	13.1

maintenance, $c_M = 0.6\ \mu g\ L^{-1}$ growth, $c_G = 0.8\ \mu g\ L^{-1}$

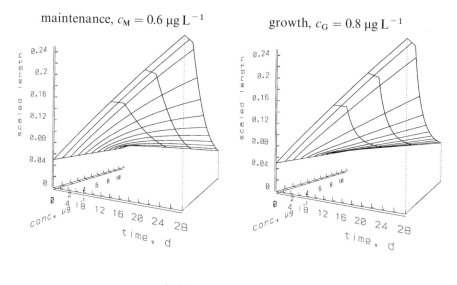

assimilation, $c_A = 10\ \mu g\ L^{-1}$

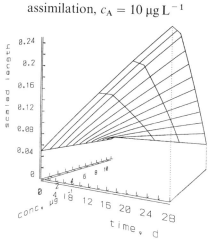

Figure 16.6 Direct and indirect effects on growth. The elimination rate is $\dot{k}_a = 0.1\ \text{day}^{-1}$ and the no-effect concentration is $c_0 = 1.5\ \mu g\ L^{-1}$. The von Bertalanffy growth rate $\dot{\gamma} = 0.008\ \text{day}^{-1}$ is typical for zebra fish at 26°C.

Toxicants sometimes stimulate reproduction at low concentrations, rather than reduce it, a phenomenon known as "hormesis". The actual cause is largely unknown and therefore difficult to model. For some compounds that showed hormesis at high feeding levels, I have been able to avoid hormesis in *Daphnia* reproduction tests by reducing feeding levels. This points to an explanation in terms of suppression of a secondary stress by the toxicant at low concentrations. It is far from obvious to what extent this explanation is general.

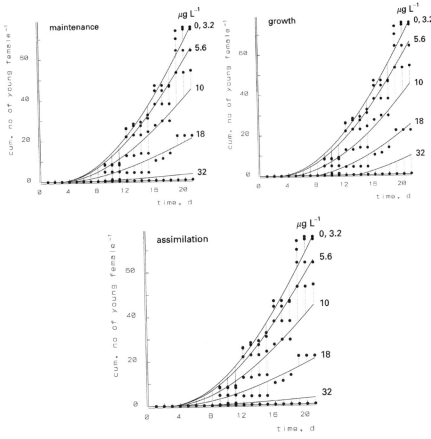

Figure 16.7 Indirect effects on *Daphnia* reproduction: 3,4-DCAN. The mean cumulated number of young per female daphnid as a function of the exposure time to several concentrations of 3,4-dichloroaniline. The fitted curves represent least squares fits of the model for effects on reproduction via maintenance, growth, and assimilation. The estimated parameters are

	$c_0, \mu g\,L^{-1}$	$c_*, \mu g\,L^{-1}$	k_a, day^{-1}	$\dot{R}_m, day^{-1}$
Maintenance	3.84	31.7	0.75	9.50
Growth	3.51	3.85	0.90	9.56
Assimilation	3.37	65.8	0.87	9.53

A wise strategy to deal with hormesis is to favor test conditions in which hormesis is avoided.

16.7 EFFECTS ON POPULATION DYNAMICS

In principle, the consequences of effects on individuals on population dynamics can be analyzed on the basis of the now rapidly developing theory of structured population dynamics (Webb 1985; Metz and Diekmann 1986; Hallam et al. 1988; Ebenman and Persson 1988; DeAngelis and Gross 1992). The mathematics of this theory is tedious, however, and computer simulation studies will remain necessary for cases that are of practical interest. In such simulation studies, each individual or group of individuals is followed in time, which requires substantial computation efforts. Ecotoxicological applications are beginning to appear (Hallam et al. 1988; Lassiter and Hallam 1990; Kooijman 1993) and point to the fact that effects of chemicals can manifest in a rather complex way at the population level. The simulation studies will hopefully help to find useful mathematical approximations that allow a qualitative analysis of population dynamics.

Direct toxicity testing for effects on the population growth rate is only feasible for very small organisms. An example is the alga growth inhibition test, where batch cultures are followed during a short exposure period (Kooijman et al. 1996). Since alga cells are so small, it seems safe to assume that the intracellular concentration of toxicant is instantaneously in equilibrium with the environment for most compounds. Three modes of action can be distinguished for the toxicant. It might increase the costs of growth in terms of the required amount of energy and/or nutrient per volume of structural biomass. It might affect the hazard rate (over the generations of cells), and it might do so for a short period only. The latter refers to the process of selection. Algal cells originate from control cultures. The transition to experimental conditions that include chemical stress might be lethal for some cells; some cells might be more sensitive than others, possibly depending on the cell cycle. In any case, it seems that some toxicants only delay population growth, but once a population begins growing, it does so at the control rate. All three modes of action lead to two toxicity parameters: the NEC and the tolerance concentration. The elimination rate is taken to be large with respect to the inverse of the interdivision interval of the cells. It therefore does not show up as an independent parameter.

The cells in the control grow exponentially as long as the environment is constant. This results from the fact that the daughter cells repeat the physiological behavior of the mother cell. For the same reason, all animal populations grow exponentially in constant environments. The fact that they rarely do this for a prolonged time in the field relates to changes in the environment, frequently due to exhaustion of the resources. The way how hazard and reproduction rate together determine the population growth rate is given by a particular result in the theory of structured population dynamics: the characteristic equation. A detailed discussion is beyond the scope of this chapter. It suffices here to note that the age-specific hazard and reproduction rates not only specify the population growth rate, but also the age structure in the population. The size distribution can be obtained from the age distribution on the basis of the DEB model, which specifies the age-specific growth rates that are necessary for this translation.

This reasoning can be used to evaluate the effect of a change in parameters on the population growth rate, which reveals that the effect of toxicants on population dynamics depends on the gross production rate of the population

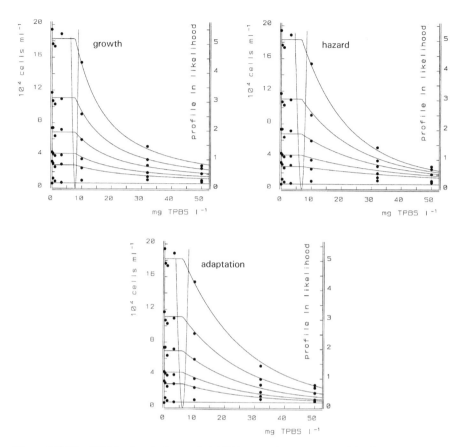

Figure 16.8 The NEC in the algal growth inhibition test is not very sensitive for errors in the selection of the mode of action of a compound. The effects of tetrapropybenzenesulfonate (TPBS) on the growth of the diatom *Stephanodiscus hantzschii* are shown. The NEC is well defined; the 99% confidence interval can be read from the figure by looking at concentrations for which the profile likelihood is below 3.317. Three different models are fitted to the same data set. Data from Kooijman et al. (1983).

Parameter	Unit	Growth	Hazard	Adaptation
c_0	mg l^{-1}	7.82	6.78	6.08
$c_G, \dot{k}_\dagger, c_H$	mgL^{-1}, L (mg d)$^{-1}$, mgL^{-1}	36.3	0.0123	20.5
$\dot{\mu}_0$	d^{-1}	0.480	0.480	0.480
N_0	10^4 cells mL^{-1}	0.655	0.633	0.696
$\hat{\sigma}$	10^4 cells mL^{-1}	0.412	0.447	0.416

(Kooijman and Metz 1983; Kooijman 1991, 1993). If food density is low, almost all food is used for maintenance. If a compound affects maintenance, this directly translates into an effect on the standing crop. If a compound affects reproduction, however, this hardly affects the population at low food densities, because of the limitation by food. These insights can easily be verified by simple experiments with batch cultures. If we supply a population of daphnids with a fixed amount of food per time, the population will rapidly grow to a plateau value, where the incoming food just suffices for the maintenance of all individuals. Since the life span of daphnids is several months, death hardly plays a role, and growth and reproduction are arrested when the population reaches its plateau value. If we add a toxicant at a concentration where it affects reproduction, it might affect the time at which the plateau is reached, but not the plateau itself. If we start harvesting the population at steady state, we will see that the biomass will settle at a value that is slightly reduced compared with the plateau value (so each individual gets more food, because the food supply remains constant), and growth and reproduction are resumed such that it matches our harvesting rate. Only now do we start to see that the toxicant affects the population biomass. If we increase the harvesting rate, we will see that the maximum sustainable harvesting rate is reduced by the toxicant. These considerations show that a toxicant at a constant concentration will have a dynamic effect on a population, even apart from problems which are physical and chemical in nature. This is one reason why the ecological consequences of effects of toxicants are difficult to analyze.

Toxicants reach organisms directly from the environment and food. In terrestrial environments this leads to an accumulation through the food chain, which makes top predators the most vulnerable. The importance of this type of accumulation in aquatic environments is still under discussion because of the more intensive direct exchange with the environment. Remarkable differences exist between various groups of animals; mammals, birds, and insects obtain most toxicants via food, even in aquatic environments, because their skin/exoskeleton is not permeable to many toxicants. Even if there is no accumulation via the food chain, we should expect that the tissue concentration increases with body size, which tends to increase towards the top of the food web. This is mainly due to the amount of food these animals eat.

16.8 TOXICITY PATTERNS

It is possible to obtain a wide range of experimental results from a simple toxicity experiment for almost any particular combination of organism and toxicant. This is due to a phenomenon known as secondary stress: the apparent toxicity is much higher if test conditions in the control are physiologically marginal. It is difficult to optimize test conditions because of the lack of knowledge about the detailed needs of organisms. This knowledge is to some extent available for a very limited set of species only. Even for these organisms, magic additives to medium and/or food have to be applied in chemically "pure" environments, where multiply distilled water is used with pure salt additions and food is cultured under equally well defined conditions. These types of problems hamper the detection of the more subtle toxicity patterns.

16.8.1 Temperature

All rates depend on temperature. The Arrhenius relationship describes how physiological rates depend on temperature: the logarithm of such rates depend linearly on the inverse of the absolute temperature within a certain tolerance range. The slope of this line, the Arrhenius temperature, is typically 12500 K. This approximately corresponds to an increase by a factor of 3 for an increase in temperature of $10°C$. Diffusion rates are proportional to absolute temperature to the power of 1.5, however. This means an almost linear increase in the range from $0°$ to $35°C$ by a factor of 1.2 only. To my knowledge, no systematic study has yet been done to relate uptake kinetics in combination with effects to temperature. I expect that uptake kinetics are physiologically controlled. Complex patterns can emerge if one process depends on temperature in a way which is different from that of other processes.

16.8.2 pH

Ionized as well as nonionized (molecular) forms of a compound are taken up at different rates, while the pH affects their relative abundance and therefore the toxicokinetics (Könemann 1980). Homeostasis, a cornerstone concept in the DEB theory, ensures that the pH inside the organism is independent of that in the environment (within a certain range of pH values). Therefore the elimination rate is independent of pH in the environment; only the uptake rate and hence the bioconcentration factor is important. The ratio of the uptake rates of the ionized and molecular forms therefore equals the ratio of the killing rates that would result if all of the compound was present in either the ionized or the molecular form. The killing rate depends on the pH in the same way as Könemann (1980) proposed for LC_{50}^{-1}, namely

$$\dot{k}_\dagger(pH) = \frac{\dot{k}_\dagger(-\infty) + \dot{k}_\dagger(\infty)K_a 10^{pH}}{1 + K_a 10^{pH}}$$

where $\dot{k}_\dagger(-\infty)$ and $\dot{k}_\dagger(\infty)$ stand for the killing rate if all of the compound was present in the molecular and the ionized form, respectively, and K_a is the dissociation coefficient $K_a \equiv [H^+][A^-][HA]^{-1}$, where $[A^-]$ and $[HA]$ stand for the concentrations of the ionized and molecular forms.

16.8.3 Body Size and Reserves

The relationship between uptake kinetics and body size has already been discussed. I mention it here just as a reminder that the uptake and elimination rates are inversely proportional to volumetric length, so that the bioconcentration factor is independent of body size, as well as the killing rate and tolerance concentrations for sublethal effects. Variations in body size among individuals directly translates into variations in response to toxicants. The fat content tends to increase with body size when we compare different species (Kooijman 1993). This is due to the fact that reserve capacity, expressed as density, is proportional to volumetric length. Part of the fat belongs to structural body mass and part to the reserves. The next subsection deals with fat content.

16.8.4 Solubility in Fat

The interest in the relationship between chemical structure and physiological effect goes back to Crum-Brown and Fraser (1868), but the first application of this relationship to ecotoxicological problems is very recent (Verhaar 1995). The focus is on general trends rather than precise predictions. The most obvious property of chemicals which is necessary for the understanding of toxicity is the n-octanol/water partition coefficient, P_{ow}, which can be estimated from chemical structure of the compound. Octanol serves as a model for typical lipids of animals. It has a density of 827 g dm^{-3} and a molecular weight of 130, so that 1 dm^3 of octanol contains 6.36 mol. Most comparisons are restricted to the interval $(10^2, 10^6)$. The size of the molecule tends to increase with P_{ow} and if it is larger than 10^6, the molecules are generally too big to enter cells easily. Relationships between toxicity measures and P_{ow} are one class of quantitative structure–activity relationships (QSARs).

The bioconcentration factor P_{aw} for fish relates to the octanol/water partition coefficient as $P_{aw} = 0.048\ P_{ow}$ (Mackay 1982). Hawker and Connell (1986) found $P_{aw} = 0.0484\ P_{ow}^{0.898}$ for daphnids and $P_{aw} = 0.0582\ P_{ow}^{0.844}$ for molluscs in the range $10^2 \leq P_{ow} \leq 10^6$. The scatter in the data is big enough for the relationship $P_{ow} = 0.02\ P_{ow}$ to apply for both daphnids and molluscs. The proportionality factor directly relates to the fat content. In general we can say that $P_{aw} = P_{oa}P_{ow}$, where P_{oa} stands for the mass-specific octanol equivalent of the organism. High correlations between P_{aw} and P_{oa} have been found, for example, for fenitrothion in a variety of algae (Kent and Currie 1995). The DEB theory predicts that P_{oa} increases with the body size of the different species of animal because the maximum reserve capacity increases with volumetric length as density and reserves are relatively rich in lipids. These are only general trends and many exceptions occur. The eel *Anguilla* is much fatter than other fish of similar size, for instance.

Hawker and Connell (1985, 1986) related the elimination rate $\dot{k}_a$ to P_{ow} and found $\dot{k}_a = 8.851\ P_{ow}^{-0.663}$ for fish, $\dot{k} = 113\ P_{ow}^{-0.507}$ for *Daphnia pulex* and $\dot{k}_a = 9.616\ P_{ow}^{-0.540}$ for molluscs. The proportionality factor is inversely proportional to the volumetric length of the animal (see above), which explains the wide range of values. The results for daphnids are most reliable because they all have the same body size, in this case, which suggests that $\dot{k}_a \propto 1\sqrt{P_{ow}}$. In view of the finding that the bioconcentration factor is proportional to P_{ow}, the uptake rate should be proportional to $\sqrt{P_{ow}}$. These relationships directly follow from a symmetry argument for the rate at which a substance moves from one matrix

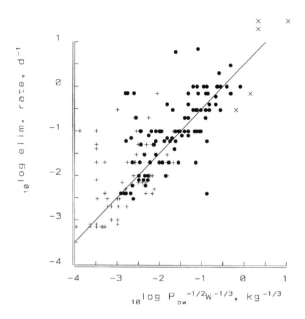

Figure 16.9 The elimination rate depends on the n-octanol/water partition coefficient P_{ow} and the weight W of an organism. It is roughly proportional to $P_{ow}^{1/2} W^{-1/3}$ with proportionality constant $\sqrt{10}\, day^{-1}\, kg^{-1/3}$ for 181 halogenated organic compounds in fish. Data compiled by Hendriks (1995). The marker codes are: $P_{ow} \leq 10^2$ ($\times$), $10^2 \leq P_{ow} < 10^6$ ($\bullet$), $10^6 \leq P_{ow} \leq 10^8$ (+). The range of fish weights is from 0.1 to 900 g. No corrections for differences in temperature have been made, nor for differences in fat content of the fish.

into another: the way that the uptake rate depends on P_{ow} and how the elimination rate depends on P_{wo} are the same, while $P_{wo} = P_{ow}^{-1}$ (Figure 16.9.) Notice that the elimination rate being proportional to $P_{ow}^{-0.5}$ implies that the accumulation time, i.e., the waiting time until a fixed percentage of the assymptotic burden is exceeded, is proportional to $P_{ow}^{0.5}$.

Since the hazard model has not yet been widely applied, there is no empirical information about the relationship between the killing rate and P_{ow} at this moment. Since the equilibrium tissue concentration is proportional to P_{ow}, we should expect to find that $\dot{k}_{\dagger} \propto P_{ow}$. The idea is that effects relate to the aqueous fraction of the tissue, which is proportional to P_{ow} because P_{aw} is proportional to P_{ow}. Similarly we should expect that NEC $\propto P_{ow}^{-1}$, which also holds for the tolerance concentrations.

Könemann (1981) observed that the 14 days LC_{50} of the guppy *Poecilia reticulata* for 50 "industrial chemicals" behaves as $LC_{50} = 0.0794\, P_{ow}^{-0.87}$ mol dm^{-3}. To understand this relationship, we have to realize that for large elimination rates, i.e., small P_{ow}, the 14 days LC_{50} is close to the ultimate value, but for large P_{ow}, the ultimate LC_{50} is much lower than the LC_{50} after 14 days. Taking these complexities into account, numerical studies confirm that $\dot{k}_{\dagger} \propto P_{ow}$ and

$\dot{k}_a \propto P_{ow}^{-0.5}$ is indeed consistent with the finding by Könemann (Figure 16.10). The LC_{50} values were obtained by Könemann by application of the standard model, rather than the hazard model. The results can probably be improved by application of the hazard model. Notice that the NEC, killing and elimination rate of the hazard model define the survival probability as a function of time and concentration. By equating the survival probability to 0.5, we can obtain a LC_{50}–time curve from the three parameters. Unfortunately, these data did not allow checking the relationship for the NEC. Although the NECs are set to zero in Figure 16.10, the function $NEC = 10P_{ow}^{-1}$ mmol dm^{-3} hardly changes the result. The conclusion is that we can now understand the QSAR for LC_{50} values from first principles. Notice that the standard QSARs for LC_{50}s require two parameters, while we need just one for killing rates.

16.8.5 Classes of Compounds

The usefulness of QSAR relationships can be greatly improved by restricting the compounds that are to be compared to compounds that have the same mode of action, physiologically. Based on the work of Veith and Broderius (1987, 1990), Hermens (1989, 1990), Lipnick (1989), Bradbury and Lipnick (1990), Verhaar et al. (1992), and Verhaar (1995) Hermens and Verhaar distinguish four main

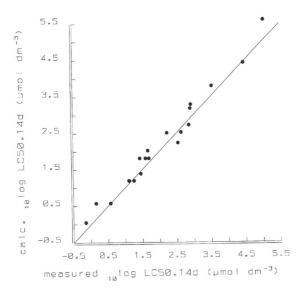

Figure 16.10 The calculated 14 days LC_{50} values as a function of "measured" values for guppies (*Poecilia reticulata*) exposed to 21 chlorinated aromatic and other some chlorinated hydrocarbons whose P_{ow} ranged from $10^{-0.22}$ (pentachlorobenzene) to $10^{5.21}$ (acetone). (Data from Könemann (1979)). The calculations are based on the assumptions that the elimination rates equal $50/\sqrt{P_{ow}^{-1}}$, the killing rates equal $10^{-6.6}P_{ow}$ d^{-1} µmol^{-1} dm^3, and the NECs are zero (see text).

classes of organic compounds that may contain nitrogen, sulfur, and/or halogens (excluding iodine), with $1 < P_{ow} < 10^6$ and a molecular mass < 600 daltons.

- Type 1: narcotic compounds with baseline toxicity.
- Type 2: less inert compounds. These include non- or weakly acidic phenols, anilines, mononitroaromatics, primary alkylamines and pyridines. These compounds may have one or two chlorine or alkyl substituents.
- Type 3: compounds with unspecific reactivity.
- Type 4: compounds with specific reactivity. Examples of this type of compound are DDT and analogues, (dithio)carbamates, organotin compounds, pyrethroids, and organophosphorothionate esters.

Each of these types have been further subdivided.

16.8.6 Sensitivity Among Species

Apart from compounds with specific reactivity, the only patterns in sensitivity among species that have been detected so far relate to fat content. Acute toxicity also involves body size and surface area (Hoekstra et al. 1994), but these quantities relate to kinetics. Van Straalen (1994) found a relationship between the sensitivity and the food spectrum of a species: the broader the spectrum, the better detoxification systems are developed and the less sensitive a species is. He found a correlation between the LC_{50} and the activity of detoxification enzymes such as cytochrome P450 and glutathione-S-transferase in terrestrial arthopods. Practical problems usually hamper the detection of such patterns. The first problem is the extremely limited ability to standardize laboratory cultures of many species. Many data relate to specimens taken from the field and tested under less than optimal conditions, which amplify scatter to an extent that sensitivity patterns disappear. Another problem is that only acute toxicity has been quantified with LC_{50} values. These data have not been corrected for differences in body size or other factors influencing effect. It should be clear by now why it will be very hard to detect patterns in toxicity measures that have such poor properties as LC_{50}s.

I have proposed a factor that can be applied to LC_{50} values of a very limited set of tested species to estimate the LC_{50} of the most sensitive species in an ecosystem (Kooijman 1987). This factor is based on the idea that the (known) LC_{50}s of a limited set of tested species as well as the (unknown) LC_{50}s of all species in the ecosystem represent random trials from the same frequency distribution. My purpose has been to illustrate that a very simple reasoning leads to the expectation that effects in the field can be expected when standard safety factors are applied. Mesocosm experiments usually fail to reveal this sensitivity of complex systems, first because of lack of power due to erratic behavior of the control (a problem similar to that connected to the NOEC) and second because of lack of detail of the observations (species are not monitored separately, but rather only together as a class of "secondary producers"). Van Straalen and Denneman (1989) and Aldenberg and Slob (1993) modified this idea and proposed an estimate for a LC_{50} value, the HC5, which is exceeded by a fixed percentage (namely 95%) of the species, rather than by all of the species. The value of the HCx rapidly increases with decreasing x. The practical application of this factor has been stimulated by the observation that it matches current practice. The problem that I have with the application is in the objectives of legislation, which should not aim at the disappearance of any percentage of the species. The reason why this practice has not yet resulted in disasters is that an increase in the variance of LC_{50}s leads to a sharp decrease HC5. As has been discussed already, many factors contribute to an increase in the variance of LC_{50}s, although this variance does not reflect a proper scatter in sensitivities among species. This is a rare example where ignorance turns out to be protective.

16.9 CONCLUSIONS AND PERSPECTIVES

The direct application of ecotoxicology in risk assessment has given standardization and purely descriptive methods priority above developing a sound scientific basis. Reliable applications are only possible if they are based on a scientific basis, however. I have shown that mechanistic underpinnings need not lead to models that are too complex to apply on a routine basis. (The software package DEBtox, as provided in Kooijman and Bedaux 1996d, can be used to perform the calculations that are discussed here for the full set of standardized aquatic toxicity tests.) Scientific progress is only possible if we develop a range of *consistent* models from simple (for routine applications) to complex (for mechanistic studies). Consistency is essential for the cross-fertilization between large scale routine applications and scientific progress. Routine test protocols should be developed such that the simplifications make sense. This means that these protocols should be subjected to a regular update procedure. The development of a database with raw data would facilitate the reexamination of old data in a newly developed scientific light. Existing data bases with just summary statistics, such as $LC_{50}/EC_{50}/NOEC$ values, decrease in value with changing scientific insight. The rapidly decreasing costs of computer memory has removed technical obstacles for such a more extensive database. International cooperation and organization are the main bottlenecks.

In my opinion, legislation should aim at no effects. Spatial and time scales should be worked out in more detail in this respect. Very local and temporary effects can, in some cases, rapidly be cancelled by immigration from the surroundings as long as the surroundings are not affected. Degradation of compounds is in such cases more important than toxicity. Immigration from unaffected surrounding areas is less likely in extensive areas with intense industrial activity that lack "virgin" sites.

Whatever the aims of legislation, ecological effects of toxicants will continue to occur. An evaluation of the ecological significance of these effects remains necessary. The development of fundamental ecology would benefit from two endeavors in ecotoxicology:

- a widening of the problem of effects of human-induced chemical disturbance to chemical disturbance in general: the understanding of the development of life in chemically changing environments
- the analysis of population dynamics in terms of the ecophysiological behavior of individuals: the further development of structured population dynamics.

Both endeavors are very complex, and progress will be slow. The reaction of organisms to changes in the chemical environment calls for a deeper understanding of physiology and molecular biology. Although the field of molecular biology is developing fast as an independent discipline, little attention is given to the relationship between the molecular and organism levels of organization. An understanding of ecosystem dynamics in terms of population dynamics is remote and it is still questionable whether it is feasible at all. Lists of more specific problems, such as the toxicity of poorly soluble compounds and mixtures of compounds, seem so incomplete that the effort is useless.

16.10 REFERENCES

Agosta, W., 1995, *Bombardier Beetles and Fever Trees*, Addison-Wesley, Reading, MA, USA

Aldenberg, T., and W. Slob, 1993, "Confidence Limits for Hazardous Concentrations Based on Logistically Distributed NOEC Data," *Ecotoxicol. Environ. Safety, 25*, 48–63

Bedaux, J. J. M., and S. A. L. M. Kooijman, 1994, "Statistical Analysis of Bioassays, Based on Hazard Modeling," *J. Environ. Ecol. Stat.*, 303–314

Bradbury, S. P., and R. L. Lipnick, 1990, "Introduction: Structural Properties for Determining Mechanisms of Toxic Action," *Environ. Health Persp., 87*, 181–182

Bruce, R. D., and D. J. Versteeg, 1992, "A Statistical Procedure for Modelling Continuous Toxicity Data," *Environ. Toxicol. Chem., 11*, 1485–1492

Crommentuijn, T., C. J. A. M. Doodeman, A. Doornekamp, J. J. C. van der Pol, J. J. M. Bedaux, and C. A. M. van Gestel, 1994, "Lethal Body Concentrations and Accumulation Patterns Determine Time-Dependent Toxicity of Cadmium in Soil Arthropods," *Environ. Toxicol. Chem., 11*, 1781–1789

Crum-Brown, A., and T. Frazer, 1868–69, "On the Connection Between Chemical Constitution and Physiological Action. Part 1. On the Physiological Action of the Ammonium Bases, Derived from Strychia, Brucia, Thebaia, Codeia, Morphia, and Nicotia," *Trans. Roy. Soc. Edinburgh, 25*, 151–203

David, H. A., and M. L. Moeschberger, 1978, *The Theory of Competing Risks*, Griffin's Statistical Monographs and Courses, Vol. 39, Griffin, London, UK

DeAngelis, D. L., and L. J. Gross, (Eds.), 1992, *Individual-Based Models and Approaches in Ecology*, Chapman & Hall, London, UK

Ebenman, B., and L. Persson, 1988, *Size-Structured Populations Ecology and Evolution*, Springer, Berlin, Germany

Hallam, T. G., R. R. Lassiter, J. Li., and W. McKinney, 1988, "Physiologically Structured Population Models in Risk Assessment," in *Biomathematics and Related Computational Problems* (L. M. Riccardi, Ed.), Kluwer Academic Press, Dordrecht, The Netherlands, pp. 197–211

Haren, R. J. F. van, H. E. Schepers, and S. A. L. M. Kooijman, 1994, "Dynamic Energy Budgets Affect Kinetics of Xenobiotics in the Marine Mussel *Mytilus edulis*," *Chemosphere, 29*, 163–189

Hawker, D. W., and D. W. Connell, 1985, "Relationships Between Partition Coefficient, Uptake Rate Constant, Clearance Rate Constant, and Time to Equilibrium for Bioaccumulation," *Chemosphere, 14*, 1205–1219

Hawker, D. W., and D. W. Connell, 1986, "Bioconcentration of Lipophilic Compounds by Some Aquatic Organisms," *Ecotoxicol. Environ. Safety, 11*, 184–197

Hendriks, A. J., 1995, "Modelling Non-Equilibrium Concentrations of Microcontaminants in Organisms: Comparative Kinetics as a Function of Species Size and Octanol–Water Partitioning," *Chemosphere, 30*, 265–292

Hermens, J. L. M., 1989, "Quantitative Structure-activity Relationships of Environmental Pollutants," in *Handbook of Environmental Chemistry*, Vol. 2E (O. Hutzinger, Ed.), Springer, Berlin, Germany, pp. 111–162

Hermens, J. L. M., 1990, "Electrophiles and Acute Toxicity to Fish," *Environ. Health Persp., 87*, 219–225

Hoekstra, J. A., 1993, "Statistics in Ecotoxicology," *PhD Thesis*, Vrije Universiteit, Amsterdam, The Netherlands

Hoekstra, J. A., and P. H. van Ewijk, 1992, "Alternatives for the No-Observed-Effect Level," *Environ. Toxicol. Chem., 12*, 187–194

Hoekstra, J. A., M. A., Vaal, J. Noteboom, and W. Slooff, 1994, "Variation in the Sensitivity of Aquatic Species to Toxicants," *Bull. Environ. Contam. Toxicol., 53*, 98–105

Hoeven, N. van der, S. A. L. M. Kooijman, and W. K. de Raat, 1990, "*Salmonella* Test: Relation Between Mutagenicity and Number of Revertant Colonies," *Mut. Res., 234*, 289–302

Hothorn, L., 1994, "Multiple Comparisons in Long-term Toxicity Studies," *Environ. Health Persp. Suppl., 102, Suppl. 1*, 33–38

Kent, R. A., and D. Currie, 1995, "Predicting Algal Sensitivity to a Pesticide Stress," *Environ. Toxicol. Chem., 14*, 983–991

Könemann, W. H., 1980, "Quantitative Structure–Activity Relationships for Kinetics and Toxicity of Aquatic Pollutants and Their Mixtures in Fish," *PhD Thesis*, Utrecht University, The Netherlands

Könemann, W. H., 1981, "Quantitative Structure–Activity Relationships in Fish Toxicity Studies. 1. Relationship for 50 Industrial Pollutants," *Toxicology, 19*, 209–221

Kooi, B. W., and S. A. L. M. Kooijman, 1994, "Existence and Stability of Microbial Prey–Predator Systems," *J. Theor. Biol., 170*, 75–85

Kooi, B. W., and S. A. L. M. Kooijman, 1994a, "The Transient Behaviour of Food Chains in Chemostats," *J. Theor. Biol., 170*, 87–94

Kooi, B. W., and S. A. L. M. Kooijman, 1995, "Many Limiting Behaviours in Microbial Food Chains," in *Conf. Proc. Third Internat. Conf. Math. Pop. Dyn., Biological Systems* (O. Arino, M. Kimmel, and D. Axelrod, Eds.), Wuerz Publ., Winnipeg, Canada, pp. 131–148

Kooijman, S. A. L. M., 1981a, "Parametric Analyses of Mortality Rates in Bioassays," *Water Res.*, *15*, 107–119

Kooijman, S. A. L. M., 1981b, "The Estimation of Mortality Rates in Toxicity Tests," *Inserm*, *106*, 467–474

Kooijman, S. A. L. M., 1983, "Statistical Aspects of the Determination of Mortality Rates in Bioassays," *Water Res.*, *17*, 749–759

Kooijman, S. A. L. M., 1987, "A Safety Factor for LC_{50} Values Allowing for Differences in Sensitivity Among Species," *Water Res.*, *21*, 269–276

Kooijman, S. A. L. M., 1988, "Strategies in Ecotoxicological Research," *Environ. Aspects Appl. Biol.*, *17* (1), 11–17

Kooijman, S. A. L. M., 1991, "Effects of Feeding Conditions on Toxicity for the Purpose of Extrapolation," *Comp. Biochem. Physiol.*, *C100* (1/2), 305–310

Kooijman, S. A. L. M., 1993, *Dynamic Energy Budgets in Biological Systems. Theory and Applications in Ecotoxicology*, Cambridge University Press, Cambridge, UK

Kooijman, S. A. L. M., 1995, "The Stoichiometry of Animal Energetics," *J. Theor. Biol.*, *177*, 139–149

Kooijman, S. A. L. M., and J. J. M. Bedaux, 1996a, "Some Statistical Properties of Estimates of No Effects Concentrations," *Water Res.*, *30*, 1724–1728

Kooijman, S. A. L. M., and J. J. M. Bedaux, 1996b, "Analysis of Toxicity Tests on *Daphnia* Survival and Reproduction," *Water Res.*, *30*, 1711–1723

Kooijman, S. A. L. M., and J. J. M. Bedaux, 1996c, "Analysis of Toxicity Tests on Fish Growth," *Water Res.*, *30*, 1633–1644

Kooijman, S. A. L. M., and J. J. M. Bedaux, 1996d, *The Analysis of Aquatic Toxicity Data.* VU University Press, Amsterdam, The Netherlands (160 pp + floppy)

Kooijman, S. A. L. M., A. O. Hanstveit, and N. van der Hoeven, 1987, "Research on the Physiological Basis of Population Dynamics in Relation to Ecotoxicology," *Water Sci. Tech.*, *19*, 21–37

Kooijman, S. A. L. M., A. O. Hanstveit, and N. Nyholm, 1996, "No-Effect Concentrations in Alga Growth Inhibition Tests," *Water Res.*, *30*, 1625–1632

Kooijman, S. A. L. M., A. O. Hanstveit, and H. Oldersma, 1983, "Parametric Analyses of Population Growth in Bioassays," *Water Res.*, *17*, 727–738

Kooijman, S. A. L. M., and R. J. F. van Haren, 1990, "Animal Energy Budgets Affect the Kinetics of Xenobiotics," *Chemosphere*, *21*, 681–693

Kooijman, S. A. L. M., B. W. Kooi, and M. P. Boer, 1995, "Rotifers Do it With Delay. The Behaviour of Reproducers vs Dividers in Chemostats," *Nonlin. World 3*, 107–128

Kooijman, S. A. L. M., and J. A. J. Metz, 1983, "On the Dynamics of Chemically Stressed Populations; The Deduction of Population Consequences from Effects on Individuals," *Ecotoxicol. Environ. Safety*, *8*, 254–274

Laskowski, R., 1995, "Some Good Reasons to Ban the Use of NOEC, LOEC and Related Concepts in Ecotoxicology," *OIKOS*, *73*, 140–144

Lassiter, R. R., and T. G. Hallam, 1990, "Survival of the Fattest: Implications for Acute Effects of Lipophilic Chemicals on Aquatic Populations," *Environ. Toxicol. Chem.*, 9, 585–595

Lipnick, R. L., 1989, "Base-line Toxicity Predicted by Quantitative Structure–Activity Relationships as a Probe for Molecular Mechanism of Toxicity," in *ACS Symposium Series*, Vol. 413 (P. S. Magee, D. R. Henry, and J. H. Block, Eds.), American Chemical Society, Washington, DC, USA, pp. 366–389

MacDonald, D., 1995, *European Mammals; Evolution and Behaviour*. Harper Collins Publishers, London, UK

Mackay, D., 1982, "Correlation of Bioconcentration Factors," *Environ. Sci. Technol.*, 16, 274–278

Metz, J. A. J., and O. Diekmann, (Eds.), 1986, *The Dynamics of Physiologically Structured Populations*, Springer Lecture Notes in Biomathematics, Springer, Berlin, Germany

Moran, N., and P. Baumann, 1994, "Phylogenetics of Cytoplasmically Inherited Microorganisms of Arthropods," *TREE*, 9, 15–20

Nyholm, N., 1985, "Response Variable in Algal Growth Inhibition Tests—Biomass of Growth Rate?," *Water Res.*, 19, 273–279

Pack, S., 1983, "A Review of Statistical Data Analysis and Experimental Design in OECD Aquatic Toxicity Test Guidelines," Technical Report, Shell Research Ltd., Sittingbourne, UK

Pack, S., 1993, "A Review of Statistical Data Analysis and Experimental Design in OECD Aquatic Toxicity Test Guidelines," Shell Research Ltd., Sittingbourne, UK

Robertson, J. R., and G. W. Salt, 1981, "Responses in Growth, Mortality, and Reproduction to Variable Food Levels by the Rotifer *Asplanchna Girodi*," *Ecology*, 62, 1585–1596

Skalski, J. R., 1981, "Statistical Inconsistencies in the Use of No-observed-effect-levels in Toxicity Testing," in *Aquatic Toxicology and Hazard Assessment* (D. R. Branson, and K. L. Dickson, Eds.), *Fourth Conf. ASTM STP 737*, American Society for Testing and Materials, Philadelphia, USA, pp. 328–338

Sprague, J. B., 1969, "Measurement of Pollutant Toxicity to Fish I. Bioassay Methods for Acute Toxicity," *Water Res.*, 3, 793–821

Stephan, C. E., and J. W. Rogers, 1985, "Advantages of Using Regression to Calculate Results of Chronic Toxicity Tests," in *Aquatic Toxicology and Hazard Assessment* (R. C. Bahner, and D. J. Hansen, Eds.), *8th Symp. ASTM STP 891*, American Society for Testing and Materials, Philadelphia, USA, pp. 328–338

Straalen, N. M. van, 1994, "Biodiversity of Ecotoxicological Responses in Animals," *Neth. J. Zool.*, 44, 112–129

Straalen, N. M. van, and C. A. J. Denneman, 1989, "Ecotoxicological Evaluation of Soil Quality Criteria," *Ecotoxicol. Environ. Safety*, 18, 241–251

Veith, G. D., and S. J. Broderius, 1987, "Structure-toxicity Relationships for Industrial Chemicals Causing Type (II) Narcosis Syndrome," in *QSAR in Environmental Toxicology II* (K. L. E. Kaiser, Ed.), Reidel, Dordrecht, The Netherlands, pp. 385–391

Veith, G. D., and S. J. Broderius, 1990, "Rules for Distinguishing Toxicants that Cause Type I and Type II Narcosis Syndromes," *Environ. Health Persp.*, 87, 207–211

Verhaar, H. J. M., 1995, "Predictive Methods in Aquatic Toxicology," *PhD Thesis*, University of Utrecht, The Netherlands

Verhaar, H. J. M., C. J. van Leeuwen, and J. L. M. Hermens, 1992, "Classifying Environmental Pollutants. Structure-activity Relationships for Prediction of Aquatic Toxicity," *Chemosphere*, *25*, 471–491

Webb, G. F., 1985, *Theory of Nonlinear Age-dependent Population Dynamics*, Marcel Dekker, New York, USA, p. 294

White, T. C. R., 1993, *The Inadequate Environment; Nitrogen and the Abundance of Animals*, Springer, Berlin, Germany, 425 pp

Yanagawa, T., Y. Kikuchi, and K. G. Brown, 1994, "Statistical Issues on the No-Observed-Adverse-Effect Level in Categorical Response," *Environ. Health Persp. Suppl.*, *102*, Suppl. *1*, 95–101

17

Cellular Response Profile to Chemical Stress

Helmut Segner (*Leipzig, Germany*)

and

Thomas Braunbeck (*Heidelberg, Germany*)

17.1 SUMMARY

Cells are the site of primary interaction between chemicals and biological systems. The importance of toxicity assessment at the cellular level in the context of ecotoxicology is based on the assumption that the cellular change can ultimately develop into ecological change. However, it has to be emphasized that the relationship between the different levels of biological organization is neither straightforward nor deterministic. This chapter discusses cellular structures and functions that are of major relevance for toxicant fate and effect, including membrane processes, biotransformation, multixenobiotic resistance, metal homeostasis, receptors for xenobiotics and endogenous compounds, nuclear and DNA changes, protective molecules such as glutathione and stress proteins, and, finally,

---

Ecotoxicology, Edited by Gerrit Schüürmann and Bernd Markert.
ISBN 0-471-17644-3   © 1998 John Wiley & Sons, Inc. and Spektrum Akademischer Verlag.

cell death. Among the examples of the application of cellular studies in ecotoxicology are the establishment of toxicity equivalency factors, identification of biomarkers, rational prognosis of mixture effects, etc. The present challenges for further extension of cell toxicological approaches in ecotoxicology are (a) the identification of the factors that decide whether a cellular effect will continue at higher levels of biological hierarchy or not, and (b) to develop an understanding which cellular changes influence the ecological success of organisms and populations.

17.2 INTRODUCTION

The initial interaction between chemicals and biological systems occurs at the cellular level, the first tier integrating all criteria of life. Thus, cells may be expected to react to chemical exposure. The cellular response is the starting point of a chain of reactions leading to ecological change, although it has to be emphasized that there is usually no straightforward, but rather a complex relationship between toxic events at different levels of biological organization. Most importantly, not every cellular reaction has a pathologic meaning; many responses are of a protective nature and may lead to acclimation to the toxic impact instead of disease and death (Figure 17.1). The regulating and modifying processes at the cellular level determine whether the interaction with a toxic chemical eventually results in consequences at higher levels of biological integration or not.

Mechanistic approaches are indispensable for a solid theoretical basis of the scientific discipline of ecotoxicology. From an evolutionary point of view, basal cellular functions and structures may be considered highly conserved biological entities, and studies on cells provide an approach for the detection of general mechanisms of toxicity. A mechanism-based classification of toxic effects—instead of the descriptive collection of isolated phenomena—is of outstanding importance for the ecotoxicologist, who, confronted with the huge variety of both living organisms and chemical compounds, strongly depends on extrapolation. In fact, the systematic classification of environmental chemicals, e.g., in studies on structure–activity relationships, is related to their cellular mode of toxic action. Examples for such a mode-of-action-based approach is the classification of chemicals as acetylcholinesterase inhibitors or uncouplers of oxidative phosphorylation.

The fact that chemical uptake, storage, and excretion, as well as adaptive or toxic action take place at the cellular level makes cellular responses suitable tools for the early and sensitive detection of chemical exposure. Actually, cellular reactions have successfully been used as biomarkers of exposure. Their use as biomarkers of effect, however, is still limited due to gaps in our knowledge of the mechanistic basis of toxicant-induced cellular changes and of the conditions under which cellular changes translate into alterations at higher biological levels. An outline of potential reactions and processes following exposure of cells to chemicals is given in Figure 17.2.

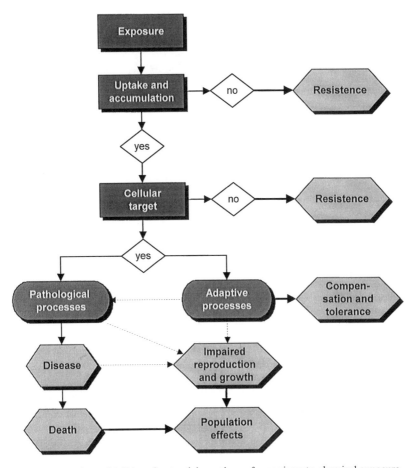

Figure 17.1 The multiplicity of potential reactions of organisms to chemical exposure.

In the subsequent sections, we discuss the following selection of examples of the cellular response profile to chemical stress:

(1) interaction with membrane processes;
(2) intracellular fate of chemicals;
(3) intracellular receptors;
(4) interaction with nuclear structures and functions;
(5) protective molecules and processes;
(6) cell injury and death.

In the final section an attempt is made to summarize the potential functions of cell toxicology in ecotoxicology.

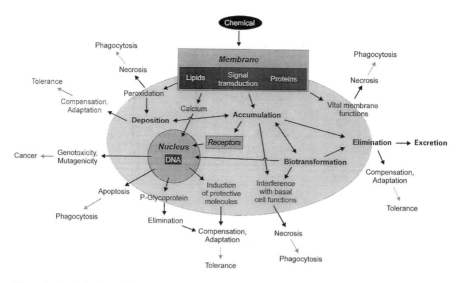

Figure 17.2 Possible reactions and processes in the interaction between chemicals and cells as well as their relation to tissue responses.

In most examples, cellular reactions are discussed with emphasis on responses in fish. However, due to the fundamental nature of most cellular processes, many conclusions can be extrapolated to other taxa. Due to the wealth of information on hepatic processes, the central position of the liver in basic metabolism and, in particular, in accumulation, biotransformation, and excretion of xenobiotics, the liver will be given special attention. In the following chapters, the terms "stress" or "stressor" refer to an environmental stimulus (rather than to the physiological response), which, by exceeding a threshold value, disturbs normal animal function (Bayne 1985). Thus, there is an equivalency with the term usually applied in ecosystem research.

17.3 INTERACTION WITH MEMBRANE PROCESSES

Animal membranes, both intracellular ones as well as the plasma membrane, consist of a phospholipid bilayer with polar head groups (phosphatidylcholine, phosphatidylethanolamine) projecting on both outer and inner surfaces of the membrane and fatty acids directed more or less perpendicularly to the membrane and filling out the inner space. The membrane lipids of aquatic animals are particularly rich in $(n-3)$polyunsaturated fatty acids (PUFAs), whereas in terrestrial organisms the $(n-6)$PUFAs are dominant (Sargent et al. 1989). In addition to lipids, proteins are inserted into the bilayer or may even span it, forming pores and channels. On the outer face of membranes, eucaryotic cells have a coat or glycocalyx consisting of glycoproteins and glycolipids. To function properly, the membranes must not be rigid, but fluid, and the maintenance of this fluid state under changing environmental conditions is critical for membrane function (concept of homeoviscosity; cf. Sargent et al. 1989).

The plasma membrane is the first structure encountered by a toxic agent upon reaching a cell. Membranes are the site for the uptake, deposition, and elimination of chemicals. The vital regulatory mechanisms inherent in membranes and their chemical composition make them a target particularly susceptible to many chemical compounds. Thus, membranes are of importance for both toxicokinetics and toxicodynamics.

A toxicant may pass the membrane by either special transport processes including active transport, facilitated transport, endocytosis, or passive diffusion (Hodgson and Levi 1987). Membrane permeability and, as a consequence, the cellular uptake and accumulation of organic molecules is closely related to the lipophilicity of compounds. Hydrophobic organic chemicals can traverse the lipid bilayer of membranes by passive diffusion. This explains the frequently observed dependency of bioaccumulation of organic chemicals on their lipophilicity, the latter being usually expressed in the form of the octanol/water partition coefficient, $\log K_{OW}$ or P_{OW} (Leo et al. 1971). The transmembrane uptake of organic compounds may be complicated by steric factors (Opperhuizen 1991; Schüürmann and Segner 1995) or the presence of charged moieties (e.g., Könneman and Musch 1981; Fent 1996), since membranes are less permeable to compounds in their ionized states.

Nonessential toxic metals such as cadmium (Cd) appear to utilize adventitiously existing pathways for essential metals to cross the membrane (Simkiss and Taylor 1989a; Roesijadi and Robinson 1995). Cadmium seems to cross cell membranes via calcium (Ca) channels. Interaction between the uptake of, for example, Cd and Ca have been reported for the cellular (e.g., Verbost et al. 1989a), organ (e.g., Pärt et al. 1984), and organismic level (e.g., Wright 1977; Markich and Jeffree 1994). In experiments with both vertebrate and invertebrate cells, antagonists of various types of calcium channels were able to consistently reduce cellular Cd uptake (Hinkle et al. 1987; Verbost et al. 1989b; Roesijadi and Unger 1993). Use of metabolic inhibitors such as sodium cyanide or rotenon (inhibitors of oxidative respiration) did not alter Cd uptake, an effect interpreted as an indication that the metal transfer across the membrane is not an energy-dependent active process (Carpene and George 1981; Stacey and Klaassen 1984). Although Roesijadi and Unger (1993) observed that 2,4-dinitrophenol, an uncoupler of oxidative phosphorylation, inhibited Cd uptake in the gills of the oyster *Crassostrea virginica*, it was speculated that ATP is not required to fuel active uptake against concentration gradients, but rather to support the function of calcium channels (Roesijadi and Robinson 1995).

In addition to being a barrier to toxicant uptake, membranes also represent a target for toxicant action. The integrity of the plasma membrane is essential for cell viability, homeostasis, and communication capabilities. By selective permeability, membranes establish an asymmetric distribution of ions, and regulate uptake and release of nutrients. Membrane-anchored molecules provide the basis for cell–cell communication; e.g., membrane-based connexins form the gap junctions responsible for intercellular communication, or membrane-based hormone receptors and transduction pathways integrate individual cells in whole organism physiology and metabolism. In the interior of cells, membrane systems guarantee for an effective separation of compartments (nucleus, mitochondria, peroxisomes, lysosomes, Golgi fields, endoplasmic reticulum, vacuoles) and carry numerous enzymes.

An example of chemicals that exert their toxicity by disturbance of membrane function are compounds which act by a narcotic mode of action. Narcosis due to environmental pollutants in aquatic organisms is defined as a nonspecific disturbance of the cell membrane caused by the accumulation of lipophilic xenobiotics in the hydrophobic phase of the membrane (Wezel and Opperhuizen 1995). Dysfunction of the membrane can ultimately result in cell and organismic death. It is estimated that about 60% of industrial chemicals entering the aquatic environment are toxic by narcosis under acute, lethal exposure (Veith et al. 1983). The membrane effects of narcotic chemicals may be caused by either direct interaction with membrane proteins (enzymes, membrane channel components, receptors, etc.; protein theory of narcosis; Franks and Lieb 1978) or hydrophobic interaction, i.e., the molar or volume accumulation of lipophilic xenobiotics in the lipid bilayer (lipoid theory of narcosis; Overton 1901; Mullins 1954). The resulting disturbance of membrane order and membrane swelling can secondarily result in impaired function of membrane-embedded proteins (Sikkema et al. 1994). Independent of the detailed mechanism, the fact that narcotic chemicals are toxic due to their accumulation in the membrane and that their accumulation depends on the $\log K_{OW}$, provides a rational explanation for the observable correlation between the acute toxicity of narcotics and their lipophilicity (Wezel and Opperhuizen 1995).

Whereas narcotic effects on the lipid layer of membranes seem to be mainly related to physical processes such as interdigitation or changes in fluidity (Wezel and Opperhuizen 1995), many toxic compounds can also modify membrane lipids chemically. An example of such a process is lipid peroxidation, i.e., the oxidative destruction of membrane PUFAs in an autocatalytic process (Cheeseman 1993; Di Giulio et al. 1995). During the inititiation of lipid peroxidation, oxyradicals such as hydroxyl (HO•), peroxyl (ROO•), or alkoxyl radicals (RO•) abstract a hydrogen atom from the target fatty acid (LH). The resulting fatty acid radical L• rapidly rearranges to a conjugated diene radical that reacts with O_2 to produce a fatty acid peroxy radical LOO•. This fatty acid peroxy radical can, in turn, react with other lipids, thus beginning a new chain of reactions (propagation phase). Termination of lipid peroxidation involves the combination of two radicals to form nonradical products. Moreover, the lipid peroxy radical can spontaneously degrade to typical products such as malondialdehyde. The result of lipid peroxidation is the destabilization of hydrophobic membrane bilayer integrity and function, thus decreasing membrane fluidity and its ability to act as a semipermeable barrier (Cheeseman 1993). Peroxidative damage to membranes appears to be a major causative factor in

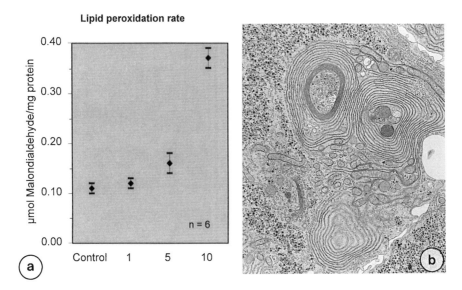

Figure 17.3 Lipid peroxidation rate and associated formation of myelin-like membrane whorls in the liver of rainbow trout (*Oncorhynchus mykiss*) after 4 weeks combined exposure to 50 ng/L endosulfan and 1, 5, and 10 ng/L of disulfoton. (a) Lipid peroxidation rate measured as micromoles of malondialdehyde formed per milligram of protein in relation to variable concentration of disulfoton (data from Arnold et al. 1995); (b) electron microscopical image of membrane whorls formed in the endoplasmic reticulum of hepatocytes of rainbow trout exposed to 50 ng/L endosulfan plus 5 µg/L disulfoton (6,500 ×).

the development of tissue necrosis (Klaassen and Eaton 1991). In morphological terms, increased lipid peroxidation has frequently been associated with the formation of membrane whorls and, eventually, with their transformation into myelinated bodies (Figure 17.3). In general, lipid peroxidation has been implicated as a cause of an extraordinarily large range of pathological processes, although the relationship between peroxidative destruction of lipid membranes and disease remains a tenuous one in the majority of cases (Dix and Aikens 1993).

The chemicals leading to lipid peroxidation can roughly be subdivided into three categories (Comporti 1993): (1) those forming reactive free radicals as a result of endogenous metabolism, e.g., carbon tetrachloride, which is metabolized by cytochrome P450 to the trichloromethyl radical; (2) those depleting cellular glutathione levels, e.g., cadmium, allyl alcohol, acetaminophen, and (3) those generating reactive oxygen species by redox cycling, e.g., paraquat and menadione. At least for some of these chemical compounds, the ability to induce lipid peroxidation has been shown in aquatic animals (Wofford and Thomas 1988; Roche and Bogé 1993). At present, however, the pathological significance or the biomarker value of lipid peroxidation in aquatic organisms is not understood at all.

Membrane-based information systems (receptors, protein channels) regulate cellular differentiation, proliferation, quiescence, metabolic cooperation, and physiological cell death (apoptosis; see below). Communication between adjacent cells occurs through membrane-anchored gap junction channels, which allow the exchange of ions and small water-soluble molecules including signal molecules such as cyclic nucleotides, calcium, and inositol triphosphate. The loss of gap junction intercellular communication plays an important role in the process of carcinogenesis (Loewenstein and Kanno 1966). Tumor cells frequently display decreased communication capacity, and, in fact, the inhibiting effect of a chemical on gap junctional exchange is taken as an indicator of tumor-promoting activity (e.g., Budunova and Williams 1994). Although to date this screening test has been almost exclusively applied in mammalian toxicology, explorative studies by Baldwin and Calabrese (1994) have already indicated its potential use in aquatic toxicology.

Disturbance and destabilization of membrane structure and function by toxic agents result in a loss of the semipermeability of the membrane, i.e., the release of cytoplasmic components to the extracellular space or the diffusion of external substances into the cell. In cytotoxicity tests, for example, this loss of membrane integrity can be measured by lactate dehydrogenase leakage or trypan blue uptake, respectively. Similarly, the disturbance of intracellular membranes leads to a loss of an essential component of cellular organization: intracellular concentration gradients and compartmentation of molecules into nucleus, organelles, and cytoplasm. Chemicals with an uncoupling mode of action illustrate the importance of membrane-mediated intracellular separation. ATP production of mitochondria depends, according to the chemiosmotic model, on the existence of a proton gradient across the intact mitochondrial membrane. Any disturbance of the membrane will discharge the electrochemical membrane potential by destruction of the proton gradient and, as a consequence, will uncouple oxidative phosphorylation (Terada 1990).

Another example of the importance of intracellular membranes in maintaining cellular homeostasis and viability is lysosomal fragility, i.e., the destabilization of lysosomal membranes under the impact of chemicals (Moore 1990). Two alternative test procedures are common to assess the integrity of lysosomal membranes: lysosomal latency and neutral red retention. Lysosomal latency has frequently been used as a biomarker for the interaction between chemicals and membranes (Moore 1976, 1992; Köhler 1991; Köhler et al. 1992; Lowe et al. 1992, 1995). It determines alterations in the permeability properties of the lysosomal membrane to the passage of substrates from the cytosol into the lysosome (Schneider et al. 1984) and subsequent release of the lysosomal enzyme N-acetyl-β-o-hexosaminidase. The neutral red retention assay is based on a reduced capability of the lysosomal membrane to retain the cationic dye neutral red trapped before within lysosomes by protonation and measures the increased efflux from damaged lysosomes into the cytoplasm (Lowe et al. 1995). The neutral red assay has routinely been applied as a cell viability test in in vitro toxicity tests (Borenfreund and Puerner 1985). As hypothesized for the neutral red assay by Lowe et al. (1992), damage of the lysosomal membrane is possibly due to impairment of the lysosomal membrane Mg^{2+}-ATP-dependent proton pump and subsequent equilibration of the proton gradient and free passage over the lysosomal membrane. Much of the damage to lysosomes in either test is a consequence of their ability to concentrate a wide range of chemicals from the environment, including lipophilic xenobiotics and metals, resulting in increased membrane permeability and loss of acid hydrolases into the cytosol, causing further cellular damage and, eventually, necrosis.

One group of chemicals that largely manifest their toxicity by interactions with cell membranes are organotins, particularly trialkyltins (Cameron et al. 1991; Raffray and Cohen 1991; Fent 1996). Although these compounds have lipophilic properties, their impact on the membrane seems not to be caused by hydrophobic interaction but by specific interactions with membrane proteins. In mammals, organotins have been shown to influence pH gradients across membranes, obviously by inhibiting membrane ATPases (Aldridge 1976; Zucker et al. 1988). They also interfere with elements of signal transduction pathways such as cyclic-AMP-dependent protein kinases (Siebenlist and Taketa 1983) or membrane-bound phospholipase A_2 and arachidonic acid metabolism (Käfer et al. 1992). A frequently observed neurotoxic effect of triorganotins—astrocytic and axonal swelling and edema—seems to be related to this inhibitory effect of organotins on membrane proteins (Kobayashi et al. 1994). In fish, tributyltin inhibition of adrenergically activated sodium/proton exchange in trout erythrocytes have been demonstrated (Virkki and Nikinmaa 1993).

In the brain of juvenile rainbow trout exposed to tributyltin oxide, massive concentration of tin in endothelial cells and myelin sheaths in the optic nerve, the stratum opticum, and stratum album centrale of the tectum opticum as revealed by energy filtering transmission electron spectroscopic imaging (Figures 17.4a, b) and electron energy loss spectroscopy (EELS) could be correlated to intramyelin vacuolization and subsequent disintegration and necrosis of optic fibers (Triebskorn et al. 1994a). Such vacuolization of the tectum opticum was also reported for other fish and mammals (Aldridge 1992; Fent and Meier 1992). As a consequence of TBTO-induced damage in the brain, vision was apparently impaired and swimming behavior was drastically altered: In contrast to control fish swimming, which was characterized by regular orientation along the edges of the circular miniaquaria, fish exposed to TBTO almost completely lacked normal orientation (Figure 17.4c). Eventually, fish displayed drastically reduced weight gain due to less effective feeding (Triebskorn et al. 1994b). Thus, changes finally relevant at the population level could be traced back to pathological alterations in definite brain areas due to the accumulation of tributyltin within membranes of the brain.

17.4 INTRACELLULAR FATE OF CHEMICALS

Having crossed the cell membrane, chemicals can undergo binding to biomolecules, storage or deposition in subcellular compartments (e.g., lysosomes, lipid droplets), metabolic transformation, or elimination (Figure 17.2). The

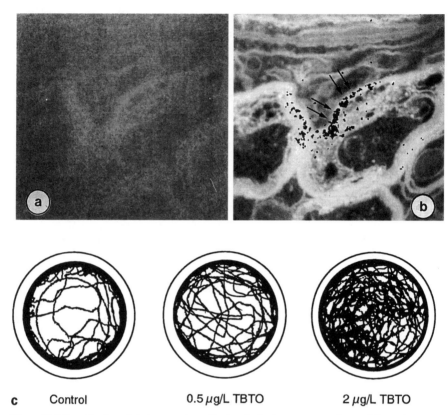

c   Control   0.5 µg/L TBTO   2 µg/L TBTO

Figure 17.4 Localization of tin in axonal myelin sheaths in the stratum opticum of the optic tectum of juvenile rainbow trout (*Oncorhynchus mykiss*) exposed to 2 µg/L bis(tri-*n*-butyltin)oxide (TBTO) for 21 days by means of electron spectroscopic imaging. Difference between inelastic dark field image at the tin-specific energy loss and the background image below the tin edge (a) and combination of tin distribution and inelastic dark field image taken at $\Delta E = 250$ eV (b; difference marked by black dots), showing localization of tin in the myelin sheath of an axon in the stratum opticum of the optic tectum (45,000 ×). TBTO accumulation could be correlated to intramyelin vacuolization and subsequent disintegration and necrosis of optic fibers, as well as impaired vision and swimming behavior. Video imaging of swimming tracks (duration: 5 min) of individual control rainbow trout, which were characterized by regular orientation along the edges of the circular miniaquaria, displayed significant differences from that of rainbow trout exposed to 0.5 or 2 µg/L TBTO, which almost completely lacked normal orientation (c). Reprinted with permission by Elsevier Science from Triebskorn et al., *Aquat. Toxicol.* 30: 189–197 and 199–213 (1994).

cellular dose of a chemical is largely determined by the rates of uptake and elimination, biotransformation, and the extent of storage and deposition.

As a determinant of intracellular xenobiotic accumulation and toxicity (and in particular, carcinogenicity; see below), metabolism of xenobiotics or "biotransformation" plays a key role, since many chemicals are not toxic, mutagenic and/or carcinogenic per se, but require metabolic activation to reactive species that can interact with target molecules in the cell. Conversely, cellular metabolism may also transform toxic chemicals to less toxic metabolites that can readily be eliminated by renal or biliary excretion. Biotransformation can usually be subdivided into two phases: in the first step (phase I), a polar functional moiety such as a hydroxyl group is introduced into the lipophilic contaminant by oxidation, epoxidation, dealkylation, dehalogenation, desamination, sulfoxidation, or desulfuration, thus giving it more hydrophilic properties. In phase II, metabolites of phase I are conjugated to various endogenous substrates such as peptides (e.g., glutathione), carbohydrates (e.g., UDP-glucuronic acid), or sulfate to further increase the water solubility of the metabolites.

The primary oxidative enzymes involved in phase I reactions belong to a large enzyme family collectively referred to as cytochrome-P450-dependent monooxygenases (mixed function oxygenases, MFO; for reviews, see Stegeman and Hahn 1994; Bucheli and Fent 1995; Goksøyr and Husoy 1997). Since cytochrome-P450-dependent monooxygenases not only metabolize xenobiotic compounds, but also endogenous substrates such as steroid hormones, prostaglandins, bile salts, and fatty acids, the interpretation of cytochrome-P450-dependent monooxygenase induction should be made with caution (Stegeman 1989). In fish, cytochrome-P450-dependent monooxygenases are most abundant in the endoplasmic reticulum of the liver and blood vessels in various organs (Miller et al. 1989; Lester et al. 1993); in crustaceans and molluscs, it is primarily located in the midgut gland (James 1989; Livingstone et al. 1989a).

Cytochrome-P450-dependent monooxygenase induction can be visualized by means of immunocytochemical techniques in native tissues (Figure 17.5; cf. Goksøyr and Husoy 1996) or by electrophoretic separation of homogenates and microsomal preparations (Figure 17.6). Induction of the cytochrome P450 system has been given particular attention as a biomarker of environmental contamination (Stegeman et al. 1992; Goksøyr et al. 1996). Cytochrome-P450-dependent monooxygenase induction, however, may not only serve as an early warning system for the exposure to specific classes of chemicals dioxins, PCBs, and PAHs, but also provide a more solid theoretical foundation for applied environmental research by elucidating fundamental principles of toxicology (Lech and Vodicnik 1985; McCarthy and Shugart 1990; Stegeman and Hahn 1994; Di Giulio et al. 1995; Rand et al. 1995).

By modifying the hydrophobicity of xenobiotics and/or conjugated metabolites, enzyme-mediated biotransformation enhances the excretion of xenobiotics from the cell. Apart from renal elimination, biliary excretion is the most important route of excretion for xenobiotics and their metabolites in aquatic vertebrates. Due to the central position of the liver in the circulatory system, hepatocytes can, thus, effectively remove chemicals from the blood. The mechanisms of transport of foreign substances from the liver cell into the bile, however, are not known with certainty, but in mammals it has been shown that the biliary membrane of hepatocytes possesses active transport systems for organic compounds (Klaassen and Watkins 1984).

Another elimination mechanism of the cell is based on a multixenobiotic transport protein in the cell membrane, the P-glycoprotein. This protein exports moderately hydrophobic chemicals from the cell cytoplasm (Kartner et al. 1983).

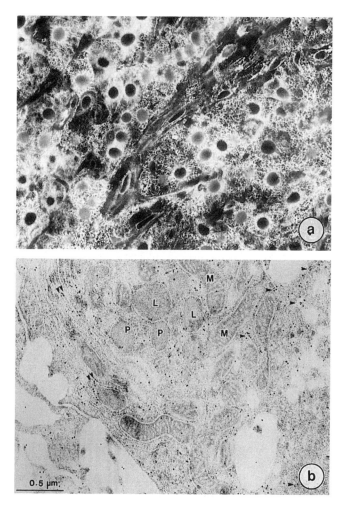

Figure 17.5 Light (a) and electron microscopical localization (b) of cytochrome P4501A1 (CYP1A) in the liver of juvenile rainbow trout (*Oncorhynchus mykiss*) given a single i.p. injection of β-naphthoflavone suspended in cod liver oil (50 μg/g; total volume of injection: 500 μL) 5 days prior to sampling. (a) In the light micrograph of glycol-methacrylate-embedded liver incubated with a monoclonal antibody (Mab 1-12-3 raised against cytochrome P-450 CYP 1A in scup, *Stenotomus versicolor*) and a fluorescent secondary antibody, nuclear envelopes and perinuclear regions within hepatocytes as well as endothelial cells are labelled. (b) In the electron micrograph, protein-G-colloidal gold (10 nm) labelling of the secondary antibody (➤) can be found on membranes of the endoplasmic reticulum (data from Lester et al. 1993). L, Lipid droplets; M, Mitochondria; P, Peroxisomes.

The phenomenon was first discovered in mammalian cancer cells that became resistant to treatment with chemotherapeutic drugs by exporting chemicals via an ATP-dependent membrane efflux pump, thus reducing intracellular drug accumulation, and, consequently, drug toxicity (Gottesman and Pasan 1993).

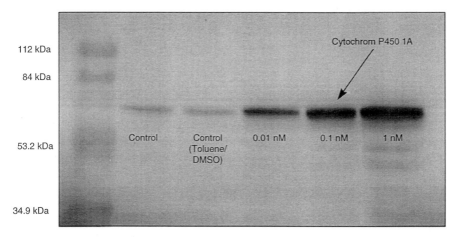

Figure 17.6 Western blot illustrating the induction of cytochrome P4501A1 (CYP1A) in primary cultures of hepatocytes isolated from the liver of rainbow trout (*Oncorhynchus mykiss*) after in vitro exposure to varying concentrations of 2,3,7,8-tetrachlorodibenzo-*p*-dioxin (TCDD). Lane 1: molecular weight standards; lane 2: hepatocytes without addition of solvent or TCDD; lane 3: solvent control (toluene/dimethylsulfoxide); lanes 4–6: hepatocytes exposed in vitro to 0.01, 0.1 and 1 nM TCDD for 3 days (courtesy of Th. Berbner, Department of Zoology, University of Heidelberg).

The broad-spectrum resistance to a wide variety of structurally unrelated chemicals sharing no common mechanisms of action was named multidrug resistance or—in the context of aquatic toxicology—multixenobiotic resistance (Ling and Gerlach 1985; Kurelec 1992). In addition to tumor cells, P-glycoprotein has been detected in cells of many nontransformed tissues, mainly in epithelia forming important physiological barriers such as those in brain, gut, or dermis (Pastan and Gottesman 1991; Lechardeur and Scherman 1995). The synthesis of P-glycoprotein is inducible by exposure of cells to xenobiotics; interestingly, there is evidence that the genes for P-glycoprotein and cytochrome-P450-dependent monooxygenases share common regulatory elements (Burt and Thorgeirsson 1988). This indicates the existence of a coordinated protective response of cells against potentially harmful xenobiotics.

Multidrug resistance and P-glycoprotein efflux pumps have been shown to be present and operative in all aquatic organisms studied so far and may partly explain how aquatic organisms can survive in a polluted environment (Kurelec 1992). However, the available data on this cellular process are still very limited for aquatic animals so that, at present, it is difficult to quantify the exact implications on xenobiotic accumulation and toxicity. Cornwall et al. (1995) speculated that elimination of chemicals by P-glycoprotein might explain discrepancies observed in the correlation between bioaccumulation of xenobiotics and lipophilicity, but experimental proof for this hypothesis is lacking.

The intracellular fate of metals is the result of the interaction between free metal ions, their binding to a diverse group of cytoplasmic, soluble ligands, and their sequestration into vesicle-bound granules and lysosomes (Viarengo 1989; Roesijadi and Robinson 1994; Di Giulio et al. 1995). These binding sites enable the cells to actively regulate the availability of both essential and nonessential metals and, as a consequence, cells are able to deal with the ambivalency of metals in biological systems—their essentiality for many biochemical reactions or metallo-enzymes, and their potential toxicity by nonspecific binding to oxygen, nitrogen, or sulfur.

Among the cytoplasmic ligands involved in metal regulation, metallothioneins have been studied most intensively. Metallothioneins represent a family of constitutively expressed low molecular mass proteins with numerous cysteinyl thiol groups, a property that allows them to bind to metals (Waalkes and Goering 1990; Roesijadi 1992). They provide a reserve pool of essential metals such as Cu and Zn for cell metabolism, but they also represent a high affinity sink for nonessential metals such as Hg and Cd. Metal exposure can lead to the induction of cellular metallothionein synthesis, with the induction being highest in tissues that are active in organismic metal uptake, storage, and excretion; in fish these organs are the gills, intestine, liver, and kidney. However, cellular metallothionein levels appear to be not always strictly correlated with metal exposure, since various endogenous factors such as hormones are also able to alter metallothionein synthesis (Burgess et al. 1991; Roesijadi 1992). The influence of such confounding factors have to be carefully considered when using cellular metallothionein levels as indicators of metal exposure.

An alternative way of intracellular metal sequestration is compartmentalization into membrane-bound vesicles (Sternlieb and Goldfischer 1976; George et al. 1982; Back and Prosi 1985; Segner 1987; Viarengo 1989; Segner and Braunbeck 1990; Nott 1991; Roesijadi and Robinson 1994; Vogt and Quinitio 1994; Cajaraville et al. 1996). Under conditions of enhanced metal burden, organisms increase the number and/or size of lysosomes (Figure 17.7). Particularly in invertebrates, irreversible metal deposition occurs in inorganic granules with high phosphate or calcium contents (Brown 1982; Simkiss and Taylor 1989b; Nott 1991). Cells containing elevated numbers of the inorganic granules can be extruded by apocrine secretion (Simkiss and Taylor 1989b; Vogt and Quinitio 1994), thus leading to decreased metal burdens of the whole organism.

17.5 INTRACELLULAR RECEPTORS

In the foregoing paragraphs of this chapter, factors influencing the cellular dose of chemicals have been discussed. In the following paragraphs, attention is given to adaptive or destructive responses of the cell to the presence of toxic compounds.

To be able to develop adaptive responses against a chemical, the cell needs a sensor to detect the intracellular presence of the compound. For this purpose, the cell applies a principle known from physiological processes, i.e., the induction of the synthesis of protective molecules via a receptor-mediated pathway.

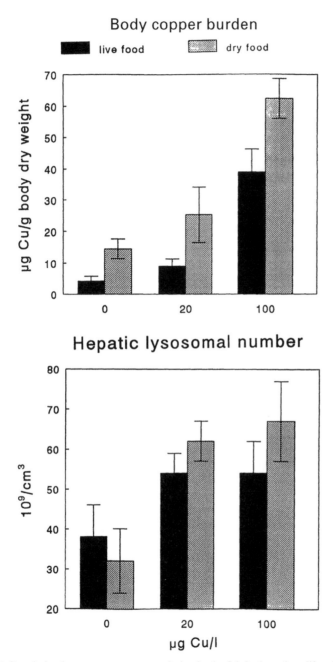

Figure 17.7 Correlation between copper accumulation in the fish body and proliferation of hepatocellular lysosomes in larvae of milkfish (*Chanos chanos*) after sublethal exposure to nominal concentrations of 20 and 100 µg Cu/L for 27 days. The copper burden was measured by atomic absorption spectroscopy; numeric density of lysosomes was evaluated by quantitative morphology, i.e., stereology.

Receptors are macromolecular components of cells with which a chemical (ligand) interacts to produce its characteristic biological activity. Receptor-ligand interactions are generally highly stereospecific (e.g., Hankinson 1995). The existence of a receptor has been described for metals (cf. Roesijadi 1992) and for substances inducing stress proteins (Sanders 1993; Di Giulio et al. 1995), but by far the best studied receptor is the aromatic hydrocarbon (Ah) receptor. The Ah receptor mediates induction of the cytochrome P450 1A system by halogenated aromatic hydrocarbons such as 2,3,7,8-tetrachlorodibenzo-p-dioxin (TCDD) and polycyclic aromatic hydrocarbons such as 3-methylcholanthrene (3-MC; Knutson and Poland 1982; Safe 1986; Okey 1990; Hankinson 1995). In the initial stage, chemicals such as TCDD and 3-MC reversibly bind to the cytosolic Ah receptor protein (Figure 17.8). The ligand–receptor complex then undergoes a process of transformation into a heterodimer consisting of the Ah-receptor–ligand binding subunit and the Ah receptor nuclear translocator

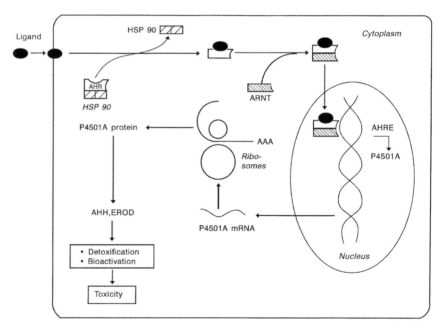

Figure 17.8 Proposed mechanism for Ah receptor-mediated changes in gene expression: After having entered the cell, ligands (chemicals) such as 2,3,7,8-tetrachlorodibenzo-p-dioxin bind to the cytoplasmic Ah receptor (AHR). The receptor subsequently undergoes a transformation including the dissociation of a 90-kDa heat shock protein (HSP 90) and the binding of the Ah receptor nuclear translocator (ARNT protein). In this form, the receptor–ligand complex is transferred into the nucleus, where it interacts with Ah receptor-responsive elements (AHRE; also known as XRE, i.e., xenobiotic responsive elements, or DRE, i.e., dioxin-responsive elements) close to the promoter region of the P4501A gene. The DNA binding initiates the transcription of P4501A1 mRNA, which in turn leads to enhanced cytoplasmic synthesis of P4501A protein and to elevated catalytic EROD and AHH activities. In addition to the P4501A gene, other genes and effector pathways appear to be under the control of the Ah receptor (Stegeman and Hahn 1994).

protein (ARNT). The dimer is translocated into the nucleus and interacts with specific Ah-receptor-responsive enhancer (AHRE) elements of the DNA that are located upstream of the CYP1A gene. The binding to AHRE leads to enhanced transcription of the CYP1A gene, elevated levels of CYP1A1 mRNA, and, subsequently, increased synthesis of CYP1A1 protein and associated catalytic activities including aryl hydrocarbon hydroxylase (AHH) and 7-ethoxy re-sorufin-O-deethylase (EROD).

In mammals, there is a close correlation between the affinity of the ligand (inducer) for the Ah receptor, its potency as an inducer of CYP1A, and several toxic endpoints such as thymic atrophy, body weight loss, and acute lethality (Safe 1986; Safe 1990; Kafafi et al. 1993). The correlations between these parameters support the hypothesis that many of the pleiotropic toxic effects of polyaromatic halogenated hydrocarbons in animals are effected via a common sequence of events initiated by ligand binding to the Ah receptor. However, although the structure–activity relationship holds well within individual classes of chemicals, e.g., within the polychlorinated biphenyls, interclass comparisons may be controversial. Riddick et al. (1994) showed that the affinity of TCDD and 3-MC for the mammalian Ah receptor is close to equal, but that TCDD is almost 30 000 times more potent as a CYP1A inducer than 3-MC. This discrepancy in receptor binding affinity and CYP1A induction potency between the two substances is obviously due to the more rapid biotransformation and, thereby, elimination of 3-MC compared with the very slow rate of TCDD metabolism. This example illustrates that cellular toxicity is not determined by one isolated event or process alone, but rather depends on the integrated cellular response.

The knowledge that many of the toxic effects of the structurally similar polychlorinated dibenzo-p-dioxin, dibenzofuran, and biphenyl congeners are mediated through the Ah receptor pathway has led to the formulation of the "toxic equivalency factor" (TEF) concept (Safe 1990). The TEF value of a chemical is an estimate of its toxic potency relative to that of TCDD, which is taken as a prototype compound. Usually, TEF values are determined by measuring the ability of a chemical to induce EROD or AHH activity relative to TCDD. The TEF approach has been proven to be particularly valuable as an integrative tool to assess the toxic potency of complex environmental mixtures containing polyaromatic halogenated hydrocarbons (e.g., Tillit et al. 1991). On the assumption that congeners which are effective via the Ah receptor act additively, the toxic potency of mixtures can be expressed as an equivalent concentration of TCDD by use of congener-specific TEF values (Safe 1990).

In aquatic animals, the presence of an Ah receptor has been demonstrated in a few species only (cf. Hahn and Stegeman 1992; Stegeman and Hahn 1994). Although there is as yet no direct evidence, it is generally assumed that CYP1A induction in aquatic animals occurs via an Ah-receptor-dependent mechanism comparable to that described above for mammals. There is a lack of direct structure–activity examinations of ligand–receptor binding in aquatic species; such studies might provide an explanation for some of the observed differences in CYP1A induction between fish and mammals. For instance, in contrast to mammals, mono*ortho*-substituted chlorinated biphenyl congeners are not active or only weakly active in fish (Gooch et al. 1989; Skaare et al. 1991).

During recent years, the TEF concept has been introduced into aquatic toxicology. Various bioassay systems for the establishment of piscine TEF values have been suggested, including embryo toxicity tests (Walker and Peterson 1991), in vivo measurements of EROD induction, (Newstedt et al. 1995), as well as in vitro EROD analysis using cell culture systems (Clemons et al. 1994). However, whereas in mammals a proven correlation exists between the EROD induction capacity of a chemical and its toxic potential, such a correlation has not yet been demonstrated unequivocally for fish (Stegeman and Hahn 1994; Newstedt et al. 1995). Thus, the use of the term "*induction* equivalency factor" (IEF) as suggested by Kennedy et al. (1996) may be more appropriate than the term "*toxicity* equivalency factor".

Binding of xenobiotics to receptors such as the Ah receptor induces a physiologic response of the cell that is intended to compensate the toxic impact of the chemical. However, xenobiotics may also bind to nontarget receptors, thus leading to toxic interference with normal ligand–receptor interaction. An example for such an effect is the binding of a variety of organic chemicals including phthalate plasticizers and organochlorine pesticides to the mammalian and piscine estradiol receptor (e.g., Soto et al. 1994; Jobling et al. 1995; Sumpter and Jobling 1996). This accidental occupation of the estradiol receptor can result in either antiestrogenic or estrogenic effects (e.g., Jobling and Sumpter 1993; Sumpter and Jobling 1996). It will be a challenging task for future studies to establish the consequences of such endocrine disruption on the reproductive output of organisms and populations.

17.6 INTERACTION WITH NUCLEAR STRUCTURES AND FUNCTIONS

Exposure to toxic chemicals not only affects cytoplasmic features, but also interferes with nuclear structure and function. With respect to nuclear morphology, common alterations after sublethal exposure are changes in nuclear size and shape and karyoplasm density. The functional background for these structural changes due to sublethal challenge is usually the need for enhanced DNA transcription—frequently initiated by receptor-mediated processes as described above—and enhanced mRNA export into the cytoplasm. More drastic changes of nuclear morphology appear during the late stages of cell death, with the two principal modes of cell death—apoptosis and necrosis—evoking distinctly different responses of nuclear morphology (Wyllie 1981, 1987).

In the previous section, the example of the Ah receptor was used to describe how chemicals may interact with nuclear DNA via a receptor molecule. This action represents a physiologically regulated process. However, chemicals can also directly interfere with DNA, thus acting as genotoxic substances. Genotoxic compounds, either directly or following biotransformation (see above), can bind to the DNA molecule in different ways. The most important sites of attack are the nucleotide bases. Planar molecules, for example, may intercalate between the bases of the DNA, or electrophilic substances may form covalent bonds with the nucleophilic bases, thus leading to DNA adducts, etc. (Walum et al. 1990; Shugart 1995). The formation of DNA adducts can result in mispairing during subsequent replication or clastogenic changes such as strand breaks. Since maintenance of the correct genetic information is essential for the organism, cells have developed a number of effective repair mechanisms which serve as protective measures against errors in the DNA (Walum et al. 1990; Maccubin 1994; Shugart 1995).

DNA adduct formation is the genotoxic endpoint that has received by far most attention in aquatic toxicology (Maccubin 1994; Lloyd-Jones 1995; Shugart 1995). Nonlethal genetic alterations in dividing somatic cells may ultimately result in the development of cancer (Maccubin 1994), and DNA adducts are thought to play a key role in initiation, early promotion (clonal expansion of transformed cells), and even in later stages of tumor progression (formation of metastases, malignancy, etc.; Harris 1991). Thus, measurement of DNA adducts has become a common tool to monitor exposure of fish to chemical carcinogens, both in carcinogenicity tests at the laboratory scale (mainly conducted with small fish species; Hawkins et al. 1995) and in field studies on epizootics in fish tumors from contaminated areas (mainly carried out with bottom-dwelling fish species; Malins et al. 1988; Harshbarger and Clark 1990). Although, in both approaches, major interest was focused on liver and skin tumors (Köhler 1991), the origin of metaplastic cells in, e.g., piscine liver neoplasia has not yet been clarified (Hinton et al. 1988a, b; Braunbeck et al. 1992; Parker et al. 1993).

The most widely used method for determination of DNA adducts is the ^{32}P-postlabelling technique, which is based on the direct detection of DNA adducts of radiolabelled chemicals (Baird 1979). Alternative methods for measuring genotoxicity are the microscopical determination of micronucleus formation, sister chromatid exchange, and chromosome aberrations (Al-Sabti 1991). More recently, the so-called comet assay (single cell or microgel electrophoresis) has been successfully adapted to fish cells. The comet assay is based on the differential electrophoretic migration of DNA fragments resulting from strand breaks induced by chemical or radiation damage to DNA (McKelvey-Martin et al. 1993; Fairbairn et al. 1995). The greater the damage to DNA, the greater the amount of DNA which migrates into the so-called comet during electrophoresis (Figure 17.9).

The comet assay can be used to discriminate between single- and double-strand DNA breaks (Singh et al. 1988), to investigate the organ specificity of genotoxic action and susceptibility to radiation, to investigate the capacity of repair mechanisms and the effectiveness of tumor therapy, as well as to study phenomena related to senescence of cells such as apoptosis, which, in contrast to

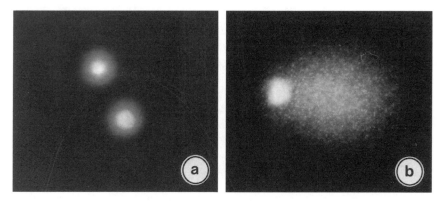

Figure 17.9 Genotoxicity of 2,3,7,8-tetrachlorodibenzo-*p*-dioxin in primary cultures of hepatocytes isolated from zebrafish (*Brachydanio rerio*). Fragments of DNA produced by *in vitro* exposure of hepatocytes show differential migration in electrophoresis primarily according to their size and form the so-called comet in the single cell gel electrophoresis according to Singh (1988). Whereas control cells are almost free of fragmented DNA extracted from the cell nucleus (a), cells exposed to ≥ 0.01 nM TCDD display a fluorescent tail after staining with ethidium bromide (courtesy of N. Rahman, Department of Zoology, University of Heidelberg).

necrosis, is characterized by early fragmentation of DNA (McKelvey-Martin et al. 1993; Fairbairn et al. 1995). The facts that the comet assay can be applied to both cells isolated from contaminated fish and cell cultures exposed to chemicals in vitro, that it requires only few cells or little tissue and seems to be very sensitive (Gedik et al. 1992), that it is independent of the cell phase, and that it is rapid and inexpensive make it particularly suitable for routine testing of chemical genotoxicity.

17.7 PROTECTIVE MOLECULES AND PROCESSES

Cells are equipped with protective mechanisms against the toxic impact of chemicals, including (a) the alteration of bioaccumulation by, e.g., biotrans-formation or P-glycoprotein activity, (b) the modification of intracellular avail-ability of toxic molecules by trapping and removal agents, e.g., metallothioneins or glutathione (GSH), and (c) the compensation of injury by repair mechanisms, e.g., excision repair of damaged DNA or restoration of denatured proteins with the help of heat shock proteins (hsp).

Metallothioneins are a good example of protective molecules. As discussed above, these proteins control, in response to metabolic needs, the availability of essential metals for specific metal-dependent functions, while they sequester nonessential metals into a nonaccessible form (Roesijadi 1992). As with many other defense mechanisms, metallothionein synthesis is inducible, i.e., it re-sponds via a receptor-mediated pathway to the elevation of intracellular metal concentrations (see above). The increasing levels of metallothioneins keep the amount of free toxic metal ions in the cell below critical levels and, thus, prevent toxicity. According to the "spill-over" hypothesis (Hodson 1988), toxicity only occurs when, with a continuously increasing metal burden, the rate of metallothionein synthesis has reached its maximum and excess free metal ions are no longer bound. Convincing evidence that metallothioneins have a protective function against metal toxicity comes from studies with yeast. Yeast strains deficient in the metallothionein gene are no longer capable of tolerating elevated levels of copper in the culture medium (Hamer et al. 1985); however, upon insertion of a mammalian metallothionein gene into the deficient yeast strains, the capability for copper tolerance reappears (Thiele et al. 1986). For aquatic organisms, it has been repeatedly shown that the induction of metallothioneins, as it occurs during exposure to sublethal levels of metals, confers increased tolerance to the toxicity of elevated metal concentrations (Hodson 1988; Roesijadi 1992). Preexposure to low doses of copper or zinc enhanced metal tolerance of fish in subsequent acute lethality tests by a factor of approximately 2 (Dixon and Sprague 1981; McCarter and Roch 1984; Klaverkamp and Duncan 1987; Hodson 1988). Congruent findings have been reported for aquatic invertebrates (Roesijadi and Fellingham 1987; Aoki et al. 1989). Additionally, induction of metallothionein by one metal such as zinc can induce cross-tolerance to another (Klaverkamp and Duncan 1987; Roesijadi and Fellingham 1987).

Another important molecule that protects cells against damage is reduced GSH. By trapping reactive electrophilic compounds, GSH prevents their binding to cell components; toxicity then depends primarily on the balance between the rates of reactive compound formation and removal by GSH.

The importance of GSH conjugation in determining xenobiotic toxicity may be exemplified by acetaminophen, which is metabolized by the cytochrome P450 system to a reactive, electrophilic intermediate; this reactive intermediate is usually removed by coupling to GSH in the glutathione-S-transferase (GST) reaction (Hodgson and Levi 1987). However, with higher doses of acetaminophen, the cellular GSH levels are progressively depleted, resulting in extensive covalent binding of the reactive metabolites to cell macromolecules with subsequent cell injury and death. Supplementation of the cells with agents that stimulate GSH synthesis is effective in preventing acetaminophen-induced cell death and liver necrosis. In contrast, depletion of cellular GSH levels increases toxicity of many xenobiotics (e.g., Gallagher and Di Giulio 1992). For instance, cyclophosphamide toxicity is partly related to its metabolic activation to acrolein. Normally, acrolein is detoxified by a reaction of the activated double bonds with GSH (Perry et al. 1995). If, however, cellular GSH contents are reduced, cyclophosphamide-induced toxicity is enhanced (Ishikawa et al. 1989).

GSH is of particular importance for the cellular defense against oxidative damage. Oxygen-derived radicals, such as the hydroxyl radical ($\cdot$OH), superoxide radical ($O_2\cdot$), and hydrogen peroxide ($H_2O_2\cdot$), can originate from the normal aerobic metabolism of the cell (Sies 1985; Winston and Di Giulio 1991;

Di Giulio et al. 1995), but they can also arise from cellular metabolism of drugs and chemicals such as quinones, biphenyls, and nitroaromatics (Borg and Schaich 1984; Kappus 1987; Washburn and Di Giulio 1988; Livingstone et al. 1989b). The activated species of oxygen, in particular the hydroxyl radical, are extremely reactive and have been implicated in oxidative cell damage and free radical pathology (Kappus 1987; Kehrer 1993). The oxyradicals attack the lipids of cell membranes resulting in lipid peroxidation (see above); they lead to mutagenesis and cancer by interacting with DNA, and they destroy cellular proteins, etc. The protective role of GSH against the deleterious action of oxyradicals is based on its action (a) as a radical scavenger (nonenzymatic protection) and (b) as a cofactor for the antioxidative enzyme glutathione peroxidase (Figure 17.10).

For aquatic animals, there are indications of an involvement of GSH in the defense against radicals, although data clearly demonstrating a protective role of GSH are scant. After exposure to oxyradical-generating compounds, both elevation and decline of cellular glutathione levels have been reported (Thomas and Wofford 1984; Babich et al. 1994; Regoli and Principato 1995).

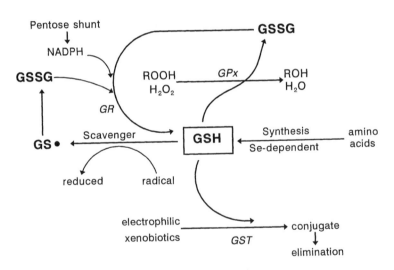

Figure 17.10 Involvement of reduced glutathione (GSH) in protective cellular reactions towards toxic compounds: GSH is synthesized from amino acids in a selenium (Se)-dependent pathway. GSH functions in biotransformation: in the glutathione-S-transferase (GST) reaction, electrophilic xenobiotics are conjugated to reduced GSH and thereby prepared for elimination from the cell. GSH protects against increased levels of peroxidants such as oxyradicals or lipid peroxy radicals from lipid peroxidation processes. The protective role of GSH in oxidative stress is based either on a direct function as a scavenger similar to, e.g., ascorbate, or on its function as a cosubstrate in the glutathione peroxidase (GPx) reaction. Oxidized glutathione (GSSG) is reduced to GSH by glutathione reductase (GR); the NADPH required for the GR reaction is usually derived from the pentose shunt.

The particular response—GSH increase or decrease—may be time dependent, with a depletion during acute exposure and an induction of synthesis during a longer exposure. In a comparative study with two different catfish species, Hasspieler et al. (1994) demonstrated reduced sensitivity to oxidative stress in species with higher GSH concentrations, a finding that indicates a protective role of GSH in fish. Similarly, Babich et al. (1994) observed in *in vitro* studies that GSH depletion of fish cell cultures results in enhanced cytotoxicity of quinone compounds. Congruent data are available with respect to the cytotoxicity of heavy metals that might induce oxidative stress by changing the oxidation status of the cells (Figure 17.11).

In field studies with fish populations, a relationship between exposure to aromatic hydrocarbons, the generation of free radicals, mutagenesis, and cancer has repeatedly been observed (Di Giulio et al. 1993; Malins and Haimanot 1993); however, the relationship between cancer etiology and oxidative stress has yet to be clarified.

Another component of the cellular stress response entails the rapid synthesis of a suite of proteins referred to as stress proteins, which are heat-inducible and

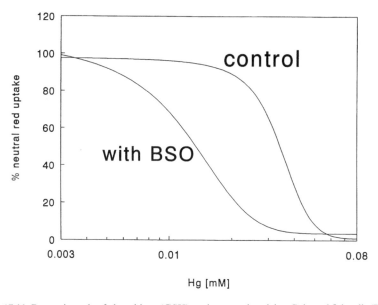

Hg [mM]

Figure 17.11 Protective role of glutathione (GSH) against metal toxicity. Cultured fish cells (RTG-2 cells derived from rainbow trout gonads) were exposed to various concentrations of mercury, and survival of the cells was assessed by means of the neutral red uptake inhibition assay. This method is based on the fact that viable, but not injured or dead cells can accumulate and retain neutral red in their lysosomes. In control cells, a mercury concentration of 0.035 mM killed 50% of the RTG-2 cells as evidenced by a 50% reduction of neutral red uptake. Preexposure of the cells to the inhibitor of GSH synthesis, buthionine sulfoximine (BSO), resulted in a 40% decrease in cellular GSH contents and a decrease in the NR_{50} value to 0.012 mM in GSH-depleted cells.

for this reason were initially referred to as heat-shock proteins (Ritossa 1962). The role of stress proteins in protein synthesis and function was first discovered during cellular reactions to heat shock, when stress proteins of different classes (hsp 90, hsp 70, hsp 60, and low molecular weight (LMW, 16–24 kDa) hsps) are induced (Sanders 1993; Di Giulio et al. 1995). Since, in addition to temperature, multiple stressors including toxic endogenous metabolites, ultraviolet light (UV), antibiotics, oxidants, viruses, organic xenobiotics, as well as heavy metals have been demonstrated to induce the synthesis of stress proteins (Kelly and Schlesinger 1978; Ashburner and Bonner 1979; Garry and Wixson 1983; Caltabiano 1986; Cochrane et al. 1991), heat-shock proteins were later termed stress proteins (Lindquist 1986). Heat-shock proteins thus represent an excellent example for the "coincidental" use of the presence of an endogenous process for protection against toxic damage. In order to establish consistency and minimize confusion in nomenclature, it was suggested that "stress-70" and "stress-90" be used for the 70- and 90-kDa families, respectively, and that the 60-kDa family be referred to as "chaperonin" (Rothman 1989; Gething and Sambrook 1992).

From a phylogenetic point of view, except for the LMW class, stress proteins are highly conserved molecules (Schlesinger 1986; Margulis et al. 1989) and apparently synthesized in almost all cell types (Gething and Sambrook 1992, Sanders 1993). Although the elucidation of the biological functions of stress proteins is still in its infancy, primary functions of stress proteins seem to be (1) in normal metabolism, the assistance in correct three- dimensional folding of proteins during their synthesis and (2) under adverse conditions, the identification and reorganization of denatured proteins and, thus, the reestablishment of normal protein function. Stress-90, stress-70, and chaperonin thus belong to a ubiquitous group of proteins called molecular chaperons, which direct protein folding and assembly (Ellis 1990; Gething and Sambrook 1992; Hartl et al. 1992; Welch 1991). Since the function and metabolism of cellular proteins frequently represent primary targets of xenobiotic toxicity ("proteotoxicity"; Hightower 1991), preservation and continuous repair of protein structure and function are of major importance for the integrity of the cell.

The accumulation of stress proteins correlates with acquired tolerance, wherein exposure to a mild stress regime confers the ability to survive a subsequent more severe stress that otherwise would be lethal to the organism (Lindquist 1986). The protective effect of stress proteins appears to involve common targets of environmentally induced damage, since tolerance is enhanced as long as stress protein levels are elevated; protection is independent of the specific chemical or physical properties of the stressor (Sanders 1993).

Stress-90, stress-70, and chaperonin, however, are also produced under normal conditions (Craig et al. 1983), since they play a pivotal role in translocation of proteins across membranes (Craig 1990) and correct folding by aggregation with nascent proteins (Rothman 1989; Langer and Neupert 1991; Martin et al. 1991; Braakman et al. 1992; Gething and Sambrook 1992; Sanders 1993). These constitutively synthesized stress proteins are frequently referred to as cognates (hsc; Sanders 1993). Under normal conditions, stress-90 forms a stable inactive complex with enzymes, hormone receptors (including, e.g., the estrogen and Ah receptors (see above); Chambraud et al. 1990; Perdew 1988), and components of the cytoskeleton (Gething and Sambrook 1992). In the case of hormone receptors, stress-90 is thought to inhibit activation and translocation into the nucleus in absence of the specific inducers. Stress-70 represents a multigene family of at least 21 proteins (Nover 1991) and prevents incorrect protein folding also in undamaged cells by binding to nascent target proteins and modulating protein folding, transport, and repair until completion of synthesis (for review, see Gething and Sambrook 1992; Sanders 1993). As a molecular chaperone, chaperonin also binds proteins to facilitate folding and assembly, although in a way different to that of stress-70 (Landry et al. 1992).

Upon exposure to stressful conditions, the synthesis of stress-90, stress-70, and chaperonin increases and redirects by mechanisms unknown in detail cellular metabolism to protect the cell from proteotoxicity and, thus, enhance tolerance. Cytoplasmic stress-70 (1) migrates into the nucleus, where it binds to preribosomes and other proteins to protect them from denaturation (Lindquist 1986; Gething and Sambrook 1992), (2) prevents formation of particularly damaging insoluble aggregates (Ellis 1990; Pelham 1990), (3) breaks up existing aggregates and, thus, allows protein refolding and reactivation (Ellis 1990; Gaitanaris et al. 1990), and (4) directs severely damaged proteins to lysosomes for elimination (Chiang et al. 1989). Stress-related functions of chaperonin are similar to those of stress-70 (Martin et al. 1992) except for the lysosomal elimination of damaged proteins and the breaking up of protein aggregates (Buchner et al. 1991). In contrast to the other classes of stress proteins, LMW stress proteins are more species-specific and have, so far, only been identified under adverse conditions.

In eukaryotic cells, environmentally stimulated transcription of heat-shock proteins is activated by binding of a heat-shock factor (HSF; Jurivich et al. 1992) to a conserved sequence in the promoter region referred to as a heat-shock element (HSE; Bienz and Pelham 1987; Morimoto et al. 1992). The activation mechanisms of HSF, however, remain obscure (Sanders 1993); according to the "abnormal protein hypothesis" (Edington et al. 1989), HSF is induced by damage to proteins (Ananthan et al. 1986). The induction of stress proteins is highly tissue- and stressor-specific, and, as a consequence, selective induction and relative concentrations of stress proteins can be used to identify tissues that are particularly vulnerable to damage by xenobiotics (e.g., Dyer et al. 1991, 1993a, b). The phosphate ester diazinon and the organochlorine insecticide lindane, for example, are known to function as inhibitors of cholinesterase and γ-aminobutyric-acid-activated chloride channels in nervous tissues, respectively; as a consequence, stress protein induction was highest in nervous tissues (Ogata et al. 1988; Dyer et al. 1993a, b). Such studies may be extended further to examine the subcellular distribution of stress proteins as a means of determining the mechanism of toxicity and specific cellular sites of stressor-induced damage.

With respect to physiological significance, stress proteins play important roles in processes that involve rapid breakdown and reorganization of tissues, such as larval settling and molting, resorption of gametes, breakdown of symbiosis, parasitic invasion of host organisms, and acclimation to changing environments (Sanders 1993). In ectothermic species, especially those living in rapidly changing habitats such as intertidal areas, shallow estuaries, or ephemeral desert ponds, acquisition of thermotolerance is of particular evolutionary and ecological importance (short-term "heat hardening" and temporally longer "resistance acclimation"; Sanders 1993). The role of stress proteins in stress-induced teratogenesis has been reviewed by Lindquist (1986), Black and Subjeck (1991), and Nover (1991).

Due to their central role in protective cellular responses, heat-shock proteins have considerable potential as diagnostic tools for detection of damage caused by multiple environmental stressors including chemical contaminants (Sanders 1990; Dyer et al. 1993b; Hightower and Schultz 1991). On the other hand, since induction of stress proteins is not exclusively linked to contaminant exposure, its use as biomarker requires thorough consideration of natural internal variables which modify stress protein levels, such as developmental and reproductive state (Lindquist 1986; Ramachandran et al. 1988), experimental conditions, and the presence of additional environmental stressors (Nover 1991; Dyer et al. 1993b).

The specific characteristics of stress protein induction (rapidity of response, relative induction of different stress protein classes, level of induction, tissue specificity) appear to be contaminant-specific (Kothary and Candido 1982; Lee et al. 1991; Sanders 1993). Induction of stress proteins by chemical contaminants, for example, seem to be generally slower than induction by heat shock (Mosser et al. 1988; Nover 1991), since the proteotoxic effect of the former is dependent on bioavailability, uptake, tissue-specific accumulation, elimination, and mechanism of toxicity. Moreover, many contaminants have been demonstrated to induce other proteins that are normally not expressed under heat-shock conditions, but might equally be termed "stress proteins" (Sanders 1990), e.g., metallothioneins and heme oxygenase (Caltabiano et al. 1988; Murphy et al. 1991). The environmental relevance of stress protein induction and tolerance acquisition has been demonstrated in numerous *in situ* studies with aquatic species (Sanders 1993).

17.8 CELL INJURY AND DEATH

Once protective measures can no longer counterbalance toxic effects of chemicals, processes of cell injury and, ultimately, cell death develop. Toxic effects are produced via alterations in normal cellular biochemistry, physiology, and morphology. The ways in which chemicals can interfere with cell structure and function are diverse, but some cellular responses appear to be of a more general nature as they occur under exposure to a wide variety of chemicals (Table 17.1).

Table 17.1 Selected general toxic effects of chemical stressors

---

Alterations in membrane functions

- Ion flux (DDT (dichlorodiphenyl trichloroethane) blocks sodium channels)
- Neurotransmitters and neuroreceptors (cholinesterase inhibitors prevent acetylcholine hydrolysis, resulting in ongoing membrane depolarization and excessive nerve excitation)
- Membrane fluidity (organic solvents and narcotics alter fluidity nonspecifically by insertion into the lipid bilayer)
- Disturbance of selective permeability (uncoupling chloro- and nitrophenols short-circuit the proton gradient across the mitochondrial membrane)
- Lipid peroxidation (e.g., carbon-tetrachloride-derived radicals attack the cell membranes as well as the membranes of lysosomes, mitochondria, and endoplasmic reticulum; the latter effect contributes to hepatic steatosis)

Interference with normal ligand–receptor interaction

- Agonistic or antagonistic effect on receptors for endogenous biomolecules (e.g., estrogenic effects of many organic chemicals)

Binding to biomolecules

- Binding to DNA (electrophilic metabolites of xenobiotics bind to nucleophilic nucleotides which can result in DNA damage; see below)
- Binding to enzymes (e.g., inhibition of glutathione-S-transferase and cytochrome P4501A1 by organotins, inhibition of urease by metals)

Accumulation of lipids

- Destruction of cytoskeleton and resulting reduction in lipoprotein excretion (e.g., phalloidin, norethandrolone)
- Impaired (apo)protein synthesis (see below)

Alterations in protein synthesis

- Destruction of the endoplasmic reticulum, the cellular site of protein synthesis (e.g., acetaminophen, ochratoxin)
- Decreased activity of RNA due to the binding of chemicals (e.g., dimethylnitrosamine)

Interference with cellular energy production

- Inhibition of electron transport chain (e.g., cyanide, 2,4-dinitrophenol)

Oxidative stress

- Redox cycling of quinones forming oxyradicals
- Metal catalysis of Haber-Weis and Fenton reaction

(Continued)

Table 17.1 (*Continued*)

Perturbation of calcium metabolism

- Disruption of Ca cycling (e.g., paraquat)
- Ca^{2+} release from mitochondria (e.g., menadione)

Neoplastic changes

- Formation of DNA adducts by electrophilic substances without metabolic transformation (direct carcinogens; e.g., N-methyl-N-nitronitrosoguanidine, methylazoxymethanol acetate)
- Adduct formation after metabolic activation (indirect carcinogens; e.g., aflatoxin B_1)
- Reaction with macromolecules other than DNA (epigenetic form of neoplastic change by nongenotoxic agents; e.g., immunosuppressants, lindane, polychlorinated biphenyls, diethylstilbestrol, 2,3,7,8-tetrachloro-p-dioxin)
- Oncogene activation (e.g., *ras* in goldfish, *c-myc* in rainbow trout)

The mechanism of toxic action of a chemical is usually not restricted to one specific process, but rather multiple sites and mechanisms are involved. For many chemicals, the cell can prevent toxicity up to a certain critical concentration; beyond this threshold, however, toxic injury and eventually cell death follow. The sequence of chemically induced cell damage frequently involves the following steps: (1) activation of the chemical to the initiating (toxic) form, (2) either detoxification of the reactive compound or early cellular changes, (3) either recovery and repair or irreversible changes, and finally (4) cell death. Although many toxic effects at the organismic level are ultimately a result of cell death, other organismic effects may be caused by structural and functional cellular changes which are not lethal to the cell. Examples are neoplastic transformation, impairment of immune function due to lowered phagocytotic activity of macrophages, and behavioral changes due to altered signal transduction in neural cells.

Recurrent reaction patterns can also be observed at the cytological level, which already integrates a multitude of biochemical responses. For instance, as shown by electron microscopy, there is a syndrome of cytopathological reactions in isolated hepatocytes from rainbow trout (*Oncorhynchus mykiss*), following exposure to chemical agents, which consists of both effects specific for particular chemicals and responses which are apparently independent of the contaminant (Table 17.2). General hepatocellular alterations include an increase in size heterogeneity and deformation of mitochondrial profiles; diminution, reduced stacking, dilation, vacuolation, and fragmentation of rough endoplasmic reticulum (RER) cisternae; transformation of RER cisternae into concentric and eventually myelin-like membrane whorls; and proliferation of lysosomes, myelinated bodies, and cytoplasmic vacuoles (Braunbeck 1993, 1994; Zahn and Braunbeck 1993, 1995; Zahn et al. 1996).

Table 17.2 Lowest observed effect concentration of dinitro-*o*-cresol (DNOC), 2,4-di-chlorophenol (2,4-DCP), 4-chloroaniline (4-CA), atrazine (ATR), disulfoton (DS), triphenyltin acetate (TPTA), and malachite green (MG) with respect to cytological alterations in isolated rainbow trout (*Oncorhynchus mykiss*) hepatocytes after in vitro exposure

	DNOC	2,4-DCP	4-CA	ATR	DS	TPTA	MG
Nucleus							
Irregular shape	7.2 (2)	—	—	—	—	—	—
Dilation of nuclear envelope	0.7 (2)	6.1 (4)	7.8 (1)	—	7.3 (3)	0.24 (1)	—
Rupture of nuclear envelope	—	—	—	2.32 (5)	—	2.44 (1)	—
Condensation of chromatin	0.7 (5)	—	—	—	—	—	—
Marginalization of heterochromatin	0.7 (5)	—	7.8 (5)	23.2 (5)	72.9 (1)	0.24 (3)	—
Proliferation of heterochromatin	—	—	—	—	—	0.24 (5)	0.029 (1)
Dense heterochromatin aggregates	—	—	—	—	—	2.44 (3)	—
Reduction of heterochromatin	—	61.3 (4)	7.8 (5)	2.32 (5)	72.9 (1)	—	0.029 (1)
Altered heterochromatin pattern	—	—	—	—	—	24.4 (3)	—
Loss of nucleolus	—	—	78.4 (5)	—	—	—	—
Mitochondria							
Heterogeneity in size	72.2 (4)	613 (2)	7.8 (3)	—	—	—	—
Proliferation of mitochondria	—	—	23.5 (1)	—	72.9 (1)	0.24 (1)	0.29 (1)
Reduction of mitochondria	—	—	—	—	—	—	—
Formation of mitochondria clusters	—	61.3 (3)	—	0.2 (1)	—	0.24 (5)	—
Deformation of mitochondria	72.2 (4)	613 (2)	7.8 (3)	0.2 (1)	72.9 (1)	0.24 (1)	0.029 (3)
Dilation of matrix	—	—	23.5 (5)	23.2 (5)	—	0.24 (1)	—
Formation of myelin-like whorls	0.7 (2)	—	7.8 (1)	23.2 (5)	—	—	—
Loss of mitochondrial cristae	—	—	23.5 (5)	—	—	—	—
Association with lipid droplets	—	—	—	0.2 (1)	—	—	0.029 (5)

(*Continued*)

Table 17.2 (Continued)

	DNOC	2,4-DCP	4-CA	ATR	DS	TPTA	MG
Peroxisomes							
Proliferation	—	—	23.5 (5)	0.2 (1)	—	0.24 (1)	0.29 (1)
Formation of peroxisome clusters	—	—	78.4 (1)	—	—	0.24 (1)	—
Rough endoplasmic reticulum							
Reduction	7.2 (3)	6.1 (4)	7.8 (3)	0.2 (3)	—	0.24 (1)	0.29 (3)
Loss of cisternal stacking	0.7 (3)	6.1 (1)	7.8 (3)	0.2 (1)	—	0.24 (1)	0.29 (3)
Proliferation of cisternae	—	—	—	—	7.3 (3)	—	—
Fenestration of cisternae	—	—	23.5 (2)	—	72.9 (1)	—	—
Vesiculation and/or fragmentation	0.7 (4)	6.1 (1)	7.8 (3)	0.2 (5)	—	0.24 (1)	0.029 (1)
Degranulation	—	—	7.8 (1)	—	7.3 (3)	—	—
Concentric arrangement (whorls)	0.7 (3)	6.1 (1)	7.8 (1)	0.2 (1)	7.3 (3)	0.24 (1)	0.029 (1)
Dilation of cisternae	7.2 (1)	6.1 (1)	—	0.2 (3)	7.3 (3)	0.24 (1)	0.29 (1)
Smooth endoplasmic reticulum							
Proliferation	—	6.1 (2)	7.8 (1)	—	—	0.24 (1)	0.029 (5)
Steatosis (microvesicular lipid)	—	—	7.8 (1)	23.2 (5)	72.9 (1)	—	—
Golgi fields							
Inactivation of VLDL secretion	—	—	7.8 (5)	—	—	—	—
Dilation of cisternae	7.2 (1)	6.1 (2)	—	2.3 (3)	—	0.24 (1)	0.029 (5)
Concentric arrangement of cisternae	7.2 (3)	—	—	23.2 (3)	—	—	—

Table 17.2 (Continued)

	DNOC	2,4-DCP	4-CA	ATR	DS	TPTA	MG
Lysosomal elements							
Proliferation of lysosomes	7.2 (1)	6.1 (2)	7.8 (1)	0.2 (3)	7.3 (3)	0.24 (1)	0.29 (1)
Formation of lysosome clusters	—	—	7.8 (3)	—	—	0.24 (1)	—
Induction of myelinated bodies	72.2 (2)	61.3 (4)	7.8 (3)	2.3 (3)	0.7 (1)	0.24 (1)	0.029 (3)
Induction of cytoplasmic vacuoles	7.2 (1)	6.1 (3)	23.5 (3)	23.2 (5)	0.7 (1)	0.24 (1)	0.029 (3)
Storage products							
Decrease in glycogen content	7.2 (5)	61.3 (4)	—	23.2 (3)	—	0.24 (1)	0.029 (5)
Formation of glycogenosomes	0.7 (2)	6.1 (1)	—	2.3 (3)	—	—	0.029 (3)
Formation of glycogen bodies	—	—	7.8 (3)	0.2 (5)	—	—	0.029 (3)
Increase in lipid droplets	0.7 (4)	6.1 (4)	—	2.3 (3)	72.9 (1)	2.44 (3)	0.29 (1)

Data are presented as lowest observed effect concentration in μM plus shortest experimental period in days, after which the alteration was recorded (in brackets).

VLDL, Very low density lipoprotein; — indicates no observed effect.

Development from initial injury to final cell death does not follow a straight line, but involves the interaction of numerous—reversible and irreversible—processes (Wyllie 1981, 1987). Two different forms of cell death can be differentiated: necrosis, the cell death induced as a consequence of cell injury, and apoptosis, the gene-directed "programmed" cell death. The processes of cell proliferation and physiological cell death (apoptosis) maintain tissue homeostasis in multicellular organisms.

The two forms of cell death can be differentiated by means of morphological and biochemical criteria. Necrotic cell death (Wyllie 1981) is initiated by swelling of the cytoplasm and organelles. This effect is due to the loss of selective permeability of the membrane. These alterations are in response to the early disappearance of the ion-pumping activities of the membrane, either directly due to membrane damage or secondary to cellular energy depletion. Further changes include rupture of cell and organelle membranes, rapid decreases in protein and RNA levels, and cleavage of DNA by lysosomal deoxyribonuclease activity into fragments displaying a continuous spectrum of sizes. Necrosis typically affects groups of contiguous cells and an inflammatory reaction usually develops.

Apoptosis is morphologically characterized by initial nuclear and cytoplasmic condensation, cytoskeletal damage, membrane blebbing, and cell shrinking (Schwartzman and Cidlowski 1993). Eventually, the cell disintegrates into a number of smaller, membrane-bound fragments containing intact organelles. The apoptotic cell or the cell fragments, are taken up by phagocytosis and rapidly degraded by neighboring cells without induction of an inflammatory response. Biochemically, one of the earliest detectable changes of apoptosis is a rapid, sustained increase in intracellular Ca^{2+} concentrations. This rise in Ca^{2+} levels may activate an endonuclease responsible for degradation of DNA into fragments of distinct length.

Chemically induced lethal impairment of vital cellular structures and functions results in necrotic cell death. Particularly under conditions of acute or high-dose exposure, necrosis is frequent. A well-studied example is necrosis of epithelial cells in fish gills during the initial shock phase (Mallat 1985; McDonald and Wood 1993). Apoptosis, on the other hand, is usually not triggered by chemicals, but by physiological factors. However, chemicals such as genotoxic agents (Ronen and Hedle 1984) or dimethylnitrosamine (Pritchard and Butler 1989) may directly or indirectly interfere with the regulatory signals for apoptosis, and thus circumstantially induce apoptosis. For instance, oxidative stress, free radicals, and DNA damage have been shown to be able to switch on the programme of apoptosis (Thompson 1995).

17.9 FUNCTION OF CELL TOXICOLOGY IN ECOTOXICOLOGY

The manifestation of ecotoxicological effects of chemicals on populations is dependent on (1) exposure and bioavailability of compounds, (2) the balance between toxic mechanisms and protective responses at the cellular level, (3) the number of individuals with alterations in population-relevant parameters as a consequence of adverse cellular change, (4) the dependence of population dynamics on the fraction of affected members of the population, and (5) other, indirect effects (e.g., alterations in the composition of the food web as a result of contamination) and factors not related to contamination such as season, temperature, changes in the physical environment, etc. (Depledge 1993). Cell toxicology investigates (2) and partly (3).

Cell toxicological studies provide ecotoxicologically relevant information in a number of fields: (1) retrospective analysis of the causative relationship between chemical exposure, primary effects, and observed ecological consequences, (2) insight into the underlying mode of toxic action as a prerequisite for extrapolation and the systematic prognosis of potential future damage, (3) an understanding of the modification of toxic challenge by protective mechanisms.

The establishment of a causative correlation between the ecological effects observed in the field and a given chemical exposure is usually extremely difficult due to the multifactorial control of ecological processes. As a rule, ecological investigations alone do not have sufficient resolving power to identify the causative agent (e.g., Adams and Ryon 1994). Similarly, a chemical analysis of toxicants alone cannot provide evidence for the toxic consequences of exposure. Rather, to identify chemical agents as the cause of ecological damage, a combination of ecological studies, chemical analytics, and toxicological investigations at the suborganismic and cellular levels appears to be a suitable approach. One example of such a strategy is the use of biomarkers in ecological surveys to verify the presence of a chemical contamination (Peakall 1994). A careful selection of the type of biomarker employed allows a first discrimination between groups of contaminants, e.g., dioxine-like compounds versus heavy metals, as well as the biological functions affected, e.g., genotoxic versus reproductive effects. The combination with chemical fractionation and analysis leads to the identification of the causative agent(s) and, thus, provides the basis for the design of remediation strategies. This philosophy is reflected in the TIE (toxicity identification evaluation) approach, i.e., the bioassay-directed determination of toxic constituents in environmental samples (e.g., Ankley et al. 1992; Burgess et al. 1994). Further, isolation of the cellular responses used for detection of specific toxicity and their integration into a biotechnological sensor system may lead to the development of environmental diagnostics, i.e., the simplified, but optimized and preferably automatized identification of chemical contamination in environmental surveillance.

Given the multitude of chemical reagents, organisms, and biological and ecological structures and functions, practical ecotoxicology will never be able to avoid extrapolations. Thus, a major task of ecotoxicology is to improve the quality of the basis for extrapolation. The better the knowledge of cellular fates and effects, the more likely it will be that the extrapolation and the resulting prognosis will be correct. The importance of mechanistic knowledge for the prediction of toxicity can be exemplified by the SAR (structure–activity relationship) approach: provided a strong correlation between a specific structural element and a certain mode of toxic action can be established, the toxicity of a new substance bearing this structural element can be predicted with reasonable certainty.

Knowledge of cellular modes of toxic action is also particularly useful for the development of a systematic categorization of mixture effects. Since in the field organisms usually experience multiple exposure, and since by far not all possible combinations can be tested, it is essential to develop tools for the systematic classification of mixture effects. For instance, Sandmann and Böger (personal communication) have demonstrated that the simultaneous presence of peroxidative agents (e.g., paraquat) results in a decrease in the toxicity of inhibitors of photosynthesis (e.g., diuron).

Another approach that uses knowledge of cellular processes to cope with the complexity of chemical mixtures in the environment is the TEF (toxicity equivalency factors) procedure, which has recently been introduced into ecotoxicology for Ah-receptor-binding chemicals. Expression of the toxicity of any polychlorinated biphenyl (PCB) congener as a fraction of the toxicity of the reference substance 2,3,7,8-tetrachlorodibenzo-p-dioxin (TCDD) allows an estimation of the combined toxicity of a PCB mixture without the necessity to consider each compound individually. However, a necessary prerequisite for the development of this approach is the knowledge that the toxic effects of PCBs and TCDD are mediated via an identical cellular route, the Ah receptor pathway (see above).

Particularly under conditions of sublethal exposure, the organism can initiate the following protective measures: (1) adaptive measures, e.g., increased metallothionein synthesis to reduce the cellular contents of free metal ions, (2) compensative measures, e.g., cellular proliferation to counteract cell death, and (3) repair measures, e.g., enzymatic restoration of chemically damaged DNA. Since the purpose of such processes is to protect the organism against pathological consequences, the toxic effect at the cell level may be not translated into an overt effect at the organism level, as long as the toxic stress does not exceed the protective capacity. For a given toxicant dose, the threshold between protection and pathology depends on a multitude of endogenous and exogenous factors including age, sex, nutritional status, social stress, parasitism, season, temperature, oxygen supply, and pH. Cell toxicology provides an insight into the extent of protective mechanisms and their modulation by confounding factors. Given the low dose exposure typical of present environmental contamination, identification of thresholds up to which cellular adaption and compensation are able to prevent organismic damage (e.g., Braunbeck and Völkl 1991, 1993) represents a major task in ecotoxicology. Under sublethal exposure conditions, it is not sufficient to distinguish between "life" and "death"; rather, a careful consideration of confounding factors is required to define the range of transition from protection to overt damage.

Cells are the site of primary interaction between chemicals and biological systems. Whether a chemically induced change in cell metabolism, physiology, and structure will develop into an adverse, toxic effect depends on the integration of multiple cellular processes. A toxic cellular effect, however, is not necessarily deterministic for toxic effects at higher levels of biological organization. Although it is widely assumed that biological hierarchy represents a continuum with linear transfer of information, factors such as the presence of compensatory mechanisms at higher organization levels, the presence of nondirect effects, etc. are likely to influence the relevance of the toxicological response of the cell for the overall ecotoxicological effect of a chemical (Underwood and Peterson 1988). It should be emphasized that in ecotoxicology, when considering the relation between primary cellular responses to toxicity and subsequent responses at higher levels of biological organization, those cellular changes which may influence population-relevant organismic parameters such as growth, health, energetics, survival, and reproduction deserve particular attention (Munkittrick and McCarty 1995). For instance, provision of vitellogenin in sufficient amounts and quality is a major prerequisite for reproductive success. In the liver of female zebrafish (*Brachydanio rerio*) exposed to sublethal concentrations of 4-chloroaniline, for example, drastic alterations in the RER indicated severe impairment of protein synthesis and reduced amounts of vitellogenin-containing Golgi vesicles gave rise to the hypothesis that reproductive success (number of eggs, survival of the offspring) might be impaired (Braunbeck et al. 1990). In fact, in a parallel study, Bresch et al. (1990) were able to reveal a decline in the number of surviving larvae in the subsequent generation.

Integrated with other disciplines such as environmental chemistry and ecology, cell toxicology provides an essential tool to understand processes in ecotoxicology. Cell toxicology not only plays its classical role in elucidating toxic modes of action, but also diagnoses existing hazards in the field and prospectively identifies biological functions at risk. The future value of cell toxicology in ecotoxicology, however, will strongly depend on the focus on population-relevant cellular parameters and a more intimate integration with ecological studies.

17.10 REFERENCES

Adams, S. M., and M. G. Ryan, 1994, "A Comparison of Health Assessment Approaches for Evaluating the Effects of Contaminant-related Stress on Fish Populations," *J. Aquat. Ecosyst. Health*, *3*, 15–25

Al-Sabti, K., 1991, *Handbook of Genotoxic Effects and Fish Chromosomes*, Ljubljana, Slovenia, p. 221

Aldridge, N. W., 1992, "Selective Neurotoxicity: Problems in Establishing the Relevance of *in vitro* Data to the *in vivo* Situation," in *Tissue-Specific Toxicity: Biochemical Mechanisms* (W. Dekant, and H. G. Neumann, Eds.), Academic Press, New York, USA.

Aldridge, W. N., 1976, "The Influence of Organtin Compounds on Mitochondrial Functions," *Adv. Chem. Ser.*, *157*, 166–179

Ananthan, J., A. L. Goldberg, and R. Voellmy, 1986, "Abnormal Proteins Serve as Eukaryotic Stress Signals and Trigger the Activation of Heat Shock Genes," *Science*, *232*, 522–524

Ankley, G. T., M. K. Schubauer-Berigan, and R. A. Hoke, 1992, "Use of Toxicity Identification Evaluation Techniques to Identify Dredged Material Disposal Options: A Proposed Approach," *Environ. Managem.*, *16*, 1–6

Aoki, Y., N. Hatakeyama, Y. Kobayashi, T. Sumi, T. Suzuki, and K. T. Suzuki, 1989, "Comparison of Cadmium-Binding Protein Induction Among Mayfly Larvae of Heavy Metal Resistant (*Baetis thermicus*) and Susceptible Species (*B. yoshinensis* and *B. sahoensis*)," *Comp. Biochem. Physiol.*, *C93*, 345–357

Arnold, H., H. J. Pluta, and T. Braunbeck, 1995, "Simultaneous Exposure of Fish to Endosulfan and Disulfoton in vivo: Ultrastructural, Stereological and Biochemical Reactions in Hepatocytes of Male in Rainbow Trout (*Oncorhynchus mykiss*) Liver," *Aquat. Toxicol.*, *33*, 17–43

Ashburner, M., and J. J. Bonner, 1979, "The Induction of Gene Activity in *Drosophila* by Heat Shock," *Cell*, *17*, 241–254

Babich, H., M. R. Palace, E. Borenfreund, and A. Stern, 1994, "Naphthoquinone Cytotoxicity to Bluegill Sunfish BF-2 Cells," *Arch. Environ. Contam. Toxicol.*, *27*, 8–13

Back, H., and F. Prosi, 1985, "Distribution of Inorganic Cations in *Limnodrilus udekemianus* (Oligochaeta, Tubificidae) Using Laser Induced Microprobe Mass Analysis, with Special Emphasis on Heavy Metals," *Micron Microsc. Acta*, *16*, 145–150

Baird, W. M., 1979, "The Use of Radioactive Carcinogens to Detect DNA Modifications," in *Chemical Carcinogens and DNA* (P. L. Grover, Ed.), CRC Press, Boca-Raton, USA, pp. 59ff

Baldwin, J. A., and E. J. Calabrese, 1994, "Gap Junction-mediated Intercellular Communication in Primary Cultures of Rainbow Trout Hepatocytes," *Ecotox. Environ. Safety*, *28*, 201–207

Bayne, B. L., 1985, "General Introduction," in *The Effects of Stress and Pollution on Marine Animals* (B. L. Bayne, D. A. Brown, K. Burns, D. R. Dixon, A. Ivanovici, D. R. Livingstone, D. M. Lowe, M. N. Moore, A. D. R. Stebbing, and J. Widdows, Eds.), Praeger Scientific, New York, USA, pp. xi–xvi

Bienz, M., and H. R. B. Pelham, 1987, "Mechanisms of Heat-shock Activation in Higher Eukaryotes," *Adv. Genet.*, *24*, 31–72

Black, A. R., and J. R. Subjeck, 1991, "The Biology and Physiology of the Heat Shock and Glucose-Regulated Stress Protein Systems," *Methods Achiev. Exp. Pathol.*, *15*, 126–166

Borenfreund, E., and J. A. Puerner, 1985, "A Simple Quantitative Procedure Using Monolayer Cultures for Cytotoxicity Assays (HTD/NR-90)," *J. Tissue Cult. Methods*, *9*, 7–9

Borg, D. C., and K. M. Schaich, 1984, "Cytotoxicity from Coupled Redox Cycling of Autooxidizing Xenobiotics and Metals," *Israel J. Chem.*, *24*, 38–53

Braakman, I., J. Helenius, and A. Helenius, 1992, "Role of ATP and Disulfide Bonds During Protein Folding in the Endoplasmic Reticulum," *Nature*, *356*, 260–262

Braunbeck, T., S. J. Teh, S. M. Lester, and D. E. Hinton, 1992, "Ultrastructural Alterations in Hepatocytes of Medaka (*Oryzias latipes*) Exposed to Diethylnitrosamine," *Toxicol. Pathol.*, *20*, 179–196

Braunbeck, T., V. Storch, and H. Bresch, 1990, "Species-Specific Reaction of Liver Ultrastructure of Zebra Fish (*Brachydanio rerio*) and Trout (*Salmo gairdneri*) after Prolonged Exposure to 4-Chloroaniline," *Arch. Environ. Contam. Toxicol.*, *19*, 405–418

Braunbeck, T., 1994a, "Cytological Alterations in Isolated Hepatocytes from Rainbow Trout (*Oncorhynchus mykiss*) Exposed to 4-Chloroaniline," *Aquat. Toxicol.*, *25*, 83–110

Braunbeck, T., 1994a, "Structural and Functional Alterations in Hepatocytes of Fish as Biomarkers for Contamination by Environmental Chemicals," *Habilitation Thesis*, University of Heidelberg, Germany, p. 214

Braunbeck, T. 1994b, "Detection of Environmentally Relevant Pesticide Concentrations Using Cytological Parameters: Pesticide Specificity in the Reaction of Rainbow Trout Liver?" in *Sublethal and Chronic Effects of Pollutants on Freshwater Fish* (R. Müller, and R. Lloyd, Eds.), Blackwell, Oxford, UK, pp. 15–29

Braunbeck, T., and A. Völkl, 1993, "Toxicant-Induced Cytological Alterations in Fish Liver as Biomarkers of Environmental Pollution? A Case-study on Hepatocellular Effects of Dinitro-*o*-cresol in Golden Ide (*Leuciscus lidus melanotus*)," in *Fish in Ecotoxicology and Ecophysiology* (T. Braunbeck, W. Hanke, and H. Segner, Eds.), Verlag Chemie, Weinheim, Germany, pp. 55–80

Bresch, H., H. Beck, D. Ehlermann, H. Schlaszus, and M. Urbanek, 1990, "A Long-Term Test Comprising Reproduction and Growth of Zebrafish with 4-Chloroaniline," *Arch. Environ. Contam. Toxicol.*, *19*, 419–427

Brown, B. E., 1982, "The Form and Function of Metal-Containing Granules in Invertebrate Tissues," *Biol. Rev.*, *57*, 621–667

Bucheli, Th., D. and K. Fent, 1995, "Induction of Cytochrome P450 as a Biomarker for Environmental Contamination in Aquatic Ecosystems," *Crit. Rev. Environ. Sci.*, *25*, 201–268

Buchner, J., M. Schmidt, M. Fuchs, R. Jaenicke, R. Rudolph, F. X. Schmidt, and T. Kiefhaber, 1991, "GroE Facilitates Refolding of Citrate Synthase by Suppressing Activation," *Biochemistry*, *30*, 1586–1591

Budunova, I. V., and G. M. Williams, 1994, "Cell Culture Assays for Chemicals With Tumour-promoting or Tumour-inhibiting Activity Based on the Modulation of Intercellular Communication," *Cell Biol. Toxicol.*, *10*, 71–116

Burgess, R. M., K. T. Ho, M. D. Tagliabue, A. Kuhn, R. Comeleo, P. Comeleo, G. Modica, and G. E. Morrison, 1995, "Toxicity Characterization of an Industrial and a Municipal Effluent Discharging to the Marine Environment," *Mar. Pollut. Bull.*, *30*, 524–535

Burgess, D., N. Frerichs, and S. George, 1993, "Control of Metallothionein Expression by Hormones and Stressors in Cultured Fish Cell Lines," *Mar. Environ. Res.*, *35*, 25–28

Burt, R. K., and S. S. Thorgeirsson, 1988, "Coinduction of MDR-1 Multidrug Resistance and Cytochrome P-450 Genes in Rat Liver by Xenobiotics," *J. Natl. Cancer Inst.*, *80*, 1383–1386

Cajaraville, M. P., I. Abascal, M. Etxeberria, and I. Marigomez, 1996, "Lysosomes as Cellular Markers of Environmental Pollution – Time-dependent and Dose-dependent Responses of the Digestive Lysosomal System of Mussels after Petroleum Hydrocarbon Exposure," *Environ. Toxicol. Water Qual.*, *10*, 1–8

Caltabiano, M. M., G. Koestler, G. Poste, and R. G. Greig, 1986, "Induction of 32 and 34 kD Stress Proteins by Sodium Arsenite, Heavy Metals, and Thiol-Reactive Agents," *J. Biol. Chem.*, *261*, 13381–13386

Caltabiano, M. M., G. Poste, and R. G. Greig, 1988, "Induction of the 32 kDHuman Stress Protein by Auranofin and Related Triethylphosphine Gold Analogs," *Biochem. Pharmacol.*, *37*, 4089–4093

Cameron, J. A., P. R. S. Kodavanti, S. N. Pentyala, and D. Desaiah, 1991, "Triorganotin Inhibition of Rat Cardiac Adenosine Triphosphatases and Catecholamine Binding," *J. Appl. Toxicol.*, *11*, 403–411

Carpene, E., and S. G. George, 1981, "Absorption of Cadmium by Gills of *Mytilus edulis* (L.)," *Mol. Physiol.*, *1*, 23–34

Chambraud, B., M. Berry, G. Redeuilh, P. Chambon, and E.-E. Baulieu, 1990, "Several Regions of the Human Estrogen Receptor are Involved in the Formation of Receptor-heat Shock Protein 90 Complexes," *J. Biol. Chem.*, *265*, 20686–20691

Cheeseman, K. H. 1993, "Mechanisms and Effects of Lipid Peroxidation," *Mol. Aspects Med.*, *14*, 191–197

Chiang, H. -L., S. R. Terlecky, C. P. Plant, and J. F. Dice, 1989, "A Role for a 70-Kilodalton Heat Shock Protein in Lysosomal Degradation of Intercellular Proteins," *Science*, *246*, 382–385

Clemons, J. H., M. R. van den Heuvel, J. J. Stegeman, D. G. Dixon, and N. C. Bols, 1994, "Comparison of Toxic Equivalent Factors for Selected Dioxin and Furan Congeners Derived Using Fish and Mammalian Liver Cell Lines," *Can. J. Fish. Aquat. Sci.*, *51*, 1577–1584

Cochrane, B. J., Irby, R. B., and T. W. Snell, 1991, "Effects of Copper and Tributyltin on Stress Protein Abundance in the Rotifer *Brachionus plicatilis*," *Comp. Biochem. Physiol.*, *C98*, 385–390

Comporti, M. 1993, "Lipid Peroxidation. Biopathological Significance," *Mol. Aspects Med.*, *14*, 199–207

Cornwall, R., B. H. Toomey, S. Bard, C. Bacon, W. M. Jarman, and D. Epel, 1995, "Characterization of Multixenobiotic/Multidrug Transport in the Gills of the Mussel *Mytilus californianus* and Identification of Environmental Substrates," *Aquat. Toxicol.*, *31*, 277–296

Craig, E. A. 1990, "Role of hsp70 in Translocation of Proteins Across the Membranes," in *Stress Proteins in Biology and Medicine* (R. I. Morimoto, A. Tissiéres, C. Georgopoilos, Eds.), Cold Spring Harbor, Laboratory Press, Cold Spring Harbor, USA, pp. 279–295

Craig, E. A., T. D. Ingolia, and L. J. Manseau, 1983, "Expression of *Drosophila* Heat Shock Cognate Genes during Heat Shock and Development," *Dev. Biol.*, *99*, 418–426

Depledge, M. H., 1993, "The Rational Basis for the Use of Biomarkers as Ecotoxicological Tools," in *Nondestructive Biomarkers in Vertebrates* (M. C. Fossi, and C. Leonzio, Eds.), Lewis Publishers, Boca Raton, USA, pp. 272–295

Di Giulio, R. T., W. H. Benson, B. M. Sanders, and P. A. Van Veld, 1995, "Biochemical Mechanisms: Metabolism, Adaptation and Toxicity," in *Aquatic Toxicology*, 2nd Edn. (G. M. Rand, Ed.), Taylor & Francis, Washington, USA, pp. 523–561

Di Giulio, R. T., C. Habig, and E. P. Gallagher, 1993, "Effects of Black Harbor Sediments on Indices of Biotransformation, Oxidative Stress, and DNA Integrity in Channel Catfish," *Aquat. Toxicol.*, *26*, 1–22

Dix, T. A., and J. Aikens, 1993, "Mechanisms and Biological Relevance of Lipid Peroxidation Initiation," *Chem. Res. Toxicol.*, *6*, 2–18

Dixon, D. G., and Sprague, J. B., 1981, "Acclimation to Copper by Rainbow Trout–Modifying Factor in Toxicity," *Can. J. Fish. Aquat. Sci.*, *38*, 880–888

Dyer, S. D., G. L. Brooks, K. L. Dickson, B. M. Sanders, and E. G. Zimmerman, 1993, "Synthesis and Accumulation of Stress Proteins in Tissues of Arsenite Exposed Fathead Minnows (*Pimephales promelas*)," *Environ. Toxicol. Chem.*, *12*, 913–924

Dyer, S. D., K. L. Dickson, and E. G. Zimmerman, 1993, "A Laboratory Evaluation of the Use of Stress Protein in Fish to Detect Changes in Water Quality," in *Environmental Toxicology and Risk Assessment* (W. G. Landis, J. J. Huges, and M. A. Lewis, Eds.), ASTM Publishers, Philadelphia, USA, pp. 273

Dyer, S. D., K. L. Dickson, E. G. Zimmerman, and B. M. Sanders, "Tissue-Specific Patterns of Heat Shock Protein Synthesis and Thermal Tolerance of the Fathead Minnow (*Pimephales promelas*)," *Can. J. Zool.*, *69*, 2021–2027

Edington, B. V., S. A. Whelan, and L. E. Hightower, 1989, "Inhibition of Heat Shock (Stress) Protein Induction by Deuterium Oxide and Glycerol: Additional Support for the Abnormal Protein Hypothesis of Induction," *J. Cell Physiol.*, *139*, 219–228

Ellis, R. J. 1990, "The Molecular Chaperon Concept," *Semin Cell Biol.*, *1*, 1–28

Fairbairn, D. W., P. L. Olive, and K. L. O'Neill, 1995, "The Comet Assay: A Comprehensive Review," *Mutat. Res.*, *339*, 37–59

Fent, K., and W. Meier, 1992, "Tributyltin-induced Effects on Early Life Stages of Minnows *Phoxinus phoxinus*," *Arch. Environ. Contam. Toxicol.*, *22*, 428–438

Fent, K., 1996, "Ecotoxicology of Organotin Compounds," *Crit. Rev. Toxicol.*, *26*, 1–117

Franks, N. P., and W. R. Lieb, 1978, "Where do General Anaesthetics Act?," *Nature*, *274*, 339–344

Gaitanaris, G. A., A. G. Papavassiliou, P. Rubock, S. J. Silverstein, and M. E. Gottesman, 1990, "Renaturation of Denatured Lambda Repressor Requires Heat Shock Proteins," *Cell*, *61*, 1013–1020

Gallagher, E. P., and R. T. Giulio, 1992, "Glutathione-Mediated Chlorthalonil Detoxification in Channel Catfish Gills," *Mar. Environ. Res.*, *34*, 221–226

Garry, R. F., and B. G. Wixson, 1983, "Induction of Stress Proteins in Sinbis Virus- and Vesicular Stomatitis Virus-Infected," *Virology*, *129*, 329–332

Gedik, C. M., S. W. Ewen, and A. R. Collins, 1992, "Single Cell Gel Electrophoresis Applied in to the Analysis of UV-C Damage and its Repair in Human Cells," *Int. J. Radiat. Biol.*, *62*, 313–320

George, S. G., T. L. Coombs, and B. J. S. Pirie, 1982, "Characterization of Metal-containing Granules from the Kidney of the Common Mussel, *Mytilus edulis*," *Biochim. Biophys. Acta*, *716*, 61–71

Gething, M. J., and J. Sambrook, 1992, "Protein Folding in the Cell," *Nature*, *355*, 33–45

Goksøyr, A., and A. M. Husøy, 1997, "Immunochemical Approaches to Studies of CYP1A Localization and Induction by Xenobiotics in Fish," in *Fish Ecotoxicology* (T. Braunbeck, D. E. Hinton, and B. Streit, Eds.), Birkhäuser, Basel, Switzerland, in press

Gooch, J. W., A. A. Elskus, P. J. Kloepper-Sams, M. E. Hahn, and J. J. Stegeman, 1989, "Effects of Ortho and Non-ortho Substituted Polychlorinated Biphenyl Congeners on the Hepatic Monooxygenase System in Scup (*Stenotomus chrysops*)," *Toxicol. Appl. Pharmacol.*, *98*, 422–428

Gottesman, M. M., and I. Pastan, 1993, "Biochemistry of Multidrug Resistance Mediated by the Multidrug Transporter," *Annu. Rev. Biochem.*, *62*, 385–427

Hahn, M. E., and J. J. Stegeman, 1992, "Phylogenetic Distribution of the Ah Receptor in Non-mammalian Species: Implications for Dioxin Toxicity and Receptor Evolution," *Chemosphere*, *25*, 931–937

Hamer, D. H., D. J. Thiele, and J. E. Lemontt, 1985, "Function and Autoregulation of Yeast Copper–Thionein," *Science*, *228*, 685–690

Hankinson, O. 1995, "The Aryl Hydrocarbon Receptor Complex," *Annu. Rev. Pharmacol. Toxicol.*, *35*, 307–340

Harris, C. C., 1991, "Chemical and Physical Carcinogensis: Advances and Perspectives for the 1990's," *Cancer Res.*, *Suppl. 51*, 5023–5044

Harshbarger, J. C., and J. B. Clark, 1990, "Epizootiology of Neoplasms in Bony Fish of North America," *Sci. Total Environ.*, *94*, 1–32

Hartl, F. U., J. Martin, and W. Neupert, 1992, "Protein Folding in the Cell: The Role of Molecular Chaperones Hsp70 and Hsp60," *Annu. Rev. Biophys. Biomol. Struct.*, *21*, 293–322

Hasspieler, B. M., J. V. Behar, D. B. Carlson, D. E. Watson, and R. T. Di Giulio, "Susceptibility of Channel Catfish (*Ictalurus punctatus*) and Brown Bullhead (*Ameirurus nebulosus*) to Oxidative Stress: A Comparative Study," *Aquat. Toxicol.*, *28*, 53–64

Hawkins, W. E., W. W. Walker, and R. M. Overstreet, 1995, "Carcinogenicity Tests Using Aquarium Fish," in *Fundamentals of Aquatic Toxicology* (G. M. Rand, Ed.), 2nd Edn., Taylor & Francis, Washington, DC, USA, pp. 523–562

Hightower, L. E., and R. J. Schultz, 1991, "Poeciliopsis: A Fish Model for Evaluating Genetically Variable Responses to Environmental Hazards," in *Biological Criteria: Research and Regulation.* U.S. Environm. Prot. Agency, Washington, DC, USA, pp. 129–152

Hightower, L. E., "Heat Shock, Stress Proteins, Chaperons, and Proteotoxicity (Meeting Review)," *Cell*, *66*, 191–197

Hinkle, P. M., P. A. Kinsella, and K. C. Ousterhoudt, 1987, "Cadmium Uptake and Toxicity via Voltage-Dependent Calcium Channels," *J. Biol. Chem.*, *262*, 16333–16337

Hinton, D. E., D. J. Laurén, and S. J. Teh, 1988, "Cellular Composition and Ultrastructure of Hepatic Neoplasms Induced by Diethylnitrosamine in *Oryzias latipes*," *Mar. Environ. Res.*, *24*, 307–310

Hinton, D. E., J. A. Couch, T. J. Swee, and L. A. Courtney, 1988, "Cytological Changes During Progression of Neoplasia in Selected Fish Species," *Aquat. Toxicol.*, *11*, 77–112

Hodgson, E., and P. E. Levi, 1987, *Modern Toxicology*, Elsevier, New York, USA

Hodson, P. V., "The Effect of Metal Metabolism on Uptake, Disposition and Toxicity in Fish," *Aquat. Toxicol.*, *11*, 3–18

Ishikawa, M., I. Sasaki, and Y. Takayanagi, 1989, "Injurious Effect of Buthionine Sulfoximine, an Inhibitor of Glutathione Biosynthesis, on the Lethality and Urotoxicity of Cyclophosphamide in Mice," *Jap. J. Pharmacol.*, *51*, 146–149

James, M. O., 1989, "Cytochrome P450 Monooxygenase in Crustaceans," *Xenobiotica*, *19*, 1063–1076

Jobling, S., and J. P. Sumpter, 1993, "Detergent Components in Sewage Effluent are Weakly Oestrogenic to Fish: An *in vitro* Study Using Rainbow Trout (*Oncorhynchus mykiss*) Hepatocytes," *Aquat. Toxicol.*, *27*, 361–372

Jobling, S., T. Reynolds, R. White, M. G. Parker, and J. P. Sumpter, 1995, "A Variety of Environmentally Persistent Chemicals, Including Some Phtalate Plasticizers, are Weakly Estrogenic," *Environ. Health Persp.*, *103*, 582–587

Jurivich, D. A., L. Sistonen, R. A. Kroes, and R. I. Morimoto, 1992, "Effect of Sodium Salicylate on the Human Heat Shock Response," *Science*, *255*, 1243–1245

Kafafi, S. A., H. Y. Afeefy, H. K. Said, and A. G. Kafafi, 1993, "Relationship Between Aryl Hydrocarbon Receptor Binding, Induction of Aryl Hydrocarbon Hydroxylase and 7-Ethoxyresorufin-O-deethylase Enzymes, and Toxic Activities of Aromatic Xenobiotics in Animals. A New Model," *Chem. Res. Toxicol.*, *6*, 328–334

Käfer, A., H. Zöltzer, and Krug, H. F. 1993, "The Stimulation of Arachidonic Acid Metabolism by Organic Lead and Tin Compounds in Human HL-60 Leukemia Cells," *Toxicol. Appl. Pharmacol.*, *116*, 125–132

Kappus, H, 1987, "Oxidative Stress in Chemical Toxicity," *Arch. Toxicol.*, *60*, 144–149

Kartner, N., J. R. Riordan, and V. Ling, 1983, "Cell Surface P-glycoprotein Associated with Multidrug Resistance in Mammalian Cell Lines," *Science*, *221*, 1285–1287

Kehrer, J. P., 1993, "Free Radicals as Mediators of Tissue Injury and Disease," *CRC Crit. Rev. Toxicol.*, *23*, 21–48

Kelly, P., and M. J. Schlesinger, 1978, "The Effect of Amino Acid Analogies and Heat Shock on Gene Expression in Chicken Embryo Fibroblasts," *Cell*, *15*, 1277–1286

Kennedy, S. W., A. Lorenzen, and R. J. Norstrom, 1996, "Chicken Embryo Hepatocyte Bioassay for Measuring Cytochrome P4501A-based 2,3,7,8-Tetrachlorodibenzo-*p*-dioxin Equivalent Concentrations in Environmental Samples," *Environ. Sci. Technol.*, *30*, 706–715

Klaassen, C. D., and J. B. Watkins, 1984, "Mechanisms of Bile Formation, Hepatic Uptake, and Biliary Excretion," *Pharmacol. Rev.*, *36*, 1–67

Klaassen, C. D., and D. L. Eaton, 1991, "Principles of Toxicology," in *Casarett and Doull's Toxicology*, 4th Edn. (M. O. Amdur, J. Doull, and C. D. Klaassen, Eds.). Pergamon Press, New York, USA, pp. 12–49

Klaverkamp, J. F., and D. A. Duncan, 1987, "Acclimation to Cadmium Toxicity by White Suckers: Cadmium Binding Capacity and Metal Distribution in Gill and Liver Cytosol," *Environ. Toxicol. Chem.*, *6*, 275–289

Knutson, J. C., and A. Poland, 1982, "Response of Murine Epidermis to 2,3,7,8-Tetra-chlorodibenzo-p-dioxin: Interaction of the *Ah* and *hr* Loci," *Cell*, *30*, 225–230

Kobayashi, H., T. Suzuki, H. Sato, and N. Matsusaka, 1994, "Neurotoxicological Aspects of Organotin and Lead Compounds on Cellular and Molecular Mechanisms," *TEN*, *1*, 23–30

Köhler, A., 1991, "Lysosomal Perturbations in Fish Liver as Indicators for Toxic Effects of Environmental Pollution," *Comp. Biochem. Physiol.*, *C100*, 123–127

Köhler, A., H. Deisenmann, and B. M. Lauritzen, 1991, "Histological and Cytochemical Indices of Toxic Injury in the Liver of Dab *Limanda limanda*," *Mar. Ecol. Progr. Ser.*, *91*, 141–153

Könnemann, H., and Musch, A. 1981, "Quantitative Structure–Activity Relationships in Fish Toxicity Studies. Part 2: The Influence of pH on the QSAR of Chlorophenols," *Toxicology*, *19*, 223–228

Kothary, R. K., and E. P. M. Candido, 1982, "Induction of a Novel Set of Polypeptides by Heat Shock or Sodium Arsenite in Cultured Cells of Rainbow Trout, *Salmo gairdnerii*," *Can. J. Biochem.*, *60*, 347–355

Kurelec, B., 1992, "The Multixenobiotic Resistance Mechanism in Aquatic Organisms," *CRC Crit. Rev. Toxicol.*, *22*, 23–43

Landry, S. J., R. Jordan, R. McMacken, and L. M. Gierasch, 1992, "Different Conformations for the Same Polypeptide Bound to Chaperons DnaK and GroEL," *Nature*, *355*, 455–457

Langer, T., and W. Neupert, 1991, "Heat Shock Proteins hsp60 and hsp70: Their Role in Folding, Assembly and Membrane Translocation of Proteins," in *Heat Shock Proteins and Immune Response* (S. H. E. Kaufmann, Ed.), Springer, Berlin, Heidelberg, New York, pp. 3–30

Lech, J. J., and M. J. Vodicnik, 1985, "Biotransformation," in *Fundamentals of Aquatic Toxicology* (G. M. Rand, and S. R. Petrocelli, Eds.), Hemisphere Publish. Corp, Cambridge, UK, pp. 527–557

Lechardeur, D., and D. Scherman, 1995, "Functional Expression of the P-Glycoprotein *mdr* in Primary Cultures of Bovine Cerebral Capillary Endothelial Cells," *Cell Biol. Toxicol.*, *11*, 283–293

Lee, Y. J., L. Curetty, and P. M. Corry, 1991, "Differences in Preferential Synthesis and Redistribution of HSP70 and HSP28 Families by Heat or Sodium Arsenite in Chinese Hamster Ovary Cells," *J. Cell Physiol.*, *149*, 77–87

Leo, A. C. Hansch, and D. Elkins, 1971, "Partition Coefficients and their Use," *Chem. Rev.*, *71*, 525–616

Lester, S. M., T. Braunbeck, S. J. Teh, J. J. Stegeman, , M. R. Miller, and D. E. Hinton, 1993, "Hepatic Cellular Distribution of Cytochrome P-450 IA1 in Rainbow Trout (*Oncorhynchus mykiss*): An Immuno- Histo- and -Cytochemical Study," *Cancer Res.*, *53*, 3700–3706

Lindquist, S., 1986, "The Heat Shock Response," *Annu. Rev. Biochem.*, *55*, 1151–1191

Ling, V., and J. Gerlach, 1985, Multidrug Resistance. *Breast Cancer Res. Treat.*, *4*, 89–94

Livingstone, D. R., P. Garcia Martinez, and G. W. Winston, 1989, "Menadione-Stimulated Oxyradical Production in Digestive Gland Microsomes of the Mussel, *Mytilus edulis*," *Aquat. Toxicol.*, *15*, 231–236

Livingstone, D. R., M. A. Kirchin, and A. Wiseman, 1989, "Cytochrome P-450 and Oxidative Metabolism in Molluscs," *Xenobiotica*, *19*, 1041–1062

Livingstone, D. R., P. Garcia Martinez, and G. W. Winston, 1989b, "Menadione-Stimulated Oxyradical Production in Digestive Gland Microsomes of the Common Mussel, *Mytilus edulis*," *Aquat. Toxicol.*, *15*, 231–236

Lloyd-Jones, G., 1995, "[32]P-Postlabeling: A Valid Biomarker for Environmental Assessment?," *TEN*, *2*, 100–104

Loewenstein, W. R., and Y. Kanno, 1966, "Intercellular Communication and the Control of Tissue Growth: Lack of Communication Between Cancer Cells," *Nature*, *209*, 1248–1249

Lowe, D. M., M. N. Morre, and B. M. Evans, 1992, "Contaminant Impact on Interactions of Molecular Probes with Lysosomes in Living Hepatocytes from Dab *Limanda limanda*," *Mar. Ecol. Progr. Ser.*, *91*, 135–140

Lowe, D. M., C. Soverchia, and Moore, M. N. 1995, "Lysosomal Membrane Responses in the Blood and Digestive Cells of Mussel Experimentally Exposed to Fluoranthene," *Aquat. Toxicol.*, *33*, 105–112

Maccubin, A. E., 1994, "DNA Adduct Analysis in Fish: Laboratory and Field Studies," in *Aquatic Toxicology* (D. C. Malins, and G. K. Ostrander, Eds.), Lewis Publishers, Boca Raton, USA, pp. 267–294

Malins, D. C., B. B. McCain, J. T. Landahl, M. S. Myers, M. M. Krahn, D. W. Brown, S. L. Chann, and W. T. Roubal, 1988, "Neoplasms and Other Diseases in Fish in Relation to Toxic Chemicals: An Overview," *Aquat. Toxicol.*, *11*, 43–52

Malins, D. C., and R. Haimanot, 1991, "The Etiology of Cancer: Hydroxyl-Radical Induced DNA Lesions in Histologically Normal Livers of Fish from a Population with Liver Tumours," *Aquat. Toxicol.*, *20*, 123–130

Mallat, J., 1985, "Fish Gill Structural Changes Induced by Toxicants and Other Irritants: A Statistical Review," *Can. J. Fish. Aquat. Sci.*, *42*, 630–648

Margulis, B. A., O. Y. Antropova, and A. D. Kharazova, "70 kD Heat Shock Proteins from Mollusc and Human Cells Have Common Structural and Functional Domains," *Comp. Biochem. Physiol.*, *B94*, 621–623

Markich, S. J., and R. A. Jeffree, 1994, "Absorption of Divalent Trace Metals as Analogues of Calcium by Australian Freshwater Bivalves: An Explanation of How Water Hardness Reduces Metal Toxicity," *Aquat. Toxicol.*, *29*, 257–290

Martin, J., T. Langer, R. Boteva, A. Schramel, A. L. Horwich, and F. U. Hatri, "Chaperonin-Mediated Folding at the Surface of GroEL Through a "Molten Globule"-Like Intermediate," *Nature*, *352*, 36–42

Martin, J., A. L. Horwich, and F. U. Hartl, 1992, "Prevention of Protein Denaturation Under Heat Stress by the Chaperonin hsp60," *Science*, *258*, 995–998

McCarter, J. A., and M. Roch, 1983, "Hepatic Metallothionein and Resistance to Copper in Juvenile Coho Salmon," *Comp. Biochem. Physiol.*, *C74*, 133–138

McCarthy, J. F., and L. R Shugart, 1990, *Biomarkers of Environmental Contamination.* CRC Press, Boca Raton, USA, 457 pages

McDonald, D. G., and C. M. Wood, 1993, "Branchial Mechanisms of Acclimation to Metals in Freshwater Fish," in *Fish Ecophysiology* (J. C. Rankin, and F. B. Jensen, Eds.), Chapman & Hall, London, UK, pp. 297–321

McKelvey-Martin, V. J., M. H. L. Green, P. Schmezer, B. L. Pool-Zobel, M. P. De Meo and A. Collins, 1993, "The Single Cell Gel Electrophoresis (Comet Assay): A European Review," *Mutat. Res.*, *288*, 47–63

Miller, M. E., D. E. Hinton, and J. J. Stegeman, 1989, "Cytochrome P450E Induction and Localization in Gill Pillar (Endothelial) Cells of Scub and Rainbow Trout," *Aquat. Toxicol.*, *14*, 307–322

Moore, M. N., 1992, "Pollutant-Induced Cell Injury in Fish Liver: Use of Fluorescent Molecular Probes in Live Hepatocytes," *Mar. Environ. Res.*, *34*, 25–31

Moore, M. N., 1976, "Cytochemical Demonstration of Latency of Lysosomal Hydrolases in the Digestive Cells of the Common Mussel," *Mytilus edulis*, and Changes Induced by Thermal Stress. *Cell Tissue Res.*, *175*, 279–287

Moore, M. N. 1990, "Lysosomal Cytochemistry in Marine Environmental Monitoring," *Histochem. J.*, *22*, 189–191

Morimoto, R. I., K. D. Sarge, and K. Abravaya, 1992, "Transcriptional Regulation of Heat Shock Genes," *J. Biol. Chem.*, *31*, 21987–21990

Mosser, D. D., and N. C. Bols, 1988, "Relationship Between Heat-Shock Protein Synthesis and Thermotolerance in Rainbow Trout Fibroblasts," *J. Comp. Physiol.*, *B158*, 457–467

Mullins, L. J., 1954, "Some Physical Mechanisms in Narcosis," *Chem. Rev.*, *54*, 289–305

Munkittrick, K. R., and L. S. McCarty, 1995, "An Integrated Approach to Aquatic Ecosystem Health: Top-down, Bottom-up, or Middle-out?" *J. Aquat. Ecosyst. Health*, *4*, 77–90

Murphy, B. J., K. R. Laderoute, S. M. Short, and R. M. Sutherland, 1991, "The Identification of Heme Oxygenase as a Major Hypoxic Stress Protein in Chinese Hamster Ovary Cells," *Br. J. Cancer*, *64*, 809–814

Newstedt, J. L., J. P. Giesy, G. T. Ankley, D. E. Tillit, R. A. Crawford, J. P. Gooch, P. D. Jones, and M. D. Denison, 1995, "Development of Toxic Equivalency Factors for PCB Congeners and the Assessment of TCDD and PCB Mixtures in Rainbow Trout," *Environ. Toxicol. Chem.*, *14*, 861–871

Nott, J. A. 1991, "Cytology of Pollutant Metals in Marine: 191—Invertebrates: A Review of Microanalytical Applications," *Scanning Microsc.*, *5*, 191–204

Nover, L., 1991, *The Heat Shock Response*, CRC Press, Boca Raton, USA

Ogata, N., S. M. Vogel, and T. Narahashi, 1988, "Lindane, but Not Deltamethrin Blocks a Component of GABA-activated Chloride Channels," *FASEB J.*, *2*, 2895–2900

Okey, A. B. 1990, "Enzyme Induction in the Cytochrome P-450 System," *Pharmacol. Ther.*, *45*, 241–2988

Opperhuizen, A., 1991, "Bioconcentration and Biomagnification: Is a Distinction Necessary?" in *Bioaccumulation in Aquatic Systems* (R. Nagel, and R. Loskill, Eds.), Verlag Chemie, Weinheim, Germany, pp. 67–80

Overton, C. E., 1901, *Studien über die Narkose, Zugleich ein Beitrag zur Allgemeinen Pharmakologie*, Gustav Fischer, Stuttgart, Germany

Parker, L. M., D. J. Laurén, B. D. Hammock, B. Winder, and D. E. Hinton, 1993, "Biochemical and Histochemical Properties of Hepatic Tumors of Rainbow Trout, *Oncorhynchus mykiss*," *Carcinogenesis*, *14*, 211–217

Pärt, P, O. Svanberg, and A. Kiessling, "The Availability of Cadmium to Perfused Rainbow Trout Gills in Different Water Qualities," *Water Res.*, *23*, 427–434

Pastan, I., and M. M. Gottesman, 1991, "Multidrug Resistance," *Annu. Rev. Med.*, *42*, 277–286

Peakall, D. B., 1994, "Biomarkers. The Way Forward in Environmental Assessment," *TEN*, *1*, 55–60

Pelham, H. R. B., 1990, "Functions of the hsp70 Protein Family: An Overview," in *Stress Proteins in Biology and Medicine* (R. I. Morimoto, A. Tissiéres, and C. Georgopoulos, Eds.), Cold Spring Harbor, Laboratory Press, Cold Spring Harbor, USA, pp. 287-ff

Perdew, G. H., 1988, "Association of the Ah Receptor with the 90-kDa Heat Shock Protein. *J. Biol. Chem.*, *263*, 13802–13805

Perry, C. S., X. Liu, C. G. Lund, C. P. Whitman, and J. P. Kehrer, 1995, "Differential Toxicities of Cyclophosphamide and its Glutathione Metabolites to A549 Cells," *Toxicol. In Vitro*, *9*, 21–26

Pritchard, D. J., and W. H. Butler, 1989, "Apoptosis—the Mechanism of Cell Death in Dimethylnitrosamine Induced Hepatotoxicity," *J. Pathol., 158*, 253–260

Raffray, M., and G. M. Cohen, 1991, "Bis(tri-n-butyltin)oxide Induces Programmed Cell Death (Apoptosis) in Immature Rat Thymocytes," *Arch. Toxicol., 65*, 135–139

Ramachandran, C., M. G. Catelli, W. Schneider, and G. Shyamala, "Estrogenic Regulation of Uterine 90-Kilodalton Heat Shock Protein," *Endocrinology, 123*, 956–961

Rand, G. M., P. G. Wells, and L. S. McCarty, 1995, "Introduction to Aquatic Toxicology," in *Fundamentals of Aquatic Toxicology*, 2nd Edn. (G. M. Rand, Ed.), Taylor & Francis, Washington, USA, p. 3

Regoli, F., and G. Principato, 1995, "Glutathione, Glutathione-dependent and Antioxidant Enzymes in Mussel, *Mytilus galloprovincalis*, Exposed to Metals under Field and Laboratory Conditions: Implications for the Use of Biomarkers," *Aquat. Toxicol., 31*, 143–164

Regoli, F., and E. Orlando, 1994, "Accumulation and Subcellular Distribution of Metals (Cu, Fe, Mn, Pb, and Zn) in the Mediterranean Mussel Mytilus Galloprovincialis during a Field Transplant Experiment," *Mar. Pollut. Bull., 28*, 592–600

Riddick, D. S., Y. Huang, P. A. Harper, and Okey, A. B. 1994, "2,3,7,8-Tetrachlorodibenzo-p-dioxin Versus 3-Methylcholanthrene: Comparative Studies of Ah Receptor Binding, Transformation and Induction of CYP1A1," *J. Biol. Chem., 269*, 12118–12128

Ritossa, F., 1962, "A New Puffing Pattern Induced by Heat Shock and DNP in Drosophila," *Experientia, 18*, 571–582

Roche, H., and G. Bogé, 1993, "Effects of Cu, Zn and Cr Salts on Antioxidant Enzyme Activities *in vitro* in Red Blood Cells of a Marine Teleost *Dicentrachus labrax*," *Toxicol. In Vitro, 7*, 623–629

Roesijadi, G., and W. E. Robinson, 1995, "Metal Regulation in Aquatic Animals: Mechanisms of Uptake, Accumulation and Release," in *Aquatic Toxicology* (D. C. Malins, and G. K. Ostrander, Eds.), Lewis Publishers, Boca Raton, USA, pp. 387–420

Roesijadi, G., and G. W. Fellingham, 1987, "Influence of Cu, Cd, and Zn Preexposure on Hg Toxicity in the Mussel, *Mytilus edulis*," *Can. J. Fish. Aquat. Sci., 44*, 680–684

Roesijadi, G., and M. E. Unger, 1993, "Cadmium Uptake in Gills of the Mollusc *Crassostrea virginica* and Inhibition by Calcium Channel Blockers," *Aquat. Toxicol., 24*, 195–206

Roesijadi, G., 1992, "Metallothioneins in Metal Regulation and Toxicity in Aquatic Animals," *Aquat. Toxicol., 22*, 81–114

Ronen, A., and J. A. Heddle, 1984, "Site-Specific Induction of Nuclear Abnormalities (Apoptotic Bodies and Micronuclei) by Carcinogens in Mice," *Cancer Res., 44*, 1536–1540

Rothman, J. E., "Polypeptide Chain Binding Proteins: Catalysts of Protein Folding and Related Processes in Cells," *Cell, 59*, 591–601

Safe, S., 1986, "Comparative Toxicology and Mechanism of Action of Polychlorinated Dibenzo-p-dioxins and Dibenzofurans," *Annu. Rev. Pharmacol. Toxicol., 22*, 517–554

Safe, S., 1990, "Polychlorinated Biphenyls (PCBs), Dibenzo-p-dioxins (PCDDs), Dibenzofurans (PCDFs) and Related Compounds: Environmental and Mechanistic Considerations Which Support the Development of Toxic Equivalency factors (TEFs)," *Crit. Rev. Toxicol., 21*, 51–88

Sanders, B. M., 1990, "Stress Proteins: Potential as Multitiered Biomarkers," in *Biomarkers of Environmental Contamination* (L. Shugart, and J. McCarthy, Eds.), Lewis Publishers, Chelsea, UK, pp. 165–192

Sanders, B. M., 1993, "Stress Proteins in Aquatic Organisms: An Environmental Perspective," *Crit. Rev. Toxicol., 23*, 49–75

Sargent, J. R., R. J. Henderson, and D. R. Tocher, 1989, "The Lipids," in *Fish Nutrition*, 2nd edn. (J. E. Halver, Ed.), Academic Press, San Diego, USA, pp. 153–218

Schlesinger, M. J., 1986, "Heat Shock Proteins: The Search for Functions," *J. Cell Biol., 103*, 321–325

Schneider, J. A., A. J. Jonas, M. L. Smith, and A. A. Greene, 1984, "Lysosomal Transport of Cystine and Other Molecules," *Biochem. Soc. Trans., 12*, 9908–9910

Schüürmann, G., and H. Segner, 1995, "Struktur-Wirkungs-Analyse von Trialkylzinnverbindungen," in *Eco- Informa'94*, Vol. 7 (K. Totsche, M. Matthies, F. Strutzenberger, W. Petek, W. Klöpffer, P. Czedik-Eysenberg, H. Meinholz, H. Fiedler, K. Alef, and O. Hutzinger, Eds.)., Umweltbundesamt, Wien, Austria, pp. 439–453

Schwartzman, R. A., and J. A. Cidlowski, 1993, "Apoptosis: The Biochemistry and Molecular Biology of Programmed Cell Death," *Endocr. Rev., 14*, 133–151

Segner, H, 1987, "Response of Fed and Starved Roach, *Rutilus rutilus*, to Sublethal Copper Contamination," *J. Fish Biol., 30*, 423–427

Segner, H., and T. Braunbeck, 1990, "Qualitative and Quantitative Assessment of the Response of Milkfish, *Chanos chanos*, Fry to Low-Level Copper Exposure," in *Pathology in Marine Science* (F. O. Perkins, and T. C. Cheng, Eds.), Academic Press, San Diego, USA, pp. 347–368

Shugart, L. R., 1995, "Environmental Genotoxicology," in *Aquatic Toxicology* (G. M. Rand, Ed.). Taylor & Francis, Washington, USA, pp. 405–420

Siebenlist, K. R., and F. Taketa, 1983, "The Effects of Triethyltin Bromide on Red Cell and Brain Cyclic AMP-Dependent Protein Kinases," *J. Biol. Chem., 258*, 11384–11390

Sies, H. (Ed.), 1985, *Oxidative Stress*. Academic Press, London, UK

Sikkema, J., J. A. M., d Bont, and B. Poolman, 1994, "Interactions of Cyclic Hydrocarbons with Biological Membranes," *J. Biol. Chem., 269*, 8022–8028

Simkiss, K., and M. G. Taylor, 1989a, "Metal Fluxes Across the Membranes of Aquatic organisms," *CRC Crit. Rev. Aquat. Sci., 1*, 173–188

Simkiss, K. and M. G. Taylor, 1989b, "Convergence of Cellular Systems of Metal Detoxification. *Mar. Environ. Res., 28*, 211–214

Singh, N. P., D. B. Danner, R. R. Tice, L. Brandt, and E. L. Schneider, 1988, "DNA Damage and Repair with Age in Individual Human Lymphocytes," *Mutat. Res., 237*, 123–130

Skaare, J. U., E. G. Jensen, A. Goksoyr, and E. Egaas, 1991, "Response of Xenobiotic Metabolizing Enzymes of Rainbow Trout (*Oncorhynchus mykiss*) to the Mono-ortho Substituted Polychlorinated PCB Congener 2,3',4,4',5-Pentachlorobiphenyl, PCB 118, Detected by Enzyme Activities and Immunochemical Methods," *Arch. Environ. Contam. Toxicol., 20*, 349–355

Soto, A. M., H. Justicia, J. W. Wray, and C. Sonnenschein, 1991, "*p*-Nonyl-phenol: An Estrogenic Xenobiotic Released from "Modified" Polystyrene," *Environ. Health Persp., 92*, 167–173

Stacey, N. H., and C. D. Klaassen, 1984, "Cadmium Uptake by Rat Hepatocytes", *Toxicol. Appl. Pharmacol.*, *55*, 448–455

Stegeman, J. J., 1989, "Cytochrome P-450 Forms in Fish: Catalytic, Immunological and Sequence Similarities," *Xenobiotica*, *19*, 1093–1110

Stegeman, J. J., M. Brouwer, R. T. DiGiulio, L. Förlin, B. A. Fowler, B. M. Sanders, and P. A. van Veld, 1992, "Enzyme and Protein Synthesis as Indicators of Contaminant Exposure and Effect," in *Biomarkers. Biochemical, Physiological, and Histological Markers of Anthropogenic Stress* (R. J. Huggett, R. A. Kimerle, P. M. Mehrle, and H. L. Bergman, Eds.), Lewis Publishers, Boca Raton, USA, pp. 235–336

Stegeman, J. J., and M. E. Hahn, 1994, "Biochemistry and Molecular Biology of Monooxygenases: Current Perspectives on Forms, Functions, and Regulation of Cytochrome P450 in Aquatic Species," in *Aquatic Toxicology* (D. C. Malins, and G. K. Ostrander, Eds.), Lewis Publishers, Boca Raton, USA, pp. 87–206

Sternlieb, I., and S. Goldfischer, "Heavy Metals and Lysosomes," in *Lysosomes in Biology and Pathology* (J. T. Dingle, and R. T. Dean, Eds.), North-Holland, Amsterdam, The Netherlands, pp. 185–200

Sumpter, J. P., and S. Jobling, 1996, "Vitellogenesis as a Biomarker for Estrogenic Contamination of the Aquatic Environment," *Environ. Health Persp. Suppl.*, 173–178

Terada, H, 1990, "Uncouplers of Oxidative Phosphorylation," *Environ. Health Persp.*, *87*, 213–217

Thiele, D. J., M. J. Walling, and D. H. Hamer, 1986, "Mammalian Metallothionein is Functional in Yeast," *Science*, *231*, 854–856

Thomas P., and H. W. Wofford, 1984, "Effects of Metals and Organic Compounds on Hepatic Glutathione, Cysteine and Acid-soluble Thiol Levels in Mullet (*Mugil cephalus L.*)," *Toxicol. Appl. Pharmacol.*, *76*, 172–182

Thompson, C. B., 1995, "Apoptosis in the Pathogenesis and Treatment of Disease," *Science*, *267*, 1456–1462

Tillit, D. E., J. P. Giesy, and G. T. Ankley, 1991, "Characterization of the H4IIE Rat Hepatoma Cell Bioassay as a Tool Assessing Toxic Potency of Planar Halogenated Hydrocarbons in Environmental Samples," *Environ. Sci. Technol.*, *25*, 87–92

Triebskorn, R., H. R. Köhler, J. Flemming, T. Braunbeck, R. D. Negele, and H. Rahmann, 1994, "Evaluation of Bis(tri-n-butyltin)oxide (TBTO) Neurotoxicity in Rainbow Trout (*Oncorhynchus mykiss*). I. Behaviour, Weight Increase, and Tin Contents," *Aquat. Toxicol.*, *30*, 189–197

Triebskorn, R., H. R. Köhler, K. H. Körtje, R. D. Negele, H. Rahmann, and T. Braunbeck, 1994, "Evaluation of Bis(tri-n-butyltin)oxide (TBTO) Neurotoxicity in Rainbow Trout (*Oncorhynchus mykiss*). II. Ultrastructural Diagnosis and Tin Localization by Energy Filtering Transmission Electron Microscopy (EFTEM)," *Aquat. Toxicol.*, *30*, 199–213

Underwood, A. J., and C. H. Peterson, 1988, "Towards an Ecological Framework for Investigating Pollution," *Mar. Ecol. Prog. Ser.*, *46*, 227–234

Veith, G. D., D. J. Call, and L. T. Brooke, 1983, "Structure-Toxicity Relationships for the Fathead Minnow, *Pimephales promelas*: Narcotic Industrial Chemicals," *Can. J. Fish. Aquat. Sci.*, *40*, 743–749

Verbost, P. M., P. M. J. van Rooij, G. Flik, R. A. C. Lock, and S. E. Wendelaar Bonga, 1989a, "The Movement of Cadmium Through Freshwater Trout Branchial Epithelium and Its Interference with Calcium Transport," *J. Exp. Biol.*, *145*, 185–197

Verbost, P. M., G. Flik, P. T. K. Pang, R. A. C. Lock, and S. E. Wendelaar Bonga, 1989b, "Cadmium Inhibition of the Erythrocyte Ca^{2+} Pump. A Molecular Interpretation," *J. Biol. Chem.*, *264*, 5613–5615

Viarengo, A., "Heavy Metals in Marine Invertebrates: Mechanisms of Regulation and Toxicity at the Cellular Level," *CRC Crit. Rev. Aquat. Sci.*, *1*, 295–317

Virkki, L., and N. Nikinmaa, 1993, "Tributyltin Inhibition of Adrenergically Activated Sodium/Proton Exchange in Erythrocytes of Rainbow Trout (*Oncorhynchus mykiss*)," *Aquat. Toxicol.*, *25*, 139–146

Vogt, G., and E. T. Quinitio, 1994, "Accumulation and Excretion of Metal Granules in the Prawn, *Penaeus monodon*, Exposed to Water-Borne Copper, Lead, Iron and Calcium," *Aquat. Toxicol.*, *28*, 223–241

Waalkes, M. P., and P. L. Goering, 1990, "Metallothionein and Other Cadmium-Binding Proteins: Recent Developments," *Chem. Res. Toxicol.*, *3*, 263–270

Walker, M. K., and R. E. Peterson, 1991, "Potencies of Polychlorinated Dibenzo-*p*-dioxin, Dibenzofuran and Biphenyl Congeners Relative to 2,3,7,8-Tetrachloro-dibenzo-*p*-Dioxin for Producing Early Life Stage Mortality in Rainbow Trout (*Oncorhynchus mykiss*)," *Aquat. Toxicol.*, *21*, 219–238

Walum, E., K. Stenberg, and D. Jenssen, 1990, "Understanding Cell Toxicology," Ellis Horwood, New York, USA

Washburn, P. C., and R. T. Di Giulio, 1988, "Nitrofurantoin-stimulated Superoxide Production by Channel Catfish (*Ictalurus punctatus*) Hepatic Microsomal and Soluble Fractions," *Toxicol. Appl. Pharmacol.*, *95*, 363–377

Welch, W. J., 1991, "The Role of Heat-Shock Proteins as Molecular Chaperones," *Curr. Opin. Cell Biol.*, *3*, 1033–1038

Wezel, A. P. van, and A. Opperhuizen, 1995, "Narcosis due to Environmental Pollutants in Aquatic Organisms: Residue-based Toxicity, Mechanisms, and Membrane Burdens," *CRC Crit. Rev. Toxicol.*, *195*, *25*, 255–279

Winston, G. W., and R. T. Di Giulio, 1991, "Prooxidant and Antioxidant Mechanisms in Aquatic Organisms," *Aquat. Toxicol.*, *19*, 137–161

Wofford, H. W., and P. Thomas, 1988, "Effect of Xenobiotics on Peroxidation of Hepatic Microsomal Lipids Striped Mullet (*Mugil cephalus*) and Atlantic Croaker (*Micropogonias undulatus*)," *Mar. Environ. Res.*, *24*, 285–289

Wright, D. A., 1977, "The Effect of Calcium on Cadmium Uptake by the Shore Crab *Carcinus maenas*," *J. Exp. Biol.*, *67*, 163–173.

Wyllie, A. H., 1977, "Cell Death: A New Classification Separating Apoptosis from Necrosis," in *Cell Death in Biology and Pathology* (I. D. Bowen, and R. A. Lockshin, Eds.), Chapman and Hall, London, UK, pp. 9–34

Wyllie, A. H. 1987, "Apoptosis: Cell Death Under Homeostatic Control," *Arch. Toxicol. Suppl.*, *11*, 3–10

Zahn, T., H. Arnold, and T. Braunbeck, 1996, "Cytological and Biochemical Response of R1 Cells and Isolated Hepatocytes from Rainbow Trout (*Oncorhynchus mykiss*) to Subacute *in vitro* Exposure to Disulfoton," *Exp. Toxicol. Pathol.*, *48*, 47–64

Zahn, T., and T. Braunbeck, 1993, "Isolated Fish Hepatocytes as a Tool in Aquatic Toxicology–Sublethal Effects of Dinitro-*o*-Cresol and 2,4-Dichlorophenol," *Sci. Total Environ. Suppl.*, 721–734

Zahn, T., and T. Braunbeck, 1996, "Acute and Sublethal Toxicity of Seepage Water from Garbage Dumps to Permanent Cell Cultures and Primary Cultures of Hepatocytes from Rainbow Trout (*Oncorhynchus mykiss*): A Novel Approach to Environmental Risk Assessment," *Zentralbl. Hyg.*, *196*, 455–479

Zucker, R. M., K. H. Elsetin, R. E. Easterling, H. P. Ting-Beall, J. W. Allis, and E. J. Massaro, "Effects of Tributyltin on Biomembranes: Alteration in Flow Cytometric Parameters and Inhibition of Na$^+$, K$^+$-ATPase Two-dimensional Crystallization," *Toxicol. Appl. Pharmacol.*, *96*, 393–403

18

Long-Term Effects of Chemicals in Aquatic Organisms

Horst Peter and Wolfgang Heger (Berlin, Germany)

Ecotoxicology, Edited by Gerrit Schüürmann and Bernd Markert.
ISBN 0-471-17644-3   © 1998 John Wiley & Sons, Inc. and Spektrum Akademischer Verlag.

18.1 SUMMARY

Different theoretical approaches deal with the prediction of long-term toxic effects of chemicals in aquatic organisms on the basis of acute test data. These methods are presented and critically discussed. For a further evaluation of these methods we investigated the relationship between acute toxicity EC/LC_{50} and measured toxic long-term no observed effect concentration (NOEC) values of substances in our database. The data were mainly from pesticide registrations and notifications according to the Chemicals Act. Since the period of exposure influences strongly the type and variety of the toxic effect, the calculated prolonged or chronic effects of chemicals in aquatic organisms may differ considerably from measured data. Despite this variability, 85% of the test data on acute and prolonged toxicity do not differ by more than a factor of 100. The only way to precisely determine prolonged or chronic toxicity is to perform tests of appropriate duration and to measure the effects. The available prolonged and chronic toxicity tests with aquatic organisms used for notifications according to the Chemicals Act and Pesticides Act are presented and discussed. Recommendations are given for choosing an appropriate test system.

18.2 INTRODUCTION

The Chemicals Act and the German Plant Protection Act prescribe eco-toxicological testing of new chemicals and pesticides prior to being placed on the market. The test results are used to estimate the hazard potential of the chemical for the environment. For risk assessment the lowest EC_{50} or LC_{50} values of the acute tests or the lowest no observed effect concentration (NOEC) of prolonged or chronic tests are divided by an assessment factor, which is between 10 and 1000, in order to calculate the predicted no effect concentration (PNEC).

If, for instance, the EC/LC_{50} values are available for the three short-term tests of the base set according to the Chemicals Act, the assessment factor of 1000 is applied to the lowest EC/LC_{50} value. An assessment factor of 10 is applied to the lowest NOEC if there are available long-term NOECs from three species across three taxonomic groups (e.g., fish, daphnia, and algae). The PNEC is compared with the calculated maximum predicted environmental concentration (PEC). If the PEC/PNEC ratio exceeds 1 a possible danger for the environment is indicated and further testing is necessary in order to confirm or refute this suspicion (Technical Guidance Document of the EU). If the concentration of a substance that is found to be toxic in a laboratory test is of the same order of magnitude as the PEC or if the PEC exceeds the effect concentration then the substance is considered to be dangerous for the environment and regulatory measures or even prohibition is proposed.

In most cases the concentrations of the substance expected in the environment are orders of magnitude lower than the experimentally determined effect concentrations. This is especially the case when considering results from short-term tests with a test duration of one to several days. Data from prolonged toxicity tests with a test duration of one or several weeks, or chronic toxicity tests which minimally include one complete life cycle, are generally much more proximate to environmental concentrations, and extrapolations to ecosystem effects can be made with greater confidence.

According to the Chemicals Act, the number and the type of ecotoxicity tests for the notification of new substances (chemicals produced or imported after January 1, 1982) depend on the production volume or the imported quantities. The majority of chemicals notified in Europe do not exceed the production or importation volume of 1 t/year and, therefore, only short-term toxicity tests had to be performed. These 'base set tests' are acute toxicity for fish, acute toxicity for daphnia, and a growth inhibition test with algae. For chemicals produced or imported before January 1, 1982 ecotoxicity data after prolonged or chronic exposure are hardly available.

Approaches have been devised to predict the prolonged or chronic toxicity for aquatic organisms from acute toxicity considering physicochemical data such as $\log p_{ow}$ (partition coefficient of a test substance between octanol and water) or water solubility, bioconcentration, and the different mechanisms of toxic action. For the Plant Protection Act active components in general have to be tested under acute *and* prolonged exposure conditions. Prolonged toxicity tests on fish and daphnia are prescribed at latest for level 1 of the Chemicals Act (production volume or imported volume > 100 t/year). Chronic tests with aquatic organisms are intended for level 2 (production volume or imported volume > 1000 t/year).

18.3 PREDICTION OF PROLONGED AND LONG-TERM EFFECTS FROM LC/EC$_{50}$ TIME CURVES

The base set tests for ecotoxicology demanded by the Chemicals Act generally have to be performed according to the test guidelines of the EU and those required by the Plant Protection Act have to be performed according to the EU or OECD test guidelines. These tests prescribe for the acute fish test a test duration of 96 h; the duration of the acute test with daphnids is normally 48 h. The used end points are 50% mortality for fish or 50% immobilization of the daphnids. Exposure of the test organisms for 48 or 96 h to the test substance makes it possible to interpret the slope of the LC_{50}/EC_{50} time curves. A flat time–effect curve indicates that prolonged toxicity—if only lethal effects are considered—is not very likely. However, other delayed effects, for instance perturbance of reproduction, cannot be excluded (Peter and Franke 1992). If the ratio LC_{50} (24 h)/LC_{50} (96 h) or the ratio EC_{50} (24 h)/EC_{50} (48 h) is > 2, prolonged exposure might give rise to higher toxicity for fish or daphnids. This mode of evaluation of the German Federal Environmental Agency is in good agreement with, e.g., the proposal of the Danish National Agency of Environmental Protection (Kristensen and Tyle 1991).

18.4 PREDICTABILITY OF PROLONGED
AND LONG-TERM NO EFFECT CONCENTRATIONS
FROM ACUTE TOXICITY DATA

Some approaches have been developed to predict prolonged or long-term (chronic) effects from acute toxicity data using special factors for LE/EC_{50} values of acute tests in order to calculate NOEC for prolonged or chronic tests. The discussion about application factors was initiated by Mount and Stephan already in 1967.

Kenaga (1982) evaluated the relationship of the acute LC_{50} of chemicals to their chronic toxicity for aquatic animals expressed as the acute chronic ratio (ACR). Calculation of ACRs was performed by many scientists using data from many different sources, species, test methods, and chemicals. Therefore, the validity of the tests and the comparability of data could not be checked. Furthermore, the test methods used today generally relate the toxicity data to the analytically determined concentrations of the test substance, whereas in the past analytical determination of real concentrations was often not performed. Because of this, the LC_{50} values evaluated in Kenaga's paper, especially for unstable and volatile compound or substances which adsorb to the glass vessel, tend to be larger and therefore produce larger ACR values.

The data include 135 different ACRs, 84 different chemicals, nine species of fish, and two species of aquatic invertebrates. The chemicals were divided into different categories of pesticides and into inorganic and industrial chemicals categories. Kenaga found that the lowest ACR was 1 and the highest, 18 000. Among the industrial organic chemicals the average ACR was 12. Of these ACR values 93% were 25 or below. Pesticides and heavy metals had a higher percentage of ACR values above 25. The application factors did not differ greatly between fish and daphnids. The ACR values did not correlate very well with the bioconcentration factors or the octanol–water partition coefficients (K_{OW}), although it was mentioned that such correlations may exist between closely related chemicals, such as homologous series. It was concluded that chronic toxicity of industrial organic compounds can be reliably estimated in about 93% of cases from acute toxicities in the respective species of aquatic organisms using an ACR of 25 or less.

A different application factor was proposed in the ECETOC (European Chemical Industry, Ecology and Toxicology Centre) technical report No. 51 (ECETOC 1993a). In this report the ratio of acute to prolonged NOEC applicable to 90% of substances was claimed to be 40. Recently, the scientific basis for the derivation of an application factor in risk assessment was newly discussed by ECETOC and a value of 28 was calculated (ECETOC 1993b). We would like to anticipate that these results are in contrast to our application factors calculated from data which had been presented to the Federal Environmental Agency especially for the purpose of the pesticide registration and the notification of new chemicals.

Table 18.1 Comparison of acute toxicity with prolonged toxicity (American flagfish)

Compound	Flow-through LC_{50} (µg/L)	LOEC µg/L	Toxic end point	ACR
Pentachlorophenol	218	75	larval survival	2.91
2,3,5,6-Tetrachlorophenol	1160	245	fry survival	4.73
2,4,6-Trichlorophenol	2207	750	fry growth	2.94
1,2,4,5-Tetrachlorobenzene	2150	85	fry growth	25.29
1,2,4-Trichlorobenzene	1217	1130	larval survival	1.08
1,4-Dichlorobenzene	2053	263	larval survival	7.81
Tetrachloroethyle	8430	3100	larval survival	2.72
Trichloroethylenene	28280	1100	larval survival	2.57
1,1,2,2-Tetrachloroethane	18480	7230	larval survival	2.56
1,1,2-Trichloroethane	45117	31200	larval survival	1.45

LOEC, lowest effect concentration; ACR, acute chronic ratio.

Smith et al. (1991) compared acute toxicity with chronic toxicity with emphasis on matched data sets, i.e., using the same chemicals and the same test facilities. They determined acute toxicity expressed as 96 h LC_{50} and chronic toxicity for 10 compounds in the American flagfish and compared these toxicity estimates with each other and with values reported in the literature for other freshwater fish species. For chronic toxicity not only lethality to fish but also egg hatchability, larval survival, and growth were considered, and the data were related to the most sensitive end point. All data were based on measured water concentrations. Table 18.1 summarizes the main results including the ACR values which we calculated from the results of Smith et al. These results show that for the chlorinated aliphates and aromates investigated, the ACR was rather low and varied from 1 to 25. In most cases larval survival was the most sensitive toxic end point.

Recently, Mayer et al. (1994) developed a method for predicting chronic lethality of chemicals for fish from acute toxicity test data. This method is based on the reasoning of Green (1965) that as the time of exposure becomes sufficiently long, the LC_{50} approaches an asymptotic value and the course of the LC_{50} curve as a function of time follows a hyperbola. Therefore, the reciprocal plot of that function is linear

$$LC_{50} = a + b(1/t), \tag{18.1}$$

and the intercept with the ordinate represents the LC_{50} over an indefinite (chronic) time of exposure ($t \Rightarrow \infty$, $1/t \Rightarrow 0$).

Instead of LC_{50} values Mayer et al. calculated the LC_0 values at the observation times 24, 48, 72, and 96 h. The data were taken from acute flow-through tests as probit percentage mortality. Extrapolating for $t \Rightarrow \infty$ from the plot

$$LC_0 = a + b(1/t) \tag{18.2}$$

they could determine the predicted no effect concentration (PNOEC) with the end point 'lethality' for chronic exposure as intercepts with the ordinate:

$$(LC_0)_{t \Rightarrow \infty} = a = PNOEC \tag{18.3}$$

The technique was applied to a data base of 18 chemicals and seven fish species and was highly accurate among various single chemicals and mixtures. This accuracy seemed to be independent of lipophilic properties of the chemicals, expressed by their $\log K_{OW}$ value. The predicted NOEC was accurate for industrial chemicals such as carbon tetrachloride or pentachlorophenol and pesticides such as endosulfan or chlordane.

18.5 EVALUATION OF DATA FROM PESTICIDE REGISTRATIONS AND NOTIFICATIONS ACCORDING TO THE CHEMICALS ACT

The differences in the proposed application factors obviously cause some uncertainty and lead to the question of whether it is scientifically justifiable to determine application factors on the basis of relatively limited data sets. Because of the relatively great number of registrations especially according to the Pesticides Act and of notifications according to the Chemicals Act we were able to collect numerous data on acute and prolonged toxicity of substances for aquatic organisms. Data on ecotoxicological properties of existing chemicals were obtained from research projects sponsored by the Federal Environmental Agency. We evaluated this data also in view of the question of whether the potential of a substance to bioaccumulate enhances its toxicity after long-term exposure (Heger et al. 1995).

As described in the introduction we defined an exposure of one week to several weeks not as a chronic, but as a prolonged exposure. Therefore, we used instead of the ACR the acute-prolonged toxicity ratio APR. Acute toxicity was expressed as LC_{50}, or, in the case of daphnids, as EC_{50}. The NOECs of the prolonged studies were related to the most sensitive end points, which could be mortality, immobilization, adverse effects on reproduction, reduced growth, reduced body length, or changes in behavior. The APR values were determined for fish and daphnids with existing chemicals, new chemicals, and active components of pesticides and are summarized in Tables 18.2–18.5. In the case of pesticides the data are subdivided into fungicides, herbicides, and insecticides. The last columns in Tables 18.4 and 18.5 also contain data sets of other pesticides such as molluscicides, acaricides, and rodenticides; therefore these columns include more data than only the sum of fungicides, herbicides, and insecticides.

As can be seen from Tables 18.2–18.5, mean and median values of APR for fish and for daphnids differ to a large extent, and the standard deviation

Table 18.2 Acute LC/EC$_{50}$ to prolonged NOEC ratio (APR) for new chemicals in fish and daphnids

	Daphnids	Fish
Data sets (n)	39	25
Mean	2793	2089
Median	33	4
STD	16003	7268
SEM	2563	1454
Minimum	0.02	1
Maximum	100000	31250
Q1[a]	4	1
Q3[a]	100	55

[a] Q1 and Q3 are the lower and upper quartiles.

STD, standard deviation; SEM, standard error of mean.

Table 18.3 Acute EC$_{50}$ to prolonged NOEC ratio (APR) for existing chemicals in daphnids

Data sets (n)	94
Mean	113
Median	25
STD	360
SEM	37
Minimum	2
Maximum	3000
Q1	8
Q3	65

Table 18.4 Acute LC$_{50}$ to prolonged NOEC ratio (APR) for pesticides in fish

	Fungicides	Herbicides	Insecticides	All pesticides
Data sets (n)	36	63	35	146
Mean	22	48	470	144
Median	8	7	6	6
STD	42	148	2602	1278
SEM	7	19	440	106
Minimum	0.01	0.1	0	0.009
Maximum	246	1120	15415	15415
Q1	3	2	3	2
Q3	29	32	25	30

considerably exceeds the mean value. This high variability was found especially with new chemicals and to a lesser extent with existing chemicals; the variability was relatively low with pesticides. Obviously the toxicity of substances with a high acute toxicity, such as pesticides and especially insecticides, is not considerably enhanced with prolonged exposure times (Tables 18.4, 18.5).

In order to facilitate the discussion of reasonable application factors for calculating prolonged toxicity from acute toxicity we divided the APR values for new and existing chemicals and for pesticides into 4 classes (Tables 18.6–18.9): > 1000, > 100–1000, > 10–100, ≤ 10.

For numerous chemicals the acute to prolonged toxicity ratio is ≤ 10 or > 10–100 according to the results of Kenaga (1982), Smith et al. (1991), and ECETOC (1993a,b). However, for some chemicals the APR exceeds 100 or even 1000. The mean values of APR vary from 0.001 to 142276. As can be seen from Table 18.10, a factor of 40 applied to a measured EC/LC_{50} value would lead to a prolonged NOEC which would be valid for about 72% of the tests. Therefore, an application factor of 100—which would lead to a protection of 85% of the test animals exposed for a prolonged period of time—is much more justified (Table 18.10).

Table 18.5 Acute EC_{50} to prolonged NOEC ratio (APR) for pesticides in crustacea[a]

	Fungicides	Herbicides	Insecticides	All pesticides
Data sets (n)	*32*	*59*	*35*	*141*
Mean	61	2676	144	1175
Median	26	9	3	9
STD	93	18 509	605	11 985
SEM	16	2410	102	1009
Minimum	0.7	0.005	0.001	0.001
Maximum	388	142 276	3540	142 276
Q1	5	3	0	2
Q3	75	75	22	54

[a] Most frequently daphnids.

Table 18.6 Classification of acute to prolonged toxicity ratios (APR) of new chemicals[a]

APR	Fish	Daphnids
> 1000	2 (8%)	2 (5%)
> 100–1000	3 (12%)	6 (15%)
> 10–100	5 (20%)	14 (36%)
≤ 10	15 (60%)	17 (44%)

[a] Absolute values; percent in parentheses.

Table 18.7 Classification of acute to prolonged toxicity ratios (APR) of existing chemicals[a]

APR	Daphnids
> 1000	3 (3%)
> 100–1000	14 (15%)
> 10–100	50 (53%)
≤ 10	27 (29%)

[a] Absolute values; percent in parentheses.

Table 18.8 Classification of acute to prolonged toxicity ratios (APR) of pesticides[a]

APR	Fungicides	Herbicides	Insecticides	Pesticides
> 1000	0	1 (2%)	1 (3%)	2 (1%)
> 100–1000	1 (3%)	7 (11%)	2 (6%)	11 (8%)
> 10–100	14 (39%)	20 (32%)	8 (23%)	43 (29%)
≤ 10	21 (58%)	35 (55%)	24 (68%)	90 (62%)

[a] Data were taken from results of fish tests; absolute values; percent in parentheses.

Table 18.9 Classification of acute to prolonged toxicity ratios (APR) of pesticides[a]

APR	Fungicides	Herbicides	Insecticides	Pesticides
> 1000	0	6 (10%)	1 (3%)	7 (1%)
> 100–1000	5 (16%)	6 (10%)	3 (8%)	15 (8%)
> 10–100	17 (53%)	15 (25%)	8 (22%)	46 (29%)
≤ 10	10 (31%)	32 (55%)	24 (67%)	73 (62%)

[a] Data were taken from results of crustacea tests; absolute values; percent in parentheses.

Table 18.10 Effectiveness of application factors in predicting prolonged effects from acute test data

Application factor	Fish	Crustacea	Fish and crustacea
40	80%	67%	72%
100	90%	82%	85%
ca. 310	95%		
ca. 970		95%	
ca. 610			95%

We would like to emphasize that the prolonged NOEC values from the data sets which we evaluated not only considered lethal effects but also others which had consequences mainly for reproduction. Therefore, discrepancies between our data and those of other authors who found a great effectiveness of lower APR (ACR) values but considered mainly or exclusively mortalities (Mayer et al. 1994) are explainable. The low ACR values from the results of Smith et al. (1991) were estimated with rather toxic substances which act acutely and are therefore not in contrast to ours which were estimated using highly acute toxic chemicals such as pesticides.

18.6 BIOACCUMULATION AND PROLONGED TOXICITY

We also investigated whether an enhanced toxicity during a prolonged exposure is related to the potential of a substance to bioaccumulate. It is generally accepted that the octanol–water partition coefficient ($\log p_{ow}$) yields at least some information which allows the prediction of bioaccumulation of water-borne chemicals (Veith et al. 1979). Therefore, the $\log p_{ow}$ value may be of interest especially for lipophilic, persistent, and poorly metabolizable chemicals for assessment of their toxicity (Koenemann 1981; Veith et al. 1983; Veith et al. 1985). Indeed, the majority of computer programs for QSAR (quantitative structure activity relationships) are based on the lipophilic properties of substances expressed by their $\log p_{ow}$ value and the equations derived by McCarty (1987):

$$\log \text{ acute toxicity} = -\log p_{ow} - 1.38.$$

$$\log \text{ chronic toxicity} = -\log p_{ow} - 2.33.$$

However, McKim and Schmieder (1991) have questioned the relationship between toxicity and bioaccumulation. They concluded in their review "Bioaccumulation: Does it Reflect Toxicity?" that whether or not a chemical is highly bioaccumulative says nothing about its toxic potency. Bioaccumulation merely informs of the activity of the chemical and how it will act kinetically. Our examinations support this conclusion concerning the relationship between prolonged toxicity and bioaccumulation. Representative for all chemicals evaluated we show in Figures 18.1 and 18.2 comparisons of APRs derived from $\log p_{ow}$ values of pesticides in fish and crustacea (Heger et al. 1995). It is clear that neither fish nor crustacea toxicity depends on the duration of exposure; hence, there is obviously no correlation between $\log p_{ow}$ and prolonged toxicity. Therefore, in our opinion, the use of QSAR for predicting prolonged toxicity based on lipophilic properties of the chemical cannot lead to reasonable results. Evidently high concentrations of lipophilic chemicals which are acumulated in lipid-rich pools of organisms are not necessarily toxic, because they do not reach a receptor or target site that elicits a toxic response. Therefore bioaccumulation or bioconcentration is not hazardous in itself (Weber et al. 1993) and should be considered as a fate and exposure-related process (Kristensen and Tyle 1991). However, the question of what happens with bioaccumulated substances which are degraded and metabolized during phases of stress or hunger remains to be answered.

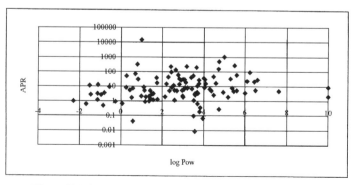

Figure 18.1 Comparison of APRs with $\log p_{ow}$ of pesticides (fish).

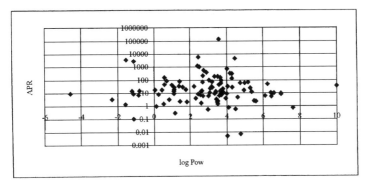

Figure 18.2 Comparison of APRs with $\log p_{ow}$ of pesticides (crustacea).

The only conclusion which can be drawn from the results referred to so far is that chemicals with a high $\log p_{ow}$ value may or may not be more toxic for aquatic organisms in comparison with their acute toxicity when exposed for a prolonged period of time. It is highly probable that acute and prolonged toxicities do not differ by more than a factor of 100. However, the only possibility to determine prolonged or chronic effects caused by chemicals is to perform tests with a prolonged or chronic duration.

18.7 PROLONGED AND CHRONIC TESTS FOR AQUATIC ORGANISMS

Long-term tests for aquatic organisms which are used for notifications according to the Chemicals Act and the Pesticides Act are presented and discussed below.

18.7.1 NOEC and EC_x

The results of long-term toxicity tests are normally expressed as NOECs, and the effects are related to the most sensitive end point. One disadvantage of this procedure is that the NOEC is dependent on the differences between the concentration steps selected, and, therefore, cannot be very exact. Furthermore, the first effective concentration above the NOEC must be by definition significantly different from the control, which often leads because of the variation between measuring points to the unreasonable result that NOEC lies in the region of the LC/EC_{50} value. Much more exact is the determination of an EC_x value by regression analysis, where x may be 10 or 15, or even 20, depending on the accuracy of the test system (number of animals). This procedure is independent of the coefficient of variation and the number of replicates and yields a maximum of information with high statistical power.

18.7.2 Algal Inhibition Test (92/69/EWG, Part C.3)

The test duration of 72 h includes several generations of algal cells, so that the test has to be considered, according to the definition given in the introduction, as a chronic test. However, for risk assessment the algal test is generally regarded as an acute test, similar to the acute tests with fish and daphnids; therefore, it should not be discussed here in relationship to long-term tests. In contrast to the level 1 tests presented in Sections 18.7.3–18.7.7, the algal test belongs to the base level tests according to the Chemicals Act.

18.7.3 Daphnia, Acute Immobilization Test, and Reproduction Test
Part II—Reproduction Test Duration of at Least 14 Days, OECD 202, Part II

The aim of the acute immobilization and reproduction test is to determine the effects of chemicals mainly on the immobilization and the reproduction capacity of daphnids. The daphnids are exposed to solutions containing the test substance in various concentrations for a period of not less than two weeks and long enough for the development of at least three broods. The mortality, the time of the first production of young, the number of young born, and the signs of intoxication observed are compared with the corresponding parameters in controls.

The main difficulty in interpreting the results in the past was the high variability of the NOEC values, so that results from different laboratories could scarcely be compared. For that reason the OECD is preparing a revised test guideline for determining long-term toxicity with daphnids, which will be more precise: the test duration will be an obligatory 21 days; the test medium must be completely defined; the test species is prescribed as *Daphnia magna*, preferentially from clone A (Baird et al. 1991). Organic solvents as solubilizer should be avoided and the test should only be performed up to concentrations representing the maximum achievable water solubility (Heger et al. 1994).

18.7.4 Fish, Prolonged Toxicity Test: 14-Day Study, OECD 204

The prolonged toxicity test is frequently performed with *Oncorhynchus mykiss* or *Brachydanio rerio*, but also other fish species can be chosen such as those recommended in the OECD Test Guideline 203 (acute toxicity for fish). The test period of at least 14 days should be extended by 1 or 2 weeks if necessary. The main aim of the test is to measure lethal effects in fish exposed to the test substance. Also other effects such as behavioral changes or reduction in body weight or body length should be considered when determining the NOEC. However, it turned out to be very difficult to determine these effects and to identify behavioral effects as sublethal effects. Therefore, in many cases mortalities had been considered exclusively, so that the prolonged toxicity test with fish was reduced to a "prolonged acute test". Recently, Lorenz et al. (1995) introduced a test system which uses the integrative whole-organism response by automatized recording and quantifying the spontaneous behavior of fish and other test organisms. This 'BehavioQuant®' system is claimed, according to the descriptions of the authors, to be able to determine NOEC or LOEC (lowest observed effect concentration) values related to the following behavioral parameters: average swimming velocity (motility), average swimming depth below water surface, inconstancy of motility, frequency of turns, standard frequency distributions of vertical and horizontal positions, and distance behavior. However, before this test system can be recommended by the competent authorities it has to be evaluated in an interlaboratory ring test.

18.7.5 Fish, Juvenile Growth Test—28 Days, OECD, Draft

Juvenile fish in exponential growth phase are exposed for 28 days to different concentrations of the test substance dissolved in water, preferably under flow-through conditions. The aim of the test is to determine the NOEC and LOEC with respect to inhibitory effects on the specific growth rate. A ring test was performed with the rainbow trout (*O. mykiss*), but experience gained from growth tests with zebra fish (*B. rerio*) has also been taken into consideration. As described by Nagel (1988), rainbow trout and zebra fish exhibit a comparable susceptibility with respect to the test parameter growth. This test parameter can easily and clearly be quantified and it seems reasonable to combine the OECD test 204 with the juvenile growth test, which would lead to a test system with high sensitivity and high ecotoxicological relevance.

18.7.6 Fish, Early-Life Stage Toxicity Test, OECD 210

The early-life stages of fish are exposed to a range of concentrations of the test substance dissolved in water, preferably under flow-through conditions. The test species recommended are zebra fish or rainbow trout. The test duration is dependent on the species used: for the zebra fish it should be 6 weeks (Nagel et al. 1991). Toxicological end points of the fish early-life stage test (FELS test) are lethality, growth, impact on embryonic development, hatching, and larval development. Thus, a multitude of diverse differentiation and development processes as well as the weaning period with the shift from utilization of endogenous reserves within the yolk sac to external feeding are included (Nagel 1993). Therefore, we recommend the FELS test as a level 1 test according to the Chemicals Act. Growth and survival rates were found to be the most sensitive end points of all parameters in the FELS test (Ensenbach and Nagel 1995).

18.7.7 Fish, Toxicity Test on Egg and Sac Fry Stages, OECD, Draft

The egg and sac fry test is a shorter version of the early-life stage test with fish (OECD 210). According to this test the life stages from the newly fertilized egg to the end of the sac fry stage are exposed to the test substance. This guideline is intended to define lethal, and to some extent, sublethal effects of chemicals on the specific stages and species tested. The test duration is dependent on the fish species used and is limited to the time when the yolk sac of all larvae in all of the test chambers has been absorbed completely. For the zebra fish *B. rerio* the test duration is 10 days. This test guideline is expected to be, in general, less sensitive than the early-life stage toxicity test. However, for many chemicals, especially those with a narcotic mode of action, the differences in sensitivity between the two tests are rather small (Kristensen 1990).

18.7.8 Chronic Toxicity Test for Fish

For level 2, according to the Chemicals Act, determination of long-term effects on aquatic organisms including effects on reproduction may be necessary for the risk assessment. Although at present no OECD guideline about chronic toxicity for fish is available, the necessity to develop such a guideline is under discussion. The early-life stage test can be considered as a part of a chronic or long-term test. However, as concluded by Suter et al. (1987) and Chorus (1987) on the basis of the data available, it is not scientifically justified to apply an early-life stage test instead of a full life-cycle study.

This conclusion was confirmed by Bresch et al. (1990), who described a long-term test over three generations with zebra fish (*B. rerio*) and with 4-chloroaniline as the test substance. The most sensitive parameter in this test was reduction in egg release by fish, with this end point being tenfold more sensitive than growth of the fish. Bresch et al. came to the conclusion that a long-term test is the most appropriate method to assess the chronic toxicity of a substance in fish, and they recommended to perform a chronic toxicity test which comprises two generations, with the zebra fish as test species. This test comprises effects of permanent exposure to the test chemical on embryos, larvae, juveniles, and adults in the F_1 generation and during embryo-larval development in the F_2 generation. Therefore, numerous processes such as development of gametes, fertilization, courtship, and mating can be studied (Nagel 1993).

18.8 CONCLUSIONS

- For risk assessment acccording to the Chemicals Act and the Pesticides Act prolonged or chronic effects of chemicals on aquatic organisms cannot be calculated on the basis of acute toxicity data, although in most cases prolonged NOEC differs from acute EC/LC_{50} by a factor of ≤ 100.
- Bioaccumulative properties of chemicals including pesticides are not necessarily related to an enhanced toxicity under conditions of prolonged exposure.
- In many cases mortality is not the most sensitive endpoint in prolonged or chronic tests with aquatic organisms, but rather growth rate or reproduction.
- From all existing prolonged toxicity tests with fish the early-life stage test (FELS test) seems to be the most sensitive and ecotoxicologically the most relevant test.

• Even more sensitive than the FELS test is a chronic fish test over, e.g., two generations, so that the FELS test cannot substitute for a chronic test.

18.9 REFERENCES

Baird, D. J., I. Barber, M. Bradley, A. M. V. M. Soares, and P. Calow, 1991, "A Comparative Study of Genotype Sensitivity to Acute Toxic Stress Using Clones of Daphnia Magna Straus," *Ecotoxicol. Environ. Safety, 21*, 257–265

Bresch, H., H. Beck, D. Ehlermann, H. Schlaszus, and M. Urbanek, 1990, "A Long-Term Toxicity Test Comprising Reproduction and Growth of Zebra Fish with 4-chloraniline," *Arch. Environ. Contam. Toxicol., 19*, 419–427

Chorus, I., 1987, "Literaturrecherche und Auswertung zur Notwendigkeit chronischer Tests – insbesondere des Reproduktionstests – am Fisch für die Stufe II nach dem Chemikaliengesetz," *UBA F&E-Vorhaben Gesch–Z.* I 4. 1-97 316/7

ECETOC, 1993a, "Environmental Hazard Assessment of Substances," *Technical Report*, No. 51

ECETOC, 1993b, "Aquatic Toxicity Data Evaluation," *Technical Report*, No. 56

Ensenbach, U., and R. Nagel, 1995, "Toxicity of Complex Chemical Mixtures: Acute and Long-Term Effects on Different Life Stages of Zebra Fish (*Brachydanio rerio*)," *Ecotoxicol. Environ. Safety, 30*, 151–157

Green, R. H., 1965, "Estimation of Tolerance Over an Indefinite Time Period," *Ecology, 46*, 887.

Heger, W., S-J. Jung, S. Martin, and H. Peter, 1995, "Acute and Prolonged Toxicity to Aquatic Organisms of New and Existing Chemicals and Pesticides. 1. Variability of the Acute to Prolonged Ratio. 2. Relation to log p_{ow} and Water Solubility," *Chemosphere, 31*, 2707–2726

Heger, W., S-J. Jung, S. Martin, U. Schiecke, H. Teichmann, and H. Peter, 1994, "Chemikaliengesetz Heft 11; Ökotoxikologische Testverfahren mit aquatischen Organismen. Bewertung von Prüfberichten durch das Umweltbundesamt im Rahmen des Vollzuges Chem G und PflSchG und Hinweise für die Durchführung der Tests," *UBA-Texte 14/94*, 2, Aufl., 1–150, Berlin, Germany

Kenaga, E. E., 1982, "Predictability of Chronic Toxicity from Acute Toxicity of Chemicals in Fish and Aquatic Invertebrates," *Environ. Toxicol. Chem., 1*, 347–358

Könemann, H., 1981, "Quantitative Structure–Activity Relationships in Fish Toxicity Studies. Part 1: Relationship for 50 Industrial Pollutants," *Toxicology, 19*, 209–221.

Kristensen, P., 1990, "Evaluation of the Sensitivity of Short Term Fish Early Life Stage Test in Relation to other FELS Test Methods," *Final Report to the Commission of the European Communities*, 60 pp, Hørsholm, Denmark

Kristensen, P., and H. Tyle, 1991, "The Assessment of Bioaccumulation," in *Bioaccumulation in Aquatic Systems* (R. Nagel, and R. Loskill, Eds.), VCH Publishers, New York, USA, pp. 189–227

Lorenz, R., O. H. Spieser, and C. Steinberg, 1995, "New Ways to Ecotoxicology: Quantitative Recording of Behaviour of Fish as Toxicity Endpoint," *Acta Hydrochim. Hydrobiol., 23*(5), 197–201

Mayer, F. L., G. F. Krause, D. R. Buckler, M. R. Ellersieck, and G. Lee, 1994, "Predicting Chronic Lethality of Chemicals in Fishes From Acute Toxicity Test Data: Concepts and Linear Regression Analysis," *Environ. Toxicol. Chem.*, *13*, 671–678

McCarty, L. S., 1987, "Relationship Between Toxicity and Bioconcentration for Some Organic Chemicals. I. Examination of the Relationship. II. Application of the Relationship," in *QSAR in Environmental Toxicity – II* (K. L. E. Kaiser, Ed.), D. Reidel Publishing, Dordrecht, The Netherlands, pp. 207–230

McKim, J. M., and P. K. Schmieder, 1991, "Bioaccumulation—Does it Reflect Toxicity," in *Bioaccumulation in Aquatic Systems* (R. Nagel, and R. Loskill, Eds.), VCH Publishers, New York, USA, pp. 161–188

Mount, D. I., and C. E. Stephan, 1967, "A Method for Establishing Acceptable Limits for Fish—Malathion and the Butoxyathanol Ester of 2,4-D," *Trans. Am. Fish Soc.*, *96*, 185–193

Nagel, R., 1988, "Fische und Umweltchemikalien – Beiträge zu einer Bewertung," Habilitationsschrift, Universität Mainz, Mainz, Germany

Nagel, R., 1993, "Fish and Environmental Chemicals – a Critical Evaluation of Tests," in *Fish–Ecotoxicology and Ecophysiology* (T. Braunbeck, W. Hanke, and H. Segner, Eds.), VCH Verlagsgesellschaft, Weinheim, Germany, 147–156

Nagel, R., H. Bresch, N. Caspers, P. D. Hansen, M. Markert, R. Munk, N. Scholz, and B. B. Ter Höfte, 1991, "Effect of 3,4-dichloroaniline on the Early Life Stages of the Zebra Fish (*Brachydanio rerio*): Results of a Comparative Laboratory Study," *Ecotoxicol. Environ. Safety*, *21*, 157–164

Peter, H., and C. Franke, 1992, "Ökotoxikologische Prüfungen nach dem Chemikaliengesetz (ChemG)–Durchführung und Bewertung", *UWSF-Z. Umweltchem. Ökotox.*, *4*, 333–338

Smith, A. D., A. Bharath, C. Mallard, D. Orr, K. Smith, J. A. Sutton, J. Vukmanich, L. S. Mccarty, and G. W. Ozburn, 1991, "The Acute and Chronic Toxicity of 10 Chlorinated Organic Compounds to the American Flagfish (*Jordanella floridae*)," *Arch. Environ. Contam. Toxicol.*, *20*, 94–102

Steinberg, C. E. W., R. Lorenz, and O. H. Spieser, 1995, "Effects of Atrazine on Swimming Behavior of Zebra Fish, *Brachydanio rerio*," *Water Res.*, *29*, 981–985

Suter, G. W., A. E. Rosen, E. Linder, and D. F. Parkhurst, 1987, "Endpoints of Responses of Fish to Chronic Toxic Exposures," *Environ. Toxicol. Chem.*, *6*, 793–809

Veith, G. D., and D. L. Defoe, 1979, "Measuring and Estimating the Bioconcentration Factor of Chemicals in Fish," *J. Fish. Res. Board Can.*, *36*, 1040–1048

Veith, G. D., D. L. Defoe, and M. Knuth, 1985, "Structure–Activity Relationships for Screening Organic Chemicals for Potential Ecotoxicity Effects," *Drug Metabolism Reviews*, *15*, 1295–1303

Veith, G. D., D. J. Call, and L. T. Brooke, 1983, "Structure–Toxicity Relationships for the Fathead Minnow, *Pimephales promelas*: Narcotic Industrial Chemicals," *Can. J. Fish Aquat. Sci.*, *40*, 743–748

Weber, H., W. Pflüger, and R. Grau, 1993, "The Use of Test Data from Registration of Agrochemicals to Determine the Toxicological and Ecological Relevance of Bioaccumulation," in *Mitteilungen aus der Biologischen Bundesanstalt für Land- und Forstwirtschaft No. 290: Evaluation and Assessment of Bioaccumulation of Active Ingredients of Plant Protection Products* (A. Wilkening, and H. Köpp, Eds.), Paul Parey, aaBerlin, Germany, pp. 142–148

19

Effects of Heavy Metals in Plants at the Cellular and Organismic Level

Wilfried H. O. Ernst (Amsterdam, The Netherlands)

19.1 SUMMARY

Ecotoxicity of heavy metals cannot be determined by the total amount present in the soil but rather only by their bioavailable fraction which is affected by environmental conditions such as pH and redox potential and by biological activities of roots and rhizosphere organisms. A surplus of heavy metals has an impact on plants at all biological organization levels.

Ecotoxicology, Edited by Gerrit Schüürmann and Bernd Markert.
ISBN 0-471-17644-3   © 1998 John Wiley & Sons, Inc. and Spektrum Akademischer Verlag.

At the cellular level the cell wall is the first contact zone with external metals. The metal-binding capacity of cell walls can modify metal transport (velocities) across the neighboring biomembranes and allow the precipitation of metals. Under natural conditions an equilibrium between cell wall adsorption and metals in the soil solution is soon reached.

The plasma membrane is the cell compartment which regulates metal entry into the cell interior; in addition its proteins, especially the SH groups may be hampered or blocked in their activity resulting in damage to membrane stability. Once metals have passed the plasma membrane, enzymes in the cytosol are the next target if the metals are not properly detoxified. Some biomarkers, i.e., phytochelatins and stress proteins may be indicators of the free metal ion concentration in the cytosol.

Cell organelles are protected from direct metal exposure by a biomembrane. Chloroplasts are the organelles most sensitive to heavy metals, at least in vitro, with photosystem I being affected. In contrast, mitochondria and nuclei seem to be the most metal resistant. Vacuoles are metal storage compartments of the cell. Judged by the concentration, Co, Ni, and Zn can be hyperaccumulated in the vacuole whereas the concentration of Cd and Cu remain low.

Survival of a metal stress at the cellular level in multicellular organisms depends on cell interactions at the tissue and whole-plant level and may change during life history. Except in the case of aerial metal fallout, roots of terrestrial plants are the first tissues which are exposed to metals. Root exudation and interaction with rhizosphere organisms become important in changing metal speciation and metal availability and in modifying metal translocation from root to shoot. Accumulation of metals in shoots helps to ensure longevity of the roots and detoxification by storage in senescent tissue. Heavy metal concentration in leaves may help to protect them from pathogenic fungi and bacteria and from herbivores.

At the reproductive stage the seed is the most protected from metal influx. At the same time a metal-poor environment creates an opportunity for metal-sensitive seed predators. After seed shedding seed contact with the soil solution does not affect seed survival; but the seedling is immediately affected.

Understanding metal toxicity and metal resistance demands a knowledge of all biological levels.

19.2 INTRODUCTION

Heavy metals occur in all ecosystems of the world. The total concentration in soils and waters, however, varies at a local, regional, and continental scale (Ernst 1974; Allan 1995). Plant species have developed often very specific demands for chemical elements including heavy metals during their evolution. Central metabolic roles are played by iron, manganese, copper, zinc, and molybdenum in all higher plants, by nickel alone in legumes, and by cobalt in the legume–rhizobium relationship. Iron has a main function in all heme groups; manganese is involved in the water-splitting systems of photosynthesis; iron and manganese act together in Fe–Mn–superoxide-dismutase; copper is necessary for cytochrome a activity; zinc is necessary for the formation of zinc-finger proteins and carboanhydrase; copper and zinc act together in Zn–Cu–superoxide-dismutase and in a lot of other systems (Marschner 1995).

The occurrence of certain species may be related to these specific demands, e.g., high sulfur content in many cruciferous plants (Ernst 1993a). In addition many plant species and genotypes of species have adapted to exceptional element concentrations in their environment: halophytes have adapted to saline soils and waters, metallophytes have adapted to a surplus of one or more heavy metals, and selenophytes have adapted to a surplus of selenium.

Anthropogenic activities have modified or often completely changed the concentration of chemical elements in the various environmental compartments. These changes can have strong impacts on the physiology and ecology of organisms adapted originally only to a specific environment. These organisms are now forced to adapt once more and often more rapidly than they adapted to changes at a geological time scale (Bradshaw 1976; Ernst 1976). In elaborating on the ecotoxity of heavy metals in plants we have to keep in mind the high geochemical variability, the high biodiversity of organisms, and the time scale, which alone and in combination are effective, and the synchronous demands for major and minor nutrients. Often a distinction is made between chronic, i.e., low level exposure over a long period of time, and acute ecotoxic effects, i.e., high level exposure for a short duration. Due to the retention of heavy metals in the soil, each acute event will persist as a chronic process, having a strong impact on the selection of species and genotypes. Therefore, this distinction between acute and chronic exposure demands a very good knowledge of genotypic metal sensitivities and is not considered in this contribution.

The best examples of this long-term adaptation and selection processes are the heavy metal vegetation on ore outcrops around the world (Ernst 1974; Brooks 1993) and the very recent and thus rapid selection processes which have occurred in plant populations and vegetation near metal-emitting point sources, constructed in formerly metal-poor environments (Ernst 1993b). As soon as the heavy metal concentration exceeds a certain level, it may have adverse effects on plant metabolism and performance. The degree of injury established by effect concentration (EC) levels will depend on metal availability and the interaction of metals with other soil factors.

19.3 METAL AVAILABILITY

The total metal concentration in an environmental compartment is biologically not interesting, although laws and guidelines for environmental protection have often taken this very simplistic approach (US EPA 1986). Considering total metal concentration alone, storage of silver, copper, and gold coins and ingots by the various National Banks renders these sites the most toxic environments on earth, even surmounting ore bodies. In judging ecotoxity only the small proportion of the total concentration of a heavy metal in the environment which is available to plants is relevant. This bioavailability is determined by abiotic and biotic factors. The pH and redox potential (Eh) of the abiotic environment influence availability in a metal-specific manner (Brookins 1988). Most heavy metals are more available at high acidity and moderate oxidation state of the environmental compartment. Even within the available fraction the proportion which is accessible to plants depends on the chemical speciation of the metal (Bourg 1988), which is governed by abiotic and biotic processes. Speciation of iron and manganese in the divalent state is preferred in an anoxic environment; reduced iron and manganese can be taken up by plants in essentially higher amounts than the same metals in a highly oxygenated environment (Ernst 1988). Plants themselves can modify the speciation. Microbial oxidation of metals, especially by sulfur- and iron-oxidating bacteria, can strongly modify metal availability (Kelley and Tuovinen 1988). Other microorganisms are involved in the methylation of heavy metals such as lead (Walton et al. 1988), mercury, and arsenic (Kersten 1988).

Microorganisms and higher plants have further possibilities to modify abiotic metal speciation by the exudation of complexing agents such as organic acids, siderophores, protons, and oxygen (Mench et al. 1988; Ernst 1996). The amount of exudates are metabolically controlled by the demand of the organism and therefore vary in a rhythm of hours and days (Marschner 1995). Seasonal processes such as litter fall supply the abiotic environment with a huge amount of complexing agents, which affect the bioavailability and thus the ecotoxicity of metals (Kuiters 1993; Ernst 1996). These high dynamics of metal speciation in nearly all environments makes it very difficult to judge the effective bioavailable metal concentration. All chemical extraction techniques for simulating metal availability to plants can only be a rough approach (Scheffer and Schach-tschabel 1992). Therefore, all calculations and models may only be specific for one plant species or only one genotype (Wu and Aasen 1994). This species-specific metal sensitivity makes it difficult to carry out an appropriate risk assessment because there is no absolute level of ecotoxicity in the long-term.

Interaction with other environmental factors can decrease or increase metal availability. Application of deicing salts in winter as calcium-magnesium acetate mobilizes lead and zinc (Amrhein et al. 1994); acetate is a well-known complex-ing agent for heavy metals and is used as an extractant for the determination of adsorbed heavy metals (Ernst 1974). A correlation between total metal concen-tration in a soil and the survival of a plant species at that site (Turner and Dickinson 1993) reveals nothing about the exposure of the plant roots to the bioavailable metal concentration.

19.4 SENSITIVITY AND RESISTANCE TO HEAVY METALS

Generally, only the free ionic metal in the abiotic environment will affect the performance of plants at the various levels of biological organizations, from the subcellular up to the ecosystem level (Ernst et al. 1992). Sensitivity and resistance depends on the genotype-specific demand, the metabolic regulation of a surplus at the various levels of an individual, and population and ecosystem processes affecting metal availability. In the context of ecotoxicology the final effect at the individual level determines the survival of the individual, the persistence of the population, and the functioning of the ecosystem. In the natural environment it is not only the impact of the metal alone which affects plant performance; the presence of other abiotic and biotic factors may modify metal sensitivity. Co-occurrence of a surplus of several metals, shortage or surplus of other major and minor nutrients, water and radiation, and the absence or presence of pathogens and herbivores determine whether the plant survives.

19.4.1 Sensitivity and Resistance at the Cellular Level

The outer part of a plant cell, the cell wall, is the first cellular compartment which is exposed to heavy metals. Most of the cell-wall-associated heavy metals are bound to polygalacturonic acids with a decreasing affinity in the order Pb > Cr > Cu > Ca > Zn. Binding of heavy metals to the cell wall has frequently been reported, first for *Silene vulgaris* (Ernst 1969) and later for many other plant species (Ernst et al. 1992). This high metal-specific binding capacity of cell walls (Figure 19.1) does not reflect the metal concentration of roots if the roots are not desorbed prior to harvesting by placing the plants in an ice-cold 5-mM Pb(NO$_3$)$_2$ solution (Harrison et al. 1979). Unfortunately in the case of lead, this procedure cannot be applied; therefore, the lead concentration in root tissue is very difficult to determine.

The metal-binding capacity of cell walls is not only species specific (high for legumes, low for grasses), but also genotype specific, as shown for the difference between British and German populations of the perennial grass *Agrostis capillaris* (Ernst 1976). In addition to adsorptive binding to polygalacturonic acid, heavy metals are also precipitated as crystal-like bodies between the lamellae

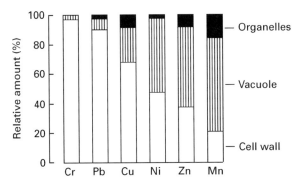

Figure 19.1 Metal-specific compartmentation in cell walls, vacuoles, and organelles of root cells of *Indigofera setiflora* (after Ernst 1972).

(Ernst and Weinert 1972; Neumann et al. 1995). These pseudocrystals of heavy metals contribute less than 10% to the tissue metal concentration (Ernst 1974). The cell wall contains several extracellular enzymes such as the acid phosphatases. The sensitivity of these enzymes to heavy metals was found to be higher in the grasses *A. capillaris* and *Anthoxanthum odoratum* from uncontaminated sites than those from metal-contaminated sites (Wainwright and Woolhouse 1975; Cox and Thurman 1978). If this finding can be generalized, then the first metabolic effect of heavy metals may be on the activity of extracellular enzymes in the cell wall.

19.4.2 The Plasma Membrane

The plasma membrane is the compartment of the cell which may at least partially regulate the entry of a heavy metal ion into the cell interior. This direct exposure may have consequences for the adaptation and selection of plant species and genotypes (Verkleij and Schat 1990; Meharg 1993). Cation and anion uptake by the cell can be actively regulated by electrogenic proton pumps (H^+-ATPase, PP_iase), transmembrane redox pumps (NAD(P)oxidase), and ion channels (Marschner 1995). Some of the processes are partially ion specific: K^+ has to compete with Na^+ and other monovalent ions in the monovalent K^+ channel; competition of Ca^{2+} and Mg^{2+} occurs with a lot of other divalent cations, such as many heavy metals, in the divalent cation channel and perhaps in the specific Ca^{2+} channel. There are specific channels for other divalent cations such as Cd^{2+}. The knowledge of the functioning of all these channels is very incomplete so that only general metabolic disturbances of the biomembranes can be judged upon. Biochemical changes of the plasma membrane of metal-sensitive plants are found with regard to lipid composition after the exposure to Cd, permeability after exposure to Cu, and Mg^{2+}-ATPase and H^+-ATPase (Veltrup 1978; Ros et al. 1990; Costa and Morel 1994). Exposure of Cu-sensitive and Cu-resistant ecotypes of *S. vulgaris* to copper did not lead to significant differences in lipid and fatty acid composition of the plasma membrane (De Vos et al. 1993). Therefore, changes in the metal-sensitive genotype of this biomembrane may involve ion channels.

Ion competition for the same channels in relation to heavy metals has only been reported for arsenic (Meharg and Macnair 1990, 1991a,b). In the arsenate-resistant grasses *A. capillaris*, *Deschampsia cespitosa*, and *Holcus lanatus*, the low phosphate-affinity system is suppressed (Meharg 1993) so that in addition to arsenate/phosphate competition other changes in the plasma membrane have taken place. A common ion channel has been suggested for Cd, Cu, and Zn (Clarkson and Luettge 1989). Plants growing on soils rich in either copper or zinc should have difficulties in acquiring the appropriate amount of the heavy metal not present in surplus. Such a deficiency, however, has not been found (Figure 19.2). There are two possible explanations (Ernst 1996): (1) Zn and Cu are not taken up by the same ion channel, and thus the hypothesis is incorrect. (2) Plants growing for a long period of time on such metal-surplus sites have

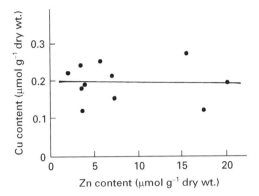

Figure 19.2 Relation between Zn content and Cu content of leaves of *Agrostis capillaris* growing on sites with increasing Zn concentration in the rhizosphere (Ernst, unpublished).

modified the affinity of the uptake system in favor of the essential heavy metals present in low amounts. At this point it should be mentioned that many studies on metal toxicities are not conclusive because the applied metal concentration was lethal and did not allow further experimentation (cf. Fodor et al. 1995). This misconception of physiologists hamper real progress in understanding metal toxicity.

Sulfhydryl-reactive heavy metals have a strong impact on the integrity of the plasma membrane. Cu and Hg cause leakage of potassium from cells exposed to increasing concentrations of these heavy metals, as originally shown for *Chlorella* (McBrien and Hassall 1965). This effect has been confirmed for different plant species: copper-sensitive and copper-resistant genotypes strongly differ in the amount of potassium leakage (Wainwright and Woolhouse 1977; De Vos et al. 1991; Strange and Macnair 1991). This K^+ leakage by the plasma membrane is a very sensitive parameter for testing deleterious effects of copper and mercury in the soil solution and in water. Exposure of plants to Zn and Cd does not result in this metabolic disturbance (Ernst et al. 1992).

The transmembrane redox pump, which is essential for the uptake of iron, by reducing Fe^{3+} to Fe^{2+} (Chaney et al. 1972), is also involved in the reduction of Cu^{2+} to the radical producing Cu^+ (Holden et al. 1995) and may possibly reduce As^{5+} to As^{3+}. The stability of Cu^+ is so low that it is immediately oxidized during membrane passage.

19.4.3 The Cytosol

As mentioned above, higher plants contain certain heavy metals necessary for normal metabolic performance. Nothing is known of how these essential heavy metals are channeled to the sites of demand in appropriate amounts and how intracellular competition is balanced. When all the demands are saturated the problem of handling the metal surplus arises.

Enzymes

As soon as heavy metals have passed the plasma membrane, they can immediately interact with all metabolic processes in the cytosol. Cytosolic enzymes of metal-sensitive and metal-resistant genotypes of the same plant species have the same sensitivity to all tested metals, as demonstrated for *S. vulgaris* (Mathys 1975). Each enzyme, however, differs in sensitivity to the various heavy metals. The decrease in activity can be used for the calculations of EC values of free metal ions in the cytosol. The most sensitive enzyme was found to be nitrate reductase (Figure 19.3), the most resistant one, peroxidase. This result makes clear that differences in metal sensitivity between genotypes are not based on the selection of genotype-specific changes in metal sensitivity of cytosolic enzymes. The absence of a difference at the enzyme level was confirmed for glutamine synthetase in Zn-sensitive and Zn-resistant ecotypes of *D. cespitosa* (Smirnoff and Stewart 1987). In vivo, a surplus of heavy metals not properly detoxified by cell metabolism will hamper the activity of these cytosolic enzymes (Mathys, 1975). In such a situation they may be relevant biomarkers for early warning systems (Ernst and Peterson 1994).

Isolation of enzymes from cells of plants grown in a metal-polluted environment may be problematic because during the isolation and purification procedure the metal compartmentation of the cells is broken down. This means that the compartmented metals may react during the isolation procedures with the enzymes to such a degree that an assessment of metal toxicity in the environment is impossible.

Phytochelatins

One way of rendering a surplus of heavy metals innocuous to cytosolic metabolism is to change their speciation. Heavy metals with a high affinity to sulfhydryl groups have immediate access to the SH in glutathione. Direct binding to this

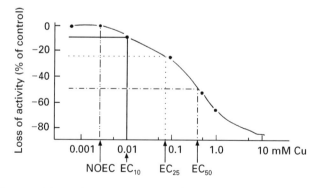

Figure 19.3 Impact of increasing copper concentration on the in vitro activity of the nitrate reductase from leaves of *Silene vulgaris*, grown in a clean soil. NOEC, EC_{10}, EC_{25}, and EC_{50} levels are indicated. Basis data from Mathys (1975).

tripeptide is obviously not a realistic solution. Therefore another system has evolved for the rapid change of speciation by metal binding. A surplus of free heavy metals in the cytosol induces the synthesis of phytochelatins (PCs; Grill et al. 1985) by activating the enzyme phytochelatin synthase (Grill et al. 1989). PCs have the general structure $(\tau\text{-Glu-Cys})_n\text{Gly}$ in most higher plants. In several grasses the glycine at the terminal position is substituted by serine, in which case the phytochelatins are called hydroxyl-phytochelatins (Klapheck et al. 1994) and have the general structure $(\gamma\text{-Glu-Cys})_n\text{-Ser}$. In Fabaceae the terminal glycine is substituted by β-alanine; here, the phytochelatins are called homophytochelatins (Grill et al. 1986) and have the general structure $(\gamma\text{-Glu-Cys})_n\text{-}\beta\text{-Ala}$. The various heavy metals have quite specific impacts on the responsiveness of the phytochelatin synthase to a surplus of free metal ions in the cytosol. At the same exposure metal concentration there is a decreasing order: Hg > Cd > As > Te > Ag > Cu > Ni > Sb > Au > Sn > Se > Bi > Pb > W > Zn (Grill et al. 1987; Ernst 1996). Therefore, the amount of PCs and the chain length may be relevant in early warning systems (Figure 19.4) because once more there is a great difference at the same external metal exposure in the synthesis of PCs between metal-sensitive and metal-resistant genotypes of higher plants (De Knecht et al. 1994). When both genotypes are compared not at the same external Cd concentration, but at the same EC_{50} level (cf. Chapter 4), i.e., at the same level of free Cd in the cytosol, the PC concentration is the same in both genotypes. Experimentally produced mutants of *Arabidopsis thaliana* which are low in glutathione, the precursor for PC synthesis, are therefore very Cd sensitive (Howden et al. 1995).

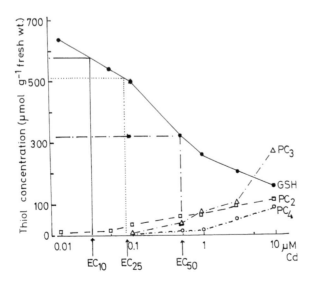

Figure 19.4 Increasing synthesis of phytochelatins (PC_2, PC_3, and PC_4) and the decrease of glutathione (GSH) in maize roots after exposure to increasing Cd concentration in the nutrient solution. EC_{10}, EC_{25}, and EC_{50} levels are indicated. Data from Tukendorf and Rauser (1990).

19.4.4 Protein Metabolism

Proline

Plants growing in copper-enriched soils (Farago 1981) and in copper mine soils (Schat et al. 1997) have an increased proline concentration. A similar effect was achieved by growing plants under conditions of increasing metal exposure (Barcelo et al. 1986). Proline accumulation is a typical symptom of water stress, independently of whether the cellular water stress is caused by drought, salinity, low temperature, or heavy metals (cf. Ernst and Peterson, 1994). However, by protecting metal-sensitive plants from exceptional transpiration, proline does not accumulate despite exposure to a surplus of heavy metals (Schat et al. 1997). This example demonstrates that a complete knowledge of a habitat is necessary to estimate the degree of stress a plant will suffer from heavy metals.

Polyamines

Other nonspecific stress metabolites belong to the polyamines. In addition to their enhanced synthesis under conditions of nutrient deficiency, salinity, acidity, and drought, exposure to Cd and Cr can stimulate polyamine synthesis (Weinstein et al. 1986). Under conditions of Cd exposure only the synthesis of putrescine was found to increase rapidly (Figure 19.5). Doubling of the putrescine concentration in oat seedlings is achieved by a very low external Cd concentration. A comparison of the polyamine concentration in *Avena* after in

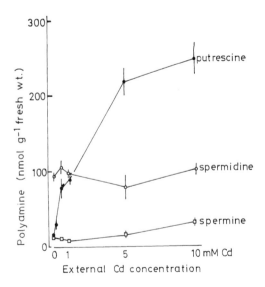

Figure 19.5 Polyamine synthesis upon Cd exposure of intact oat seedlings. Data from Weinstein et al. (1986).

vitro and in vivo exposure highlights the dependency of this stress reaction on the general environmental conditions. The physiological importance may be based on a series of metabolic protection steps, i.e., stabilization of membrane structures and delaying of the process of senescence.

Heat Shock Proteins

High temperature can induce synthesis of the mRNAs for so-called heat shock proteins (HSPs). The various HSP families are involved in protein assembly and folding, and in membrane stabilization. After short-term exposure to heat followed by an exposure to Cd, the plasma membrane of tomato plants (*Lycopersicon peruvianum*) were less sensitive to Cd (Neumann et al. 1994). It is doubtful whether HSPs can play a role in the natural situation. To be effective the heat shock should be applied to the plant roots, which are the first metal targets. The condition for such an event is an open vegetation on a relatively dry soil. Plant species or ecotypes growing at such sites, such as the vegetation on copper schists, are already adapted to incidental high temperature, comparable to their adaptation to drought with a structurally enhanced proline level (Schat et al. 1997). Therefore, it is conceivable that HSPs do not play a role in modifying metal sensitivity in nature; in any case they do not induce heavy-metal tolerance.

19.4.5 Cell Organelles

Photosynthesis

Chloroplasts are the cell organelles which are responsible for the fixation of solar energy. They are protected from a direct confrontation of a surplus of heavy metals in the cells by (1) a special biomembrane and (2) by their position in the cell, i.e., surrounded by the cytosol. For an appropriate functioning the chloroplast demands several heavy metals as micronutrients, i.e., Cu, Fe, Mn, and Zn. Mechanisms of metal homeostasis in photosynthetically active cells are necessary for an adequate metal supply at the required enzymatic site and for a protection from a metal surplus. Chloroplasts have their own cpDNA. Although chloroplast multiplication is more rapid than cellular division theoretically it is conceivable that cpDNA allows the evolution of metal-resistant chloroplasts in a metal-sensitive cytosol. However, there is no evidence that cpDNA can play a role in the evolution of metal resistance, perhaps because metal exposure of chloroplasts is a cellular third-order event.

A lot of research has been carried out with in vitro systems, i.e., isolated chloroplasts, thus excluding all regulatory mechanisms of the cytosol. Isolated chloroplasts can be affected by heavy metals (Krupa and Baszynski 1995). Due to the instability of isolated biological systems high exposure concentrations and short exposure periods have been used. A further draw-back of in vitro studies is the lack of adequate measurements of metal availability and metal

speciation in the assay (e.g., Hampp et al. 1973). Therefore, the in vitro experiments can be summarized as follows: (1) In vitro studies established that
heavy metals have a direct inhibitory effect on photosynthetic electron transport reactions, especially on photosystem II, both at the oxidizing (donor)
and reducing (acceptor) sites (for a review, see Krupa and Baszynski 1995).
(2) However, the results of in vitro studies are not relevant for in vivo heavy
metal exposure; they have no relevance for ecotoxicology.

In vivo experiments have the advantage that the exposure of the chloroplasts
to heavy metals occurs at the end of a whole sequence from uptake by roots,
translocation to the leaves, and passage across the plasma membrane and
through the cytosol. The complexity of the process makes it more difficult to
achieve clear injury symptoms. Certainly, the first interaction of SH-reactive
metals occurs at the chloroplast membranes similar to that described for the
plasma membrane, i.e., lipid peroxidation (Sandmann and Boeger 1980; Maksymiec et al. 1992). One of the problems in the evaluation of most water-culture
experiments is the insufficient information on the composition of the nutrient
solution, especially on the metal speciation (e.g., Tukendorf 1989). Short exposure times and high exposure concentrations make experiments ecologically and
ecotoxicologically not relevant (e.g., Angelov et al. 1993). On the other hand,
a fluorescence measurement of in vivo chlorophyll as a general stress parameter
(Lichtenthaler and Rinderle 1988) has successfully been applied to plants growing in nature (Lanaras et al. 1993) and in water culture (Krupa et al. 1992;
Ouzounidou 1994).

Krupa and Baszynski (1995) have proposed a very plausible hypothesis for
the impact of metal on chloroplasts (Figure 19.6). A surplus of Cd and Cu
initiates premature senescence through increased activity of galactolipase, which

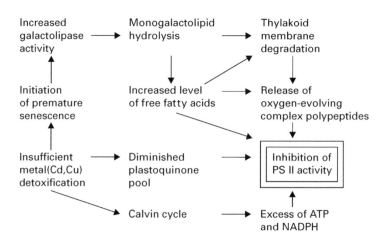

Figure 19.6 Model of the impact of heavy metals on in vivo photosynthesis. Modified after Krupa
and Bazynski (1995).

causes the degradation of the thylakoid membrane and, thus, finally inhibits photosystem II (PS II) activity. Another mechanism of metal injury is a direct effect on the Calvin cycle (Greger and Oegren 1991), which results in excess levels of ATP and NADPH and finally downregulates PS II. At the moment the number of exposure concentrations are too low to allow calculations of EC and NOEC (no observed effect concentration) values. Other effects giving rise to chlorosis may involve ion competition at the root surface; this is discussed in Section 19.5.4.

Respiration

Mitochondria are the cell organelles which are responsible for the production of energy by respiration using compounds rich in chemical energy. The impact of heavy metals on mitochondria is obviously less than that on chloroplasts, although it is a similar third-order event. Micromorphological changes can occur after exposure to lead and zinc (Silverberg 1975; Davies et al. 1995). Swelling of mitochondria, a general stress reaction known from NaCl and anoxia exposure (Smith et al. 1982; Vartapetian and Zakhmilova 1990), is more extensive in Zn-sensitive than Zn-resistant genotypes of *Festuca rubra* in in vivo experiments.

In vitro studies on mitochondria isolated from maize confirmed the low sensitivity to heavy metals (Bittel et al. 1974). Cd was the most effective: the EC_{25} for succinate oxidation was around 30 µM. Phosphorylation was uncoupled only at 500 µM Cd. This low in vitro injury to respiration is supported by in vivo respiration measurements of metal-exposed roots of *S. vulgaris*.

Nucleus and Cell Division

Growth of plants is very dependent on cell division. The root elongation test is principally based on root growth, and thus cell division. In the perennial grass *F. rubra* exposure of the Zn-sensitive ecotype to increased Zn levels (3 µM, 4 days) decreased the nuclear volume by 30% and doubled the length of the cell cycle (Powell et al. 1986). The cell cycle of the Zn-resistant ecotype, however, was not affected, although their nuclear volume increased by 50%. If metals can interact with the nucleus, a deregulation of the mitotic cycle may be expected, as has also been demonstrated for copper (Liu et al. 1994). In the latter case the EC_{25} value (more than 1 mM Cu) was much greater than the EC_{100} value of the plasma membrane. The fixation techniques for studying morphological cell parameters may interfere with the metal exposure; therefore, in ecotoxicological studies changes in the morphology of the nucleus by metals are not of primary importance.

Vacuoles

Since the first metal compartmentation study (Ernst 1969) the accumulation of heavy metals in the vacuole has been generally accepted as a cellular process

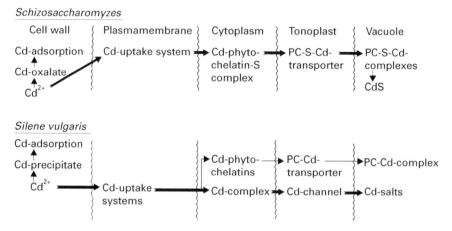

Figure 19.7 Model of cellular metal compartmentation for the fungus *Schizosaccharomyces* and *Silene vulgaris* (Ernst, unpublished).

(Figure 19.7). The presence of heavy metals in the vacuoles has been demonstrated by using various physical and chemical techniques. Compartmentation of heavy metals into the vacuole is the principle of all hypotheses explaining metal tolerance in plants (Mathys 1975; Ernst et al. 1992). The open questions concern the transporter(s) across the tonoplast: metal ion, metal complexes such as the Zn-malate complex (Mathys 1975), or metal peptides such as the Cd-phytochelatin complex (Voegeli-Lange and Wagner 1990).

To ensure the stability of metal compartmentation heavy metals can be precipitated in the vacuole, either as metal oxalates (Ernst and Weinert 1972), Zn-phytate (Van Steveninck et al. 1990) or as other metal salts (Ernst et al. 1992). Complexation with anthocyanins, organic acids, and glucosinolates (Mathys 1977) is another possible mechanism for the stable compartmentation of heavy metals in the vacuole. An understanding of the different degrees of metal toxicity requires further knowledge of the specificity and velocity of metal transport across the tonoplast.

19.5 REACTION OF PLANT PARTS TO A SURPLUS OF HEAVY METALS

An exposure to a surplus of heavy metals can occur during various stages of the life history of a plant, from germination through vegetative growth up to reproduction. Within a life phase different plant parts can be affected, such as bark, leaves and flowers, or roots, depending on the polluted environmental compartment. The whole plant has to try to repair the injury suffered by one plant part. The exposure time, even with a permanent general exposure of the plant, will depend on the lifespan of the plant part, varying from very short in pistil and pollen up to very long in bark and root cells.

19.5.1 Seed and Germination

Starting with the colonization of a polluted site by a seed, at the time of germination the seedling will be confronted with the heavy metal. The seed seems to be the least metal-sensitive stage since imbibition of a seed in a metal-containing solution does not damage it. Obviously the metal adsorption to the seed (fruit) coat keeps the metal outside the embryo; in addition, the micropyle may operate as a metal barrier. Therefore, up to now no negative effects of heavy metals are reported from the imbibition phase (Table 19.1). As soon as the radicula penetrates the seed coat, the root takes up water and with it the dissolved heavy metals directly from the surrounding soil solution. The first injury may occur in a metal-specific manner (see Section 19.4.2).

19.5.2 Seedling Stage

Seedling development comprises the elongation and development of a root system, the elongation of the hypocotyl, and the activation of the enzymes for the mobilization of stored energy and nutrients and for photosynthesis of the cotyledons. The exposure of the developing seedlings to a metal surplus in the soil will affect plant metabolism and morphogenesis. The direct observation of morphological changes of the root will be hampered in soil in contrast to water culture (Table 19.1). At a sufficiently high metal concentration root growth will be inhibited; nevertheless, the development of cotyledons may continue (Gries 1965). The photosynthetic apparatus has been observed to be the least affected during the first phase of metal exposure of seedlings.

Table 19.1 Exposure of seeds and seedlings from a zinc-sensitive and a zinc-resistant population of _Silene vulgaris_ to medium and high zinc levels for 10 days

Population	Zinc concentration (mM)		
	0	3.8	15.3
Zn-sensitive germination (%)	96	92	88
Unfolded cotyledons			
(% of germinated seeds)	91.7	13.0	4.5
Root length (mm)	3.2 ± 0.5	0.2 ± 0.55	0.2 ± 0.05
Zn-resistant germination (%)	96	96	100
Unfolded cotyledons			
(% of germinated seeds)	91.7	87.5	20.0
Root length (mm)	3.2 ± 0.5	0.3 ± 0.1	0.2 ± 0.05

Source: Gries (1965).

When the lifespan of cotyledons lasts more than 1 week, a surplus of heavy metals will cause discoloration of the cotyledons, hamper their expansion and cause malformations such as curling and dwarfism. The hypocotyl is often more affected by metal exposure than the cotyledons. Copper pollution gives rise to seedlings with very stunted hypocotyls, whereas the impact of zinc and cadmium is not so dramatic. At very high metal exposure concentration the cotyledons are shed earlier than those of nonexposed plants.

The OECD guideline (208 GTP) for testing heavy metal toxicity of soils by seedlings is inappropriate: it prescribes to finalize the test 14 days after 50% seed germination has occurred. In a very homogenous population, i.e., a highly selected cultivar with long-living cotyledons, this period will be 15 days after sowing. In a seed collection of wild plant species this period may last for months resulting in a completely different statement on soil metal toxicity. Therefore the OECD guideline has to be highly specified with regard to genotype, cultivar or wild population, the seedling reserve, the applied genotype-specific temperature, the metal speciation including the effects of root exudates, etc.

19.5.3 Root Growth and Metal Exposure

Root performance strongly depends on the way the root environment was contaminated. If metal pollution occurs at a later stage of a plant's life history, i.e., aerial deposition from a smelter fallout on an established vegetation (Ernst 1973; Dueck et al. 1984), or fertilization with metal-polluted sewage sludge, only the upper roots will be exposed. After years, decades, or even after a century the deeper roots may still exploit a noncontaminated soil if the metals are fixed in the humic layer. As a consequence deep-rooting plants will not be exposed to selection in contrast to shallow-rooting plants. In such a contaminated ecosystem metal-resistant plants will coexist with metal-sensitive plants as demonstrated for the shallow-rooted, highly zinc-resistant grass *Agrostis capillaris* and the deep-rooting, still zinc-sensitive grass *Molinia caerulea* in the vicinity of a zinc smelter (Figure 19.8). Plants growing in a deeply polluted soil or in a soil over natural ore outcrops will be strongly selected. In the medium term these populations and the whole vegetation will consist only of metal-resistant plants.

The direct exposure of roots to heavy metals results in a rapid accumulation of these elements in the roots (Figure 19.9). A steep metal gradient can build up between the rhizodermis and the xylem and phloem of the roots (Ernst 1974). The latter tissues have to function throughout the life cycle of a plant whereas the rhizodermis is substituted by a periderm; therefore, xylem and phloem are highly protected from metal accumulation. The visible effect of a surplus of metals in the rhizosphere is a stunted root growth. Differences in root elongation is one of the best established parameters for screening metal resistance in plants. The metabolic injury is based on metal accumulation in the cytosol and an insufficient detoxification of free metal ions. Metal stress results in an increase in phytochelatin levels (Figure 19.10), but due to the limitations of this

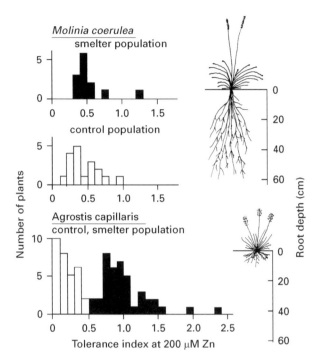

Figure 19.8 Long-term exposure of two grass species with different rooting strategies to metal fallout and the selection of zinc-resistant genotypes in the shallow-rooting *Agrostis capillaris* and the deep-rooting *Molinia caerulea* (Ernst, unpublished, and data from Dueck et al. 1984).

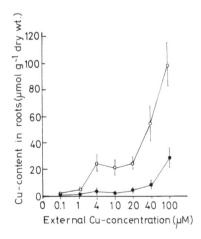

Figure 19.9 Copper accumulation in roots of two genotypes of *Silene vulgaris* in relation to the external exposure concentration (after Lolkema et al. 1984).

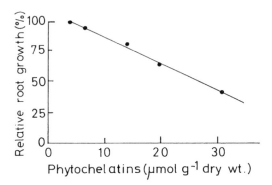

Figure 19.10 Relation between root elongation and the phytochelatin content of root tips of *Silene vulgaris* after a 3-day exposure to Cd (after De Knecht 1994).

decontamination process, cell elongation and cell division are hampered in a genotype-specific way.

Plants can respond to the metal concentration of the soil by changes in root physiology due to selection. One of the reaction patterns to a metal surplus is a change of the uptake system ("avoidance" sensu Levitt 1980; "excluder" sensu Baker 1981). This avoidance may be realized by a change in metal affinity as known for *Saccharomyces cerevisiae* and *Neocosmospora vasinfecta* with long-term Cd exposure and for *S. vulgaris* and *Minuartia hirsuta* with long-term Cu exposure. Another adaptation is the rapid translocation of heavy metals from root to shoot, as found in several metal-sensitive genotypes of *S. vulgaris* (Lolkema et al. 1984) and maize (Florijn 1993).

Root exudates can reduce metal toxicity in plants of all degrees of metal sensitivity (Mench et al. 1988). Phytosiderophores are synthesized by grasses primarily to mobilize iron from the rhizosphere during iron deficiency of the shoot. However, they can also biomobilize other heavy metals depending on the ion competition in the rhizosphere and thus enhance metal toxicity (Roemheld 1991). Other quite common root exudates with metal-complexing properties are citric and oxalic acid and many phenolic acids. They all affect metal uptake and metal accumulation in a metal-specific manner (Kuiters and Mulder 1993).

Rhizophere Symbionts

The interaction of the roots with the soil is mostly modified by symbionts. Roots of most higher plants live in a symbiosis with mycorrhizal fungi, which enhance uptake of phosphorus and micronutrients and improve the water supply (Tinker and Gildon 1983). The association may involve endomycorrhizal, especially vesicular-arbuscular mycorrhizal (VAM), or ectomycorrhizal fungi. Spore germination and root colonization by vesicular-arbuscular mycorrhizal fungi may be severely affected by elevated concentrations of heavy metals (Weissenhorn 1994). Metal exposure can diminish the degree of mycorrhization of roots if the mycorrhizal fungi are not selected for metal resistance (Galli et al. 1995). As with amycorrhizal roots mycorrhizal fungi can exudate siderophores which change metal speciation and bioavailability to the host plant. In the case of *Calluna vulgaris* the affinity of the siderophores for iron seems to be so high that the plant is protected from a surplus of other heavy metals in the presence of ericoid mycorrhizal fungi.

With long-term metal exposure VAM and ectomycorrhizal fungi contain sufficient genetic potential to be selected for metal resistance, which is as high or even higher than that of higher plant species (Ietswaart et al. 1992; Weissenhorn 1994; Colpaert and van Asche 1992). The mycelia of the mycorrhizal fungi can exploit more soil than the higher plant. This opens the danger that the toxicity of a moderately contaminated soil may be more toxic to a mycorrhized higher plant than to a nonmycorrhized one. An experiment in the metal-contaminated area near Overpeelt (Belgium) has indeed established metal transport from a contaminated soil via ectomycorrhizal fungi to trees in a noncontaminated soil with lethal effects to trees after a period of 30 years. The metal concentration in tree species which are restricted to moderately polluted soils may strongly depend on their symbiosis with ectomycorrhizal fungi. As long as the metal concentration is below an EC_{25} level the adsorption of heavy metal to the cell wall of the mycorrhizal fungus will diminish the translocation to the host tree as demonstrated for Cd and the *Laccaria laccata/Picea abies* symbiosis (Galli et al. 1993). Ectomycorrhizal fungi have an additional organ for metal accumulation, the carpophore (Figure 19.11). It can accumulate high concentrations of heavy metals (Lepsova and Kral 1988), surpassing often the critical value for human consumption. The Chernobyl catastrophe in particular has shown the accumulation potential of basidiomycetes for chemical elements (Ernst and van Rooij 1987).

In addition to VAM fungi, legumes exist symbiotically with nitrogen-fixing bacteria, which may be particularly sensitive to heavy metals (McGrath et al. 1988). This sensitivity of nitrogen-fixing bacteria contrasts with the high insensitivity of soil bacteria in general (Angle et al. 1993). Long-term exposure leads to the selection of metal-resistant *Rhizobium* bacteria (Wu 1989) so that finally a completely metal-resistant rhizosphere develops.

Ecotoxicity tests with naturally occurring amycorrhizal plant species such as *S. vulgaris* (Ernst et al. 1990) or cruciferous species have an advantage compared to the same tests with mycorrhizal grasses. A relevant testing of metal toxicity for grasses demands the addition of appropriate mycorrhizal strains; the testing of metal toxicity for legumes requires the addition of an appropriate *Rhizobium* strain. The OECD guidelines for metal toxicity testing are very inadequate in this area. TGL 208 does not specify, for example the mycorrhizal state of a plant, the equilibrium time after metal addition, soil pH, water content, the quality of the organic matter, and its metal-complexing properties.

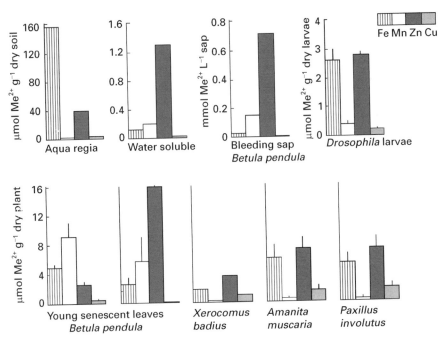

Figure 19.11 Metal accumulation in the carpophores of ectomycorrhizal fungi of *Betula pendula* (*Amanita muscaria, Paxillus involutus, Xerocomus badius*) in comparison with the soil metal content (aqua regia and water extracts), the metal content of birch leaves and birch bleeding sap, and the metal content of *Drosophila* larvae feeding in the carpophores.

19.5.4 Shoot Growth and Leaf Response to a Surplus of Metals

The toxicity of a metal to the shoot is determined by the translocation rate from the root to the shoot, which is very genotype specific. Cu-sensitive genotypes of *S. vulgaris* translocate more rapidly the copper from the shoot to the root than Cu-resistant genotypes (Lolkema et al. 1986). Similar differences have been reported for maize cultivars with regard to Cd (Florijn 1993). As mentioned above, stunted hypocotyls indicate that shoot morphology may suffer from a surplus of heavy metals.

Most interest, however, has been focused on metal accumulation in leaves because most of the heavy metals are translocated from the stalk into the leaves. The preference of metal accumulation in the leaves is conditioned by life-history. A stalk has to support all leaves and later on the reproductive tissues, whereas leaves can be substituted by younger leaves higher up the stalk or by regrowth from dormant buds. Metal-contaminated plants show a strong correlation between age gradients of leaves and metal concentration (Ernst 1974; Ernst et al. 1992). In addition to the metal supply of the shoot by the roots from the pedosphere leaves can absorb heavy metals from wet and dry deposition of the atmosphere (Ernst 1973) and from the hydrosphere (aquatic plants; Marquenie-van der

Werff and Ernst 1979). Metal uptake by leaves is widespread due to the spraying of metal fertilizer on plants. Metal storage in the leaves has a further advantage: in leaf cells a great proportion of the cell volume is occupied by vacuole(s), the well-known final deposit site for many waste products of plant metabolism. At senescence metal-loaded leaves can be shed without diminishing the survival of the plant.

In the natural vegetation metal-loaded leaves protect the plants from herbivory and parasitism. The more metal in the leaf the fewer are the attacks by herbivores as shown for the caterpillars of *Parasemia plantaginis* on leaves of Zn-accumulating *S. vulgaris* (Ernst 1985) and for the caterpillars of *Pieris rapae* on leaves of Ni-accumulating *Streptanthus polygaloides* (Boyd and Martens 1992). Leaves will only be protected by heavy metal accumulation from herbivores if the degree of accumulation is high and if the metal is present in an ionic or easily biodegradable complex. Metals bound to phytochelatins and other less degradable compounds are less available to animals (McKenna et al. 1992). The metal defense strategy does not work in the case of phloem-feeding insects. Aphids sucking phloem sap are exposed to enhanced metal concentrations, but they show different metal-specific accumulation patterns, being high for cadmium and zinc and low for copper (Ernst et al. 1990; Crawford et al. 1995). Despite their often high metal concentration, aphids transferred from metal-poor plants to metal-enriched plants have no problems with survival and reproduction (Ernst et al. 1990). A similar metal-resistant herbivore is the larvae of *Drosophila* species which feed on the carpophores of metal-polluted ectomycorrhizal fungi (Figure 19.10).

High metal concentrations are also effective in the defense against pathogenic mildew *Erysiphe polygoni* and the bacteria *Xanthomonas campestris* of *S. polygaloides* (Boyd et al. 1994). The protective capacity of heavy metals against pathogens is commercially used for the management of pesticides. Copper-based protectants, e.g., the Bordeaux mixture, are applied to vineyards, coffee stands, and hop cultures, where they can cause soil and plant toxicity if applied for long periods (Rieder and Shcwertmann 1972; Magalhaes et al. 1985; Lepp and Dickinson 1994).

The presence of metal in leaves can hamper plant metabolism even at very low concentrations in metal-sensitive plants in contrast to metal-resistant ones (De Vos et al. 1991; De Knecht et al. 1994). All aspects of the cellular impact of heavy metals described may come to full expression in intact leaves. Often chlorosis, necrosis, and the synthesis of anthocyanins are the first visible symptoms of metal toxicity. Chlorosis can be caused by metals via two pathways: (1) direct inhibition of enzymes involved in chlorophyll synthesis and (2) competition in the translocation of iron and magnesium with other heavy metals and interaction at the reactive sites. In the case of chlorosis the enzyme δ-aminolaevulinic acid (ALA) dehydratase plays a key role in chlorophyll biosynthesis, because it synthesizes porphobilinogene from δ-ALA. Heavy metals with a high affinity for sulfhydryl groups, i.e., As, Cd, Cu, Hg, and Pb, can block the enzyme and thus hamper chlorophyll biosynthesis (Stobart et al. 1985). In the chloroplast heavy metals can have an impact on the enzyme ribulose-1,5-biphosphate carbo-xylase/oxygenase. The activity of the enzyme depends on two SH groups in the active center and on the formation of a complex between Mg, CO_2, and the enzyme. Substitution of Mg by Co, Ni, Mn, and Zn can diminish the carboxyla-tion (Van Assche et al. 1980). Even metal-tolerant plants cannot withstand all metal concentrations and they become chlorotic at very extreme metal concen-trations in the soil (Ernst 1974).

A permanent production of new leaves creates new disposal sites and allows detoxification up to the individual level. If the balance between new and dying tissue is in favor of the new tissue, the plant will have the opportunity to survive up to seed production and maturity. As a consequence the oldest leaves of metal-exposed plants exhibit the highest metal concentration (Ernst 1974). The longer leaves are exposed to a metal surplus, the more morphological changes become visible; one of these symptoms is dwarfism, as observed in coniferous trees. At the same time an increasing metal concentration diminishes the life expectancy of the leaves (Figure 19.12).

The leaf is the plant organ which plays a predominant role in water economy. Wilting of plants is a first sign of disturbed water balance and is described for many plant species at heavy metal exposure, especially with cadmium (Page et al. 1972; Lamoureux and Chaney 1977; Barcelo et al. 1986). The reason for this wilting is the stimulation of the transpiration rate (Paul and de Foster 1981; Greger and Johansson 1992). The already mentioned enhanced synthesis of proline is a typical symptom of water stress caused by a disturbance of water transport and transpiration by heavy metals. As a metabolic consequence abscisic acid levels may be enhanced, but this reaction is too nonspecific to be of value as a biomarker (Poschenrieder et al. 1989).

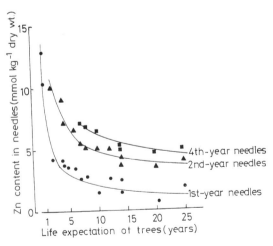

Figure 19.12 Zinc content in different needle age classes of Scotch pine in relation to the life expectancy of the tree (after Ernst 1985).

19.5.5 Reproductive Growth and Metal Exposure

Metal-sensitive plants upon metal exposure have great difficulties in reaching the reproductive phase. Aerial metal fallout may interact with the normal processes of flowering and seed development. Plants which are metal resistant enough to complete one life cycle in a metal contaminated soil are excellent subjects with which to study allocation and impact of heavy metals on the reproductive system. In contrast to the calyx, petals of flowers are relatively short-living. Their metal concentration is therefore low. Within the inflorescence and infructescence the calyx lives the longest, often up to seed ripeness, and exhibits the highest metal concentration of the reproductive organ (Ernst et al. 1990).

Accumulation of metals on the pollen due to aerial pollution, especially by leaded fuel, does not affect pollen fertility, but can have a negative impact on animal and human food chains (Ernst and Bast-Cramer 1980). Adsorption of heavy metal to and metal accumulation in the stigma have greater adverse effects. Low concentrations of Cd (0.2-0.45 µM) hamper pollen growth in *Petunia* and *Pinus* (Kapur and Malik 1976; Strickland and Chaney 1979). The injury to species with such a great taxonomic distance as that between angiosperms and gymnosperms may indicate a general sensitivity of pollen and stigma to a surplus of heavy metals. Nevertheless, the general principle of species- and genotype-specific reaction to such a surplus is applicable to the reproductive tissue. In *Mimulus guttatus* the in vitro vitality of pollen is strongly correlated with the copper resistance of the plant (Figure 19.13). Comparable to the situation of vegetative growth, pollen germination in the presence of enhanced copper levels is more hampered in Cu-sensitive than in Cu-resistant genotypes. Similar differences in the pollen germination have been observed between

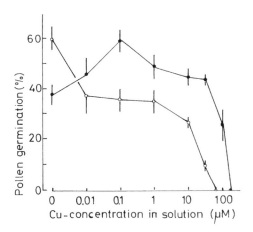

Figure 19.13 Germination of pollen of a Cu-resistant (closed circles) and a Cu-sensitive (open circles) genotype of *Mimulus guttatus* at increasing external Cu concentrations (after Searcy and Mulcahy 1985).

Zn-sensitive and Zn-resistant genotypes of *Silene dioica* (Searcy and Mulcahy 1985a,b). To date, measurements of metal concentration in the pistil up to the ovaries have not been carried out. Therefore, it is an open question how important differences in pollen sensitivity to heavy metals are in relation to the genetic isolation between the offspring of tolerant and sensitive plants. Plants producing seeds in a metal-affected environment can help to understand the transfer of metals to the seeds. During seed development heavy metals have to pass the chalaza before entering the embryo. The seeds of most plant species growing on heavy-metal-enriched soils have very low metal concentrations (Ernst 1974; Lepp and Dickinson 1994). This low load has an ecological advantage for a seedling establishing in a metal-enriched soil. In contrast to the seed, the fruit is not protected from metal accumulation; from an evolutionary point of view it belongs to the leaf structure. Therefore metal concentrations of fruits increase with metal exposure, as shown for Zn and Cd concentration in grain after long-term sewage sludge application (Juste and Mench 1992) and increased Cu concentration in wine and coffee berries (Lep and Dickinson 1994). The uptake of lead from traffic emission by wine (Lobinski et al. 1994) underlines the physiological nature of fruit as leaf structure. Therefore, the metal burden of fruit-consuming organisms will increase with the contamination of the food.

At the same time seed as a metal-poor food in a metal-enriched environment is a welcome opportunity for seed predators. Seeds in capsules of *S. vulgaris* are heavily predated by the caterpillars of *Hadena* moths (*Hadena cucubali*, *Hadena bicruris*), often destroying the whole seed crop (Ernst et al. 1990). Seeds of *Cardaminopsis halleri*, a hyperaccumulator of zinc (Ernst 1974) are predated by the larvae of the curculionid beetles *Ceutorrhynchus* spp. Both examples, the caterpillar and the beetle larvae, extend the theory of safe niches to ecotoxicology.

Fruit and seed production is diminished with increasing metal excess. Application of sewage sludge (100 t ha^{-1}) to maize over a long period reduced the final grain harvest by more than 20% (Juste and Mench 1992). Even metal-resistant plants suffer from increasing amounts of available metals in their reproduction. Zn-resistant genotypes of *S. vulgaris* are stimulated by increased levels of zinc, not only in parameters of vegetative growth (Ernst 1968, 1974; Mathys 1975), but also in flower and seed production (Table 19.2). Surpassing a critical Zn concentration results in diminished growth and the lack of flower production.

19.6 THE INDIVIDUAL

Performance and survival of a plant up to reproduction involves more than the sum of the resistance of plant parts and metal compartmentation at the level of cells and organs. Most experiments exposing plants to a surplus of heavy metals allow only a determination of EC concentrations. These experiments cannot predict the acceptable concentration throughout the life cycle.

Table 19.2 Performance of vegetative and generative growth of Zn-resistant plants of *Silene vulgaris* at various Zn levels throughout a life cycle period

Ionic Zn in the solution (mM)	Biomass production (mg dry mass per plant)	Number of flowers per plant	Number of seeds per plant
0.15	1790 ± 52	4 ± 2	161 ± 23
0.76	2670 ± 15	8 ± 1	245 ± 47
1.15	1990 ± 93	1 ± 1	7 ± 9
1.53	352 ± 123	0	0

Source: Data on vegetative growth from Ernst (1968); those on reproductive growth are unpublished data from the same experiment.

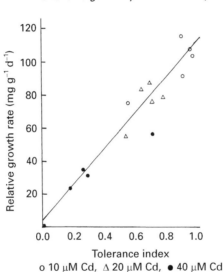

Silene vulgaris $y = 3.8 + 108.9x$; $r = 0.97$

Figure 19.14 Correlation between relative growth rate, index of Cd tolerance, and Cd exposure concentration for *Silene vulgaris*. Data from Verkleij and Prast (1989).

Relative growth rate is a parameter which integrates the uptake of heavy metals and other nutrients by roots, their incorporation, detoxification, and translocation to the shoot, and the allocation, incorporation, and detoxification within the shoot. A comparison between exposure concentrations (at identical exposure time), metal tolerance indices and relative growth rates (Verkleij and Prast 1989) showed that the highest relative growth rate is found at the lowest exposure concentration of nonessential heavy metals such as Cd (Figure 19.4). As a rule low exposure concentration at long exposure periods of time will guarantee the survival of a plant up to reproduction (cf. Chapter 4).

Another aspect at the level of the individual is the demand for the various mineral nutrients in relation to the surplus of the heavy metals and the co-occurrence of several heavy metals (multiple stress) at enhanced bioavailability. The data base is too small for generalizations. Ernst (1968) has shown that an increasing amount of phosphorus interferes in such a way with the availability of iron that the toxicity of Zn is enhanced. Fertilizing a zinc-mine soil stimulates the growth of well-adapted plant species but does not have positive effects on metal-sensitive species. The often told story that an improved water supply in metal-contaminated sites stimulates metal-sensitive species is not substantiated by thorough experiments which consider the dilution of the metal in the soil.

Changing environmental conditions, i.e., in this context increasing external metal concentrations in the rhizosphere, will enhance the selection of the most metal-resistant individuals if the first reproduction can be realized during the first growing season after germination. Many of the perennial plants growing in soils naturally or anthropogenically enriched by heavy metals reproduce in the first year. Only a limited supply of major nutrients can delay flowering and reproduction. Therefore, it is a misinterpretation of ecological reality when Turner (1994, p. 156) states: "most of the research.....has been concerned with short lived species". The herb *S. vulgaris* is as long-living as the tree *Acer pseudoplatanus*. The decisive factor is not plant age, but rather the time up to the first reproduction. In this respect trees have a great disadvantage. Birches, one of the few species growing in slightly contaminated soils, reach the stage of reproduction after at least 5 years of growth. Although they have some potential for the evolution of metal resistance, birches as all other tree species have not been able to colonize metal-enriched soils during the past 12 000 years despite a high annual seed production.

Judging by the ecotoxicity of a surplus of the various heavy metals, those with a high affinity to SH groups will primarily injure the root system; if the root system survives it will then be the decisive target for As, Cu, Hg, and partially Cd. For a surplus of other heavy metals such as Co, Ni, and Zn, the shoot is the primary target. In this case plants with a high shoot productivity may have a better chance of survival than those with a low shoot productivity. Despite the different target organs it is surprising that plants growing at metal-enriched sites belong to the same species, but not to the same genotypes. An improved knowledge of the genetic background of metal resistance will enhance the understanding of metal sensitivity (Schat et al. 1996).

19.7 CONCLUSIONS

A surplus of heavy metals can affect plants at all levels of biological organization, from the subcellular through the individual plant levels up to populations and ecosystems. It can cause changes in metabolism ranging from a slight injury to lethality. The degree of ecotoxicity of a surplus of heavy metals is the result of a combination of abiotic and biotic factors: the bioavailability and the speciation of the heavy metals, the fertility of the environment, the genetic constitution of the plant genotypes with regard to tolerance and resistance to heavy metals, and other factors.

Mycorrhization may confer an advantage to the plant with regard to supply with water and phosphorus, but an exploitation of the heavy metals by the hyphae may counteract the positive contribution of mycorrhizal fungi.

For risk assessment of the heavy metal load in the environment there are sufficient techniques available: specific (PCs) and aspecific metabolic biomarkers, and parameters of plant growth and reproduction. A comparison of different metal-sensitive genotypes of the same plant species will further the understanding of metal ecotoxicity; the overall differences with regard to other metabolic processes within one species are small compared with differences between species.

19.8 REFERENCES

Allan, R. J. (1995) "Impact of Mining Activities on the Terrestrial and Aquatic Environment with Emphasis on Mitigation and Remedial Measures." In *Heavy Metals. Problems and Solutions.* (W. Salomons, U. Foerstner, and P. Mader, Eds.), Springer, Berlin, Germany, pp. 119–140

Angelov, M., T. Tsonev, A. Uzunova, and K. Gaidardjieva, 1993, "Cu^{2+} Effect upon Photosynthesis, Chloroplast Structure, RNA and Protein Synthesis of Pea Plants." *Photosynthetica, 28,* 341–350

Angle, J. S., R. L. Chaney, and D. Rhee, 1993, "Bacterial Resistance to Heavy Metals Related to the Extractable and Total Metal Concentration in Soil and Media," *Soil Biol. Biochem., 25,* 1443–1446

Baker, A. J. M. 1981, "Accumulators and Excluders—Strategies in the Response of Plants to Heavy Metals," *J. Plant Nutrition, 3,* 643–654

Barcelo, J., Ch. Poschenrieder, J. Andreu, and B. Gunse, 1986, "Cadmium-Induced Decrease of Water Stress Resistance in Bush Bean Plants (*Phaseolus vulgaris* L. cv. Contender) I. Effects of Cd on Water Potential, Relative Water Content, and Cell Wall Elasticity," *J. Plant Physiol., 125,* 17–25

Bittel, J. E., D. E. Koeppe, and J. E. Miller, 1974, "Sorption of Heavy Metal Cations by Corn Mitochondria and the Effects of Electron and Energy Transfer Reaction," *Physiologia Plantarum, 30,* 226–230

Bourg, A. C. M., 1988, "Metals in Aquatic and Terrestrial Sytems: Sorption, Speciation, and Mobilization," in *Chemistry and Biology of Solid Waste* (W. Salomons, U. Foerstner, Eds.), Springer, Berlin, Germany, pp. 3–32

Boyd, R. S., and S. C. Martens, 1992, The Raison d'être of Metal Hyperaccumulation by Plants," in *The Vegetation of Ultramafic (Serpentine) Soils* (A. J. M. Baker, J. Proctor, R. D. Reeves, Eds.), Intercept Ltd., Andover, UK, pp. 279–289

Boyd, R. S., J. J. Shaw, and S. C. Martens, 1994, "Nickel Hyperaccumulation Defends *Streptanthus polygaloides* (Brassicaceae) against Pathogens," *Am. J. Bot., 81,* 294–300

Bradshaw, A. D. 1976, "Pollution and Evolution." In: *Effects of Air Pollutants on Plants* (T. A. Mansfield, ed.), Cambridge University Press, Cambridge, UK, pp. 135–159

Brookins, D. G., 1988, *Eh-pH Diagrams for Geochemistry*, Springer, Berlin, Germany

Brooks, R. R., 1993, "Geobotanical and Biogeochemical Methods for Detecting Mineralization and Pollution from Heavy Metals in Oceania, Asia and the Americas," in *Plants as Biomonitors. Indicators for Heavy Metals in the Terrestrial Environment* (B. Markert, Ed.), VCH Verlagsgesellschaft, Weinheim, Germany, pp. 127–153

Chaney, R. L., J. C. Brown, and L. O. Tiffin, 1972, "Obligatory Reduction of Ferric Chelates in Iron Uptake by Soybeans," *Plant Physiol., 50*, 208–213

Clarkson, D. T., and U. Luettge, 1989, "Divalent Cations, Transport and Compartmentation," *Progress in Botany, 51*, 93–112

Colpaert, J. V., and J. A. Van Assche, 1992, "The Effects of Cadmium and the Cadmium-Zinc Interaction on the Axenic Growth of Ectomycorrhizal Fungi, *Plant Soil, 145*, 237–243

Costa, G., and J. L. Morel, 1994, "Efficiency of H^+-ATPase Activity on Cadmium Uptake by Four Cultivars of Lettuce," *J. Plant Nutr., 17*, 627–637

Cox, M. R., and D. A. Thurman, 1978, "Inhibition by Zinc of Soluble and Cell Wall Acid Phosphatases of Zinc-Tolerant and Non-Tolerant Clones of Anthoxanthum Odoratum," *New Phytol., 80*, 17–32

Crawford, L. A., I. D. Hodkinson, and N. W. Lepp, 1995, "The Effects of Elevated Host-Plant Cadmium and Copper on the Performance of the Aphid *Aphis fabae* (Homoptera: Aphididae)," *J. Appl. Ecol., 32*, 528–535

Davies, K. L., M. S. Davies, and D. Francis, 1995, "The Effects of Zinc on Cell Viability and on Mitochondrial Structure in Contrasting Cultivars of *Festuca rubra* L. – A Rapid Test for Zinc Tolerance," *Environ. Pollut., 88*, 109–113

De Knecht, J. A., M. Van Dillen, P. L. M. Koevoets, H. Schat, J. A. C. Verkleij, and W. H. O. Ernst, 1994, "Phytochelatins in Cadmium-Sensitive and Cadmium-Tolerant *Silene vulgaris*," *Plant Physiol., 104*, 255–261

De Vos, R. C. H., H. Schat, M. A. M. and De Waal, R. Voijs, and W. H. O. Ernst, 1991, "Increased Resistance to Copper-Induced Damage of the Root Cell Plasmalemma in Copper-Tolerant *Silene cucubalus*," *Physiol. Plant., 82*, 523–528

De Vos, C. H. R., W. M. Ten Bookum, R. Vooijs, H. Schat, and L. J. De Kok, 1993, "Effect of Copper on Fatty Acid Composition and Peroxidation of Lipids in the Roots of Copper Tolerant and Sensitive *Silene cucubalus*," *Plant Physiol. Biochem., 31*, 151–158

Dueck, T. A., W. H. O. Ernst, Faber, J. and F. Pasman, 1984, "Heavy Metal Emission and Genetic Constitution of Plant Populations in the Vicinity of Two Metal Emission Sources," *Angew. Bot., 58*, 47–59

Ernst, W., 1968, "Der Einfluss der Phosphatversorgung sowie die Wirkung von ionogenem und chelatisiertem Zink auf die Zink- und Phosphataufnahme einiger Schwermetallpflanzen," *Physiol. Plant., 21*, 323–333

Ernst, W., 1969, "Zur Physiologie der Schwermetallpflanzen – Subzelluläre Speicherungsorte des Zinks," *Ber. Deutsch. Bot. Ges., 82*, 161–164

Ernst, W. H. O., 1973, "Zink- und Cadmium-Immissionen auf Böden und Pflanzen in der Umgebung einer Zinkhütte," *Ber. Deutsch. Bot. Ges., 85*, 295–300

Ernst, W. H. O., 1974, *Schwermetallvegetation der Erde*," Stuttgart: G. Fischer Verlag

Ernst, W. H. O., 1976, "Physiological and Biochemical Aspects of Metal Tolerance," in *Effects of Air Pollutants in Plants* (T. A. Mansfield, Ed.), Cambridge University Press, Cambridge, pp. 115–133

Ernst, W. H. O., 1985, "Schwermetallimmissionen – oekophysiologische und populations-genetische Aspekte," *Geobotanisches Colloquium Düsseldorf, 2*, 43–57

Ernst, W. H. O., 1988, "Response of Plants and Vegetation to Mine Tailings and Dredged Materials," in *Chemistry and Biology of Solid Waste* (W. Salomons, and U. Foerstner, Eds.), Springer, Berlin, pp. 54–69

Ernst, W. H. O., 1993a, "Ecological Aspects of Sulfur Nutrition in Higher Plants: The Impact of SO_2 and the Evolution of the Biosynthesis of Organic Sulfur Compounds on Populations and Ecosystems," in *Sulfur Nutrition and Assimilation in Higher Plants. Regulatory, Agricultural and Environmental Aspects* (L. J. De Kok, I. Stulen, H. Rennenberg, C. Brunold, and W. E. Rauser, Eds.), SPB Academic Publishing, The Hague, The Netherlands, pp. 295–313

Ernst, W. H. O., 1993b, "Geobotanical and Biogeochemical Prospecting for Heavy Metal Deposits in Europe and Africa," in *Plants as Biomarkers. Indicators for Heavy Metals in the Terrestrial Environment* (B. Markert, Ed.), VCH Verlagsgesellschaft, Weinheim, Germany, pp. 107–126

Ernst, W. H. O., 1996, "Schwermetalle," in *Stress bei Pflanzen* (C. Brunold, R., Braendle, A. Rüegsegger, Eds.), Ulmer Verlag (UTB), Stuttgart, Germany, pp. 191–219

Ernst, W. H. O., and W. B. Bast-Cramer, 1980, "The Effect of Lead Contamination of Soils and Air on its Accumulation in Pollen," *Plant Soil, 57*, 491–496

Ernst, W. H. O., and P. J. Peterson, 1994, "The Role of Biomarkers in Environmental Assessment. (4) Terrestrial Plants," *Ecotoxicology, 3*, 180–192

Ernst, W. H. O., and Van Roij, 1987, "[134/137]Cs Fall-Out from Chernobyl in Dutch Forest," in *Heavy Metals in the Environment* (S. E. Lindberg, T. C. Hutchinson, Eds.), CEP Consultants, Edinburgh, UK, pp. 284–286

Ernst, W., and H. Weinert, 1972, "Lokalisation von Zink in den Blättern von *Silene cucubalus* Wib," *Z. Pflanzenphysiol., 66*, 258–264

Ernst, W. H. O., H. Schat, and J. A. C. Verkleij, 1990, "Evolutionary Biology of Metal Resistance in *Silene vulgaris*," *Evol. Trends Plants, 4*, 45–51

Ernst, W. H. O., J. A. C. Verkleij, and H. Schat, 1992, "Metal Tolerance in Plants," *Acta Bot. Neerl., 41*, 229–248

Farago, M. E., 1981, "Metal Tolerant Plants," *Coordination Chem. Rev., 36*, 155–182

Florijn, P. J., 1993, "Differential Distribution of Cadmium in Lettuce (*Lactuca sativa* L.) and Maize (*Zea mays* L.)," *Doctorate Thesis*, Agricultural University, Wageningen, The Netherlands

Fodor, E., A. Szabo-Nagy, and L. Erdei, 1995, "The Effects of Cadmium on the Fluidity and H_+-ATPase Activity of Plasma Membrane from Sunflower and Wheat Roots," *J. Plant Physiol., 147*, 87–92

Galli, U., M. Meier, and C. Brunold, 1993, "Effects of Cadmium on Non-mycorrhizal Norway Spruce Seedlings [*Picea abies* (L.) Karst.] and its Ectomycorrhizal Fungus *Laccaria laccata* (Scop. ex Fr.) Bk. and Br.: Sulphate Reduction, Thiols and Distribution of the Heavy Metal," *New Phytol., 125*, 837–843

Galli, U., H. Schueepp, and C. Brunold, 1995, "Thiols of Cu-treated Maize Plants Inoculated with the Arbuscular-mycorrhizal Fungus Glomus Intraradius," *Physiol. Plant.*, *94*, 247–253

Greger, M., and M. Johansson, 1992, "Cadmium Effects on Leaf Transpiration of Sugar Beet (*Beta vulgaris*)," *Physiol. Plant.*, *86*, 465–473

Greger, M., and E. Oegren, 1991, "Direct and Indirect Effects of Cd^{2+} on Photosynthesis in Sugar Beet (*Beta vulgaris*)," *Physiol. Plant.*, *83*, 129–135

Gries, B., 1965, "Zellphysiologische Untersuchungen über die Zinkresistenz bei Galmeiökotypen und Normalformen von *Silene cucubalus* Wib," *Dissertation*, Universitaet Münster, Germany

Grill, E., E. L. Winnacker, and M. H. Zenk, 1985, "Phytochelatins: The Principal Heavymetal Complexing Peptides of Higher Plants," *Science*, *230*, 674–676

Grill, E., W. Gekeler, E. L. Winnacker, and M. H. Zenk, 1986, "Homo-Phytochelatins are the Heavy Metal-binding Peptides of Homo-Glutathione Containing Fabaceae," *FEBS*, *205*, 47–50

Grill, E., E. L. Winnacker, and M. H. Zenk, 1987, "Phytochelatins, a Class of Heavymetal-binding Peptides from Plants, are Functionally Analogous to Metallothioneins," *Proc. Natl. Acad. Sci. USA*, *84*, 439–443

Grill, E., S. Loeffler, E. L. Winnacker, and M. H. Zenk, 1989, "Phytochelatins, the Heavy-Metal-binding Peptides of Plants, are Synthesized from Glutathione by a Specific γ-glutamylcysteine Dipeptidyl Transpeptidase (Phytochelatin Synthase)," *Proc. Natl. Acad. Sci. USA*, *86*, 6838–6842

Hampp, R., H. Ziegler, and I. Ziegler, 1973, "Die Wirkung von Bleiionen auf die CO_2-Fixierung und die ATP-Bildung von Spinatchloroplasten," *Biochem. Physiol. Pflanzen*, *124*, 588–595

Harrison, S. I., N. W. Lepp, and Phipps, 1979, "Uptake of Copper by Excised Roots. II. Copper Desorption from the Free Space," *Z. Pflanzenphysiol.*, *94*, 27–34

Holden, M. J., T. J. Crimmins, and R. L. Chaney, 1995, "Cu^{2+} Reduction by Tomato Root Plasma Membrane Vesicles," *Plant Physiol.*, *108*, 1093–1098

Howden, R., C. R. Andersen, P. B. Goldsbrough, and C. S. Cobbett,. 1995, "A Cadmium-Sensitive, Glutathione-Deficient Mutant of *Arabidopsis Thaliana*," *Plant Physiol.*, *107*, 1067–1073

Ietswaart, J. H., W. A. J. Griffioen, and W. H. O. Ernst, 1992, "Seasonality of VAM Infection in Three Populations of *Agrostis capillaris* (Gramineae) on Soil With or Without Heavy Metal Enrichment," *Plant Soil*, *139*, 67–73

Juste, C., and M. Mench, 1992, "Long-Term Application of Sewage Sludge and its Effects on Metal Uptake by Crops," in *Biogeochemistry of Trace Metals* (D. C. Adriano, Ed.), Lewis Publishers, Boca Raton, USA, pp. 159–193

Kapur, A., and C. P. Malik, 1976, "Effect of Metabolic Inhibitions on Pollen Germination and Pollen Tube Growth of *Petunia alba*," *Plant Sci. (Lucknow)*, *8*, 26–27

Kelley, B. C., and O. H. Tuovinen, 1988, "Microbiological Oxidations of Minerals in Mine Tailings," in *Chemistry and Biology of Solid Waste* (W. Salomons, and U. Foerstner, Eds.), Springer, Berlin, Germany, pp. 33–53

Kersten, M., 1988, "Geochemistry of Priority Pollutants in Anoxic Sludges: Cadmium, Arsenic, Methyl Mercury, and Chlorinated Organics," in *Chemistry and Biology of Solid Waste* (W. Salomons, and U. Foerstner, Eds.), Springer, Berlin, Germany, pp. 170–213

Klapheck, S., W. Fliegner, and I. Zimmer, 1994, "Hydroxymethyl-Phytochelatins [(γ-glutamylcysteine)$_n$-serine] are Metal-induced Peptides of the Poaceae," *Plant Physiol.*, *104*, 1325–1332

Krupa, Z., and T. Baszynski, 1995, "Some Aspects of Heavy Metals Toxicity Towards Photosynthetic Apparatus – Direct and Indirect Effects on Light and Dark Reactions," *Acta Physiol. Plant.*, *17*, 177–190

Kuiters, A. T., 1993, "Dissolved Organic Matter in Forest Soils: Sources, Complexing Properties and Action on Herbaceous Plants," *Chem. Ecol.*, *8*, 171–184

Kuiters, A. T., and W. Mulders, 1993, "Water-Soluble Matter in Forest Soils. II. Interference with Plant Cation Uptake," *Plant Soil*, *152*, 225–235

Lamoureux, R. J., and W. R. Chaney, 1977, "Growth and Water Movement in Silver Maple Seedlings," *J. Environ. Quality*, *6*, 201–205

Lanaras, T., M. Moustakas, L. Symeonidis, S. Diamantoglou, and S. Karataglis, 1993, "Plant Metal Content, Growth Responses and Some Photosynthetic Measurements on Field-cultivated Wheat Growing on Ore Bodies Enriched in Cu," *Physiol. Plant.*, *88*, 307–314

Lepp, N. W., and N. M. Dickinson, 1994, "Fungicide-Derived Copper in Tropical Plantation Crops," in *Toxic Metals in Soil-Plant Systems* (S. M. Ross, Ed.), Wiley, Chichester, UK, pp. 367–393

Lepsova, A., and R. Kral, 1988, "Lead and Cadmium in Fruiting Bodies of Macrofungi in the Vicinity of a Lead Smelter," *Sci. Total Environ.*, *76*, 129–138

Levitt, J., 1980, *Response of Plants to Environmental Stress* [2nd Edn.] Academic Press, New York, USA

Lichtenthaler, H. K., and U. Rinderle, 1988, "The Role of Chlorophyll Fluorescence in the Detection of Stress Conditions in Plants," *CRC Crit. Rev. Anal. Chem.*, *19*, S29–S85

Liu, D., W. Jiang, C. Lu, F. Zhao, Y. Hao, and L. Guo, 1994, "Effects of Copper Sulfate on the Nucleolus of *Allium cepa* Root-Tip Cells," *Hereditas*, *120*, 87–90

Lobinski, R., C. Witte, F. C. Adams, P. L. Teissedre, J. C., Cabanis, and C. F. Boutron, 1994, "Organolead in Wine," *Nature*, *370*, 24

Magalhaes, M. J., E. M. Sequeira, and D. Lucas, 1985, "Copper and Zinc in Vine Yards of Central Portugal," *Water, Air, Soil Pollut.*, *26*, 1–17

Maksymiec, W., R. Russa, T. Urbanik-Sypniewska, and T. Baszynski, 1992, "Changes in Acyl Lipid and Fatty Acid Composition in Thylakoids of Copper Non-tolerant Spinach Exposed to Excess Copper," *J. Plant Physiol.*, *140*, 52–55

Marschner, H., 1995, *Mineral Nutrition of Higher Plants* [2nd Edn.] Academic Press, London, UK

Mathys, W., 1975, "Enzymes of Heavy Metal-Resistant and Non-Resistant Populations of *Silene cucubalus* and Their Interaction with Some Heavy Metals In Vitro and In Vivo," *Physiol. Plant.*, *33*, 161–165

Mathys, W., 1977, "The Role of Malate, Oxalate, and Mustard Oil Glucosides in the Evolution of Zinc-Resistance in Herbage Plants," *Physiol. Plant.*, *40*, 130–136

McBrien, D. C. H., and R. A. Hassall, 1965, "Loss of Cell Potassium by *Chlorella vulgaris* After Contact with Toxic Amounts of Copper Sulphate," *Physiol. Plant.*, *18*, 1053–1065

McGrath, S. P., P. C. Brookes, and K. E. Giller, 1988, "Effect of Potentially Toxic Elements in Soil Derived from Application of Sewage Sludge on Nitrogen Fixation by *Trifolium repens* L," *Soil Biol. Biochem.*, *20*, 415–424

McKenna, I. M., R. L. Chaney, S. H. Tao, R. M. Leach, and F. M. Williams, 1992, "Interactions of Plant Zinc and Plant Species on the Availability of Plant Cadmium to Japanese Quail Fed Lettuce and Spinach," *Environ. Res.*, *57*, 73–87

Meharg, A. A., 1993, "The Role of Plasmalemma in Metal Tolerance in Angiosperms," *Physiol. Plant.*, *88*, 191–198

Meharg, A. A., and M. R. Macnair, 1990, "An Altered Phosphate Uptake System in Arsenate-tolerant *Holcus lanatus* L," *New Phytol.*, *116*, 29–35

Meharg, A. A., and M. R. Macnair, 1991a, "Uptake, Accumulation and Translocation in Arsenate Tolerant and Non-tolerant *Holcus lanatus* L," *New Phytol.*, *117*, 225–231

Meharg, A. A., M. R. Macnair, 1991b, "The Mechanism of Arsenate Tolerance in *Deschampsia cespitosa* (L.) Beauv. and *Agrostis capillaris* L. Adaptation of the Arsenate Uptake System," *New Phytol.*, *119*, 291–297

Mench, M., J. L. Morel, A. Guckert, and B. Guillet, 1988, "Metal Binding with Root Exudates of Low Molecular Weight," *J. Soil Sci.*, *39*, 521–527

Neumann, D., O. Lichtenberger, D. Günther, K. Tschiersch, and L. Nover, 1994, "Heat-Shock Proteins Induce Heavy-Metal Tolerance in Higher Plants," *Planta, 194*, 360–367

Neumann, D., U. Zur Nieden, O. Lichtenberger, and I. Leopold, 1995, "How does *Armeria maritima* Tolerate High Heavy Metal Concentrations?" *J. Plant Physiol.*, *146*, 704–717

OECD, 1984, "Guideline for Testing Chemicals no. 208. Terrestrial Plants, Growth Test," Organization for Economic Cooperation and Development, Paris, France

Ouzounidou, G., 1994, "Copper-Induced Changes on Growth, Metal Content and Photosynthetic Function of *Alyssum montanum* L. Plants," *Environ. Exp. Bot.*, *34*, 165–172

Page, A. L., F. T. Bingham, and C. Nelson, 1972, "Cadmium Absorption and Growth of Various Plant Species as Influenced by Solution Cadmium Concentration," *J. Environ. Quality*, *1*, 288–291

Paul, R., and E. de Foresta, 1981, "Effects du cadmium sur la transpiration de plantes," *Bulletin de la Recherche Agronomique Gembloux*, *16*, 371–378

Poschenrieder, C., B. Gunse, and J. Barcelo, 1989, "Influence of Cadmium on Water Relations, Stomatal Resistance, and Abscisic Acid Content in Expanding Bean Leaves," *Plant Physiol.*, *90*, 1365–1371

Powell, M. J., M. S. Davies, and D. Francis, 1986, "Effects of Zinc on Cell, Nuclear and Nuclear Size, and on RNA and Protein Content in the Root Meristem of a Zinc-Tolerant and a Non-tolerant Cultivar of *Festuca rubra* L," *New Phytol.*, *104*, 671–679

Rieder, W., and U. Schwertmann, 1972, "Kupferanreicherung in hopfengenutzten Böden der Hallertau," *Landwirtsch. Forsch.*, *25*, 170–177

Roemheld, V., 1991, "The Role of Phytosiderophores in Acquisition of Iron and Other Micronutrients in Graminaceous Species: An Ecological Approach," *Plant Soil, 130,* 127–134

Ros, R., D. T. Cooke, R. S. Burden, and C. S. James, 1990, "Effects of the Herbicide MCPA, and the Heavy Metals, Cadmium and Nickel on the Lipid Composition, Mg_{2+}-ATPase Activity and Fluidity of Plasma Membranes from Rice, *Oryza sativa* (cv. Bahia) Shoots," *J. Exp. Bot., 41,* 457–462

Sandmann, G., and P. Boeger, 1980, "Copper-Mediated Lipid Peroxidation Processes in Photosynthetic Membranes," *Plant Physiol., 66,* 797–800

Schat, H., R. Vooijs, and E. Kuiper, 1996, "Identical Major Gene Loci for Heavy Metal Tolerances that Have Independently Evolved in Different Local Populations and Subspecies of *Silene vulgaris,*" *Evolution, 50,* 1888–1895

Schat, H., S. S. Sharma, and R. Vooijs, 1997, "Heavy Metal-Induced Accumulation of Free Proline in a Metal-Tolerant and a Non-Tolerant Ecotype of *Silene vulgaris,*" *Physiol. Plant.,* in press

Searcy, K. B., and D. C. Mulcahy, 1985a, "Pollen Tube Competition and Selection for Metal Tolerance in *Silene dioica* (Caryophyllaceae) and *Mimulus guttatus* (Scrophulariaceae)," *Am. J. Bot., 72,* 1695–1699

Searcy, K. B., and D. C. Mulcahy, 1985b, "Pollen Selection and the Gametic Expression of Metal Tolerance in *Silene dioica* (Caryophyllaceae) and *Mimulus guttatus* (Scrophulariaceae)," *Am. J. Bot., 72,* 700–706

Scheffer, F., and P. Schachtschabel, 1992, *Lehrbuch der Bodenkunde* [13th Edn.], F Enke Verlag, Stuttgart, Germany

Silverberg, B. A., 1975, "Ultrastructural Localization of Lead in Stigeoclonium tenue (Chlorophyceae, Ulotrichales) as Demonstrated by Cytochemical and X-ray Micro-analysis," *Phycologia, 14,* 265–274

Smirnoff, N., and G. R. Stewart, 1987, "Glutamine Synthetase and Ammonium Assimilation in Roots of Zinc-Tolerant and Non-tolerant Clones of *Deschampsia cespitosa* (L.) *Beauv.* and *Anthoxanthum odoratum* L," *New Phytol., 107,* 659–670

Smith, M. M., M. J. Hodson, H. Opik, and S. J. Wainwright, 1982, "Salt-Induced Ultrastructural Damage to Mitochondria in Root Tips of Salt-Sensitive Ecotypes of *Agrostis Stolonifera,*" *J. Exp. Bot., 33,* 886–895

Stobart, A. K., W. T. Griffiths, K. Ameen-Bukhari, and R. P. Sherwood, 1985, "The Effect of Cd^{2+} on the Biosynthesis of Chlorophyll in Leaves of Barley," *Physiol. Plant., 63,* 293–298

Strange, J., and M. R. Macnair, 1991, "Evidence for a Role for the Cell Membrane in Copper Tolerance in *Mimulus guttatus* Fischer ex DC," *New Phytol., 119,* 383–388

Strickland R.C., and R. L. Chaney, 1979, "Cadmium Influence on Respiratory Gas Exchange of *Pinus resinosa* Pollen," *Physiol. Plant., 47,* 129–133

Tinker, P. B., and A. Gildon, 1983, "Mycorrhizal Fungi and Ion Uptake," in *Metals and Micronutrients. Uptake and Utilization of Metals by Plants,* (A. D. Robb, and W. S. Pierpoint, Eds.), Academic Press, London, UK, pp. 21–32

Tukendorf, A., 1989, "Characteristics of Copper-Binding Proteins in Chloroplasts of Spinach Tolerant to Excess Copper," *J. Plant Physiol., 135,* 280–284

Tukendorf, A., and W. E. Rauser, 1990, "Changes in Glutathione and Phytochelatin in Roots of Maize Seedlings Exposed to Cadmium," *Plant Sci., 70,* 155–166

Turner, A. P., 1994, "The Response of Plants to Heavy Metals," in *Toxic Metals in Plant-Soil Systems* (S. M. Ross, Ed.), Wiley, Chichester, UK, pp. 153–187

Turner, A. P., and N. M. Dickinson, 1993, "Survival of *Acer pseudoplatanus* L. (Sycamore) Seedlings on Metalliferous Soils," *New Phytol., 123,* 509–521

US EPA, 1986, "Acid Digestion of Sediment, Sludge and Soils," in *Test Methods for Evaluation Solid Wastes,* EPA SW-846, US Government Printing Office, Washington, DC, USA

Van Assche, F., R. Ceulemans, and H. Clijsters, 1980, "Zinc Mediated Effects on Leaf CO_2 Diffusion Conductances and Net Photosynthesis in *Phaseolus vulgaris* L.," *Photosynthesis Res., 1,* 171–180

Van Steveninck, R. F. M., M. E. Van Steveninck, D. R. Fernando, W. J. Horst, and W. J. Marschner, 1990, "Deposition of Zinc Phytate in Globular Bodies in Roots of Deschampsia Caespitosa Ecotypes; a Detoxification Mechanism?" *J. Plant Physiol., 131,* 247–257

Vartapetian, B. B., and N. A. Zakhmilova, 1990, "Ultrastructure of Wheat Seedling Mitochondria Under Anoxia and Postanoxia," *Protoplasma, 156,* 39–44

Veltrup, W., 1978, "The Effect of Cu^{2+} and Mg^{2+} on the Activity of ATPases from the Roots of Barley," *Biochem. Physiol. Pflanzen, 173,* 17–22

Verkleij, J. A. C., J. E. Prast, 1989, "Cadmium Tolerance and Co-Tolerance in *Silene vulgaris* (Moench.) Garcke [= S. cucubalus (L.) Wib.]," *New Phytol., 111,* 637–645

Verkleij, J. A. C., and H. Schat, 1990, "Mechanisms of Metal Tolerance in Plants," in *Heavy Metal Tolerance in Plants: Evolutionary Aspects* (A. J. Shaw, Ed.), CRC Press, Boca Raton, USA, pp. 179–193

Voegeli-Lange, R., and G. J. Wagner, 1990, "Subcellular Localization of Cadmium and Cadmium-Binding Peptides in Tobacco Leaves. Implications of a Transport Function for Cadmium-Binding Peptides," *Plant Physiol., 92,* 1086–1093

Wainwright, S. J., and H. W. Woolhouse, 1975, "Physiological Mechanisms of Heavy Metal Tolerance," in *The Ecology of Resource Degradation and Renewal* (M. J. Chadwick. G. T. Goodman, Eds.), Blackwell Scientific Publications, Oxford, UK, pp. 231–259

Wainwright, S. J., and H. W. Woolhouse, 1977, "Some Physiological Aspects of Copper and Zinc Tolerance in *Agrostis tenuis* Sibth.: Cell Elongation and Membrane Damage," *J. Exp. Bot., 28,* 1029–1036

Walton, A. P., L. Ebdon, and G. E. Millward, 1988, "Methylation of Inorganic Lead by Tamar Estuary (UK) Sediments," *Appl. Organometallic Chem., 2,* 87–90

Weinstein, L. H., R. Kaur-Sawhney, M. Venkat Rajam, S. C. Wettlaufer, and A. W. Galston, 1986, "Cadmium-Induced Accumulation of Putrescine in Oat and Bean Leaves," *Plant Physiol., 82,* 641–645

Weissenhorn, I., 1994, "Les Mycorhizes à Arbuscules dans des Sols Pollus par des Métaux Lourds," *Dissertation,* Université de Nantes, France

Wu, L., 1989, "Colonization and Establishment of Plants in Contaminated Sites," in *Metal Tolerance in Plants: Evolutionary Aspects* (A. J. Shaw, Ed.), CRC Press, Boca Raton, USA, pp. 269–284

Wu, X., and I. Aasen, 1994, "Models for Predicting Soil Zinc Availability for Barley," *Plant Soil, 163,* 279–285

20

Assessment of Ecotoxicity at the Population Level using Demographic Parameters

*Nico M. van Straalen (**Amsterdam, The Netherlands**)*

and

*Jan E. Kammenga (**Wageningen, The Netherlands**)*

20.1 SUMMARY

Ecotoxicity of chemicals at the population level may be assessed on the basis of experiments in which test organisms are exposed for the whole length of their life-cycle, and observations are made for growth, development, reproduction and survival. The data obtained from these experiments may be analysed using demographic techniques that allow population parameters, such as the intrinsic rate of increase and the biomass turnover rate, to be estimated. Age-structure and stage-structure models, using either continuous time or discrete time, may be used for the analysis of ecotoxicological data. This chapter evaluates the application of demographic approaches in the ecotoxicological literature; a total of 29 studies are reviewed. Short-lived aquatic invertebrates are the

Ecotoxicology, Edited by Gerrit Schüürmann and Bernd Markert.
ISBN 0-471-17644-3   © 1998 John Wiley & Sons, Inc. and Spektrum Akademischer Verlag.

most popular test organisms. However, more recently, soil invertebrates have been increasingly used. The studies show that, in contrast to some suggestions in the literature, population parameters are not necessarily more sensitive than individual parameters and the most sensitive individual endpoint may not be the most relevant in terms of population level effects. The use of population parameters lies in their ecological relevance, and the possibility of summarizing a variety of effects in the life-cycle using a single endpoint. An "elasticity index of toxicant exposure" is proposed to measure the change of a life-cycle variable, and its effects on fitness, with a change of the exposure concentration. Although theory shows that the trade-off structure among life-cycle variables is important, empirical studies have not been able to reveal these trade-offs clearly. The application of demographic methods in ecotoxicology may contribute to improving the ecological relevance of ecotoxicological risk assessment.

20.2 INTRODUCTION

A complete impression of the ecotoxicity of a chemical to a single species is obtained from life-cycle experiments. In these experiments individuals or groups of individuals are exposed to a series of constant concentrations, and observations are made on various life-history events, such as (depending on the species) hatching, germination, growth, moulting, flowering, seed formation, egg laying, development, survival. Each of these observations is a potential endpoint in which the toxicity of the chemical may be expressed. Often one will find that the endpoints differ considerably in their response to toxicants. This leaves the risk assessor with the question: which criterion should be chosen for the protection of the species?

The dilemma of sensitivity is illustrated in a theoretical way in Figure 20.1a. Three concentration–response relationships are shown, for different life-history traits, each with its own EC_{50} (50% effect concentration) and NEL (no effect level). Usually survival is a less sensitive endpoint than body-growth or reproduction. However, this depends on the mode of action of the chemical and the physiological response of the species. To protect the species, should one take the most sensitive criterion (in this case NEL for body-growth), or is it better to focus on one of the criteria, e.g. reproduction? A similar problem may arise when two species are compared (Figure 20.1b). If the two species differ in their basic reproductive capacity, but have the same EC_{50}, will the toxicant affect the two species in a similar manner?

Population ecology, specifically life-history theory, offers a straightforward answer to the questions posed above: one should take the criterion that has the greatest influence on the intrinsic rate of population increase, or even better: one should integrate all life-history traits into this single parameter. Various ecotoxicological studies have followed this approach, often referred to as "life-table evaluation". This chapter will review these studies, discuss some related approaches, and evaluate their contribution to the improvement of ecotoxicological risk assessment.

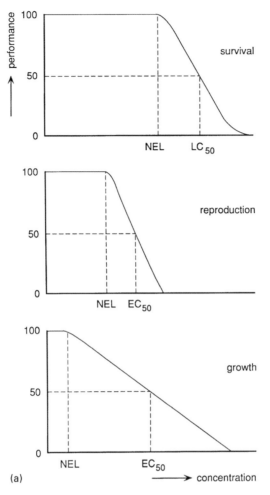

Figure 20.1 Theoretical concentration–response relationships for one species with three endpoints differing in sensitivity (a), and two species (A and B) with the same toxicological sensitivity but a difference in control reproduction (b). NEL = no-effect level, LC_{50} = median lethal concentration, EC_{50} = 50% effect concentration.

The question may be raised as to whether the study of population parameters might provide endpoints that are more sensitive than individual parameters (Suter and Donker 1993). This argument was initially formulated by Halbach et al. (1983), who called it the "looking glass effect": small, hardly perceptible, effects on individuals are magnified by the great number of individuals constituting a population, and so become more easily visible. Recent literature, however, seems to agree on the fact that population parameters cannot be more sensitive than individual parameters, and, in fact, are sometimes less sensitive (Day and Kaushik 1987; Meyer et al. 1987; Barbour et al. 1989; Crommentuijn et al. 1993).

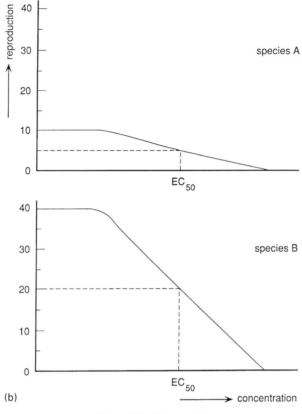

Figure 20.1 (*Continued*)

Hence, the greatest value of the use of population parameters does not lie in their assumed sensitivity, but in their ecological relevance, and the possibility of summarizing the variety of possible effects in the course of a life–cycle by a single measure.

20.3 DEMOGRAPHIC TECHNIQUES

The classic demographic methodology starts with a summary of the life-history using two functions, $l(x)$ and $m(x)$, that define survival (fraction still alive, relative to birth), and fertility (no. offspring per time unit per parent) as a function of age, x. From these two functions, the Malthusian parameter, r, may be calculated by means of Euler's equation:

$$\int_{x=0}^{\infty} l(x)\, m(x)\, e^{-rx}\, dx = 1 \qquad (20.1)$$

This equation relates to the basic demographic theorem (see, e.g., Keyfitz 1968) that may be formulated as follows:

Any age-structured population with time-invariant $l(x)$ and $m(x)$ schedules will approach a stable age distribution, independent of the initial conditions; the population will then increase at an exponential rate r (per time unit).

From an ecological point of view, the relevance of equation 20.1 is contained in Fisher's fundamental theorem of natural selection, which states that, in a population composed of genotypes with different life-histories, the types with the highest Malthusian parameter (r) tend to displace others from the population, and that the rate of change of the population growth rate is approximately equal to the additive genetic variance of r (Charlesworth 1980). In other words, r is a measure of fitness (Kozlowski 1993). This theorem forms the basis for the study of life-history evolution, in which an analysis is made of the changes in the life-history that maximize fitness, given certain trade-off patterns between life-history parameters (Stearns 1992). In an ecotoxicological context, the relevance of equation 20.1 may be summarized in the following statement:

The population consequences of the effects of a toxicant on survival and fertility must be judged by the extent to which a reduction in survival or fertility affects the Malthusian parameter of the life-cycle.

Despite its obvious ecological relevance, the use of r has a drawback in that it only measures the numerical aspects of the population (size in numbers, mortality and fertility in numbers per time), not its qualitative aspects. At the individual level body-weight is another common endpoint, and its analogue at the population level is total biomass. Van Straalen (1985a) and Aldenberg (1986) discussed the concepts of biomass and productivity, as defined for populations structured by body-size instead of age. Following from this theory, Van Straalen and De Goede (1987) proposed the intrinsic biomass turnover rate, P/B, as a new population performance index that includes possible effects of a toxicant on body-growth. The intrinsic biomass turnover measures the productivity of the population, that is the rate at which biomass is produced, relative to the biomass present. Production of biomass may be a more relevant measure for the performance of a natural population than production of numbers, especially in relation to its functioning in trophic chains. The parameter P/B may be calculated from the age-specific survival rate, $l(x)$, the age-specific fertility, $m(x)$, and the age-specific body weight, $w(x)$, by means of the following equation:

$$\frac{P}{B} = \frac{\int_0^\infty -\frac{dl}{dx} w(x) e^{-rx} dx}{\int_0^\infty l(x) w(x) e^{-rx} dx} + r \qquad (20.2)$$

The derivation of this equation, given in Van Straalen and De Goede (1987), is based on what may be called the weight-extended demographic theorem:

Any structured population with time-invariant survival, fertility and body-growth will approach a stable weight distribution, independent of the initial conditions. The total biomass of the population (B), and its total rate of production (P) will then both increase exponentially at a rate r, while their quotient (P/B) will approach a constant.

The calculation of r from real data arranged in a life-table is explained in Stearns (1992). It requires the use of a (simple) computer programme, because of the implicit expression of r in equation 20.1. In a similar manner, P/B may be calculated from a life-table supplemented with body-growth data (Van Straalen 1985a; Van Straalen and De Goede 1987).

The reference to "time invariance" in the above statements may be considered a serious limitation of the application of the theory, because the vital statistics of a population often show considerable variation, in response to changing environmental conditions. The value of the statements should therefore not be judged by the extent to which r and P/B are able to provide estimates for the real rates of increase and productivity in a population under field conditions. Rather, they must be considered instantaneous values that summarize the fitness of a species in a particular environment and indicate the potential for population growth in the future, *if* the present conditions persist. So it is more appropriate to interpret r and P/B as instantaneous performance indices, rather than as real predictions of future behavior.

When temperature is the only variable affecting the changes in vital rates, the time-invariant approach may still be used after a suitable transformation of the time-scale. Van Straalen (1983) argued that, under certain conditions, any process whose rate changes with time due to temperature changes, can be made to proceed at a constant rate by warping the time-scale. The new time-scale, referred to as "physiological time", can be defined in terms of the temperature responses of the specific organism and the change in temperature with time. This approach has proven useful when modelling population development of small ectotherms, for which temperature is the dominating factor determining life-history events (Van Straalen 1985b; Stamou et al. 1993; Diekkrüger and Röske 1995; Axelsen 1997).

The mathematical theory underlying the derivation of equations 20.1 and 20.2 considers time and age as continuous variables. The techniques used are those of calculus algebra. Alternatively, age may be considered a discrete variable. All events are then assumed to take place at certain moments in time. At regular intervals (of length T) individuals are assumed to die, produce offspring, or jump from one class to the next, while nothing happens in between. The mathematical techniques used in this approach are those of matrix algebra. Development of the population may be described by a matrix equation in which the age distribution at a certain moment (considered a column vector) is multiplied by a matrix L, to obtain the age distribution one time step later. The population projection matrix L is usually referred to as Leslie's matrix, and has a typical configuration, as follows:

$$
L = \begin{bmatrix}
0 & f_1 & f_2 & . & . & f_n \\
p_0 & 0 & 0 & . & . & 0 \\
0 & p_1 & 0 & . & . & 0 \\
0 & 0 & p_2 & . & . & 0 \\
. & . & . & . & . & . \\
0 & 0 & 0 & . & p_{n-1} & 0
\end{bmatrix}
\qquad (20.3)
$$

All the elements of the projection matrix are zero, except for the off–diagonal elements p_i, which represent the probabilities of survival from age-class i to age class $i + 1$, and some of the elements in the first row, f_i, which represent the number of offspring produced at age i and surviving to the next time step, when they enter the zeroth age class. Demographic theory shows that the projection matrix has one real dominant eigenvalue, usually denoted by λ. This parameter represents the (dimensionless) factor by which the population is multiplied per time interval T, when it is in a stable age distribution and growing exponentially. The parameter λ is also referred to as the "finite rate of increase" and it is related to r according to $r\,T = \ln \lambda$, where T is the time step considered (usually unit of time).

The matrix description of population development has the advantage that it is more flexible towards incorporating time-dependent processes. Projection matrix elements may be given different values for different time steps (e.g., a stop on reproduction in the winter period). An annual projection matrix may be composed of twelve monthly matrices, each describing the life-history traits in that month. This approach is therefore more suitable for predicting population changes under field conditions. The matrix approach is easy to formulate in computer programmes and is often used in simulation models.

The Leslie matrix approach may be extended to include stage-structured populations rather than populations structured by age only. Stages may be real biological stages, such as the instars of arthropods, or they may be arbitrary classes of body-weight. Lefkovitch (1965), Vandermeer (1975), Longstaff (1977) and Van Groenendael et al. (1988) considered projection matrices for stage-structured populations. This theory is briefly summarized here.

Compared to the classical Leslie approach, the matrix description of stage-structure changes poses a problem because the stages are of unequal duration and the individuals in the population do not all jump from one stage to another after each time interval. After each time step T there are three possibilities for an individual in stage i: to die, to stay in stage i or to jump to stage $i + 1$. Transition rates are denoted by $q_{i,i}$ (probability to stay in stage i) and $q_{i,i+1}$ (probability to jump to stage $i + 1$). The projection matrix, M, then has the following configuration:

$$
M = \begin{bmatrix}
q_{0,0} & f_1 & f_2 & \cdot & \cdot & f_n \\
q_{0,1} & q_{1,1} & 0 & \cdot & \cdot & 0 \\
0 & q_{1,2} & q_{2,2} & \cdot & \cdot & 0 \\
0 & 0 & q_{2,3} & \cdot & \cdot & 0 \\
\cdot & \cdot & \cdot & \cdot & \cdot & \cdot \\
0 & 0 & 0 & \cdot & p_{n-1,n} & q_{n,n}
\end{bmatrix}
\tag{20.4}
$$

In contrast to the Leslie matrix, the survival probabilities per time interval T are now not equal to the off-diagonal elements $q_{i,i+1}$ but to $q_{i,i} + q_{i,i+1}$. Schobben and Van Straalen (1987) showed that the matrix elements of the stage-structure projection matrix may be given the following biological interpretation:

$$
q_{i,i+1} = \frac{p_i T}{d_i} \quad \text{and} \quad q_{i,i} = \frac{p_i (d_i - T)}{d_i}
\tag{20.5}
$$

where p_i is the survival rate per time interval T, as above, and d_i is the duration of stage i. The Leslie matrix L may be considered as a special case of the stage-structure projection matrix M. It can be seen from equations 20.5 that if all stage durations d_i are the same and equal to the time interval T, the probability of staying in a stage, $q_{i,i}$, equals zero, $q_{i,i+1}$ equals p_i, and the two matrices are identical. So, the description of stage–structured populations does not necessarily require the completely filled projection matrices proposed by Lefkovitch (1965). If it is assumed that individuals do not skip stages (that is, do not jump from stage i to stage $i + 2$ during time interval T), only one extra diagonal of elements is necessary to extend the Leslie matrix to stage–structured populations. The prohibition to skip stages can easily be met by making the time interval T shorter than the shortest stage duration.

In addition to the parameters r, P/B, and λ, other population characteristics have been used in ecotoxicological studies for describing certain aspects of population change. Gentile et al. (1982) proposed the use of reproductive value, $v(x)$, as a measure of the expected contribution of a female to future population growth. Various authors have used the carrying capacity, K, of the population, usually estimated after fitting the logistic growth model to experimental data. Halbach et al. (1983) introduced two other parameters, the frequency of oscillation (f), and the "pregnancy", p, of population fluctuations. The latter index is a measure of the persistence of oscillations when these are analysed using autocorrelation functions. The indices K, f, and p do not have a strong theoretical underpinning, they are mere descriptors of population fluctuations.

20.4 POPULATION PERFORMANCE INDICES IN ECOTOXICOLOGICAL STUDIES

The demographic techniques reviewed above and elsewhere (Caswell 1996; Kareiva et al. 1996) demonstrate that there is a good theoretical framework for the application of the life-history approach in ecotoxicology. To exemplify the approach, some examples from the literature will be reviewed in this section.

Demographic analysis of ecotoxicity at the population level has been applied to a variety of different species. Table 20.1 lists studies reported in the literature, with a specification of the species and chemicals investigated. Among the species, freshwater invertebrates with short life-cycles are represented well, presumably because these species can easily be cultured and observations on life-history traits can be made within a relatively short experimental period. The most widely used population performance index is r, the intrinsic rate of population increase, derived from Equation 20.1, or estimated directly from populations growing exponentially in laboratory cultures. More recently, terrestrial ecotoxicology studies have also considered the demographic approach.

By way of example, Figure 20.2 demonstrates the use of the intrinsic rate of population increase for populations of rotifers, cladocerans and algae, exposed to various concentrations of 3,4-dichloroaniline (Seitz and Ratte 1991; Ratte et al. 1992). The response curves clearly show the differential effects of the toxicant on the species investigated. The use of the intrinsic rate of population increase in this case made it possible to express toxicity in terms of an endpoint that is relevant to all five species. Although the toxicant may affect different life-history traits in different species, the effects can be compared on a common basis by using r.

Another lesson to be learnt from the studies listed in Table 20.1 is that the population effects of a toxicant, particularly when they reduce juvenile survival, greatly depend on food availability. This was shown in a study by Kooijman (1985). The toxicity of vanadate (affecting survival) to the rotifer *Brachionus rubens*, expressed in terms of r, was hardly visible at high food concentrations, but became clearly apparent at low food availability. Toxicity of 3,4-dichloroaniline (affecting reproduction) was, however, independent of food supply. These observations are in line with predictions from a dynamic energy budget model (Kooijman and Metz 1984; Kooijman 1993).

An example of the matrix approach in ecotoxicology can be found in Schobben and Van Straalen (1987). In this study, a model was constructed for the population dynamics of the oribatid mite *Platynothrus peltifer*, which develops in six discrete stages: egg, larva, three nymphal stages, and the adult stage. Based on life-history information obtained from field studies, survival probabilities, stage durations, and reproduction activities were specified for each stage, and were assumed to vary per month. The annual population projection matrix was obtained by subsequently multiplying the monthly projection matrices. The eigenvalue of the annual projection matrix (λ) represents the factor by which the

Table 20.1 A historically ordered listing of studies in which demographic techniques were applied to analyse the population effects of toxicants, specifying species, chemicals, and population endpoints considered

Reference	Species	Group	System	Chemical	Population endpoint[1]
Winner et al. (1977)	*Daphnia magna*	Cladocera	freshwater	copper	r
Marshall (1978a, b)	*Daphnia galeata*	Cladocera	freshwater	cadmium	r, K
Daniels and Allan (1981)	*Eurytemora affinis*	Copepoda	marine	dieldrin	r
	Daphnia pulex	Cladocera	freshwater	dieldrin	r
Allan and Daniels (1982)	*Eurytemora affinis*	Copepoda	marine	kepone	r
Gentile et al. (1982)	*Mysidopsis bahia*	Mysidacea	marine	mercury, nickel	$r, v(x)$
Halbach et al. (1983, 1984)	*Brachionus rubens*	Rotifera	freshwater	pentachlorophenol	r, K, f, p
Kooijman (1985)	*Brachionus rubens*	Rotifera	freshwater	3,4-dichloroaniline, vanadate	r
Bengtsson et al. (1985)	*Onychiurus armatus*	Collembola	soil	copper, lead	r
Van Leeuwen et al. (1985)	*Daphnia magna*	Cladocera	freshwater	cadmium	r, K
Van Leeuwen et al. (1986)	*Daphnia magna*	Cladocera	freshwater	bromide	r, K
Van Leeuwen et al. (1987)	*Daphnia magna*	Cladocera	freshwater	cadmium, maneb, dichromate, vanadate, 2,4-dichloroaniline, pentachlorophenol, TPBS, bromide, pentachlorobenzene	r, K
Van Straalen and De Goede (1987)	*Orchesella cincta*	Collembola	soil	cadmium	P/B
Day and Kaushik (1987)	*Daphnia galeata*	Cladocera	freshwater	fenvalerate	r
Meyer et al. (1987)	*Daphnia pulex*	Cladocera	freshwater	cadmium, copper	r
Schobben and Van Straalen (1987)	*Platynothrus peltifer*	Oribatida	soil	cadmium	λ
Barbour et al. (1989)	*Ceriodaphnia*	Cladocera	freshwater	complex effluents	r
Van Straalen et al. (1989)	*Orchesella cincta, Platynothrus peltifer*	Collembola, Oribatida	soil	cadmium	r

Table 20.1 (*Continued*)

Reference	Species	Group	System	Chemical	Population endpoint [1]
Van der Hoeven (1990)	*Daphnia magna*	Cladocera	freshwater	3,4-dichloroaniline, metavanadate	r
Wong and Wong (1990)	*Moina macrocopa*	Cladocera	freshwater	cadmium	r
Seitz and Ratte (1991)	*Daphnia, Chydorus, Ceriodaphnia, Keratella*	Cladocera Rotifera	freshwater	3,4-dichloroaniline	r
Crommentuijn et al. (1993)	*Folsomia candida*	Collembola	soil	cadmium	r
Enserink et al. (1993)	*Daphnia magna*	Cladocera	freshwater	cadmium, chromium	r
Bechman (1994)	*Tisbe furcata*	Copepoda	marine	copper	r
Klüttgen and Ratte (1994)	*Daphnia magna*	Cladocera	freshwater	cadmium	r
Kammenga et al. (1996)	*Plectus acuminatus*	Nematoda	soil	cadmium	r
Klok and de Roos (1996), Klok et al. (1997)	*Lumbricus rubellus*	Lumbricidae	soil	copper	r
Hendriks and Enserink (1996)	*Phalacrocorax carbo*	Aves	freshwater	PCB-153	r
Crommentuÿn et al. (1997)	*Folsomia candida*	Collembola	soil	cadmium, chlorpyrifos, triphenyltin	r
Laskowski (1997)	*Helix aspersa, Lithobius mutabilis*	Gastropoda Chilopoda	terrestrial	cadmium, copper, zinc, lead	λ

[1] r = intrinsic rate of increase, λ = population multiplication factor ($= e^r$), P/B = intrinsic rate of biomass turnover, $v(x)$ = reproductive value, K = carrying capacity, f = oscillation frequency, p = "pregnancy" of population fluctuations with time. See text for explanation.

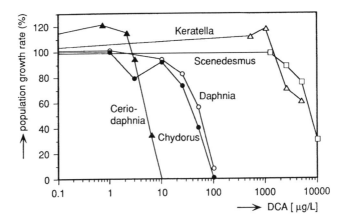

Figure 20.2 Concentration–response relationships for the effect of 3,4-dichloroaniline (DCA) on the population growth rate (r) of five species of aquatic organisms. For each species, r is expressed as a percentage of the value in the blank treatment. After Seitz and Ratte (1991).

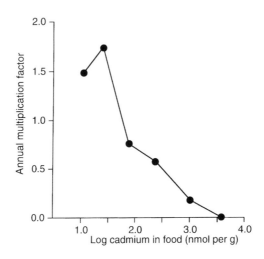

Figure 20.3 Calculated population increase, expressed as the annual replacement factor, λ, of the oribatid mite *Platynothrus peltifer*, exposed to cadmium-polluted food. The factor λ was estimated as the eigenvalue of the annual projection matrix (obtained after subsequently multiplying 12 monthly projection matrices), in which the life-cycle in the field was combined with laboratory observations on cadmium toxicity. After Schobben and Van Straalen (1987).

population is multiplied yearly. To maintain a stable population over the years, λ should equal unity.

Figure 20.3 illustrates some of the results obtained when the matrix model for *P. peltifer* was combined with observations for the effects of cadmium on reproduction, stage duration and survival. It may be seen that cadmium has a rather drastic effect on this species, λ is reduced below unity already by the third exposure concentration. At this and all higher concentrations the population will become extinct ($\lambda < 1$ implies $r < 0$). The use of the matrix model in this case provided a way to summarize a variety of complicated ecological phenomena (development through stages of unequal duration, time-varying fertility, etc.) as a single measure for the performance of the population. It also illustrated that the effect of cadmium on λ was more easily visible, compared to the effects on individual reproduction (cf. Van Straalen et al. 1989).

20.5 SENSITIVITY ANALYSIS BASED ON LIFE-HISTORY RESPONSES TO TOXICANTS

The concept of sensitivity is not an easy one. Meyer et al. (1987) discussed various aspects relating to the question of how to decide whether one variable responds more sensitively to a toxicant than another variable. From a toxicological point of view, sensitivity may be discussed in terms of the exposure–response graph. An endpoint is considered to be more sensitive if its toxicological effect criterion (e.g., EC_{50}) is positioned more to the left on the exposure axis. In practice, however, sensitivity is also determined by the possibility of demonstrating an effect emerging above the noise in the data. It would be possible, in principle, to have a variable that is very sensitive in a toxicological sense, but that has such a high degree of statistical uncertainty that it requires relatively high exposure levels to demonstrate a response differing significantly from the control. This is particularly relevant in the case of secondary parameters, such as the intrinsic rate of population increase, which are usually estimated from different sources of information, each contributing to increase the error. The statistical aspects of estimating confidence intervals for r, using numerical techniques, were discussed by Meyer et al. (1986). Although we recognize the practical importance of the statistical concept of sensitivity, it will be ignored for the purpose of this paper and sensitivity will be discussed in the classical toxicological sense.

When the life-cycle of an organism is specified in demographic terms, and equation 20.1, or its discrete time analog, is used to estimate r or λ, it becomes possible to analyse how the value of r (or λ) changes with changes in life-history traits. In an ecological context, this analysis simulates the effects of gene substitutions that affect a certain life-history trait. The response of r tells us how strong the selection pressure is on that particular trait. This type of sensitivity analysis dates back to Cole (1954), and has been elaborated intensively in the context of life-history theory (Charlesworth 1980; Stearns 1992; Caswell 1996). Caswell (1978) approached the problem by considering the relative change in λ brought about by a small relative change in one of the matrix elements of the population projection matrix, if all the other matrix elements were held constant. In mathematical terms, this means that one can consider the partial derivative of ln λ with respect to the natural logarithm of the matrix element as a measure of the "elasticity" of that matrix element (see also Van Groenendael et al. 1988). Elasticity of the a_{ij} element is denoted as e_{ij} and is defined as follows:

$$e_{ij} = \frac{\partial \ln \lambda}{\partial \ln a_{ij}} \tag{20.6}$$

This type of analysis has led to the conclusion that fitness is sensitive to different life-history traits under different rates of population growth. In a rapidly expanding population, there is a strong selection pressure on age at maturity, while in a stable population, survival and fertility rates are more important variables (Stearns 1992).

The sensitivity analysis from life-history theory was applied in an ecotoxicological context by Kammenga (1995) and Kammenga et al. (1996). Figure 20.4 shows a theoretical illustration of the argument. In the right part of the figure, a normal concentration–response relationship is seen, for two different species, 1 and 2. The effects of the toxicant are expressed in terms of some life-history trait (e.g., reproduction, survival, development). At a certain concentration of the toxicant, species 2 is affected to a greater extent than species 1. In the left part of the curve, graphs are shown that relate the intrinsic rate of population increase, r, to the life-history trait considered. The theoretical example is chosen such that the r value of species 1 reacts much more strongly to a decrease in the life-history trait than the r value of species 2. The result is that r of species 1 is affected to a greater extent than r of species 2, even though the two species are exposed to the same exposure concentration and species 2 is more sensitive with respect to the life-history trait considered.

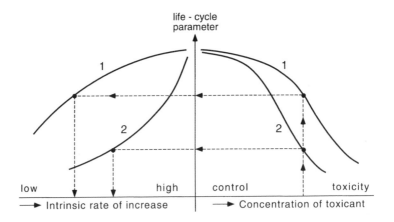

Figure 20.4 Theoretical response curves for the effect of a toxicant on a life-history parameter (right) and the influence of this parameter on the intrinsic rate of population increase (left). Curves are shown for two species, 1 and 2, that are assumed to differ in sensitivity.

A sensitivity analysis of the type outlined above was applied by Kammenga et al. (1996) in experiments using the soil nematode *Plectus acuminatus* exposed to cadmium. Their analysis challenged the generally held opinion that population level effects of toxicants are determined by the effects on the most sensitive life-history trait. This is, for example, the rationale behind the development of the early life-stage test for fish (McKim 1977). However, Kammenga et al. (1995) showed that in the nematode experiments the effect of cadmium on r was not determined by the effect on the most sensitive life-history component, the reproductive period. In fact, a reduction of the reproductive period by 45% did not reduce fitness to any extent, while a prolongation of the juvenile period, the least sensitive trait for cadmium toxicity, had a clear effect on r.

Similar analyses of sensitivity to toxicants have been applied to bird populations, using demographic models (Young 1968; Emlen and Pikitch 1989; Meyer and Boyce 1994). These studies have emphasized the importance of different life-history traits to population growth rates, depending on the life-time of the bird species. In addition, Meyer and Boyce (1994) showed that to evaluate the effect of pesticides on avian populations, it is essential to estimate the variability in survival and fertility.

The fitness effects of toxicants may be analysed by means of elasticity indices that measure not only the change of fitness with a small change of a life-history variable, as in equation 20.6, but also the response of fitness to a change in the exposure concentration (Kammenga 1995). In analogy to equation 20.6, one may define a new elasticity index, "toxicant exposure elasticity", as the partial derivative of ln λ with respect to the exposure concentration. Because concentrations act on a multiplicative scale rather than on a linear scale it is logical to consider a change in the logarithm of the concentration (ln c), and define the toxicant exposure elasticity index, k, as:

$$k = \frac{\partial \ln \lambda}{\partial \ln a_{ij}} \frac{\partial \ln a_{ij}}{\partial \ln c} \tag{20.7}$$

This index would be a direct measure of the change in fitness that is experienced by a species when the concentration of a toxicant in the environment increases by a small amount, relative to the concentration present ($\partial \ln c = (1/c)\partial c$).

Another approach to sensitivity analysis was developed by Crommentuijn et al. (1995). Following Daniels and Allan (1981), Crommentuijn et al. (1995) considered the concentration–response relationships of survival and sublethal characteristics, such as reproduction, in a single graph (Figure 20.5). The ratio between the LC_{50} and the threshold concentration for reproduction was defined as the "sublethal sensitivity index", abbreviated SSI. A large value for this index indicates that reproduction is inhibited at a level far below the LC_{50} (a type III response in the terminology of Daniels and Allan 1981), while a low value indicates that the organism maintains reproduction until it dies (type I response according to Daniels and Allan 1981). An inventory of SSI-values for several

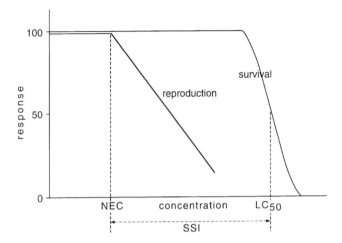

Figure 20.5 Showing hypothetical concentration–response relationships for survival and reproduction of the same species, from which the sublethal sensitivity index, SSI, according to Crommentuijn et al. (1995), may be derived as the quotient of LC_{50} and the no-effect concentration (NEC) for reproduction.

invertebrates exposed to cadmium showed that it varied between 280 and 3.1 (Crommentuijn et al. 1995). SSI may be considered a measure of an organism's priorities when, under toxicant stress, it is forced to allocate energy either to reproduction or survival. It can be an interesting tool for the analysis of species-specific responses to toxicants, as related to the life-history strategy. Previous analyses have taken the r and K selection theory as a basis for classifying life-history related responses to toxicants, but with limited success (Neuhold 1987).

20.6 TRADE-OFF RELATIONSHIPS AMONG DEMOGRAPHIC CHARACTERS

In the previous section, the question of sensitivity was approached on a character-by-character basis, that is, the influence of one life-history trait on r or λ was considered under the condition that all other traits remained the same. This is a rather unrealistic situation, because it is well known that a change in one variable will often evoke a correlated response in another variable. When the secondary response can be considered as (partly) compensatory to the first, the relationship between the two variables is called a trade-off-relationship. For example, a reduction in fertility may increase the probability of adult survival. In these cases, the question of sensitivity cannot be solved for one isolated character, but the life-history must be seen as a whole.

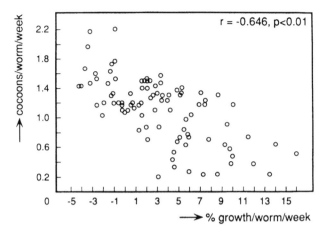

Figure 20.6 Correlation between cocoon production and body growth in experiments using the earthworm *Eisenia andrei*, cultured for 3 weeks in OECD artificial soil substrate. The data are from 18 separate experiments (all without toxicants). After Van Gestel et al. (1992).

As an example of a trade-off mechanism, Figure 20.6 gives the results of a series of experiments using the earthworm *Eisenia andrei* in an artificial soil substrate (Van Gestel et al. 1992). The production of cocoons in the controls of ecotoxicity experiments was plotted against the increase in body mass in the same group of animals. Figure 20.6 clearly shows that there is a negative correlation between these two variables: in experiments where the worms produced many cocoons, their body growth was slow, sometimes even negative. Conversely, toxicants that inhibit reproduction may increase the weight gain of these animals. A similar trade-off mechanism was noted by Crommentuijn et al. (1997) in experiments where the collembolan *Folsomia candida* was exposed to cadmium and triphenyltin. Both toxicants reduced body growth significantly, with a correlated reduction of egg output, but adult survival time increased with a decrease in growth.

The trade-off structure among life-cycle variables was the basis for a theory formulated by Sibly and Calow (1989). This theory attempts to link the physiology of an organism with its life-history, and consequently with its population growth. Calow and Sibly (1990) visualized the theory by defining the state of an organism in a "physiological state space" diagram. Production and metabolism were considered the most important variables for characterizing the physiological state (Figure 20.7). In the diagram, isoclines were drawn to indicate all combinations of production and metabolism that have equal values for the intrinsic rate of increase, *r*. Three different lay-outs of the *r* isoclines were proposed, corresponding to a stepwise dependence of production on metabolism (left), a linear decrease (middle) and a curvilinear decrease (right). When a toxicant moves the physiological state of an organism in a certain direction, the consequences for the population are determined by the trajectory it takes in the

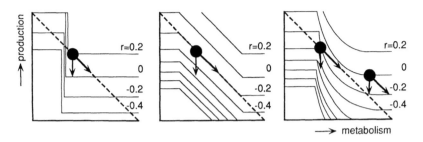

Figure 20.7 Physiological state space diagrams, showing the theoretical relationship between metabolism and production, according to Calow and Sibly (1990). The isoclines indicate all combinations of metabolism and production that have equal values for the intrinsic rate of population increase, r. The arrows indicate possible trajectories when the organism is exposed to a toxicant.

physiological state space. When the trajectory coincides with the local direction of the r isocline, there is no net effect on fitness (Figure 20.7).

The theory of Sibly and Calow (1989) leads to the conclusion that the trade-off structure among life-history traits and physiological variables is of utmost importance for predicting effects of toxicants. Unfortunately, these trade-offs are not easily measured in practice, and so a test of the theory is difficult to make at the moment. There are many examples in the literature in which an expected negative correlation between life-history traits could not be demonstrated empirically (e.g., Van Dijk 1979; Högstedt 1981; Meyer 1986). Ernsting and Isaaks (1991) showed that the costs of reproduction in the carabid beetle *Notiophilus biguttatus* depended on the way in which reproduction was manipulated (by temperature variation or by food availability). The analysis is further complicated by the fact that there may be differences between trade-offs at the physiological level and the genetic level. In the context of life-history evolution, it is the genetic correlation between two characters that matters. However, the existence of a negative genetic correlation between two traits does not preclude the existence of a positive phenotypic correlation between the same traits (Van Noordwijk and De Jong 1986). For the direct population effects of toxicants, the phenotypic trade-offs seem to be most relevant. However, for the selective pressure exerted by a toxicant, genetic correlations between life-history variables and toxicant tolerance are decisive (Posthuma et al. 1993).

20.7 CONCLUSIONS

The present chapter has emphasized the demographic approach towards ecotoxicity assessment at the population level. Other population approaches in ecotoxicology have not been addressed. Among these are the use of physiologically-based population models (Hallam et al. 1990; Kooijman 1993; Klok and de Roos 1996), the application of approaches developed for exploited populations (Jensen 1984) and for populations of protected wildlife species (Grant et al. 1984; Emlen 1989), population analysis in multispecies test systems (Woltering 1985; Ratte et al. 1992), and the incorporation of a spatial component into population level effects (Jepson and Thacker 1990; Sherratt and Jepson 1993).

The material reviewed above has, nevertheless, illustrated the considerable progress recently made by the demographic approach. In several examples the study of population responses has provided a sharper picture of the effects of chemicals, although the looking glass effect mentioned in the introduction does not seem to have general validity. An important observation is that demographic analysis may help to identify the ecological relevance of effects observed in individuals. The fact that the most sensitive variable is not necessarily the most relevant one in terms of r indicates the presence of plasticity and compensation mechanisms. These may be considered as the population analogue of the "functional redundancy" concept often discussed in an ecosystem context (see Levine 1989). The application of demographic analysis in an ecotoxicological context is a new area of research, and it opens up an interesting perspective for improving the ecological relevance of ecotoxicology.

Acknowledgements. The authors acknowledge comments made by Dr. J. S. Meyer (Wyoming University), Dr. C. Klok (IBN, Arnhem) and Dr. C. A. M. van Gestel (VU, Amsterdam) regarding an earlier version of this paper. Mrs. Nan Kasanpawiro and Mrs. Thea Laan helped with word processing and Mr. L. Sanna prepared the figures.

20.8 REFERENCES

Aldenberg, T., 1986, "Biological Productivity in Relation to Cohort and Continuous Structured Population Models," in *The Dynamics of Physiologically Structured Populations* (J. A. J. Metz, and O. Diekmann, Eds.), Lecture Notes in Biomathematics, Vol. 68, Springer, Berlin, Germany, pp. 409–428

Allan, J. D., and R. E. Daniels, 1982, "Life Table Evaluation of Chronic Exposure of *Eurytemora affinis* (Copepoda) to Kepone," *Marine Biol.*, 66, 179–184

Axelsen, J. A., 1997, "A Physiologically Driven Mathematical Simulation Model as a Tool for Extension of Results from Laboratory Tests to Ecosystem Effects," in *Ecological Risk Assessment of Contaminants in Soil* (N. M. van Straalen and H. Løkke, Eds.), Chapman and Hall, London, UK, pp. 233–250

Barbour, M. T., C. G. Growes, and W. L. McCulloch, 1989, "Evaluation of the Intrinsic Rate of Increase as an Endpoint for *Ceriodaphnia* Chronic Tests," in *Aquatic Toxicology and Environmental Fate: Eleventh Volume* (G. W. Suter II, and M. A. Lewis, Eds.), American Society for Testing and Materials, Philadelphia, USA, pp. 273–288

Bechman, R. K., 1994, "Use of Life Tables and LC_{50} Tests to Evaluate Chronic and Acute Toxicity Effects of Copper on the Marine Copepod *Tisbe furcata* (Baird)," *Environ. Toxicol. Chem.*, 13, 1509–1517

Bengtsson, G., T. Gunnarsson, and S. Rundgren, 1985, "Influence of Metals on Reproduction, Mortality and Population Growth in *Onychiurus armatus* (Collembola)," *J. Appl. Ecol.*, *22*, 967–978

Calow, P., and R. M. Sibly, 1990, "A Physiological Basis of Population Processes: Ecotoxicological Implications," *Funct. Ecol.*, *4*, 283–288

Caswell, H., 1978 "A General Formula for the Sensitivity of Population Growth Rate to Changes in Life History Parameters," *Theor. Populat. Biol.*, *14*, 215–230

Caswell, H., 1996, "Demography Meets Ecotoxicology: Untangling the Population Level Effects of Toxic Substances," in *Ecotoxicology. A Hierarchical Treatment* (M. C. Newman, and C. H. Jagoe, Eds.), Lewis Publishers, Boca Raton, USA, pp. 255–282

Charlesworth, B., 1980, *Evolution in Age-structured Populations.* Cambridge University Press, Cambridge, UK

Cole, L. C., 1954, "The Population Consequences of Life History Phenomena," *Q. Rev. Biol.*, *29*, 103–137

Crommentuijn, T., C. J. A. M. Doodeman, A. Doornekamp, and C. A. M. van Gestel, 1997, "Life Table Study with the Springtail *Folsomia candida* (Willem) Exposed to Cadmium, Chlorpyrifos and Triphenyltin Hydroxide," in *Ecological Risk Assessment of Contaminants in Soil* (N. M. van Straalen, and H. Løkke, Eds.), Chapman and Hall, London, UK, pp. 275–291

Crommentuijn, T., C. J. A. M. Doodeman, J. J. C. van der Pol, A. Doornekamp, M. C. J. Rademaker, and C. A. M. van Gestel, 1995, "Sublethal Sensitivity Index as an Ecotoxicity Parameter Measuring Energy Allocation under Toxicant Stress: Application to Cadmium in Soil Arthropods," *Ecotoxicol. Environ. Saf.*, *31*, 192–200

Crommentuijn, T., J. Brils, and N. M. Van Straalen, 1993, "Influence of Cadmium on Life-history Characteristics of *Folsomia candida* (Willem)," *Ecotoxicol. Environ. Saf.*, *26*, 216–227

Daniels, R. E. and J. D. Allan, 1981, "Life Table Evaluation of Chronic Exposure to a Pesticide," *Can. J. Fish. Aquat. Sci.*, *38*, 485–494

Day, K., and N. K. Kaushik, 1987, "An Assessment of the Chronic Toxicity of the Synthetic Pyrethroid, Fenvalerate to *Daphnia galeata mendotae*, using Life Tables," *Environ. Pollut.*, *44*, 13–26

Diekkrüger, B., and H. Röske, 1995, "Modelling the Dynamics of *Isotoma notabilis* (Collembola) on Sites of Different Agricultural Usage," *Pedobiologia*, *39*, 58–73

Emlen, J. M., 1989, "Terrestrial Population Models for Ecological Risk Assessment: A State-of-the-art review," *Environ. Toxicol. Chem.*, *8*, 831–842

Emlen, J. M., and E. K. Pikitch, 1989, "Animal Population Dynamics: Identification of Critical Components," *Ecol. Modell.*, *44*, 253–273

Enserink, L., M. De la Haye, and H. Maas-Diepeveen, 1993, "Reproductive Strategy of *Daphnia magna*: Implications for Chronic Toxicity Tests," *Aquat. Toxicol.*, *25*, 111–124

Ernsting, G. and J. A. Isaaks, 1991, "Accelerated Aging: A Cost of Reproduction in the Carabid Beetle *Notiophilus biguttatus F*," *Funct. Ecol.*, *5*, 299–303

Gentile, J. H., S. M. Gentile, N. G. Hairston Jr., and B. K. Sullivan, 1982 "The Use of Life-Tables for Evaluating the Chronic Toxicity of Pollutants to *Mysidopsis bahia*," *Hydrobiologia*, *93*, 179–187

Grant, W. E., S. O. Fraser, and K. G. Isakson, 1984, "Effect of Vertebrate Pesticides on Non-Target Wildlife Populations: Evaluation Through Modelling," *Ecol. Model., 21*, 85–108

Halbach, U., 1984, "Population Dynamics of Rotifers and its Consequences for Ecotoxicology," *Hydrobiologia, 109*, 79–96

Halbach, U., M. Siebert, M. Westermayer, and C. Wissel, 1983, "Population Ecology of Rotifers as a Bioassay Tool for Ecotoxicological Tests in Aquatic Environments," *Ecotoxicol. Environ. Saf., 7*, 484–513

Hallam, T. G., R. R. Lassiter, J. Li, and W. McKinney, 1990, "Toxicant-Induced Mortality in Models of *Daphnia* populations," *Environ. Toxicol. Chem., 9*, 597–621

Hendriks, A. J., and E. L. Enserink, 1996, "Modeling Responses of Single Species Populations to Microcontaminants as a Function of Species Size, with Examples for Water-Fleas (*Daphnia Magna*) and Cormorants (*Phalacrocorax Carbo*)," *Ecol. Model., 88*, 247–262

Högstedt, G., 1981, "Should there be a Positive or a Negative Correlation Between Survival of Adults in a Bird Population and Their Clutch Size?," *Am. Naturalist, 118*, 568–571

Jensen, A. L., 1984, "Assessing Environmental Impact on Mass Balance, Carrying Capacity and Growth of Exploited Populations," *Environ. Pollut. Ser. A., 36*, 133–145

Jepson, P. C., and J. R. M. Thacker, 1990, "Analysis of the Spatial Component of Pesticide Side-Effects on Non-Target Invertebrate Populations and its Relevance to Hazard Analysis," *Funct. Ecol., 4*, 349–355

Kammenga, J. E., 1995, "Phenotypic Plasticity and Fitness Consequences in Nematodes Exposed to Toxicants," *Ph.D. Thesis*, Landbouwuniversiteit, Wageningen, The Netherlands

Kammenga, J. E., M. Busschers, N. M. van Straalen, P. C. Jepson, and J. Bakker, 1996, "Stress-Induced Fitness Reduction is not Determined by the Most Sensitive Life-Cycle Trait," *Funct. Ecol., 10*, 106–111

Kareiva, P., J. Stark, and U. Wennergren, 1996, "Using Demographic Theory, Community Ecology and Spatial Models to Illuminate Ecotoxicology," *Ecotoxicology: Ecological Dimensions* (D. J. Baird, L. Maltby, P. W. Greig-Smith, and P. E. T. Douben, Eds.), Chapman and Hall, London, UK, pp. 13–23

Keyfitz, N., 1968, *Introduction to the Mathematics of Population.* Addison-Wesley Publishing Company, Massachusetts, USA

Klok, C., and A. M. de Roos, 1996, "Population Level Consequences of Toxicological Influences on Individual Growth and Reproduction in *Lumbricus rubellus* (Lumbricidae, Oligochaeta)," *Ecotoxicol. Environ. Saf., 33*, 118–127

Klok, C., A. M. de Roos, J. C. Y. Marinissen, H. M. Baveco, and W. C. Ma, 1997, "Assessing the Effects of Abiotic Environmental Stress on Population Growth in *Lumbricus rubellus* (Lumbricidae, Oligochaeta)," *Soil Biol. Biochem., 29*, 287–293

Klüttgen, B., and H. T. Ratte, 1994, "Effects of Different Food Doses on Cadmium Toxicity to *Daphnia magna*," *Environ. Toxicol. Chem., 13*, 1619–1627

Kooijman, S. A. L. M., 1985, "Toxicity at Population Level," in *Multispecies Toxicity Testing* (J. Cairns Jr., Ed.), Pergamon Press, New York, USA, pp. 143–164

Kooijman, S. A. L. M., 1993, *Dynamic Energy Budgets in Biological Systems. Theory and Applications in Ecotoxicology*, Cambridge University Press, Cambridge, UK

Kooijman, S. A. L. M., and J. A. J. Metz, 1984, "On the Dynamics of Chemically Stressed Populations: The Deduction of Population Consequences from Effects on Individuals," *Ecotoxicol. Environ. Saf.*, 8, 254–274

Kozlowski, J., 1993, "Measuring Fitness in Life-history Studies," *Trends Ecol. Evolut.*, 8, 84–85

Laskowski, R., 1997, "Estimating Fitness Costs of Pollution in Iteroparous Invertebrates," *Ecological Principles for Risk Assessment of Contaminants in Soil* (N. M. van Straalen, and H. Løkke, Eds.), Chapman and Hall, London, UK, pp. 305–319

Lefkovitch, L. P., 1965, "The Study of Population Growth in Organisms Grouped by Stages," *Biometrics*, 21, 1–18

Levine, S. N., 1989, "Theoretical and Methodological Reasons for Variability in the Responses of Aquatic Ecosystem Processes to Chemical Stresses," *Ecotoxicology: Problems and Approaches* (S. A. Levin, M. A. Harwell, J. R. Kelly, and K. D. Kimball, Eds.), Springer, New York, USA, pp. 145–179

Longstaff, B. C., 1977, "The Dynamics of Collembolan Populations: A Matrix Model of Single Species Growth," *Can. J. of Zool.*, 55, 314–324

Marshall, J. S., 1978a, "Population Dynamics of *Daphnia galeata mendotae* as Modified by Chronic Cadmium Stress," *J. Fish. Res. Board Canada*, 35, 461–469

Marshall, J. S., 1978b, "Field Verification of Cadmium Toxicity to Laboratory *Daphnia* Populations," *Bull. of Environ. Contam. Toxicol.*, 20, 387–393

McKim, J. M., 1977, "Evaluation of Tests with Early Life Stages of Fish for Predicting Long-Term Toxicity," *J. Fish. Res. Board Canada*, 34, 1148–1154

Meyer, J. S., 1986, "Variability and Bias in Estimating Cladoceran Population Growth Rates," *Ph.D. Thesis*, University of Wyoming, Laramie, USA

Meyer, J. S., and M. S. Boyce, 1994, "Life Historical Consequences of Pesticides and Other Insults to Vital Rates," in *Wildlife Toxicology and Population Modeling* (R. J. Kendall, and T. E. Lacher Jr., Eds), Lewis Publishers, Boca Raton, USA, pp. 349–363

Meyer, J. S., C. G. Ingersoll, and L. L. McDonald, 1987, "Sensitivity Analysis of Population Growth Rates Estimated from Cladoceran Chronic Toxicity Tests," *Environ. Toxicol. Chem.*, 6, 115–126

Meyer, J. S., C. G. Ingersoll, L. L. McDonald, and M. S. Boyce, 1986, "Estimating Uncertainty in Population Growth Rates: Jackknife vs. Bootstrap Techniques," *Ecology*, 67, 1156–1166

Neuhold, J. M., 1987, "The Relationship of Life History Attributes to Toxicant Tolerance in Fishes," *Environ. Toxicol. Chem.*, 6, 709–716

Posthuma, L., R. F. Hogervorst, E. N. G. Joosse, and N. M. van Straalen, 1993, "Genetic Variation and Covariation for Characteristics Associated with Cadmium Tolerance in Natural Populations of the Springtail *Orchesella cincta* (L.)," *Evolution*, 47, 619–631

Ratte, H. T., U. Dülmer, B. Klüttgen, and M. Pelzer, 1992, "Investigation and Modelling of Primary and Secondary Effects of 3,4-dichloroaniline in Experimental Aquatic Laboratory Systems and Mesocosms," *GSF-Bericht*, 1, 49–55, Germany

Schobben, J. H. M., and N. M. van Straalen, 1987, "A Model for the Calculation of the Intrinsic Population Growth Rate for a Stage-structured Population in Ecotoxicology: Applied to *Platynothrus peltifer*," *Report* (in Dutch), Department of Ecology and Ecotoxicology, Vrije Universiteit, Amsterdam, The Netherlands

Seitz, A., and H. T. Ratte, 1991, "Aquatic Ecotoxicology: On the Problems of Extrapolation from Laboratory Experiments with Individuals and Populations to Community Effects in the Field," *Comparative Biochem. Physiol., C100,* 301–304

Sherratt, T. N., and P. C. Jepson, 1993, "A Metapopulation Approach to Modelling the Long-Term Impact of Pesticides on Invertebrates," *J. Appl. Ecol., 30,* 696–705

Sibly, R. M., and P. Calow, 1989 "A Life-Cycle Theory of Responses to Stress," *Biol. J. Linnean Soc., 37,* 101–116

Stamou, G. P., M. D. Asikidis, M. D. Argyropoulou, and S. P. Sgardelis, 1993, "Ecological Time Versus Standard Clock Time: The Asymmetry of Phenologies and the Life History Strategies of Some Soil Arthropods from Mediterranean Ecosystems," *Oikos, 66,* 27–35

Stearns, S. C., 1992, *The Evolution of Life Histories.* Oxford University Press, Oxford, UK

Suter, II, G. W., and M. H. Donker, 1993, "Parameters for Population Effects of Chemicals," *Sci. Total Environ., Suppl,* 1793–1797

Van der Hoeven, N., 1990, "Effect of 3,4-dichloroaniline and Metavanadate on *Daphnia* Populations," *Ecotoxicol. Environ., Saf., 20,* 53–70

Van Dijk, Th.S., 1979, "On the Relationship Between Reproduction, Age and Survival in Two Carabid Beetles: *Calathus melanocephalus* L. and *Pterostichus coerulescens* L. (Coleoptera, Carabidae)," *Oecologia, 40,* 63–80

Van Gestel, C. A. M., E. M. Dirven-van Breemen, and R. Baerselman, 1992, "Influence of Environmental Conditions on the Growth and Reproduction of the Earthworm *Eisenia andrei* in an Artificial Soil Substrate," *Pedobiologia, 36,* 109–120

Van Groenendael, J, H. de Kroon, and H. Caswell, 1988, "Projection Matrices in Population Ecology," *Trends Ecol. Evolut., 3,* 264–269

Van Leeuwen, C. J., G. Niebeek, and M. Rijkeboer, 1987, "Effects of Chemical Stress on the Population Dynamics of *Daphnia magna*: A Comparison of Two Test Procedures," *Ecotoxicol. Environ. Saf., 14,* 1–11

Van Leeuwen, C. J., M. Rijkeboer, and G. Niebeek, 1986, "Population Dynamics of *Daphnia magna* as Modified by Chronic Bromide Stress," *Hydrobiologia, 133,* 277–285

Van Leeuwen, C. J., W. J. Luttmer, and P. S. Griffioen, 1985, "The Use of Cohorts and Populations in Chronic Toxicity Studies with *Daphnia magna*: A Cadmium Example," *Ecotoxicol. Environ. Saf., 9,* 26–39

Van Noordwijk, A. J. and G. de Jong, 1986, "Acquisition and Allocation of Resources: Their Influence on Variation in Life History Tactics," *Am. Naturalist, 128,* 137–142

Van Straalen, N. M., 1983, "Physiological Time and Time-Invariance," *J. Theor. Biol., 104,* 349–357

Van Straalen, N. M., 1985a, "Production and Biomass Turnover in Stationary Stage-Structured Populations," *J. Theor. Biol., 113,* 331–352

Van Straalen, N. M., 1985b, "Comparative Demography of Forest Floor Collembola Populations," *Oikos, 45,* 253–265

Van Straalen, N. M. and R. G. M. de Goede, 1987, "Productivity as a Population Performance Index in Life-Cycle Toxicity Tests," *Water Sci. Technol., 19,* 13–20

Van Straalen, N. M., J. H. M. Schobben, and R. G. M. de Goede, 1989, "Population Consequences of Cadmium Toxicity in Soil Microarthropods," *Ecotoxicol. Environ. Saf., 17,* 190–204

Vandermeer, J. H., 1975, "On the Construction of the Population Projection Matrix for a Population Grouped in Unequal Stages," *Biometrics, 31*, 239–242

Winner, R. W., T. Keeling, R. Yeager, and M. P. Farrell, 1977, "Effect of Food Type on the Acute and Chronic Toxicity of Copper to *Daphnia magna*," *Freshwater Biology, 7*, 343–349

Woltering, D. M., 1985, "Population Responses to Chemical Exposure in Aquatic Multispecies Systems," in *Multispecies Toxicity Testing* (J. Cairns, Jr., Ed.), Pergamon Press, New York, USA, pp. 61–75

Wong, C. K., and P. K. Wong, 1990, "Life Table Evaluation of the Effects of Cadmium Exposure on the Freshwater Cladoceran, *Moina Macrocopa*," *Bull. Environ. Contam. Toxicol., 44*, 135–141

Young, H., 1968, "A Consideration of Insecticide Effects on Hypothetical Avian Populations," *Ecology, 49*, 991–993

21

Effects of Pollutants on Soil Invertebrates: Links between Levels

*Jason M. Weeks (**Huntingdon, UK**)*

21.1 SUMMARY

This section describes the difficulties encountered when one attempts to extrapolate measurements of effects between different biological levels. It also considers the complexity of natural ecosystems, and their likely responses to natural and anthropogenic perturbations. It is essential that whole system responses to pollutants are considered during ecological risk assessment protocols, but how this should be undertaken is not clear. It is apparent that more emphasis must be placed on community-level responses

Ecotoxicology, Edited by Gerrit Schüürmann and Bernd Markert.
ISBN 0-471-17644-3   © 1998 John Wiley & Sons, Inc. and Spektrum Akademischer Verlag.

and bioassays, but this requires a fine balancing act between simplifying assessment and obtaining sufficient detail but at an affordable price.

21.2 INTRODUCTION

Most research in environmental toxicology focuses on the understanding of effects at lower levels of biological organization (such as molecules, cells and individuals), and usually such work is accomplished in the laboratory. However, contaminants exert their effects at all levels of biological organization, from molecules to ecosystems. Although effects at the lower levels may influence populations, communities and ecosystems, surprisingly little research has been conducted to demonstrate a causal relationship between effects at different levels of organization. Molecular endpoints and other biomarkers, including other sublethal measures, may be effective as endpoints to assess contaminant exposure, or may provide good early warning signs of potential contaminant impacts. However, the ecological consequences of exposure to contaminants are probably best examined at higher levels of organization. This chapter serves to address the problems of establishing links between levels, in an effort to avoid the frequently encountered "so what?" questions.

21.3 WHY SOIL INVERTEBRATES?

Due to their essential functions in ecosystems, soil invertebrates are useful for assessing the ecological effects of chemicals. Soil invertebrate tests measure acute toxicity at the species level (e.g. earthworm test, OECD 1984) and the population and community levels (e.g. soil insect tests, OECD 1984). The preliminary endpoints of these tests are survival, growth (measured as biomass), reproductive success and behavioral changes. This chapter considers the merits of attempting to link together the observed effects of pollutants between different levels of biological organization and soil invertebrate functions, and in particular considers the enhanced value (if any) of evaluating the significance of any effects of pollutants on ecosystems.

The true nature of ecosystems is complex, consisting of many communities and assemblages of thousands of species of plant and animal each in dynamic equilibrium and each interacting with the complex physico-chemical components of their ecosystems. The grouped biota of an ecosystem play major structuring roles in the maintenance of ecological processes, and modify its physical and chemical environment. The ecosystem, in turn, influences the composition and overall diversity of the communities that contribute to it. Thus, it may be considered important in terms of ecological risk assessment that an effort is made to appraise the effects of chemical stressors on the community structure and dynamics of an ecosystem. This has been the ultimate goal of much of the current research undertaken in the field of ecotoxicology.

However, the belief by many scientists that if one were to reduce the properties of an ecosystem to its individual components the total of the separated parts would be equal to the sum of the original properties is misleading and hinders the development of appropriate technologies. Were such a concept robust then the isolated use of bioindicator and biomonitoring techniques would be sufficient. However, such a hypothesis may be argued against for many reasons. It is known that the detrimental effects of many chemicals occur at concentrations well below those that are lethal—particularly effects on reduced feeding, growth and fitness and reductions in reproductive performance. Furthermore, a comprehensive knowledge of the toxicity of a single chemical to a particular population (e.g. from short-term laboratory toxicity testing) will be insufficient to predict the characteristic toxicity that may be manifest at the level of the whole ecosystem. For these and numerous other reasons, the approach of ecological risk assessment must be replaced by a paradigm that focuses on the toxicity of pollutants throughout entire ecosystems. This can only be undertaken with the development and validation of new techniques and methodologies that more accurately assess the impacts of chemicals at the ecosystem level. Such methodologies require a strong ecological basis and approach.

21.4 STRUCTURAL AND FUNCTIONAL CHARACTERISTICS REQUIRED TO EVALUATE THE CONDITION OF ECOSYSTEMS

In order to evaluate the effects of chemical (and other) stress on an ecosystem it is necessary to differentiate between and examine the structural and functional properties of that ecosystem. Structural properties (e.g. productivity, species composition and demographic descriptions) and functional parameters (e.g. nutrient cycling and trophic structure) are interdependent, yet still have distinct characteristics. Structural parameters describe the parts that make up an ecosystem, i.e. what is there. Functional parameters describe the actions or innerworkings of a system, i.e. how the components work. Generally, structural properties of a more descriptive nature are much easier to assess or measure because fewer parameters can be measured over a shorter time period. Structural properties correlate with functional properties that are expressed as rates and must then be measured in a more complicated manner. For example, measuring a functional property such as decomposition of plant matter on a forest floor requires taking complex, multiple measurements over time. Conversely, describing the composition of the microbial biomass is all the more complicated and time-consuming.

Additionally, it is possible that functional redundancies can buffer ecosystems from the effects of chemical perturbations within specific populations of soil invertebrates. Thus, if groups of organisms are killed their function(s) may be replaced by other groups of organisms. The overall effect may thus be that the functioning of that particular ecosystem remains unchanged, whilst the structure of the ecosystem changes considerably. Populations within ecosystems are further able to resist stress-induced changes through compensatory (often deleterious) alterations in, for example, growth and reproduction, which may in turn alter the structural characteristics of the ecosystem. Ultimately, if the magnitude or duration of such a stressor persists for any prolonged period the limitations of the compensatory mechanisms will be exceeded, and consequently the condition of the ecosystem may decline. Environmental management tends to favour the sustaining of an ecosystem function at the expense of the ecosystem structure. The danger here is, of course, that in a chemically stressed ecosystem many of the functions may be performed by a severely diminished community, and may ultimately be more susceptible to further future stress (either anthropogenic or not).

Many functional properties of ecosystems can be monitored to evaluate ecosystem conditions, for example, primary production, biotic factors and trophic interactions, rates of decomposition and mineralization. Structural properties may also be useful indicators of ecosystem conditions. These may include changes in species composition and abundance, reduced biodiversity, less complex food webs and many more. It was recently suggested in the Scientific Group on Methodologies for the Safety Evaluation of Chemicals (SCOPE 53 1995) that many of these parameters may be monitored to evaluate changes in ecosystems or to assess trends that may be a cause for concern. However, the assessment of environmental impacts of chemicals on ecosystems at the ecosystem level raises considerable theoretical and methodological problems. The ultimate protection of an ecosystem from an ecological point of view should be the preservation of and prevention from extinction of each individual species. Due to the difficulties of operating at the ecosystem level, and given the time scale, it is difficult to measure extinction. At the ecosystem level, gradual (and potentially unnoticed) extinction of interconnecting species will result in potential malfunctioning of the system. Testing of ecosystems by the compilation of data collected at the lower levels of biological organization may well prove more useful. Testing at the cell and tissue level, primarily aimed at biochemical and physiological processes for soil invertebrates, often termed biomarkers, has received growing interest recently. For studies in natural habitats, biomarkers have been developed to improve the estimation of sublethal exposure on populations of critical species in the wild (Peakall 1992).

21.5 LEVELS OF ORGANIZATION FOR EVALUATING THE EFFECTS OF CHEMICALS

As alluded to in the previous section, the methods available for evaluating the effects of chemicals range from the molecular to the ecosystem level. Evaluating effects becomes more difficult as one moves toward the more complex organizational level. One may start at either end, namely the molecular or the ecosystem.

If one starts at the lower levels the difficulty begins with the need to extrapolate laboratory data or limited field trials to effects at the ecosystem level. Conversely, starting at a higher tier within this scale, say at the individual or physiological level, effects are surmised by examining contaminated ecosystems. Ideally, both approaches should be adopted to determine how chemicals are affecting the ecosystem at all levels. Each approach, simple or complex, has advantages and disadvantages. The first approach has the clear advantage of establishing a laboratory dose/response scenario. However, this approach then suffers from the increasing degree of complexity in extrapolating results and observations at the lower levels to higher organizational levels, whereas the second approach requires that a well-defined contaminated ecosystem be studied. As stated earlier, ecosystems are not the simple sum of their components, but encompass several ecosystem functions that are the products of interactions of the species. Measuring ecosystem changes directly does have the advantage of allowing us to determine the effects of chemicals on these systems, and hence extrapolation is not required. However, the lack of appropriate technologies and the costs in time and money may well be prohibitive to the measurement of the effects of even one chemical on all aspects of the structure and function of a single ecosystem. Thus, for ecosystems one must be selective in the choice of suitable indicators of structures and functions, such a choice is problematic.

The effects of chemicals on ecosystem functions can be assessed only by studying these functions, and cannot be directly measured or estimated by merely examining individual species. As with measuring effects on species, measuring effects on ecosystem functions requires indicators for endpoints. These indicators should be sufficiently sensitive to provide early warnings, distributed over a wide range of stresses, independent of sample size and cost-effective (Sheehan 1984a; Noss 1990). Ecosystem indicators must be ecologically significant phenomena (Sheehan 1984b). Indicators for ecosystems might include indices of species diversity, relative species abundance, indices of species richness and landscape parameters. Ramade (1995) provides a brief review of methods for estimating damage to terrestrial ecosystems. He concluded that current knowledge of ecosystems was insufficient to identify key species that may affect either structure or function, and to improve the knowledge of species that may serve as bioindicators for monitoring the effects of chemicals on communities. In an ideal world laboratory studies should predict ecosystem effects. Such extrapolation should be possible as one learns to use micro- and mesocosm studies to duplicate the structure and function of ecosystems.

21.6 THE VALUE OF MESOCOSM EXPERIMENTS AS LINKS TO REAL ECOSYSTEMS

Microcosm and mesocosm tests are useful intermediates between bioassays and ecosystem experiments. They provide controlled experimental conditions in the laboratory or field for studying changes at any level (population, community or ecosystem) of a chemical or other stressor. Microcosm studies are usually small, contain a few species, and are generally conducted indoors, whereas mesocosm studies are normally relatively large, contain many species from an ecosystem and are usually conducted in outdoor settings.

Microcosm studies offer several advantages over mesocosm studies and field surveys. Multispecies microcosm studies allow greater realism than can be achieved with more traditional single species toxicity tests. Microcosms allow the researcher to observe the integrated effects of contaminants on community and ecosystem functions and pathways. Data obtained in such studies, however, must be interpreted with some caution due to the limitations of such studies. Most limitations result from the fact that the microcosm is really a very simplified representation of a real ecosystem. More specific limitations include the difficulties of extrapolating observations to the broader environment, and the use of small population sizes, which may lead to chance extinctions.

The importance of earthworms in testing the adverse effects of chemicals in the soil environment has been recognized by various environmental organizations and has resulted in rigid test guidelines (e.g. OECD 1994). However, the ecotoxicological testing and risk assessment of the effects of chemicals on earthworms has historically been based on short-term laboratory experiments. During such tests animals are exposed to constant and typically extreme concentrations of xenobiotics that are usually in a bioavailable form. The laboratory test can usually be regarded as the "worst case" situation, and thus appears to be the most realistic option which enables the screening and ranking of the acute toxicity of various compounds under controlled conditions. Such information is clearly necessary and important in basic toxicological research as well as for initial chemical legislation. However, under field conditions the actual exposure to chemicals may be lowered since the chemical may not be distributed uniformly throughout the soil and its bioavailability will often be decreased due to various sorption or binding processes. This makes the extrapolation from laboratory studies to the field situation very difficult, necessitating field testing in order to validate the predictions of environmental effects that are made on the basis of standard laboratory tests (Christensen and Mather 1994). Disappointingly, if one reviews the many field experiments with earthworms so far reported in the literature, it will prove extremely difficult to reliably assess the relative toxicity of different chemicals, primarily due to the considerable variability between sites, soils, formulations, doses and methods of application adopted (Greig-Smith et al. 1992). Furthermore, although the reported field experiments have been adequate for identifying chemicals that are extremely toxic to earthworms, they have failed to accurately identify moderately or only slightly toxic compounds (Lofs-Holmin and Boström 1988; Edwards 1992; Edwards and Bohlen 1992).

Laboratory-based tests, on the one hand, are carried out under artificial conditions and disregard any ecological interaction between different species, which makes realistic extrapolation to the field situation uncertain. On the other hand, results from field-based experiments typically show such large inherent variability, so that the results serve little purpose in ecotoxicological testing. As a compromise model ecosystems have been designed that capture certain aspects of real ecosystems but at the same time are simple enough to enable a sensible interpretation of the results. Such systems are often called mesocosms.

Mesocosm experiments generally make an attempt to isolate a small part of an ecosystem to permit some degree of control of the physico-chemical conditions and composition of the biota (Steel 1979; Menzel and Case 1977). Thus they facilitate the measurement of physical and chemical parameters and permit population level effects to be investigated under near natural conditions for prolonged periods (Donaghay 1984; Snatschi et al. 1984).

Several terrestrial model ecosystems have been described without any attempt at standardization (Anderson 1978; Teuben and Roelofsma 1990). Usually the system takes the form of a cylinder with an encased soil column, and often has various devices which collect leachate from the column. Some authors have taken intact core samples from natural habitats to make up such mesocosms (Tolle et al. 1985; Tolle et al. 1995; Zwick et al. 1984), others have reconstituted a column from more or less standardized materials (e.g. Bond et al. 1976). The vast majority of these studies have involved bringing the core into the laboratory for experimentation and have included various pretreatments of the soil column.

The major disadvantage in using any type of mesocosm is that normally they are very costly, both in true economic terms and in relation to manpower. Previously this has meant that few mesocosm experiments have been undertaken and, when adopted, they were often statistically poorly designed and not repeated often enough. This, together with the large inherent variability of these more complex systems, has caused several large, expensive studies to produce inconclusive results (Donaghay 1984; Lofs-Holmin and Boström 1988; Steel 1979). This problem can be addressed from several fronts: First, by decreasing the variation caused by non-experimental factors (i.e. various soil gradients) by providing more standardized conditions and second, by making the experimental set-up more economically practicable to enable greater replication.

21.7 BIOMARKERS: ADVANTAGES AND DISADVANTAGES

In recent years there has been an increasing interest in the use of biomarkers for the assessment of the potential adverse effects of chemicals on the environment. Peakall (1994) defined a biomarker as "a biological response to a chemical or chemicals that gives a measure of exposure and sometimes also, of toxic effect". According to Peakall (1994) the major strength of biomarkers lies in their potential to circumvent the serious limitations of the classical approach to environmental toxicology; that of measuring the residue of a chemical in either the organism or the environment and relating it to acute effects, through toxicity studies. Furthermore, in soils there is the question of the bioavailability of chemicals which are known to vary widely with different soil characteristics (Ma et al. 1983; Corp and Morgan 1991).

In theory the types of biological response that could be considered as biomarkers range from the molecular to effects on the intact organism, the population or community structure and perhaps also, the structure and function of ecosystems. Three statements can be made about this continuum (Peakall 1994). Firstly, the time scale of response increases, with progress along it, moving from seconds or minutes to years or even decades. Secondly, the degree of importance increases (we would perhaps be more concerned with the impairment of the function of ecosystems than with subtle molecular changes) and finally, it becomes increasingly difficult to relate effects to a single cause as one moves through this continuum. For these reasons, if biomarkers are to be used for anything other than stating the obvious (i.e., the species is likely to become extinct or a forest area is in decay) they must be applied at the lower end of the continuum, where responses are sensitive and rapid, whilst being reasonably easy to interpret (Huggett et al. 1992; Peakall 1994).

Biomarkers have been used extensively to document and quantify both exposure to and effects of environmental pollutants. As monitors for exposure, biomarkers have the advantage of quantifying only biologically available pollutants. There is a vast amount of literature on biomarkers (see reviews by McCarthy and Shugart 1990 and Huggett et al. 1992). However, there is a general paucity of biomarkers for use with terrestrial soil invertebrates. Earthworms are virtually ubiquitous and are ecologically important soil organisms. They are valuable in situ sentinels for assessing biological risks from hazardous and toxic compounds in terrestrial environments. As such they have been used in numerous laboratory, toxicity and bioaccumulation studies (see, for example, van Gestel et al. 1992). The development of a series of earthworm immune-based biomarkers has been documented by Rodriguez-Graun et al. (1989) and measurements of lysozyme activity in earthworm coelomic fluid have indicated that such measurements have potential as a biomarker for assaying the immunotoxicity of metals (Goven et al. 1994).

The field of biomarkers has evolved rapidly and significantly in the last 10 years. The word biomarker, however, evokes different passions within different people. A true definition of the term is lost within a convoluted pathway of misnomers and it often represents a simple substitution of a new encompassing name for the more classic biochemical measures of exposure, effective dose, response or susceptibility. For example, blood lead levels, serum ALA dehydrase, serum enzyme markers of hepatotoxicity and so on, are now encompassed within this broad definition of biomarkers. In general a biomarker may be considered as a biological response to a chemical or chemicals that gives a measure of exposure and sometimes also of the toxic effect. One question worthy of greater consideration, however, is the validity of a predictive role for biomarkers in the assessment of risks to individual soil organisms and the potential for the circumvention of lasting damage to their populations. The field of biomarkers is large and continually expanding. However, this section hopes to provide some incentive for establishing comprehensive links between biomarkers and soil ecology.

The increasing development of new indices (markers) for measuring exposure, effective dose, responses and susceptibility has been brought forward by the application of new analytical techniques, mainly based on the tools of molecular biology. Such developments have opened new avenues of research in medical toxicology using cellular and animal models in human populations. For example, the concept of using specific DNA adducts as markers of biological effective dose has shifted the focus of dose response investigations on to more relevant measures of dose which have direct analogues in humans. Thus medically applied biomarkers have a proven and valid role in human toxicological risk evaluation. Perhaps, fallen by the wayside, however, is the ecotoxicological role of biomarkers in estimating or predicting ecological harm at higher levels of organization than as a simple biochemical measure of response. The step of measuring a selective response in a single representative organism and extrapolating from this simple biochemical change to be able to understand the ecological significance of a slight increase or decrease of a measured parameter is a major hurdle. However, if it is ever brought to a satisfactory conclusion it will provide a potentially powerful legislative tool that would enable objective definitions of emissions and the environmental effects that these could (but not necessarily will) have.

Environmental (rather than medical) biomarkers may be used to investigate uncertainties that arise in preliminary hazard assessments. This objective may need to be carefully refined for specific chemicals and for specific end users. A link must be established between exposure and biological effect. This is the area of greatest uncertainty in hazard assessment. Is exposure always conducive of the same effect, if so at what dose, and is it specific to that compound, if not, does it matter? Furthermore, mathematical models can be used to validate such test systems, and once developed and accepted can be used to extrapolate from relatively simple laboratory experiments to other more complex systems and possibly higher organizational levels. However, extrapolation of exposure to real world scenarios is fraught with problems, many of which are recognized. Nevertheless, such mathematical models are widely accepted and utilized to their fullest capacity in other fields. Why then the reluctance or even shying away from the attempted development of similar models to establish such links between biomarkers and higher order ecological perturbations? It is clearly apparent that such links have been made and are being readily applied to human toxicology and the solution of epidemiological problems. We trust such systems for our own health, why then not for the health of our environment?

One stalling point may be the difficulties encountered when one attempts to apply some meaningful interpretation in terms of risk. It is now possible to detect potentially toxic compounds in biological samples at the level of one DNA adduct per cell. Similarly, biological responses may be measured in terms of an increase or decrease in enzyme activity or levels of proteins. Yet our ability to translate this knowledge into reasonable risk factors is relatively poor. This problem has arisen perhaps because of the lack of establishing links between observed or measured biomarker responses to changes in the higher orders of the ecosystem, i.e. changes in sex ratios, body weights, population structure, etc. These are all measurable factors that have a weighting system applicable to risk assessment. The mere presence of a chemical or the detection of a biochemical change is not necessarily indicative of a hazard, and therefore that the particular individual or population is at risk. No doubt, with the ever increasing technical sensitivity for the detection of chemicals in biological systems, novel biomarkers will invariably become irrelevant to direct toxic effects, or indeed be too sensitive to provide any meaningful interpretation, such as the natural induction, often within minutes, of elevated stress protein levels in organisms brought about by natural environmental changes.

Perhaps an even greater obstacle to interpreting results of studies at higher levels of organization is the difficulty in determining a cause–and–effect relationship between levels of contaminants and the degree of change in community and ecosystem level endpoints. It is also important that we consider the questions of the sensitivity and specificity of any biomarker response—if a biomarker is too sensitive or is non-specific it may detect effects which are not toxicologically significant or indeed relevant. Non-specificity, however, may be useful if it is not known what the potential toxicant is. It may also be advantageous when applying a broad spectrum survey to a new locality, with the proviso that the non-specificity is related to a toxicant, rather than to other natural physiological perturbations.

Part of the problem in the routine application and acceptance of biomarkers lies in the difficulties in distinguishing between measured alterations that are adaptive and reversible and those that are pathological and irreversible. This problem may be resolved by the adoption of a range of different types of biomarker which are used, where appropriate, in conjunction. It is important to emphasize that in risk assessment scenarios it is usually essential to use more than one biomarker for detecting biological effects, as it is rare for one marker to be sufficiently reliable and specific to be used alone.

21.8 BIOMARKERS AND LINKS TO ECOLOGICAL EFFECTS: LABORATORY VERSUS FIELD STUDIES

One of the ultimate goals of the science of ecotoxicology is to predict or assess the risk of ecological effects occurring due to environmental contamination. The environmental risk assessment processes combine exposure assessment and effects assessment, respectively (Løkke 1994). The largest source of uncertainty involved in this assessment process results from the extrapolation of laboratory results to the field. Such an extrapolation seems theoretically impossible, as the essential characteristics of an ecosystem are not incorporated in a single species laboratory test (Van Straalen et al. 1994). To meet this challenge it is important to establish clear links of ecotoxicological significance between biomarkers and ecological endpoints such as lethality, reproductivity, growth impairment or other ecologically important functions.

Some differences in the characteristics of laboratory and field tests are widely applicable but more so to earthworm studies (see reviews by Greig-Smith et al. 1992; Van Gestel 1992). In the laboratory one or more (not necessarily field-relevant) species, with all individuals of the same age and health, are exposed to chemicals that are mixed homogeneously into the soil for a short time under optimal conditions (temperature, pH, soil moisture content, etc.) (OECD 1994).

In direct contrast, in the field many different earthworm species, each represented by several age classes with great variations in physical condition, will be exposed under fluctuating environmental conditions to chronic, possibly heterogeneously distributed chemicals or often even mixtures of chemicals. Furthermore, the bioavailability of these substances may differ considerably from that in laboratory tests (Spurgeon et al. 1994; Spurgeon and Hopkin 1995). Additionally, organisms exposed in the field may be further subjected to stress factors, such as predation, competition or habitat stress, which may make them more susceptible, potentially increasing their sensitivity to chemical stress (Bayne et al. 1985).

To undertake a comprehensive environmental risk assessment by way of a full-scale field study is impossible (Edwards 1992). Such studies are inherently expensive, the risk assessment would become an incredibly slow process, results would often be overshadowed by natural variation and data influenced by widely fluctuating environmental conditions.

Biomarkers, especially physiological and non-specific markers, have been used extensively in the laboratory to document and quantify both exposure to and the effects of environmental pollutants (Mayer et al. 1992). For the monitoring of exposure, biomarkers have the advantages of reacting rapidly to exposure and quantifying only the biologically available pollutants, and as monitors of effects they are able to integrate the effects of multiple stressors (Peakall 1994). The utility of many biomarkers in the field is unknown since testing and extensive evaluation of the methods are lacking (Huggett et al 1992). While field research is more difficult, as stated above, the rewards of having a biomarker (the response of which ideally has been related to ecologically relevant effects, and further evaluated and validated in the field) would be more beneficial. There would be potential for using such a biomarker in ecological risk assessment. This would enable more flexible and sensitive risk assessment which could be performed both more rapidly and at lower cost (Dickerson et al. 1994).

A simple biomarker, neutral-red retention by earthworm coelomocyte cell lysosomes (Weeks and Svendsen 1996), has been successfully compared with the responses of ecological parameters in order to link effects at different levels of organization in a laboratory study (Svendsen and Weeks 1997a). Exposure of the earthworm *Eisenia andrei* to an increasing range of soil copper concentrations in the laboratory showed a threshold for the neutral-red retention assay which coincided with changes in the earthworms' regulation of tissue copper concentrations (Svendsen and Weeks 1997a). Effects at the individual and population levels were observed to appear only at soil exposure concentrations beyond that of the biomarker threshold, thus enabling differentiation between exposure and exposures causing detrimental effects to the earthworm (Svendsen and Weeks 1997b).

However, the ultimate challenge of ecotoxicological relevance to be met by a biomarker is to enable judgment and subsequent interpretation to be placed upon a simple measurement, not in terms of the consequences for any particular individual earthworm, but in terms of the consequences for the population from which that individual was taken (Weeks 1995). This provides an early warning about the risk of ecological effects occurring on a longer-term basis.

21.9 AN EXAMPLE OF THE USE OF SOIL INVERTEBRATE BIOMARKERS TO ASSESS ECOLOGICAL DAMAGE TO A FOREST ECOSYSTEM

Chemically contaminated terrestrial habitats are evaluated in much the same manner as aquatic habitats. Essentially, three types of data can be obtained and subsequently integrated: those from chemical analyses of samples, toxicity tests (both in the laboratory and in situ), and field or ecological surveys of biota. In addition, studies on biomarkers may be useful or controlled ecosystem studies appropriate when diagnozing the causes of observed effects on terrestrial habitats and establishing causal relationships between contaminants and observed changes. This section describes the field application of a non-specific and non-destructive biomarker technique that has been developed for use with earthworms (Weeks and Svendsen 1996). The technique makes use of changes in the lysosomal membrane stability of earthworm coelomocyte cells as a measure of subcellular, pollution-induced stress.

Industrial chemical accidents, such as fires, are relatively common (Meharg 1994). Nevertheless, each incident is unique in terms of the environment polluted, the prevailing weather conditions and the chemicals and materials involved. The uniqueness and unexpected nature of accidents (and the scarce background data on the specific areas affected) makes assessment of the ecological impact difficult. Just such a fire in October 1991 at a plastics recycling factory in Thetford, Norfolk in the United Kingdom involved 600 t of polyvinylchloride (PVC), 200 t of polyethylene (PE) and 200 t of a range of other plastics (Meharg 1994). To appreciate the potential environmental impact of these quantities of plastic, Meharg (1994) estimated, using the figures of Tamaddon and Hogland (1993), that 6–18 t of Cd alone was contained within the plastic stored on site at the time of the fire. The fire blazed for three days and severely contaminated a discrete area of a neighbouring Scots pine and beech stand with molten pyrolyzed plastic. Since the fire (3 years ago), a soil and leaf litter layer has developed overlying the now friable plastic layer.

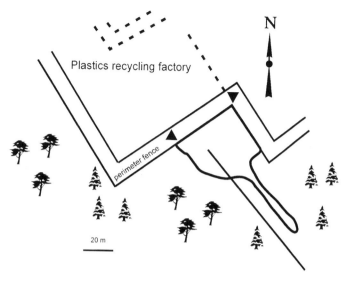

Figure 21.1 The extent of the plastic mantel (bold irregular line) at the fire site in Thetford (UK), with the original molten plastic runoff points (▼) and the sampling transect indicated (after Svendsen et al. 1996).

Svendsen et al. (1996) collected earthworms of the species *Lumbricus castaneus* (Savigny), and soil samples along a 200 m transect leading south-east from the factory perimeter fence, over a layer of molten plastic impregnated soil and into the surrounding forest (see Figure 21.1). Coelomic fluid extracted from the earthworms was dye-loaded with neutral-red and lysosomal leaking observed.

Metal residues in soil and earthworms were found to be highly elevated close to the factory perimeter and to rapidly drop to background levels within the first 50 metres of the transect. Coelomocyte cells taken from earthworms adjacent to the factory perimeter showed the shortest period of neutral-red retention (2 min); cells taken from worms further into the surrounding forest had a longer retention time (12 min), whilst cells taken from worms from a control site showed even greater retention times (25 min). Thus, the neutral-red retention times correlated negatively with measured residues of heavy metals in the earthworms, the higher the body metal concentration the shorter the retention time. Pollutants are known to induce cellular damage (Moore 1985). It follows that effects measured at the cellular or subcellular level would be expected to occur before or at lower contamination levels than effects measured at the physiological level. The fact that shorter neutral-red retention times were found at distances of 140 and 200 m from the source, whereas elevated metal concentrations in the worms were more or less confined to the first 30 m of the transect, lends support to this concept.

The major benefit gained by using a good biomarker is that the results obtained reflect the effects of toxicants experienced by the biota, thus circumventing the problems of assessing bioavailability. The question of bioavailability is a problem when evaluating measurements of single substances in the environment against fixed criteria. One advantage gained by using a non-specific biomarker, such as the neutral-red retention assay, is that it is sensitive to a wide spectra of pollutants and can integrate the combined effects of a complex pollution matrix even if all the individual components are below detection limits. This study has therefore clearly validated the applicability of the neutral-red retention assay as a cheap and time-efficient field tool capable of indicating the spatial distribution of biological effects resulting from a pollution incident. However, there is still no information or evidence to support the claim that such effects may be linked to higher level effects.

21.10 DIAGNOSIS AND ESTABLISHMENT OF A LINK TO CAUSATION

This chapter has attempted to briefly review some of the techniques or practices available to enable scientists to define the effects of soil contamination on soil invertebrates and to establish links between effects measured at different levels of biological complexity and observed adverse effects on the ecosystem. These tools include chemical analyses, toxicity tests, field trials, field surveys and special studies, such as biomarkers and simulated mesocosm studies. Each tool provides specific information, which if used alone is certainly useful. However, the greatest value would be obtained by using all the tools together, each one providing a vital piece of the overall environmental puzzle. Once the tools have been deployed and the results assessed, the final step is to demonstrate that the contaminants (if any) have caused the damage. In most cases the extent and nature of the toxic effects do not unequivocally demonstrate which agent was responsible for the injury. Correlations between a substance and the type of damage may be obvious, but causation is most difficult to prove, and can often be established only indirectly.

The use of micro- and mesocosms and field experiments has improved our ability to predict the effects of contaminants on higher levels of biological organization. The use of field manipulations has truly begun in the areas of hazard and risk assessment and will greatly improve the credibility of our interpretations. Such experimental approaches allow researchers to establish cause and effect relationships between contaminants and ecological endpoints, and permit the separation of direct and indirect effects. Manipulative studies remain the best way of establishing a causal relationship between contaminants and responses at higher levels of organization. Furthermore, due to the inconsistencies among various approaches, it is also important (particularly at moderately polluted sites) to develop integrated approaches for addressing contaminant impacts.

In conclusion, although research at lower levels of organization improves our understanding of mechanisms of toxic action and exposure assessment, they must ultimately be linked with changes in communities and ecosystems. The difficulty remains in interpreting the ecological significance of sublethal effects. Alternative methods, such as those described for the earthworms, need to be developed that will more easily allow the translation of lower level effects into value or benefits for the overall assessment process at higher biological levels of organization.

21.11 REFERENCES

Anderson, J. M., 1978, "Competition between Two Unrelated Species of Soil Cryptostigmata (Acari) in Experimental Mesocosms," *J. Anim. Ecol., 47*, 787–803

Bayne, B. L., D. A. Brown, K. Burns, D. R. Dixon, A. Ivanovici, D. R. Livingstone, D. M. Lowe, M. N. Moore, A. R. D. Stebbing, and J. Widdows, 1985, *The Effects of Stress and Pollution on Marine Animals*, Praeger Publishers, New York, USA, 384 pp

Bond, H., L. Lighthart, R. Shimabuku, and L. Russell, 1976, "Some Effects of Cadmium on Coniferous Forest Soils and Litter Microcosms," *Soil Sci., 121*, 278–287

Christensen, O. M., and J. G. Mather, 1994, *Earthworms as Ecotoxicological Test Organisms*. Ministry of the Environment, Danish Environmental Protection Agency, Denmark

Corp, N., and A. J. Morgan, 1991, "Accumulation of Heavy Metals from Polluted Soils by the Earthworm, *Lumbricus rubellus*: Can Laboratory Exposure of "Control" Worms Reduce Biomonitoring Problems?," *Environ. Pollut., 74*, 39–52

Dickerson, R. L., M. J. Hooper, N. W. Gard, G. P. Cobb, and R. J. Kendall, 1994, "Toxicological Foundations of Ecological Risk Assessment: Biomarker Development and Interpretation Based on Laboratory and Wildlife Species," *Environ. Health Perspectives, 102*, 65–69

Donaghay, P. L., 1984, "Utility of Mesocosms to Assess Marine Pollution," in *Concepts in Marine Pollution Measurements* (H. White, Ed.), College Park, MD, USA, pp. 589–620

Edwards, C. A., 1992, "Testing the Effects of Chemicals on Earthworms: The Advantages and Limitations of Field Testing," in *Ecotoxicology of Earthworms* (P. W. Greig-Smith, H. Becker, P. J. Edwards, and F. Heimbach, Eds.), Intercept Ltd., Hampshire, UK, pp. 75–85

Edwards, C. A. and P. J. Bohlen, 1992, "The Effects of Toxic Chemicals on Earthworms," *Rev. Environ. Contam. Toxicol., 125*, 23–100

Goven, A. J., S. C. Chen, L. C. Fitzpatrick, and B. J. Venables, 1994, "Lysozyme Activity in Earthworm (*Lumbricus terrestris*) Coelomic Fluid and Coelomocytes: Enzyme Assay for Immunotoxicity of Xenobiotics," *Environ. Toxicol. Chem., 13*, 607–613

Greig-Smith, P. W., H. Becker, P. J. Edwards, and F. Heimbach, 1992, *Ecotoxicology of Earthworms*, Intercept Ltd., Hampshire, UK

Huggett, R. J., R. A. Kimerle, P. M. Mehrle Jr., and H. L. Bergman, 1992, *Biomarkers. Biochemical, Physiological, and Histological Markers of Anthropogenic Stress*, Lewis Publishers, Boca Raton, FL, USA

Lofs-Holmin, A., and U. Boström, 1988, "The Use of Earthworms and Other Soil Animals in Pesticide Testing," in *Earthworms in Waste and Environmental Management* (C. A. Edwards, and E. F. Neuhauser, Eds.), SPB Academic Publ., Hague, The Netherlands, pp. 303–313

Løkke, H., 1994, "Ecotoxicological Extrapolation: Tool or Toy?" in *Ecotoxicology of Soil Organisms* (M. H. Donker, H. Eijsackers, and F. Heimbach, Eds.), CRC Press, Boca Raton, Florida, USA, pp. 412–425

McCarthy, J. F., and L. R. Shugart, 1990, *Biomarkers of Environmental Contamination*, Lewis Publishers, Boca Raton, FL, USA

Ma, W. C., T. Edelman, I. V. Beersum, and T. Jans, 1983, "Uptake of Cadmium, Zinc, Lead and Copper by Earthworms Near a Zinc-Smelting Complex: Influence of Soil pH and Organic Matter," *Bull. Environ. Contam. Toxicol.*, *30*, 424–427

Mayer, F. L., D. J. Versteeg, M. J. McKee, L. C. Folmar, R. L.Graney, D. C. McCume, and B. R. Rattner, 1992, "Physiological and Nonspecific Biomarkers," in *Biomarkers. Biochemical, Physiological, and Histological Markers of Anthropogenic Stress* (R. J. Huggett, R. A. Kimerle, P. M. Mehrle Jr., and H. L. Bergman, Eds.), Lewis Publishers, Boca Raton, FL, USA, pp. 5–85

Meharg, A. A., 1994, "Inputs of Pollutants into the Environment from Large-Scale Plastics Fires," *Toxicol. Ecotoxicol. News*, *1*, 117–122

Menzel, D. W., and J. Case, 1977, "Concept and Design: Controlled Ecosystem Pollution Experiments," *Bull. Marine Sci.*, *27*, 1–7

Moore, M. N., 1985, "Cellular Response to Pollutants," *Marine Pollut. Bull.*, *16*, 134–139

Noss, R., 1990, "Indicator for Monitoring Biodiversity: A Hierarchial Approach," *Con. Biol.*, *4*, 355–364

OECD, 1994, OECD Guidelines for the Testing of Chemicals (No. 207). Earthworm, Acute Toxicity Tests. Organisation for Economic Co-Operation and Development (OECD), Paris, France

Peakall, D., 1992, *Animal Biomarkers as Pollution Indicators*, Chapman and Hall, London, UK, p. 291

Peakall, D. B., 1994, "The Role of Biomarkers in Environmental Assessment. 1. Introduction," *Ecotoxicology*, *3*, 157–160

Ramade, F., 1995, "Estimation of Damage to Ecosystems," in *Methods to Assess the Effects of Chemicals on Ecosystems* (R. A. Linthurst, P. Bourdeau, and R. G. Tardiff, Eds.), SCOPE no. 53, Wiley, Chichester, UK

Rodriguez-Grau, J., B. J. Venables, L. C. Fitzpatrick, A. J. Goven, and E. L. Cooper, 1989, "Suppression of Secretory Rosette Formation by PCBs in Lumbricus Terrestris: An Earthworm Assay for Humoral Immunotoxicity of Xenobiotics," *Environ. Toxicol. Chem.*, *8*, 1201–1207

SCOPE no. 53, 1995, *Methods to Assess the Effects of Chemicals on Ecosystems* (R. A. Linthurst, P. Bourdeau, and R. G. Tardiff, Eds.), Wiley, Chichester, UK

Sheehan, P. J., 1984a, "Functional Changes in Ecosystems," in *Effects of Pollutants at Ecosystem Level* (P. J. Sheehan, D. R. Miller, G. C. Butler, and P. Bourdeau, Eds.), SCOPE no. 22, Wiley, Chichester, UK, pp. 101–146

Sheehan, P. J., 1984b, "Effects on Community and Ecosystem Structure and Dynamics," in *Effects of Pollutants at Ecosystem Level* (P. J. Sheehan, D. R. Miller, G. C. Butler, and P. Bourdeau, Eds), SCOPE no. 22, Wiley, Chichester, UK, pp. 51–99

Snatschi, P. H., U. Nyffeler, R. Anderson, and S. Schiff, 1984, "The Enclosure as a Tool for the Assessment of Transport and Effects of Pollutants in Lakes," in *Concepts in Marine Pollution Measurements* (H. White, Ed.), College Park, MD, USA

Spurgeon, D. J., S. P. Hopkin, and D. T. Jones, 1994, "Effects of Cadmium, Copper, Lead and Zinc on Growth, Reproduction and Survival of the Earthworm *Eisenia Fetida* (Savigny): Assessing the Environmental Impact of Point-Source Metal Contamination in Terrestrial Ecosystems," *Environ. Pollut., 84,* 123–130

Spurgeon, D. J., and S. P. Hopkin, 1995, "Extrapolation of the Laboratory-Based OECD Earthworm Toxicity Test to Metal-Contaminated Field Sites," *Ecotoxicology, 4,* 190–205

Steel, J. H., 1979, "The Use of Experimental Ecosystems," *Phil. Trans. Roy. Soc. London B, 286,* 583–595

Svendsen, C. and J. M. Weeks, 1995, "The Use of a Lysosome Assay for the Rapid Assessment of Cellular Stress from Copper to the Freshwater Snail *Viviparus contectus* (Millet)," *Marine Pollut. Bull., 31,* 139–142

Svendsen, C., A. A. Meharg, P. Freestone, and J. M. Weeks, 1996, "Use of an Earthworm Lysosomal Biomarker for the Ecological Assessment of Pollution from an Industrial Plastics Fire," *Appl. Soil Ecology, 3,* 99–107

Svendsen, C., and J. M. Weeks, 1997a, "Relevance and Applicability of a Simple Earthworm Biomarker of Copper Exposure: I. Links to Ecological Effects in a Laboratory Study with *Eisenia andrei*," *Ecotoxicol. Environ. Saf., 36,* 72–79

Svendsen, C., and J. M. Weeks, 1997b, "Relevance and Applicability of a Simple Earthworm Biomarker of Copper Exposure: II Validation and Applicability under Field Conditions in Mesocosm Experiment with *Lumbricus rubellus*," *Ecotoxicol. Environ. Saf., 36,* 80–88

Tamaddon, F., and W. Hogland, 1993, "Review of Cadmium in Plastic Waste in Sweden," *Water Management Res., 11,* 287–295

Teuben, A., and T. A. P. J. Roelofsma, 1990, "Dynamic Interactions Between Functional Groups of Soil Arthropods and Microorganisms during Decomposition of Coniferous Litter in Microcosm Experiments," *Biol. Fertil. Soils, 9,* 145–151

Tolle, D. A., M. F. Arthur, and J. Chesson, 1985, "Comparison of Pots Versus Microcosms for Predicting Agroecosystem Effects due to Waste Amendment," *Environ. Toxicol. Chem., 4,* 501–509

Tolle, D. A., C. L. Frye, R. G. Lehmann, and T. C. Zwick, 1995, "Ecological Effects of PDMS-augmented Sludge Amended to Agricultural Microcosms," *Sci. Total Environ., 162,* 193–207

Van Gestel, C. A. M., E. M. D. V. Breemen, R. Baerselman, H. J. B. Emans, J. A. M. Janssen, R. Postuma, and P. J. M. V. Vliet, 1992, "Comparison of Sublethal and Lethal Criteria for Nine Different Chemicals in Standardized Toxicity Tests using the Earthworm *Eisenia andrei*," *Ecotoxicol. Environ. Saf., 23,* 206–220

Van Straalen, N. M., P. Leeuwangh, and P. B. M. Stortelder, 1994, "Progressing Limits for Soil Ecotoxicological Risk Assessment," in *Ecotoxicology of Soil Organisms* (M. H. Donker, H. Eijsackers, and F. Heimbach, Eds.), CRC Press, Boca Raton, FL, USA, pp. 397–409

Weeks, J. M., 1995, "The Value of Biomarkers for Ecological Risk Assessment, Academic Toys or Legislative Tools," *Appl. Soil Ecology, 2,* 215–216

Weeks, J. M., and C. Svendsen, 1996, "Neutral-Red Retention by Lysosomes from Earthworm (*Lumbricus Rubellus*) Coelomocytes: A Simple Biomarker of Exposure to Soil Copper," *Environ. Toxicol. Chem.*, *15*, 1801–1805

Zwick, T. C., M. F. Arthur, and D. A. Tolle, 1984, "A Unique Laboratory Method for Evaluating Agro-Ecosystem Effects of an Industrial Waste Product," *Plant and Soil*, *77*, 395–399

PART 4

Contributions to an Ecological Risk Assessment

22

Ecotoxic Modes of Action of Chemical Substances

Gerrit Schüürmann (Leipzig, Germany)

Ecotoxicology, Edited by Gerrit Schüürmann and Bernd Markert.
ISBN 0-471-17644-3   © 1998 John Wiley & Sons, Inc. and Spektrum Akademischer Verlag.

22.1 SUMMARY

Anthropogenic chemicals affect biological systems at various levels of organization. Their fate and bioavailability in the environment is governed by physicochemical properties and chemical reactivity, which also determine their ability to bioaccumulate and exert certain modes of toxic action in organisms. Chemical contaminants act primarily on suborganismic entities, such as molecules and cells, which are common features of all organisms. These provide a suitable level for classifying effects according to characteristic features. The main focus of the present chapter is a mechanistic analysis of such ecotoxic effects on aquatic organisms. They are discussed with a view to identifying relationships between structural features and ecotoxic profiles of chemicals. It thus belongs to the substance-oriented branch of ecotoxicology, which is complementary to system-oriented approaches discussed in other chapters of the book.

After an illustration of causative links between chemical exposure and system-level effects in the field, an outline of the biochemical background of chemical toxicity is presented, and subsequently current theoretical methods for quantifying relevant physico-chemical properties and reactivity aspects of compounds are reviewed. Application of these techniques for deriving structure–activity relationships and identifying structural determinants of ecotoxic bioactivity is then demonstrated with five case studies. This main section of the chapter starts with aromatic phosphorothionates, followed by phenols, alkylphenol ethoxylates, triorganometals, and unsaturated alcohols. Each case study includes general aspects of the environmental relevance, chemistry and fate of the compound class under investigation, and with each example particular approaches for analyzing activity profiles of chemicals through application of theoretical techniques are introduced. The general discussion also addresses the problem as to how concepts could be developed to translate ecotoxic effects and modes of action on suborganismic and organismic levels into effects on populations, communities and ecosystems.

22.2 INTRODUCTION

A major concern of current ecotoxicology is the question of whether and how information about chemical stress on individuals can be used for understanding and predicting hazardous effects of contaminants on higher levels of biological organization. Strategies for protecting the environment from chemical contamination require consideration of ecosystem structures and functions, whereas the vast majority of established ecotoxicological test procedures focus on direct chemical effects at the organismic or even suborganismic level. As compared to more complex test systems, such as microcosm and mesocosm studies or experiments in the field, single-species tests have the advantage of a generally high standardizability and reproducability, and explicit laboratory control of relevant system variables, such as exposure time and species condition, enables derivation of well-defined cause–effect relationships. On the other hand, single-species tests do not address indirect ecotoxicological effects and their propagation mechanisms, which form an important feature of system-level responses to chemical stress.

Only a few years after the introduction of ecotoxicology as a scientific discipline (Truhaut 1977), profound claims for more ecosystem-oriented work in this field have been made (Cairns 1985; Kimball and Levin 1985; Levin et al. 1989; Levin and Kimball 1984; Sheehan et al. 1984). Nonetheless, most ecotoxicological studies up till now have been devoted to short-term direct effects of chemicals on individual organisms (Clements and Kiffney 1994; Maltby and Calow 1989), and acute toxicity constitutes an important criterion for regulatory decisions in the notification process of chemical substances. This apparent conflict between the scope of ecotoxicology and the limits of its current methodological equipment has lead to an ongoing debate about the value of single-species tests as indicators of adverse effects on populations and beyond, which is reflected in recent book publications (Bartell et al. 1992; Cairns and Niederlehner 1995; Forbes and Forbes 1993; Suter II 1993) as well as in a recent series of journal editorials (Brown and Reinert 1992; Cairns 1992; Calow 1994; Clements and Kiffney 1994; Hope 1993).

Three simple model examples of effects at the population and community level will illustrate the principal problem in extrapolating from toxicity results with single species to situations in the field: Species with similar individual sensitivity to toxic exposure of chemicals may differ substantially in their population dynamics and resultant recruitment performance. The latter is a critical factor which determines whether the population will survive or decline under chemical stress. In communities, predator species with relatively high tolerance towards chemical exposure may suffer from selective eradication of more sensitive prey species (provided that compensation through invasion from neighbouring locations is not possible or sufficient), which would result in alterations of the food web. Finally, habitats with high levels of chemical contamination are expected to show shifts in the community structure to greater abundancies of more tolerant species and decreases in species diversity, with the associated consequences for ecosystem functions, such as nutrient cycling, productivity and genetic development.

22.2.1 Two Branches of Ecotoxicology

Generally, xenobiotics will have both direct and indirect effects on biological systems, and it seems quite obvious that ecotoxicological test systems should include and study system-level information in order to enable prediction of such events in the field. However, system theory has brought up the notions of the uniqueness and (almost) infinite variety of ecosystems. It would mean in the strict sense, that even experimental studies with the highest level of complexity would not be able to (precisely) predict the actual response of a given ecosystem to a given chemical stress.

A potential solution to this apparent dilemma of ecotoxicology starts with a reconsideration of the fundamental difference between organisms and more complex ecological structures. The former are characterized by a genetic disposition, which corresponds to a predefined range of responses to external stress factors, such as exposure to chemical contaminants. In contrast, the present status and future development of ecological structures depend heavily on their individual history, as was demonstrated in recent aquatic microcosm studies (Landis et al. 1996; Matthews et al. 1996). This means that besides the structural and functional inventory of ecosystems, their *historical conditioning* is crucial for the further development, both without and under the influence of chemical stress. A concrete example was presented in the investigations just mentioned, where treated microcosms with significant responses, mainly from daphnids, continued to differ from their untreated counterparts, even after complete disappearance of the chemical contaminants and subsequent recovery of the daphnids. Analysis suggested that detritus from unconsumed algae (during temporary decline of the daphnids) differed substantially from detritus coming from partially digested algae, such that temporary intoxication lead to ongoing alterations of the energy flow from primary producers to grazers (Matthews et al. 1996). From this viewpoint, the often quoted stochastic nature of more complex systems, such as ecological structures, has two components: One is the generally large variety of possibilities for further development that are all compatible with the current internal and external conditions (the *multidimensional component*), and the other is the storage of information of prior stressor events that conditions the future dynamics of the system (the *historical component*). Thus it will be generally (almost) impossible to predict an ecosystem response to chemical stress in a deterministic sense. Instead, a probabilistic prediction of events most likely to occur after chemical intoxication should be feasible, provided that sufficient knowledge about ecological structures in general and about chemical effects on species and abiotic compartments has been developed.

From this reasoning it follows that both single-species tests and experiments at higher levels of organization contribute profoundly to ecotoxicological research: The former allow the ranking of chemicals according to their toxicity to selected species, and enable elucidation of underlying modes of action and the establishment of direct relationships between chemical exposure and biological effect. This approach may be called the *substance-oriented branch of ecotoxicology*. In contrast, the *system-oriented branch of ecotoxicology* aims at identifying and characterizing indirect effects and their principal propagation mechanisms, using microcosms, mesocosms and field studies. Here, historical conditioning puts principal constraints on the reproducability and standardizability, and consequently the aim of this branch is to develop knowledge about the impact of outer and inner parameters on the probability of system-level events and the responses that occur.

22.2.2 Outline of Subsequent Sections

The model examples outlined above illustrate that knowledge about system-level properties and dynamics is needed for a sound understanding and assessment of direct and indirect ecotoxicological effects in the environment. On the other hand, environmental fate, bioavailability and bioaccumulation of chemical contaminants, as well as their disposition to direct ecotoxic effects on individual organisms, are driven by their physicochemical properties and reactivity characteristics. In the present chapter, the focus will be on the substance-oriented branch of ecotoxicology, and particular emphasis will be given to quantitative structure–activity relationships (QSARs) as a tool for the identification and characterization of molecular modes of ecotoxic action.

The potential impact of chemical structure on higher-level effects in the field will be illustrated with three examples in the next section, where causative relationships can be shown to exist between anthropogenically caused exposure to chemicals and damage to populations, communities and environmental systems. An outline of the biochemical background of the interaction between chemicals and biota follows, and the principal modes of bioactivity will be analyzed using classification criteria at the molecular and phenomenological level. Subsequently, theoretical methods for quantifying physicochemical properties and chemical reactivity will be reviewed. This section includes a somewhat more extended discussion of quantum chemical parameters, which describe local or global aspects of the electronic structure of molecules, such as site-specific or overall electrophilicity and nucleophilicity. All major reactivity parameters are specified in mathematical form to enable independent application without reference to the primary literature, and it is suggested that readers with less interest in the details may concentrate on the concepts of chemical reactivity and the resulting possibilities for characterizing reactivity-driven modes of toxic action. After this presentation of relevant theoretical methods, the subsequent section will present five application examples in the form of separate case studies. This main part of the chapter will illustrate how molecular descriptors for physicochemical properties and chemical reactivity can be used to analyze activity profiles of chemicals from a mechanistic viewpoint. The concluding paragraph will discuss the potential scope of ecotoxic modes of action for addressing effects on higher levels of biological organization.

22.3 CAUSE-EFFECT RELATIONSHIPS IN THE FIELD

In view of the complexity of ecological structures it would seem almost impossible to trace biological effects in environmental systems to the presence of certain chemicals with specific property and activity profiles. However, there are prominent examples of distinct chemical pollution in the field with subsequent effects on population and community levels, which illustrate that a proper understanding of such events needs information about system-level properties, modes of action at the organismic or suborganismic level, compound properties and reactivity characteristics of the chemical toxicant.

22.3.1 DDT

The perhaps most widely discussed case of biological damage as a result of anthropogenic chemicals is the relationship between application of p,p'-DDT (dichlorodiphenyltrichloroethane; systematic name: 1,1,1-trichloro-2,2-di-(4-chlorophenyl)-ethane) as insecticide and a dramatic population decline of the peregrine falcon (*Falco peregrinus*) in the 1950s and 1960s in the USA. DDT was first synthesized in 1874 by the German chemist Othmar Zeidler, and its insecticidal activity was detected in 1939 by Paul Müller, who was honoured for this finding with the nobel prize for medicine in 1946. From 1940 to 1970, DDT was used in high volumes throughout the world and proved its merits in the control of malaria and typhus. It had reached annual levels of around 100,000 t in 1963 (Metcalf 1973). One major cause for the observed population decline of the falcons was their reduced breeding success, which could be traced back to a significant eggshell thinning caused by DDT and its metabolite DDE in the bird. Uptake of DDT occurred with contaminated prey species, illustrating that persistent insecticides may well reach and affect non-target animals at higher trophic levels through bioaccumulation along the food chain.

The relatively high persistence of DDT is associated with its lipophilicity, which in turn enables the compound to penetrate insect cuticles much better than animal skins and thus makes it appear safe and target-specific for in-field applications. DDT is neurotoxic and interferes with the axonic membrane of the nervous system, and DDT as well as DDE were shown to impair the thyroid function (Jefferies and French 1971, 1972). Analysis of falcon eggs revealed substantial concentrations of DDE, and correlations between eggshell thickness and DDE burdens suggest that this metabolite is the active agent. There is evidence that the eggshell thinning is caused by inhibition of the enzyme Ca^{2+}-ATPase (Cooke 1973, 1979), but the molecular level of the toxic mechanism has not been fully unravelled (Moriarty et al. 1986).

Interestingly, bird sensitivity and response to DDT and DDE differ markedly: Thinning of eggshells of falcons and mallard ducks (Hodgson and Levi 1987) contrasts with thickening of eggshells of the bengalese finch (*Lonchura striata*) (Jefferies 1969), and eggshells of gallinaceous species, such as domestic hen, pheasant and quail, remain essentially unaffected by DDT (Cooke 1973). In any case, recovery of the peregrine population started during the mid-1970s, which is in qualitative agreement with a decrease in DDE residues in eggs of this species (Peakall and Tucker 1985). In 1972, DDT production and application were banned in the USA and Germany, and nowadays DDT is still being used in developing countries in estimated annual volumes of 40,000 t. Due to its high stability under environmental conditions, substantial levels of DDT and DDE can be found in environmental compartments all over the world, and recent investigations suggest DDE to act as a potent antagonist of the androgen receptor (Kelce et al. 1995). The DDT story with the peregrine falcon thus illustrates how complex and unexpected the ecotoxicological effect of xenobiotics may be, and how much information on various levels of biological organization and about the toxic agent itself is necessary to detect and understand such events. It also shows how difficult it is to develop ecotoxicological knowledge for a predictive assessment of potential environmental damage due to xenobiotics.

22.3.2 Acid Rain

Another well-known example of ecotoxicological effects with a clear chemical cause at the system level is given by the acidification of lakes by acid rain. Major sources of acid rain are anthropogenic emission of sulfur oxides and nitrogen oxides, which undergo hydrolyzation and oxidation in the troposphere and return to the ground as H_2SO_3, H_2SO_4 and HNO_3, respectively, dissolved in raindrops. Uncontaminated rain already has a pH of ca. 5.6 as a result of dissolution of background, airborne CO_2, and acid rain may reach pH values of 4 and below.

In Scandinavian lakes and those of the Great Lakes area of Canada and the USA, acidic deposition has lead to death of aquatic organisms and corresponding reductions in species diversity, disruption of normal food–chain relations, and shifts to greater abundances of acidophilic species. A common geogenic feature of these lakes is their relatively low buffer capacity, which makes them particularly susceptible to a drop in pH due to the uptake of airborne acids. The resultant ecotoxicological effects could be traced back to (at least) two components: Acidic water is directly toxic for biota, perturbating the osmoregulation caused by disruption of the transepithelial electrochemical gradient, and acidification of lake water leads to mobilization of heavy metals with subsequent uptake and more specific toxic effects in aquatic organisms. A typical example is the generation of toxic Al^{3+} by acidic hydrolysis of $Al(OH)_3$, a natural component of sediments and soils. In contrast to the DDT example outlined above, the primary action of the chemical contaminant (the acid) is the alteration of an important physicochemical characteristic (the pH) of the abiotic environmental compartment, and the associated ecotoxicological effect results from a superposition of chemical stress by the primary agent as well as by other mobilized toxicants.

Much work has been devoted to identifying ecological consequences in the aquatic communities of acidified lakes. These form model cases for aquatic ecosystem responses to chemical stress on local to regional scales. In the northwestern area of Ontario in Canada, the effect of pH lowering on aquatic communities has been studied in long-term ecosystem-level experiments with poorly buffered small lakes (Schindler 1987; Schindler et al. 1985). Artificial acidification was achieved by gradual addition of sulfuric acid to decrease the lake pH to 5.0 over a period of eight years. Small species with high reproduction rates and wide dispersal powers, such as phytoplankton, turned out to be most sensitive at the early stage of acidification, and other early indicators of acidic stress were provided by morphological abnormalities in benthic invertebrates.

On the other hand, global ecosystem functions, such as primary production, nutrient cycling and respiration, remained essentially unaffected by acidification of the lake water.

As a further interesting finding, comparison of microcosm and mesocosm experiments suggested that both of them are still too small and not sufficiently complex to mimic results with real ecosystems (Schindler 1987). In particular, mesocosm experiments with a longer duration seemed less adequate for simulating selected features of ecosystems. This observation is not too surprising in view of the above-mentioned hypothesis that the individual history of a system conditions its future dynamics: As a consequence of the historical conditioning, greater divergence between initially similar (but not identical) systems would be expected with an increasing development time. On the other hand, indirect ecotoxicological effects through interactions between the populations could of course not be predicted from results with single-species tests. These, however, could yield valuable information about principal differences in species sensitivity and in the mode of response to acids and heavy metals in aqueous solution. As noted above, a predictive characterization of the typical or most likely scenarios of a lake during acidification would need knowledge about relevant ecological structures and dynamics of the aquatic ecosystem as well as about response characteristics at the level of individuals.

22.3.3 Chlorofluorocarbons

A third and somewhat different example of a causative relationship between chemical pollution and environmental damage in the field is the depletion of stratospheric ozone by anthropogenically emitted chlorofluorocarbons (CFCs). Manufacture of CFCs started in the 1930s, and their physicochemical property profile and lack of toxicity made them highly suitable as refrigerants. Annual worldwide production reached levels of around 1.2 million tons in the 1980s, and it can be assumed that most anthropogenic CFCs migrate upwards into the stratosphere within 10 years of their initial emission. First warnings about a potential depletion of the ozone layer came in the 1970s, and since the late 1980s there is convincing evidence that stratospheric ozone loss beyond diurnal and seasonal variation patterns has indeed occurred (McElroy and Salawitch 1989; McFarland 1989; McFarland and Kaye 1992; Stolarski 1988).

The mechanism of ozone removal by CFCs starts with photolytic generation of chlorine radicals, which then participate in the catalytic degradation cycle of stratospheric ozone. Typical residence times for the relevant CFCs of around 50 to 100 years indicate that even after a complete stop of CFC release into the atmosphere, reduction of the ozone layer would continue for years. One important function of stratospheric ozone is its absorption of UV-B radiation (290–320 nm wavelength), through which it protects humans from skin cancer, and animals and crops from general UV damage. It follows that increased transmission of the UV-B component of sunlight due to depletion of the ozone layer is likely to affect a variety of species, including microorganisms.

Current discussions focus on potential deleterious effects on krill in the Antarctic with subsequent implications for associated food chains. Another matter of concern is the potential sensitivity of crop plants to increased UV-B radiation, which may also make the plant more susceptible to other natural and anthropogenic stressors. Finally, depletion of stratospheric ozone might lead to climatic changes with a corresponding impact on the global ecosystem. These scenarios demonstrate that although CFCs do not act directly on biological systems, their interference with the ozone layer may well have dramatic indirect effects on the population, community and ecosystem level. Because the initial stressor would be electromagnetic radiation and thus non-chemical, the CFC problem would not fall within the focus of classical ecotoxicology. However, the potentially hazardous consequences for biological systems could still be traced back to the environmental fate and reactivity of anthropogenically emitted chemicals.

22.4 BIOCHEMICAL BACKGROUND AND MODES OF TOXIC ACTION

The DDT example in the previous section has illustrated two important aspects of the toxic action of chemicals: Biological sensitivity to contaminant concentrations may vary substantially from species to species, and the mode of toxic action may be highly selective and impair specific biological functions of the organism. Despite the great variety of hazardous effects in biota exerted by chemicals, systematic relationships exist between the chemical structure of compounds and their potential for toxic activity. The latter has been advanced by development of molecular toxicology as a relatively new sub-discipline of toxicology. Here, toxic action is analyzed by distinguishing between several stages of the overall process including uptake, metabolism, interaction with endogenous sites, biological repair processes and elimination.

22.4.1 Biotransformation

A simplified scheme of the in vivo fate of chemicals is given in Figure 22.1, emphasizing the two phases of toxicokinetics and toxicodynamics in the course of a toxic process. After uptake into the organism, the chemical has to pass further biological membranes to reach internal compartments. With organic substances, a certain range of lipophilicity is usually a prerequisite for efficient membrane transfer, which indicates the great importance of this physicochemical property for the toxicological fate of compounds (cf. Section *Compound Properties and Molecular Descriptors* below). Elimination can occur by passive diffusion of the initial compound as well as of (usually more polar) metabolites, and biotransformation can increase or decrease its toxicity directly or by affecting its ability to pass further membranes and reach the site of action. In this

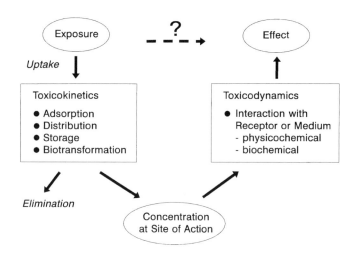

Figure 22.1 Fate and effect of chemicals in biological organisms.

context, the dual role of biotransformation should be stressed: In addition to its natural task of protecting the organism against chemical stress, it may also generate more toxic metabolites, as will be discussed in more detail below with the class of proelectrophiles and their modes of toxic action (see case study *Unsaturated Alcohols*). Other prominent examples of toxication through biotransformation are given by benzo[a]pyrene-type polycyclic aromatic hydrocarbons (containing the so-called bay region), which are activated in vivo to carcinogenic diol epoxides by stepwise cyctochrome P-450 oxidation, by dehalogenation of tetrachloromethane (CCl_4) to the trichloromethyl radical with subsequent peroxidation of lipid tissues, and by phosphorothionate insecticides with oxidative activation to oxon derivatives as potent acetylcholinesterase inhibitors (see case study *Aromatic Phosphorothionates* below).

The metabolism of xenobiotics during the toxicokinetic phase is again separated into two general phases. In phase I, polar functional groups are introduced into the molecule, making it more water-soluble than the parent compound and generally more suitable for excretion or further conjugation with endogenous substrates. A major phase I biotransformation is oxidation, which may be mediated by cytochrome P-450 (mixed function oxidases), alcohol dehydrogenase, aldehyde dehydrogenase or monoamine oxidase. Reduction and hydrolysis are further important phase I reactions listed in Table 22.1. Generally, phase I metabolism of a chemical results from competition between different functionalization reactions, which in turn affects its overall toxicological profile.

In phase II, conjugation with endogenous molecules further increases polarity and water solubility, and correspondingly decreases the compound's ability to passively diffuse through cellular membranes. The most important phase II biotransformations are summarized in Table 22.2, and the general effect is

Table 22.1 Phase I transformations

General type	Example	Xenobiotic substrate	Functional group of metabolite
Oxidation	Hydroxylation	Aliphatics, aromatics	-OH
	Epoxidation	Alkenes, alkynes, aromatics	Epoxide
	O-, N-, S-dealkylation	Amines, ethers, thioethers	-OH, -NHR, -NH$_2$, -SH
	N-, S-, P-oxidation	Amines	-NH-OH, -NR-OH, -NO$_2$, R$_3$N$^+$-O$^-$
	S-, P-oxidation	Thioethers, phosphines	-S(=O)-, -S(=O)(=O)-, R$_3$P=O
	Desulfuration	Phosphorothionates	R$_3$P=O
	Transformation mediated by alcohol dehydrogenase	Aliphatic and aromatic alcohols	-CHO, -C(=O)-
	Transformation mediated by aldehyde dehydrogenase	Aliphatic and aromatic aldehydes	-COOH
	Deamination	Aliphatic and aromatic amines	-C(=O)-
	Aliphatic dehydration	Aliphatic hydrocarbon moieties	-CH = CH-
Reduction	Ketone reduction	Ketones	-OH
	Nitro and azo reduction	Aromatic nitro and azo compounds	-NH-NH, -NH$_2$
	Dehalogenation	Halogenated hydrocarbons	-CHR$_2$, -CH$_2$R
	Hydration	Alkene moiety	Alkane moiety
	Disulfide reduction	Disulfides	-SH
Hydrolysis	Epoxide hydrolysis	Epoxides	Diol
	Ester hydrolysis	Carboxylic esters, phosphates	-COOH and -OH, (RO)$_2$P(=O)OH and OH
	Carboxyamide hydrolysis	Amides	-COOH, -NHR
	Carboxythioester hydrolysis	Thioesters	-COOH, -SH

Table 22.2 Phase II conjugations

Endogenous agent	Xenobiotic substrate	Type of conjugation
UDPGA (uridine diphosphate glucuronic acid)	Alcohols, phenols, thiols, carboxylic acids, amines	Glucuronide formation
PAPS (3'-phosphoadenosine-5'-phosphosulfate)	Alcohols, phenols, aromatic amines	Sulfate conjugation
SAM (S-adenosyl methionine)	Phenols, catechols, thiols, aliphatic and aromatic amines	Methylation
Acetyl coenzyme A	Amines, hydrazines, amino acids	Acetylation
Glycine	Aliphatic and aromatic carboxylic acids	Hippuric acid formation
Glutathione (γ-L-glutamyl-L-cysteinylglycine)	Electrophiles (epoxides, alkenes, halides etc.)	Mercapturic acid formation

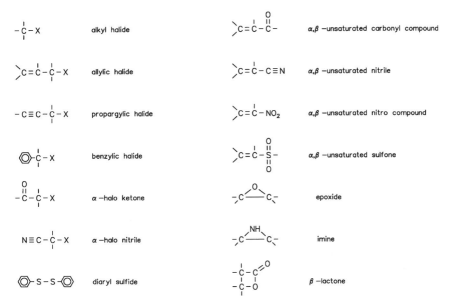

Figure 22.2 Electrophilic structures in chemical compounds.

a substantially enhanced rate of elimination of the (chemically modified) xenobiotic. For electrophilic compounds with the ability to attack basic sites of endogenous macromolecules, the tripeptide glutathione forms a particularly important conjugating agent. Due to the ubiquitous presence of nucleophilic sites, such as -SH, -NH$_2$ and -OH (ordered in increasing hardness) in endogenous molecules, electrophilicity is an important general basis of chemical toxicity. As a rule, compounds with intermediate electrophilicity and lipophilicity are expected to show the greatest toxic potency due to their ability to pass cellular membranes and interact with critical endogenous sites. Alkylhalogenides, α,β-unsaturated carbonyl compounds and epoxides are examples of electrophilic chemicals which may be deactivated by conjugation with glutathione. A more extended list of chemical structures with electrophilic character and corresponding toxicological relevance is given in Figure 22.2. However, conjugation by glutathione may eventually be reversible and enable translocation to intracellular sites by temporarily masking electrophilic reactivity, which is one example of possible toxication processes as a result of phase II metabolism (cf. Baillie and Slatter 1991).

22.4.2 Classification of Chemical Toxicity

Electrophilic toxicity is an example of an adverse effect resulting from covalent interaction between the xenobiotic chemical and essential compounds of the cellular metabolism. At the organismic and sub-organismic level, chemical toxicity is generally caused by impairment of biochemical and physiological processes. As noted above, the chemical structure of a compound encodes a disposition for its reactivity and toxicity profile, and the actual in vivo fate of the compound after uptake depends on both chemical properties and the physiology and metabolism of the organism. From the molecular viewpoint, the great variety of overall toxic effects can be traced back to a few general mechanisms. A corresponding tentative list of basic molecular mechanisms of chemical toxicity is given in Table 22.3. According to this classification, chemical burn through hydrolysis of body fluids and subsequent disruption of proteins and tissues, and impairment of membrane fluidity through hydrophobic association form two principal types of unspecific xenobiotic action in organisms. Alteration of cellular transport systems, such as ion channels, interference with normal receptor-ligand functions, impairment of enzymatic catalysis and covalent binding to other endogenous sites of essential macromolecules, form another four principal mechanisms of toxic action, which may lead to a variety of different effects, depending on the location, type and function of the affected site. At this level of classification, no distinction between different types of chemical reactions, such as, for example, nucleophilic displacement or a radical reaction, is made, and the targets are characterized only by their principal type and biochemical function in the organism.

When distinguishing more specifically between different endogenous targets and their biochemical or physiological functions, a much greater list of distinct modes of toxic action is achieved. An example is given by specifying different types of cytotoxicity: Impairment of cellular vitality can originate from hydrophobic association of chemicals at the membrane surface with subsequent modifications of membrane structure and fluidity. In an alternative explanation

Table 22.3 Mechanisms of primary molecular interaction of chemical toxicity

Toxic mechanism	Type of chemical agents
Hydrolysis of compartments and macromolecules	
Dissolution of tissues, denaturation of proteins	Strong acids and bases
Interference with membrane functions	
Activation or deactivation of ion channels	Sodium antagonists
Alteration of cellular transport systems and membrane fluidity	Narcotics, diuretics, moderate and weak acids and bases
Interaction with endogenous molecules	
Interference with normal receptor-ligand interactions	Agonists and antagonists like neurotoxicants or β-blockers
Activation or deactivation of enzymes	AChE inhibitors
Covalent bond formation with proteins, DNA, RNA or tissue lipids	Alkylating agents, lipid peroxidizing agents

model, unspecific membrane irritation results from hydrophobic binding of chemicals to protein pockets, and the true mode of narcotic action is still a matter of discussion (Franks and Lieb 1982, 1984, 1985, 1986). Other modes of cytotoxicity include alkaline or acidic hydrolysis, peroxidation of membrane lipids by chemical radicals, and alkylation of membrane proteins, peptides, RNA or DNA by electrophilic compounds (s.a.). Correspondingly, other targets or compartments may be affected by a variety of mechanisms and on different time scales. It is an ongoing task of ecotoxicology to identify and characterize the mechanistic basis of toxicity exerted by chemicals which are considered to have substantial relevance for aquatic and terrestrial species under environmental conditions. Examples of such molecular modes of ecotoxic action with potential environmental relevance have been collected from the literature (Corbett et al. 1984; Draber and Fujita 1992; Könemann 1981; Schüürmann 1991; Veith et al. 1983; Veith and Broderius 1990) and are listed in Table 22.4. Some of these will be discussed in more detail in the Section *Molecular Mode-of-Action Analysis of Compound Toxicity* below.

At this stage, however, a short comment should be made about the alternative notions of toxic mechanism and toxic modes of action. The former is usually understood as indicating precise knowledge of the relevant processes at the molecular and possibly sub-molecular level, whereas the latter is often used in cases where the underlying mechanism is not definitely disclosed, but rather characterized at a more phenomenological level. In practice, the differentiation between these two terms is not always that strict, and in Table 22.4 the category "mode of action" includes both phenomenologically and more distinctly defined characterizations of toxic processes.

Narcosis is a good example of an ecotoxic mode of action which has developed as a useful concept in aquatic toxicology for grouping different chemicals under the same type of overall toxic effect. For a given species, narcosis may be exerted through different molecular mechanisms as outlined above. With all of these possible mechanisms, however, narcosis is governed by hydrophobicity or lipophilicity (or other related physicochemical properties) of the xenobiotic compound, which has enabled the derivation of simple quantitative structure–activity relationships (QSARs) for narcotic-type toxicity of various classes of chemicals in aquatic organisms (Könemann 1981; Veith et al. 1983; Veith and Broderius 1990; Schüürmann 1991). An important application of these baseline QSARs is their use as (more or less empirical) frames of reference for identifying more specifically acting xenobiotics by comparing their experimental toxicity with the prediction derived from narcosis theory (Lipnick 1995). This approach has proven useful in detecting and characterizing different ecotoxic modes of action and in linking them to distinct structural features and chemical reactivities of the compounds. Prominent examples include phosphorothionate insecticides, phenolic oxidative uncouplers, organotins, polyoxyethylene surfactants and proelectrophilic unsaturated alcohols, which are all discussed in more detail in the remainder of this chapter. Here, theoretical analysis of more specific interactions with endogenous sites needs consideration of suitable parameters

Table 22.4 Examples of molecular modes of ecotoxic action of chemicals

Mode of action	Site of action	Example chemicals
Nonpolar narcosis[a]	Cell membrane	Octanol, acetone, chlorobenzene
Polar narcosis[a]	Cell membrane	Phenol, 1-naphthol, aniline, 4-toluidine
Ethoxylate narcosis[a]	Cell membrane	Alkylphenol ethoxylate surfactants
Electrophilic cytotoxicity	Membrane-bound macromolecules	Acrylnitrile, acrolein, thiiran, epoxide
Peroxidating cytotoxicity	Membrane-bound lipids	Tetrachloromethane, hydroquinone
Respiratory electron transport inhibition	Various sites of respiratory chain	Rotenone, dibutylchloromethyl tin, cyanide
Inhibition of oxidative phosphorylation	ATP synthetase of respiratory chain	Oligomycin, bis(tributyltin) oxide, tetrasul
Inhibition of respiratory ATP carriage	ATP-ADP translocase of respiratory chain	Atractylate
Oxidative uncoupling of respiration	Mitochondrial membranes	2,4-dinitrophenol, pentachlorophenol
Photosystem I electron transport inhibition	Thylakoid membranes of chloroplasts	Diquat, paraquat
Photosystem II electron transport inhibition	Thylakoid membranes of chloroplasts	Atrazine, diuron, linuron, benzthiazuron
Amino acid biosynthesis inhibition	PSCV transferase in cytoplasma	Glyphosate
Amino acid biosynthesis inhibition	Glutamine synthetase in chloroplast stroma	Bialaphos, glufosinate
Acetylcholinesterase inhibition	Postsynaptic membrane of nerve cells	Parathion, bromophos, diazinon, sarin
Mutagenicity and genotoxic carcinogenicity	DNA	Aziridine, dimethylsulfate, mustard gas
Epigenetic carcinogenicity	Various cellular sites	Purine analogs, catechol, phorbol ester

[a] The notion narcosis as introduced in aquatic toxicology refers to general membrane irritation through hydrophobic association of the xenobiotic with membrane-bound lipids or proteins.

for characterizing and quantifying relevant reactivity aspects of chemical structures. This will be outlined in the next section.

22.5 COMPOUND PROPERTIES AND MOLECULAR DESCRIPTORS

The environmental fate and ecotoxicological profile of organic xenobiotics are largely determined by physicochemical properties (cf. Chapter 8). After anthropogenic input into the environment or mobilization from matrices, passive transfer of the chemicals between different abiotic compartments may lead to local, regional or global dispersion and accumulation in remote ecosystems as well as in biological organisms. Variation in only a few physicochemical properties results in completely different dispersion and accumulation patterns. Degradation by photolysis or hydrolysis also depends on distinct structural features of the compounds (cf. Chapter 9). Similarly, the ecotoxic profile of xenobiotics can often be traced back to relatively few molecular properties and chemical reactivity characteristics.

22.5.1 Octanol/Water Partition Coefficient

According to findings in the last three decades, one of the key properties for the environmental and ecotoxicological behaviour of organic contaminants is their hydrophobicity, which is often quantified as octanol/water partition coefficient in logarithmic form, $\log K_{OW}$. Sorption to soils and sediments, bioaccumulation and baseline toxicity to aquatic organisms are often related to $\log K_{OW}$. A particularly important area of application of this free-energy parameter has been and still is its use for quantitative structure–activity relationships (Dearden 1985; Lipnick 1995), where it provides both baseline predictions of toxicity and a frame of reference for identifying and characterizing more specifically acting chemicals with toxic potencies greater than narcotic-type isolipophilic compounds (s.a.).

Apart from experimental techniques for determining $\log K_{OW}$, calculation methods based on additive-constitutive increment schemes have been developed which are also available in computerized form (Klopman and Wang 1991, Leo et al. 1971; Leo 1993; Meylan and Howard 1995). Despite a generally wide application range and remarkable prediction capability of these tools for $\log K_{OW}$ calculations, there is still room for improvement with certain compound classes and more complex chemical structures (Lyman et al. 1990; Schüürmann et al. 1995). In addition to the classical increment schemes, completely different calculation techniques have been advanced: one method uses Monte Carlo simulations of molecular systems containing the solute and a large cluster of solvent molecules (Jorgensen et al. 1990), which is computationally quite demanding. A more recent approach is based on quantum chemical continuum-solvation methods for calculating solvation energy and molecular contact surface area as parameters for polar and non-polar solute-solvent interactions (Schüürmann 1995). Furthermore, hydrophobicity can be estimated from water solubility, which in turn can also be approximately calculated using increment methods (Klopman et al. 1992; Kühne et al. 1995).

A general shortcoming of current $\log K_{OW}$ estimation methods is the lack of proper fragment values for ionic and dissociated structures. This has also not yet been studied in depth with quantum chemical methods. As discussed below in the case study *Chlorophenols and nitrophenols*, the degree of ionization affects phase transfer, partitioning, bioavailability and bioaccumulation as well as toxicity, and further investigation is needed to evaluate the usefulness of $\log K_{OW}$ as a proper surrogate parameter for the bioconcentration and narcotic-type toxicity of anions and cations.

22.5.2 Dissociation Constant in Aqueous Solution

For an ionizing compound, the degree of dissociation (or protonation) in aqueous solution is usually quantified by the pK_a value of the acidic form, for which there are different calculation techniques available. The classical approach is based on the Hammett equation (s.b.; Perrin et al. 1981) and also has the form of a fragment method. Provided that the compound of interest can be separated into parent structure and substituents with known increment values, calculation of pK_a is straightforward and has been used routinely in structure–activity analyses to evaluate the impact of ionization on bioaccumulation or toxicity (cf. case study *Chlorophenols and Nitrophenols* below). The method has also been implemented in computerized form, and its overall success is reflected by statistics of $r^2 = 0.78$ and a standard deviation of 1.43 pK_a units with 1401 organic compounds (Hunter 1988). However, application of this classical method is principally restricted to certain types of parent structures and substituents, which may not apply to the compound under investigation. Consequently, there is a need for alternative methods for calculating pK_a, and recent findings with two simple compound classes suggest that continuum-solvation models of semiempirical quantum chemical schemes may be suited for providing reasonable pK_a estimates after proper scaling of the results (Schüürmann 1996). Further improvement can be expected by using corresponding ab initio methods to calculate the free energy of aqueous solvation (which is directly related to pK_a), and additional experience is needed to evaluate the performance of this more fundamental approach to more complex and diverse structures of ionizing compounds.

As indicated above, both $\log K_{OW}$ and pK_a (if relevant) affect the environmental behaviour of chemical xenobiotics and their toxicokinetic and toxicodynamic phase after uptake in biota. However, a more distinct evaluation of the in vivo fate of chemicals needs characterization of their biochemical reactivity, which in turn depends on their electronic structure. In organisms, the chemical may interact with endogenous macromolecules through formation of covalent bonds via substitution or addition reactions. Here, polar functionalities of the macromolecules, such as amino, sulfhydryl, hydroxyl and carboxyl groups form typical sites of attack. The electron-rich character of these sites makes them susceptible to electrophiles, an important general basis of chemical toxicity as noted in the previous section.

Due to the variety of chemical reaction mechanisms and pathways, an adequate characterization of molecular reactivity does not seem straightforward at first sight. However, physical, organic and theoretical chemistry have provided several concepts for addressing the molecular disposition for certain electronic interactions in a general way. The following discussion will focus on linear free energy relationships and three approaches derived from quantum chemistry, all of which have already proven useful for the analysis and understanding of specific modes of toxic action.

22.5.3 Hammett Equation

In the 1930s, Hammett analyzed dissociation of meta- and para-substituted benzoic acids in aqueous solution and hydrolysis of the corresponding esters for a large number of compounds. Parallel trends between the acid dissociation equilibria, K, and the hydrolysis rate constants, k, lead to the derivation of a linear free-energy relationship,

$$\log \frac{k}{k_0} = \rho \cdot \sigma \qquad (22.1)$$

where the reaction constant ρ depends on the parent structure of the chemical compounds, and the substituent constant σ is defined as

$$\sigma = \log \frac{K}{K_0} = pK_{a(0)} - pK_a \qquad (22.2)$$

with the index 0 denoting the dissociation constant and reaction rate of unsubstituted benzoic acid and ester. Using Equation 22.1, which is known as the Hammett equation, electronic substituent effects can be analyzed in a systematic manner, and corresponding linear free-energy relationships have been observed with varying precision for hundreds of reactions. The substituent constant σ has positive values for π-acceptors, such as NO_2, CN, halogen, COOR, SOR and SO_2R groups, and thus decreases electron density at the reaction site and makes it more susceptible for attacking nucleophiles. In contrast, π-donors, such as OH, NH_2 and SH have negative σ values, supporting electrophilic attack at the reaction site. Extensive compilations of Hammett constants are available in the literature (Hansch et al. 1991; Hansch et al. 1995). Halogens form a particularly interesting group of substituents, where the mesomeric effect (π-donor with a negative σ constant) is opposed to the electron-attracting inductive effect. With chlorine, both effects are of comparable strength, which allows the substituent to accommodate for quite different electronic demands. This may partially explain its well-known impact on the chemical stability and toxicity of organic compounds.

The absolute size of the reaction constant ρ reflects the sensitivity of the reaction to polar substituent effects, and the sign indicates the type of electronic demand along the reaction path: positive ρ corresponds to electrophilic reaction sites (e.g. ester carbon) with an excess charge in the transition state, and negative ρ indicates attack on a nucleophilic reaction centre. After introduction of the Hansch approach 30 years ago, electronic substituent constants and pK_a values have often been used in combination with hydrophobic parameters to derive and analyze structure-activity relationships for a diverse range of compounds and modes of action (Hansch and Fujita 1964; Hansch and Leo 1995; Hermens and Opperhuizen 1990; Kaiser 1984 and 1987; Lipnick 1995; Schüürmann 1992; Turner et al. 1988).

Despite the fundamental importance and success of the Hammett equation for quantifying electronic interactions within molecules and their implications for molecular reactivity and biological activity, a principal shortcoming is the need for suitable substituent constants which may be lacking for the compound of interest (cf. discussion with pK_a above). More generally, this classical approach requires the compound to be structurally related to some parent compound in such a way that substituent constants can be (and have been) defined. In contrast, quantum chemical methods allow, in a more general way, the quantification of various aspects of the electronic structure of molecules and its response to perturbations. For a given molecule, the electronic structure can be analyzed at all sites of interest and at various levels of complexity, and its reactivity to a toxicologically relevant (model) agent can also be studied using explicit calculations of energy barriers along reaction paths of the interacting system.

22.5.4 Ground-State Global Molecular Reactivity

Chemical reactions are intimately related to the gain and loss of electrons by the interacting species. From this viewpoint (ignoring steric effects), differences in molecular reactivity are driven by differences in the ability of the interacting compounds to exchange electron charge. Both the extent and ease of electronic deformation may vary from compound to compound as well as from site to site within a given molecule. These considerations form the basis of the concept of characterizing chemical reactivity by molecular electronegativity and hardness (Parr and Pearson 1983). Electronegativity, EN, characterizes the electron attraction tendency of atoms or molecules. Generally, electron charge is transferred from lower EN (donor molecule) to higher EN (acceptor molecule). Thus EN determines the direction of net electron transfer between interacting molecules, whereas hardness, η, is a measure of how easily molecular electron density can be changed. The harder the molecule, the more it will resist accepting or donating electron charge. EN and η are related to the ionization potential, IP, and electron affinity, EA, according to

$$EN = \tfrac{1}{2}(IP + EA) \tag{22.3}$$

$$\chi = \tfrac{1}{2}(IP - EA) \tag{22.4}$$

With closed–shell systems (i.e. if the molecule is not a radical), IP and EA can be approximated using the negative energies of the HOMO (highest occupied molecular orbital) and LUMO (lowest unoccupied molecular orbital),

$$IP = -\varepsilon_{\mathrm{HOMO}} \tag{22.5}$$

$$EA = -\varepsilon_{\mathrm{LUMO}} \tag{22.6}$$

thus making EN and η readily accessible by routine quantum chemical calculations. Although both electronegativity and electron affinity characterize the electron attraction tendency of molecules, they differ not only mathematically, but also conceptually, as can be seen from the original rationalization of EN by Mulliken (1934). Two chemical species X and Y have equal electronegativities if the energy changes ΔE associated with the two disproportionation reactions

$$X + Y \rightarrow X^+ + Y^- \tag{22.7}$$

$$X + Y \rightarrow X^- + Y^+ \tag{22.8}$$

are equal. From

$$IP(X) = \Delta E(X \rightarrow X^+ + e^-) \tag{22.9}$$

$$EA(X) = -\Delta E(X + e^- \rightarrow X^-) \tag{22.10}$$

it follows that the first reaction energy is $IP(X) - EA(Y)$, and the second is $IP(Y) - EA(X)$; equality implies that $IP(X) + EA(X) = IP(Y) + EA(Y)$. Thus EN is defined by chemical interaction, whereas EA refers to isolated chemical species. The contribution of IP to EN can be explained by the fact that increasing IP decreases donor capability and thus indirectly increases the acceptor capability of molecules.

Analysis of the disproportionation of chemical species X

$$X + X \rightarrow X^+ + X^- \tag{22.11}$$

leads to a corresponding rationalization of the hardness η: The reaction energy, referring to two X molecules, is $IP(X) - EA(X)$, thus leading to Equation 22.4 for hardness. This shows that hardness can be identified as the resistance to changing the number of electrons. The combined influence of electronegativity and hardness on the interaction between two molecules X and Y is summarized by the following formula for the fractional number of transferred electrons, ΔN (Parr and Pearson 1983):

$$\Delta N(X) = \frac{EN(X) - EN(Y)}{2[\chi(X) + \chi(Y)]} = -\Delta N(Y) \qquad (22.12)$$

It is seen that electron charge is increased in the molecule with greater EN, with increasing EN differences driving electron transfer, and that hardness resists the exchange of electrons.

For practical purposes, all four parameters, IP, EA, EN, and η are useful entities for characterizing the general disposition of molecular structures to undergo changes in their electronic structure in the context of structure–property and structure–activity relationships (cf. Mekenyan et al. 1994; Purdy 1988; Schüürmann 1990a and 1996, Schüürmann and Röderer 1988; Veith and Mekenyan 1993; Verhaar et al. 1994). It should be noted, however, that these parameters describe global molecular reactivity, which may not necessarily be relevant for the reactivity behaviour at specific sites. Consequently, there should be room for improvement by focusing on corresponding local parameters which address more directly the biochemical disposition of contaminants for specific modes of toxic action. These parameters should properly reflect the energies of atoms in molecules.

22.5.5 Ground-State Local Molecular Reactivity

A more detailed treatment of incipient interactions between molecules is based on perturbation theory. The starting point is the consideration that in the initial stage of a chemical reaction, the interaction can be described in terms of the mutual perturbation of the electronic structures of the reactants. Here, polarization of the molecules may have two physically different origins: Atomic sites may undergo attractive or repulsive Coulomb interaction according to their local charge densities (and depending on the interatomic distances), and a further attractive interaction may result from partial delocalization of electrons between sites on the reacting molecules with a possible net transfer of electron charge. According to Klopman (1968), the energy change associated with the initial interaction between the atomic site r of the donor molecule X (nucleophile) and the site s of the acceptor molecule Y (electrophile) at an interatomic distance d_{rs} can be written in a molecular orbital (MO) treatment (neglecting solvation effects) as

$$\Delta E_{rs} = -\frac{q_r q_s}{d_{rs}} + 2 \sum_i^{occ} \sum_k^{vac} \sum_{\mu(r)} \sum_{\nu(s)} \frac{c_{\mu i}^2 c_{\nu k}^2}{\varepsilon_i^* - \varepsilon_k^*} \beta_{\mu\nu} \qquad (22.13)$$

The first term represents the Coulomb interaction between the net atomic charges q_r and q_s, and the second term represents the orbital interaction with contributions from all occupied MOs i of the nucleophile X and all vacant MOs k of the electrophile Y. Summation takes into account all atomic orbitals (AOs) μ of r and ν of s according to their expansion coefficients $c_{\mu i}$ and $c_{\nu i}$ (these are often called LCAO-MO coefficients because they specify how the MOs are derived from linear combinations of AOs), and the terms are weighted by the energy differences of the perturbed MO energies ε_i^* and ε_k^* and the resonance integrals $\beta_{\mu\nu}$. For qualitative applications in practice, ε_i^* and ε_k^* are replaced by the MO energies ε_i and ε_k of the unperturbed molecules, and d_{rs} as well as $\beta_{\mu\nu}$ (which are parameters anyway) are typically omitted. The net atomic charges in the Coulomb interaction term are often calculated using semiempirical quantum chemical schemes in the zero-differential-overlap (ZDO) approximation according to

$$q_r = Z_r - 2 \sum_i^{occ} \sum_{\mu(r)} c_{\mu i}^2 \tag{22.14}$$

with Z_r denoting the core charge of the atomic centre r.

Great reactivity between r of X and s of Y is indicated by large values for the perturbational interaction energy, which can be driven in two ways (Klopman 1968): Large net atomic charges of opposite sign lead to a substantial electrostatic contribution and (if dominating) to a charge-controlled type of reaction, which can be shown to explain the preference of hard-hard over hard-soft interactions between chemical species. Alternatively, substantial contributions from pairs of occupied and vacant MOs with small energy gaps and large LCAO-MO coefficients at the interacting sites r and s may lead to orbital-controlled reactions, which correspond to a preference for soft-soft interactions. Of particular interest is the special case where the sum is dominated by the HOMO-LUMO interaction and may be reduced to the corresponding term. Thus Equation 22.13 provides qualitative explanations both for Pearson's principle of hard and soft acids and bases and their interaction features, and for Fukui's concept of frontier-orbital interactions.

In the general case, however, direct application of Equation 22.13 for the elucidation of the impact of chemical reactivity on toxicity is not straightforward, because the (potentially) relevant endogenous receptor cannot be sufficiently specified in terms of the relevant atomic charges, MO energies and LCAO-MO coefficients. It follows that for typical QSAR applications, analysis will often concentrate on the properties of the molecules alone. Electrostatic influence on reactivity can be tested by examining the variation of site-specific atomic charge within the compound series of interest. An assessment of the potential relevance of soft (orbital-related) reactivity requires further simplifications. Here, a possible candidate is Fukui's delocalizability, which was introduced as a measure for the site-specific energy stabilization due to fractional electron transfer to or from a reagent, corresponding to an electron delocalization between the interacting molecules (Fukui 1970). Interestingly, replacement of the orbital interaction term in Equation 22.13 by such a delocalizability parameter for structure-property studies was proposed already 25 years ago (Cammarata 1971).

The susceptibility of a molecule to nucleophilic attack at its atomic centre r is characterized by its nucleophilic delocalizability (D_r^N)

$$D_r^N = 2 \sum_{k}^{vac} \sum_{\mu(r)} \frac{c_{\mu k}^2}{\alpha - \varepsilon_k} \tag{22.15}$$

where summation includes all vacant MOs assessing the readiness of r to accommodate excess electron charge. Electrophilic delocalizability represents the corresponding energy stabilization resulting from electrophilic attack at r (D_r^E)

$$D_r^E = 2 \sum_{i}^{occ} \sum_{\mu(r)} \frac{c_{\mu i}^2}{\varepsilon_i - \alpha} \tag{22.16}$$

and involves occupied MOs and their contribution to electron density at the atomic site. In these formulae, the inner sums account for all AOs μ belonging to the centre r of interest, and α represents the relevant orbital energy of the unspecified reaction partner, which would be -IP for a nucleophilic receptor, and -EA for an electrophilic receptor. In practice, α is often skipped or used as an arbitrarily selected value to shift the MO energies, which may be needed to avoid accidentally dominant contributions from MO energies close to (or even identical to) zero. Practical applications of nucleophilic and electrophilic delocalizability for structure–activity studies go back to the 1960s, and various levels of semiempirical quantum chemistry have been employed for their calculation (Cammarata 1968; Lewis 1988 and 1989a; Mekenyan et al. 1993; Purdy 1991, Schüürmann 1990a; Verhaar et al. 1994). In addition, structure-property analyses for calculating log K_{OW} and other physicochemical compound properties have made use of these electronic parameters (Cammarata 1971; Lewis 1987; Schüürmann 1990b).

The concept of delocalizability was originally developed for π-electron systems, where the corresponding parameters were called superdelocalizability. A similar approach, also based on perturbation theory, leads to the atom polarizability, which characterizes the response of the molecule to local changes in electron energy. In the initial phase of intermolecular interaction, deformation of electronic structure will be generally easier at atomic sites with greater polarizability, which reflects the potential suitability of this parameter for quantifying local molecular reactivity. Self-atom polarizability π_{rr} characterizes the readiness of the atomic site r to accommodate local changes in electron energy

$$\pi_{rr} = 4 \sum_{i}^{occ} \sum_{k}^{vac} \sum_{\mu(r)} \frac{c_{\mu i}^2 c_{\mu k}^2}{\varepsilon_i - \varepsilon_k} \tag{22.17}$$

and atom polarizability π_{rs} quantifies the corresponding molecular response due to local energy changes at two (typically neighbouring) sites r and s:

$$\pi_{rs} = 4 \sum_i^{occ} \sum_k^{vac} \sum_{\mu(r)} \sum_{\nu(s)} \frac{c_{\mu i} c_{\nu i} c_{\mu k} c_{\nu k}}{\varepsilon_i - \varepsilon_k} \tag{22.18}$$

Generally, π_{rr} may be interpreted as indicating softness at the atomic site r within the molecule, thus yielding a measure of local soft reactivity. In an analogous way, π_{rs} may be understood as reflecting softness of the bond between r and s, which can be useful for characterizing the readiness to undergo bond cleavage under the influence of an approaching reagent.

An alternative quantification of covalent bond strength is given by the bond order p_{rs} according to Wiberg (1968),

$$p_{rs} = \sum_{\mu(r)} \sum_{\nu(s)} \left(2 \sum_i^{occ} c_{\mu i} c_{\nu i} \right)^2 \tag{22.19}$$

and finally the charge density difference between neighboring atoms, Δq_{rs}, may also be used to characterize the susceptibility of intramolecular bonds to heterolytic cleavage:

$$\Delta q_{rs} = |q_r - q_s| \tag{22.20}$$

Applications of self-atom polarizability, π_{rr}, and the parameters for covalent and ionic bond strength (Equations 22.17, 22.19 and 22.20) can also be found in the QSAR literature (Dearden and Nicholson 1986; Lewis 1989b; Schüürmann 1990a,b).

The reactivity indices presented so far show that for a given type of reaction and associated toxic effect there may be several parameters with varying or even equivalent suitability for the derivation and analysis of structure–activity relationships. Furthermore, the performance may vary with the type and level of quantum chemical calculation as well as with the particular series of compounds under investigation. If soft and hard reactivity are involved, molecular parameters for both aspects may be included, as was discussed already with the Klopman equation above. Interestingly, a corresponding equation for the specific case of electron structure perturbations in π-electron systems resulting from changes in local one-electron energies α_r at the atomic site r has been discussed already in the mid 1960s:

$$\Delta E_r = q_r \delta \alpha_r + \tfrac{1}{2} \pi_{rr} \delta \alpha_r^2 \tag{22.21}$$

Equation 22.21 shows that in π-electron theory, q_r and π_{rr} give the first- and second-order change in molecular energy due to a local energy shift $\delta \alpha_r$. It also shows that in the general case, consideration of both hard (electrostatic) and soft (orbital-related) reactivity may be needed to characterize the energy of attraction with which some reagent would adhere at a specific site on the molecule (Greenwood and McWeeny 1966).

22.5.6 Transition-State Theory

A more explicit investigation of molecular reactivity is possible by using transition-state theory, which assumes that thermodynamic properties, such as enthalpy and entropy, can be ascribed to molecular species both in the ground state and transition state of a reaction path connecting reagents and products. The relevant energies are, in principle, available with theoretical methods which provide the energy of a chemical system as a function of nuclear coordinates. However, of course, the accuracy depends on the actual (or possible) level of calculation. A chemical reaction corresponds to a rearrangement of atoms, where the energy function shows minima for the initial and final stable points, and a saddle point for the transition state (which is a maximum with respect to the reaction coordinate, and a minimum for all remaining coordinates).

The basic equations are illustrated for a replacement reaction, where a reagent Y attacks a molecule R–X:

$$Y + R - X \leftrightarrows [Y \ldots R \ldots X]^{\neq} \leftrightarrows Y - R + X \qquad (22.22)$$

The initial ground state and the transition state are linked by an equilibrium constant $K^{\neq}$, which is related to the free energy of activation, $\Delta G^{\neq}$, and its enthalpic and entropic contributions according to

$$\log K^{\neq} = -\frac{\Delta G^{\neq}}{2.3\,RT} = -\frac{\Delta H^{\neq}}{2.3\,RT} + \frac{\Delta S^{\neq}}{2.3\,R} \qquad (22.23)$$

(cf. textbooks). The total reaction rate constant k is directly proportional to $K^{\neq}$, which finally gives the Eyring equation in logarithmic form

$$\log k = \log \frac{k_B \cdot T}{h} + \frac{\Delta S^{\neq}}{R} - \frac{\Delta H^{\neq}}{RT} \qquad (22.24)$$

with Boltzmann's constant k_B and Planck's constant h. As can be seen from Equation 22.24, the rate of a certain reaction can be calculated from the differences in enthalpy and entropy between the transition state and the initial ground state of the chemical system, which in turn can be obtained (approximately) by applying quantum chemical schemes. It follows that for a series of compounds and a model reagent to mimic the relevant chemical functionality of an endogenous receptor, the associated reaction rates can be calculated, and the differences between these reactions rates can be used to explain differences in the reactivity-driven toxicity of the compounds.

Application of Equation 22.24 for a series of compounds assumes kinetic control of the differences in reactivity. Alternatively, the corresponding energy and entropy values can also be calculated for the reactants and products, yielding the free energy of reaction, ΔG_r, according to

$$\Delta G_r = \Delta H_r - T \cdot \Delta S_r \qquad (22.25)$$

The entropy terms in Equations 22.24 and 22.25 need particular attention. Recent experience when estimating compound acidity from molecular structure suggests that with semiempirical schemes it may be preferable to confine the calculations to the relevant enthalpies, probably because the underlying method parameters are fitted to macroscopic properties in order to best reproduce heats of formation rather than free energy terms (Schüürmann 1996). In contrast, with ab initio methods of quantum chemistry, explicit inclusion of the entropy term is generally recommended.

A final note will be given to a more general shortcoming of many quantum chemical methods for quantifying molecular reactivity. Most calculations of molecular structure and properties have been performed in the gas phase, thus deliberately ignoring solvation effects and their impact on the structure and reactivity of the compounds. With the advance of continuum-solvation models, approximate inclusion of aqueous solvation effects has been made possible at both semiempirical and ab initio levels of theory (Cramer and Truhlar 1992 and 1995; Klamt and Schüürmann 1993; Miertus et al. 1981; Tomasi and Persico 1994). Up till now, these methods have been used mostly to quantify the free energy or enthalpy of solvation, and only few applications included compound properties with direct environmental relevance, such as dissociation constant and Henry's law constant (Lim et al. 1991; Schüürmann 1995 and 1996). Further studies are needed to assess the performance of solution-phase reactivity parameters and their suitability for providing improved structure–property and structure–activity relationships.

22.6 MOLECULAR MODE-OF-ACTION ANALYSIS OF COMPOUND TOXICITY

Chemical disposition to bioactivity depends on physicochemical properties and on reactivity characteristics. As outlined above, both aspects can be addressed using a number of theoretical methods which allow mechanistic hypotheses about the impact of molecular structure on compound toxicity to be tested by calculating the relevant parameters and comparing with observed trends in biotests. Experimental analysis of the ability of a molecule to donate or accept electrons, protons or greater structural fragments at arbitrary sites would not be an easy task in the general case, which shows the great merit of quantum chemical and other theoretical methods for the discussion and quantification of global as well as local chemical reactivity. On the other hand, however, many chemicals can undergo various primary interactions with endogenous molecules, and the disposition for a specific biotransformation does not necessarily indicate the relevant process behind the prevailing toxicity observed in a specific assay.

In the following, five case studies will be presented that demonstrate the use of structure–activity analyses to elucidate molecular modes of ecotoxic action exerted by chemical contaminants. The examples also demonstrate how application of theoretical methods can alert experimentalists to traits on which to focus their attention.

22.6.1 Aromatic Phosphorothionates

Organophosphorus esters (OPs) form an important class of agrochemicals. They account for ca. 45% of the insecticide world market. In addition, they also have important industrial applications as oil additives, flame retardants and plasticizers (Muir 1988). Most of the commercially relevant bioactive OPs have sulfur directly attached to the central pentavalent phosphorus as shown in Figure 22.3, with their insect : mammal toxicity typically being greater than that of corresponding non-sulfur analogues. For convenience, some important OP compound classes are given in Figure 22.4 together with their generic names.

In contrast to organochlorine insecticides, OPs do not generally bioaccumulate but undergo hydrolysis and biodegradation under environmental conditions. Hydrolytic degradation may proceed by two different mechanisms: S_N2 reaction on the central phosphorus results in cleavage of a P–O bond, and nucleophilic attack on the ester carbon leads to dealkylation of the OP by C–O bond fission (Figure 22.5).

Aromatic phosphorothionates as a subgroup of bioactive OPs are known to inhibit acetylcholinesterase (AChE) in mammals and insects. Prominent examples are given by parathion (*O,O*-diethyl *O*-paranitrophenol phosphorothionate; E 605), methyl parathion (*O,O*-dimethyl *O*-paranitrophenyl phosphorothionate) and diazinon (*O,O*-diethyl *O*-(2-isopropyl-4-methyl pyrimid-6-yl) phosphorothionate), with rat LD_{50} (acute oral toxicity) values of

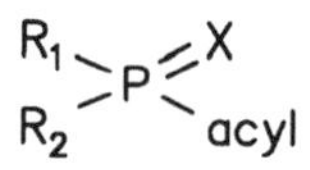

$$R_1 \diagdown P \diagup X$$
$$R_2 \diagup \diagdown acyl$$

R_1 , R_2 = alkoxy, alkyl, alkylamino

acyl = acidic group

(anion of organic or inorganic acid)

X = O, S

Figure 22.3 Schrader formula of bioactive organophosphorus esters.

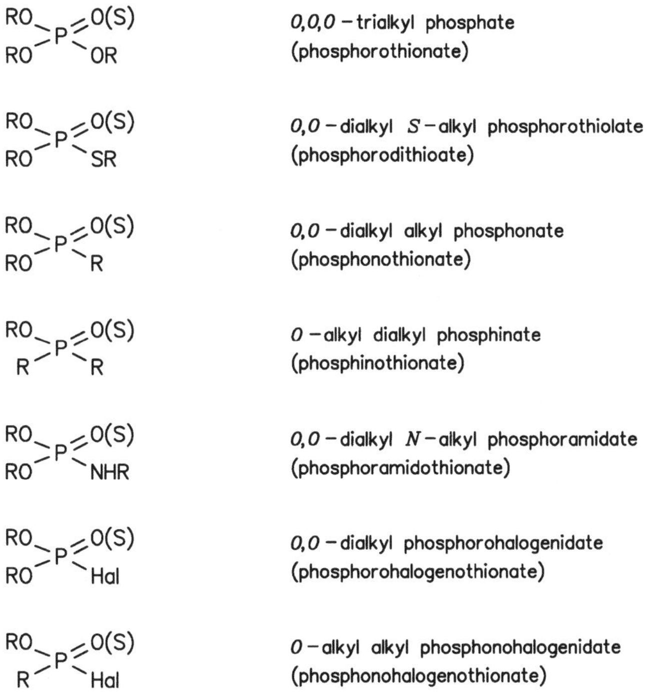

RO $\diagdown_{P}\diagup^{O(S)}$
RO $\diagup$ $\diagdown$ OR

0,0,0 – trialkyl phosphate
(phosphorothionate)

RO $\diagdown_{P}\diagup^{O(S)}$
RO $\diagup$ $\diagdown$ SR

0,0 – dialkyl S – alkyl phosphorothiolate
(phosphorodithioate)

RO $\diagdown_{P}\diagup^{O(S)}$
RO $\diagup$ $\diagdown$ R

0,0 – dialkyl alkyl phosphonate
(phosphonothionate)

RO $\diagdown_{P}\diagup^{O(S)}$
R $\diagup$ $\diagdown$ R

0 – alkyl dialkyl phosphinate
(phosphinothionate)

RO $\diagdown_{P}\diagup^{O(S)}$
RO $\diagup$ $\diagdown$ NHR

0,0 – dialkyl N – alkyl phosphoramidate
(phosphoramidothionate)

RO $\diagdown_{P}\diagup^{O(S)}$
RO $\diagup$ $\diagdown$ Hal

0,0 – dialkyl phosphorohalogenidate
(phosphorohalogenothionate)

RO $\diagdown_{P}\diagup^{O(S)}$
R $\diagup$ $\diagdown$ Hal

0 – alkyl alkyl phosphonohalogenidate
(phosphonohalogenothionate)

Figur-e 22.4 Generic structures and names of common organophosphorus compound classes.

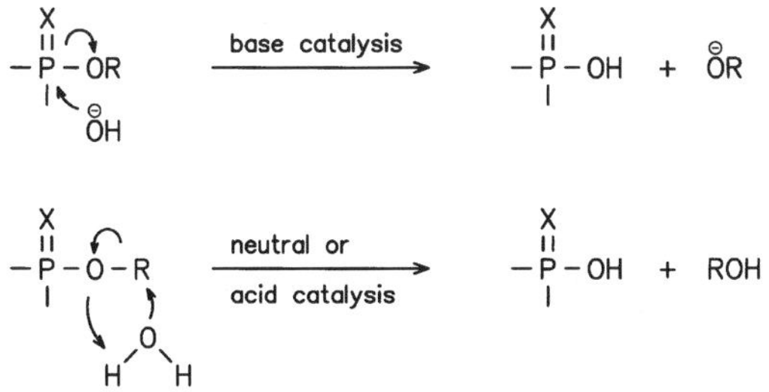

Figure 22.5 Hydrolytic degradation pathways of organophosphorus esters: P–O bond cleavage (top) and C–O bond cleavage (bottom).

6.4 mg/kg, 15–20 mg/kg and 108 mg/kg, respectively (Fest and Schmidt 1977). The aromatic moiety of this substance class allows minor modifications of the chemical structure by variation of substituents, which forms a suitable basis for analyses of relationships between structural features and chemical reactivity or biological activity of the compounds.

Principal Mechanisms of Bioactivity

For target as well as non-target organisms, inhibition of acetylcholinesterase (AChE) appears to be the major toxicity mechanism of Ops: Electrophilic phosphorus attacks AChE at a serine OH group, which is probably activated by H-bond interaction with the imidazole nitrogen of a histidine unit. Cleavage of the acidic OP leaving group enables formation of a covalent P–O(serine) bond, thus inhibiting enzymatic hydrolysis of the natural substrate acetylcholine, which would proceed via a reaction between serine OH and the electrophilic carbonyl carbon of the neurotransmitter. Phosphorothionates (esters containing the thiono group P=S) are less effective as AChE inhibitors than their oxon metabolites (with P=S being replaced by P=O), which are formed in vivo by cytochrome P-450 oxidation. This has often been explained by the fact that phosphorus has a greater positive charge in P=O than in P=S due to the greater electronegativity of oxygen compared to sulfur, leading to a greater electrophilicity and related reactivity of phosphorus in phosphate esters. Although this interpretation gets support from both semiempirical PM3 results and ab initio calculations on the polarized double-zeta level, experimental as well as calculated ^{31}P NMR shifts seem to indicate the opposite. P=S units in aromatic phosphorothionates and phosphorodithioates show downfield shifts for δ^{31}P of ca. 70 ppm to 100 ppm compared to the P=O unit in phosphates and phosphorothiolates (cf. Figure 22.4), and in thioester functions R–S–P of aromatic OPs, δ^{31}P is shifted downfield by ca. 30 ppm compared to R–O–P ester functions (Schüürmann 1992; Schüürmann and Schindler 1993). More positive δ^{31}P values in P=S and P–S units would thus indicate an increased deshielding of the central phosphorus, which, according to the usual interpretation, would correspond to greater positive charge and associated electrophilicity. Consequently, the observed differences in the δ^{31}P NMR shifts are in apparent conflict with experimental observations of differences in AChE inhibition potency and with direct charge density calculations. This surprising feature of the electronic structure of OPs indicates that classical concepts for explaining molecular reactivity and the related toxicity may be difficult to apply to compound classes with atoms from higher rows of the periodic system.

As a general but not strict rule, bioactivity of OPs is increased with increasing AChE inhibition potency and increasing stability towards hydrolysis or dealkylation, which represent two different detoxification pathways. Phosphorothionates are less reactive to hydrolysis and more lipophilic than phosphates, which makes them more efficient in reaching critical endogenous sites in sufficiently effective concentrations, with cytochrome P-450 oxidation to oxon metabolites forming a major toxification step in vivo (s.a.). Whereas both AChE inhibition and hydrolysis proceed via hard-hard interactions between central phosphorus and hydroxyl oxygen, dealkylation requires soft nucleophilic attack on the ester carbon. For the latter, the endogenous tripeptide glutathione (γ-L-glutamyl-L-cysteinylglycine; GSH) as an important non-protein thiol is a major candidate. However, GSH conjugation may also be reversible, in which case it

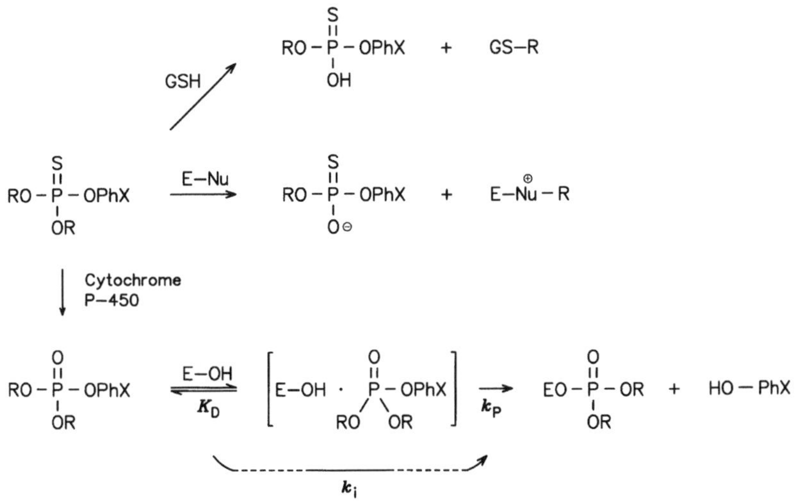

Figure 22.6 Biotransformation and toxicity of aromatic phosphorothionates. Detoxification through conjugation with glutathione (GSH) and alkylation of endogenous macromolecules (E-Nu) are shown in the top and middle, and inhibition of acetylcholinesterase (E-OH) after oxon formation is shown in the bottom.

could even enable in vivo transport of reactive electrophiles to intracellular sites (Baillie and Slatter 1991). A further OP toxicity pathway, which apparently has not been discussed so far, may result from the alkylation potency of the ester carbons, which could attack nucleophilic sites of critical macromolecules without interfering with AChE or GSH. A summary of major biotransformations related to the toxicity profile of OPs is given in Figure 22.6.

Aquatic Toxicity Profile

Despite an abundance of literature about OPs in general and aromatic phosphorothionates in particular, very few investigations have addressed their ecotoxic potential for aquatic organisms. In this context, a question of particular interest is whether acute fish toxicity would also be caused by the AChE inhibition potency of the compounds. More recently, a thorough experimental study of 12 aromatic phosphorothionates became available, in which various aspects of their in vitro and in vivo fate and their toxicity towards the guppy (*Poecilia reticulata*) were analyzed (de Bruijn 1991). This set of compounds with a characterization of suborganismic and organismic aspects of their ecotoxic profile will be used as an example to show how theoretical methods can be applied to identify and test potentially relevant biotransformations as causes for specific modes of toxic action.

The molecular structures of the 12 aromatic phosphorothionates are specified in Table 22.5, and the relevant biological data are summarized in Table 22.6.

Table 22.5 Substitution pattern of 12 aromatic phosphorothionates used for the PLS study[a]

Compound	Substituents at phenolic moiety		
	R_1 (ortho)	R_3 (para)	R_4 (meta)
Cyanophos	H	CN	H
SV5	H	H	H
Methylparathion	H	NO_2	H
Fenitrothion	H	NO_2	CH_3
Methylisocyanthion	CN	Cl	H
Chlorothion	H	NO_2	Cl
Dicapton	Cl	NO_2	H
Fenthion-S2145	H	SCH_3	H
Fenthion	H	SCH_3	CH_3
Ronnel	Cl	Cl	Cl
Bromophos	Cl	Br	Cl
Iodofenphos	Cl	I	Cl

[a] The general molecular structure is $(MeO)_2P(=S)OPhX$ (cf. Figure 22.4).

Aqueous exposure of the compounds leads to uptake and elimination with rate constants k_1 and k_2. If partitioning has reached equilibrium, body burden (dose) nd exposure concentration are related by the bioconcentration factor $BCF = k_1/k_2$, which in turn allows estimates of LD_{50} (lethal dose 50%) values from exposure-related LC_{50} (lethal concentration 50%) values according to

$$LD_{50} = LC_{50} \cdot BCF \tag{22.26}$$

(McCarty 1986). To analyze the influence of oxon formation by cytochrome P-450 on toxicity, OP clearance rate constants from in vitro oxidation have been determined in NADPH-enriched rainbow trout (*Salmo gairdneri*) liver homogenate (k_{NADPH}). Correspondingly, in vitro dealkylation in GSH-enriched rainbow trout liver homogenate has been quantified to address the detoxification pathway via GSH conjugation (k_{GSH}). Following Figure 22.6, the toxicokinetic phase of the OPs would depend on both k_{NADPH} (activation for AChE inhibition) and k_{GSH} (deactivation by formation of dealkylated metabolites).

The toxicodynamic phase at the site of action (i.e. at serine OH of AChE) can be divided into two steps (cf. Figure 22.6): Complex formation as a still reversible binding between the attacking OP and the enzyme AChE, which depends on the affinity of the inhibitor for the active site and can be characterized by the dissociation constant K_D, and subsequent phosphorylation through elimination of the acidic moiety of the OP. The latter is quantified by the rate constant k_p. Both constants have been determined in vitro with AChE purified from electric eel (*Electrophorus electricus*), and the overall AChE inhibition potency (k_i) can be characterized by the bimolecular rate constant $k_i = k_p/K_D$ (de Bruijn 1991).

Table 22.6 Ecotoxic profile of aromatic phosphorothionates[a]

Compound	log k_{NADPH}	log k_{GSH}	log BCF	log k_1	log k_2	log K_D	log k_p	log k_i	log LC_{50}	log LD_{50}
Cyanophos	−3.89	−3.96	2.62	3.11	0.62	2.67	0.45	3.78	−1.25	1.37
SV5	−4.40	−4.15	2.95	3.41	0.19	n.a.	n.a.	1.43	−1.32	1.63
Methylparathion	−3.92	−3.96	2.98	3.41	0.38	2.16	1.44	5.28	−2.39	0.59
Fenitrothion	−3.55	−3.82	3.36	3.59	0.05	2.19	0.84	4.65	−2.00	1.36
Methylisocyanothion	−3.89	−3.21	3.39	3.61	0.07	2.05	0.22	4.16	−2.77	0.62
Chlorothion	−3.54	−2.69	2.61	3.31	0.84	1.98	1.24	5.26	−3.19	−0.58
Dicapton	−3.54	−3.07	2.95	3.49	0.41	2.07	1.01	4.95	−2.57	0.38
Fenthion-S2145	−2.80	−4.10	3.84	3.70	0.04	2.01	1.18	5.17	−1.60	2.24
Fenthion	−3.19	−3.92	4.22	3.94	−0.22	2.10	0.91	4.81	−2.11	2.11
Ronnel	−3.74	−3.15	4.64	4.13	−0.42	2.07	−0.36	3.58	−3.00	1.64
Bromophos	−3.82	−3.29	4.65	4.12	−0.48	1.77	−0.22	4.01	−2.91	1.74
Iodofenphos	−3.68	−3.47	4.68	3.86	−0.44	1.45	−0.22	4.32	−2.68	2.00

[a] Experimental data were taken from de Bruijn 1991 and refer to the guppy (k_1, k_2, BCF, LC_{50}), to rainbow trout liver homogenate fortified with NADPH-generating cofactors or with glutathione (k_{NADPH}, k_{GSH}), and to electric eel acetylcholinesterase (K_D, k_p, k_i). LD_{50} was calculated from LC_{50} and BCF according to Equation 22.26 and has the unit mmol $* kg^{-1}$. The other data refer to the following units: k_{NADPH}: 1/(min $*$ mg protein), k_{GSH}: 1/(min $*$ mg protein), k_1: mL /g $*$ day), k_2: 1/day, K_D: µmol, k_p: 1/min, k_i: 1/(min $*$ mol).

The 10 biological endpoints summarized in Table 22.6 form a suitable test case for the question as to whether and how an ecotoxic profile of chemicals rather than a single effect can be traced back to compound properties and molecular reactivity. Provided that suitable parameters for quantifying relevant structural features can be found, the statistical method of choice would be the multivariate PLS (partial least-squares) regression method (cf. Geladi and Kowalski 1986; Wold 1995). A corresponding PLS analysis of the 12 OPs and the endpoints of Table 22.6 [except the (calculated) LD_{50}] showed that for most of the nine endpoints the major variation within the OP series could be explained by a total of 10 molecular descriptors including $\log K_{OW}$ and reactivity parameters calculated using semiempirical quantum chemical methods (Verhaar et al. 1994). For the present discussion, we will include all 10 biological endpoints and derive a simplified PLS model based on the following three molecular descriptors (cf. Section *Compound Properties and Molecular Descriptors*): hydrophobicity $\log K_{OW}$, electron affinity EA (Equation 22.6) and molecular hardness η (Equation 22.5). The last two descriptors are calculated using the semiempirical AM1 scheme (Dewar et al. 1985). The relevant descriptor values are shown in Table 22.7.

With PLS regression, linear relationships between biological endpoints (**y**) and molecular descriptors (**x**) are obtained which are usually considered to be statistically more robust than with ordinary multilinear regression. In matrix notation, with **X** and **Y** representing the blocks of descriptors and endpoints, the result can be written as

$$\mathbf{Y} = \mathbf{X} \cdot \mathbf{B}' + \mathbf{F} \tag{22.27}$$

Table 22.7 Molecular descriptors of aromatic phosphorothionates[a]

Compound	$\log K_{OW}$	Electron affinity	Hardness
Cyanophos	2.71	1.84	4.14
SV5	3.00	1.57	4.14
Methylparathion	3.04	2.07	4.16
Fenitrothion	3.47	2.03	4.16
Methylisocyanthion	3.58	1.86	4.10
Chlorothion	3.63	2.12	4.16
Dicapton	3.72	2.12	4.14
Fenthion-S2145	3.74	1.55	3.42
Fenthion	4.17	1.53	3.42
Ronnel	5.07	1.81	4.01
Bromophos	5.21	1.84	4.01
Iodofenphos	5.51	1.84	4.03

[a] Experimental $\log K_{OW}$ is taken from de Bruijn 1991; electron affinity [eV] and hardness [eV] are calculated according to the semiempirical AM1 scheme including geometry optimization (Dewar et al. 1985) using Equations 22.6 and 22.4, respectively.

with the residual matrix $\mathbf{F}$ representing the noise component of the PLS model, and $\mathbf{B}$ specifying the relationships between $\mathbf{y}$ and $\mathbf{x}$ variables, as with ordinary multilinear regression. For the set of 12 aromatic phosphorothionates (Table 22.5), with $\mathbf{Y}$ and $\mathbf{X}$ being specified in Tables 22.6 and 22.7, respectively, the PLS result is summarized in Table 22.8. According to this model, 73% (adjusted for degrees of freedom) of the overall variation in the 10 endpoints is explained by corresponding variations in the three molecular descriptors. However, there are substantial differences in performance for the individual endpoints, with variances ranging from 43% for $\log k_{GSH}$ to 92% for $\log k_p$.

Uptake ($\log k_1$) and bioconcentration ($\log$ BCF) are primarily governed by compound hydrophobicity ($\log K_{OW}$), as is also observed with many other classes of chemicals (cf. Chapter 14). Elimination ($\log k_2$) gives greater PLS regression coefficients for both $\log K_{OW}$ (-0.79) and electron affinity EA (0.47). The former represents passive excretion according to physicochemical partitioning, whereas the latter reveals that with this set of compounds, biotransformation forms an important second route of elimination (cf. Verhaar et al. 1994). Both coefficient signs are as expected: Increasing hydrophobicity decreases physicochemical elimination, and increasing electron affinity increases the probability of biotransformation via chemical reaction with endogenous nucleophiles, leading to phase II metabolites with greater water solubility and enhanced excretion (cf. Section *Biochemical Background and Modes of Toxic Action* above). Accordingly, biotransformation decreases overall bioconcentration, as can also be seen by the negative contribution of EA to $\log$ BCF (-0.25).

Table 22.8 PLS relationship between biological endpoints and molecular descriptors of aromatic phosphorothionates[a]

Biological end point	$\log K_{OW}$	Electron affinity	Hardness	Explained variance [%]
$\log k_{NADPH}$	0.00	0.80	-1.35	85
$\log k_{GSH}$	0.49	0.64	0.01	43
$\log$ BCF	0.86	-0.25	-0.08	90
$\log k_1$	0.82	-0.14	-0.16	77
$\log k_2$	-0.79	0.47	-0.14	73
$\log K_D$	-0.77	-0.48	0.36	59
$\log k_p$	-0.88	0.91	-1.12	81
$\log k_i$	-0.05	1.36	-1.26	92
$\log$ LC$_{50}$	-0.66	-0.69	0.13	63
$\log$ LD$_{50}$	0.30	-0.78	0.02	62

[a] Columns 2, 3 and 4 specify the relationship between the 10 biological endpoints Y given in column 1 (cf. Table 22.6) and the three molecular descriptors X given in the top row (cf. Table 22.7) in terms of weighted B coefficients according to Equation 22.27, calculated by the partial least-squares (PLS) method (Wold 1995). The overall explained variance is 73%.

It follows that for the assessment of the bioconcentration potential of aromatic phosphorothionates, biotransformation must be considered. This has been discussed earlier for chloroanilines, chloroanisols, dibenzodioxins and dibenzofurans (de Wolf et al. 1992; Opperhuizen and Voors 1987; Opperhuizen and Sijm 1990; Sijm and Opperhuizen 1988) and probably holds true more generally for reactive electrophilic xenobiotics.

As noted above, the toxicokinetic phase of the OPs after incorporation includes two major metabolic pathways: cytochrome P-450 oxidation to oxon analogs, and dealkylation via GSH conjugation (cf. Figure 22.6). The latter (quantified as $\log k_{GSH}$ in vitro, s.a.) is not well accounted for in the present PLS model, which may indicate the need for more site-specific descriptors. On the other hand, the oxidative clearance rate, $\log k_{NADPH}$ (again quantified in vitro), can be traced back to molecular reactivity in terms of electron affinity and hardness. Again, the coefficient signs are significant: both increasing electron affinity and increasing softness (decreasing hardness) make the compounds more susceptible to oxidative metabolism.

PLS analysis of the interaction with AChE in terms of $\log K_D$ (dissociation constant), $\log k_p$ (phosphorylation constant) and $\log k_i$ (bimolecular inhibition constant) reveals other interesting features. First, OP variation in the dissociation constant is modelled by only 59%, which indicates that inclusion of more distinct structural features (e.g., electrostatic interactions) will be necessary to characterize the affinity of OPs adequately for the active site. Nonetheless, the negative contribution of $\log K_{OW}$ suggests, in agreement with conventional considerations, that more hydrophobic substrates would form more stable complexes with AChE. Following Table 22.8, phosphorylation efficiency increases with increasing electron affinity and softness of the compounds. The latter is somewhat surprising, as AChE phosphorylation is usually considered as a hard-hard interaction between the central phosphorus and the serine hydroxyl oxygen (s.a.). This aspect needs further clarification. A potential difficulty may be that decreasing values of η would also indicate increasing ability to undergo deformation of the electron shell and hence a general increase in chemical reactivity. As $\log K_D$ and $\log k_p$ have very similar PLS contributions from hydrophobicity, the overall inhibition constant $\log k_i$ is almost independent of $\log K_{OW}$. Both EA and η show significant contributions with significant coefficient signs.

The PLS results summarized in Table 22.8 show that the general characteristics of OP bioconcentration, biotransformation and AChE inhibition can be traced back to compound properties and molecular reactivity parameters. However, the moderate performance of some of the endpoints, including acute toxicity, indicates that additional and more specific reactivity aspects are relevant which are apparently not accounted for by electron affinity and hardness as global electronic descriptors. A similar situation was achieved with a larger set of molecular descriptors, including site-specific delocalizabilities as reported previously (Verhaar et al. 1994). This suggests that the acute toxicity profile of phosphorothionates is more complex than would be expected from their general AChE inhibition potency.

Acute Fish Toxicity and Dealkylation

In the following, a separate structure-activity analysis of acute fish toxicity ($\log LC_{50}$) will illustrate that the mechanistic understanding can be improved by applying advanced modelling techniques. As outlined in Figure 22.6, a second mode of phosphorothionate toxicity in vivo is alkylation of endogenous nucleophilic sites. Here, the active site of the toxicant is its ester function, and the reaction proceeds through cleavage of the bond between the ester oxygen (attached to the central phosphorus) and the ester carbon. This suggests that (besides general membrane irritation) toxicity may result from both AChE inhibition and alkylation of critical macromolecules, with relative contributions varying according to the site-specific reactivities of the phosphorothionate compounds.

The critical step in AChE inhibition is assumed to be a nucleophilic substitution reaction on the central phosphorus of the oxon derivative, where previous formation of the active metabolite through cytochrome P-450 oxidation is not considered to be rate-limiting (s.a.). An approximate characterization of the relevant reactivity of the organophosphorus compounds can be achieved by computational analysis of a suitable model reaction, focussing on the release of the aromatic moiety while the compound approaches a nucleophilic site. A corresponding study was performed for the present set of 12 aromatic phosphorothionates (cf. Table 22.5) at the semiempirical PM3 level of quantum chemistry (Stewart 1989). For each compound, the model system consisted of the oxon metabolite, a methoxy anion (MeO^-) as (certainly crude) model nucleophile interacting with the central phosphorus, and a water molecule solvating the phenolic oxygen during bond dissociation. Geometry optimizations gave stationary ground states for the molecular complexes, and the (pentacoordinated) transition states of the above-mentioned substitution reaction were localized using reaction path calculations and subsequent optimization with the eigenvector routine as implemented in the MOPAC package (MOPAC93 1993; Stewart 1990). The resulting differences between transition state and ground state heats of formation, $\Delta H^{\neq}$, represent reaction barriers and thus provide local measures of the phosphorylation ability of the oxon metabolites. It should be noted that this approach is based on the Eyring equation of transition state theory as outlined above (cf. Equation 22.24 in the previous section).

Linear regression of $\log LC_{50}$ on $\Delta H^{\neq}$ gives an explained variance (adjusted for degrees of freedom) of 53% (48%), which increases to 67% (59%) when $\log K_{OW}$ is included as a second descriptor. These only moderate results support the above-mentioned hypothesis that phosphorothionate fish toxicity cannot be fully explained by a superposition of narcotic-type membrane irritation and AChE inhibition. A test of the potential relevance of dealkylation as a further, more specific toxicity component needs quantification of the corresponding reactivity of the compounds, bearing in mind that both the phosphorothionates and the oxon metabolites may have potential for alkylating electron-rich sites of critical macromolecules. A simple way to derive reactivity descriptors is the calculation of the differences in heat of formation between the organophosphorus compounds and the respective dealkylated anions (cf. Equation 22.25), which are formed by abstraction of one ester methyl cation. These calculations have been performed at the PM3 level for both the phosphorothionates and the oxon analogs, and the resulting ΔH_f and ΔH_f^{oxon} values are approximate measures of the relevant enthalpies of dealkylation (the latter would also include the constant heat of formation for the methyl cation). Multilinear regression of $\log LC_{50}$ on $\log K_{OW}$, $\Delta H^{\neq}$ and ΔH_f gives the following equation:

$$\log LC_{50} = -1.14 \log K_{OW} - 0.154 \Delta H^{\neq} + 0.0918 \Delta H_f + 30.6$$
(22.28)
$$r_{adj}^2 = 0.88, \quad r^2 = 0.91, \quad SD = 0.23, \quad F_{3,8} = 27, \quad n = 12$$

Now around 90% of the intraseries variation in toxicity can be explained using molecular parameters, which suggests that the observed fish toxicity of the organophosphorus compounds is indeed partly related to their potential for alkylating endogenous sites. Interestingly, replacement of ΔH_f by ΔH_f^{oxon} lowers r_{adj}^2 to 0.82, which would mean that in vivo dealkylation of the phosphorothionates is toxicologically more relevant than dealkylation of the oxon metabolites. However, the small number of compounds allows only tentative conclusions at this stage of analysis.

The signs of the regression coefficients for $\log K_{OW}$ and ΔH_f are as expected: Increasing lipophilicity decreases $\log LC_{50}$ and thus increases toxicity, and toxicity is also increased by greater negative dealkylation enthalpies (ΔH_f has calculated values ranging from -307.5 kJ/mol for dicapton to -242.0 kJ/mol for SV5). Surprisingly, according to Equation 22.28 toxicity is also increased with greater reaction barriers (and thus smaller reactivity) for phosphorylation, $\Delta H^{\neq}$ (with calculated values ranging from 5.2 kJ/mol to 39.1 kJ/mol for the oxon derivatives of dicapton and SV5, respectively). An attempt to rationalize this unexpected result would be the consideration that, for some reason, AChE inhibition is less toxic for the guppy than alkylation of critical macromolecules. The situation is further complicated by the relatively high intercorrelation between $\Delta H^{\neq}$ and ΔH_f of 79%, which allows only a tentative interpretation of the regression coefficients. An indirect test of potential statistical problems was performed through application of PLS, which can handle correlated data without restriction; inclusion of a number of additional descriptors yields an almost equivalent PLS result based on three latent variables with the same coefficient signs (according to the B matrix, cf. Equation 22.27) for $\log K_{OW}$, $\Delta H^{\neq}$ and ΔH_f like in Equation 22.28.

In any case, the present structure–activity analysis points to a second mode of toxic action of aromatic phosphorothionates in fish, which will have to be analyzed in more detail using experimental studies. This shows that QSAR results can point the way for future experimental investigations for a mechanistic understanding of the in vivo fate and toxicity of xenobiotics.

22.6.2 Chlorophenols and Nitrophenols

Chlorinated phenols have antiseptic properties and a wide spectrum of applications as fungicides and bactericides (cf. Ahlborg and Thunberg 1980). The estimated global production volume amounts to 200,000 t per year. Congeners with large commercial importance include pentachlorophenol, 2,4,5-trichlorophenol and 2,4-dichlorophenol. In addition, 2,3,4,6-tetrachlorophenol is a major by-product (up to 10–20%) of technical pentachlorophenol. Di- and trichlorophenol isomers play an important role in the industrial production of chlorinated phenoxyacetic acid herbicides, from which the chlorophenol precursors can be recovered by hydrolysis under environmental conditions. Evaporation from agricultural areas forms an important non-point source and dispersion pathway. On the other hand, airborne chlorophenols can undergo abiotic degradation by direct photolysis as well as by attack from OH radicals (Bunce et al. 1991).

In the 1960s and early 1970s, pentachlorophenol was used in Japan as a major herbicide for rice fields (Goto 1971), and an important application internationally in the 1980s was as a fungicide in wood preservatives. In Germany, the commercial distribution and use of pentachlorophenol was prohibited in late 1989. From the ecotoxicological viewpoint, polychlorinated dibenzodioxins are the most important contaminants of pentachlorophenol, with concentrations of typically 10 ppm and above. It follows that a sound assessment of toxic effects associated with commercial formulations must include consideration of these and other by-products.

Nitrophenols have phytotoxic properties and are used as herbicides and insecticides. They are intermediate products in the industrial synthesis of pesticides and dyes. In the environment, a non-point source may be their tropospheric formation from monoaromatic chemicals, which has been discussed in the context of forest decline and the occurrence of substantial amounts of nitrophenols in rain. Analysis of rain at urban sites in Switzerland in 1985 gave concentrations of 2,4-dinitro-6-methylphenol (DNOC) and other nitrated phenols of up to 15 nM (Leuenberger et al. 1988), and in Germany analysis of rain showed concentrations of up to 10 nM for 2-nitrophenol and up to 173 nM for 4-nitrophenol (Rippen et al. 1987). Known biological effects include perturbation of cell metabolism at doses of around 10 µM and below as well as uncoupling of oxidative phosphorylation, which is also observed for pentachlorophenol and other weak organic acids (McKim et al. 1987; Terada 1990; see also below).

For a given aqueous ionic strength, the dissociation of phenols in water depends on the relationship between pH and pK_a, with the latter being determined by the number, type and position of substituents (cf. Section *Compound Properties and Molecular Descriptors* above). The speciation of hydrophobic ionogenic organics in aqueous and lipophilic solutions under environmental or physiological conditions has gained much attention, but the relevance of ionic organic species for aqueous bioconcentration and toxicity has not yet been clarified satisfactorily. More recent experimental studies have shown that for the octanol/water distribution of organic acids and bases, ion-pair formation with inorganic counterions, such as Cl^-, may contribute substantially to the overall concentration in the octanol phase (Jafvert et al. 1990; Johnson and Westall 1990; Lee et al. 1990; Westall et al. 1985). In addition, comparison of octanol/water distribution with distribution between lipid membranes (egg phosphatidylcholine) and water has revealed that the latter is several hundred times greater for pentachlorophenolate than for pentachlorophenol (Smejtek and Wang 1993).

Impact of Dissociation on Partitioning and Bioavailability

In the following, the basic equations for the dependence of dissociation or protonation on pH and pK_a will be recalled. These are important for an understanding of partitioning, bioconcentration and associated ecotoxic effects of ionogenic compounds as outlined subsequently. Although the present discussion focusses on phenols, the relevant formulae are given for both acids and bases in order to enable more general applications.

For given pK_a and pH values, the unionized and ionized fractions f_u and f_i can be calculated for acids and bases according to

$$f_u^{acid} = f_i^{base} = \frac{1}{1 + 10^{pH - pK_a}} \qquad (22.29)$$

$$f_i^{acid} = f_u^{base} = \frac{1}{1 + 10^{pK_a - pH}} \qquad (22.30)$$

where the pK_a of bases is understood as the pK_a of their conjugate acids (e.g. the pK_a of NH_3 means the pK_a of its conjugate acid NH_4^+, which is 9.25). The dependence of overall partitioning on dissociation or protonation is now illustrated with K_{OW}, which is known as an important descriptor for the bioconcentration of organic chemicals (cf. Chapter 14). The possibility of ion-pair formation in the octanol phase is included by the assumption that the octanol/water system contains small amounts of inorganic salts. Following Figure 22.7, the distribution of an ionogenic compound between octanol and water can be written as

$$D_{OW} = K_{OW} \cdot f_u + (K_i + K_{ip}) \cdot f_i \qquad (22.31)$$

where K_i denotes the octanol/water partitioning of the ionized organic species, and K_{ip} the distribution between ion-pairs in octanol and the respective organic and inorganic ions in water. Combination with Equations 22.29 and 22.30 and rearrangement gives the overall distribution coefficients D_{OW} for acids and bases, from which the more familiar approximate terms can be derived by

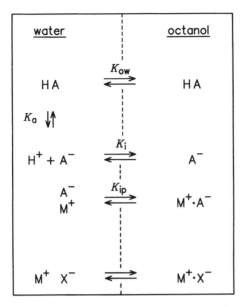

Figure 22.7 Distribution of an ionogenic compound HA between octanol and water in the presence of inorganic salts MX. K_a denotes the acid dissociation constant of HA, K_{OW} and K_i the octanol/water partition coefficients of HA and A$^-$, and K_{ip} the corresponding distribution coefficient of the ion pair A$^-$ + M$^+$.

neglecting the ion partitioning (K_i) and ion-pair distribution (K_{ip}):

$$D_{OW}^{acid} = \frac{K_{OW} + (K_i + K_{ip}) \cdot 10^{pH - pK_a}}{1 + 10^{pH - pK_a}} \geq \frac{K_{OW}}{1 + 10^{pH - pK_a}} \qquad (22.32)$$

$$D_{OW}^{base} = \frac{K_{OW} + (K_i + K_{ip}) \cdot 10^{pK_a - pH}}{1 + 10^{pK_a - pH}} \geq \frac{K_{OW}}{1 + 10^{pK_a - pH}} \qquad (22.33)$$

Use of the expressions on the right-hand side of Equations 22.32 and 22.33 implies that the contribution from K_i and K_{ip} is small compared to K_{OW}, which also depends on the pH–pK_a relationship.

The situation is illustrated by three examples (experimental data taken from Jafvert et al. 1990): 2,3,4,5-tetrachlorophenol with $pK_a = 6.35$, $\log K_{OW} = 4.87$, $\log K_i = -2.05$ and $\log K_{ip} = 1.81$ (referring to 0.1 M aqueous NaCl) gives $\log D_{OW} = 4.13$ at pH 7 both with the exact and the simplified formula of Equation 22.32. The more acidic and more hydrophobic pentachlorophenol ($pK_a = 4.83$, $\log K_{OW} = 5.09$, $\log K_i = -1.62$ and $\log K_{ip} = 3.10$ at 0.1 M aqueous LiCl) leads to $\log D_{OW} = 3.32$ at pH 7, as opposed to the simplified calculation of 2.91 (neglecting K_i and K_{ip}). Finally, 2,4,5-trichlorophenoxy acetic acid ($\log K_i = -3.36$, $\log K_{ip} = 1.25$ with 0.1 M aqueous LiCl) is less hydrophobic ($\log K_{OW} = 4.64$) but considerably more acidic ($pK_a = 2.83$), yielding exact and approximate $\log D_{OW}$ values at pH 7 of 1.25 and -0.86, respectively. This shows that ion-pair contribution to D_{OW} becomes increasingly important for increasing acidity or basicity and decreasing hydrophobicity, and that neglection of this term may lead to serious calculation errors of two or more orders of magnitude if the difference between pH and pK_a is close to or even greater than $\log K_{OW}$.

On the other hand, the approximate formulae of the right-hand sides of Equations 22.32 and 22.33 can be further simplified in the case of weak hydrophobic acids or bases with pK_a and pH differing by more than 1 (Howard and Scherrer 1977):

$$\log D_{OW}^{acid} \approx \log K_{OW} + pK_a - pH \qquad (22.34)$$

$$\log D_{OW}^{base} \approx \log K_{OW} - pK_a + pH \qquad (22.35)$$

Equation 22.34 gives $\log D_{OW}$ values of 4.22, 2.92 and -0.86 for the three compounds mentioned above, which are in fact close to the results from the right-hand side formula of Equation 22.32.

The example calculations with tetra- and pentachlorophenol might suggest that with chlorophenols the ionized fraction taken up in lipophilic phases would be too small to contribute substantially to their aquatic bioconcentration and toxicity. It would follow that for structure–activity analyses, the conventional $\log K_{OW}$ scale should be replaced by the $\log D_{OW}$ scale. Although this is supported by a number of cases (Dearden et al. 1995; Saarikoski and Viluksela 1982; Scherrer and Howard 1977), chlorophenol bioconcentration and acute aquatic toxicity often correlate very well with $\log K_{OW}$ (Butte et al. 1987; Könemann and Musch 1981; Saito et al. 1991; Schüürmann et al. 1996 and 1997; Schultz 1987; Smith et al. 1994; Veith and Broderius 1990). Interestingly, $\log D_{OW}$ is indeed often inferior to $\log K_{OW}$ as a descriptor for bioconcentration and narcotic-type effects, as will also be demonstrated below.

A possible explanation for this observation includes three aspects: First, uptake and bioavailability of organic ions from aqueous exposure is more efficient due to ion-pair formation than expected from the correction for ionization referring to a pure octanol/water system. Secondly, lipid membranes apparently have greater affinity for organic ions than (water-saturated) octanol (s.a.). Thirdly, organic ions probably differ in their toxicokinetics and internal bioavailability from their neutral counterparts, and even small ion concentrations could well contribute substantially to overall toxicity. Very few investigations have addressed this aspect, which is now discussed in more detail using experimental data from the literature (Könemann and Musch 1981).

Aquatic Toxicity from Unionized and Ionized Compound Fractions

Generally, both neutral and ionic forms of a chemical contribute to aquatic toxicity, but to a different extent depending on the amounts of unionized and ionized fraction as well as on intrinsic differences in bioavailability and efficacy. Both chemical forms can be transformed into each other, and a simple assumption would be that their individual toxicities add up to the overall toxicity. Without further specification, however, this assumption would still leave room for investigation to see, which of the classical concepts of additivity, concentration addition (Loewe and Muischnek 1926) or effect summation (Plackett and Hewlett 1948), would be the most suitable approach for assessing the mixture toxicity of the neutral and ionic form of ionogenic compounds. In the following, the concept of concentration addition will be used to interpret acute toxicity towards the guppy of phenols taken from the literature (Könemann and Musch 1981).

Concentration addition implies that the individual toxic potencies of the unionized and ionized fractions, $P_u = 1/LC_{50u}$ and $P_i = 1/LC_{50i}$, contribute to the overall effect proportional to their concentrations c_u and c_i, such that

$$1 = \frac{c_u}{LC_{50u}} + \frac{c_i}{LC_{50i}} = P_u \cdot c_u + P_i \cdot c_i \qquad (22.36)$$

holds true. For a given LC_{50} at a given pH, the concentrations of the unionized and ionized compound fractions, c_u and c_i, can be calculated through application of Equations 22.29 and 22.30:

$$c_u = f_u \cdot LC_{50} \qquad (22.37)$$

$$c_i = f_i \cdot LC_{50} \qquad (22.38)$$

Introduction into Equation 22.36 leads to

$$\frac{1}{LC_{50}} = P_u \cdot f_u + P_i \cdot f_i \qquad (22.39)$$

which can be rearranged to

$$\frac{1}{LC_{50}} = P_u + (P_i - P_u) \cdot f_i \qquad (22.40)$$

For an ionogenic compound, the toxic potencies P_u and P_i can be determined from the intercept and slope of a linear regression plot of $1/LC_{50}$ for different pH values against f_i (Equation 22.29), the pH-dependent ionized compound fraction (Könemann and Musch 1981). This approach is based on the further assumption that P_u and P_i are constant over the pH range analyzed.

The corresponding analysis for phenol and 10 chlorophenols is summarized in Table 22.9. At pH 7.3, the ionized compound fraction f_i varies from 0.2% (phenol) to 99.7% (pentachlorophenol). Similar ranges are found for the relative toxicity contributions $(P_u * f_u)/P$ and $(P_i * f_i)/P$, where $P = 1/LC_{50}$ denotes the

Table 22.9 Toxicity contributions from unionized and ionized compound fractions of phenols[a]

Compound	log LC$_{50}$ [mM]	pK$_a$	f_u [%]	f_u*P_u/P [%]	f_i*P_i/P [%]
Phenol	− 0.50	9.92	99.8	89	11
2-chlorophenol	− 1.06	8.52	94.3	140	− 40
3-chlorophenol	− 1.30	8.97	97.9	100	0
2,4-dichlorophenol	− 1.59	7.90	79.9	103	− 3
3,5-dichlorophenol	− 1.78	8.25	89.9	99	1
2,3,5-trichlorophenol	− 2.10	6.43	11.9	29	71
2,3,6-trichlorophenol	− 1.59	5.80	3.1	46	54
3,4,5-trichlorophenol	− 2.24	7.55	64.0	69	31
2,3,4,5-tetrachlorophenol	− 2.48	5.64	2.1	11	89
2,3,5,6-tetrachlorophenol	− 2.23	5.03	0.5	20	80
Pentachlorophenol	− 2.85	4.74	0.3	15	85

[a] Experimental LC$_{50}$ towards the guppy (*Poecilia reticulata*) at pH 7.3, $P_u = 1/LC_{50u}$ and pK$_a$ are taken from Könemann and Musch 1981. Unionized and ionized compound fractions f_u and f_i are calculated according to Equations 22.29 and 22.30, respectively, and their relative toxicity contributions are calculated according to Equation 22.39 with $P = 1/LC_{50}$ and $P_i = 1/LC_{50i}$, assuming concentration addition for the mixture effect.

overall potency. With 2,3,4,5-tetrachlorophenol and pentachlorophenol, 89% and 85% of the total toxicity are calculated to come from the ionized fraction, whereas with 3-chlorophenol toxicity is attributed solely to the unionized fraction. In two cases, the unionized fraction is calculated to contribute more than 100% of the total toxicity, which consequently leaves a negative contribution for the ionized fraction. This suggests limitations of the concentration addition model for the interpretation of the present data set, perhaps due to minor antagonistic effects of unionized and ionized forms in certain cases and under certain conditions. The overall picture, however, clearly indicates that the ionized fractions may contribute substantially to the aquatic toxicity of phenols. It follows that neglection of this compound fraction by simplified calculation of D_{ow} may well be inappropriate, as was observed with a number of chlorophenol data sets.

As dissociation or protonation certainly affect partitioning (s.a.), it is striking that with substituted phenols conventional K_{ow} performs relatively well in rationalizing bioconcentration from aqueous exposure. This is illustrated for pentachlorophenol in Table 22.10, where literature data for bioconcentration factors normalized to lipid content, BCF$_L$, are compared with K_{ow} and D_{ow}. The latter is again calculated according to the right-hand side of Equation 22.32 with neglection of the (unknown) contributions from ion partitioning and ion-pair distribution and refers to a pH of 7. As can be seen from the table, lipid-normalized bioconcentration factors of pentachlorophenol vary from 1,600 to

24,400 for four species, thus being considerably smaller than K_{OW} (123 000) but clearly greater than D_{OW} (813). According to the classical bioconcentration paradigm, BCF_L would be expected to roughly equal K_{OW} in the case of non-metabolizing (and non-dissociating) hydrophobic compounds (cf. Chapter 14). It follows that metabolizing compounds, such as substituted phenols, should bioconcentrate less due to enhanced excretion (cf. corresponding discussion with *Aromatic phosphorothionates* above). For pentachlorophenol, phase II biotransformation is known to play an important role, and with zebra fish 73% of the initial compound was metabolized to pentachlorophenyl glucuronide and sulphate within 48 h of exposure time (Kasokat et al. 1986). It can be concluded that the observed BCF_L/K_{OW} ratios from 1% to 20% in Table 22.10 are in qualitative agreement with expectations, provided that the four fish species differ in their metabolic activity. Here, the results with zebra fish are particularly striking: The theoretical bioconcentration yield of 20% (referring to the K_{OW} model) fits almost perfectly to the experimental biotransformation yield of 73% just mentioned. In contrast, BCF_L/D_{OW} ratios of 200–3000% suggest that D_{OW} derived from simplified calculation is not an appropriate descriptor for bioconcentration, and that future investigations should also focus on possible differences between octanol/water and lipid/water distribution of ionized organics (cf. Smejtek and Wang 1993, s.a.).

For the ecotoxic evaluation of phenols it follows that bioaccumulation and hazardous effects depend on both intrinsic compound properties (log K_{OW}, pK_a) and on the aqueous milieu (pH and buffer conditions). This will be of particular relevance when studying and assessing contaminations in the field, where different pH levels may lead to substantial differences in the resulting hazardous potential. Apart from chemical contamination, pH ranges in aquatic ecosystems also form selection criteria for the fitness of communities. Lower pH values favour survival and reproduction of acidophilic species compared to higher pH values; this is used for a long time for biological monitoring of water quality. On the other hand, more detailed knowledge about differences in species sensitivity to chemical stress is less well developed. Although experience has lead to the characterization of certain species as being more or less tolerant, it is often unclear to which compound classes, external conditions and life stages these characterizations would apply, and whether the underlying cause is at the physiological level of individual organisms or (if referring to situations in the field) is linked to indirect effects by interaction between different levels of biological organization (cf. *Introduction* and Chapters 16, 20, 21, 23, and 24).

Chemical Toxicity and Biological Response Profiles

At the organismic level, species response to a xenobiotic chemical will depend in particular on as to whether and to what extent certain routes of metabolic conversion and modes of toxic action can take place. Such knowledge is also valuable from the practical viewpoint of implementing and surveying water quality criteria, as different test species may differ in their sensitivity to specific

Table 22.10 Lipid-normalized bioconcentration factors of pentachlorophenol with aquatic species[a]

Species	BCF_L		BCF_L/K_{ow}	BCF_L/D_{ow} at pH 7
American flagfish (*Jordanella floridae*)	1600	(Smith et al. 1990)	1.3%	200%
Rainbow trout (*Oncorhynchus mykiss*)	4600–7600	(van den Heuvel et al. 1991)	3.7%–6.2%	570%–930%
Guppy (*Poecilia reticulata*)	7200	(Saarikoski and Viluksela 1982)	5.9%	890%
Zebra fish (*Brachydanio rerio*)	24400	(Butte et al. 1987)	20%	3000%

[a] BCF_L denotes the lipid-normalized bioconcentration factor, and K_{ow} the octanol-water partition coefficient of pentachlorophenol. BCF_L/K_{ow} refers to $K_{ow} = 123000$ (log $K_{ow} = 5.09$), and BCF_L/D_{ow} to $D_{ow} = 813$ (log $D_{ow} = 2.91$) calculated according to the right-hand side of Equation 22.32.

modes of toxic action. This means that different biological test systems will, in general, respond to different parts of the activity profile of chemicals, which will also limit the possibility of extrapolating toxic responses between different species. In the following, this is illustrated by a comparative analysis of the acute toxicity of eight phenols to 10 different species, showing how variation of toxic profiles with both chemical structure and biological organism can be studied and evaluated in a systematic way. A more detailed discussion of this matter is given elsewhere (Schüürmann et al. 1997).

Table 22.11 contains acute toxicity data from phenol, five chlorophenols (including pentachlorophenol) and two nitrophenols (including 2,4-dinitrophenol). Chlorophenols are known to exert polar narcosis or uncoupling of oxidative phosphorylation (cf. Section *Biochemical Background and Modes of Toxic Action*), depending on the compound acidity. The latter is increased by π-acceptor substituents, such as -CN, -NO$_2$ and -Hal, and decreased by π-donor substituents, such as -NH$_2$, -OH and -SH (cf. Section *Compound Properties and Molecular Descriptors*). Pentachlorophenol and 2,4-dinitrophenol are prominent examples of lipid-soluble weak organic acids that act as oxidative uncouplers. It follows that the phenol set comprises (at least) two different modes of action and appears to form a suitable test case for the present question of identifying systematic differences in the toxic response between the species.

The biological test systems listed in Table 22.11 include fish, daphnids, ciliates, bacteria and plant pollen, thus representing both eucaryotic and procaryotic species. It should be recalled that there are competing models which can be used to understand the mechanism of oxidative uncoupling, which in turn give different expectations as regards the response of eucaryotic and procaryotic species to this mode of action. According to the classic chemiosmotic model, oxidative uncouplers destroy the electrochemical membrane potential by carrying protons into mitochondria, thus depleting the energy needed for ATP formation. Procaryotic cells, such as bacteria, do not have mitochondria as the primary target membrane and thus would be considered (at least) less sensitive to oxidative uncoupling. In contrast, the receptor model suggests that the critical step is binding of the xenobiotic substrate to a specific protein site on the inner membrane of the cell, impairing the protein function required for oxidative phosphorylation. In this latter model, both eucaryotic and procaryotic cells would be potentially sensitive to this mode of action, but only the former would have the uncoupler binding site located in mitochondria. The data in Table 22.11 now lead to the following two questions:

- Do the response profiles of the test species, as characterized by toxicity results for all eight compounds, indicate systematic differences in their sensitivities to different groups of chemicals and associated modes of bioactivity?
- Does comparison of the ecotoxic profiles of the chemicals, as characterized by the 10 biological endpoints, reveal a systematic grouping reflecting different modes of action?

Table 22.11 Acute toxicity of 8 phenols towards 10 biological species[a]

Compound	Bluegill 96h-LC_{50}	Fathead minnow 96h-LC_{50}	Rainbow trout 96h-LC_{50}	Daphnia magna 24h-IC_{50}	Tetrahymena pyriformis 48h-IC_{50}	Vibrio fischeri LUMIStox 30min-EC_{50}	Vibrio fischeri Microtox 30min-EC_{50}	Bluegill cell line BF-2 24h-NR_{50}	Rainbow trout cell line R1 24h-MTT	Tobacco plant assay PTG 24h-IC_{50}
Phenol	− 0.84	− 0.409	− 1.02	− 0.47	0.431	− 0.43	− 0.42	0.84	1.05	0.59
2-Cl	− 1.29	− 0.969	− 1.69	− 1.15	− 0.277	− 0.53	− 0.58	0.38	0.61	0.20
2,4-Cl	− 1.92	− 1.323	− 1.80	− 1.43	− 1.036	− 1.44	− 1.51	− 0.29	− 0.09	− 0.90
2,4,6-Cl	− 2.80	− 1.334	− 2.43	− 2.10	− 1.55[d]	− 1.66	− 1.38	− 0.32	− 0.26	− 2.29
Tetra-Cl[b]	− 3.15	− 2.752	− 3.05	− 2.16	− 2.712	− 1.95	− 2.26	− 0.57	− 0.70	− 3.57
Penta-Cl	− 3.70	− 3.045	− 3.71	− 2.57	− 2.568	− 2.17	− 2.72	− 0.77	− 0.89	− 3.75
4-NO_2	− 1.22	− 0.493	− 1.25	− 1.20[c]	− 1.426	− 1.18	− 1.02	0.63	0.29	− 0.97
2,4-NO_2	− 2.52	− 1.229	− 2.20	− 2.05[c]	− 1.096	− 1.28	− 1.24	0.61	− 0.12	− 1.65

[a] Experimental data are given in decadic logarithms and were taken from various sources of literature (cf. Schüürmann et al. 1997).
[b] Covers the following isomers: 2,3,4,6-Cl (Daphnia magna, LUMIStox, Microtox, R1, PTG); 2,3,5,6-Cl (bluegill, BF-2); 2,3,4,5-Cl (rainbow trout, fathead minnow, Tetrahymena pyriformis).
[c] Extrapolation from 48h to 24h by subtraction of 0.40, which is the corresponding average difference for 2-Cl, 4-Cl, 2,4-Cl and 2,4,6-Cl.
[d] Average of two different literature values.

Both questions can be addressed using principal component analysis (PCA), which is a multivariate chemometric technique for studying the correlation structure of data matrices. It should be noted, however, that PCA requires autoscaling of all endpoint variables to a mean of zero and a standard deviation of 1, so that only similarities and differences in relative trends can be identified, differences in absolute toxicities are ignored. For a detailed description of PCA, the reader is referred to the literature (Aries et al. 1991; Geladi and Kowalski 1986).

Application of PCA to the toxicity data shows that 93% of the correlation structure can be represented in two dimensions, with 90% being given by the first principal component. The results are summarized in Figure 22.8. In the graphs, endpoints with higher intercorrelation are located more closely to each other, whereas greater distances indicate greater differences in the response pattern. The loadings plot on the left reveals similarities as well as differences between the response patterns of the 10 test systems, with the major variation being introduced by the second (vertical) principal component. The bluegill BF-2 fish cell line in the top left corner apparently shows the greatest difference in its response profile to all other test systems. Rainbow trout and the bacteria *Vibrio fischeri* in the Microtox assay give quite similar toxicity characterizations as regards the relative trend along the phenol series. In the lower right corner,

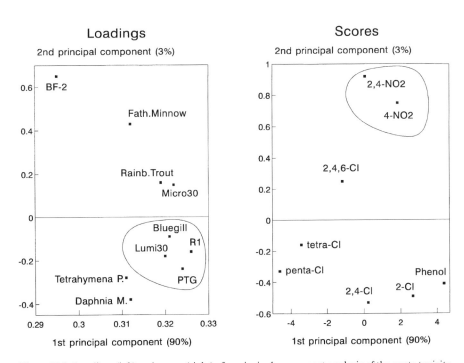

Figure 22.8 Loadings (left) and scores (right) of a principal component analysis of the acute toxicity of eight phenols towards 10 biological species. The respective toxicity data are given in Table 22.11.

four different species with relatively high similarity in their response pattern can be identified: the fish bluegill, the bacteria *Vibrio fischeri* in the LUMIStox assay, the rainbow trout cell line R1, and the pollen tube growth test PTG. In this context, the substantial difference between the two fish cell lines R1 and BF-2 is particularly striking. Furthermore, the difference in biological response between Microtox and LUMIStox is somewhat surprising, as both assays use the same species and endpoints and differ only in the preconditioning and perhaps slightly in the milieu.

Although the PCA loadings plot shows how the response patterns of the different endpoints are related to each other, it does not indicate the underlying cause for this correlation structure. This leads to the second question raised above, which is addressed with the scores plot on the right of Figure 22.8. Separate location of both nitrophenols indicates a systematic difference in their response profile from that of phenol and the five chlorophenols. Even more striking is the finding that the two oxidative uncouplers, pentachlorophenol and nitrophenol, are not grouped together. In contrast, acute toxicity exerted by pentachlorophenol, as characterized by the 10 endpoints of Table 22.11, is more closely related to that of other chlorophenols which act as polar narcotics. This suggests that the acute toxicity of pentachlorophenol is driven mainly by its narcotic component, and that uncoupling of oxidative phosphorylation does not yield a substantial enhancement of its acute aquatic toxicity. On the other hand, the graphical location of 4-nitrophenol suggests a toxic component from oxidative uncoupling, as is known for 2,4-nitrophenol.

The dependence of toxicity on underlying modes of action is more clearly seen in Figure 22.9, where $\log LC_{50}$ is plotted against $\log K_{OW}$ for four different species (fish, daphnid, ciliates, plant pollen) and compared with corresponding linear regression relationships for the subgroup of phenol and the five chlorophenols. With all four species, 2,4-dinitrophenol is more toxic than would be expected for isolipophilic chlorophenols. This suggests that the excess toxicity is due to oxidative uncoupling, which is also observed for all other test systems except the BF-2 cell line. The deviation of 4-nitrophenol toxicity from the chlorophenol regression line for some of the species is probably also due to oxidative uncoupling as an additional specific toxicity component. In contrast, pentachlorophenol nicely fits narcotic-type regressions on $\log K_{OW}$ with all species under investigation.

Differences in species sensitivity to specific modes of action can be seen with the response to 4-nitrophenol. Small excess toxicity with bluegill (Figure 22.9, top left) contrasts with the greatest excess toxicity for the ciliate *Tetrahymena pyriformis* (Figure 22.9, top right), and intermediate toxicity enhancement is observed with the waterflea *Daphnia magna* and the marine bacteria assay LUMIStox (Figure 22.9, bottom left and right). In addition, intermediate excess toxicities for 4-nitrophenol are also found with the rainbow trout fish cell line R1 and the pollen tube growth test PTG, whereas the two fish fathead minnow and rainbow trout as well as the bluegill cell line BF-2 show acute responses equivalent to isolipophilic chlorophenols (no graphs given). This demonstrates

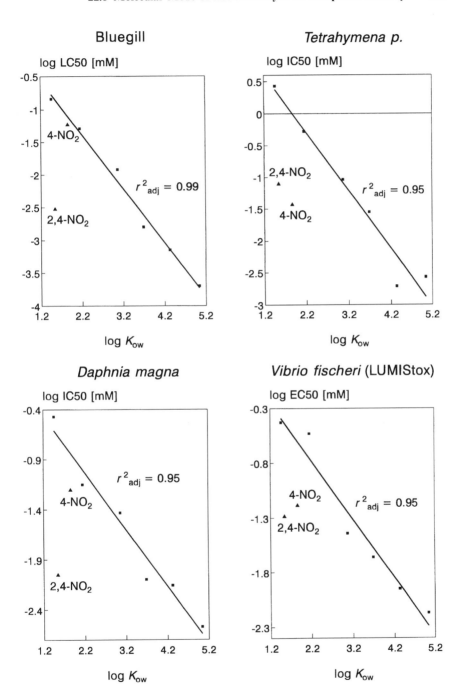

Figure 22.9 Acute toxicity vs. hydrophobicity ($\log K_{ow}$) of eight phenols for four different species (cf. Table 22.11). The regression lines refer to the subgroup of phenol and five chlorophenols.

that species responses to specific modes of toxic action may differ substantially, and that relevant knowledge will be needed to assess the ecotoxicological effects of contaminants on different species in food webs and communities. Future investigations should centre on similar analyses with other groups of chemicals and their associated modes of action. In addition, research should be devoted to the important question as to whether and how such systematic differences in species response to toxicants can be found for sublethal and chronic doses as well as for higher levels of biological organization.

Shift in Aqueous pH Due to Compound Ionization

A final note will be given for the design and interpretation of laboratory experiments with ionogenic compounds. As is known from textbooks on chemistry, pH of unbuffered solutions depends directly on pK_a and the total concentration of the acid or base. On the one hand, natural waters usually contain inorganic buffers in concentrations well above highly toxic contaminants, such as pentachlorophenol. On the other hand, however, it may still be possible that in certain biological experiments toxicity is caused both directly by the ionogenic compound and indirectly by acid stress due to pH lowering.

A useful orientation about upper limits of potential pH shifts can be derived from formulae for the extreme case of a truly unbuffered (i.e. pure) aqueous solution of the contaminant. For acids and bases with a molar concentration, c, of above $10^{-6.5}$ M (s.b.), neglection of H^+ and OH^- contributions from autoprotolysis of water leads to the following expressions for the pH shift due to contaminant dissociation or protonation:

$$pH^{acid} = pK_a - \log \left(\sqrt{1 + \frac{4c^{acid}}{K_a}} - 1 \right) + 0.301 \qquad (22.41)$$

$$pH^{base} = pK_a + \log \left(\sqrt{1 + \frac{4c^{base}K_a}{10^{-14}}} - 1 \right) - 0.301 \qquad (22.42)$$

(with constant $0.301 = -\log 0.5$). Application of Equation 22.41 for pentachlorophenol shows that aqueous concentrations of 1 mM and 1 µM would result in pH values of 4 and 6, respectively (if the solution is unbuffered). The latter concentration is close to the fathead minnow LC_{50} of 0.9 µM (s.b.), illustrating the potential relevance of evaluating the actual speciation of xenobiotic acids and bases. The error introduced by the above-mentioned simplification of the proton balance (i.e. assuming that in acidic solutions all H^+ comes from the contaminant dissociation, and in basic solutions all OH^- is caused by contaminant protonation) can be limited to 10% if molar concentration and acidity comply with the following condition (cf. Pankow 1991):

$$K_a \cdot (c^{acid} - 10^{-6.5}) \geq 10^{-13} \tag{22.43}$$

$$\frac{10^{-14}}{K_a} \cdot (c^{base} - 10^{-6.5}) \geq 10^{-13} \tag{22.44}$$

Thus for increasingly dilute contaminants, increasingly strong acids or bases are needed to allow calculation of the resultant pH using Equations 22.41 and 22.42. The lower limit would be a xenobiotic concentration of $10^{-6.5}$ M, which shows that our example with pentachlorophenol is indeed within the boundary conditions of the approximation. In this context it should be recalled that the more familiar textbook equations

$$pH^{acid} = \frac{1}{2}(pK_a - \log c^{acid}) \tag{22.45}$$

$$pH^{base} = \frac{1}{2}(14 - pK_a - \log c^{base}) \tag{22.46}$$

rely on the additional assumption that the resultant H^+ concentration is small compared to the total acid or base concentration. This approximation is valid for weak acids and bases at concentrations around 0.1 to 0.01 M, but generally less appropriate for contaminant loads typical for environmental systems. Comparative calculation of the pH for 1 μM acid with a pK_a varying from 1 to 6 gives pH values between 6.0 and 6.2 according to Equation 22.41, as opposed to pH values between 3.5 and 6.0 according to Equation 22.45. This shows that application of the former equations to derive upper limits for potential pH shifts is preferable for decreasing concentration and increasing acidity or basicity of the contaminant.

22.6.3 Alkylphenol Ethoxylate Surfactants

Organic surface-active agents are compounds containing hydrophobic and hydrophilic moieties. The hydrophobic part is usually composed of alkyl or alkylphenol groups, and the hydrophilic part consists of ionic or polar neutral groups. This forms the basis for a classification into anionic, cationic, amphoteric and non-ionic surfactants. Industrial use includes application as cleaners in various areas of production and manufacturing, as emulsifiers for creams and salves in the pharmaceutical industry, as additives for insecticides, herbicides and agricultural fertilizers, and as auxiliaries for textiles and fibres. Major domestic use covers consumer products such as detergents, dishwashing agents and cleaning agents as well as personal products (Piorr 1987). The high commercial relevance is reflected by marketing levels of 140,000 t per year for alkylphenol polyethoxylates alone in the USA in the early 1980s (Haupt 1983), of 220,000 t per year for non-ionic surfactants in former West Germany around

1985 (Gerike 1987), and of ca. 1 Mio and 0.5 Mio tons per year for linear alkylbenzene sulfonates (LAS) and linear alcohol ethoxylates (AEO) as the most common anionic and non-ionic surfactants in the USA, Western Europe and Japan in the late 1980s (Richtler and Knaut 1988). More recent estimates suggest an annual worldwide total of 7 Mio tons of synthetic surfactants (Stache and Kosswig 1990). As a consequence, surfactants have been (and are being) subjected to various ecotoxicological studies to characterize and assess their potential impact on aquatic and terrestrial environments. These studies examine bioaccumulation (Ekelund et al. 1990; Tolls and Sijm 1994; Wakabayashi et al. 1987), biodegradation (Ahel et al. 1994a and 1994b; Gerike 1987; Giger et al. 1984; Knaebel et al. 1990; Roch and Alexander 1994; Trehy et al. 1996), aquatic toxicity (Baillie et al. 1989; Gerike 1987; Hall et al. 1989; Roberts 1991), estrogenic activity of alkylphenol polyethoxylate metabolites (Routledge and Sumpter 1996; White et al. 1994), and general human safety (Bartnik and Künstler 1987).

The potential ecotoxicological concern is already apparent from model estimates for average concentrations of surfactants in German rivers, based on simple mass-balance calculations with the assumption of no biodegradation (Gerike 1987). This conservative (and, in fact, unrealistic) scenario leads to theoretical exposure levels of around 0.6–1.2 mg/L for non-ionic surfactants, which are not too far from laboratory-derived acutely toxic concentrations (LC_{50}) between 0.25 and 100 mg/L using various fish. With anionic surfactants, predicted exposure levels of 1.30–2.6 mg/L (again assuming no biodegradation) are even more critical compared to fish LC_{50} values of 1–20 mg/L (Gerike 1987). This illustrates that without surfactant biodegradation in efficiently operating wastewater treatment plants as well as under natural conditions, the discharge of such large amounts of surfactants would indeed have the most serious effects on aquatic communities in the freshwaters of industrialized countries.

Narcotic-type Mode of Aquatic Toxicity

Polyethoxylate derivatives of alkylphenols, fatty alcohols and fatty acids with the general structure

$$R\text{-}O\text{-}(CH_2CH_2O)_nH$$

represent major compound classes among the non-ionic surfactants (s.a.). This is partly due to the fact that both the hydrophobic (alkyl) and hydrophilic (oxyethylene) moieties can be varied to adapt the compound properties to specific needs. As a general observation, water solubility increases with increasing ethoxylate chain length, reflecting the net hydrophilic character of the oxyethylene unit, which is probably due to its propensity to accept hydrogen bonds in aqueous solutions. Interestingly, oxymethylene ($-CH_2O-$) derivatives with an even higher oxygen to carbon ratio show decreasing water solubility with increasing chain length, which is also known for oxypropylene ($-CH_2CH_2CH_2O-$) and isooxyethlylene ($-CH(CH_3)O-$) congeners.

Investigations of non-ionic surfactant toxicity to aquatic organisms suggest a direct relationship with the compound's ability to partition into membranes and subsequently disrupt membrane functions (Baillie et al. 1989). It would follow that, as with narcotics, a dependence of acute toxicity on lipophilicity in terms of $\log K_{OW}$ could be expected. Unfortunately, experimental $\log K_{OW}$ data for non-ionic surfactants, such as polyethoxylates, are at best scarce, and application of the prominent CLOGP calculation scheme (Daylight 1994; Leo 1993) does not yield increasing hydrophilicity with increasing ethoxylate chain length. In addition, octanol/water partitioning of surface-active agents may well depend on their concentrations and thus violate the conditions of Nernst's law due to an enhanced mixing between the two phases caused by solubilization.

However, the rationale that polyethoxylate toxicity is driven (mainly) by the compound's lipophilicity can be tested indirectly, if two further assumptions are made (Schüürmann 1990c): Both hydrophilic and hydrophobic moieties of the polyethoxylate contribute to acute toxicity proportional to their lipophilicity, and the (unknown) $\log K_{OW}$ of the ethoxylate moiety is additive with respect to the number of ethoxylate (EO) units. Subdivision of the compounds R-O-$(EO)_nH$ into the monooxyethylene congener R-O-$(EO)_1$-H and the remaining EO chain $(EO)_{n-1}$ leads to the following regression model for toxicity:

$$\log Tox = a_1 \cdot (\#EO\text{-}1) + a_2 \cdot \log K_{OW}(\text{R-O-EO-H}) + constant \quad (22.47)$$

Based on the assumption of lipophilicity-driven toxicity, the first term can be transformed to $\log K_{OW}$ units using

$$a_1 \cdot (\#EO\text{-}1) = a_2 \cdot \log K_{OW}((EO)_{\#-1})$$
$$= a_2 \cdot (\#EO\text{-}1) \cdot \log K_{OW}(EO) \quad (22.48)$$

which makes use of the above-mentioned (assumed) additivity of $\log K_{OW}$ with respect to EO units. Comparison of the left- and right-hand sides of Equation 22.48 gives $\log K_{OW}$ as a ratio of the two regression coefficients:

$$\log K_{OW}(EO) = \frac{a_1}{a_2} \quad (22.49)$$

It follows that polyethoxylate toxicity, if driven mainly by compound partitioning into membranes, would give a regression model according to Equation 22.47, which can be converted to a $\log K_{OW}$ scale by using Equation 22.49. A corresponding analysis of eight alkylphenol ethoxylates with 48-hr LC_{50} data for the marine crustacean *Mysidopsis bahia* (taken from Hall et al. 1989) lead to a tentative $\log K_{OW}$ increment of -0.10 per ethoxylate unit and a narcotic-type regression model

$$\log LC_{50} = -0.739 \cdot \log K_{OW} - 2.01 \quad (22.50)$$

(LC_{50} expressed in mol/L) with slope and intercept between the corresponding values for nonpolar and polar narcosis (Schüürmann 1990c, 1991). This illustrates again that QSAR techniques can be used to derive and develop mechanistic hypotheses about ecotoxic modes of action. In the present case, Equation 22.50 suggests a narcotic-type toxicity for non-ionic surfactants, which is probably exerted by disruption of cell membranes or intracellular organelles (cf. Section *Biochemical Background and Modes of Toxic Action* above).

A comparison of various narcosis models of aquatic toxicology covering different species and different compound classes is given in Table 22.12 and shown in Figure 22.10, with hydrophobicity in terms of log K_{OW} as the common molecular descriptor. Whereas prediction of bioconcentration using log K_{OW} usually gives upper limits (s.a.), prediction of toxicity using log K_{OW} is considered to yield lower limits, the so-called baseline toxicity. In both cases the underlying reason is implicit neglect of biotransformation, which may yield enhanced elimination as well as toxification, as was seen above with the oxidative metabolization of the aromatic phosphorothionates. Furthermore, it has already been demonstrated using the example of oxidative uncouplers, that prediction of baseline toxicity may provide a useful criterion for detecting compounds that are more toxic than their isolipophilic counterparts, which is likely to indicate a more specific mode of toxic action. Under certain (and probably not uncommon) circumstances, however, narcosis models may also hide some average biotransformation activity. This could be the case, if biotransformation has no or only a minor effect on overall toxicity, if it were relatively similar for the compounds under investigation, or if it correlated with log K_{OW} for the given compound set. This suggests that corresponding situations as well as variations in species sensitivity account for the observed differences in slope and intercept among different narcotic-type regression relationships. A typical example is the comparison between nonpolar and polar narcosis (cf. Table 22.12 and Figure 22.10), where a greater intercept and smaller slope for the latter indicate a greater average toxicity and weaker dependence on log K_{OW}, probably due to a generally higher participation of biotransformation.

Estimation of log K_{OW}

A final note should be given to the tentative log K_{OW} increment of -0.10 for the ethoxylate unit, which was derived from alkylphenol polyethoxylates with at least three monomer units (Schüürmann 1990c). The same increment value appears to explain log K_{OW} differences between short-chain alcohol ethoxylates $HO(EO)_nH$ with n varying from 1 to 4 (Roberts 1991). However, analysis of experimental log K_{OW} data for n-alkyl mono- and diethoxylates (Funasaki et al. 1984) suggests a more hydrophilic EO increment of around -0.17 to -0.19, and later a substantially more hydrophilic log K_{OW} increment of -0.34 was proposed based on comparison with partition coefficients for the iso-octane/water system (Ahel et al. 1993). However, correspondingly large decreases in

Table 22.12 QSAR models for narcotic-type modes of action[a]

Mode of action	Species	Endpoint log [mol/L]	log K_{OW} coefficient	Constant	CBR [mmol/kg]	log K_{OW} range	n	r^2	SD	Reference
Nonpolar narcotic										
	Guppy	14d-LC$_{50}$	-0.87	-1.13	3.7	$-1.3..5.7$	50	0.98	0.24	Könemann 1981
	Daphnia magna	48h-IC$_{50}$	-0.91	-1.28	2.6	$-1.3..5.7$	19	0.98	0.24	Hermens et al. 1984
	Tetrahymena pyriformis	48h-IC$_{50}$	-0.80	-0.83	7.4	$-0.8..5.7$	20	0.99	0.14	Schultz et al. 1990
	Vibrio fischeri	5min-EC$_{50}$	-1.17	-0.52	15	$-0.8..4.2$	12	0.97	0.38	Schultz et al. 1990
Ethoxylate narcotic										
	Crustacean	48h-LC$_{50}$	-0.74	-2.01	0.48	$0.8..5.1$	8	0.98	0.18	Schüürmann 1991
Polar narcotic										
	Fathead minnow	96h-LC$_{50}$	-0.65	-2.29	0.26	$1.3..6.4$	39	0.90	n.d.	Veith and Broderius 1990
	Tetrahymena pyriformis	48h-IC$_{50}$	-0.63	-1.95	0.56	$0.5..5.3$	36	0.91	0.24	Schultz et al. 1989
Polar narcotic-type										
	Pollen tube	18h-IC$_{50}$	-1.74	0.98	480	$2.2..3.8$	10	0.89	0.35	Schüürmann et al. 1996

[a] The QSAR models are linear regression equations of the general form $\log TOX = a \cdot \log K_{OW} + constant$. CBR denotes the critical body residue, which is calculated from the QSAR intercept according to Schüürmann et al. 1996b.

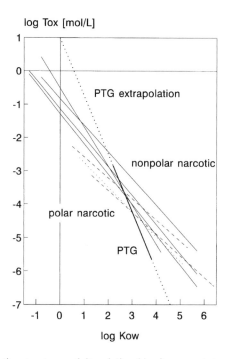

Figure 22.10 Quantitative structure-activity relationships for narcosis-type toxicity. The respective regression equations are listed in Table 22.12. Solid lines represent nonpolar narcosis, broken lines polar narcosis, one dotted line ethoxylate narcosis, and one solid line with dotted linear extrapolations a narcosis-type toxicity with the pollen tube growth (PTG) test (cf. Schüürmann et al. 1996).

hydrophobicity are not observed in the series of nonylphenol mono-, di- and triethoxylate with almost identical $\log K_{ow}$ values of 4.17, 4.21 and 4.20, and application of -0.34 as the $\log K_{ow}$ increment for an ethoxylate unit leads to apparently unrealistic $\log K_{ow}$ values for polyethoxylate congeners, which are more than three orders of magnitude below corresponding estimates from linear correlation with water solubility (Ahel et al. 1993). The latter suggests only small decreases in $\log K_{ow}$ with an increase in the chain length of below or equal to -0.10 per EO unit, which is qualitatively in agreement with the QSAR-based estimate mentioned above. It follows that with polyethoxylates, partitioning between iso-octane and water is probably not a good surrogate for octanol/water partitioning (i.e. the Collander relationship is not applicable in this case). On the other hand, $\log K_{ow}$ estimates based on incremental values for the ethoxylate unit also depend on the (experimental or calculated) value for the monoethoxylate congener. With extended hydrocarbon moieties as short-chain parent compounds, application of current calculation schemes may well yield estimates which are too hydrophobic. This needs further attention in future investigations of the hydrophobicity of polyethoxylates.

22.6.4 Triorganotins and other Organic Heavy Metals

Organotin compounds show biocidal activity and have gained high commercial relevance. This is reflected by an increase in the global annual marketing level from 50 t in 1950 to 30,000 t in 1980. The German production volume was estimated as 11,000 t per year in 1990 (Besler and Laschka 1991). Industrial use of di- and triorgano derivatives includes applications in anti-fouling paints, as fungicides in agricultural production, as stabilizers for PVC and other plastics, and as preservatives for wood, textiles and other materials. In general, organo-metal compounds are more toxic to a variety of organisms of all phyla than their inorganic counterparts. Among the tetravalent heavy metals germanium, tin and lead, the triorgano form shows the highest biological activity compared to mono-, di- and tetra-organo analogs (Saxena 1987). In this context, delayed toxicity of tetraorganotins, R_4Sn, has been explained by in vivo dealkylation to the triorgano form, R_3Sn^+, which was demonstrated to take place in the liver of vertebrates and in the midgut glands of invertebrates in both in vitro and in vivo studies (Blunden and Chapman 1986; Lee 1991).

Bioconcentration in the Field

Use pattern and physicochemical compound properties of organotins have lead to substantial contamination levels in aquatic environments. Besides direct input, there are important secondary intake pathways, such as, for example, run-off from agricultural applications. A recent analysis of Dutch rainwater revealed 16 ng/L (as Sn) of tributyltin (TBT) (Stäb et al. 1994). With TBT, surveys in French mediterranean coastal waters gave exposure concentrations of between 4 and 833 ng/L (Alzieu et al. 1987), around 150 ng/L in Rhine estuaries (Quevauviller and Donard 1990), ca. 400 ng/L in freshwater boat harbors of the Swiss Lake Lucerne in 1988 (Fent and Hunn 1995), and 34–367 ng/L in Chesapeake bay in the USA (Matthias et al. 1988). The greatest TBT concentrations in freshwater of 3000 ng/L were found in yacht harbours of the Canadian lake St. Clair, and in Europe the top TBT concentrations reported were around 1600 ng/L (Hall and Pinkney 1985).

The lipophilic character of organometals leads to substantial accumulation in sediments (Espourteille et al. 1993; Fent and Hunn 1991) with subsequent uptake by benthic organisms (Mercier et al. 1994). Correspondingly, significant bioconcentration of waterborne organometals in aquatic organisms may take place. With bivalves, TBT doses of 40–1160 ng/g dry weight were found on the coasts of the USA (Uhler et al. 1993), and body burdens of 210–1400 ng/g wet weight of triphenyltin (TPT) as co-toxicant of TBT were reported for mussels (*Mytilus edulis*) in Japanese harbors (Shiraishi and Soma 1992). Bioconcentration factors of TBT and TPT in zebra mussels (*Dreissena polymorpha*) are typically around 10,000 to 70,000 (Fent and Hunn 1991, 1995), but may be as high as 900,000 as reported for TBT in a study of freshwater mussels taken from harbors of Lake Geneva (Becker van Slooten and Tarradellas 1994). Here, body residues were up to 13,000 ng/g dry weight TBT and 4,000 ng/g dry weight dibutyltin (DBT), with elimination half-lives of ca. 26 days. In a Dutch freshwater survey, triorganotin contents in zebra mussels showed similar top levels in ng/g dry weight for TBT, DBT and TPT of up to 11,500, 1,740 and 3,200, respectively, with apparent half-lives of around 100 to 200 days in the field (Stäb et al. 1995). With various fish species, TBT and TPT bioconcentration factors range from 90 to 5,000 and 180 to 10,000, respectively, and body residues in the field from 0.01 to 0.7 nmol/g come surprisingly close to lethal body burdens of 20 nmol/g, determined in laboratory experiments with guppy (Tas 1993). Generally, the bioconcentration potential of TBT in freshwaters is substantially greater than in marine waters (Tsuda et al. 1990).

The widespread use of TBT, mainly as an anti-fouling agent on ship hulls, has lead to deleterious effects on oyster cultures, and sublethal toxicity concentrations of as low as 1–10 ng/L with various non-target organisms in freshwater and marine environments suggest that TBT is among the most toxic synthetic chemicals ever introduced to aquatic systems (Alzieu 1991; Hall and Pinkney 1985; Maguire 1987; Mercier 1994). In 1982, France was the first country to introduce legislative regulation of the use of TBT in anti-fouling paints, and bans on retail sale and use for ships below 25 m overall length were implemented in the UK in 1987, followed by the USA in 1988, Canada in 1989 and Switzerland between 1988 (Lake Constance) and 1990 (whole Switzerland; Becker van Slooten and Tarradellas 1994; Huggett et al. 1992; Tester et al. 1996). However, reduction in organotin concentrations is apparently slow, and harbor sediments, as long-term reservoirs, may form current sources of ongoing contamination of water and molluscs (cf. Fent and Hunn 1995; Uhler et al. 1993). This also suggests that in-field degradation by consecutive dealkylation (Bock 1981; Freitag and Bock 1974) is rather slow.

Modes of Toxic Action

The biological activity of triorganotins includes different modes of toxic action. One major cause is thought to be the impairment of mitochondrial functions, which may proceed in several ways (Aldrige 1987; Blunden and Chapman 1986; Reader et al. 1993; Virkki and Nikinmaa 1993; Yatome 1993):

- inhibition of oxidative phosphorylation (i.e. oxygen uptake from coupled mitochondria) and ATP synthesis;
- gross swelling of mitochondria due to interaction with their membranes;
- perturbation of ion-channel functions;
- perturbation of signal transduction processes.

In addition, neurotoxic, teratogenic and androgenic effects have been reported (Aschner and Aschner 1992; Gibbs et al. 1990; Kobayashi et al. 1994; Tester et al. 1996; Yonemoto et al. 1993). The latter may affect reproduction at the population level, and effects such as impaired recruitment have been reported (Bryan et al. 1986; Laughlin et al. 1989). In addition, it has been shown that tributyltin exposure leads to a time- and concentration-dependent decrease in total microsomal P-450 content (Fent and Stegeman 1991). Tributyltin interacts strongly with this enzyme causing its destruction and inhibition of MFO activities. Comparable results were obtained for molluscs, where tributyltin acts as an endocrine disruptor which disturbs steroid metabolism, namely the aromatization of androgens to estrogens by the P-450 depended aromatase (Oehlmann et al. 1993). In these animals TBT exposure in the lower ng/L range leads to a virilization and finally a sex change from females to males with deleterious effects not only on reproduction but also on sexual differentiation. These examples illustrate how ecotoxic modes of action at the organismic level may translate to effects at higher levels of biological organization. This also holds true for xenobiotic impact on the behaviour of individual organisms, e.g., swimming, foraging and predator avoidance, which will affect their life-history functions and thus the abundance of the respective population (Semlitsch et al. 1995).

Organometal salts of the general structure $R_{3-n}MeX_n$ possess lipophilic and ionic properties. Toxicity depends both on the structure of the organic residues R (alkyl or aryl), and on the central metal cation Me^+ (Ge^+, Sn^+ or Pb^+ in the case of tetravalent metals of group IV). On the other hand, the type of anionic residue X^- (e.g. halogenide or acetate) has only minor influence on toxicity, except if X is biologically active by itself or through specific circumstances associated with the actual mode of action. The latter could lead to synergistic or antagonistic effects and would, for example, explain why with certain triorganotins an oligomycin-type inhibition of oxidative phosphorylation was observed only in the absence of Cl^- residues (Aldrige 1987).

Non-Linear Dependence of Toxicity on the Alkyl Chain Length

For symmetric tri-*n*-alkylmetals, activity varies with the alkyl chain length, and the most active congener depends on the species. This is illustrated by organotin salts in Table 22.13, which shows that trimethyl and triethyl forms, for example, are particularly toxic to insects and mammals, respectively, and that tri-*n*-butyltin shows the highest toxicity for plants, fungi and some fish species. These general findings suggest that for a given organism and a congeneric series of tri-*n*-alkylmetal salts, toxicity will show a parabolic-type dependence on alkyl chain length: increase of alkyl chain length will first increase and, having passed the congener with maximum activity, then decrease toxicity. However, there are a few apparent exceptions with linear dependencies of organotin toxicity on parameters related to alkyl chain length (Eng et al. 1991; Laughlin et al. 1985; Vighi and Calamari 1985).

A possible explanation for the non-linear toxicity dependence of alkyl chain length has been developed using quantitative structure-activity analyses with various species (fungus *Botrytis allii*, rainbow trout cell line R1, marine bacteria *Vibrio fischeri*) and triorgano salts of tin, lead and germanium (Kaars Sijpesteijn

Table 22.13 Species specificity of triorganotin toxicity[a]

Organismic class	R of most toxic R_3SnX
Insects	CH_3
Mammals	C_2H_5
Fish cell line	C_2H_5
Gram-negative bacteria	$n\text{-}C_3H_7$
Gram-positive bacteria, fish, fungi, molluscs, plants, human erythrocytes	$n\text{-}C_4H_9$
Fish, fungi, molluscs	C_6H_5
Fish, mites	$cyclo\text{-}C_6H_{11}$

[a] Modified after Blunden and Chapman 1986.

et al. 1962; Schüürmann and Röderer 1988; Schüürmann and Segner 1994). The data sets are summarized in Figure 22.11, and show parabolic trends of toxicity in all cases.

Application of semiempirical quantum chemistry to quantify molecular hardness, electronegativity and enthalpy of solvation of triorganotins as well as the charge density of the central tin cation (cf. Equations 22.3, 22.4 and 22.14 of Section *Compound Properties and Molecular Descriptors*) revealed that none of these reactivity descriptors showed a parabolic trend (Schüürmann and Röderer 1988; Schüürmann and Segner 1994). However, with all calculated parameters the difference was greatest between trimethyltin and triethyltin, and higher-alkylated congeners showed an apparent saturation of reactivity. Similarly, the contact surface area of the central tin (which quantifies the portion of the tin surface area accessible to van der Waals interactions with water molecules; cf. Schüürmann 1995) was also greatest for trimethyltin and almost constant for all other congeners. This shows that except for trimethyltin there is no substantial variation in the potentially relevant aspects of molecular reactivity or in the steric situation at the central cation within the congeneric compound series. As a consequence, observed non-linear toxicity is not likely to be driven by local reactivity variation at the central tin.

Two-Component Model of Organometal Toxicity

An increase in the alkyl chain length yields two different trends in the property profile of the compounds (Schüürmann and Segner 1994): On the one hand,

Figure 22.11 Toxicity of tri-*n*-alkyl metals against alkyl chain length ($\# C$) for three species. Top: Complete growth inhibition within 72 h of the fungus *Botrytis allii* for lead, tin and germanium. Middle: Effective luminiscence inhibition of the marine bacterium *Vibrio fischeri* in the LUMIStox assay through tri-*n*-alkyl tins, using exposure times of 5, 15 and 30 min. Bottom: Inhibition of cell attachment within 24 h of the R1 cell line through tri-*n*-alkyl tins, with quantification of the total cell protein within each well through staining with crystal violet (CV) solution (Lenz et al. 1993).

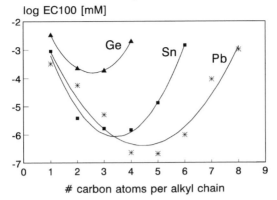

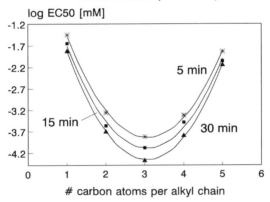

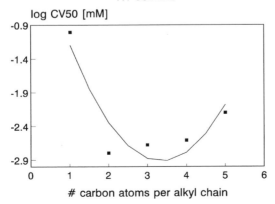

greater alkyl moieties make the associated hydrophobicity greater, and on the other hand electronic reactivity first declines and then is almost constant for the tripropyl form and higher congeners. This suggests that the parabolic course of overall toxicity can be understood as a superposition of both trends. According to this hypothesis, organometal toxicity has two basic components: Impairment of membrane functions of cells or intracellular organelles through hydrophobic association of the organic residues (hydrophobic component), and specific interaction with intracellular sites, e.g. enzymes or proteins (polar component).

Relative trends in triorganometal hydrophobicity may be quantified by (calculated) contact surface area and polar reactivity at the central cation using the (calculated) energy of aqueous solvation (Schüürmann and Segner 1994). Corresponding bilinear regression analyses of the above-mentioned data sets with triorganotins (cf. Figure 22.11) gave explained variances (adjusted for degrees of freedom) of 77% (53%) for the marine bacteria *Vibrio fischeri*, 79% (66%) for the fungus *Botrytis allii*, and 99.6% (99%) for the rainbow trout cell line R1. Interestingly, in all cases triethyltin has the greatest calculated toxicity, although this is only true for the fish cell line. Detailed inspection of the regression results reveals that with both of the other two test systems, the only moderate statistics are largely due to the wrong prediction for this compound. With the other compounds, the calculated results are in qualitative agreement with the above-mentioned hypothesis that trialkyltin toxicity results from both hydrophobic association with membrane surfaces and polar interaction with intracellular sites.

Following this model, trimethyltin with the greatest reactivity but smallest hydrophobicity is less toxic than triethyltin because of the lower probability of it reaching critical intracellular sites. Greater hydrophobicity due to greater alkyl chain length facilitates membrane penetration, but above the optimum hydrophobicity increasing association with membrane surfaces decreases successful penetration and thus the probability of subsequent polar interaction with the endogenous receptor. NMR spectroscopic analysis of the interaction of triorganotins with synthetic phospholipid membranes (vesicles) revealed only weak association at the membrane surface as indicated by the very small chemical shift of phosphatidylcholine phosphorus and tin. Furthermore, investigation of the membrane structure with electron microscopy showed substantially greater damage through tributyltin than through trimethyltin (Heywood et al. 1989). These experimental findings support the interpretation that association of triorganotins with membranes is hydrophobic in nature and not caused by polar interaction with phosphate groups, which probably also holds true for triorgano salts of other heavy metal cations. The combination of biochemical reactivity and hydrophobicity with maximum overall toxicity depends on the physiology and metabolism of the species, thus offering a qualitative explanation for the findings summarized in Table 22.13.

A further interesting aspect is the observation that greater absolute toxicity tends to shift the toxicity maximum to greater alkyl chain lengths. Figure 22.11 contains different examples of this phenomenon: In the top plot, variation of growth inhibition efficacy within 72 h of the fungus *Botrytis allii* with alkyl chain length is shown for a congeneric series of tri-*n*-alkyl germanium, tin and lead. The toxicity maximum is between four and five alkyl chain carbons for the generally most toxic lead derivatives, between three and four carbons for the tins, and between two and three carbons for the (much shorter) germanium series. The situation is similar when comparing responses of different species to one congeneric compound series, as with tri-*n*-alkyl tins and the species fungus *Botrytis allii*, rainbow trout cell line R1 (bottom plot in Figure 22.11), and the marine bacterium *Vibrio fischeri* (LUMIStox bioassay, middle plot in Figure 22.11). Most sensitive fungi show the highest toxicity for tri-*n*-butyl tin, LUMIStox with intermediate sensitivity gives maximum response to the tri-*n*-propyl congener for all three exposure times of 5, 15 and 30 min, and the least sensitive R1 cell line shows maximum toxicity for triethyltin (as noted earlier) in the 24 h assay to quantify inhibition of the cell attachment. One possible cause could be that with triorganotins greater toxicity on the exposure scale corresponds to greater intrinsic toxicity and thus smaller body burdens for exerting the toxic effect. Smaller toxic body burdens at the polar intracellular site of action would then still be reached with greater compound hydrophobicity, so that maximum activity would occur with greater chain lengths. This would give a qualitative explanation for the observed trend that greater absolute toxicity is accompanied by a shift of the toxicity maximum to greater hydrophobicity.

However, the present hypothesis is clearly tentative and needs further experimental investigation as well as a more distinct specification of the endpoints and modes of action to which it may be applied. Nonetheless, it provides a qualitative frame of reference for the further elucidation of differences in triorganometal sensitivity between different species as well as of differences in hydrophobicity for the most toxic derivative within congeneric series of compounds. Similarly, the breakdown of overall triorganometal toxicity into hydrophobic and polar components offers distinct criteria for further investigations of the underlying modes of action associated with this class of xenobiotics.

22.6.5 Unsaturated Alcohols

The aquatic toxicity of saturated aliphatic alcohols is characterized by a narcotic-type mode of action. If expressed in terms of exposure concentrations (e.g. LC_{50}), it is correlated with compound hydrophobicity as quantified by $\log K_{OW}$ (Könemann 1981; Veith et al. 1983; Lipnick et al. 1985). Enhanced toxicity is observed for substituted or unsaturated alcohols, which may be identified by comparing actual toxicity and theoretical toxicity predicted from baseline narcosis as noted above. With reference to LC_{50} values, excess toxicity T_e can be quantified according to

$$T_e = \frac{LC_{50\,predicted}}{LC_{50\,observed}} \tag{22.51}$$

(Lipnick 1989a), where $LC_{50predicted}$ is taken from a suitable baseline quantitative structure–activity relationship.

Prominent examples among industrial alcohols with substantial excess toxicities are ethylene chlorohydrin (2-chloroethanol), allyl alcohol (2-propen-1-ol) and propargyl alcohol (2-propyn-1-ol). The first is used as a solvent for cellulose acetate and organic dyes, and is a by-product of ethylene oxide and ethylene glycol. Furthermore, ethylene chlorohydrin occurs as an intermediate in the industrial synthesis of insecticides, anaesthetics and softening agents for plastics. The known alkylating activity of this compound can be explained by analyzing its chemical structure: The chlorine substituent on the β carbon increases the electrophilicity of the α carbon carrying the hydroxyl group, thus enabling electrophilic attack at endogenous nucleophilic sites, such as amino or sulfuhydryl groups. In addition, ethylene chlorohydrin is biotransformed to chloroacetaldehyde and chloroacetic acid, metabolites with greater electrophilicity. A survey of the toxicological relevance of electrophilic features in chemical structures is available in the recent literature (Lipnick 1995), and some examples of general electrophilic structures are collected in Figure 22.2 of the section *Biochemical Background and Modes of Toxic Action*.

Allylic alcohols have a double bond in the α,β-position (i.e. adjacent to the carbon bearing the hydroxyl group), and propargylic alcohols contain a corresponding triple bond. The generic structures of these compounds as well as of homopropargylic alcohols are shown in Figure 22.12. The parent compounds allyl alcohol and propargyl alcohol have aquatic excess toxicities of up to 16,000 and ca. 4,500, respectively (Lipnick 1989b; Veith et al. 1989). They are completely miscible with water and are used as solvents. They occur as intermediates in the industrial synthesis of pharmaceuticals and agricultural chemicals, and

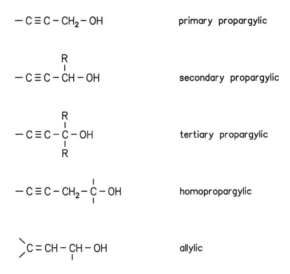

Figure 22.12 Molecular structures of propargylic, homopropargylic and allylic alcohols.

a further application of propargyl alcohol is to protect steel against corrosion and attack by mineral acids.

Proelectrophilic Mechanism of Chemical Toxicity

The high excess toxicities of these compounds and related derivatives can be traced back to a proelectrophilic mechanism (Lipnick et al. 1989a): The overall toxic potency of proelectrophiles is driven by the formation of electrophilic metabolites, which may well include several biotransformation steps to yield the relevant agent with substantially increased biological activity. The relevant metabolic activation mechanism depends on the chemical structure of the proelectrophile as outlined in Figure 22.13: Enzymatic oxidation of α,β-unsaturated alcohols by alcohol dehydrogenase yields the corresponding aldehydes or ketones, which may act as Michael-type electrophiles and form adducts with endogenous macromolecules. With homopropargylic alcohols containing an additional methylene group between the acetylenic and hydroxyl moieties, the corresponding carbonyl metabolite is not a Michael-type agent, but can undergo further biotransformation through temporary loss of an acidic proton and subsequent protonation to an allenic aldehyde or ketone (with two adjacent double bonds conjugated to the carbonyl group) as the final electrophilic metabolite (Veith et al. 1989).

The enzyme monooxygenase is involved in another more complex in vivo toxification mechanism as proposed for the proelectrophile pentaerythritol triallyl ether (Lipnick et al. 1987). Initial hydrogen abstraction leads to an allylic

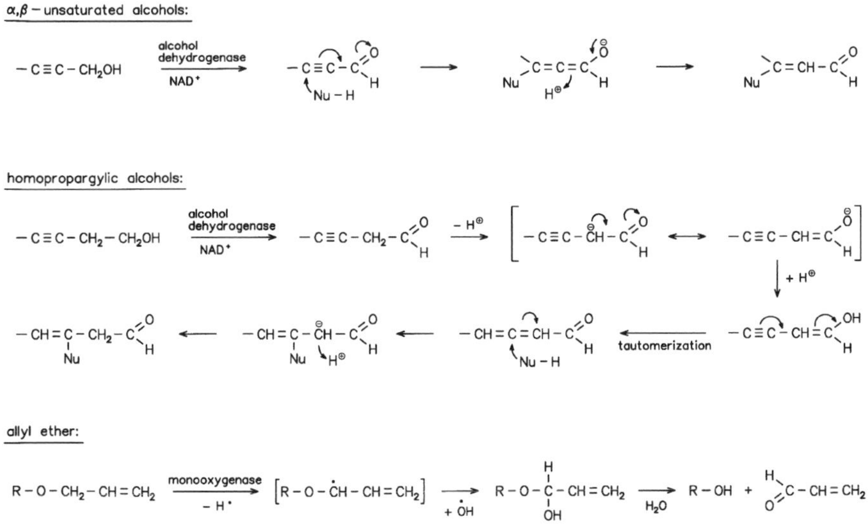

Figure 22.13 Proelectrophilic mechanisms of toxicity of unsaturated alcohols and allylic ethers.

radical with subsequent perhydroxyl radical attack to form a labile hemiacetal (with two hydroxyl groups attached to one carbon). The latter decomposes to give the final electrophile acrolein (cf. Figure 22.13). Theoretically, one mole of triallyl ether could yield up to three moles of acrolein. However, the latter (if applied directly) exerts an excess toxicity of 81,000 as compared to 18,000 for pentaerythritol triallyl ether. This suggests that the actual biotransformation performance of the triallyl ether must be much less efficient and may also include detoxification pathways.

A further proelectrophilic mechanism is associated with enzymatic toxification of the compound by reaction with glutathione S-transferase (GSH). As noted in previous sections, GSH conjugation does not always lead to an enhanced excretion of the xenobiotic (cf. Baillie and Slatter 1991). Besides the possibility to temporarily mask electrophilic chemicals and thus enable transport through cellular membranes to potential sites of interaction with biological macromolecules, conjugation of electrophilic compounds with GSH may also lead to metabolites with greater electrophilicity. A corresponding example is given by 1,3-dibromopropane, whose mutagenic activity has been traced back to metabolic formation of a four-membered sulfonium ring, an intermediate with increased electrophilicity which attacks endogenic sites by nucleophilic displacement reactions. This activation mechanism would also explain its excess toxicity of 87 for fish with some evidence from a corresponding metabolic study (Lipnick 1991).

These examples demonstrate the usefulness of a qualitative analysis of chemical reactivity for an assessment of potential in vivo activation and deactivation mechanisms and associated effects on the overall toxicity of compounds. For a given mode of action and biological organism, the variation in toxicity between structurally related chemicals should then be associated with differences in reactivity, which in turn could be tested by analyzing the suitable reactivity descriptors for the rate-determining step of biotransformation. This is now illustrated with the acute fish toxicity of various types of unsaturated alcohols, where initial observation of distinct excess toxicities lead to the hypothesis of corresponding proelectrophilic modes of action with subsequent confirmation by systematic structure-activity studies (Lipnick et al. 1985; Veith et al. 1989; Mekenyan et al. 1993).

Structure–Activity Model for Propargylic and Allylic Alcohols

Analysis of propargylic and homopropargylic alcohols as well as of some alkenols (cf. Figure 22.12) revealed that only tertiary derivatives of propargyl alcohol and α,β-alkenols are acutely toxic for fathead minnow (*Pimephales promelas*) with potencies comparable to isolipophilic polar narcotics. In contrast, substantial excess toxicities of primary and secondary propargylic and homopropargylic alcohols could be traced back to proelectrophilic mechanisms as outlined above, and acute toxicities of allylic (α,β-unsaturated) alcohols were also consistent with proelectrophilic activation. Within each subclass of these propargylic or allylic alcohols, toxicity was shown to increase with increasing $\log K_{OW}$ (Veith et al. 1989), indicating that both partitioning and chemical reactivity have an impact on the overall toxic effect.

In a subsequent structure-activity study, nucleophilic delocalizability (cf. Equation 22.15) at the β-carbon of the carbonyl metabolites formed was used to model their soft electrophilicity for covalent bond formation with electron-rich sites of endogenous macromolecules. This approach allowed a nice discrimination between narcotic-type and proelectrophilic alcohols (Mekenyan et al. 1993). However, tertiary alcohols cannot be metabolized to aldehydes or ketones, and with homopropargylic alcohols the β-carbon becomes electrophilic only after further biotransformation of the carbonyl metabolite to the respective allenic form. This suggests that a separate treatment of unsaturated alcohols according to different activation mechanisms is preferable to explain and predict proelectrophilic excess toxicity in terms of the intrinsic reactivity of the relevant active metabolites. From this viewpoint, non-tertiary propargylic and allylic alcohols would form one class of proelectrophilic compounds, which can all be metabolized by alcohol dehydrogenase to active aldehydes or ketones. In the following it will be demonstrated that detailed consideration of the relevant metabolic steps leads to a simple quantitative structure–activity relationship for this subset of unsaturated alcohols.

As outlined above, the rate-determining step in this toxicity mechanism is assumed to be the electrophilic attack of the carbonyl metabolites at electron-rich sites of biological macromolecules. A simplified evaluation of the associated reactivities can be achieved by analyzing the corresponding reaction with a model nucleophile, provided that the precise nature of the endogenous site and associated stereochemical parameters do not play a major role. Here, nucleophilic addition of NH_3 at the β-carbon of the propargylic or allylic carbonyl metabolite was selected as a simple model reaction for the final toxic step. The respective reaction enthalpies can be further simplified by omitting the constant heat of formation of NH_3. The remaining differences in heat of formation between NH_3-adducts and preceding carbonyl metabolites, ΔH_f, have been calculated with the MNDO model (Dewar and Thiel 1977) as implemented in the MOPAC package (Stewart 1990; MOPAC93 1993) and are listed in Table 22.14 together with the set of compounds, associated $\log LC_{50}$ values for fathead minnow (Veith et al. 1989) and further descriptor values. Multilinear regression of $\log LC_{50}$ on $\log K_{OW}$ and ΔH_f yields an explained variance (adjusted for degrees of freedom) of 84%, which can be improved to 89% by additional inclusion of the electronegativity (*EN*, cf. Equation 22.3) of the unsaturated alcohols:

$$\log LC_{50} = -0.43 \log K_{OW} + 0.019 \, \Delta H_f + 3.86 \, EN + 15.7$$

$$(22.52)$$

$$r_{adj}^2 = 0.89, \quad r^2 = 0.93, \quad SD = 0.27, \quad F_{3,5} = 24, \quad n = 9$$

Table 22.14 QSAR data set for proelectrophilic toxicity of non-tertiary propargylic and allylic alcohols[a]

Compound	log LC_{50} [mM]	Excess toxicity	log K_{OW}	ΔH_f [kJ/mol]	EN [eV]
3-butyn-2-ol	− 0.78	383	− 0.06	− 168.8	4.60
1-heptyn-3-ol	− 1.80	134	1.52	− 167.8	4.59
1-octyn-3-ol	− 2.48	214	2.05	− 168.1	4.59
2-propyn-1-ol	− 1.59	4880	− 0.37	− 169.0	4.46
2-butyn-1-ol	− 0.84	276	0.16	− 124.3	4.47
2-butyn-1,4-diol	− 0.21	4740	− 1.83	− 130.3	4.50
2-decyn-1-ol	− 2.16	7.7	3.33	− 120.1	4.45
1,5-hexadien-3-ol	− 0.41	42.2	0.51	− 92.4	4.51
1-hexen-3-ol	− 0.52	16.5	1.12	− 94.5	4.51

[a] Experimental toxicity (LC_{50}) towards fathead minnow (*Pimephales promelas*), calculated log K_{OW} and excess toxicity (cf. Equation 22.51) are taken from Mekenyan et al. 1993 and Veith et al. 1989. ΔH_f denotes the calculated difference in heat of formation between carbonyl metabolites after and before nucleophilic addition of NH_3, and EN the molecular electronegativity (cf. Equation 22.3) of the respective unsaturated alcohols (cf. text), using the semiempirical MNDO model (Dewar et al. 1977).

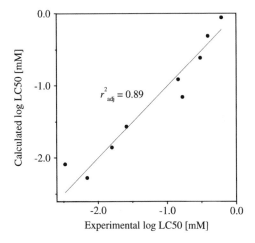

Figure 22.14 QSAR data distribution of the acute fish toxicity of non-tertiary propargylic and allylic alcohols according to Equation 22.52 and Table 22.14.

This suggests that overall toxicity depends on three factors: partitioning from aqueous exposure into the organisms (log K_{OW}), ease of oxidation to the carbonyl metabolites (EN), and electrophilic reactivity of the latter to attack endogenous macromolecules (ΔH_f). The coefficient signs of the regression equation are in agreement with this interpretation: Toxicity increases (log LC_{50}

decreases) with increasing $\log K_{OW}$, with decreasing EN and with decreasing ΔH_f. The latter reflects the fact that electrophilic reactivity is greater for more negative ΔH_f values, and that oxidation to carbonyl derivatives is easier with less electronegative alcohols. The resulting data distribution is shown in Figure 22.12.

It should be kept in mind that these results are based on a rather small set of compounds, and that they rely on further inherent approximations, such as complete neglection of solvation effects and the assumption that enzymatic catalysis by alcohol dehydrogenase would not change relative differences in the oxidative reactivity of the compounds. However, such assumptions have often proven useful in qualitative and quantitative structure–activity studies, and the example demonstrates again how mechanistic hypotheses about ecotoxic modes of action can be analyzed and evaluated using suitable reactivity descriptors calculated from quantum chemical methods.

22.7 CONCLUDING REMARKS

The ecotoxic modes of action discussed in this chapter address biological effects of chemicals at the organismic or suborganismic level. They lead to the important question as to whether criteria can be developed to recognize which types of individual-level effects would translate under which conditions and through which propagation mechanisms to higher levels of biological organization, such as populations, communities and ecosystems. Although a profound analysis of this problem is beyond the scope of the present chapter, a few aspects merit further consideration. First of all there is no doubt that connections do exist between processes operating at the level of individual organisms and at levels of more complex ecological systems (Koehl 1989). This is reflected by the rise of individual-based models in theoretical ecology, which have shown great potential as a primary modelling tool for the emergence of system patterns from interactions between individual organisms (DeAngelis and Gross 1992; Huston et al. 1988; Judson 1994). It also seems obvious that the ecological relevance of individual-level effects will vary substantially for different organismic endpoints, and that both organismic sensitivity and potential effect propagation to higher levels will also depend on the life stage of the species. Consequently, the potential ecological relevance of single-species tests is likely to have room for improvement if a proper selection of endpoints (including specification of species and the associated life stage) is used.

A possible step in this direction may be the focus on chemical impairment of physiological processes, such as growth and reproduction of individuals with evident relevance for population dynamics (cf. Chapters 16, 20, and 24). According to models of physiological energy balance in organisms, there are tradeoffs between energetic costs for metabolic activity and for somatic growth and reproduction. The latter two are the components of the scope for production

(S_p), which is the net energy available for production of biomass (Brett 1976; Calow 1988; Calow and Sibly 1990; Forbes and Forbes 1994; Koehn and Bayne 1989). More precisely, scope for somatic growth and reproduction ($S_g + S_r$) competes with metabolic costs for organismic maintenance (C_m) and production (C_p) at the expenditure of absorbed energy (A):

$$S_p = S_g + S_r = A - (C_m + C_p) \tag{22.53}$$

Physiological endpoints, such as growth and reproduction, integrate a variety of processes, each of which may be affected in different ways by chemical stress. Furthermore, interference of xenobiotics with endogenous molecules may provoke increased metabolic activity to initiate defense and repair mechanisms, which in turn can reduce the energy budget for production and other natural functions of the organism. Although all species have fundamental similarities in suborganismic structures and functions, differences in body size, constitution and metabolism cause differences in biological response on the physiological level. On the other hand, the chemical structure of contaminants conditions their overall bioactivity profile, as was demonstrated using structure–activity relationships and classification schemes for ecotoxic modes of action.

It follows that future ecotoxicological research should address the question, as to whether and how the presently discussed individual-level modes of biological action of contaminants translate into physiological endpoints with accepted relevance for population dynamics. Here, theoretical methods, such as structure–property and structure–activity relationships and chemometric design could provide valuable criteria for proper selections of test compounds to represent certain groups of chemicals and types of biological effects, and to enable mechanistic interpretation of the results both from the chemical and biological side.

Acknowledgements. The author thanks Dr. Rolf Altenburger and Dr. Helmut Segner for critically reading the manuscript and for helpful discussions, and Dr. Wolfgang Bödeker for helpful suggestions. All figures with chemical structures and reaction mechanisms have been drawn by Mrs. Elke Büttner, which is gratefully acknowledged.

22.8 REFERENCES

Ahel, M., and W. Giger, 1993, "Partitioning of Alkylphenols and Alkylphenol Polyethoxylates between Water and Organic Solvents," *Chemosphere*, 26, 1471–1478

Ahel, M., W. Giger, and C. Schaffner, 1994a, "Behaviour of Alkylphenol Polyethoxylate Surfactants in the Aquatic Environment. I. Occurrence and Transformation in Sewage Treatment," *Water Res.*, 28, 1131–1142

Ahel, M., W. Giger, and C. Schaffner, 1994b, "Behaviour of Alkylphenol Polyethoxylate Surfactants in the Aquatic Environment. II. Occurrence and Transformation in Rivers," *Water Res.*, 28, 1143–1152

Ahlborg, U. G., and T. Thunberg, 1980, "Chlorinated Phenols: Occurrence, Toxicity, Metabolism, and Environmental Impact," *CRC Crit. Rev. Toxicol., 7*, 1–35

Aldrige, W. N., B. W. Street, and D. N. Skilleter, 1987, "Oxidative Phosphorylation: Halide-Dependent and Halide-Independent Effects of Triorganotin and Triorgano-lead Compounds on Mitochondrial Functions," *Biochem. J., 168*, 353–364

Alzieu, C., 1991, "Environmental Problems caused by TBT in France: Assessment, Regulations, Prospects," *Mar. Environ. Res., 32*, 7–17

Aries, R. E., D. P. Lidiard, and R. A. Spragg, 1991, "Principal Component Analysis," *Chem. Br.*, 821–824

Baillie, A. J., H. Al-Assadi, and A. T. Florence, 1989, "Influence of Non-Ionic Surfactant Structure on Motility Inhibition of Tetrahymena Elliotti: A Model For Surfactant–Membrane Interactions," *Int. J. Pharm., 53*, 241–248

Baillie, T. A., and J. G. Slatter, 1991, "Glutathione: A Vehicle for the Transport of Chemically Reactive Metabolites in vivo," *Acc. Chem. Res., 24*, 264–270

Bartell, S. M., R. H. Gardner, and R. V. O'Neill, 1992, *Ecological Risk Assessment*, Lewis Publishers, Chelsea, MI, USA

Becker van Slooten, K., and J. Tarradellas, 1994, "Accumulation, Depuration and Growth Effects of Tributyltin in the Freshwater Bivalve *Dreissena polymorpha* under Field Conditions," *Environ. Toxicol. Chem., 13*, 755–762

Besler, W., and Laschka, 1991, "Umweltbundesamt (German Environmental Protection Agency)," Forschungsbericht 102 05 138, Berlin, Germany

Blunden, S. J., and A. Chapman, 1986, "Organotin Compounds in the Environment," in *Organometallic Compounds in the Environment*. Principles and Reactions (P. J. Craig, Ed.), Longman, Burnt Mill, Harlow, UK, pp. 111–159

Bock, R., 1981, "Triphenyltin Compounds and their Degradation Products," *Residue Rev., 79*, 1–270

Brett, J. R., 1976, "Scope for Metabolism and Growth of Sockey Salmon, *Oncorhynchus nerka*, and some Related Energetics," *J. Fish. Res. Board Can., 33*, 307–313

Brown, S. S., and K. H. Reinert, 1992, "A Conceptual Framework for Ecological Risk Assessment," *Environ. Toxicol. Chem., 11*, 143–144

Bunce, N. J., J. S. Nakai, and M. Yawching, 1991, "Estimates of the Tropospheric Lifetimes of Short- and Long-Lived Atmospheric Pollutants," *J. Photochem. Photobiol. (A), 57*, 429–440

Butte, W., A. Willig, and G.-P. Zauke, 1987, "Bioaccumulation of Phenols in Zebra Fish Determined by a Dynamic Flow-through Test," in *QSAR in Environmental Toxicology – II* (K. L. E. Kaiser, Ed.), D. Reidel, Dordrecht, The Netherlands, pp. 43–53

Cairns Jr. J., 1992, "Paradigms Flossed: The Coming Age of Environmental Toxicology," *Environ. Toxicol. Chem., 11*, 285–287

Cairns Jr. J., (Ed.), 1985, *Multispecies Toxicity Testing*, Pergamon Press, Headington Hill Hall, Oxford, UK

Cairns Jr. J., and B. R. Niederlehner, (Eds.), 1995, *Ecological Toxicity Testing*, Lewis Publishers, Boca Raton, Florida, USA

Calow, P., 1988, "Physiological Ecotoxicology: Theory, Practice and Application," in *Proc. 1st European Conf. on Ecotoxicology* (H. Løkke, H. Tyle, and F. Bro-Rasmussen Eds.), Copenhagen, Denmark, October 17–19, Conference Organizing Committee, Lyngby, Denmark, pp. 23–35

Calow, P., 1994, "Ecotoxicology: What are we Trying to Protect?" *Environ. Toxicol. Chem.*, *13*, 1549

Calow, P., and R. M. Sibly, 1990, "A Physiological Basis of Population Processes: Ecotoxicological Implications," *Funct. Ecol.*, *4*, 283–288

Cammarata, A., 1968, "Some Electronic Factors in Drug-Receptor Interactions," *J. Med. Chem.*, *11*, 1111–1115

Cammarata, A., 1971, "Electronic Representation of the Lipophilic Parameter π," *J. Med. Chem.*, *14*, 269–274

Cooke, A. S., 1973, "Shell Thinning in Avian Eggs by Environmental Pollutants," *Environ. Pollut.*, *4*, 85–152

Cooke, A. S., 1979, "Changes in Egg Shell Characteristics of the Sparrowhawk (*Accipiter nisus*) and Peregrine (*Falco peregrinus*) Associated with Exposure to Environmental Pollutants During Recent Decades," *J. Zool. London*, *187*, 245–263

Corbett, J. R., K. Wright, and A. C. Baillie, 1984, *The Biochemical Mode of Action of Pesticides*, Academic Press, London, UK

Cramer, C. J., and D. G. Truhlar, 1992, "An SCF Solvation Model for the Hydrophobic Effect and Absolute Free Energies of Solvation," *Science*, *256*, 213–217

Cramer, C. J., and D. G. Truhlar, 1995, "Continuum Solvation Models: Classical and Quantum Mechanical Implementations," in *Reviews in Computational Chemistry* (K. B. Lipkowitz and D. B. Boyd, Eds.), Vol. 6, VCH Publishers, New York, USA, pp. 1–72

Daylight Software 4.34, 1994, Daylight Chemical Information Systems, Irvine, CA, USA

DeAngelis, D. L., and L. J. Gross, 1992, (Eds.), *Individual-Based Models and Approaches in Ecology*, Chapman and Hall, New York, USA

Dearden, J. C., 1985, "Partitioning and Lipophilicity in Quantitative Structure–Activity Relationships," *Environ. Health Perspect.*, *61*, 203–228

Dearden, J. C., M. T. D. Cronin, T. W. Schultz, and D. T. Lin, 1995, "QSAR Study of the Toxicity of Nitrobenzenes to Tetrahymena Pyriformis," *Quant. Struct.–Act. Relat.*, *14*, 427–432

Dearden J. C., and R. M. Nicholson, 1986, "The Prediction of Biodegradability by the Use of Quantitative Structure–Activity Relationships: Correlation of Biological Oxygen Demand with Atomic Charge Difference," *Pestic. Sci.*, *17*, 305–310

De Bruijn, J., 1991, "Hydrophobicity, Biokinetics and Toxicity of Environmental Pollutants: A Structure–Activity approach, *Ph.D. Thesis*, University of Utrecht, Utrecht, The Netherlands

Dewar M. J. S., E. G. Zoebisch, E. F. Healy, and J. J. P. Stewart, 1985, "AM1: A New General Purpose Quantum Mechanical Molecular Model," *J. Am. Chem. Soc.*, *107*, 3902–3909

De Wolf, W., J. H. M. de Bruijn, W. Seinen, and J. L. M. Hermens, 1992, "Influence of Biotransformation on the Relationship between Bioconcentration Factors and Octanol–Water Partition Coefficients," *Environ. Sci. Technol.*, *26*, 1197–1201

Draber. W., and T. Fujita, (Eds.), 1992, *Rational Approaches to Structure, Activity and Ecotoxicology of Agrochemicals*, CRC Press, Boca Raton, Florida, USA

Ekelund, R., A. Bergman, A. Granmo, and M. Berggren, 1990, "Bioaccumulation of 4-Nonylphenol in Marine Animals – a Reevaluation," *Environ. Pollut. 64*, 107–120

Eng G., E. J. Tierney , G. J. Olson, F. E. Brinckman, and J. M. Bellama, 1991, "Total Surface Areas of group IV Organometallic Compounds: Predictors of Toxicity to Algae and Bacteria," *Appl. Organomet. Chem. 5*, 33–37

Espourteille, F. A., J. Greaves, and R. J. Hugget, 1993, "Measurement of Tributyltin Contamination of Sediments and *Crassostrea virginica* in the Southern Chesapeake Bay," *Environ. Toxicol. Chem., 12*, 305–314

Fent, K., and J. Hunn, 1991, "Phenyltins in Water, Sediment, and Biota of Freshwater Marinas," *Environ. Sci. Technol., 25*, 956–963

Fent, K., and J. Hunn, 1995, "Organotins in Freshwater Harbors and Rivers: Temporal Distribution, Annual Trends and Fate," *Environ. Toxicol. Chem., 14*, 1123–1132

Fent, K., and J. J. Stegeman, 1991, "Effects of Tributyltin Chloride in vitro on the hepatic Microsomal Monooxygenase System in the Fish *Stenotomus chrysops*," *Aquat. Toxicol., 20*, 159–168

Fest, C., and K.-J. Schmidt, 1977, "Insektizide Phosphorsäureester," in *Pflanzenschutz und Schädlingsbekämpfung* (K. H. Büchel, Ed.), Thieme, Stuttgart, Germany, pp. 22–60

Forbes, V. E., and T. L. Forbes, 1994, *Ecotoxicology in Theory and Practice: A Critique of Current Approaches*, Chapman and Hall, London, UK

Franks, N. P., and W. R. Lieb, 1982, "Molecular Mechanisms of General Anaesthesia," *Nature, 300*, 487–493

Franks, N. P., and W. R. Lieb, 1984, "Do General Anaesthetics Act by Competitive Binding to Specific Receptors?" *Nature, 310*, 599–601

Franks, N. P., and W. R. Lieb, 1985, "Mapping of General Anaesthetic Target Sites Provides a Molecular Basis of Cutoff Effects," *Nature, 316*, 349–351

Franks, N. P., and W. R. Lieb, 1986, "The Pharmacology of Simple Molecules," *Arch. Toxicol. Suppl., 9*, 27–37

Freitag, K.-D., and R. Bock, 1974, "Degradation of Triphenyltin Chloride on Sugar Beet Plants and in Rats," *Pestic. Sci., 5*, 731–739

Fukui, K., 1970, "Theory of Orientation and Stereoselection," *Top. Curr. Chem., 15*, 1–85

Funasaki, N., S. Hada, S. Neya, and K. Machida, 1984, "Intramolecular Hydrophobic Association of Two Alkyl Chains of Oligoethylene Glycol Diethers and Diesters in Water," *J. Phys. Chem., 88*, 5786–5790

Geladi, P., and B. R. Kowalski, 1986, "Partial Least-Squares Regression: A Tutorial," *Anal. Chem., 185*, 1–17

Gerike, P., 1987, "Environmental Impact," in *Surfactants in Consumer Products*, (J. Falbe, Ed.), Springer, Heidelberg, Germany, pp. 450–474

Gibbs, P. E., G. W. Bryan, P. L. Pascoe, and G. R. Burt, 1990, "Reproductive Abnormalities in Female *Ocenebra erinacea* (Gastropoda) Resulting from Tributyltin-Induced Imposex," *J. Mar. Biol. Assoc. U.K. 70*, 639–656

Giger, W., P. H. Brunner, and C. Schaffner, 1984, "4-Nonylphenol in Sewage Sludge: Accumulation of Toxic Metabolites from Nonionic Surfactants," *Science, 225*, 623–625

Goto, M., 1971, "Organochlorine Compounds in the Environment of Japan. Int. Symp. on Pesticide Terminal Residues," *Pure Appl. Chem. Suppl.*, 105–110

Greenwood, H. H., and R. McWeeny, 1966, "Reactivity Indices in Conjugated Molecules: The Present Position," *Adv. Phys. Org. Chem., 4*, 73–145

Hall, L. W., and A. E. Pinkney, 1985, "Acute and Sublethal Effects of Organotin Compounds on Aquatic Biota: An Interpretative Literature Evaluation," *CRC Crit. Rev. Toxicol., 14*, 159–209

Hall, W. S., J. B. Patoczka, R. J. Mirenda, B. A. Porter, and E. Miller, 1989, "Acute Toxicity of Industrial Surfactants to *Mysidopsis bahia*," *Arch. Environ. Contam. Toxicol., 18*, 765–772

Hansch, C., and T. Fujita, 1964, "ρ-σ-π Analysis. A Method for the Correlation of Biological Activity and Chemical Structure," *J. Am. Chem. Soc., 86*, 1616–1626

Hansch, C., and A. Leo, 1995, *Exploring QSAR. Fundamentals and Applications in Chemistry and Biology*, American Chemical Society, Washington DC, USA, 557 pp

Hansch, C., A. Leo, and D. Hoekman, 1995, *Exploring QSAR. Hydrophobic, Electronic and Steric Constants*, American Chemical Society, Washington DC, USA, 348 pp

Hansch, C., A. Leo, and R. W. Taft, 1991, "A Survey of Hammett Substituent Constants and Resonance and Field Parameters," *Chem. Rev., 91*, 165–195

Hermens, J., 1990, "Electrophiles and Acute Toxicity to Fish," *Environ. Health Perspect., 87*, 219–225

Hermens, J., H. Canton, P. Janssen, and R. de Jong, 1984, "Quantitative Structure–Activity Relationships and Toxicity Studies of Mixtures of Chemicals with Anaesthetic Potency: Acute Lethal and Sublethal Toxicity to *Daphnia magna*," *Aquat. Toxicol., 5*, 143–154

Hermens, J. L. M., and A. Opperhuizen, (Eds.), 1991, *QSAR in Environmental Toxicology – IV*, Elsevier, Amsterdam, The Netherlands, 705 pp

Heywood, B. R., K. C. Molloy, and P. C. Waterfield, 1989, "Organotin Biocides XV: Modelling the Interactions of Triorganotins with Cell Membranes," *Appl. Organomet. Chem., 3*, 443–450

Hodgson, E., and P. E. Levi, 1987, *A Textbook of Modern Toxicology*, Elsevier, New York, USA, 93 pp

Hope, B. K., 1993, "Ecological Considerations in the Practice of Ecotoxicology," *Environ. Toxicol. Chem., 12*, 205–206

Hugget, R. J., M. A. Unger, P. F. Seligman, and A. O. Valkirs, 1992, "The marine biocide tributyltin," *Environ. Sci. Technol., 26*, 232–237

Hunter, R. S., 1988, "Computer Calculation of pK_a," in *QSAR 88, Proc. 3rd Int Workshop on Quantitative Structure–Activity Relationships in Environmental Toxicology* (J. E. Turner, M. W. England, T. W. Schultz, N. J. Kwaak, Eds.), National Technical Information Service, U.S. Department of Commerce, Springfield, Virginia, USA, pp. 33–41

Huston, M., D. DeAngelis, and W. Post, 1988, "New Computer Models Unify Ecological Theory," *BioScience, 38*, 683–691

Jafvert, C. T., J. C. Westall, E. Grieder, and R. P. Schwarzenbach, 1990, "Distribution of Hydrophobic Ionogenic Organic Compounds between Octanol and Water: Organic Acids," *Environ. Sci. Technol., 24*, 1795–1803

Jefferies, D. J., 1969, "Introduction of Apparent Hyperthyroidism in Birds Fed DDT," *Nature, 222*, 578–579

Jefferies, D. J., and M. C. French, 1971, "Hyper- and Hypothyroidism in Pigeons Fed DDT: An Explanation for the "Thin Eggshell Phenomenon," *Environ. Pollut., 1*, 235–242

Jefferies, D. J., and M. C. French, 1972, "Changes Induced in the Pigeon Thyroid by p,p'-DDE and Dieldrin," *J. Wildlife Managem.*, *36*, 24–30

Johnson, C. A., and J. C. Westall, 1990, "Effect of pH and KCl Concentration on the Octanol–Water Distribution of Methylanilines," *Environ. Sci. Technol.*, *24*, 1869–1875

Jorgensen, W. L., J. M. Briggs, and M. L. Contreras, 1990, "Relative Partition Coefficients for Organic Solutes from Fluid Simulations," *J. Phys. Chem.*, *94*, 1683–1686

Judson, O. P., 1994, "The Rise of the Individual–Based Model in Ecology," *TREE*, *9*, 9–14

Kaars Sijpestein, A., F. Fijkens, J. G. A. Luijten, and L. C. Willemsen, 1962, "On the Antifungal and Antibacterial Activity of Some Trisubstituted Organogermanium, Organotin and Organolead Compounds," *J. Microbiol. Serol.*, *28*, 346–356

Kaiser, K. L. E., (Ed.), 1984, *QSAR in Environmental Toxicology*, D. Reidel, Dordrecht, The Netherlands, 406 pp

Kaiser, K. L. E., (Ed.), 1987, *QSAR in Environmental Toxicology* – II. D. Reidel, Dordrecht, The Netherlands, 465 pp

Kasokat, T., R. Nagel, and K. Urich, 1986, "The Metabolism of Phenol and Substituted Phenols in Zebra Fish," *Xenobiotica*, *17*, 1215–1221

Kelce, W. R., C. R. Stone, S. C. Laws, L. E. Gray, J. A. Kemppainen, and E. M. Wilson, 1995, "Persistent DDT Metabolite p,p'-DDE is a Potent Androgen Receptor Antagonist," *Nature*, *375*, 581–585

Kimball, K. D., and S. A. Levin, 1985, "Limitations of Laboratory Bioassays: The Need for Ecosystem-level Testing," *BioScience*, *35*, 165–171

Klamt, A., and G. Schüürmann, 1993, "COSMO: A New Approach to Dielectric Screening in Solvents with Explicit Expressions for the Screening Energy and its Gradient," *J. Chem. Soc. Perkin Trans. 2*, 799–805

Klopman, G., 1968, "Chemical Reactivity and the Concept of Charge- and Frontier-Controlled Reactions," *J. Am. Chem. Soc.*, *90*, 223–234

Klopman, G., and S. Wang, 1991, "A Computer Automated Structure Evaluation (CASE) Approach to Calculation of Partition Coefficient," *J. Comput. Chem.*, *12*, 1025–1032

Klopman, G., S. Wang, and D. M. Balthasar, 1992, "Estimation of Aqueous Solubility of Organic Molecules by the Group Contribution Approach. Application to the Study of Biodegradation," *J. Chem. Inf. Comput. Sci.*, *32*, 474–482

Knaebel, D. B., T. W. Federle, and J. R. Vestal, 1990, "Mineralization of Linear Alkylbenzene Sulfonate (LAS) and Linear Alcohol Ethoxylate (LAE) in 11 Contrasting Soils," *Environ. Toxicol. Chem.*, *9*, 981–988

Kobayashi, H., T. Suzuki, I. Sato, and N. Matsusaka, 1994, "Neurotoxicological Aspects of Organotin and Lead Compounds on Cellular and Molecular Mechanisms," *Toxicol. Ecotoxicol. News*, *1*, 23–30

Koehl, M. A. R., 1989, "Discussion: From Individuals to Populations" in *Perspectives in Ecological Theory* (J. Roughgarden, R. M. May, and S. A. Levin, Eds.), Princeton University Press, Princeton, NJ, USA, pp. 39–53

Koehn, R. K., and B. L. Bayne, 1989, "Towards a Physiological and Genetic Understanding of the Energetics of the Stress Response," *Biol. J. Linn. Soc.*, *37*, 157–171

Könemann, H., 1981, "Quantitative Structure–Activity Relationships in Fish Toxicity Studies: Part 1. Relationships for 50 Industrial Pollutants," *Toxicology*, *19*, 209–221

Könemann, H., and A. Musch, 1981, "Quantitative Structure–Activity Relationships in Fish Toxicity Studies. Part 2: The Influence of pH on the QSAR of Chlorophenols," *Toxicology*, *19*, 223–228

Kristen, U., and R. Kappler, 1995, "The Pollen Tube Growth Test," in *In Vitro Toxicity Testing Protocols. Methods in Molecular Biology*, (S. O'Hare, C. K. Atterwill, Eds.), Vol 43, Humana Press, Totowa, NJ, USA, pp. 189–198

Kühne, R., R.-U. Ebert, F. Kleint, G. Schmidt, and G. Schüürmann, 1995, "Group Contribution Methods to Estimate Solubility of Organic Chemicals," *Chemosphere*, *30*, 2061–2077

Landis, W. G., R. A. Matthews, and G. B. Matthews, 1996, "The Layered and Historical Nature of Ecological Systems and the Risk Assessment of Pesticides," *Environ. Toxicol. Chem.*, *15*, 432–440

Laughlin, Jr. R. B., R. B. Johannesen, W. French, H. Guard, and F. E. Brinckman, 1985, "Structure–Activity Relationships for Organotin Compounds," *Environ. Toxicol. Chem.*, *4*, 343–351

Lee, L. S., P. S. C. Rao, P. Nkedi-Kizza, and J. J. Delfino, 1990, "Influence of Solvent and Sorbent Characteristics on Distribution of Pentachlorophenol in Octanol/Water and Soil–Water Systems," *Environ. Sci. Technol.*, *24*, 654–661

Lee, R. F., 1991, "Metabolism of Tributyltin by Marine Animals and Possible Linkages to Effects," *Mar. Environ. Res.*, *32*, 29–36

Lenz, D., H. Segner, and W. Hanke, 1993, "Comparison of Different Endpoint Methods for Acute Cytotoxicity Tests with the R1 Cell Line." in *Fish – Ecotoxicology and Ecophysiology* (T. Braunbeck, W. Hanke, and H. Segner, Eds.), VCH Verlagsgesellschaft, Weinheim, Germany, pp. 93–102

Leo, A. J., C. Hansch, and D. Elkins, 1971, "Partition Coefficients and their Uses," *Chem. Rev.*, *71*, 525–616

Leo, A. J., 1993, "Calculating log P_{oct} from Structures," *Chem. Rev.*, *93*, 1281–1306

Leuenberger, C., J. Czuczwa, J. Tremp, and W. Giger, 1988, "Nitrated Phenols in Rain: Atmospheric Occurrence of Phytotoxic Pollutants," *Chemosphere*, *17*, 511–515

Levin, S. A., M. A. Harwell, J. R. Kelly, and K. D. Kimball, (Eds.), 1989, *Ecotoxicology: Problems and Approaches*, Springer, New York, USA

Levin, S. A., and K. D. Kimball, 1984, "New Perspectives in Ecotoxicology," *Environ. Managem. 8*, 375–442

Lewis, D. F. V., 1987, "Molecular Orbital Calculations on Solvents and Other Small Molecules: Correlations Between Electronic and Molecular Properties μ, α_{MOL}, π^*, and β," *J. Comput. Chem.*, *8*, 1084–1089

Lewis, D. F. V, 1988, "Molecular Orbital Calculations on Tumor-Inhibitory Nitrosoureas: QSARs," *Int. J. Quant. Chem.*, *33*, 305–321

Lewis, D. F. V., 1989a, "Molecular Orbital Calculations on Tumor-Inhibitory Aniline Mustards: QSARs," *Xenobiotica*, *19*, 243–251

Lewis, D. F. V., 1989b, "The Calculation of Molar Polarizabilities by the CNDO/2 Method: Correlation with the Hydrophobic Parameter, log P," *J. Comput. Chem.*, *10*, 145–151

Lim, C., D. Bashford, and M. Karplus, 1991, "Absolute pK_a Calculations with Continuum Dielectric Methods," *J. Phys. Chem.*, *95*, 5610–5620

Lipnick, R. L., 1989a, "Narcosis, Electrophile and Proelectrophile Toxicity Mechanisms: Application of SAR and QSAR," *Environ. Toxicol. Chem., 8*, 1–12

Lipnick, R. L., 1989b, "Base-Line Toxicity Predicted by Quantitative Structure–Activity Relationships as a Probe for Molecular Mechanisms of Toxicity" in *Probing Bioactive Mechanisms* (P. S. Magee, D. R. Henry, and J. H. Block, Eds.), ACS Symposium Series 413, pp. 366–389

Lipnick, R. L., 1991, "Outliers: Their Origin and Use in the Classification of Molecular Mechanisms of Toxicity," *Sci. Total Environ., 109/110*, 131–153

Lipnick, R. L., 1995, "Structure–Activity Relationships," in *Fundamentals of Aquatic Toxicology* (G. R. Rand, Ed.), 2nd edn., Taylor & Francis, London, UK, pp. 609–655

Lipnick, R. L., D. E. Johnson, J. H. Gilford, C. K. Bickings, and L. D. Newsome, 1985, "Comparison of Fish Toxicity Screening Data for 55 Alcohols with the Quantitative Structure–Activity Relationship Predictions of Minimum Toxicity for Nonreactive Nonelectrolyte Organic Compounds," *Environ. Toxicol. Chem., 4*, 281–296

Lipnick, R. L., K. R. Watson, and A. K. Strausz, 1987, "A QSAR Study of the Acute Toxicity of Some Industrial Organic Chemicals to Goldfish. Narcosis, Electrophile and Proelectrophile Mechanisms," *Xenobiotica, 17*, 1011–1025

Loewe, S., and H. Muischnek, 1926, "Über Kombinationswirkungen. I. Mitteilung: Hilfsmittel der Fragestellung," *Naunym-Schmiedebergs Arch. Exp. Pathol. Pharmakol., 114*, 313–326

Lyman, W. J., W. F. Reehl, and D. H. Rosenblatt, (Eds.), 1990, *Handbook of Chemical Property Estimation Methods*, American Chemical Society, Washington DC, USA

Maguire, R. J., 1987, "Environmental Aspects of Tributyltin." *Appl. Organometal. Chem., 1*, 475–498

Maltby, L., and P. Calow, 1989, "The Application of Bioassays in the Resolution of Environmental Problems; Past, Present and Future," *Hydrobiologia, 188/189*, 65–76

Matthews, R. A., W. G. Landis, and G. B. Matthews, 1996, "The Community Conditioning Hypothesis and its Application to Environmental Toxicology," *Environ. Toxicol. Chem., 15*, 597–603

Matthias, C. L., S. J. Bushong, L. W. Hall, J. M. Bellama, and F. E. Brinckman, 1988, "Simultaneous Butyltin Determinations in the Microlayer, Water Column and Sediment of a Northern Chesapeake Bay Marina and Receiving System," *Appl. Organomet. Chem., 2*, 547–552

McCarty, L. S., 1986, "The Relationship Between Aquatic Toxicity QSARs and Bioconcentration for Some Organic Chemicals," *Environ. Toxicol. Chem., 5*, 1071–1080

McElroy, M. B., and R. J. Salawitch, 1989, "Changing Composition of the Global Stratosphere," *Science, 243*, 763–770

McFarland, M., 1989, "Chlorofluorocarbons and Ozone," *Environ. Sci. Technol., 23*, 1203–1207

McFarland, M., and J. Kay, 1992, "Chlorofluorocarbons and Ozone," *Photochem. Photobiol., 55*, 911–929

McKim, J. M., P. C. Schmieder, R. W. Carlson, E. P. Hunt, and G. J. Niemi, 1987, "Use of Respiratory–Cardiovascular Responses of Rainbow Trout (*Salmo gairdneri*) in Identifying Fish Acute Toxicity Syndromes. Part I. Pentachlorophenol, 2,4-Dinitrophenol, Tricaine Methanesulfonate, and 1-Octanol," *Environ. Toxicol. Chem., 6*, 295–312

MedChem 94b Database, 1994, Daylight Chemical Information Systems, Irvine, CA

Mekenyan, O. G., J. M. Ivanov, G. D. Veith, and S. P. Bradbury, 1994, "Dynamic QSAR: A New Search for Active Conformations and Significant Stereoelectronic Indices," *Quant. Struct.–Act. Relat., 13,* 302–307

Mekenyan, O. G., G. D. Veith, S. P. Bradbury, and C. L. Russom, 1993, "Structure–Toxicity Relationships for α,β-Unsaturated Alcohols in Fish," *Quant. Struct.–Act. Relat.. 12,* 132–136

Mercier, A., E. Pelletier, and J. F. Hamel, 1994, "Metabolism and Subtle Toxic Effects of Butyltin Compounds in Starfish," *Aquat. Toxicol., 28,* 259–273

Metcalf, R. L., 1973, "A Century of DDT," *J. Agric. Food Chem., 21,* 511–551

Meylan, W. M., and P. H. Howard, 1995, "Atom/Fragment Contribution Method for Estimating Octanol–Water Partition Coefficients," *J. Pharm. Sci., 84,* 83–92

Miertus, S., E. Scrocco, and J. Tomasi, 1981, "Electrostatic Interaction of a Solute with a Continuum. A Direct Utilization of Ab Initio Molecular Potentials for the Provision of Solvent Effects," *Chem. Phys., 55,* 117–129

MOPAC93 1993, Fujitsu Limited, 9-3, Nakase 1-Chome, Mihama-ku, Chiba-city, Chiba 261, Japan, and Stewart Computational Chemistry, 15210 Paddington Circle, Colorado Springs, Colorado, USA

Moriarty, F., A. A. Bell, and H. Hanson, 1986, "Does p,p'–DDE Thin Eggshells?" *Environ. Pollut., A 40,* 257–286

Müller, K., 1980, "Reaktionswege auf mehrdimensionalen Energiehyperflächen," *Angew. Chem., 92,* 1–14

Muir, D. C. G., 1988, "Phosphate esters," in *The Handbook of Environmental Chemistry,* (O. Hutzinger, Ed.), Vol 3, Part C: Anthropogenic Compounds, Springer, New York, USA, pp. 41–66

Mulliken, R. S., 1934, "A New Electroaffinity Scale, Together with Data on Valence States and on Valence Ionization Potentials and Electron Affinities," *J. Chem. Phys., 2,* 782–793

Oehlmann, J., E. Stroben, C. Bettin, and P. Fioroni, 1993, "Hormonal Disorders and TBT-induced Imposex in Marine Snails," in *Quantified Phenotypic Responses in Morphology and Physiology* (J. C. Aldrich Ed.), Proc. 27th European Marine Biology Symp., Dublin, Ireland, 7–11 September 1992, JAPAGA, Ashford, pp. 301–305

Opperhuizen, A., and D. T. H. M. Sijm, 1990, "Bioaccumulation and Biotransformation of Polychlorinated Dibenzo-p-Dioxins and Dibenzofurans in Fish," *Environ. Toxicol. Chem., 9,* 175–186

Opperhuizen, A., P. I. Voors, 1987, "Uptake and Elimination of Polychlorinated Aromatic Ethers by Fish: Chloroanisols," *Chemosphere, 16,* 953–962

Pankow, J. F., 1991, *Aquatic Chemistry Concepts,* Lewis, Chelsea, Michigan, USA, 683 pp

Parr, R. G., and R. G. Pearson, 1983, "Absolute Hardness: Companion Parameter to Absolute Electronegativity," *J. Am. Chem. Soc., 105,* 7512–7516

Plackett, R. L., and P. S. Hewlett, 1948, "Statistical Aspects of the Independent Joint Action of Poisons, Particularly Insecticides. I. The Toxicity of a Mixture of Poisons," *Ann. Appl. Biol. 35,* 347–358

Peakall, D. B., and R. K. Tucker, 1985, "Extrapolation from Single Species Studies to Populations, Communities and Ecosystems," in *Methods for Estimating the Risk of Chemical Injury: Human and Non-Human Biota and Ecosystems* (V. B. Vouk, G. C. Butler, D. G. Hoel, and D. G. Peakall, Eds.), Wiley, New York, Chichester, pp. 613–629

Perrin, D. D., B. Dempsey, and E. P. Serjeant, 1981, *pK*$_a$ *Prediction for Organic Acids and Bases*. Chapman and Hall, London, UK, 146 pp

Piorr, R., 1987, "Structure and Application of Surfactants," in *Surfactants in Consumer Products*, J. Falbe, (ed.), Springer, Heidelberg, Germany, pp. 5–22

Purdy, R., 1988, "Quantitative Structure Relationships for Predicting Toxicity of Nitrobenzenes, Phenols, Anilines, and Alkylamines to Fathead Minnows," in *QSAR 88, Proc. of the 3rd Int. Workshop on Quantitative Structure–Activity Relationships in Environmental Toxicology* (J. E. Turner, M. W. England, T. W. Schultz, and N. J. Kwaak, Eds.), National Technical Information Service, U.S. Department of Commerce, Springfield, Virginia, USA, pp. 99–110

Purdy, R., 1991, "The Utility of Computed Superdelocalizability for Predicting the LC_{50} Values of Epoxides to Guppies," *Sci. Total Environ., 109/110*, 553–556

Quevauviller, P., and O. F. X. Donard, 1990, "Variability of Butyltin Determination in Water and Sediment Samples from European Coastal Environments," *Appl. Organomet. Chem. 4*, 363–367

Reader, S., M. Marion, and F. Denizeau, 1993, "Flow Cytometric Analysis of the Effects of Tri-n-Butyltin Chloride on Cytosolic Free Calcium and Thiol Levels in Isolated Rainbow Trout Hepatocytes," *Toxicology, 80*, 117–129

Richtler, H. J., and J. Knaut, 1988, *World Prospects for Surfactants*, Proc. 2nd World Surfactant Cong., Paris, France, 24–27 May, pp. 3–58

Rippen, G., E. Zietz, R. Frank, T. Knacker, and W. Klöpffer, 1987, "Do Airborne Nitrophenols Contribute to Forest Decline?" *Environ. Technol. Lett., 8*, 475–482

Roberts, D. W., 1989, "Acute Lethal Toxicity Quantitative Structure–Activity Relationships for Electrophiles and Pro–Electrophiles: Mechanistic and Toxicokinetic Principles," in (G. W. Suter II, M. A. Lewis, Eds.), *Aquatic Toxicology and Environmental Fate*, 11th Vol, ASTM STP 1007, Ann Arbor, MI, USA, pp. 490–506

Roberts, D. W., 1991, "QSAR Issues in Aquatic Toxicity of Surfactants," *Sci. Total Environ., 109/110*, 557–568

Saarikoski, J., and M. Viluksela, 1982, "Relation Between Physicochemical Properties of Phenols and their Toxicity and Accumulation in Fish," *Ecotoxicol. Environ. Saf., 6*, 501–512

Saito, H., M. Sudo, T. Shigoka, and F. Yamauchi, 1991, "In Vitro Cytotoxicity of Chlorophenols to Goldfish GF–Scale (GFS) Cells and Quantitative Structure–Activity Relationships," *Environ. Toxicol. Chem., 10*, 235–241

Saxena, A. K., 1987, "Organotin Compounds: Toxicology and Biomedicinal Applications," *Appl. Organomet. Chem., 1*, 39–56

Scherrer, R. A., and S. M. Howard, 1977, "Use of Distribution Coefficients in Quantitative Structure–Activity Relationships," *J. Med. Chem., 20*, 53–58

Schindler, D. W., 1987, "Detecting Ecosystem Responses to Anthropogenic Stress," *Can. J. Fish. Aquat. Sci., 44* (Suppl. 1), 6–25

Schindler, D. W., K. H. Mills, D. F. Malley, D. L. Findlay, J. A. Shearer, I. J. Davies, M. A. Turner, G. A. Linsay, and D. R. Cruikshrank, 1985, "Long-Term Ecosystem Stress: The Effects of Years of Experimental Acidification on a Small Lake," *Science, 228*, 1395–1401

Schüürmann, G., 1990a, "QSAR Analysis of the Acute Toxicity of Organic Phosphorothionates Using Theoretically Derived Molecular Descriptors," *Environ. Toxicol. Chem., 9*, 417–428

Schüürmann, G., 1990b, "Quantitative Structure–Property Relationships for the Polarizability, Solvatochromic Parameters and Lipophilicity," *Quant. Struct.–Act. Relat., 9*, 326–333

Schüürmann, G., 1990c, "QSAR Analysis of the Acute Toxicity of Oxyethylated Surfactants," *Chemosphere, 21*, 467–478

Schüürmann, G., 1991, "Acute Aquatic Toxicity of Alkyl Phenol Ethoxylates," *Ecotoxicol. Environ. Saf., 21*, 227–233

Schüürmann, G., 1992, "Ecotoxicology and Structure–Activity Studies of Organophosphorus Compounds," in *Rational Approaches to Structure, Activity and Ecotoxicology of Agrochemicals* (W. Draber, T. Fujita, Eds.), CRC Press, Boca Raton, Florida, USA, pp. 485–541

Schüürmann, G., 1995, "Quantum Chemical Approach to Estimate Physicochemical Compound Properties: Application to Substituted Benzenes," *Environ. Toxicol. Chem., 14*, 2067–2076

Schüürmann, G., 1996, "Modelling pK_a of Carboxylic Acids and Chlorinated Phenols," *Quant. Struct.–Act. Relat., 15*, 121–132

Schüürmann, G., R. Kühne, R.-U. Ebert, and F. Kleint, 1995, "Multivariate Error Analysis of Increment Methods for Calculating the Octanol/Water Partition Coefficient," *Fresenius Environ. Bull., 4*, 13–18

Schüürmann, G., and G. Röderer, 1988, "Acute Toxicity of Organotin Compounds and its Predictability by Quantitative Structure–Activity Relationships (QSARs)," in *Heavy Metals in the Hydrological Cycle* (M. Astruc, and J. N. Lester, Eds.), Selper Ltd., London, UK, pp. 433–440

Schüürmann, G., and M. Schindler, 1993, "Fish Toxicity and Dealkylation of Aromatic Phosphorothionates – QSAR Analysis using NMR Shifts Calculated by the IGLO Method," *J. Environ. Sci. Health, A28*, 899–921

Schüürmann, G., H. Segner, 1994, "Struktur-Wirkungs-Analyse von Trialkylzinnverbindungen," in *ECOINFORMA '94*, (K. Totsche, M. Matthies, F. Strutzenberger, W. Petek, W. Klöpffer, P. Czedik-Eysenberg, H. Meinholz, H. Fiedler, K. Alef, and O. Hutzinger, Eds.), Volume 7, Umweltbundesamt Wien, Austria, pp. 439–453

Schüürmann, G., H. Segner, and K. Jung, 1997, "Multivariate Mode-of-Action Analysis of Acute Toxicity of Phenols," *Aquat. Toxicol., 38*, 277–296

Schüürmann, G., R. K. Somashekar, and U. Kristen, 1996, "Structure–Activity Relationships for Chloro- and Nitrophenol Toxicity in the Pollen Tube Growth Test," *Environ. Toxicol. Chem., 15*, 1702–1708

Schultz, T. W., 1987, "The Use of the Ionization Constant (pK_a) in Selecting Models of Toxicity in Phenols," *Ecotoxicol. Environ. Saf., 14*, 178–183

Schultz, T. W., M. Cajina-Quezada, and S. K. Wesley, 1989, "Structure–Toxicity Relationships for Mono Alkyl- or Halogen-Substituted Anilines," *Bull. Environ. Contam. Toxicol., 43*, 564–569

Schultz, T. W., G. W. Holcombe, and G. L. Phipps, 1986, "Relationships of Quantitative Structure–Activity to Comparative Toxicity of Selected Phenols in the *Pimephales promelas* and *Tetrahymena pyriformis* Test Systems," *Ecotoxicol. Environ. Saf., 12*, 146–153

Schultz, T. W., N. L. Wyatt, and D. T. Lin, 1990, "Structure–Toxicity Relationships for Nonpolar Narcotics: A Comparison of Data from the *Tetrahymena, Photobacterium* and *Pimephales* Systems," *Bull. Environ. Contam. Toxicol., 44,* 67–72

Semlitsch, R. D., M. Foglia, A. Mueller, I. Steiner, E. Fioramonti, and K. Fent, 1995, "Short–Term Exposure to Triphenyltin Affects the Swimming and Feeding Behavior of Tadpoles," *Environ. Toxicol. Chem., 14,* 1419–1423

Sheehan, P. J., D. R. Miller, G. C. Butler, and P. Bourdeau, (Eds.), 1984, "Effects of Pollutants at the Ecosystem Level," *Scientific Committee on Problems of the Environment (SCOPE) 22,* Anchor Brendon, Tiptree, Essex, UK

Shiraishi, H., and M. Soma, 1992, "Triphenyltin Compounds in Mussels in Tokyo Bay after Restriction of Use in Japan," *Chemosphere, 24,* 1103–1109

Sijm, D. T. H. M., and A. Opperhuizen, 1988, "Biotransformation, Bioaccumulation and Lethality of 2,8- Dichlorodibenzo-*p*-Dioxin: A Proposal to Explain the Biotic Fate and Toxicity of PCDDs and PCDFs," *Chemosphere, 17,* 83–99

Smejtek, P., and S. Wang, 1993, "Distribution of Hydrophobic Ionizable Xenobiotics Between Water and Lipid Membranes: Pentachlorophenol and Pentachlorophenolate. A Comparison with Octanol/Water Partition," *Arch. Environ. Contam. Toxicol., 25,* 394–404

Smith, A. D., A. Bharath, C. Mallard, D. Orr, L. S. McCarty, and G. W. Ozburn, 1990, "Bioconcentration Kinetics of Some Chlorinated Benzenes and Chlorinated Phenols in American Flagfish, *Jordanella floridae* (Goode and Bean)," *Chemosphere, 20,* 379–386

Smith, S., V. J. Furay, P. J. Layiwola, and J. A. Menezes-Filho, 1994, "Evaluation of the Toxicity and Quantitative Structure–Activity Relationships (QSARs) of Chlorophenols to the Copepopid Stage of a Marine Copepod (*Tisbe battagliai*) and Two Species of Benthic Flatfish, the Flounder (*Platichthys flesus*) and Sole (*Solea solea*)," *Chemosphere, 28,* 825–836

Stache, H., and K. Kosswig, (Eds.), 1990, *Tensid Taschenbuch,* 3rd Edn., Hanser, München, Germany

Stäb, J. A., W. P. Cofino, and B. van Hattum, 1994, "Assessment of Transport Routes of Triphenyltin used in Potato Culture in The Netherlands," *Anal. Chim. Acta, 286,* 335–341

Stäb, J. A., M. Frenay, I. L. Freriks, U. A. Th. Brinkman, and W. P. Cofino, 1995, "Survey of Nine Organotin Compounds in the Netherlands Using the Zebra Mussel (*Dreissena polymorpha*) as Biomonitor," *Environ. Toxicol. Chem., 14,* 2023–2032

Stewart, J. J. P., 1989, "Optimization of Parameters for Semiempirical Methods II. Applications," *J. Comput. Chem., 10,* 221–264

Stewart, J. J. P., 1990, "MOPAC: A Semiempirical Molecular Orbital Program," *J. Comput.-Aided Mol. Des., 4,* 1–105

Stolarski, R. S., 1988, "The Antarctic Ozone Hole," *Sci. Am., 258,* 30

Suter II, G. W., (ed.), 1993, "*Ecological Risk Assessment,* Lewis, Chelsea, Michigan, USA

Tas, J. W., 1993, "Fate and Effects of Triorganotins in the Aqueous Environment. Bioconcentration Kinetics, Lethal Body Burdens, Sorption and Physicochemical Properties," *Ph.D. Thesis,* University of Utrecht, Utrecht, The Netherlands

Terada, H., 1990, "Uncouplers of Oxidative Phosphorylation," *Environ. Health Perspect., 87,* 213–218

Tester, M., D. V. Ellis, and J. A. J. Thompson, 1996, "Neogastropod Imposex for Monitoring Recovery from Marine TBT Contamination," *Environ. Toxicol. Chem.*, 15, 560–567

Tolls, J., and D. T. H. M. Sijm, 1995, "A Preliminary Evaluation of the Relationship Between Bioconcentration and Hydrophobicity for Surfactants," *Environ. Toxicol. Chem.*, 14, 1675–1685

Tomasi, J., and M. Persico, 1994, "Molecular Interactions in Solution: An Overview of Methods Based on Continous Distributions of the Solvent," *Chem. Rev.*, 94, 2027–2094

Trehy, M. L., W. E. Gledhill, J. P. Mieure, J. E. Adamove, A. M. Nielsen, H. O. Perkins, and W. S. Eckhoff, 1996, "Environmental Monitoring for Linear Alkylbenzene Sulfonates, Dialkyltetralin Sulfonates and their Biodegradation Intermediates," *Environ. Toxicol. Chem.*, 15, 233–240

Truhaut, R., 1977, "Ecotoxicology: Objectives, Principles and Perspectives," *Ecotoxicol. Environ. Saf.*, 1, 151–173

Tsuda, T., S. Aoki, M. Kojima, and H. Harada, 1990, "Differences Between Freshwater and Seawater-Acclimated Guppies in the Accumulation and Excretion of Tri-n-butyltin and Triphenyltin Chloride," *Water Res.*, 24, 1373–1376

Turner, J. E., M. W. England, T. W. Schultz, and N. J. Kwaak, (Eds.), 1988, "*QSAR 88, Proc. 3rd Int. Workshop on Quantitative Structure–Activity Relationships in Environmental Toxicology*," National Technical Information Service, U.S. Department of Commerce, Springfield, Virginia, USA, 228 pp

Uhler, A. D., G. S. Durell, W. G. Steinhauser, and A. M. Spellacy, 1993, "TBT Levels in Bivalve Molluscs from the East and West Coast of the U.S.: Results from the 1988–1990 National Status and Trends Mussel Watch Project," *Environ. Toxicol. Chem.*, 12, 139–153

Van den Heuvel, M. R., L. S. McCarty, R. P. Lanno, B. E. Hickie, and D. G. Dixon, 1991, "Effect of Total Body Lipid on the Toxicity and Toxicokinetics of Pentachlorophenol in Rainbow Trout (*Oncorhynchus mykiss*)," *Aquat. Toxicol.*, 20, 235–252

Veith, G. D., and S. J. Broderius, 1990, "Rules for Distinguishing Toxicants That Cause Type I and Type II Narcosis Syndromes," *Environ. Health Perspect.*, 87, 207–211

Veith, G. D., D. J. Call, and L. T. Brooke, 1983, "Structure–Toxicity Relationships for the Fathead Minnow, *Pimephales promelas*: Narcotic Industrial Chemicals," *Can. J. Fish. Aquat. Sci.*, 40, 743–748

Veith, G. D., R. L. Lipnick, and C. L. Russom, 1989, "The Toxicity of Acetylenic Alcohols to the Fathead Minnow, *Pimephales promelas*: Narcosis and Proelectrophile Activation," *Xenobiotica*, 19, 555–565

Veith, G. D., and O. G. Mekenyan, 1993, "A QSAR Approach for Estimating the Aquatic Toxicity of Soft Electrophiles," *Quant. Struct.–Act. Relat.*, 12, 349–356

Verhaar, H. J. M., L. Eriksson, M. Sjöström, G. Schüürmann, W. Seinen, and J. Hermens, 1994, "Modelling the Toxicity of Organophosphates: A Comparison of the Multiple Linear Regression and PLS Regression Methods," *Quant. Struct.–Act. Relat.*, 13, 133–143

Vighi, M., and D. Calamari, 1985, "QSARs for Organotin Compounds on *Daphnia magna*," *Chemosphere*, 14, 1925–1932

Virkki, L., and M. Nikinmaa, 1993, "Tributyltin Inhibition of Adrenergically Activated Sodium/Proton Exchange in Erythrocytes of Rainbow Trout (*Oncorhynchus mykiss*)," *Aquat. Toxicol.*, *25*, 139–146

Wakabayashi, M., M. Kikuchi, A. Sato, and T. Yoshida, 1987, "Bioconcentration of Alcohol Ethoxylates in Carp (*Cyprinus carpio*)," *Ecotoxicol. Environ. Saf.*, *13*, 148–163

Westall, J. C., C. Leuenberger, and R. P. Schwarzenbach, 1985, "Influence of pH and Ionic Strength on the Aqueous–Nonaqueous Distribution of Chlorinated Phenols," *Environ. Sci. Technol.*, *19*, 193–198

White, R., S. Jobling, S. A. Hoare, J. P. Sumpter, and M. G. Parker, 1994, "Environmentally Persistent Alkylphenolic Compounds are Estrogenic," *Endocrinology*, *135*, 175–182

Wiberg, K. B., 1968, "Application of the Pople–Santry–Segal Complete Neglect of Differential Overlap Method to the Cyclopropyl–Carbinyl and Cyclobutyl Cation and to Bicyclobutane," *Tetrahedron*, *24*, 1083–1096

Wold, S., 1995, "PLS for Multivariate Linear Modelling," in *Chemometric Methods in Molecular Design* (H. Van de Waterbeemd, Ed.), VCH Weinheim, Germany, pp. 195–218

Yatome, C, Y. Yamauchi, and T. Ogawa, 1993, "Effects of Organotin Compounds on the Synthesis of Nucleic Acids and ATP in the Growth of Bacillus Subtilis," *Bull. Env. Contam. Toxicol.*, *51*, 234–240

23

Endpoints and Thresholds in Ecotoxicology

John Cairns, Jr. (Blacksburg, USA)

23.1 SUMMARY

The widespread interest in sustainable use of the planet will almost certainly cause a major paradigm shift in ecotoxicology from predominately single species tests to tests at higher levels of biological organization, with particular emphasis on ecosystem health. Ecosystem services (those ecological functions, such as maintaining atmospheric gas balance, regarded as valuable services to human society) will become important ecotoxicological end points. This new array of end points based on ecosystems and landscapes will increase both spatial and temporal scales for determining and predicting the effects of toxicants. This chapter discusses the ways these issues may well reshape the field of ecotoxicology.

If in the past few years you haven't discarded a major opinion or acquired a new one, check your pulse. You may be dead.   — Gelett Burgess

Ecotoxicology, Edited by Gerrit Schüürmann and Bernd Markert.
ISBN 0-471-17644-3   © 1998 John Wiley & Sons, Inc. and Spektrum Akademischer Verlag.

23.2 THE ECOLOGICAL PARADIGM SHIFT

Kuhn (1970) described a paradigm as a theory so widely accepted as an accurate description of nature that failure of an experiment to yield the result deduced from the theory leads not to rejection of the theory but rather to attempts to fault the deductive logic or experimental procedure, or simply to willful suspension of belief in the experimental result. Major forces are influencing the field of ecotoxicology—forces that are almost certainly more forceful than those of the relatively recent past. Despite the fact that these are simultaneous forces, they are listed here in order of degree of public awareness. In the professional community, these forces are probably regarded as equally important and definitely not isolated from each other.

1. *Interest in developing a strategy for sustainable use of the planet.* Generally, the term *sustainable growth* is used possibly because human society globally has a fixation that failure to grow means stagnation. Growth generally means increased affluence as measured by gross national product or gross consumer product, accompanied by increased per capita use of energy and increased per capita use of resources. Continuing this growth may be beyond the earth's capacity to deliver on a long-term basis if future generations are to have at least the same quality of life as the present generation. Human society's present paradigm is that continued growth is essential.

2. *A shift from using thresholds based on survival, reproductive function, locomotor activity, and the like, at the single species level of biological organization to ecosystem health at a much higher level of biological organization.* This follows the trend in human medicine, which first focused on absence of symptoms of disease and then, in recent years, has focused on optimal condition or health.

3. *An increased realization that human society's life support system is both technological and ecological and that natural systems provide services (i.e., those ecosystem functions deemed useful by human society) to human society.* The loss of these services, such as the regulation of atmospheric gas balance, can be appreciated by nearly everyone.

4. *Increased geographic and temporal scales for determining effects of toxicants.* Although there are no currently accepted standard methods, or even provisional methods, for determining ecotoxicological effects at the landscape level, toxicants do have major effects at this level of biological organization and extrapolations of probable effects from single species are not effective.

23.3 ECOTOXICOLOGY AND SUSTAINABLE USE OF THE PLANET

Sustainable human development is characterized by economic growth that emphasizes the quality rather than the quantity of that growth (Malone, 1994).

In other words, development, in this context, is really a strategy for sustainable use. The implementation of the sustainable use concept will require a major paradigm shift in the field of ecotoxicology. Essentially, human society is removing materials from natural systems, substantially processing them, and then discarding the waste. Sustainable use requires attention to all steps in the life history of a product: (1) extraction of raw materials, (2) processing of raw materials into a product and the inevitable production of waste, (3) use of the product by the consumer, and (4) reincorporation of the material in the discarded item into natural systems in such a way in which the lost material has been ecologically replaced. Holl et al. (1994) call this "recycling by design." The premise of this approach is that recycling must be considered in the design phase of new products in order to facilitate the separation and ecological reutilization of materials at the end of a product's life. This process of adopting a long-term view in order to optimize resource management and recycling systems will require changes in attitudes and practices on the part of individuals, the educational system, industry, and government and coordination between these different components of society (Holl et al. 1994). Recycling has been taken to mean reuse of discarded materials by society. Ecological recycling means reincorporation of the materials into natural systems in a way that does not damage the integrity of these systems and optimally improves their health. Such a goal requires testing that, in the broadest sense, is ecotoxicological—i.e., the endpoints of interest are at the ecosystem level. In this case, instead of selecting endpoints that provide no evidence of deleterious effects, the endpoints will require reincorporation into the ecosystem in ways that contribute to its well being.

There is another aspect to the sustainable use concept in which ecotoxicology will arguably play an equally important role. As the human population expands (as is predicted for at least the first half of the next century) and as expectations of increased per capita affluence increase, some of the areas now devoted to the storage of hazardous wastes, or damaged by hazardous wastes by improper storage, should be returned to a condition of greater ecological integrity. This will allow them to be used both for generation of ecosystem services and for reincorporation of materials no longer of use for society as just described. This will require the association of two fields not now closely linked—restoration ecology and ecotoxicology (e.g., Cairns 1991, 1995a). Clearly, following these strategies will require different endpoints and thresholds than are now commonly used. The question of whether present environmental toxicological endpoints can be successfully extrapolated to endpoints and thresholds at higher levels of biological and ecological organization has been addressed theoretically by Cairns (1983a, 1988), and substantial evidence has been collected on this point by a number of other investigators, such as Mayer and Ellersieck (1986). Since it is a *sine qua non* that sustainable use of the planet will not be possible if the ecosystems are not in good condition, the endpoints and thresholds suitable for sustainable use will be discussed in the section on ecosystem health that follows.

23.4 ECOTOXICOLOGY AND ECOSYSTEM HEALTH

There is some question about whether thresholds actually exist in ecosystems or whether they are artifacts of the measuring process (e.g., Woodwell 1974). Others, such as May (1977), believe that there are a number of thresholds and breakpoints in ecosystems. Proponents of both views have grounds for their beliefs. Woodwell, who uses evidence from radiation studies, finds a continuum with no dramatic thresholds for radiation effects on living material. May, basing his analysis on other attributes, finds a series of workable thresholds, even though they may be somewhat arbitrary. This is analogous to determining human health—e.g., a range of acceptable cholesterol levels in humans has been established, but there is still some dispute as to the risks involved at some of the higher levels, particularly in older persons. A more recent discussion of threshold issues may be found in Cairns (1992). This publication notes that the most commonly used thresholds in environmental toxicology are the LC_{50} (or modifications thereof) where 50% of the organisms die or are otherwise affected at a certain concentration of a chemical for a particular time of exposure under specified environmental conditions. Generally, this particular threshold is derived from single species laboratory tests low in environmental realism. Cairns (1992) notes that, if the field of ecotoxicology wishes to examine the effects of chemicals on ecosystems, serious consideration must be given to thresholds other than those commonly used in the field of ecotoxicology because attributes at the community and ecosystem level of organization are not demonstrated at lower levels of biological organization (for example, energy flow nutrient spiraling). Even if thresholds and endpoints are artifacts of testing procedures, society still must make management decisions about the risk with available methods (as is done with human health).

As noted earlier, it seems unlikely that sustainable use of the planet is possible without giving substantial attention to ecosystem health. There are two aspects to the use of the term *ecosystem health*. The first is a common dictionary definition of "flourishing condition or well-being." Narrowly used, the word *health* is restricted to the condition of an organism. However, the word can also be used in a variety of other contexts including "economic health." Socrates defined the term saying what it was not—health is not disease, weakness, or deformity. Plato used the word as a metaphor for an appropriate condition of internal order. He extended the metaphor from the individual to society by saying that justice is the healthy state of the body politic. Basically, the idea of health has been used to convey a sense of well-being, optimal condition, and, most importantly, something valuable for its own sake as well as for its more practical benefits. Present society's increased attention to human physical fitness, including diet and exercise, is just one of the many manifestations of the increased regard for health. Cairns et al. (1993) developed a suite of indicators to monitor regional environmental health similar in concept to management use of leading economic indicators. Linkages between human activities and well-being and the state of the environment are considered, in that discussion, essential to the evaluation of general environmental health. Cairns (1995b) carries the linkage between human society and ecosystem condition further by indicating that it is probably easier to measure ecological integrity by looking at the practices of human society than by looking at ecosystems themselves. From a decision-making standpoint, desirable properties of indicators of environmental health will necessarily vary with their specific management use. For example, different indicators are required when gathering data to determine the adequacy of the environmental monitoring trends over time, to provide early warning of environmental degradation, or to diagnose the cause of an existing problem. Referring again to the parallels with the econometric models, tradeoffs between desirable environmental characteristics, costs to society, and quality of information are commonplace when selecting indicators of ecosystem health for management purposes. Decisions about what information to collect for each purpose can be made more rationally when available indicators of ecosystem health are characterized and matched to management goals. Table 23.1 (from Cairns et al. 1993) provides an illustrative list of desirable characteristics of indicators for different purposes; numbers indicate their relative value for that particular purpose. Table 23.2 (from Cairns et al. 1993) provides a potential list of indicators of the response of human health to environmental degradation. Table 23.3 (from Cairns et al. 1993) lists potential indicators of the responses of ecosystems to environmental degradation.

Table 23.1 Desirable characteristics of indicators for different purposes. Table entries are on a scale of importance from 1 to 3, where 1 indicates lower importance and 3 indicates an essential attribute. Characteristics that are universally desirable and do not differ between purposes are marked with an asterisk

Characteristic of indicator	Purpose of indicator				
	Assessment	Trends	Early warning	Diagnostic	Linkages
Biologically relevant	3	3	2	2	2
Socially relevant	3	3	2	2	2
Sensitive	*	*	*	*	*
Broadly applicable	2	2	2	1	1
Diagnostic	1	1	1	3	1
Measurable	*	*	*	*	*
Interpretable	3	3	2	1	1
Cost-effective	*	*	*	*	*
Integrative	2	2	1	1	2
Historical data	*	*	*	*	*
Anticipatory	1	1	3	1	2
Nondestructive	*	*	*	*	*
Continuity	2	3	1	1	1
Appropriate scale	*	*	*	*	*
Lack of redundance	*	*	*	*	*
Timeliness	2	2	3	3	2

Table 23.2 Potential indicators of the response of human health to environmental degradation

A. Study designs—Assessment approaches with different receptor organisms
  1. Epidemiological studies on exposed human populations (see Marsh and Caplan 1987)
     a. Environmental studies
     b. Case control studies
     c. Cohort studies
  2. Studies on sentinel species of exposed feral animals (see Gilbertson 1988; Colburn 1990)
     a. Mammals: minks, voles
     b. Birds: herring gulls, Foster's terns, eagles
     c. Fish: spottail shiners, brown bullhead
  3. Studies on surrogate species of exposed laboratory animals (see Lave et al. 1988)
     a. Mammals: mice, rats
     b. Nonmammalian systems: tissue culture, bacteria (Ames assays), planaria, hydra, water fleas, frogs, fathead minnows
B. Categories of indicators
  1. Neurotoxicity (see Marsh and Caplan 1987)
     a. in vivo
        – regional incidence rates for multiple sclerosis, Parkinsons, amyotrophic lateral sclerosis
        – behavioral assays: infant cognitive function, speech, gait, visual disturbance, headaches, memory function
        – biomarkers: biopsy and histopathology, visual-evoked response, electroencephalo-gram positron emission tomography, CAT scan, electromyography
     b. in vitro
        – cell culture excitability, synaptic potential, repetitive firing properties, nerve conduction velocity
  2. Reproductive toxicity (see Marsh and Caplan 1987)
     in vivo
        – regional incidence rates for birth defects, infertility, miscarriage, stillbirth, birth weight
        – biomarkers: sister chromatid exchanges, sperm counts, motility, and morphological abnormality
  3. Carcinogenicity/Mutagenicity/Genotoxicity (see Sandu and Lower 1987; Wang et al. 1987; Colburn 1990; Marsh and Caplan 1987)
     a. in vivo
        – regional incidence rates
        – biomarkers: DNA adducts, sister chromatid exchange, DNA unwinding, histo-pathology
     b. in vitro
        – histopathology of tissue cultures
        – Ames mutagenicity tests
  4. Cardiovascular disease
     in vivo
        – regional incidence rates
  5. Immunocompetency
     in vivo
        – blood cell counts

Table 23.3 Potential indicators of ecosystem response to environmental degradation

A. Commercial fisheries (see Bird and Rapport 1986)
   1. Quantity
      – stock, harvesting, recruitment estimates
   2. Quality
      – presence of preferred species
      – restrictions on consumption
      – incidence of training, deformities
   3. Valuation
      – shadow pricing; farm reared vs feral fish
      – employment and payroll
   4. Management costs
      – stocking
      – lamprey control
B. Drinking water (see Wentworth et al. 1986)
   1. Quantity
      – stock, withdrawal, replenishment estimates
   2. Quality
      – treatment costs
      – chemical and bacterial standard violations
      – restrictions on consumption
      – reported acute illness
      – user satisfaction*
   3. Valuation
      – contingent valuation: willingness to pay and compensation for damage*
   4. Management costs
      – treatment costs
C. Recreation (see Hunsaker and Carpenter 1990; Lichtkopper and Hushak 1989)
   1. Quantity
      – visit counts: sport fishing, swimming, boating, birdwatching, bird hunting
      – boat registration
      – marina and beach counts
      – marina vacancy rates
   2. Quality
      – incidence of fish consumption restrictions
      – incidence of contact sport restrictions
      – incidence of fish deformities or tainting
      – catch per unit effort
   3. Valuation
      – employment and payroll
      – marina sales
      – admission fees
      – shadow valuation; pool construction vs beach use
   4. Management costs
      – stocking

(Continued)

Table 23.3 (*Continued*)

D. Industrial, energy, and agricultural water use
  1. Quantity
     – stock, withdrawal, replenishment rates
  2. Quality
     – productivity, crop, livestock losses attributable to water quality problems
     – costs of pre–use treatment; descaling, defouling
  3. Valuation
     – compensation for loss of use
     – increased product cost due to degradation
  4. Management costs
     – costs of post-use treatment
E. Aesthetics
  1. Quantity
     – subjective satisfaction*
     – miles of shoreline in parks
  2. Quality
     – incidence of objectionable odor*
     – incidence of turbidity
     – incidence of algal blooms
  3. Valuation
     – shadow valuation; water-view vs interior real estate
     – contingent valuation; willingness to pay and compensation for loss*
  4. Management costs
     – landscape planning
F. Transportation water use
  1. Quantity
     – water levels
  2. Quality
  3. Valuation
     – employment and payroll
  4. Management costs
     – income lost due to restrictions on dredging
     – costs of disposal for contaminated dredge spoils
     – costs of pollution controls
     – costs of control of nuisance growths: macrophytes, zebra mussels
G. Support of general economic well-being of region—traditional economic indicators (GNP, unemployment, income class distribution, etc.)
H. Future uses (genetic pool for pharmaceuticals, genetic engineering, temperature buffer in global warming)

---

* Subjective evaluations dependent on survey of shareholders

23.5 CONCEPTS OF ECOSYSTEM HEALTH: UTILITARIAN VS PHILOSOPHICAL

Suter (1993) feels that ecosystem health is a metaphor, not an observable property. Calow (1993) suggests that "healthy" ecosystems might be defined as those that persist through time. This is a useful operational definition for this discussion because it is compatible with the concept of sustainable use. Earlier, Calow (1992) notes that this definition related to stable states, but not ones achieved in the *active* cybernetic sense. Calow (1993) further notes that ecosystem "health criteria" might then be specified in terms of the properties and processes that are associated with these states. Again, this is congruent with the thresholds and attributes for ecotoxicology espoused in this discussion.

Neither ecosystem health nor sustainable use (often termed *sustainable growth*) are concepts for which there are consensus definitions. It seems counterintuitive to develop policies, goals, methodologies, and concepts before achieving some degree of unanimity. However, events are moving so rapidly that to wait until this happens will almost certainly dramatically reduce societal options. As a consequence, we must learn by doing, hoping that a consensus will develop in the process. However, the concept of ecosystem health is a useful communication bridge between scientists and non-scientists, if the metaphor is not pushed too far. In general, human society, at least in the United States and many other countries as well, regards health as good in itself, even though it also has economic value—healthy people are less likely to miss work due to illness, are likely to perform better when on the job, and, perhaps most importantly these days, will have lower health care costs. Some notable parallels exist between the practice of medicine and the maintenance of healthy ecosystems. In fact, the term *ecosystem medicine* has been used regularly over a substantial period of time (Rapport et al. 1979; Rapport 1989).

Because of the short time that ecosystem health has received even modest attention, diagnostic tests of a comparable reliability to those widely used for determining human health are lacking. Nevertheless, a number of quite reliable potential indicators of ecosystem health have been discussed already. The issue being addressed here is the criteria for selecting indicators of ecosystem health. Indicators of ecosystem health should be (Cairns et al. 1993):

(1) biologically relevant, i.e., important in maintaining a balanced community,

(2) socially relevant, i.e., of obvious value to and observable by human society or predictive of a measure that is,

(3) sensitive to stressors without an all-or-none response or extreme natural variability,

(4) broadly applicable to many stressors and sites,

(5) diagnostic of the particular stressor causing the problem,

(6) measurable, i.e., capable of being operationally defined and measured using a standard procedure with documented performance and low measurement error,

(7) interpretable, i.e., capable of distinguishing acceptable from unacceptable conditions in a scientifically and legally defensible way,

(8) cost effective, i.e., inexpensive to measure, providing the maximum amount of information per unit effort,

(9) integrative, i.e., summarizing information from many unmeasured indicators,

(10) historical data are available to define nominative variability, trends, and possibly acceptable and unacceptable conditions,

(11) anticipatory, i.e., capable of providing an indication of degradation before serious harm has occurred—early warning,

(12) nondestructive of the ecosystem,

(13) potential for continuity is measurement over time,

(14) of an appropriate scale to the management problem being addressed,

(15) not redundant with other measured indicators, i.e., providing unique information,

(16) timely, i.e., providing information quickly enough to initiate effective management action before unacceptable damage has occurred.

23.6 ECOTOXICOLOGY AT THE LANDSCAPE LEVEL

Cairns (1989) defines an *ecotoxicologist* as an individual who uses ecological parameters to access toxicity. Ecotoxicology would, by this definition, merge the fields of ecology and toxicology. The endpoints, or parameters used to determine a response, would have some ecological significance at some high levels of biological organization. Calow (e.g., 1976, 1992) argues that the concept of health as applied to organisms involves different principles from the concept of health as applied to ecosystems. Ecotoxicology must move beyond, but not discard, single species toxicity tests. If ecotoxicology is to be a field measuring the impacts of anthropogenic activities on natural processes, inevitably some studies at the landscape level will be appropriate. The strength of the field of environmental toxicology has been its ability to estimate or predict effects of chemicals and other ecological stressors in natural systems before they are actually permitted to reach these systems. In other words, ecotoxicology should prevent damage to natural systems, not merely explain–after the fact–why the damage occurred.

Although the term *landscape ecology* was first used in Europe in 1939, only in recent years has a significant trend developed among ecologists to study natural processes, not just at the individual or ecosystem level but all over the entire landscape (e.g., Forman and Godron 1986). The field of landscape ecology focuses on (1) landscape structure—the spatial arrangement of ecosystems within landscapes, (2) landscape function—the interaction among the component ecosystems through flow of energy, materials, and organisms, and (3) alterations of this structure and function by natural or anthropogenic stress or natural, successional processes, etc. (e.g., Forman and Godron 1986; Risser 1987). Allen and Starr (1982) and O'Neill et al. (1986) emphasize that the development of the field of landscape ecology has brought increasing recognition that ecosystem processes occur within a hierarchy of different spatial and temporal scales. O'Neill et al. (1986) also emphasize that ecological systems are constrained both by the range of potential behavior of lower scales and the environmental limits of higher scales. Forman and Godron (1986) have hypothesized that measurements of ecological structure and function must be taken out of the scale appropriate to the process being observed. The significance of these publications for this discussion is illustrated by the important question: If the basic assumption is that the field of ecotoxicology should represent a synthesis involving both ecology and toxicology, what effects will this newly emerging field of landscape ecology have on the field of ecotoxicology? Landscape ecotoxicology is in such an embryonic state of development that Cairns (1993) asked "Will there ever be a field of landscape toxicology?" There is already persuasive evidence of the need for landscape ecotoxicology. Many chemical stressors are either emitted or spread, soon after release, over immense areas. Kern (1979) hypothesized that the haze in the Arctic results from industrial smog spreading from Europe. Hirao and Pattern (1974) have found that damage from car exhaust fumes originating in the coastal Californian cities is detectable in the Sierra Nevada mountains in the eastern part of the state. Spencer and Cliath (1990) provided evidence that over 90% of a number of toxic pesticides volatilized into the atmosphere within a week, and were presumably then able to affect areas quite distant from their area of application. Kurtz and Atlas (1990) noted that volatilization and atmospheric transport of hexachlorocyclohexanes resulted in the contamination of surface water throughout the Pacific Ocean. Hunsaker et al. (1990) and Suter (1990) have noted that what appear to be small-scale environmental impacts can become landscape stresses when similar impacts are repeated over a large area.

Unquestionably, the field of ecology is expanding its study framework scale both spatially and temporally. The question is: How will toxicologists respond to this expanding synthesis?

23.7 OBSTACLES TO THE DEVELOPMENT OF LANDSCAPE ECOTOXICOLOGY

Some earlier publications (Kelly and Harwell 1989; Harwell and Harwell 1989) have addressed obstacles to the development of landscape ecotoxicology very satisfactorily and in much more detail than this discussion will permit. These are, of course, scientific problems, which are arguably more amenable to resolution than some of the human problems that are addressed later.

Possibly the most important obstacle is the small amount of in-depth data covering large spatial and temporal scales for a large number of species or functional groups of species. If the understanding of undisturbed systems is inadequate and the normal variability is not determined with robust information, it is difficult to distinguish between natural oscillations and deleterious effects caused by toxicants. Most toxicological studies have been at lower levels of biological organization, which does not provide adequate information on the responses of higher levels of organization, such as communities, ecosystems, and landscapes (e.g., Cairns 1983a; Levin et al. 1984). Along the same lines is the redundancy of data (e.g., Kaesler et al. 1974), which can be viewed in two ways: (1) redundant information can be used as multiple confirmations or (2) redundant information is a waste of money because no new information has been generated. Most commonly, given the complexity of the problem, redundancy is viewed as undesirable, except in the most contentious situations.

Another problem is that, although the effects of pollutants can be detected at different levels of organization (e.g., Levin et al. 1984; Hunsaker et al. 1990), it is extremely difficult at the present state of knowledge to identify the levels at which the toxic stress is occurring or, if occurring at more than one level, how the different levels are interacting. This is exceedingly important because Patten et al. (1976) have shown that effects at one level will frequently be propagated to either lower or higher levels, or both.

The third, and perhaps most intractable problem from a scientific standpoint, is that it is arguably impossible to find an ecosystem anywhere on the planet that has not been affected to a significant degree by anthropogenic activities. Generally, the impacts are cumulative, so it is difficult to determine which specific toxicant or combination of toxicants is eliciting a particular ecosystem response. Additionally, even if a determination were possible, it would be exceedingly difficult to determine the source of the toxicant. Superimposed on these problems are physical disruptions of ecosystems, such as clearing forests for agriculture, runoff from impervious surfaces such as parking lots and shopping malls, and the like. Urban runoff not only changes the shape of the runoff curve from the natural curve, thereby increasing the amplitude and decreasing the duration of the pulse, but it is also a source of deleterious chemicals. Osborne and Wiley (1988) carried out a study in central Illinois that provided persuasive evidence that increased phosphorus levels in aquatic systems resulted not from nonpoint source argicultural runoff as commonly assumed, but rather from urban runoff.

23.8 BALANCING OUR TECHNOLOGICAL AND ECOLOGICAL LIFE SUPPORT SYSTEM

For most of human society's existence on the planet, the life support system has been entirely ecological. Practically everything needed for the human species, especially food and fiber, were furnished by ecosystems. Natural processes kept the atmosphere breathable, except for episodic events such as volcanic eruptions, and even furnished some degree of shelter, such as caves or the materials for shelter. Most of the material removed for human use was readily reincorporated into natural systems, which is why so few artifacts are found today.

However, the planet experienced the agricultural revolution, followed by the industrial revolution. These two events made the support of much larger populations of humans possible, increased their per capita affluence, and dramatically altered their distribution. The process of urbanization (still underway) has been made possible by the enormous increase in per capita agricultural productivity. Thus one farmer can feed hundreds of people, a transportation system can move food from agricultural to densely populated areas, and refrigeration, canning, and other means of preserving and storing food can move perishable food longer distances and store it for future use. A concomitant feature was the enormous per capita expenditure of energy for transportation, heating, refrigeration, and light. In addition, another form of transportation, the elevator, made increased population density possible.

The present human population size, its enormous density in some locations, and the present level of comparative affluence would not be possible without technological systems, including agricultural ones. Consequently, there is a technological component to human society's life support system, although this development is, in geological time, a relatively recent development. As earthquakes, paralyzing ice storms, hurricanes, and the like have illustrated, even modest and relatively quickly repaired damage to a fraction of the technological life support systems can cause enormous hardship on that portion of human society dependent on the disrupted systems.

During all these shifts in population size, density, and level of affluence, much attention has been given to the well-being of the technological life support system. Some systems are becoming depleted, such as the water supply delivery systems of many of the world's major cities (e.g., Okun 1991), but most are maintained and repaired.

However, the rapid development of technological systems has caused major changes in the world's ecosystems and the supply of ecosystem services. Biodiversity appears to be markedly down and may be even more seriously threatened in the future (e.g., Wilson 1988). True wild lands are few and far apart and are decreasing in size and quality from year to year. The April 1994 issue of *Scientific American* has a discussion of the global decline of amphibians, probably due to a variety of causes, such as habitat destruction, increased ultraviolet radiation, and pollution.

Despite all of these notable changes, for the average citizen it is still the technological component of human society's life support system that appears to need tender loving care—not the ecological component of the life support system. The primary reason for this is that there has been no readily apparent change in the delivery of ecosystem services. The atmospheric gas balance, although changed, has not affected humans directly, and, while fossil water is not being recharged at the rate it should be, other sources of water are often available. Ecosystem services (i.e., ecosystem functions perceived as beneficial to human society) appear to be in such oversupply that the amount of environmental degradation and habitat destruction thus far has not placed the delivery of services below that necessary for the continuance of human society. In addition, many of the services of natural systems may be partly provided by agricultural systems, managed forests, and even grass on golf courses.

Another reason for a casual view toward ecosystem services by the average citizen is the possibility that the lag time is so great that it will be years, or perhaps even decades, before the failure of any major service component is evident. However, at some point, the activities of human society and its technological developments may threaten the ecological component of the life support system. It seems highly improbable that the entire planet can be converted to agricultural and industrial systems together with human settlements. Of course, over half the planet is covered by oceans, which undoubtedly contribute a substantial portion of the overall ecosystem services. It is not, however, impossible to impair the delivery of these ecosystem services even though no substantial permanent human settlements exist on them.

Some balance must be reached between the technological and ecological components of the human life support system (e.g., Cairns 1996). Even if managed ecosystems provide all the ecosystem services needed, some attention will have to be given to the delivery of services from them. *Ultimately, ecological thresholds, below which the delivery of ecosystem services is unsatisfactory, will have to be identified.* In order to do this, services must be inventoried, quantified, and the effects of human society upon their delivery determined. *In short, except for the unusual level of biological organization and temporal and spatial scales, this becomes a standard toxicological problem.* It is, as are most toxicological problems in complex multivariate systems, difficult to measure and predict because of cumulative effects, synergistic interactions, lag time between exposure and response, and a variety of other variables. It is unlikely that standard methods, comparable to those developed for single species toxicity tests, will ever exist. Since ecosystems vary and ecological landscapes are a mosaic of habitat types, much more professional judgement will be required. Since these are dynamic systems, uncertainties may be much higher than in single species laboratory toxicity tests with low environmental realism.

Traditional research methodology for both environmental toxicology and ecology is difficult to use at the landscape scale. Simulations of ecosystems (e.g., microcosms and mesocosms) have limited utility in extrapolating to ecosystems or, even more importantly, landscapes (e.g., Cairns 1983b; Levin et al. 1984). Of course, field enclosures provide another alternative for studying at least certain cause–effect pathways, but these still suffer from the same limitations as microcosms and mesocosms and have the additional problem of difficulty in replication. The now classic studies of Bormann and Likens (1979) on the Hubbard Brook ecosystem, those of Schindler (1990), and of Watras and Frost (1989) provide examples of large-scale manipulations in ecosystems themselves. Much useful information can be derived in this way, but the effort is enormous compared to a single species laboratory toxicity test. In addition, determining the correspondence between disturbed and undisturbed landscape level systems requires approximations, because undisturbed reference sites either do not exist or are exceedingly rare (e.g., Hunsaker et al. 1990). Computer simulation models and geographic information systems (GIS) have been used to predict the effects of toxicants on ecosystems. Relatively recently, GIS have also been used to integrate small-scale models over large areas. Costanza et al. (1988) modeled nutrient flows between a succession of 2479 cells, each representing 1 km² of Louisiana marshland, in order to assess the long-term effects of certain management practices. Limitations of models have been fairly well discussed, but, in the case of landscape ecotoxicology, there seems to be no better alternative. Alternatively, the landscape itself might be used as an indicator, but this is a reactive rather than a predictive system. Useful indices have been provided by O'Neill et al. (1988). Although these were not specifically designed for ecotoxicology, they should be useful models to explore as landscape-level ecotoxicological indices are developed.

23.9 THRESHOLDS, PROCESS ENDPOINTS, AND ECOSYSTEM SERVICE DELIVERY

The most important threshold for landscape-level ecotoxicology is likely to be that level of exposure below which desirable ecosystem services are delivered and above which (in terms of toxicant concentration) the services are either not delivered or seriously impaired. This would fit the standard environmental toxicological models that are presently in use, except the thresholds will be based on ecosystem-level responses and, moreover, those ecosystem functional attributes of interest to human society. If there are some ecosystems or landscapes that provide desirable services from a small fraction of the habitat mosaic, this might focus excessive attention on that particular part of the mosaic, even though the entire landscape is essential to the well-being of its component parts. While this may seem intuitively reasonable to scientists, it may not be particularly apparent to policymakers and decision makers, or, as is often the case, there may be sufficient disagreement among scientists to cause decision makers to ignore conflicting advice entirely.

The development of landscape-level ecotoxicological thresholds is essential to the implementation of sustainable use strategies. Most of these thresholds should involve delivery of ecosystem services.

Acknowledgements. I am indebted to Brenda Chandler for transcribing the dictation of this manuscript and to Darla Donald, my editorial assistant, for putting this manuscript in the form required for publication. I am indebted to my colleague B. R. Niederlehner for comments on a draft of this manuscript. Many of the ideas in this manuscript had their genesis in an earlier paper (Cairns et al. 1993) and book chapter (Holl and Cairns 1995).

23.10 REFERENCES

Allen, T. F. H., and T. B. Starr, 1982, *Hierarchy: Perspectives in Ecological Complexity*, University of Chicago, Chicago, IL, USA

Bird, P., and D. Rapport, 1986, State of the Environment Report for Canada, Minister of the Environment, Ottawa, Canada

Bormann, F. H., and G. E. Likens, 1979, *Pattern and Process in a Forested Ecosystem*, Springer, New York, USA

Cairns, J., Jr., 1983a, "The Case for Testing at Different Levels of Biological Organization," in *Aquatic Toxicology and Hazard Assessment: Sixth Symposium* (W. E. Bishop, R. D. Cardwell, and B. B. Heidolph, Eds.), American Society for Testing and Materials, Philadelphia, USA, STP802, pp. 111–127

Cairns, J., Jr., 1983b, "Are Single Species Toxicity Tests Alone Adequate for Estimating Hazard?," *Hydrobiologia*, *100*, 47–57

Cairns, J., Jr., 1988, "Putting the Eco in Ecotoxicology," *Regulatory Toxicol. Pharmacol.*, *8*, 226–238

Cairns, J., Jr., 1989, "Will the Real Ecotoxicologist Please Stand Up?" *Environ. Toxicol. Chem.*, *8*, 843–844

Cairns, J., Jr., 1991, "Restoration Ecology: A Major Opportunity for Ecotoxicologists," *Environ. Toxicol. Chem.*, *10*, 429–432

Cairns, J., Jr., 1992, "The Threshold Problem in Ecotoxicology," *Ecotoxicology*, *1*, 3–16

Cairns, J., Jr., 1993, "Will There Ever be a Field of Landscape Toxicology?" *Environ. Toxicol. Chem.*, *12*, 609–610

Cairns, J., Jr., 1995a, "Restoration Ecology and Ecotoxicology," in *Handbook of Ecotoxicology* (D. J. H. Hoffman, B. A. Rattner, A. G. Burton, Jr., and J. Cairns, Jr., Eds.), Lewis Publishers, Boca Raton, FL, USA, pp. 717–731

Cairns, J., Jr., 1995b, "Ecological Integrity of Aquatic Systems," *Regulated Rivers: Research and Management*, *11*, 313–323

Cairns, J., Jr., 1996, "Determining the Balance between Technological and Ecosystem Services," in *Engineering Within Ecological Constraints* (P. C. Schulze, Ed.), National Academy Press, Washington, DC, USA, pp. 13–30

Cairns, J., Jr., P. V. McCormick, and B. R. Niederlehner, 1993, "A Proposed Framework for Developing Indicators of Ecosystem Health," *Hydrobiologia*, *263* (1), 1–44

Calow, P., 1976, *Biological Machines: A Cybernetic Approach to Life*, Edward Arnold Publishers, London, UK

Calow, P., 1992, "Can Ecosystems be Healthy?: Critical Considerations of Concepts," *J. Aquat. Ecosystem Health*, *1*, 15–24

Calow, P., 1993, "Comment: Ecosystems Not Optimized," *J. Aquat. Ecosystem Health*, *2*, 55

Colburn, T., 1990, *Innovative Approaches for Evaluating Human Health in the Great Lakes Basin Using Wildlife Toxicology and Ecology*, International Joint Commission, Ontario, Canada

Costanza, R., F. H. Sklar, M. L. White, and J. W. Day, Jr., 1988, "A Dynamic Spatial Simulation Model of Land Loss and Marsh Succession in Coastal Louisiana," in *Wetland Modeling* (W. J. Mitsch, M. Straskraba, and S. E. Jørgensen, Eds.), Elsevier, Amsterdam, The Netherlands, 99 pp

Forman, R. T. T., and M. Godron, 1986, *Landscape Ecology*, Wiley, New York, USA

Gilbertson, M., 1988, "Epidemics in Birds and Mammals Caused by Chemicals in the Great Lakes," in *Toxic Contaminants and Ecosystem Health* (M. S. Evans, Ed.), Wiley, New York, USA, pp. 133–152

Harwell, M. A., and C. C. Harwell, 1989, "Environmental Decision Making in the Presence of Uncertainty," in *Ecotoxicology: Problems and Approaches* (S. A. Levin, M. A. Harwell, J. R. Kelly, and K. D. Kimball, Eds.), Springer, New York, USA

Hirao, Y., and C. C. Patterson, 1974, "Lead Aerosol Pollution in the High Sierra Overrides Natural Mechanisms which Exclude Lead from Food Chains," *Science*, *184*, 989

Holl, K. D., and J. Cairns, Jr., 1995, "Landscape Indicators in Ecotoxicology," in *Handbook of Ecotoxicology* (D. J. Hoffman, B. A. Rattner, A. G. Burton, Jr., and J. Cairns, Jr., Eds.), Lewis Publishers, Boca Raton, FL, USA, pp. 667–680

Holl, K. D., J. Cairns, Jr., and T. Rattray, 1994, "Recycling by Design," *Spec. Sci. Technol.*, *17* (2), 129–134

Hunsaker, C. T., and D. E. Carpenter, 1990, (Eds.), "Environmental Monitoring and Assessment Program: Ecological Indicators," Office of Research and Development, U.S. Environmental Protection Agency, Research Triangle Park, NC, USA

Hunsaker, C. T., R. L. Graham, G. W. II, Suter, R. V. O'Neill, L. W. Barnthouse, and R. H. Gardner, 1990, "Assessing Ecological Risk on a Regional Scale," *Environ. Managem.*, *14*, 325–332

Kaesler, R. L., J. Cairns, Jr., and J. S. Crossman, 1974, "Redundancy in Data from Stream Systems," *Water Res.*, *8*, 637

Kelly, J. R., and M. A. Harwell, 1989, "Indicators of Ecosystem Response and Recovery," in *Ecotoxicology: Problems and Approaches* (S. A. Levin, M. A. Harwell, J. R. Kelly, and K. D. Kimball, Eds.), Springer, New York, USA

Kern, R. A., 1979, "Arctic Haze Actually Industrial Smog?" *Science*, *205*, 290

Kuhn, T. S., 1970, *The Structure of Scientific Revolutions*, 2nd edn. University of Chicago Press, Chicago, IL, USA

Kurtz, D. A., and E. L. Atlas, 1990, "Distribution of Hexachlorocyclohexanes in the Pacific Ocean, Air and Water," in *Long-Range Transport of Pesticides* (D. A. Kurtz, Ed.), Lewis Publishers, Chelsea, MI, USA, pp. 143–160

Lave, L., F. Enhever, A. Rosenkranz, and G. Omenn, 1988, "Information Value of the Rodent Bioassay," *Nature*, *336*, 631–633

Levin, S. A., K. D. Kimball, W. H. McDowell, and S. F. Kimball, 1984, "New Perspectives in Ecotoxicology," *Environ. Managem.*, *8*, 375

Lichtkopper, F., and L. Hushak, 1989, "Characteristics of Ohio's Lake Erie Recreational Marina," *J. Great Lakes Res.*, *15*, 418–426

Malone, T. F., 1994, "Sustainable Human Development: A Paradigm for the 21st Century," A White Paper for the National Association of State Universities and Land-Grant Colleges, Research Triangle Park, North Carolina, USA

Marsh, G., and R. Caplan, 1987, "Evaluating Health Effects of Exposure at Hazardous Waste Sites," in *Health Effects from Hazardous Waste Sites* (J. Andelman, and D. Underhill, Eds.), Lewis Publishers, Chelsea, MI, USA, pp. 3–80

May, R. M., 1977, "Thresholds and Breakpoints in Ecosystems with a Multiplicity of Stable States," *Nature*, *269*, 471–477

Mayer, F. L., and M. R. Ellersieck, 1986, "Manual of Acute Toxicity: Interpretation and Data Base for 410 Chemicals and 66 Species of Freshwater Animals," Resource Publication 160, U.S. Department of Interior, Washington, DC, USA

Okun, D. A., 1991, "Meeting the Need for Water and Sanitation for Urban Populations," Abel Wolman Distinguished Lecture, National Research Council, Washington, DC, USA

O'Neill, R. V., D. K. De Angelis, J. B. Waide, and T. F. H. Allen, 1986, *A Hierarchical Concept of Ecosystems*, Princeton University, NJ, USA

O'Neill, R. V., J. R. Krummel, R. H. Gardner, G. Sugihara, B. Jackson, D. L. DeAngelis, B. T. Milne, M. G. Turner, B. Zygmunt, S. Christensen, V. H. Dale, and R. L. Graham, 1988, "Indices of Landscape Pattern," *Landscape Ecol.*, *1*, 153

Osborne, L. L., and M. J. Wiley, 1988, "Empirical Relationships Between Land Use/Cover and Stream Water Quality in an Agricultural Watershed," *J. Environ. Managem.*, *26*, 9

Patten, B. C., R. W. Bosserman, J. T. Finn, and W. G. Gale, 1976, "Propagation of Cause in Ecosystems," in *Systems Analysis and Ecosystem Ecology*, Vol. IV (B. C. Patten, Ed.), Academic Press, New York, USA

Rapport, D. J., 1989, "What Constitutes Ecosystem Health?" *Perspect. Biol. Medicine*, *33*, 120–132

Rapport, D. J., C. Thorpe, and H. A. Regier, 1979, "Ecosystem Medicine," *Bull. Ecol. Soc. Amer.*, *60*, 180–182

Risser, P. G., 1987, "Landscape Ecology: State of the Art," in *Landscape Heterogeneity and Disturbance* (M. G. Turner, Ed.), Springer, New York, USA, pp. 1–14

Sandhu, S. S., and W. R. Lower, 1987, "In Situ Monitoring of Environmental Genotoxins," in *Short-Term Bioassay in the Analysis of Complex Environmental Mixtures V* (S. S. Sandhu, D. M. DeMarini, M. J. Mass, M. M. Moore, and J. L. Mumford, Eds.), Plenum Press, New York, USA, pp. 145–160

Schindler, D. W., 1990, "Experimental Perturbations of Whole Lakes as Tests of Hypotheses Concerning Ecosystem Structures and Function," *Oikos*, *57*, 25

Spencer, W. F., and M. M. Cliath, 1990, "Movement of Pesticides from the Soil to the Atmosphere," in *Long-Range Transport of Pesticides* (D. A. Kurtz, Ed.), Lewis Publishers, Chelsea, MI, USA, pp. 1–16

Suter, G. W., II, 1990, "Endpoints for Regional Ecological Risk Assessment, *Environ. Managem.*, *14*, 9–23

Suter, G. W., II, 1993, "A Critique of Ecosystem Health Concepts and Indices, *Environ. Toxicol. Chem.*, *12*, 1533–1539

Wang, Y. Y., C. P. Flessel, L. R. Williams, K. Chang, M. J. DiBartolomeis, B. Simmons, H. Singer, and S. Sun, 1987, "Evaluation of Guidelines for Preparing Wastewater Samples for Ames Testing," in *Short-Term Bioassays in the Analysis of Complex Environmental Mixtures V* (S. S. Sandhu, D. M. DeMarini, J. J. Mass, M. M. Moore, and J. L. Mumford, Eds.), Plenum Press, New York, USA, pp. 67–87

Watras, C. J., and T. M. Frost, 1989, "Little Rock Lake (Wisconsin): Perspectives on an Experimental Ecosystem Approach to Seepage Lake Acidification," *Arch. Environ. Contam. Toxicol.*, *18*, 157

Wentworth, N., J. Westrick, and K. Wang, 1986, "Drinking Water Quality Data Bases," in *Environmental Epidemiology* (F. Kopfler, and G. Graun, Eds.), Lewis Publishers, Chelsea, MI, USA, pp. 131–140

Wilson, E. O., 1988, (Ed.), *Biodiversity*, National Academy Press, Washington, DC, USA

Woodwell, G. M., 1974, "The Threshold Problem in Ecosystems," in *Ecosystem Analysis and Predictions* (S. A. Levin, Ed.), SIAM Institute for Mathematical Society, Alta, VT, USA, pp. 9–21

24

Modeling Ecological Risks of Pesticides: A Review of Available Approaches

Lawrence W. Barnthouse (Oak Ridge, USA)

24.1 SUMMARY

In 1992 an Ecological Fate and Effects Task Force within the EPA Office of Pesticide Programs issued a paper that substantially revised the agency's approach to pesticide risk assessment. Among the most important recommendations from the Task Force was a recommendation to eliminate the field test as a routine component of the pesticide risk assessment process. As an alternative, the Task force recommended that initial decisions

Ecotoxicology, Edited by Gerrit Schüürmann and Bernd Markert.
ISBN 0-471-17644-3   © 1998 John Wiley & Sons, Inc. and Spektrum Akademischer Verlag.

be based solely on laboratory test data. Where initial test data showed a potential for long-term effects, a refined assessment would be performed. Mitigating measures, such as reduced application rates, mandatory tillage practices, or changes in formulation, could be implemented depending on the outcome of the refined risk assessment.

Modeling of ecological effects was specifically identified by the Task Force as a research need, to "predict and/or characterize adverse effects in non-target organisms, their populations and communities." This paper provides a review of available modeling approaches that could meet this need, with an emphasis on new approaches to population modeling published in the scientific literature within the past five years. Four model types were reviewed: age/stage-structured models, individual-based models, metapopulation models, and spatially explicit models. Four criteria were used to evaluate the modeling approaches:

- ability to characterize ecologically relevant effects (e.g., the abundance and/or persistence of populations)
- ability to characterize the spatial and temporal distributions of exposures and effects
- applicability to a variety of types of biota and exposure situations, and
- current degree of acceptance within the scientific community.

All four modeling approaches were found to be applicable within one or more phases of the New Paradigm. Age/stage-structured models appear most likely to be useful for initial registration decisions. For refined assessments (e.g., special reviews, regional assessments) the other three modeling approaches may be more appropriate, because of their greater flexibility and ability to realistically simulate pesticide application regimes.

Age/stage-structured models and metapopulation models are already extensively used in natural resource management and no further scientific development is needed by the EPA. Individual-based models and spatially explicit models have appeared in the refereed scientific literature only within the past five years and have had few management applications. Although, in the long run, these approaches may well be the most useful, substantial research and development is still needed to support regulatory use by the Agency.

Four steps were identified that could be taken to increase the use of population models in pesticide risk assessment:

- Expand the information used for decisions concerning pesticides to include measured effects on populations
- Develop reference population data sets for representative species and local environments
- Demonstrate new models by applying them in actual assessments
- Fund research in individual-based and spatially explicit modeling, including the support of graduate and postgraduate fellowships
- Train agency staff in the theory and application of population models.

24.2 INTRODUCTION

In 1992 an Ecological Fate and Effects Task Force within the EPA Office of Pesticide Programs issued a paper that substantially revised the agency's approach to pesticide risk assessment. Prior to 1986, the agency had operated under a tiered approach that relied heavily on mesocosms and field tests (Urban and Cook 1986). The first tiers involved simple acute toxicity tests and exposure criteria using the "quotient approach." Estimates of standard toxicity test endpoints, such as LC_{50}s and LD_{50}s, were compared directly to environmental concentrations derived from standardized environmental fate models. If the quotients exceeded specified "triggers," then more extensive testing (more species, full life-cycle tests, reproduction tests) would be required. If these tests showed the pesticide to be safe, then no further action was taken. If there was still uncertainty about the effects of the pesticide, then mesocosm or field tests could be required prior to decision making.

In theory, mesocosm and field tests were supposed to provide information on long-term, chronic, population and ecosystem-level effects of pesticide use under realistic application conditions. In practice, data obtained from these tests proved to be of limited value. The systems were difficult to standardize, expensive to use, and impossible to replicate. The Task Force found that the data were adding little to the information available for decision making, and were adding greatly to the time and money required to make a decision. Consequently, the Task Force recommended eliminating the field test as a routine component of the pesticide risk assessment process. As an alternative, they recommended that initial decisions be based solely on laboratory test data. Where initial test data showed a potential for long-term effects, mitigating measures, such as reduced application rates, mandatory tillage practices, or changes in formulation, would be implemented.

The Task Force's recommendations were formalized in an "Implementation Paper for the New Paradigm" issued in 1993. This document defined the paradigm as consisting of the following components (SETAC 1994, Appendix B):

- *Risk assessment* is the scientific phase of the overall process and consists of hazard identification, and exposure assessment, ultimately integrating hazard and exposure to characterize risk.
- *Risk mitigation* involves mitigation measures to reduce or eliminate source contamination and adverse environmental impact.
- *Risk management* is a policy-based activity that defines risk assessment questions and endpoints to protect human health and ecological systems. It takes the scientific risk assessment and incorporates social, economic, political, and legal factors, which impinge on or influence the final decision, and selects regulatory actions.

The New Paradigm emphasizes early decision making and mitigation. For new registrations, the paradigm employs the same tier-1 Levels of Concern that were used in the old assessment scheme. If these levels are not exceeded, then registration may proceed. If they *are* exceeded, then a "refined risk assessment" must be performed, and potential mitigation measures must be identified. Costs and benefits of the mitigation measures must be evaluated.

Specific procedures for performing refined risk assessments and evaluating benefits of mitigation were not specified in the implementation paper and are yet to be defined. Modeling of ecological effects was specifically identified as a research need, to "predict and/or characterize adverse effects in non-target organisms, their populations and communities." This paper provides a review of available modeling approaches that could meet this need, with an emphasis on new approaches published in the scientific literature within the past five years.

24.2.1 Endpoints

FIFRA requires decisions to incorporate analysis of costs and benefits of regulatory decisions. If a pesticide is to be restricted, it must be shown that the benefits to ecosystems will be significant compared with the economic costs of the restriction. Ecological endpoints in the past have been limited to observed fish and bird kill incidents. These were often adequate to support regulatory decisions. However, scientifically speaking, they are not really adequate for a full benefits determination. All organisms die. Except in the case of threatened or endangered species, the abundance and persistence of populations is a more relevant endpoint. Ecological risk assessments for pesticides would be more useful and scientifically credible if it were possible to base decisions on risks to populations rather than risks to individuals, and to consider both *spatial scale* and *temporal scale* in the assessment. If only a small fraction of a population is exposed, or if the population recovers rapidly after exposure events, risks associated with pesticide use may be small even if lethal exposures occasionally occur. Some pesticides are acutely toxic (e.g., neurotoxins in birds) but not persistent in the environment. Such substances are arguably preferable to persistent pesticides that have long-term chronic effects on avian reproduction that might only be detected after substantial environmental damage has occurred. If it were possible to estimate the spatial variations in pesticide exposure and to evaluate the rate of recovery of exposed populations, this information could be used to design pesticide application regimes that would minimize ecological risks.

24.2.2 Model Evaluation Criteria

None of the above considerations are included in the tier-1 assessment criteria. However, they could be considered in "refined risk assessments," if appropriate models could be found that extrapolate effects on individual organisms (mortality, reproduction, etc.) to effects on populations and ecosystems. Are there models available that can, in principle, be used for this purpose? What additional research or demonstration is required before they can be used within the new paradigm? The following criteria were used in this paper to evaluate the potential utility of the modeling approaches reviewed:

Endpoints: ability to characterize ecologically relevant effects, i.e., the abundance and/or persistence of populations.

Spatiotemporal resolution: ability to characterize the spatial distribution of exposures and effects; ability to account for variations in temporal exposure over a period of days, weeks, or months.

Generality: applicability to a variety of types of biota and exposure situations relevant to pesticide risk assessment.

Current degree of acceptance: degree of acceptance within the scientific community, as proved by the number of successful applications (e.g., in resource management or conservation biology), and the number of refereed publications that employ the approach.

"Data requirements" and "data availability" were not included as evaluation criteria, because within each of the major categories of modeling approaches one can find models possessing a wide range of data requirements and availabilities. It is almost always possible to adjust a model to fit the available data.

24.2.3 Past Reviews

There have been several recent reviews of models which are potentially useful in ecological risk assessment for chemicals and pesticides. Barnthouse et al. (1986) reviewed the general history of successes and failures of population and ecosystem theory in resource management and environmental impact assessment. Emlen (1989) reviewed general types of theoretical population models with regard to applicability to terrestrial ecological risk assessments. Both of these reviews concluded that age-structured population models, i.e., models of populations that categorize organisms in terms of age and reproductive status, have been widely used in fish and wildlife management, have been validated under many circumstances, and are ready for use in ecological risk assessment of chemicals and pesticides. The widely used population modeling program RAMAS (Ferson and Akcakaya 1989; Ferson 1990) is an age-structured population model. Barnthouse (1993) provided a discussion of the historical development and basic principles of age-structured population models, with a review of the literature and specific examples of applications to toxic chemicals. Barnthouse (1992) highlighted several new developments, specifically in landscape modeling and individual-based modeling, that appeared potentially useful for ecological risk assessment.

There is no need to repeat these reviews. Most of the literature reviewed was published prior to 1990. The age/stage-structured models emphasized by Emlen (1989) and Barnthouse (1993) are refinements and applications of a theory that is more than 50 years old. These reviews should be consulted for detailed accounts of the historical development and underlying theory of population models. The following kinds of models are included in this review:

- *Age/stage-structured population models*: These models subdivide populations into discrete classes based on age, size, sex, or reproductive status. The best-known model of this type is the Leslie Matrix (Leslie 1945). In this model, information on the age-specific reproductive rates and probabilities of survival can be used to project the future growth or decline of the population and to show how the future status of a population should change in response to changes in survival and reproduction. Variants on age-structured models have been principal tools in natural resource management, especially for fish, since the 1950s. The matrix representation of population dynamics is highly flexible and has been variously modified to accommodate stochastic environmental variation, density-dependent survival and reproduction, and other biological or physical processes. The model can be formulated in terms of the size rather than the age of organisms. Caswell (1989) provided a thorough discussion of the theoretical development of these models. Applications in 1990 were reviewed by Barnthouse (1993). The discussion in this paper will be limited to a few innovations that have appeared in the literature since 1990.

- *Individual-based models*: Individual-based models are models that characterize the dynamics of populations in terms of the physiological, behavioral, or other relevant properties of the individual organisms. The "core" of an individual-based population model is a model of the organism, including its physiology, behavior, reproduction, spatial location, or any other relevant property. For some simple models, the population-level consequences of individual properties can be generated analytically (e.g., Kooijman and Metz 1984; Hallam et al. 1990). For more complex organisms or realistic environmental scenarios, these properties are calculated by numerical simulation: a fixed number of individuals are simulated day-by-day or week-by-week and quantities such as abundance, spatial distribution, or probability of extinction are generated by tabulating the numbers and distributions of organisms. Most of the published examples of individual-based organisms involve forest composition (Huston and Smith 1987; Shugart 1984; Dale and Gardner 1987), Cladocera (McCauley et al. 1990; Gurney et al. 1990; Hallam et al. 1990) or fish (Beyer and Laurence 1980; DeAngelis et al. 1991; Madenjian and Carpenter 1991; Rose and Cowan 1993). More recently, models that simulate the behavior and distribution of animals moving over a complex landscape have been developed (Loza et al. 1992; Liu 1993).

- *Metapopulation models*: As discussed by Hanski and Gilpin (1991), a metapopulation can be defined as a "set of populations which interact via individuals moving among populations." Levins (1969) performed the first quantitative analysis of conditions under which a species consisting of many populations could remain extant even though individual populations were frequently fluctuating and going extinct. Many metapopulation models, including complex ones involving interacting species (e.g., hosts and parasites, predators and prey, plants and herbivores) have been developed for use in biological pest control studies (Murdoch et al. 1985). Many recent applications are in conservation biology, most notably in studies of the Northern spotted owl (Lande 1987; Lamberson et al. 1994) and other endangered species with fragmented spatial distributions (Lindenmayer and Lacy 1995). Hanski and Gilpin (1991) provided a good recent review.

• *Spatially explicit models*: Spatially explicit models are models that incorporate realistic features of landscape structure. These representations can range from idealized arrangements of "patches" of suitable and unsuitable habitat (Lamberson et al. 1994) to vegetation maps generated by Geographic Information Systems (GIS) (Pulliam et al. 1992; Liu 1993; Turner et al. 1993). These models can be thought of as extensions of the metapopulation and individual-based modeling concepts to complex spatial environments.

24.3 DESCRIPTION OF MODELING APPROACHES

This section provides detailed descriptions of a few recent models of each type, drawn from the recent, peer-reviewed literature. The objective of these descriptions is to provide a foundation for evaluating the consistency of each approach with the evaluation criteria and for determining its potential applicability within the New Paradigm.

24.3.1 Age/Stage-Structured Models

Both Emlen (1989) and Barnthouse (1993) noted that age/stage-structured population models, i.e., models in which all organisms belonging to a population are classified into groups according to age, size, and reproductive status, have a long history of application in natural resource management. These models are the most readily available quantitative methods for assessing risks of toxic chemicals to populations. The most simple model assumes that (1) all organisms of the same age are identical, (2) all rates of birth and death are constant and independent of environmental variation, and (3) the rate of population growth is independent of population size:

$$\sum_{x=1}^{n} \lambda^{-x} l_x m_x = 1 \qquad (24.1)$$

where

l_x = fraction of organisms surviving from birth to age x
m_x = fecundity of individuals at age x
λ = finite rate of natural increase

Provided that l_x and m_x remain constant, any population that is growing according to Equation 24.1 will assume a stable age distribution in which the fraction of organisms in each age class x will remain the same from each generation to the next. Once the stable age distribution is achieved, the population will either grow or decline exponentially according to the following equation:

$$N_t = N_0 \lambda^t \tag{24.2}$$

where

N_0 = population size at time 0

N_t = population size at time t

The above model is biologically unrealistic in many ways, but has been successfully applied to many population management problems and is still the most widely applied approach to assessment of the impacts of human activities on the abundance and persistence of fish and wildlife populations.

Leslie (1945) developed a matrix form of Equation 24.2 that permits detailed analysis of the influence of age-specific survival and reproduction rates on the rate of population growth:

$$N(t) = LN(t - 1) \tag{24.3}$$

where $N(t)$ and $N(t - 1)$ are vectors containing the numbers of organisms in each age class $(N_0, \ldots, N_k)$ and L is the matrix defined by

$$L = \begin{bmatrix} s_0 f_1 & s_1 f_2 & s_2 f_3 & \cdots & s_{k-1} f_k & 0 \\ s_0 & 0 & 0 & \cdots & 0 & 0 \\ 0 & s_1 & 0 & \cdots & 0 & 0 \\ 0 & 0 & s_2 & \cdots & 0 & 0 \\ \cdots & \cdots & \cdots & \cdots & \cdots & \cdots \\ 0 & 0 & 0 & \cdots & s_k & 0 \end{bmatrix} \tag{24.4}$$

where

s_k = age-specific probability of surviving from one time interval to the next

and

f_k = average fecundity of an organism of age k

As discussed by Barnthouse (1993), a wide variety of population models can be derived from Equation 24.4 by making the survival and reproduction parameters random variables or functions of environmental parameters or population size. The model can be defined in terms of sizes or life stages rather than ages. Caswell (1989) presents a detailed discussion of the mathematical properties of all of these models. Barnthouse (1993) and Emlen (1989) described the range of resource management and risk assessment applications to which age/stage-structured models have been applied. Some of these applications involve pesticides and toxic chemicals (e.g., Tipton et al. 1980; Samuels and Ladino 1983; Barnthouse et al. 1990). The literature search performed for this paper revealed no qualitatively new types of age/stage-structured models.

Some new research has, however, been published concerning methods for comparing the influence of different life-history characteristics on the rate of population growth (λ). The term *elasticity* (e_{ij}) (deKroon et al. 1986) has been applied as a measure of the proportional sensitivity of λ to each element of the population transition matrix (a_{ij}):

$$e_{ij} = (a_{ij}/\lambda)(\partial\lambda/\partial a_{ij}) \qquad (24.5)$$

Elasticities, as defined by Equation 24.5 measure the relative contribution of each life-cycle element to the population growth rate. They have been found useful in theoretical studies of the relative fitness of different life-history strategies, especially for organisms such as plants that have extremely complex life histories compared to most vertebrate animals. Refinements of the basic methodology have been described by van Groenendael et al. (1994) and van Tienderen (1995). Meyer and Boyce (1994) used elasticity to compare the influence of changes in fecundity and survival due to hypothetical pesticide exposures in bird populations with different age and size structures.

The elasticity methodology itself does not provide any new approaches for modeling impacts of pesticides on populations. It may, however, be useful in the design of model-based assessment schemes. Models representative of a range of life-history types (small, fast-growing, short life-span vs. large, slow-growing, long life-span) would be developed. Elasticity analyses would be used to identify the life-cycle stages most strongly influencing the long-term population growth rate. These are the life stages in which pesticide exposure would be likely to have the greatest impacts.

24.3.2 Individual-Based Models

As the term implies, individual-based models provide the opportunity to evaluate the influence of characteristics of individual organisms on the abundance of whole populations. Age- and stage-based models already account for the influence of age and (for some stage-based models) size on population dynamics, but individual-based models can expand the list of characteristics considered to *any* aspect of organismal biology believed to be relevant. For the purposes of pesticide risk assessment, the most relevant of these appear to be physiology and behavior. The general procedure is to develop a model of the individual organism to whatever level of detail is required, and then to infer the properties of the population as a whole either by analytical solution of equations or by numerical simulation of the activities of hundreds or thousands of individual organisms.

Physiological characteristics included in individual-based models have emphasized metabolism, growth and contaminant pharmacodynamics. Work on metabolism was pioneered by Kooijman and Metz (1984), who examined the influence of contaminants on metabolism and population growth using *Daphnia* as a model organism. Hallam and Lassiter (Hallam et al. 1990; Lassiter and Hallam 1990) extended this approach to include (1) a thermodynamically based model of the uptake of contaminants from aqueous media and (2) a definition of death in terms of the internal dissolved contaminant concentration within an organism. McCauley et al. (1990) and Gurney et al. (1990) developed a model of *Daphnia* growth and reproduction based on energetics and used the model to predict time-dependent changes in the age and size structure of *Daphnia* populations in response to changes in food availability.

All of the above models were developed for aquatic organisms with relatively simple life-cycles. The emphasis in model analysis was on evaluation of general properties of the models by analytical investigation of the equations. DeAngelis et al. (1991) and Rose and Cowan (1993) developed models of fish populations that include metabolism, growth, foraging behavior, and prey selection as functions of the life stage and age of the fish. DeAngelis et al. (1991) even included the nesting and nest defense behavior of male smallmouth bass. The approach followed in developing both of these models was to use the existing extensive theoretical literature on bioenergetics, reproduction, and foraging of individual fish, coupled with exhaustive evaluation of the life history of specific fish species, to develop detailed models of each life-stage from egg through to reproductive adult. Population-level consequences of changes in the physiology, behavior, or reproduction of individual fish are inferred by brute-force simulation of the birth, growth, and death of hundreds or thousands of individual fish. The models are calibrated to extensive data sets collected for specific fish populations.

The model of DeAngelis et al. (1991), for example, simulates the spawning, growth, and survival of a year-class of smallmouth bass (*Micropterus dolomeui*) in Lake Opeongo, Ontario. It is structured as a set of discrete submodels that simulate the daily activities and physiological condition of each individual fish in the model population. Reproductive behavior in smallmouth bass is quite complex. Adult males excavate nests, and after eggs are deposited the males defend the nests from predators for several weeks while the eggs hatch and develop through their early larval stages. Spawning behavior and spawning success have been shown to be both temperature- and size-dependent. Larger males spawn earlier; larvae that are spawned early have a size advantage over larvae that are spawned late. Success in rearing a brood has also been shown to be related to the size and condition of the guarding males. The males do not feed during the brood period; smaller males and males in poorer condition at spawning abandon their nests much more frequently than do larger, healthier males. Environmental conditions also have an important influence on spawning success because storm events that cause water temperatures to fall below a critical threshold cause the males that have spawned to abandon their nests.

The model of DeAngelis et al. (1991) simulates all of these processes. Given an initial size/condition distribution of males and a specified daily temperature regime, the model simulates the nesting and brood-rearing success of each male over the course of the reproductive season. The probability of successfully rearing a brood and the number of young fish produced from each brood are defined as probabilistic functions of the size and condition of the male at the time of nesting.

Growth and survival of the young-of-the-year fish is simulated on a daily time-step. The model incorporates well-established models of fish bioenergetics and foraging. Prey abundances and size distributions are specified in the model. The prey selectivities and feeding successes of each fish are specified by established predation models that account for the swim speed and visual acuity of the fish (both dependent on size) and on the frequency of encounter with prey of appropriate sizes. All of these processes are stochastic. Rather than producing single numbers for the abundance and average size of fish, the model produces, each day, a size and age (in days) distribution of fish. It is possible to calculate the relative probability of survival of larvae spawned on any given day or having any given initial size.

The objective of DeAngelis et al. (1991) was to understand and explain the processes responsible for recruitment success in smallmouth bass: why more fish are produced in some years than in others and why large males spawn earlier than small males in spite of the increased risk of low-temperature events. Because of the physiological detail included in the model, however, it would be relatively easy to modify it to include lethal and sublethal effects of toxic chemicals.

Individual-based models have also been applied to terrestrial biota. In many cases the emphasis has been on behavior rather than metabolism and physiology. Pulliam et al. (1992) developed a model of the Bachmann's sparrow population on the Savannah River Site, South Carolina, derived from the individual foraging behavior and habitat selection of the birds (this model is discussed in detail in section 24.3.4).

Lacy (1993) described a generalized computer program (VORTEX) that simulates the local population dynamics of terrestrial vertebrate populations. VORTEX was intended for use in the management of small populations threatened by habitat loss, environmental variability, and loss of genetic variation. The core of VORTEX is a stochastic model of the birth, growth, reproduction, and death of each individual animal. At the start of simulation, an initial number of animals, age/sex structure, genetic composition, and carrying capacity are specified. At each subsequent time-step (normally one year, but modifiable by the user), mature animals mate and produce young. The percentage of mature females producing young in a given year, and the number of young produced by each reproducing female, are drawn from probability distributions. Several different options are available to specify mate selection and, thus, inbreeding effects. Following reproduction, the individual animals are randomly "killed" according to numbers drawn from age-specific probability distributions. Both reproduction and survival are assumed to be density-dependent, with expected values changing depending on the relationship between the current population size and the carrying capacity. Reproduction and survival are subject to two sources of environmental variation: "normal" annual variation specified by a binomial probability distribution, and "catastrophic" variation that occurs randomly according to a uniform probability distribution. When catastrophes occur, survival and reproduction rates of all animals are reduced by a constant fraction specified by the user.

Each simulation "run" produces a single stochastically determined time-series of population sizes and age/sex/genetic compositions. The simulated population either persists throughout the simulation or goes extinct at some point during the simulation. Estimates of the probability of persistence and expected time-to-extinction of the simulated populations are obtained by performing multiple runs (hundreds or thousands). The results of these simulations can be summarized as (1) probability of extinction as a function of simulation time, (2) median and mean time to extinction of populations that go extinct, and (3) mean size and genetic variation within populations that survive.

VORTEX, in the version described by Lacy (1993), can simulate up to 20 populations, between which immigration and emigration can occur. In this form it can be viewed as a metapopulation model (see section 24.3.3). VORTEX has been applied to a variety of endangered bird and mammal species (Lacy et al. 1989; Seal and Foose 1989; Seal and Lacy 1989; Lacy and Clark 1989; Maguire et al. 1990; Foose et al. 1992; Lindenmayer et al. 1993). VORTEX, unlike the model of DeAngelis et al. (1991) does not explicitly simulate ecological or physiological processes relevant to pesticide exposure and effects assessment. However, information concerning (1) the distribution of doses within an exposed population and (2) dose-response relationships for reproduction and mortality could be used to modify the survival and reproduction functions used in VORTEX.

24.3.3 Metapopulation Models

Most species do not exist as continuous interbreeding populations. They consist of subpopulations inhabiting patches of suitable habitat mixed in with patches or regions of unsuitable habitat. All are subject to environmental variability that may be either large or small. Small populations frequently go extinct, but habitat patches are then recolonized by colonists arriving from other patches. This view of species as "metapopulations" was first formalized by Andrewartha and Birch (1954), although they did not use the term. The first quantitative studies of metapopulation biology were published in the 1960s by MacArthur and Wilson (1967) and den Boer (1968). Levins (1969) is credited with developing the first formal model to specifically address the central question in metapopulation biology: how are (1) the fraction of occupied patches and (2) the expected time to extinction of the species as a whole affected by the probabilities of extinction and recolonization of individual patches?

Levins (1969) formulated a simple relationship between the fraction of habitat patches occupied by a species at any given time ($p(t)$), the rate of extinction of occupied patches (e), and the rate of production of propagules from each occupied patch (m). He reasoned that at any time t, mp propagules would be produced. Assuming equal probability of dispersal to occupied and unoccupied patches, a fraction equal to ($1 - p$) of these would colonize unoccupied patches. At the same time, a total number of patches equal to ep would become extinct. The rate of change in p at any time would be determined by the equation:

$$\mathrm{d}p/\mathrm{d}t = mp(1 - p) - ep \tag{24.6}$$

It follows from this equation that the equilibrium frequency of occupied patches (p^*) is determined by the ratio of the extinction (e) and colonization (m) rates:

$$p^* = 1 - e/m \tag{24.7}$$

It is intuitively obvious, even without Levins' model, that if extinction is more likely than dispersal (i.e., e is larger than m) the species must become extinct. It is not obvious, however, that if these two parameters are similar in magnitude the fraction of occupied patches can be expected to be very small, even if the rate of dispersal of propagules from occupied patches is very high. Levins also investigated the influence of temporal variation in extinction and colonization rates on the size and probability of persistence of species subdivided into local populations.

The above model is clearly too simplistic to be of much value in the management of real populations. Many subsequent authors (see review by Hanski 1991) have replaced Levins' simple assumptions with more biologically realistic representations of both the dispersal of organisms between habitat patches and of the local dynamics of populations within patches. However, the fundamental processes and variables of interest, i.e., dispersal, extinction, percent occupancy of available habitat, and metapopulation persistence, have not changed.

Most early work involving metapopulation models was concerned with insect populations, either with understanding the reasons for persistence of insect species subject to wide fluctuations in local population abundance (den Boer 1968) or with designing control strategies to reduce the frequency of widespread pest outbreaks (Levins 1969). In the 1980s conservation biologists turned to metapopulation theory as a means of designing preservation strategies for vertebrate species that, although once widespread, were becoming restricted to isolated subpopulations because of increasing habitat fragmentation. The early models were extended to include influences of local population size (Hanski 1985), local population structure (Lande 1987) and spatial dispersal patterns (Ray and Gilpin 1991). The theory has also been extended to include predator-prey and host-parasite dynamics (Murdoch et al. 1985; Sabelis et al. 1991).

The relevance of this work to pesticide risk assessment comes from the observations that many wildlife species of management interest are, effectively, metapopulations. Their distribution patterns have been changed by decades of habitat conversion as the original forests and prairies of North America have been transformed into a mosaic of agricultural, urban/suburban, and successional landscapes. Pesticides of equal toxicity will have differential impacts on wildlife species depending on patterns of habitat utilization, degree of population isolation, dispersal ability, and other aspects of population biology included in metapopulation models.

Lande (1987) formulated a model of extinction and persistence in territorial populations that is a direct descendent of Levins' (1969) original model. Subsequently, other authors have developed much more detailed models of the Northern Spotted Owl, intended for use in predicting the influence of specific habitat management regimes on the recovery of this endangered subspecies. Lamberson et al. (1992) described a metapopulation model of the Northern Spotted Owl (*Strix occidentalis caurina*). In their model habitat patches are defined as nesting territories and local populations are defined as nesting pairs. A nesting pair annually produces young according to either a fixed fecundity rate or a randomly varying fecundity rate. The juvenile birds disperse at the end of each breeding season, with juvenile males seeking an unoccupied nesting territory and juvenile females seeking a site occupied by a solitary male. The probability that a dispersing juvenile finds a suitable site before it dies is determined by the fraction of the total landscape that consists of suitable sites, the fraction of those sites that is already occupied by nesting pairs, and the number of sites a juvenile can search before it dies. Adult birds are subjected to annual mortality, and nesting sites are subjected to disturbance by timber harvesting. Adults nesting on a harvested site must disperse and locate new, unoccupied sites. The outputs from the model, which are updated annually, include the total number of suitable nesting sites, the number of sites that are occupied by nesting pairs, and the number of sites occupied by single males.

Lamberson et al. (1992) used the model to evaluate the influence of initial population size, the proportion of the landscape suitable for occupancy by spotted owls, and the degree of interannual variability in fecundity (reflecting variability in food supply). Several interesting effects were found. First, because the sexes were modeled separately and colonization of a suitable site by both a male *and* a female is required to establish a nesting pair, the entire metapopulation invariably declined to extinction if the number of nesting pairs fell below a threshold determined by the searching abilities of the dispersing juveniles and the proportion of suitable habitat. This well-known phenomenon is termed by population biologists the "Allee effect" after the biologist who first described it. Second, Lamberson et al. (1992) investigated the influence of habitat availability on the probability of survival of the metapopulation under conditions of no, low, and high environmental variability. They found that, for the case of no environmental variability, extinction always occurred if less than a fixed percentage of the landscape (determined by dispersal ability) was suitable, and that extinction never occurred if the percentage of suitable habitat was greater than the threshold. Environmental variability had the effect of smoothing the transition from inevitable extinction to indefinite persistence, so that there was a small probability of persistence for habitat suitabilities slightly below the deterministic threshold and a small probability of extinction for suitabilities somewhat higher than the threshold.

By examining a range of parameter values consistent with the current state of knowledge of spotted owl population biology, the authors found that the effective persistence threshold for the metapopulation lies somewhere between 10% and 25% suitability of available habitat. Lamberson et al. (1992) also simulated a potential future timber harvesting regime in which the availability of nesting habitat was reduced gradually for 20 years and then stabilized at a level of 20%. They found that, although the simulated owl metapopulation eventually stabilized, there was significant time lag during which the total population of adult owls and the percent occupancy of suitable sites was relatively constant. The implication of this result, according to the authors, is that it would be relatively difficult to use population monitoring data collected during the harvest period to determine the ultimate response of the owl metapopulation to habitat reduction.

Lamberson et al. (1994) used a different metapopulation model to evaluate the influence of patch size and spacing on the viability of the Northern Spotted Owl. The landscape was portrayed as a rectangular array of identical circular clusters containing potential owl habitats. Each cluster consisted of a collection of territories, some or all of which were assumed to be suitable as nesting sites. All of the space between clusters was assumed to be unsuitable habitat. This idealized landscape was intended to approximate the real landscape inhabited by spotted owls, which consists of patches of old-growth forest with differing abilities to support spotted owls, separated by areas of cut forest. Within each site, owl reproduction, survival, and dispersal were modeled in the same way as Lamberson et al. (1992), except that only females were considered. Dispersal was assumed to be successful if a juvenile female found an unoccupied but suitable territory within a specified number of searches.

Dispersal within clusters was simulated using a "random walk": starting from the territory of birth, a dispersing juvenile female was assumed to have an equal probability of searching any adjacent territory. If that territory was suitable and unoccupied, it became occupied by the dispersing bird. Otherwise, the bird moved, again in a random direction, to another adjacent territory. After a fixed number of unsuccessful searches, the dispersing juvenile was assumed to exit the cluster in a random direction. Two sources of mortality were imposed on juvenile females dispersing between clusters: first, a juvenile was assumed to die if a straight line in the selected direction did not intersect another cluster. If a line in the selected direction did intersect a cluster, then the probability of survival during transit between the clusters was assumed to decline exponentially with the distance between clusters. A female successfully arriving at a new cluster searched it in the same way as her natal cluster and continued to search until she either died during transit, found an unoccupied suitable territory, or searched a total of 22 sites. All birds that searched 22 sites without success were assumed to die.

Landscape parameters investigated by the authors included the percentage of the total landscape included within habitat clusters, the number of sites within each cluster, the percentage of sites within each cluster suitable for nesting, the fraction of sites within a cluster searched prior to exiting, and the rate of mortality during dispersal. The authors evaluated the influence of different reserve design patterns on the mean occupancy of nesting sites, defined as the fraction of suitable sites occupied by nesting females.

According to the authors, field studies suggest that approximately 60% of the forested area within the range of the Northern Spotted Owl provides suitable nesting habitat. Assuming that 60% of the sites within each cluster are suitable and that a maximum of 22 sites can be searched by dispersing juveniles, simulated spotted owl metapopulations did not achieve a stable mean occupancy unless the average cluster contained at least 15 sites. Using a mean suitability of 80%, a stable population could be achieved at a smaller mean cluster size but mean occupancy declined rapidly at cluster sizes below 10 sites per cluster. For any fixed total reserve size and percent suitability, the equilibrium mean occupancy was found to increase with the size of the clusters. Lamberson et al. (1994) also investigated the influence of (1) the fraction of the total landscape occupied by clusters and (2) the distance between clusters on mean occupancy. They found that for small cluster sizes (i.e., 10 or fewer sites per cluster) increasing the fraction of the landscape in clusters and decreasing the distance between clusters both significantly increased mean occupancy; for cluster sizes greater than 25 there was little effect.

The authors used their results to evaluate the adequacy of the reserve design proposed for the Northern Spotted Owl. They concluded that, in general, the proposed sizes and spatial distributions of proposed "Habitat Conservation Areas" is adequate, provided that the recovery of currently degraded habitat within the HCAs is rapid.

Lindenmayer and Lacy (1995a, b) used the multipopulation version of VORTEX (Section 24.3.2) to evaluate the metapopulation stability (expressed as probability of persistence in the metapopulation as a whole and the inter-annual variability in abundance of local populations) of Leadbeater's Possum (*Gymnobelidius leadbeateri*) in fragmented Australian old-growth forests. Effects of patch size and number on stability were simulated by varying the carrying capacities and number of local populations; no attempt was made to simulate the influence of inter-patch distance or spatial distribution. The authors found, like Lamberson et al., that increasing the size of patches enhanced the stability of the metapopulation as a whole. When all patch sizes were small, metapopulation extinction rates were invariably high and emigration actually *decreased* metapopulation stability.

Lande's (1987) and Lindenmayer and Lacy's (1995a, b) models are more obviously relevant to pesticide risk assessment problems than is the more species-specific model of Lamberson et al. Neither, however, may provide sufficient biological realism to support pesticide regulation. In particular, neither provides for explicit consideration of local habitat requirements and distributions within agricultural landscapes. By following the example of Lamberson et al. (1992, 1994) population biologists could develop models specifically tailored to species and exposure regimes of interest in pesticide regulation. Such models could provide useful information for risk assessment if (1) the species of interest is restricted to relatively isolated subpopulations between which dispersal and recolonization occur, (2) pesticide applications have the potential to increase the risk of extinction of local populations, and (3) it is the persistence of the species as a whole, not the persistence of individual local populations, that is the regulatory endpoint of interest.

24.3.4 Spatially Explicit Models

Spatially explicit models may be thought of as extensions of individual-based or metapopulation models in which the organisms or subpopulations are distributed over a realistic rather than an idealized landscape (Dunning et al. 1995). "Suitable" and "unsuitable" habitat types can be defined explicitly in terms of vegetation, topography, or soil type. Temporal changes in habitat suitability can readily be simulated. For management applications, spatially explicit models can utilize landscape maps derived from aerial surveys and remote sensing. For obvious reasons, growth of research and application of these models did not really begin until the late 1980s and the majority of the research performed using this approach has been published within the last five years.

The approach appears especially suited to the study of mobile animal populations that forage and disperse over large, heterogeneous areas. The spatially explicit approach permits ecologists to integrate theory and observation on foraging behavior and reproduction in individual animals, relate these to specific measurable habitat characteristics, and infer influences of habitat change on populations. As noted by Pulliam (1994), information on environmental contaminant distributions and effects can easily be integrated into the same framework. Because spatially explicit models often deal with individuals, the full array of individual physiological characteristics can also be incorporated. Such models can be thought of simply as individual–based models in which the location and directional movement of the organism are included as additional characteristics.

The most thoroughly explored and tested models of this type have been developed for populations of ungulates foraging in Yellowstone National Park (Turner et al. 1993, 1994) and for the population of Bachmann's Sparrow nesting on the U.S. Department of Energy Savannah River Site (Pulliam et al. 1992). Turner et al. (1993) simulated the influence of landscape heterogeneity on winter grazing in "generic" ungulates. A standard energetics model was used to simulate the daily foraging intake and energy balance of an animal as a function of body weight, forage availability, and activity level. The authors then investigated the influence of different ungulate movement "rules" and patterns of forage availability on the energy balance and survival of model populations. Landscapes in which resource patches (sagebrush–grassland communities) were randomly distributed across the landscape were compared with landscapes derived directly from vegetation maps for Yellowstone. During each time step of the simulation, an animal feeds on resources within the patch it occupies. It may move to another patch; the probability of movement increases as it feeds and depletes the forage at its current location. While an animal is moving between patches it cannot feed.

Results obtained from the model generally supported previous theoretical predictions that (1) when resources are abundant, landscape pattern and movement rules should have no influence on weight maintenance and survival, (2) when resources are scarce, aggregated resources (i.e., the real Yellowstone landscape) should support more animals than randomly dispersed resources, and (3) when resources are scarce, behavioral rules that allow the animals to discern resource abundance at distant sites or to move over greater distances should improve survival.

Turner et al. (1994) extended their original model and used it to explore the effects of fire on free-ranging elk (*Cervus elaphus*) and bison (*Bison bison*) populations in northern Yellowstone Park. In the new analysis, the authors derived a six-category habitat map from GIS data maintained by the National Park Service, and assigned to each category a winter forage abundance derived from actual field measurements (available separately for unburned sites and for sites burned during the 1988 fires). The foraging rule used assumed that each animal visually searches within a circle around its current location and moves to the site with the highest quality. It may continue searching and moving until it either obtains its maximum daily intake or reaches its maximum daily movement distance. Because snow conditions are an important determinant of winter ungulate survival, snow was simulated in the model. A snow subroutine assigned monthly snow depth values to each grid cell based on observed data and on known influences of topography on snow depth. Foraging behavior and energetic costs were both assumed to be affected by snow depth.

The authors were able to calibrate and test their model using observed data collected both before and after the 1988 fire. For all three years, data were available on winter precipitation, fall elk/bison count, and overwintering elk/bison survival. After model parameters were calibrated so that overwintering survival during these three years matched the available data, simulation experiments were performed to evaluate the influence of winter severity, fire size, and fire pattern on ungulate survival. Observed snowfall during the most severe and most mild winters recorded in this century at Yellowstone were used to evaluate the influence of winter severity. Three levels of fire severity, expressed as the percentage of the study area burned, were examined. A range of alternative fire patterns was evaluated: a fragmented burn was simulated by distributing burned grid cells at random over the whole map; a clumped burn was simulated by generating a single patch of burned cells centered on an arbitrary location.

Several intermediate patch distributions were also evaluated, including the actual observed burn distribution of the 1988 fire. In all, 24 different scenarios were evaluated. The authors found that winter snow was the most important determinant of ungulate survival. Fire severity and pattern influenced survival only during average and severe winters. Provided winters were mild or average, large fires actually produced better long-term survival than small fires due to their stimulating effect on forage availability during post-fire winters. For small to moderate fires, ungulate survival was greater with clumped than fragmented fire patterns. The authors concluded that fires and spatial fire patterns have an important influence on ungulate population dynamics in Yellowstone only if severe winter conditions occur in the post-fire winter.

Pulliam et al. (1992) described a generalized spatially explicit population model for bird dispersal, applied to the Bachmann's sparrow (*Aimophila aestivalis*). The objective of the model was to describe influences of spatial variation in habitat suitability on the abundance and persistence of sparrow populations in a managed pine plantation. The model as described in the original paper is closely similar to the spotted owl model of Lamberson et al. (1994), discussed above. Only female birds are included, and the only life-history characteristic simulated is dispersal. The principal differences between the models are in descriptions attached to the grid cells. In the model of Pulliam et al., grid cells are identified with pine stands of different ages. Bachman's sparrows nest only in young (5 years old or less) or mature (> 80 years old) pine stands. The simulated plantation consists of a number of tracts of different ages. As the simulation proceeds, newly-seeded tracts become suitable nesting sites and previously suitable tracts age and become unsuitable. Trees are harvested on a 21-year rotation, except for a certain number of tracts of mature pine forest that provided a stable source of dispersing birds. The authors evaluated the influence of different model parameters on the abundance and persistence of populations simulated for 100 years (five rotations). They found that parameters relating to mortality and reproduction were more important than those relating to dispersal (site selectivity, dispersal mortality). Population size increased linearly with the number of tracts left in mature forest, but mature forest was not required to maintain viable sparrow populations.

Liu (1993) extended the BACHMAP model in two significant ways. First, he modified it to accept landscape classification information from a GIS. Second, he developed an economics subroutine that calculates growth, yield, income, cost, and net-present-value estimates for each tract. The extended model is coded in an object–oriented programming language, so that it is modular and can easily be adapted to different species or landscape types. Results of actual management applications are not yet published, as of the date of preparation of this paper.

Other recently published spatially explicit models simulate physiology as well as behavior. Loza et al. (1992) described a model of cattle grazing on open rangeland that simulates the influence of physiological status (energy and water balance) on the grazing behavior and land use of grazing animals. Jager et al. (1993) described a spatially explicit version of the smallmouth bass model of DeAngelis et al. (1991) that simulates reproduction, foraging, and growth in a riverine population of smallmouth bass.

The principal advantages of spatially explicit models include flexibility and realism, especially realism with respect to spatial representation of the environment. Virtually any physical or biological process can be included in such models, provided a model of that process can be developed. Both short-term and long-term events can be simulated. Extremely detailed representations of the landscape, including direct interfacing with GIS systems, is possible. The principal disadvantages are complexity of some of the models (a disadvantage if there are no data) and, especially, the relative immaturity of the applications. The vast majority of the published models are very recent. Few applications utilizing GIS technology have yet been published, although several are currently being developed. Given the flexibility of the approach, specific applications for pesticides should be straightforward and specify the spatial scale (local or regional), the species and landscape types of interest, and the pesticide application scenarios. The object-oriented programming approach described by Liu (1993) appears to provide an important advance in modeling techniques because it permits a generalized model structure to be specifically tailored to a variety of risk assessment scenarios.

24.4 INTEGRATION OF ECOLOGICAL MODELS INTO THE NEW PARADIGM

According to this New Paradigm, if the initial screening of a pesticide shows potential exceedence of Levels of Concern, then a "refined risk assessment" may be performed. The Avian Effects Dialogue Group (1994) identified fourteen specific kinds of information that could contribute to these refined risk assessments. The modeling approaches discussed in this paper could contribute to these refined assessments by (1) allowing effects to be expressed in terms of population abundance and persistence rather than as fractions of an LC_{50} or other laboratory test endpoint, (2) integrating information on lethal and sublethal effects (including behavioral effects), and (3) accounting for spatiotemporal variations in exposure.

The Avian Effects Dialogue Group discussed simulation modeling as a potential approach to risk characterization, but provided no specific recommendations. This review demonstrates that progress in some types of ecological modeling has been significant in recent years. All four types of models discussed in this paper could be used to implement the New Paradigm.

Viewed within the Framework for Ecological Risk Assessment (U.S. Environmental Protection Agency, 1992), all of these models are risk characterization techniques: they express effects of pesticides or other stressors in terms that are understandable to decision makers and are compatible with cost-benefit analyses or other management-related evaluations.

24.4.1 Types of Model Applications

There are at least three possible uses of models within the New Paradigm:

Initial registration. Initial pesticide registration is a predictive activity involving the estimation of potential changes in population size or reproductive success from standard toxicity test data. Applications of age-structured models to predictive assessment problems are described by Barnthouse et al. (1990), Emlen (1990), and Barnthouse (1993). Inter-species and inter-life stage extrapolation methods described by Barnthouse et al. (1990) can be used to quantify prediction uncertainty and express risks in terms of ranges of potential effects associated with a given exposure or ranges of exposure associated with given effects levels. As suggested by Emlen and Pikitch (1989), a few generic population models can be made to represent major life-history types of organisms, especially vertebrates (mouse, moose, sparrow, eagle). Such models could be used to illustrate the consequences of contaminant exposure for populations of organisms having different life-history characteristics.

Because of the inherent limitations of age/stage-structured models, the above approach cannot accommodate spatial or temporal variations in exposure and thus cannot be used to evaluate the influence of application regimes or site-specific environmental variability on the effects of pesticides. Such information may, however, not be needed for initial decision making. If predicted effects using conservative exposure scenarios are shown to be inconsequential, then no additional analyses may be necessary.

Another way to introduce population-level phenomena into initial registration decisions would be to expand the concept of reference environments, as it is already used in exposure assessment, to include reference populations. Descriptions of a reference field could include, in addition to estimates of soil type, slope, rainfall, and other physical parameters used in pesticide fate models, estimates of the seasonal distribution and foraging characteristics of a reference ground-feeding bird. An individual-based model would use this information to assess the impact of pesticide application on a reference bird population. Besides permitting detailed evaluation of alternative application patterns, this approach would facilitate assessments of pesticides with high potential toxicity but low environmental persistence.

Special Review. Substantial quantities of field data are often available for pesticides undergoing special reviews. For these cases, site-specific population models may aid in interpreting the results of field studies. In the case of granular carbofuran (Houseknecht 1993), for example, information on acute mortality of birds due to carbofuran exposure was available both from incidental observations and from field experiments. The value of these data for risk assessment was limited, however, because pesticide effects could only be quantified in terms of numbers of dead birds. An individual-based model of a foraging bird population, developed using the approach described by Pulliam (1994), could have been used to estimate the distributions of exposures within a local population of birds, the fraction of those birds likely to have received a lethal dose of carbofuran, and the long-term consequences of continued carbofuran application for the abundance and persistence of a reference population. Such model applications are especially feasible for birds because of the well-developed state of avian foraging theory and the relative ease of validating species-specific avian foraging models (Pulliam 1994).

Regional assessment. There is increasing interest within the EPA in performing regional, or "place-based" assessments that integrate a variety of kinds of environmental hazards and can be used to prioritize regulatory activities for specific watersheds or regions. For example, pesticide use patterns and ecosystem types present in the Midwest corn belt differ substantially from those present in the Gulf Coast region. If the ecological importance of these differences could be incorporated in risk assessments, then region-specific restrictions or labeling requirements could be designed. Issues that could be addressed might include the cumulative impacts of runoff from a region within which large quantities of a few pesticides are used, or impacts on wide-ranging species that forage over a mosaic of landscape types. For these applications, depending on scale, spatially explicit models or metapopulation models would be appropriate. Such models can accommodate land classification data that are already available from many sources. Turner et al. (1995) and Liu et al. (in press) have already shown that such applications are technically feasible.

Table 24.1 summarizes the types of models appropriate for each of the above applications. For initial registration decisions, little or no site or region-specific information is likely to be available. Relatively little toxicity data may be available as well. Thus, the models with the fewest information requirements, i.e., age/stage-based models, are the most likely to be useful. Provided that (1) reference population descriptions have been developed, and (2) some

Table 24.1 Modeling approaches suitable for three types of pesticide risk assessment applications

Applications	Models			
	Age/stage-based	Individual-based	Meta-population	Spatially explicit
Initial registration	X	X		
Special review	X	X	X	
Regional assessment			X	X

information about physiology or time-dependent pesticide fate patterns is available, then individual-based models could enable consideration of more complex phenomena. The same types of models may be used for special reviews. However, for a special review more information is likely to be available, and spatial patterns of use may have been established. If local or regional extinctions are identified as an ecological issue of concern, then metapopulation models may be useful. For regional assessments, it is likely that spatiotemporal pesticide use patterns and explicit characteristics of the landscape will be relevant. Models that cannot accommodate space, i.e., age/stage-based models and individual-based models, will not be able to address the relevant questions. Metapopulation models, and, especially, spatially explicit models, would appear to be the best choices in principle.

24.4.2 State of Development of Modeling Approaches

Table 24.2 compares the four modeling approaches discussed in this paper with respect to the evaluation criteria discussed in Section 24.2.2. All four approaches are highly flexible in form and can represent a wide range of populations of interest. Less data would in general be required to implement age/stage-structured models than to implement the other model types. However, as noted in Section 24.3, all four approaches can encompass a range of complexity so that complexity *per se* is not a useful evaluation criterion.

The four approaches differ significantly with respect to the other three evaluation criteria. Age/stage-structured models are, according to the definitions used in this paper, spatially homogeneous (spatially structured variants of this model type would be classified as metapopulation models). As normally applied, age/structured models are used to characterize the long-term or steady-state behavior of populations and cannot directly address phenomena (e.g., transient pulses of contaminant exposure) that are short with respect to the lifetime of a single organism. Individual-based models and spatially explicit models, in contrast, have arbitrarily high degrees of resolution. The activities of individual organisms can be simulated on any time-scale; the size of cells in

Table 24.2   Comparative evaluation of modeling approaches

Criteria	Models			
	Age/stage-based	Individual-based	Meta-population	Spatially explicit
Endpoints	H	H	H	H
Resolution	L	H	M	H
Generality	H	M	H	L
Acceptance	H	M	M	L

spatially distributed models can be made arbitrarily large or small. In both cases the resolution of the available data and the needs of the assessment are the limiting factors. Metapopulation models are intermediate in resolution: space can be at least implicitly represented in terms of immigration/emigration/extinction processes. Like age/stage-structured models, however, metapopulation models are generally best suited to addressing effects of long-term exposure.

As noted by Levins (1966), a tradeoff can usually be expected between generality and spatiotemporal resolution in models. Age/stage-structured models have been developed for virtually every type of living organism. The versatility of the metapopulation approach, at least when applied to vertebrates, is demonstrated by the number of species for which the multipopulation version of the VORTEX model has been implemented. In contrast, physiologically and behaviorally oriented individual-based models, such as those of DeAngelis et al. (1991) and Pulliam et al. (1992), are highly specific. The underlying theories of foraging, bioenergetics, and reproduction are quite general, however, the number of species-specific parameters needed to implement an individual-based model can be quite large. Spatially explicit models require, in addition, site-specific data on landcover, weather, and other environmental influences on the activities of the organisms being modeled.

With respect to the degree of acceptance by the scientific community, age/stage-based models are by far the best-developed type discussed in this paper. They are the type most people immediately think of when they hear the term "population model". The basic Leslie matrix and its variants are the backbone of quantitative fisheries assessment, with literally hundreds of applications over 50 years. User-friendly modeling software is widely available. The more general stage-based models have been less widely used in management, although they are common in plant demography and applied entomology.

Metapopulation models have a much shorter history, but have become very widely used in conservation biology over the last decade. The more complex models, such as the multipopulation version of VORTEX, provide a useful extension of age/stage-based models for situations in which differential exposures to isolated subpopulations are important. This will often be the case for rare or endangered species, which are restricted to specific habitat types. Like age/stage-structured models, they can be relatively general and applicable to a variety of different life-history types, as has been shown by the ease with which the VORTEX model has been adapted to a variety of mammalian species.

Despite its formulation in terms of individual organisms, VORTEX is in principle equivalent to a classic age/stage-structured model. Caswell and John (1992) showed that all matrix-type population models can be derived from models of the birth, reproduction, and death processes of individual organisms; for large population sizes VORTEX and its matrix equivalent would provide identical results. Models that incorporate physiological or behavioral influences on individuals, or that involve explicit simulation of interactions between organisms and their surrounding landscape, are fundamentally distinct from any previous approach to population modeling. Such models represent the future rather than the present of population biology. The majority of the published accounts of these kinds of models has appeared in the peer–reviewed literature only within the past three years. No standardized modeling software is available to support their development. Only a few experts, primarily associated with the authors of the papers cited in this paper, have had any significant experience in developing and applying these models. The object-oriented software described by Liu (1993) provides a general framework that would, if widely adopted, significantly simplify the programming aspect of model development. However, developing a sound biological content for the models will still be a major undertaking.

24.5 CONCLUSIONS AND RECOMMENDATIONS

1. The state-of-the-science in population biology is sufficient to support the development and use in risk assessment of models that express risks to aquatic and terrestrial biota from pesticide exposure in terms of population-level rather than individual-level endpoints.

2. Age/stage-structured models and metapopulation models can be used to qualitatively describe the influences of specific levels of mortality and reproduction on organisms with different life-history types and distributional patterns. These approaches to modeling are extensively documented in the scientific literature and are widely used in resource management. No further scientific development is needed to support their use.

3. Individual-based models and spatially explicit models can quantitatively describe effects of lethal and sublethal exposures to pesticides for specific target species in specific environments. These approaches are new and there have been few management applications. Implementation of these models requires substantial information concerning (1) the modes of action and environmental fate of the pesticides being assessed, and (2) the behavior, spatial distribution, and population dynamics of the species of interest.

4. Steps that can be taken to implement these models include:
   Broadening the management basis for decision making concerning pesticides. As documented by Troyer and Brody (1994), there are no consistent ecological assessment endpoints within the U.S. Environmental Protection Agency. Visible kills of birds and evidence of toxicity in the laboratory have been used as a basis for decision making but predicted or observed effects on populations have not. Specific population-level assessment endpoints and characteristic assessment scales, consistent with the types of decisions made at each stage of the pesticide assessment process, should be established.
   Developing reference population data sets—Data sets should be developed for representative species and local environments that can be used in the same way as the "reference environments" used in pesticide fate modeling.

These can then be used to parameterize age/stage based and individual-based models.

Demonstrating models through actual applications. Application to a high-profile assessment problem is the best way to demonstrate the value of any assessment methodology. The ideal candidate studies would be special reviews or other assessments expected to involve sensitive ecological resources and high-value pesticides. Such assessments might be expected to take several years to complete and to justify significant expenditure of funds for laboratory and field data. The availability of time and data would, in turn, allow the development of credible models.

Funding research—Individual-based models and spatially explicit models in particular lack a solid scientific foundation and, perhaps even more important, lack a corps of experienced practitioners. The best way to increase both the quality of the science and the number of practitioners is to fund research. Support of graduate and postgraduate fellowships in particular would be a highly cost-effective approach.

Training agency staff—Agency technical staff will not use a new methodology if they are unfamiliar with it and are not confident that it will improve their work. Training classes, and hands-on experience will be necessary.

Acknowledgements. Research sponsored by the U.S. Environmental Protection Agency, Office of Pesticide Programs, under InterAgency Agreement No. 1824-D073-A1 with the U.S. Department of Energy.

24.6 REFERENCES

Andrewartha, H. G., and L. C. Birch, 1954, *The Distribution and Abundance of Animals,* University of Chicago Press, Chicago, IL, USA

Avian Effects Dialogue Group, 1994, *Assessing Pesticide Impacts on Birds,* RESOLVE, Inc., Washington, DC, USA

Barnthouse, L. W., R. V. O'Neill, S. M. Bartell, and G. W. Suter II, 1986, "Population and Ecosystem Theory in Ecological Risk Assessment," in *Aquatic Toxicology and Environmental Fate,* Vol. 9 (T. M. Poston, and R. Purdy, Eds.), ASTM STP 921, American Society for Testing and Materials, Philadelphia, PA, USA, pp. 82–96

Barnthouse, L. W., 1992, "The Role of Models in Ecological Risk Assessment: A 1990's Perspective," *Environ. Toxicol. Chem., 11,* 1751–1760

Barnthouse, L. W., 1993, "Population-Level Effects," in *Ecological Risk Assessment,* (G. W. Suter II, Ed.), Lewis Publishers, Chelsea, MI, USA, pp. 247–274

Barnthouse, L. W., G. W. Suter, II, and A. E. Rosen, 1990, "Risks of Toxic Contaminants to Exploited Fish Population: Influence of Life History, Data Uncertainty, and Exploitation Intensity," *Environ. Toxicol. Chem., 9,* 297–311

Beyer, J. E., and G. C. Laurence, 1980, "A Stochastic Model of Larval Fish Growth," *Ecol. Model., 8,* 109–132

Caswell, H., 1989, *Matrix Population Models: Construction, Analysis, and Interpretation,* Sinauer, Sunderland, MA, USA, 328 pp

Caswell, H., and A. M. John, 1992, "From the Individual to the Population in Demo-graphic Models," in *Individual-Based Models and Approaches in Ecology* (D. L. DeAngelis, and L. J. Gross, Eds.), Chapman and Hall, New York, USA, pp. 36–61

Dale, V. I., and R. H. Gardner, 1987, "Assessing Regional Impacts of Growth Declines Using a Forest Succession Model," *J. Environ. Managem.*, *24*, 83–93

DeAngelis, D. L., L. Godbout, and B. J. Shuter, 1991, "An Individual-Based Approach to Predicting Density-Dependent Compensation in Smallmouth Bass Populations," *Ecol. Model*, *57*, 91–115

denBoer, P. J., 1968, "Spreading of Risk and Stabilization of Animal Numbers," *Acta Biotheoretica*, *18*, 165–194

deKroon, H., A. Plaisier, J. Van Groenendael, and H. Caswell, 1986, "Elasticity: The Relative Contribution of Demographic Parameters to Population Growth Rate," *Ecology*, *67*, 1427–1431

Dunning, J. B., D. J. Stewart, B. J. Danielson, B. R. Noon, T. L. Root, R. H. Lamberson, and E. E. Stevens, 1995, "Spatially Explicit Population Models: Current Forms and Future Uses," *Ecol. Appl.*, *5*, 3–11

Emlen, J. M., 1989, "Terrestrial Population Models for Ecological Risk Assessment: A State-of-the-art Review," *Environ. Toxicol. Chem.*, *8*, 831–842

Emlen, J. M., and E. K. Pikitch, 1989, "Animal Population Dynamics: Identification of Critical Components," *Ecol. Model.*, *44*, 253–273

Ferson, S., 1990, "RAMAS/Stage. Generalized Stage-Based Modeling for Population Dynamics," Applied Biomathematics, Setauket, New York, USA

Ferson, S., and H. R. Akcakaya, 1989, "RAMAS/Age User Manual. Modeling Fluctu-ations in Age-Structured Populations," Applied Biomathematics, Setauket, New York, USA

Foose, T. J., R. C. Lacy, R. Brett, and U. S. Seal, 1992, "Kenya Black Rhinoceros Population and Habitat Viability Assessment," International Union for the Conserva-tion of Nature and Natural Resources, Captive Breeding Specialist Group, Apple Valley, Minnesota, USA

Gurney, W. S. C., E. McCauley, R. M. Nisbet, and W. W. Murdoch, 1990, "The Physiological Ecology of *Daphnia*: A Dynamic Model of Growth and Reproduction," *Ecology*, *71*, 703–715

Hallam, T. G., R. R. Lassiter, J. Li, and W. McKinney, 1990, "Toxicant-Induced Mortality in Models of Daphnia Populations," *Environ. Toxicol. Chem.*, *9*, 597–621

Hanski, I., 1985, "Single-Species Spatial Dynamics May Contribute to Long-Term Rarity and Commonness," *Ecology*, *66*, 335–343

Hanski, I., 1991, "Single-Species Metapopulation Dynamics: Concepts, Models and Observations," *Biol. J. Linnaean Soc.*, *42*, 17–38

Hanski, I., and M. Gilpin, 1991, "Metapopulation Dynamics: A Brief History and Conceptual Domain," *Biol. J. Linnaean Soc.*, *42*, 3–16

Houseknecht, C. R., 1993, "Ecological Risk Assessment Case Study: Special Review of the Granular Formulations of Carbofuran Based on Adverse Effects on Birds. Section 3," in *A Review of Ecological Assessment Case Studies from a Risk Assessment Perspective.* EPA/630/R-92/005, U.S. Environmental Protection Agency, Washington, DC, USA

Huston, M. A., and T. M. Smith, 1987, "Plant Succession: Life History and Competition," *Am. Naturalist*, *130*, 168–198

Jager, H. I., D. L. DeAngelis, M. J. Sale, W. Van Winkle, D. D. Schmoyer, M. J. Sabo, D. J. Orth, and J. A. Lukas, 1993, "An Individual-Based Model for Smallmouth Bass Reproduction and Young-of-Year Dynamics in Streams," *Rivers*, *4*, 91–113

Kooijman, S. A. L. M., and J. A. J. Metz, 1984, "On the Dynamics of Chemically Stressed Populations: The Deduction of Population Consequences from Effects on Individuals," *Ecotoxicol. Environ. Saf.*, *8*, 254–274

Lacy, R. C., and T. W. Clark, 1990, "Population Viability Assessment of the Eastern Barred Bandicoot in Victoria," in *The Management and Conservation of Small Populations* (T. W. Clark, and J. H. Seebeck, Eds.), Chicago Zoological Society, Brookfield, IL, USA, pp. 131–146

Lacy, R. C., 1993, "VORTEX: A Computer Simulation Model for Use in Population Viability Analysis," *Wildlife Res.*, *20*, 45–65

Lacy, R. C., N. R. Flesness, and U. S. Seal, 1989, "Puerto Rican Parrot Population Viability Analysis," International Union for the Conservation of Nature and Natural Resources, Species Survival Commission, Captive Breeding Specialist Group, Apple Valley, Minnesota, USA

Lamberson, R. H., R. McKelvey, B. R. Noon, and C. Voss, 1992, "The Effects of Varying Dispersal Capabilities on the Population Dynamics of the Northern Spotted Owl," *Conservation Biol.*, *6*, 505–512

Lamberson, R. H., R. McKelvey, B. R. Noon, and C. Voss, 1994, "Reserve Design for Territorial Species: The Effects of Patch Size and Spacing on the Viability of the Northern Spotted Owl," *Conservation Biol.*, *8*, 185–195

Lande, R., 1987, "Extinction Thresholds in Demographic Models of Terrestrial Populations," *Am. Naturalist*, *130*, 624–635

Lassiter, R. R., and T. G. Hallam, 1990, "Survival of the Fattest: Implications for Acute Effects of Lipophilic Chemicals on Aquatic Populations," *Environ. Toxicol. Chem.*, *9*, 585–596

Leslie, P. H., 1945, "On the Use of Matrices in Certain Population Mathematics," *Biometrika*, *33*, 183–212

Levins, R., 1966, "The Strategy of Model Building in Population Biology," *Am. Scientist*, *54*, 421–431

Levins, R., 1969, "Some Demographic and Genetic Consequences of Environmental Heterogeneity for Biological Control," *Bull. Entomol. Soc. America*, *15*, 237–240

Lindenmayer, D. B., T. W. Clark, R. C. Lacy, and V. C., Thomas, 1993, "Population Viability Analysis as a Tool in Wildlife Management: A Review with Reference to Australia," *Environ. Management*, *17*, 745–758

Lindenmayer, D. B., and R. C. Lacy, 1995a, "Metapopulation Viability of Leadbeater's Possum, *Gymnobelideus Leadbeateri*, in Fragmented Old-growth Forests," *Ecol. Appl.*, *5*, 164–182

Lindenmayer, D. B., and R. C. Lacy, 1995b, "Metapopulation Viability of Arboreal Marsupials in Fragmented Old-growth Forests: Comparison Among Species," *Ecol. Appl.*, *5*, 183–199

Liu, J., 1993, "An Introduction to ECOLECON: A Spatially Explicit Model for ECOLogical ECONomics of Species Conservation in Complex Forest Landscapes," *Ecol. Model.*, *70*, 63–87

Liu, J., J. B. Dunning, and H. R. Pulliam, "Assessing Alternative Management Strategies: An Example Coupling GIS with Spatially Explicit Models," *Conservation Biol.,* in press

Loza, H. J., W. E. Grant, J. W. Stuth, and T. D. A. Forbes, 1992, "Physiologically Based Landscape Use Model for Large Herbivores," *Ecol. Model., 61,* 227–252

MacArthur, R. H., and E. O. Wilson, 1967, *The Theory of Island Biogeography,* Princeton University Press, Princeton, New Jersey, USA

Madenjian, C. P., and S. R. Carpenter, 1991, "Individual-Based Model for Growth of Young-of-the-Year Walleye: A Piece of the Recruitment Puzzle," *Ecol. Appl., 1,* 268–278

Maguire, L. A., R. C. Lacy, R. J. Begg, and T. W. Clark, 1990, "An Analysis of Alternative Strategies for Recovering the Eastern Barred Bandicoot," in *The Management and Conservation of Small Populations* (T. W. Clark, and J. H. Seebeck, Eds.), Chicago Zoological Society, Brookfield, IL, USA, pp. 147–164

McCauley, E., W. W. Murdoch, R. M. Nisbet, and W. S. C. Gurney, 1990, "The Physiological Ecology of *Daphnia*: A Dynamic Model of Growth and Reproduction," *Ecology, 71,* 716–732

Meyer, J. S., and M. S. Boyce, "Life Historical Consequences of Pesticides and Other Insults to Vital Rates," in *Wildlife Toxicology and Population Modeling* (R. J. Kendall, and T. E. Lacher, Jr., Eds.), Lewis Publishers, Boca Raton, FL, USA

Murdoch, W. W., J. Chessson, and P. Chesson, 1985, "Biological Control in Theory and Practice," *Am. Naturalist, 125,* 344–366

Pulliam, H. R., J. B. Dunning, Jr., and J. Liu, 1992, "Population Dynamics in Complex Landscapes: A Case Study," *Ecol. Appl., 2,* 165–167

Pulliam, H. R., 1994, "Incorporating Concepts from Population and Behavioral Ecology into Models of Exposure to Toxins and Risk Assessment," in *Wildlife Toxicology and Population Modeling* (R. J. Kendall, and T. E. Lacher, Jr., Eds.), Lewis Publishers, Boca Raton, FL, USA, pp. 13–26

Ray, C., and M. E. Gilpin, 1991, "The Effect of Conspecific Attraction on Metapopulation Dynamics," *Biol. J. Linnaean Soc., 42,* 123–134

Rose, K. P., and J. H. Cowan Jr., 1993, "Individual-based Model of Young-of-the-Year Striped Bass Population Dynamics. I. Model Description and Baseline Simulations," *Trans. Am. Fish. Soc., 122,* 415–438

Sabelis, M. W., O. Diekman, and V. A. A. Jansen, 1991, "Metapopulation Persistence Despite Local Extinction: Predator-Prey Patch Models of the Lotka-Volterra Type," *Biol. J. Linnaean Soc., 42,* 267–283

Samuels, W. B., and A. Ladino, 1983, "Calculations of Seabird Population Recovery from Potential Oil-Spills in the Mid-Atlantic Region of the United States," *Ecol. Model., 21,* 63–84

Seal, U. S., and T. J. Foose, 1989, "Javan Rhinoceros Population Viability Analysis and Recommendations," International Union for the Conservation of Nature and Natural Resources, Captive Breeding Specialist Group, Apple Valley, Minnesota, USA

Seal, U. S., and R. C. Lacy, 1989, "Florida Panther Population Viability Analysis," International Union for the Conservation of Nature and Natural Resources, Captive Breeding Specialist Group, Apple Valley, Minnesota, USA

Society of Environmental Toxicology and Chemistry (SETAC), 1994, "Aquatic Dialogue Group: Pesticide Risk Assessment and Mitigation," SETAC Press, Pensacola, FL, USA

Shugart, H. H., 1984, *A Theory of Forest Dynamics*, Springer, New York, USA

Tipton, A. R., R. J. Kendall, J. F. Coyle, and P. F. Scanlon, 1980, "A Model of the Impact of Methyl Parathion Spraying on a Quail Population," *Bull. Environ. Contam. Toxicol.*, *25*, 586–593

Troyer, M. E., and M. S. Brody, 1994, "Managing Ecological Risks at EPA: Issues and Recommendations for Progress," EPA/600/R-94/183, U.S. Environmental Protection Agency, Washington, DC, USA

Turner, M. G., Y. Wu, W. H. Romme, and L. L. Wallace, 1993, "A Landscape Simulation Model of Winter Foraging by Large Ungulates," *Ecol. Model.*, *69*, 163–184

Turner, M. G., Y. Wu, W. H. Romme, L. L. Wallace, and A. Brenkert, "Simulating Winter Interactions among Ungulates, Vegetation, and Fire in Northern Yellowstone Park," *Ecol. Appl.*, *4*, 472–496

Turner, M. G., G. J. Arthaud, R. T. Engstrom, S. J. Hejl, J. Liu, S. Loeb, and K. McKelvey, 1995, "Usefulness of Spatially Explicit Population Models in Land Management," *Ecol. Appl.*, *5*, 12–16

Urban, D. L., and N. J. Cook, 1986, "Hazard Evaluation, Standard Evaluation Procedure, Ecological Risk Assessment," EPA-540/9-85-001, U.S. Environmental Protection Agency, Washington, DC, USA

U.S. Environmental Protection Agency, 1992, "Framework for Ecological Risk Assessment," EPA/630/R-92/001, U.S. Environmental Protection Agency, Washington, DC, USA

van Groenendael, J., H. DeKroon, S. Kalisz, and S. Tuljapurkar, 1994, "Loop Analysis: Evaluating Life History Pathways in Population Projection Matrices," *Ecology*, *75*, 2410–2415

van Tienderen. P. H., 1995, "Life Cycle Trade-Offs in Matrix Population Models," *Ecology*, *76*, 2482–2489

25

Current and Future Test Strategies in Terrestrial Ecotoxicology

Reinhard Debus (Schmallenberg, Germany)

25.1 SUMMARY

This article describes the current approaches for hazard assessments of substances in the terrestrial environment. Substance-related evaluations of environmental hazard potentials and evaluations of contaminated soils have to be based on different strategies. The approaches referring to soils and substance-related approaches are therefore presented in separate sections.

The section "Evaluation of Substances" gives an example related to the German Chemical Law indicating the data required for assessments of the terrestrial environment. Contrary to aquatic compartments, important representative organisms are still unconsidered for the terrestrial environment. The problem in which additional soil-living organisms should be taken into consideration in the scope of substance testing is presently discussed on an international level (e.g. within the OECD).

National and EU-wide discussions are presented with emphasis on attempts to find out to what extent tests with aquatic organisms in aqueous soil extracts are suitable for soil assessments with respect to the ecological soil functions. Furthermore, test strategies for the determination of effects potentials in soils that enable a quality control of soils and substrates are presented. They are based on data obtained from chemical and ecotoxicological analyses.

Ecotoxicology, Edited by Gerrit Schüürmann and Bernd Markert.
ISBN 0-471-17644-3   © 1998 John Wiley & Sons, Inc. and Spektrum Akademischer Verlag.

The topic "substance-related soil assessment" describes procedures aiming at the determination of substance concentrations at which a risk for soil-living organisms is not expected. The assessment of effects potentials should include the aspects of land use, the object to be protected and the exposure.

25.2 TERRESTRIAL ECOTOXICOLOGY

The following presentations describe approaches in terrestrial ecotoxicology as presently practised and discussed in Germany and the European Union (EU). Emphasis is put on the description of current practice oriented according to German laws. International relevance is assured in so far as most methods and strategies have been the subject of international discussions.

As in the aquatic compartment, one of the main tasks in terrestrial ecotoxicology is the (prospective) evaluation of substances. Tests of terrestrial organisms are required, for example, in stage 1 within the framework of the German chemical law. These terrestrial tests are the earthworm test and the plant test. Besides the evaluation of substances, terrestrial ecotoxicology is also of interest in the context of the draft of the Federal Soil Conservation Law (Bundes-Bodenschutzgesetz, BBodSchG, 1995). Additional strategies are necessary here to derive soil test values and to access soil qualities. Various methods are in use for investigating what soil bioremediation is necessary and for documenting the bioremediation success (e.g., LfU 1994). Organizations such as DECHEMA (1995) have prepared instructions on how to go about determining soil quality depending on soil use which combine both practicability and effectiveness. However, compared with the procedure for the aquatic compartment there are still no uniform and generally recognized strategies. The fact that such strategies are now being demanded is a consequence of the change in the law. The lack of standards and methods is, on the one hand, a result of the new situation; on the other hand, it is due to the complex matrix of soil being difficult to access with chemical and biological tests.

In the soil there are a number of very different exposure paths, such as pore water, soil air, contact with contaminated particles, as well as the uptake of polluted foods, or those which are attached to polluted soil components. It is difficult to quantify contamination via the above exposure paths, because the bio–available amount of a contaminating substance often cannot be adequately determined compared to the total amount. To minimize this difficulty the soil analysis strategies are ecotoxicological as well as chemical; the latter were the common strategies up till now. Moreover, extraction methods and agents are being developed and tested which will be able to detect as accurately as possible the available and consequently the active amount of substances. In the following, the current strategies for evaluating substances, the procedure for obtaining soil limiting values and the present methods for evaluating soil quality are described.

25.3 EVALUATION OF SUBSTANCES

Terrestrial tests are required within the framework of the application procedure in stage 1 according to German chemical law. According to Rudolph and Boje (1986) data from the initial stage are suitable for evaluating the environmental presence of substances and any suspicion of the possibility of environmental pollution. In such an evaluation physicochemical data, data on the mobility and distribution as well as on degradation and test results of the toxicity to aquatic organisms are referred to. If this is confirmed, the determination of NOECs will include tests on terrestrial organisms in addition to long-term tests with fish and daphnids (Table 25.1). These terrestrial tests are the plant test according to the OECD guideline 208, and the earthworm test according to the OECD guideline 207.

Besides other higher plants, oats (*Avena sativa*) and a turnip variety (*Brassica rapa*) are also tested. In the test a substance is applied once to a defined soil and the effect on the early stages of growth is investigated. The EC_{50} is determined after 14 days with respect to growth inhibition of the biomass above ground.

The test organism *Eisenia fetida* is used for determining earthworm toxicity. The lethal effect (LC_{50}) of a substance applied once to the test soil is investigated after 14 days.

It is generally agreed that this test set is incomplete for evaluating the ecotoxicity of substances in the terrestrial compartment. For instance, certain groups of organisms are not considered even though they are involved in important processes in the soil. During an international workshop on the *Environmental Assessment of Pesticides* (17 to 19 March 1993, Fraunhofer Institute, Schmallenberg) a test set of organism groups was worked out by officials and presented to the OECD. This test set comprised birds, beetles, spiders, amphibians, Collembola, Nematodes, earthworms and higher plants.

It will not be possible in the near future to establish ecotoxicological tests with all these organisms for testing substances. However, it is now generally accepted that organisms which, to a large extent, support and maintain important soil processes must be considered within a test strategy. Among these are, for example, members of the meiofauna (GDCh 1993). Within the working group for the *Further Development of Ecological Test Methods*, managed by the Federal Ministry for Environmental Protection, criteria are currently being discussed which should be applied when selecting test organisms belonging to the meiofauna. These selection criteria include the trophic level, the relationship to functional groups, the main exposure path, the representivity for different terrestrial ecosystems as well as the cultivateability and the practicability when testing.

The objective is to identify an additional test organism. Possible candidates are members of the organism groups or classes of the Collembola, Enchytraea and Nematodes. It is anticipated that a Collembola test will soon be established as an ISO standard.

Table 25.1 Ecotoxicological tests of the Initial Stage and Stage 1 according to chemical law (modified after Rudolph and Boje 1986)

Test method	Test parameter	Test organism	Test result
Initial stage			
Acute toxicity to a fish species	Lethal effect of substance after a single application of 48 to 96 hours	Blue-striped zebra fish *Brachydanio rerio* secondary consumer	LC_{50}
Acute toxicity to a species of water flea	Inhibition of swimming ability after a single application of 24 to 48 hours	Large water flea *Daphnia magna* primary consumer	EC_{50}
Algae toxicity	Inhibition of cell multiplication (growth) after a single application of 72 hours	Algae *Scenedesmus subspicatus* primary producer	EC_{50}
Stage 1			
Long-term daphnid toxicity	Determination of the no-observed-effect concentration and the threshold concentration with respect to inhibition of reproduction and lethal effect after repeated (semistatic) or continuous (flow-through system) application for at least 21 days	Large water flea *Daphnia magna* primary consumer	NOEC
Long-term fish toxicity	Determination of the no-observed-effect concentration and the threshold concentration with respect to the lethal and sublethal effect after repeated (semistatic) or continuous (flow-through system) application for 14 to 28 days	Blue-striped zebra fish *Brachydanio rerio* secondary consumer	NOEC

Table 25.1 (*Continued*)

Test method	Test parameter	Test organism	Test result
Plant toxicity	Inhibition of seed growth (reduction of biomass) for 14 days after a single application	Oats *Avena sativa* Turnip *Brassica rapa* primary producer	EC_{50}
Earthworm toxicity	Lethal effect of substance 14 days after a single application	Earthworm *Eisenia fetida* secondary destruent	LC_{50}

Stage 2
Separate substance-specific test programme

Collembola are not only directly involved in the creation of humus from litter, fungi and soil algae, but also in the degradation of excrement from larger animals. They therefore make an important initial contribution to the subsequent decomposition processes. Furthermore, they play an important part in breaking up the soil. Collembola occur predominantly in soils with a high C/N content, which is usually the case in forest soils (Kampmann and Funke 1987). In cultivated land Collembola occur more often when the ecological balance is not intact (Spahr 1981). Compared to other arthropods, such as mites, Collembola have, owing to their thin cuticula, close contact with substances adsorbed to soil particles and dissolved in soil water. Food is the main exposure path for Collembola because they live mainly in the air-filled pores of the upper soil.

Conversely, Nematodes live in the water-filled soil pores. Their entire body surfaces are therefore exposed to dissolved and potentially available contaminants. Nematodes are most abundant particularly in agricultural soils. The class of Nematodes includes different feeding types, such as predators, bacteria-, fungi-, and algae feeders, plant- and zooparasites as well as saprophytes. With respect to handling, certain species of Nematodes prove to be easy to breed. In addition to determining the mortality, the inhibition of development and growth during juvenile phases are parameters which are easy to assess.

Like earthworms the Enchytraea belong to the order of Oligochaeta. Terrestrial species occur mostly in soils that are not too dry. Enchytraea are saprophytes that consume not only litter but also soil particles and, depending on the species, excrement from other animals. The exposure of these organisms is thus through the uptake of contaminated food and through the body surface (Römbke 1995).

In addition to the demands for further ecotoxicological tests with terrestrial organisms, more extensive recommendations were formulated at the workshop on *Soil biology and pollutants—Derivation of soil values* (Sept. 1994, Ministry for Environmental Protection, Berlin):

"In order to perform a comprehensive and practicable substance evaluation it is necessary to clarify the details of exposure, because in the soil considerable differences frequently exist between the analytically measured contamination and the exposure. By exposure we mean the behaviour of the substance in the soil with respect to time as well as spatially, i.e., its application, transformation, degradation and (bio)availability for the various soil organisms and plants."

Moreover, it is necessary to develop, similar to substance evaluation in the aquatic compartment, a tiered concept with clearly defined decision criteria.

25.4 SUBSTANCE-RELATED EVALUATION AND LIMITING VALUES FOR SOILS

By substance-related soil evaluation we mean the determination of those concentrations of a substance at which the risk for organisms living in the soil can generally be neglected. Ideally this evaluation should refer to the preservation of the structure and function of the animal and plant organisms including microorganisms. However, according to the draft of the Federal Soil Conservation Law (1995) the emphasis must be on the protection of soil functions. The protection of soil organisms is only significant in so far as they support the soil function. Consequently individuals or species of organisms are not normally the primary aim of protection. The decisive point is whether the function, and of secondary importance, the structure is maintained within acceptable limits of the natural status.

Substance-related evaluation is necessary when estimating the risk of contaminated sites. The reference for this evaluation is pollutant concentrations which lead to action being taken if certain threshold values are exceeded. At present, however, there are only few generally agreed threshold concentrations for contaminants which—if exceeded—result in action such as a change of use, protective measures or bioremediation. These threshold concentrations are those below which the ecological soil functions stay intact. Consequently, criteria, strategies and guidance documents must be referred to that were introduced for other purposes. The limited amount of data available for the complex soil compartment is probably mainly a result of there being only three tests available as guidelines. It is not only necessary to establish further tests with organisms living in the soil, but also to create test strategies and the assessment of the results such as those that exist for the aquatic compartment. During the workshop on *Soil biology and pollutants—Derivation of soil values* (Sept. 1994, Ministry for Environmental Protection, Berlin) the following was unanimously agreed:

- "A total of 5 different species should be tested to derive test values: At least 3 single species tests should be carried out on terrestrial organisms of different trophic levels supplemented by 2 functional parameters."
- Consideration of the exposure is a prerequisite for chemical evaluation.
- A tiered concept with defined decision criteria should be developed for evaluating substances and preparing soil test values which are independent of the location.

In general the potential effect of a substance should be evaluated in relation to land use, the object to be protected and exposure. Exposure in particular should be considered closely. According to Samsoe-Petersen and Petersen (1995) organisms living in the soil can be contaminated by the soil solution, soil air, oral uptake of contaminated food and particles as well as through direct contact with the soil. Thus it is necessary to determine the specific exposure paths to enable an exposure scenario to be defined that fits the object to be protected. Substance evaluation methods as laid down in chemical and pesticide laws are used for elaborating effect data. For generating LC (lethal concentration) and EC (concentration causing effects) values not only standardized tests are used.

It is possible to make a risk evaluation using extrapolation techniques, for instance. The Environmental Protection Agency in the USA (EPA 1984), for example, applies different safety factors depending on the data available (Table 25.2). Using this approach a 'concern level' is determined that indicates the substance concentration at which free-living organism populations are affected. For this purpose it is possible to use data (NOEC values and acute data EC/LC_{50} values) from all tests, single species as well as field tests, and also results of QSAR estimates (Quantitative Structure Effect Relationships). Safety factors are assigned depending on the amount and quality of data. The concentrations determined in this way are then compared with the exposure data. The aim is to identify substances that must be subjected to further ecotoxicological testing. This is particularly important when the Predicted Environmental Concentration (PEC) exceeds the concern level, i.e., it is above the Predicted Concentration of No Environmental Concern (PNEC).

Table 25.2 Assignment of safety factors according to the data available (EPA, 1984)

Data selection		Safety factor
≥ 3 NOEC values	10	on the lowest NOEC value
≥ 3 LC_{50}/EC_{50} values from species of different taxonomic groups	100	on the lowest LC_{50}/EC_{50} value
≥ 5 LC_{50}/EC_{50} values for members of two taxonomic groups	100	on the lowest LC_{50}/EC_{50} value
1 LC_{50}/EC_{50} values	1000	
LC_{50} from QSAR estimate	1000	

This type of approach aims at, for example, the determination of test values and levels for taking action. By test values we mean the substance concentrations in the soil related to land use and objectives of protection which if exceeded result in a risk evaluation of the site. When the level to take action is exceeded, measures are initiated to combat the risk associated with a contaminated site. Such measures include changing the land use, protective measures and remediation. It must be assumed in this context that test values and levels for taking action are to be defined related to the land use, the exposure and the object to be protected for every substance that comes under consideration (Bachmann 1994).

To evaluate the environmental hazard potential of substances to the soil, strategies are needed which are based on a tiered procedure containing decision making criteria. A proposal of a three-stage system is shown in Table 25.3.

The highest test level in this proposed system is stage 3 in which model ecosystems are requested. Among others van Straalen and Gestel (1993) argue the case for using terrestrial microcosms not only to detect the effect of substances under competitive conditions, but also to be able to prove a possible increased sensitivity owing to interactions between species. The sometimes very complex species structure in the soil, however, necessitates the investigation of representative organisms that belong to different trophic levels.

There is still a need to develop the framework for terrestrial microcosms because, as opposed to the aquatic compartment, little agreement exists about its design and the test parameters. For instance, details are necessary on the size of microcosm and type of filling, either with intact soil columns or disturbed soil, the end points and the application of the substances. Besides investigating the effect of the applied substances it is possible, if intact soil columns are used, to determine the spatial distribution over time of the substance including the point of entry into the leachate. With respect to the effect of the substances the test program must include pragmatically selected structure and function parameters. In addition to determining the structure for testing the habitat function using selected representative organisms, effects must be investigated regarding other ecological soil functions. These include the regulatory functions especially, such as degradation, buffer and filter capacity. When collating results to determine the habitat function it is not practicable to prepare species lists that are difficult to interpret. It is of much greater importance to quantify the effects on representative populations and species with respect to defined end points. Initial work has been carried out on this topic, for example by Debus and Hund (1995).

25.5 EVALUATION OF SOIL QUALITY

According to the draft of the Federal Soil Conservation Law under clause 5 (3) (BBodSchG, 1995): persons responsible for land, and landowners as well as users, are obliged to remediate the soil and contaminated sites so that

Table 25.3 Strategy for determining the environmental hazard potential of substances (modified after Römbke et al. 1995):

Tests recommended for stage 1

Prerequisite:	Data from PC tests as well as data about exposure path and environmental concentration
	Degradation in soil
Distribution tests:	Adsorption/desorption
	? Plant bioaccumulation
Effect tests:	Dehydrogenase activity
	Plant test
	Earthworm reproduction

Tests recommended for stage 2

Prerequisite:	Data from stage 1
Distribution tests:	Leaching potential
	? Volatility
	? Animal bioaccumulation
Effect tests:	Collembola reproduction
	Staphylinidae generation
	? Plant generation
	? Microflora short-term respiration
	? Enchytraea reproduction
	? Acute Carabidae
	? Carabidae larvae

Tests recommended for stage 3

Prerequisite:	Data from stages 1 and 2
Tests:	Terrestrial model ecosystems

? = Possible alternative test or supplement

individuals or the general public are not exposed to lasting risks, significant disadvantages or pollution. Therefore it is necessary to perform a quality control on soil and substrates after bioremediation, for example in a bioremediation facility, if it is intended for landfilling.

Implementation regulations regarding the draft of the Federal Soil Conservation Law are still in preparation, so that at present there are no uniform concepts and methods for evaluating contaminated and cleaned soil. It is now generally recognized that a characterization based solely on chemical analysis is inadequate for making a quality control of the soil and estimating the potential risk.

Tests for soil pollution are frequently based on the determination of the "total contents," with substances being analysed that are expected to occur in the soil. Subsequently a comparison is made with toxicological data. This procedure is the basis for estimating the risk to man which is presented by a polluted area. Misinterpretations occur as a result of it being fundamentally impossible to analyse the entire pool of pollutants in the soil, including the metabolites and degradation products that have been formed. Furthermore some doubt exists as to whether estimates of the potential risk of soil contamination based on human toxicological criteria are conservative for ecological soil functions. Regarding the organisms living in the soil, not only the total contents but also the mobile and available portions of substances should be measured. Initial work on extraction methods to achieve this has been completed. If these methods prove successful for determining the (bio)available parts of a contamination, then soil-independent results will be available for estimating the hazard potential, because soil-specific sorption characteristics have already been determined. The ratio of total content of a substance to its available portion allows conclusions about the mobility of pollutants (Kördel 1995).

It has been found that ecotoxicological tests are advantageous, as they react with the available portions of the substances, i.e., those parts which cause the effects. Moreover, these tests integrate the effects over the entire available portions of substances. This ensures detection of synergistic effects on the one hand, and of the effects of substances on the other, such as unknown degradation products, not recognized in chemical analysis. According to DECHEMA (1995), ecotoxicological tests are necessary or useful in respect to the following aspects amongst others:

- Supplementary risk assessment of contaminated soil,
- Assessment of the extractability of contaminants with biological effects in cases where the soil/refill can affect the groundwater (particularly uncovered areas and deep refills),
- Screening of biologically effective substances which were not detected or characterized by the chemical analysis.

At present, ecotoxicological tests with soil extracts, mainly water extracts, are a practicable way of proving the toxic potential of soils. Aquatic test organisms are also used in this respect. When these test data are applied to the soil, it is assumed that organisms living in the water have a sensitivity similar to that of organisms living in the soil. This should be the case especially when the organisms live in the pore water of the soil. In this case, the same exposure conditions can be assumed.

Figure 25.1 shows the basic strategy for estimating the environmental hazard potential that can originate from contaminated soil, from soil after bioremediation, as well as from substrates coming out of a remediation facility. From these soils and substrates, suitable extracts should be prepared, subjected to chemical

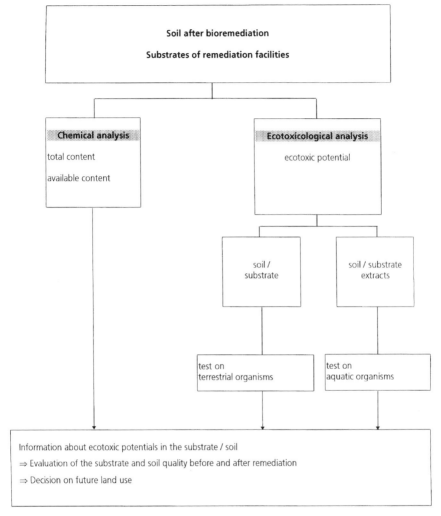

Figure 25.1 Test strategy for chemical/ecotoxicological evaluation of soils and substrates (Quality control).

analysis and examined in various ecotoxicological tests. In addition, tests should be directly carried out in the soil and substrate. Terrestrial tests provide important information concerning the reusability and landfill possibilities of the substrates.

The following points summarize the above:

1. For evaluating soils and substrates ecotoxicological tests provide
   - in aqueous soil extracts information on:

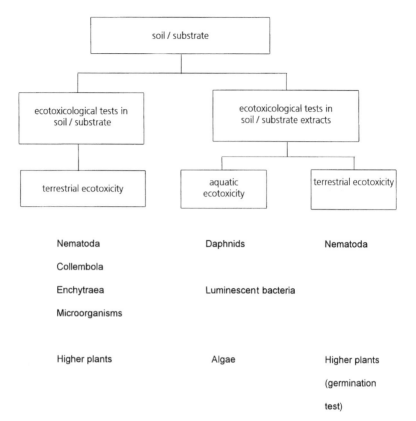

Figure 25.2. Test organisms suitable for determining ecotoxic potential in soils and substrates.

  – the existence of toxic substances
  – the retention capacity of soil/substrate or the mobility of the pollutants
• in terrestrial test systems information on:
  – the habitat function of the soil/substrate.
2. Ecotoxicological tests are used for:
  – clarifying the need for remediation of an area
  – controlling the (microbiological) remediation process
  – controlling the results of remediation
  – helping to decide how to use the soil/substrate after remediation.

Figure 25.2 shows ecotoxicological tests that can be applied to the ecotoxic potential of soils and substrates. Aquatic and terrestrial organisms are used to:

• detect ecotoxic potential of soils and substrates
• provide information about the habitat function of the soil and substrate
• estimate the leaching from the substrate.

Present findings indicate that ecological tests, even those with aquatic organisms, are suitable for identifying ecotoxic potentials in soils. This means that effect-related ecotoxicological analyses can supplement chemical analyses. Results of aquatic tests provide further information as to the extent to which substances leaching from soils can present a danger to aquatic organisms, and whether pollution of the ground water is to be expected. However, if no effect is observed in the extract it is still possible that a toxic potential exists in the soil. This is the case when sparingly soluble substances are present in the soil that can only be extracted in concentrations below the level at which they have an effect (Debus and Hund 1995). For this reason it is necessary to carry out tests directly in the soil or substrate in addition to ecotoxicological tests in extracts.

25.6 REFERENCES

Bachmann, G., 1994, "Vordringliche Anforderungen des Bodenschutzes an die Bewertung bestehender Schadstoffkonzentrationen in Böden—Thesen [Priority Requirements of Soil Protection for the Assessment of Existing Pollutant Concentrations in Soils—Theses]," Workshop-Dokument Bodenbiologie und Schadstoffe—Ableitung von Bodenwerten, 27.09.1994 im Umweltbundesamt Berlin, Germany

Bundes-Bodenschutzgesetz, 1995, Gesetz zum Schutz vor schädlichen Bodenveränderungen und zur Sanierung von Altlasten [Law on the Protection of Soils against Adverse Soil Alterations and on the Remediation of Hazardous Wastes], Referentenentwurf vom 18.08.1995, Bonn, Germany

Debus, R., and K. Hund, 1995, "Entwicklung analytischer Methoden zur Erfassung biologisch relevanter Belastungen von Böden—Ökotoxikologische Analytik im Boden und in Extrakten [Elaboration of Analytical Methods to Determine Biologically Relevant Impacts on Soils—Ecotoxicological Analyses in Soil and Soil Extracts]," in Statusseminar zum Förderschwerpunkt Ökotoxikologie des BMBF (M. Kirchner, and H. Bauer, Eds.), Forschungsbericht des Projektträgers 2/95, pp. 215–225, München, Germany

DECHEMA, 1995, "Bioassays for Soils. 4," Report of the Interdisciplinary Committee "Environmental Biotechnology-Soil". Ad-hoc-Committee "Methods for Toxicological/Ecotoxicological Assessment of Soils," Frankfurt, Germany

Environmental Protection Agency (U.S. EPA), 1984, Estimating Concern Levels for Concentrations of Chemical Substances in the Environment, Washington DC, USA

GDCh (Gesellschaft Deutscher Chemiker), 1993, "Grundsätze der ökotoxikologischen Bewertung von Chemikalien und Sachstand der heutigen Praxis [Principles in the Ecotoxicological Assessment of Chemicals and State of the Art in Today's Practice]," Positionspapier des Arbeitskreises Chemikalienbewertung in der Fachgruppe Umweltchemie und Ökotoxikologie, Frankfurt, Germany

Kampmann, S., and R. Funke, 1987, "Epigäische Collembolen mitteleuropäischer Wälder [Epigeal collembola of Central European forests]," *Verh. Ges. Ökologie, 15,* 341–350

Kördel, W., 1995, "Erfassung biologisch relevanter Bodenbelastungen [Determination of biologically relevant soil impacts]," *Mitteilungsblatt der Gesellschaft Deutscher Chemiker, Fachgruppe Umweltchemie und Ökotoxikologie*, 2/95, Frankfurt, Germany, pp. 4–10

LfU, 1994, *Handbuch Altlasten und Grundwasserschadensfälle Altlastenerkundung mit biologischen Methoden* [Handbook on Hazardous Wastes and Cases of Damage to Groundwater, Investigation of Contaminated Sites with Biological Methods], Band 13, Landesanstalt für Umweltschutz, Baden-Württemberg; Karlsruhe, Germany

Römbke, J., 1994, "Enchytraeen (Oligochaeta) als Bioindikatoren [Enchytraea (Oligochaeta) as bioindicators]," *UWSF-Z. Umweltchem. Ökotox.*, 7, 246–249

Römbke, J., C. Bauer, and A. Marschner, 1996, "Entwicklung einer Teststrategie zur Bewertung des Umweltgefährlichkeitspotentials von Umweltchemikalien im Kompartiment Boden [Development of a Test Strategy to Assess the Environmental Hazard Potential of Environmental Chemicals in the Compartment Soil]," *UWSF-Z. Umweltchem. Ökotox.*, 8, 158–166

Rudolph, P., and R. Boje, 1986, *Ökotoxikologie, Grundlagen für die ökotoxikologische Bewertung von Umweltchemikalien nach dem Chemikaliengesetz* [Ecotoxicology, Principles for the Ecotoxicological Assessment of Environmental Chemicals according to the German Chemical Law], ecomed-Verlag, Landsberg, München, Germany

Samsoe-Petersen, L., and F. Petersen, 1994, "Regarding Guidance for Terrestrial Effects Assessment," prepared for OECD, Water Quality Institute, Denmark

Spahr, H. J., 1981, "Die bodenbiologische Bedeutung von Collembolen und ihre Eignung als Testorganismen für die Ökotoxikologie [The Soil Biological Importance of Collembola and their Suitability as Test Organisms in Ecotoxicology]," *Mitt. dtsch. Ges. allg. angew. Ent.*, 3, 141

van Straalen, N. M., and C. A. M. van Gestel, 1993, "Soil Invertebrates and Microorganisms," in *Handbook of Ecotoxicology*, Vol. 1 (P. Calow, Ed.), Blackwell, Oxford, UK, pp. 251–277

26

Alternative Assays for Routine Toxicity Assessments: A Review

Colin Janssen (Ghent, Belgium)

26.1 SUMMARY

The number of available aquatic toxicity tests is large and increases every year. For certain types of applications such as the routine screening of large numbers of environmental samples, standard conventional tests may be perceived as being unpractical. This has lead to the development of a (increasing) number of alternative small-scale toxicity tests also called microbiotests. The present paper reviews the potential use and limitations of a number of these alternative assays. For bacteria, algae and invertebrates, a brief description of conventional assays is given followed by an analysis of the sensitivity, reproducibility, cost-efficiency and application potential of freshwater microbiotests.

Ecotoxicology, Edited by Gerrit Schüürmann and Bernd Markert.
ISBN 0-471-17644-3   © 1998 John Wiley & Sons, Inc. and Spektrum Akademischer Verlag.

In addition, several case studies are discussed in which alternative toxicity tests were evaluated as routine environmental monitoring tools.

From this review it can be concluded that microbiotests have an important role to play in the screening and ranking of the hazards of chemicals and environmental wastes. However, except for a few bacterial and invertebrate tests, there is a lack of published information on the various aspects of test development and especially on the application of these alternative assays. For most microbiotests, further research and test evaluation is needed before these methods will gain widespread acceptance as valid alternatives to the currently used 'conventional' test procedures.

26.2 INTRODUCTION

Since the 1960s and especially during the last decade, the development and application of aquatic toxicity tests has increased significantly. As there are literally hundreds of toxicity test methods described in the literature, the selection of the *best* test system might be conceived as looking for the proverbial *needle in the haystack*. Although it might be argued that most of these tests can provide useful information on the toxicity of chemicals for the environment, important aspects are often overlooked when selecting toxicity tests and/or developing new toxicity assessment methods. First, when using ecotoxicity tests one should keep the (envisaged) purpose in mind. Toxicity tests for screening purposes, regulatory requirements, and predictive hazard assessment each have a different set of requirements as to test precision, test organism choice, exposure time and cost (Persoone and Janssen 1994). Second, the Rs of ecotoxicity tests: relevance (ecological), reproducibility, reliability, robustness and repeatability/ sensitivity should be established within the context of the purpose of the method (Calow 1993). For an in-depth discussion on the prerequisites of a *good* toxicity test the readers are referred to Tebo (1985) and Cairns and Pratt (1989).

Numerous authors have argued that conventional aquatic toxicity tests, like the fish and cladoceran assays used in various regulatory frameworks, are unpractical for routine environmental screening (a.o. Blaise et al. 1986; Blaise 1991; Dutka et al. 1991; Persoone 1991; Willemsen et al. 1995). The main criticisms cited are that (1) the culturing of and testing with these standard species is expensive and laborious, and (2) the tests are time- and space-consuming. Indeed, for screening purposes, toxicity tests must be sufficiently simple and standardized that they can be carried out routinely by laboratories (governmental, academic and private) with widely varying capabilities.

In view of these drawbacks of conventional assays, research efforts have been directed towards the development of alternative test systems. The development of toxicity tests based on biochemical, physiological and histological test criteria—generally referred to as biomarkers—is an important research area which addresses this issue. An in-depth discussion of the recent advances in the development of these types of alternative test systems is, however, beyond the scope of the present paper. Comprehensive reviews on the use of biomarkers for the assessment of environmental contamination are given in Hugget et al. (1992) and Peakall et al. (1992).

Another research activity which has increased significantly in recent years is the development and application of alternative small-scale toxicity tests—also termed microbiotests*—for the routine screening of chemicals, effluents and other environmental wastes (Blaise 1991; Persoone 1992). In general, these assays use small-sized test species and are designed to be relatively rapid, simple, low-cost and often do not require culturing of organisms. Because of these characteristics, microbiotests are well suited for incorporation into a (multitrophic) test battery for the routine screening of environmental contamination.

The aim of this contribution is to evaluate the potential use of a number of alternative, freshwater toxicity assays. For each group of test organisms considered in this paper, a brief description will be given of the conventional assays followed by detailed analysis of a number of alternative microbiotests. This review is limited to toxicity tests with bacteria, algae and invertebrates; assays with fish will not be discussed. Finally, some case studies which illustrate the applications of some of these alternative test systems will be presented. This review is not meant to be exhaustive; it should, however, give insight into the recent developments and application of these types of assays and provide some guidance on their potential use in environmental screening programs.

26.3 MICROBIAL TEST SYSTEMS

26.3.1 General

Various microbial tests, using different test criteria and species, have been proposed. The most commonly applied methods—proposed by several national/international standardization and/or regulatory bodies—use the growth rates, biomass or number of cells of (pure or mixed) bacterial cultures as toxicity assessment endpoints. Typically, bacteria are inoculated under appropriate growth conditions into a nutrient medium, the toxicant is added at various concentrations, the flasks are incubated and at regular intervals samples are removed for the (spectrophotometric or turbidimetric) determination of biomass. Frequently used test organisms with this procedure are *Pseudomonas fluorescens* and *P. putida* (Dutka and Kwan 1982; ISO 1995).

Next to the direct measurement of toxicity by growth rate or cell viability, various microbial test methods are based on the assessment of biochemical properties of the bacteria. These biochemical indicators of toxic insults include the ATP concentration, the ATPase activity, the dehydrogenase activity and the

---

* Definition of a microbiotest: According to Blaise (1991) a microbiotest can be defined as the exposure of a unicellular or small multicellular organism to a liquid sample in order to measure a specific effect. Willemsen et al. (1995) specified the term microbiotest as an assay: (1) requiring a test volume of less than 100 ml, (2) preferably available in kit-form, (3) with an exposure period < 24 h and (4) which can be performed with standard laboratory equipment.

activity of other enzymes, such as phosphatases, urease and esterases. Changes in the ATP concentration in various species, such as *P. aeruginosa, Enterobacter aerogenes, Escherichia coli* and *Nitrosomonas europaea* exposed to toxic stress, have been used extensively as a toxicity test criterion (Parker and Pribyl 1984; Denyer 1989; Stanley 1989). Another frequently used sublethal endpoint is the dehydrogenase enzyme activity which can be easily quantified with tetrazolium salts. The use of these reduction indicators in bacterial toxicity assays has been demonstrated by Packard et al. (1983), Trevors (1984 and 1986), and Bitton and Koopman (1986). The application of resazurin, an oxido-reduction indicator, for rapid toxicity assessment with activated sludge and bacterial cultures has been reported by Liu (1981) and Liu and Thomson (1983). The quantification of the reduction of biochemical activity due to toxic stress is, for most of the assays described above, performed with the aid of a luminometer, spectrophotometer or fluorometer. The convenience of measurement together with the rapid response times of these organisms (minutes to hours) have made these bacterial assays attractive tools for toxicity screening purposes. Indeed, several bacterial tests have recently become commercially available (see below).

Bacterial assays based on the inhibition of the respiratory activity have also been applied for the assessment of toxic insult in activated sludge samples. A simple version of this type of respirometric assay is a modification of the 5-day BOD test in which 5 ml of sewage (approximately 1000 microorganisms/ml), together with a nutrient solution, is added to dilutions of the test substance. By measuring the respiratory activity of the bacteria at the beginning and at the end of the 5-day test period, the effect of the toxicant on respiration can then be calculated (King and Dutka 1986). A behavioral assay based on the changes or inhibition of the motility of a large bacterium (*Spirillum volutans*) was described by Kreig et al. (1967), Dutka (1980), and others.

26.3.2 Microbiotests

A summary of important aspects of various bacterial microbiotests is given in Table 26.1. The Microtox® (Microbics, Carlsbat, USA) bioluminescence test with the marine bacterium *Photobacterium phosphoreum* (renamed *Vibrio fisheri*) is probably the most widely applied bacterial microbiotest (Mayfield 1994). This assay, which is based on the bacteria's ability to emit light via a reaction involving the luciferase enzyme coupled to respiration (via NADH and a flavine nucleotide), measures the inhibition of the metabolic rate of the organism under toxic stress. One of the reasons for the widespread application of this assay is the availability of the bacteria in freeze-dried form, which eliminates the need for culturing the test organisms (Bulich and Greene 1979; Bulich and Isenberg 1980; Bulich 1982). The operational procedure for the Microtox assay can be summarized as follows: (1) reconstitution (rehydration) of the lyophilized bacteria (5 minutes), (2) measurement of the light output of the bacteria in the Microtox apparatus (photometer), (3) exposure of the bacteria to a dilution series of the sample for a period of 5, 15 or 30 minutes, (4) measurement of the light output of

Table 26.1 Overview of microbiotests with bacteria, algae and invertebrates

Test organism/Assay name	Test criterion/measurement	Exposure	Bioindicator material	Reference
Bacteria				
Commercially available assays				
Photobacterium phosphoreum	enzymatic activity bioluminescence	5–30 min	lyophilized	Bulich et al. 1981
Microtox				
Microtox Solid Phase				
Escherichia coli	β-galactosidase activity/ colorimetric	60 min	lyophilized	Reinhartz et al. 1987
Toxi-chromotest				
Sediment chromotest				
Escherichia coli	β-galactosidase activity/ colorimetric	2 hours	lyophilized (mutant)	Bitton et al. 1992, 1994
METpad				
METplate				
Bacillus spec.	dehydrogenase activity/ colorimetric	24 hours	impregnated dipstick	Dutka and Gorie 1988
ECHA biomonitor				
Other assays				
Vibrio harveyi	enzymatic activity bioluminescence	60 min	culturing required	Thomulka et al. 1992
Escherichia coli (or other bacterial species)	ATP concentration luminescence	5 hours	culturing required	Xu and Dutka 1987
Bacillus subtilis	size of bacteria laser technique	66 min	lyophilized	Felkner et al. 1988

Table 26.1 (Continued)

Test organism/Assay name	Test criterion/measurement	Exposure	Bioindicator material	Reference
Algae				
Commercially available assays				
Selenastrum capricornutum Algotoxkit F	growth inhibition absorbance	72 hours	immobilized algal cells	Persoone 1996
Other assays				
Selenastrum capricornutum Microplate assay	growth inhibition cell count or absorbance	72–96 hours	culturing required	Blaise et al. 1986
	ATP concentration luminescence	4 hours	culturing required	
Invertebrates				
Commercially available assays				
Brachionus calyciflorus Rotoxkit F	mortality visual	24 hours	resting eggs (cysts)	Snell et al. 1991a
Thamnocephalus platyurus Thamnotoxkit F	mortality visual	24 hours	resting eggs (cysts)	Centeno et al. 1995
Daphnia magna (and D. pulex) Daphtoxkit F	mortality visual	24–48 hours	resting eggs (ephippia)	Persoone 1996
Daphnia magna (and other invertebrates)	enzymatic inhibition/ feeding activity visual or fluorometer	60 min	culturing required	Janssen and Persoone 1993; Hayes et al. 1993
Other assays				
Brachionus calyciflorus	feeding activity esterase activity	60 min	resting eggs (cysts)	Juchelka and Snell 1994
		60 min	resting eggs (cysts)	Burbank and Snell 1995
Ceriodaphnia dubia Ceriofast™	feeding activity	60 min	culturing required	Bitton et al. 1996
Daphnia magna	feeding activity	60 min	culturing required	De Coen et al. 1995

Table 26.2 Applications of microbiotests in a battery of tests for screening environmental samples (B: bacteria; C: crustaceans; R: rotifers)

Reference	Type of alternative tests used in the study
Effluents	
Persoone et al. 1993	Microtox (B), Rotoxkit F (R), Strepto/Thamnotoxkit F (C) and the conventional acute *D. magna* test (C).
Latif et al. 1995	Rotoxkit F, Strepto/Thamnotoxkit F and the conventional acute *D. magna* test.
Costan et al. 1993	Microtox, SOS chromotest, microplate algal growth test and the conventional 7d reproduction test with *C. dubia* (C).
Espiritu 1994	Microtox, Rotoxkit F, Thamnotoxkit F, 1 h enzymatic inhibition test with *D. magna* (C) and the conventional acute *D. magna* test.

Other studies: Dutka et al. 1991; Van de Wielen *et al.* 1993; Rodrigue et al. 1995

Sediments	
Vangheluwe et al. 1995	4 microbiotests (B, C and R) and 8 conventional assays on pore waters and bulk sediments.
Persoone et al. 1993	Microtox, Rotoxkit F, Strepto/Thamnotoxkit F and the conventional acute *D. magna* test.

Other studies: Dutka and Gorrie 1989; Kwan et al. 1990; Dutka et al. 1991

Landfill elutriates	
Persoone et al. 1993	Microtox, Rotoxkit F, Strepto/Thamnotoxkit F and the conventional acute *D. magna* test.
Clement et al. 1995	Microtox, 3 types of Toxkits, acute *D. magna* and *C. dubia* assays, conventional algal growth test and the duckweed assay.

Other studies: Espiritu 1994

the exposed bacteria and calculation of the EC_x (the concentration producing x% inhibition of the light output). Details of the test procedure(s) and adaptions for working with turbid and coloured samples are reported in Bulich (1982, 1986), Bulich et al. (1981), Ribo and Kaiser (1987). The relationship between Microtox assay results and those of other bacterial tests and conventional acute toxicity tests has been investigated by Dukta and Kwan (1981, 1982, 1984); Dutka et al. (1983); Lesback et al. (1981); Qureshi et al. (1982); De Zwart and Sloof (1983); Ribo and Kaiser (1983); Vasseur et al. (1984); Bazin et al. (1987) and Retuna et al. (1989). In general, high correlation coefficients (linear regression), ranging from 0.7 to 0.9, are obtained for the Microtox-fish and the Microtox-Daphnia comparison.

Using the same apparatus, whole sediment toxicity tests can be performed with the Microtox Solid Phase assay (Brouwer et al. 1990; Kwan and Dutka 1992a). In this assay, bacteria are exposed to the sediment suspension for 15 minutes, after which they are separated from the sediment and their (reduction in) light emmision is measured.

A non-commercial bioluminescence test with *Vibrio harveyi*, a marine bacterium which can survive in oxygen-depleted environments, has recently been described by Thomulka et al. (1992).

Another commercially available bacterial microbiotest is the Toxi-chromo-test which uses lyophilized *Escherichia coli* (EBPI Inc., Brampton, Canada). This assay is based on an assessment of the inhibition of β-galactosidase activity, measured using a chromogenic substrate and a colorimeter. If less accuracy is needed, the assay (color intensity) can also be scored visually (Reinhartz et al. 1987). Recently, a modified version of this assay has been proposed for the toxicity screening of sediments (Kwan and Dutka 1992b; Kwan 1992).

MetPAD™ and MetPLATE™, two colorimetric assays available in kit-form (Group 206 Technologies, Gainesville, USA), are mainly aimed at detecting heavy metal contamination in aqueous samples (Bitton et al. 1992 a and b). Like the Toxi-chromotest, these assays use freeze-dried *E. coli* and measure the inhibition of β-galactosidase activity. The main difference between the Toxi-chromotest and the MetPAD/MetPLATE assays is that in the latter, the enzyme is already present in the bacteria and is not induced during the test. Consequently, these assays only measure the specific enzymatic inhibition and not the overall cell vigor, as is the case with the Toxi-chromotest (Willemsen et al. 1995). In the MetPAD assay, bacteria are rehydrated and added to the test solution. After 90 minutes a buffer is added and drops of the bacterial suspension are transferred to a special pad, 30 minutes later the (purple) coloration is assessed visually by comparing with the control. The MetPLATE test, although based on the same principle, allows for a quantitative assessment of the effect (colorimetric measurements in multi-well plates).

Perhaps the simplest toxicity screening procedure with bacteria is the ECHA dipstick test with *Bacillus spec.* (Dutka and Gorrie 1989; *B. cereus* in Blaise 1991). Basically this assay consists of a dipstick (impregnated with the bacteria and a tetrazolium salt), which is exposed to the toxicant for 10 seconds and then incubated for 24 hours. Toxicity is qualitatively assessed by comparing the intensity of the color with a color chart (the dipstick turns red under non-toxic conditions).

Other bacterial microbiotests for 'general' toxicity screening include:

- the ATP-tox system (with *E. coli* or local flora), a luminescence technique based on the quantification of the ATP content (after exposure) of the bacteria (Xu and Dukta 1987);
- the laser-microbe test with *Bacillus subtillis* which measures changes in the bacterial growth under toxic stress with the aid of laser techniques (Flekner et al. 1989).

Several bacterial assays have also been developed for the detection of mutagenic effects. A full discussion of these types of tests is beyond the scope of the present paper. A brief description together with some key references of the main mutagenic assays is given below:

- The Ames test with *Salmonella typhimurium* is the first and probably most widely used *in vitro* mutagenic assay (Ames et al. 1975, Maron and Ames 1983). This assay has become a standard test for mutagenicity. However, due to its long exposure time it is difficult to classify this test as a microbiotest (cf. criteria of Willemsen et al. 1995).
- The SOS chromotest with *E. coli* is based on the colorimetric measurement of the induction of the SOS repair system which is activated when DNA damage occurs (Xu et al. 1987 and 1989; Quillardet and Hofnung 1993). This assay is presently commercially available in kit form (EPBI Inc., Brampton, Canada).
- The Mutatox assay is based on bioluminesence measurements (of the reversion back to a normal light-emitting bacterium) of a dark mutant of *Photobacterium phosphoreum* M169 (Kwan and Dutka 1990; Kwan et al. 1990, Johnson 1993). This test has been commercialized and can be performed with an adapted Microtox reader.

Of all bacterial microbiotests described above, the Microtox assay is the only one for which the intra- and inter-laboratory variability and reproducibility of the test results has been extensively evaluated. In reviewing the replicability and test precision of this assay, Mayfield (1993) concluded that, in general, the Microtox test is reproducible with typical (intra- and inter-laboratory) coefficients of variation (CV) ranging from 3 to 20%. Greene et al. (1985), on the other hand, reported an inter-laboratory variability (CV) of 16.5 to 113%.

Willemsen et al. (1995) reviewed the comparative sensitivity of 28 microbiotests (bacterial and others) based on toxicity data for 202 compounds. However, for most assays considered, toxicity data were very scarce, consequently the sensitivity comparisons were extremely difficult. Based on toxicity data for 118 of the 202 compounds in the database, these authors concluded that the Microtox assay was (in general) not sensitive to metals but sensitive to organics, especially phenols. The *Vibrio harveyi* assay, on the other hand, seemed to be sensitive to metals and some organics. These authors also concluded that the Toxi-chromotest did not exhibit a high sensitivity to any of the 30 compounds tested. For the MetPAD/MetPlate data on only seven metals and a few organics the following was found: the former assay was not sensitive to the metals while the MetPLATE test was among the most sensitive for some metals. Both tests proved too insensitive to the tested organics. Very little information is available on the sensitivity towards individual compounds of the ECHA biomonitor. For Hg and di-nitrophenol the Microtox test was more sensitive than the ECHA assay. The ATP-tox test was not sensitive to any of the 14 compounds in the database. The laser-microbe assay, on the other hand, seemed to be extremely sensitive to 7 compounds evaluated (3 metals and 4 organics).

One of the major advantages of these microbiotests (except the *V. harveyi* assay) over conventional assays is that no culturing is required. Test organisms are indeed stored in a lyophilized form, thus eliminating the need for the maintenance and/or culturing of live organisms which, according to Persoone (1992), is one of the main bottlenecks in aquatic toxicology. Another attractive feature of these types of bacterial tests is their short exposure time, which ranges from < 30 minutes for the Microtox to over 90 minutes for the Toxichromo-test and MetPAD/MetPLATE assays to 24 hours for the ECHA monitor. Most of these bacterial assays, except the MetPAD- and the Toxichromo-test, require either a lumino- or colorimeter (which may mean a considerable initial investment). For the Microtox apparatus for example, Willemsen et al. (1995) reported a cost of US$ 18,000. The cost of an individual assay (without the depreciation of the equipment) ranges from US$ 5–15 for a Microtox test to US$ 54 for a Toxichromo- or MetPAD/MetPLATE assay.

Although all these microbiotests may potentially be attractive for the rapid toxicity screening of chemicals and environmental wastes, only the Microtox assay meets all the criteria (for a good assay) set out by Tebo (1985). For all other bacterial microbiotests described in this paper more research is needed on their comparative sensitivity, reproducibility, precision, and use for screening environmental samples. Based on the small number of reports retrieved in the limited (but representative) literature search conducted for this paper, one has to conclude that these assays have progressed little beyond their initial development.

26.4 ALGAL TOXICITY TESTS

26.4.1 General

Standard test procedures for conducting toxicity tests with micro algae have been published by numerous organizations, such as the American Public Health Association (APHA 1989), the American Society for Testing and Materials (ASTM 1990), the International Organization for Testing and Materials (ISO 1987) and the Organization for Economic Cooperation and Development (OECD 1984a). With these standard methods, which are very similar, the effect of a toxicant is determined on a rapidly growing algal population in a nutrient-enriched medium for 3 or 4 days (Lewis 1993). Algal tests are usually conducted in Erlenmeyer flasks (5 test concentrations and a control, in triplicate) on a shaker under constant illumination. Algal biomass is measured daily and/or at the end of the test and the results are expressed as a function of the reduction in growth rate or biomass (standing crop). Various algal species have been recommended. In practice, however, *Selenastrum capricornutum* (renamed *Raphidocelis subcapitata*) and *Scenedesmus subspicatus* are the most frequently used. A review of various aspects of conventional toxicity testing with algae, such as culture techniques, media, test species sensitivity, test conditions and endpoints are given in Lewis (1993).

26.4.2 Microbiotests

Not all algal toxicity tests are based on the generalized procedure described above. Indeed various methods have been described using alternative test end points, such as photosynthesis measured by C^{14} assimilation (Giddings et al. 1983; Versteeg 1990), oxygen evolution (Turbak et al. 1986) or *in vivo* chlorophyll fluorescence (Samuelson and Oquist 1977). Other methods use criteria such as nutrient uptake (Peterson et al. 1984; Peterson and Healey 1985) or enzyme inhibition (De Filippis and Ziegler, 1993). Research has been aimed at increasing the simplicity and cost efficiency of algal tests (Radetski et al. 1995; Persoone 1996a). These procedures include the application of flow cytometry, microplate techniques and immobilized algae (Blaise et al. 1986; Bozeman et al. 1989; Gala and Geisy 1990; Wren and Carroll 1990; Ampaarado 1995; Persoone 1996a).

A miniaturized version of the conventional flask method with *S. capricornutum* was developed by Blaise et al. (1986). In this assay the algae are exposed to the toxicant dilution series in 96-well microplates for a period of 96 hours, after which the cell density can be determined using a hemocytometer or an electronic particle counter. Other test criteria which can also be used in this test procedure are ATP content measurements (Blaise et al. 1986) or chlorophyll fluorescence (Caux et al. 1992 in Willemsen et al. 1995). Compared to the flask method, the main advantages of this microbiotest are: (a) the small sample volumes and reduced bench space requirements, (b) the use of disposable materials, (c) the large number of replicates which can be used and (d) the potential for automation of the test set-up and scoring (Blaise et al. 1986, 1988; Blaise 1991). St-Laurent et al. (1992) report a good correlation between results of the conventional flask assay and the microplate test for phenol, some metals and herbicides. The intra- and inter-laboratory reproducibility (CVs of 11–41% and 25%, respectively) of the results reported by Blaise et al. (1986) and Thellen et al. (1989) indicate that this assay is well standardized. Radetski et al. (1995) recently described a 72 h semi-static microplate test in which the algae can be transferred to fresh test medium daily.

In their review, Willemsen et al. (1995) summarized toxicity data for 25 compounds and concluded that, compared to the other (bacterial and invertebrate) microbiotests considered, the microplate algal test is very sensitive to metals and oxidizers but, in general, not very sensitive to organics (except herbicides). The application of this microbiotest as part of a battery of toxicity tests for the screening of industrial effluents has been demonstrated by Costan et al. (1993).

Another alternative algal assay with *S. capricornutum* which has recently been developed is the Algaltoxkit[TM] F (Persoone 1996a). One of the main features of this commercially available kit-test is that there is no pre-test culturing of algae required. The algae in the toxicity test kit are supplied in the form of algal beads which can be stored for several months. Although the immobilization of algal cells on alginate beads has been reported by various authors (Chibata and Wingard 1983; Vojisek and Jirku 1983; Lukavsky et al. 1986; Lukavsky 1992) the use of algae freed from (stored) beads for toxicity testing was only recently investigated (Amparado 1995; Persoone 1996a). Bozeman et al. (1989), on the other hand, used the immobilized cells (in the beads) to perform algal toxicity tests.

The Algaltoxkit test design, which adheres to test requirements set out by standard procedures, such as the OECD guideline 201 (OECD 1984a), has been developed to simplify and increase the cost-efficiency of these procedures. The test containers, for example, are (30 ml) disposable cuvettes, which allow direct measurement of the cell density using a spectrophotometer. Persoone (1996a) compared the results of toxicity tests on 13 inorganic and organic chemicals obtained using this new microbiotest with the EC50s resulting from (concurrently run) conventional "flask" tests. The difference between the two results was less than 20% for 9 of the 13 chemicals and less than 40% for all compounds tested, indicating the very good correlation between the results obtained with these two test methods ($r^2 = 0.987$). Additionally, the (intra-lab) repeatability of this algal microbiotest is reported to be very good (C.V. = 8%).

A comprehensive review of the recent development and applications of alternative microscale toxicity tests with algae is given by Blaise et al. (1996).

26.5 INVERTEBRATES

26.5.1 General

In their review on the type of assays used in aquatic toxicity assessments, Maltby and Calow (1989) concluded that (1) invertebrates were the most commonly used test organisms (75% of all papers included in the review) and, (2) 80 to 90% of those used were short-term assays. Additionally, of the 18 taxonomic groups of invertebrates considered, cladocerans were the most frequently used test animals. This is reflected in the fact that acute and chronic tests with cladocerans are currently the only frequently used freshwater invertebrate tests endorsed by most international organizations, such as the US Environmental Protection Agency (EPA), OECD and ISO. An overview of the acute and chronic standard test procedures with daphnids (and reference to the main literature) is given by Persoone and Janssen (1993).

In the standard acute test with *Daphnia magna* (or *D. pulex*), juvenile daphnids are exposed to different concentrations of the toxicant in a standard test system for a period of 24 or 48 hours. At the end of the test period the number of dead or immobilized organisms is counted and the LC_{50} or EC_{50} calculated. The two most frequently used chronic tests with invertebrates are the 7-day survival and reproduction test with *Ceriodaphnia dubia* and the 21-day reproduction assay with *D. magna* (Horning and Weber 1985; ASTM 1987; OECD 1984a). In both assays juvenile organisms are exposed to (dilutions of) the test material in a static renewal system. Test results are based on a comparison of the number of offspring (and the survival for *C. dubia*) in the toxicant treatments with the reproduction (and survival) in the controls.

Other invertebrate taxa with which various types of freshwater toxicity tests have been developed are summarized in Pascoe and Edwards (1989) and Persoone and Janssen (1993).

26.5.2 Microbiotests

In his review on the use of microbiotests in aquatic toxicology, Blaise (1991) lists 10 invertebrate assays. Among these are 3 protozoan tests (one with *Colpidium campylum* and 2 with *Tetrahymena pyriformis*), 2 rotifer tests (with *Brachionus calyciflorus* and *B. plicatilis*), 3 crustacean tests (*D. magna, C. dubia* and *Artemia salina*), one test with the nematode *Panagrellus redivivus* and a teratogenicity test with the hydrozoan *Hydra attenuata*. Using their definition of microbiotests, Willemsen et al. (1995) listed 7 assays: 2 lethality tests in kit-form with rotifers (same species as above), 2 lethality tests in kit-form with crustaceans (*Thamnocephalus platyurus* and *A. salina*), an enzymatic test with *D. magna* and an assay based on the feeding activity of rotifers.

One type of invertebrate microbiotest which is increasingly being used in environmental monitoring programs is the Toxkits™ (Creasel Ltd., Deinze, Belgium) which contain the preserved bioindicator, experimental vessels, and reagents. These assays are based on the use of cryptobiotic stages (resting eggs – cysts) of various invertebrate species from which the test organisms can be hatched when needed. The test protocol of this type of assay is simple and consists of: (1) hatching of the test animals which takes 24 to 72 h (depending on the species), (2) exposure of the hatched organisms to a toxicant dilution series for 24 to 48 h (depending on the species), (3) scoring of the mortality and calculation of the LC_{50}. For the freshwater environment, cyst-based toxicity assays have been developed with the following species: the rotifers *Brachionus rubens* and *B. calyciflorus*—Rotoxkit F (Snell and Persoone 1989a; Snell et al. 1991a), the anostracan crustaceans *Streptocephalus proboscideus* and *T. platyurus*—Thamnotoxkit F (Centeno et al. 1993a and b, 1994a and b). Recently, a microbiotest based on the Toxkit-principle was developed with the cladocerans *D. magna* and *D. pulex*—Daphtoxkit F (Persoone 1996b). The test procedures with rotifers have been accepted as a standard guideline by ASTM (ASTM 1991). The Toxkit procedure developed with the daphnids, on the other hand, fully adheres to the acute test method proposed by OECD (1984b). For estuarine/marine environments similar assays have been described with *B. plicatilis* and *A. salina* (renamed *A. fransiscana*) (Snell and Persoone 1989b; Snell et al. 1991b; Van Steertegem and Persoone 1993). Overviews of microscale toxicity testing with rotifers are given by Snell and Janssen (1995, 1996). Details on development and application of Toxkits are reported in Persoone (1996b).

Janssen and Persoone (1993) proposed a new concept in rapid toxicity screening with invertebrates which is based on the in vivo inhibition of enzymatic activity and/or feeding activity. In this type of assay, test organisms are exposed to the toxicant for one hour after which a fluorescent substrate is added, 15 minutes later the test response is scored visually (absence or presence of fluorescence in the organism) or measured with a fluorometer (Janssen et al. 1993; Espritu 1994). Under non-toxic conditions the enzymatic substrate (4-methylumbelliferyl-β-galactoside), which is non-fluorescent when added to the test medium, is taken up and metabolized by the (healthy) test organisms yielding brightly fluorescent animals. Organisms exposed to toxic stress do not feed and/or metabolize the substrate and consequently are not fluorescent. Various studies with several invertebrate species have shown that the 1h EC_{50}s obtained with this test procedure correlate well ($r^2 = 0.88$ to 0.96) with the 24- and 48-h LC_{50} obtained with acute toxicity tests (for the same species) (Janssen and Persoone 1993; Espiritu et al. 1995). Research on the automated scoring of the tests using a fluorometer (plate-reader) is reported by Espiritu (1994). This test concept has been used in toxicity assays with freshwater organisms such as *D. magna*, *C. dubia*, *S. proboscideus* and the marine crustaceans *A. salina* and *Mysidopsis bahia* (Janssen and Persoone 1993; Hayes et al. 1993; Janssen et al. 1993; Espiritu 1994; Espiritu et al. 1995). Test-kits containing the enzymatic substrate and the exposure vessel (but not the test animals) are commercially available under the tradename IQ test™ (Aqua Survey Inc., Flemington, USA). A similar test procedure, based on the esterase activity was developed with rotifers (Burbank and Snell 1994; Moffat and Snell 1995).

A new rapid acute toxicity test (Ceriofast™) based on the suppression of the feeding activity of *C. dubia* in the presence of toxicants, was proposed by Bitton et al. (1995, 1996). In this assay neonate daphnids are exposed to the toxicant for one hour after which they are allowed to feed on fluorescent-stained yeast cells for 20 minutes. The test is scored by observing the presence or absence of fluorescence in the daphnids gut with the aid of an epifluorescence microscope. Based on the results of toxicity assays with 5 heavy metals and 5 organics, Bitton et al. (1995) demonstrated that Ceriofast EC_{50}s correlated well with the 48-h EC_{50}s obtained with the acute *C. dubia* test. A similar one hour test procedure, using rotifers (*B. calyciflorus and B. plicatilis*) feeding on fluorescent beads, was proposed by Juchelka and Snell (1994, 1995). In this test, the ingestion rate is quantified using image analysis. Similar work with *D. magna* was recently reported by De Coen et al. (1995).

Of the invertebrate microbiotests described in this paper, the Toxkits are the best documented in most aspects of test standardization (intra- and inter-laboratory reproducibility, influence of environmental factors) and validation (comparative sensitivity, applications).

Based on a large intercalibration with the 2 rotifer tests (Rotoxkit F and M) and the *Artemia* test (Artoxkit M), inter-laboratory reproducibility was found to be quite satisfactory with CVs ranging from 25 to 68% (Persoone et al. 1992). Intra-laboratory variability is reported to be similar to that of the conventional

D. magna test (CV = 20%) (Persoone 1992). Modifications to the test protocol and kit materials have further improved the inter-laboratory reproducibility of these microbiotests (Persoone, pers. comm.).

In a ring-test with the IQ test™ with *D. magna*, intra- and inter-laboratory coefficients of variation of 21 and 31%, respectively, were observed (Aqua Survey 1993 in Willemsen et al. 1995). No data on the reproducibility of the feeding test with *C. dubia* and *B. calyciflorus* are available.

Willemsen et al. (1995) summarized the toxicity data for 77 and 40 compounds for the Rotoxkit F and the Thamnotoxkit F, respectively, and concluded that in general, the latter seemed to be the most sensitive. An overview of 24 h LC_{50}s obtained with rotifers is given by Snell and Janssen (1995). Persoone et al. (1995) reported on the comparative sensitivity of the crustacean microbiotests with the anostracans (*S. proboscideus* and *T. platyurus*) and the standard acute test with *D. magna*. Linear regression analysis based on the 146 data pairs included in this study (chemical compounds and environmental samples) resulted in regression coefficients ranging from 0.84 to 0.92. Furthermore, in the majority of the cases the effect ratios between the conventional tests and the microbiotests were within a factor of 2. The application of Toxkit assays and the Microtox test for screening various types of environmental samples (effluents, solid waste elutriates, monitoring wells and sediment pore waters) is reported by Persoone et al. (1993). Of the samples which were toxic to one or more test organisms, 50% were detected by the conventional acute *D. magna* test (which was run concurrently), while 60, 40 and 80% were detected by the anostracan (Thamnotoxkit/ Streptoxkit F), rotifer (Rotoxkit F) and the bacterial (Microtox) assays, respectively. A comparison of intensity of the toxicity signal showed that the *D. magna* test was the most sensitive assay in 19% of the cases, while the anostracan, rotifer and bacteria assays were the most sensitive in 39, 19 and 48% of the samples. Similar evaluations of invertebrate microbiotests for screening environmental wastes are given by Van der Wielen et al. (1993), Espiritu (1995), Clement et al. (1995), and Latif et al. (1995).

Toxkit microbiotests have also been used as part of a battery of screening assays to determine the toxicity of the 50 priority chemicals of the Multicentre Evaluation of in vitro Cytotoxicity (MEIC) programme. In a series of papers, Calleja and co-workers report on the predictive power of these screening tests for evaluating human toxicity of various classes of chemicals (Calleja et al. 1993a and b; Calleja et al. 1994a and b). Lindgren et al. (1996) used the freshwater rotifer and anostracan kit-assay in a QSAR study with non-ionic surfactants.

26.6 THE USE OF ALTERNATIVE TOXICITY TESTS IN A BATTERY OF SCREENING ASSAYS

With the number of standard and alternative ecotoxicity assays increasing every year, the problem of selecting the most appropriate test battery, which will generate the required data, has increased proportionally. Several authors have tried to address this issue through the objective evaluation and comparison of the different aspects of the various test systems.

Keddy et al. (1995) for example, critically evaluated various whole organism bioassays for the assessment of soil, freshwater sediment, and freshwater quality for their application in the Canadian 'National Contaminated Sites Remediation Program' (CCME, 1991). Using 3 essential and 12 desirable selection criteria, the toxicity tests were categorized as 'currently usable', 'prototype', or 'under development'. Based on further considerations related to bioassay application, a battery of usable screening and definitive tests was recommended for each assessment type (soil, sediment and freshwater). Of the 25 toxicity tests (most of which were conventional assays) considered for freshwater quality assessment, 3 tests (the 72-h algal growth inhibition test with *S. capricornutum*, the 48-h survival test with *D. magna* and the 5–15-min bacterial test with *P. phosphoreum*) were selected for the recommended screening battery. The 24-h acute toxicity test with *B. calyciflorus* was, together with the 3 above-mentioned tests, suggested for the augmented screening battery. For the definitive testing battery the 3 recommended screening tests mentioned above plus an acute fish test were selected; the augmented battery again included the acute rotifer assay.

An in-depth analysis of the potential use of 26 microbiotests (11 bacterial assays, 2 algal tests, 10 invertebrate microbiotests, 1 vertebrate teratogenicity test and 2 bacterial genotoxicity tests) was performed by Willemsen et al. (1995). For every assay an assessment was made of the following aspects: test convenience, availability of literature, costs, reproducibility, standardization, influence of experimental conditions, sensitivity (to single compounds and environmental samples). Based on this literature study the authors proposed a test battery designed to "detect most classes of toxicants at low concentrations with high reliability." This battery includes: the Microtox assay, the algal microplate assay and the crustacean microbiotest Thamnotoxkit F. The battery may be augmented with the Mutatox assay if genotoxicity screening is required.

Johnson (1995) made a comprehensive evaluation of conventional (higher organism) tests and alternative rapid screening toxicity assays for assessing effluent toxicity. In this study, an objective procedure was developed to select methods for specific operational roles. The usefulness of the candidate methods was assessed using a set of 17 criteria—most of which are similar to those reported above. For each criterion a score ranging from 1 to 10 was assigned by reference to guidelines and each criterion was weighted depending on the envisaged operational role of the assay. Based on the total score of all criteria a battery of rapid screening tests and a battery of confirmatory higher organism assays were selected. In total 51 screening tests and 32 sublethal and lethal confirmatory assays were evaluated. The selected battery of screening tests comprised (1) the Microtox assay, (2) the ECLOX test (Enhanced Chemical Luminescence Oxygen Reaction), (3) an in vivo enzymatic inhibition test with invertebrates (e.g., *D. magna*) and (4) a plant fluorescence test. The recommended confirmatory higher organism tests were: (1) algal growth inhibition tests, (2) growth, reproduction and lethality tests with invertebrates (e.g. *D. magna*) and (4) growth and lethality tests with indigenous fish.

Rodrigue et al. (1995) reported the results of an experimental study in which a comparison of the performance of microbiotests and standard acute assays was made for the assessment of the toxicity of mining effluents in Canada. The cost, sensitivity, speed, accuracy, applicability and reproducibility of 5 commercially available microbiotests (the Microtox-, Rotoxkit F-, Thamnotoxkit F-, Toxichromo- and *D. magna* IQ assays) were compared to 2 regulatory, conventional tests (the 96-h rainbow trout and the 48-h *D. magna* tests). The main conclusion of this (Canmet-)study can be summarized as follows:

1. not one toxicity test corresponded directly with the rainbow trout test for both comparability of toxicity response and correlation of endpoint results with chemistry;

2. based on the evaluation criteria, the "best" toxicity assay varied depending on the mine (effluent) type. From all methods evaluated, the Thamnotoxkit F, the *D. magna* IQ and the acute *D. magna* assays were selected as the 'best' alternative tests (to the rainbow trout assay).

Toussaint et al. (1995) compared the relative sensitivity and costs of 5 rapid alternative assays to that of 5 standard acute toxicity tests through literature review and testing. The alternative assays, which were selected because they do not require culturing prior to testing, were: the 24-h mortality tests with the freshwater rotifer *B. calyciflorus* and the marine crustacean *A. salina*, the 96-h root elongation test with lettuce (*Lactuca sativa*), the 5-min Microtox test and the 21-min Polytox test (respiratory activity of a bacterial blend). The standard acute toxicity tests were: the 48-h mortality tests with *D. magna* and *C. dubia*, the 96-h algal growth inhibition test with *S. capricornutum* and the 96-h mortality test with fish (*Pimephales promelas*). Based on the toxicity data from 11 chemicals, it was found that the rotifer and lettuce tests ranked closest in sensitivity to the standard tests. Microtox was slightly less sensitive and fell just outside the range of sensitivities represented by the group of standard tests. The two other assays were, for most compounds, one or more orders of magnitude less sensitive than the standard assays. The calculated costs of the alternative tests ranged from US$ 62 (Microtox, instrument purchase not included) to US$ 407 (Polytox), while the cost of the standard assays varied from US$ 703 (*D. magna* and *C. dubia*) to US$ 1280 (*S. capricornutum*).

In conclusion, Toussaint et al. (1995) suggested a battery consisting of the lettuce, rotifer and Microtox assays for the preliminary toxicity screening of chemicals. Compared to the standard tests these alternative assays do not only have a similar sensitivity, but all 3 tests can also be run for less than the cost of one standard test.

26.7 CONCLUSIONS AND RESEARCH PERSPECTIVES

From this review it is clear that alternative toxicity tests in general, and microbiotests in particular, have an important role to play in the screening and ranking of the potential hazards of chemicals and environmental wastes. The increasing application of toxicity assays in environmental assessment schemes combined with the budgetary restraints that most agencies and companies are facing, has resulted in a growing demand for alternative toxicity tests with increased cost-efficiency and diagnostic capacity. However, this review has also shown that for the majority of the alternative assays considered there is a lack of published information on various aspects of the test development and/or application. In this context Willemsen et al. (1995) state "although there are many potentially useful tests, most have not yet been studied sufficiently." Indeed, looking at the number of papers published on the various methods since their initial development, one has to conclude that many assays have not been fully evaluated and abandoned (by their developers) soon after their conception. It seems that most scientists developing new toxicity tests are not willing (or able) to pursue the further development, standardization and validation of developed methods. This is, however, crucial if the new assay is intended to be used to solve real environmental problems. Important aspects which should be considered during the initial development of a (alternative) toxicity test are: (1) the envisaged purpose of the assay (e.g. screening, regulatory), (2) standardization of the assay (precision, reproducibility), (3) sensitivity, (4) cost-efficiency and (5) the evaluation of the potential applications (e.g. screening effluents). Additionally, it is essential that the results obtained with alternative test systems are compared to those of regulatory assays, as this will help to understand the sensitivity of the new assay and the degree of agreement between the two types of test results. This is especially important when alternative assays with new endpoints (e.g. biomarkers) are developed. Analysis of the relationship between the new endpoint (e.g. feeding activity, enzymatic activity, respiration) and conventional criteria, such as mortality, growth and reproduction, will help to establish the ecological relevance and the potential applications of the new assay. As noted before, one or more of these essential aspects in the development of new toxicity assays are often overlooked.

With the recent advances in instrumental technology (affordable spectrophotometric and fluorometric equipment, image analysis, PCR and biotechnology techniques), research efforts will surely be directed towards the development of alternative assays based on physiological, biochemical, enzymatic and immunological criteria. This research would respond to the currrent demands for increased rapidity and automation of screening tests. Additional research into lyophilization and cryo-preservation techniques, or other methods for storing test organisms, will lead to the development and increased use of culture/maintenance-free screening assays.

However, as this new research unfolds and more efficient environmental screening tools become available, let us not forget—in the midst of all this technology—the ultimate goal of ecotoxicology, i.e. the protection of *ecological* systems from the adverse effects of xenobiotic chemicals.

26.8 REFERENCES

American Society for Testing and Materials (ASTM), 1987, Standard Guide for Conducting Renewal Life-Cycle Toxicity Tests with *Daphnia magna*, E-1193, ASTM, Philadelphia, USA

American Society for Testing and Materials (ASTM), 1990, Standard Guide for Conducting Static 96 h Toxicity Tests with Microalgae, E-1218-90. ASTM, Philadelphia, USA

American Society for Testing and Materials (ASTM), 1991, Standard Guide for Acute Toxicity Tests with the Rotifer *Brachionus*, E-1440, ASTM, Philadelphia, USA

American Public Health Association (APHA), 1989, "Toxicity Testing with Phytoplankton," *Standard Methods for the Examination of Water and Wastewater*, 17th Edn., APHA, Washington, USA

Ames, B. N., J. McCann, and E. Yamasaki, 1975, "Methods for Detecting Carcinogens and Mutagens with *Salmonella*/Mammalian-Microsome Mutagenicity Test," *Mutat. Res., 41*, 4192–4203

Amparado, R. F., 1995, "Development and Application of a Cost-Effective Algal Growth Inhibition Test with the Green Alga *Selenastrum Capricornutum* (Printz)," *Ph.D. Thesis*, University of Ghent, Belgium, 217 pp

Bazin, C., P. Chambon, M. Bonnefille, and G. Larbaight, 1987, "Comparaison de sensibilités du test de luminiscence bactérienne (*Photobacterium phosporeum*) et du test Daphnie (*Daphnia magna*) pour 14 substances à risque toxique élevé, *Sci. de l'Eau*," 6, 403–413

Bitton, G., M. Campell, and B. Koopman, B. 1992a, "MetPAD: A Bioassay Kit for the Specific Determination of Heavy Metal Toxicity in Sediments from Hazardous Waste Sites," *Environ. Toxicol. Water Qual. Int., 7*, 323–328

Bitton, G., and B. Koopman, 1986, "Biochemical Tests for Toxicity Screening," in *Toxicity Testing Using Microorganisms* (G. Bitton, and B. J. Dutka, Eds.), CRC Press, Boca Raton, FL, USA

Bitton, G., B. Koopman, and O. Agami, 1992b, "MetPAD: A Bioassays for Rapid Assessment of Heavy Metal Toxicity in Wastewater," *Arch. Environ. Contam., Toxicol., 64*, 834–836

Bitton, G., K. Rhodes, and B. Koopman, 1996, "Ceriofast™: An Acute Toxicity Test Based on *Ceriodaphnia dubia* Feeding Behavior," *Environ. Toxicol. Chem., 15* (2), 123–125

Bitton, G., K. Rhodes, B. Koopman, and M. Cornejo, 1995, "Short-Term Toxicity Assay Based on Daphnid Feeding Behavior," *Water Environ. Res., 67*, 290–293

Blaise, C., 1991, "Microbiotests in Aquatic Ecotoxicology: Characteristics, Utility and Prospects," *Environ. Toxicol. Water. Qual. Int. J., 6*, 145–156

Blaise, C., J.-F. Férard, and P. Vasseur, 1996, "Microplate Toxicity Tests with Microalgae: A review," in *Microscale Toxicology, Advances, Techniques and Practice* (P. G. Wells, K. Lee, and C. Blaise, Eds.), CRC Press, Boca Raton, USA, in press

Blaise, C., R. Legault, N. Bermingham, R. Van Coillie, and P. Vasseur, 1986, "A Simple Microplate Algal Assay for Aquatic Toxicity Assessment," *Tox. Assess. Int. J.*, *1*, 261–281

Blaise, C., G. Sergy, N. Bermingham, and R. Van Coillie, 1988, "Biological Testing— Development, Application and Trends in Canadian Environmental Protection Laboratories," *Tox. Assess. Int. J.*, *3*, 385–406

Bozeman, J., K. Koopman, and G. Bitton, 1989, "Toxicity Testing using Immobilized Algae," *Aquat. Toxicol.*, *14*, 345–352

Brouwer, H., T. Murphy, and L. McArdle, 1990, "A Sediment-Contact Bioassay with *Photobacterium Phosphoreum*," *Environ. Toxicol. Chem.*, *9*, 1353–1358

Bulich, A. A., 1982, "A Practical and Reliable Method for Monitoring the Toxicity of Aquatic Samples," *Proc. Biochem.*, *17*, 45–57

Bulich, A. A., 1986, "Bioluminescence Assays," in *Toxicity Testing Using Microorganisms*, (G. Bitton, and B. J. Dutka, Eds.), CRC Press, Boca Raton, USA, pp. 57–74

Bulich, A. A., and M. W. Greene, 1979, "The Use of Luminescent Bacteria for Biological Monitoring of Water Quality," in *Proc. Int. Symp. on the Analysis and Application of Bioluminescence and Chemiluminescence* (E. Schram, and S. Philip, Eds.), State Printing and Publ. Inc., USA, pp. 193–211

Bulich, A. A., M. W. Greene, and D. L. Isenberg, 1981, "Reliability of the Bacterial Luminescence Assay for the Determination of Pure Compounds and Complex Effluents," in *Aquatic Toxicology and Hazard Assessment, Fourth Conference*, ASTM STP 737 (D. R. Branson, and K. L. Dickson, Eds.), American Society for Testing and Materials, Philadelphia, USA, pp. 338–347

Bulich, A. A., and D. L. Isenberg, 1980, "Use of the Luminescent Bacteria Systems for Rapid Assessment in Aquatic Toxicology," *Adv. Instrument.*, *35*, 35–40

Burbank, S. E., and T. W. Snell, 1994, "Rapid Toxicity Assessment Using Esterase Biomarkers," in *Brachionus calyciflorus* (Rotifera), *Environ. Toxicol. Water Qual.*, *9*, 171–178

Cairns, J., Jr., and J. R. Pratt, 1989, "The Scientific Basis of Bioassays," in *Environmental Bioassay Techniques and their Application*, (M. Munawar, G. Dixon, C. I. Mayfield, T. Reynoldson, and M. H. Sadar, Eds.), Hydrobiologia, 188/189, Kluwer Academic Publishers, Dordrecht, The Netherlands, pp. 2–20

Calleja, M., G. Persoone, and P. Geladi, 1993a, "The Predictive Potential of a Battery of Ecotoxicological Tests for Human Acute Toxicity as Evaluated with the First 50 MEIC Chemicals," *ATLA*, *21*, 330–349

Calleja, M., G. Persoone, and P. Geladi, 1993b, "Comparative Acute Toxicity of the First 50 MEIC Chemicals to Aquatic Non-vertebrates," *Arch. Environ. Contam. Toxicol.*, *1*, 69–78

Calleja, M., G. Persoone, and P. Geladi, 1994a, "Human Acute Toxicity Prediction of the First 50 MEIC Chemicals by a Battery of Ecotoxicological Tests and Physical–Chemical Properties," *Fd. Chem. Toxicol.*, *32*, 173–187

Calleja, M., G. Persoone, and P. Geladi, 1994b, "QSAR Models for Predicting the Acute Toxicity of Selected Organic Chemicals with Divers Structures to Aquatic Non-vertebrates and Humans," *SAR and QSAR Environ. Res.*, *2*, 193–234

Calow, P., 1993, *Handbook of Ecotoxicology*, (P. Calow, Ed.), Blackwell Scientific Pub., Oxford, UK, 478

Caux, P. Y., C. Blaise, P. Le Blanc, and M. Tache, 1992, "A Phytoassay Procedure using Fluorescence Induction," *Environ. Toxicol. Chem.*, *11*, 549–557

CCME (Canadian Council of Ministers of the Environment), 1991, "Interim Canadian Environmental Quality Criteria for Contaminated Sites," Report CCME, EPC-CS34, Environment Canada, Ottowa

Centeno, M. D., L. Brendonck, and G. Persoone, 1993a, "Cyst-Based Toxicity Tests: XI. Influence of Production, Processing and Storage Conditions of Resting Eggs of *Streptocephalus Proboscideus* (Crustacea: Branchiopoda: Anostraca) on the Sensitivity of Larvae to Selected Reference Toxicants," *Bull. Environ. Contam. Toxicol.*, *51*, 927–934

Centeno, M. D., L. Brendonck, and G. Persoone, 1993b, "Cyst-Based Toxicity Tests: III. Development and Standardization of an Acute Toxicity Test with the Freshwater Anostracan Crustacean *Streptocephalus Proboscideus*," in *Progress in Standardization of Aquatic Toxicity Tests* (A. M. V. M. Soares, and P. Calow, Eds.), Lewis Publishers, Boca Raton, USA, pp. 37–55

Centeno, M. D., G. Persoone, and M. Goyvaerts, 1995, "Cyst-based Toxicity Tests: IX. The Potential of *Thamnocephalus Platyurus* as Test Species in Comparison with *Streptocephalus Proboscideus* (Crustacea, Branchiopoda, Anostraca)," *Environ. Toxicol. Water Qual.*, *10*, 275–282

Chibata, I., L. B. Wingard, 1983, (Eds.), Immobilized Microbial Cells, *Appl. Biochem. Bioeng.*, *4*, 344

Clément, B., G. Persoone, C. Janssen, and A. Le Dû-Delepierre, 1996, "Estimation of the Hazard of Landfills through Toxicity Testing of Leachates. I. Determination of Leachate Toxicity with a Battery of Acute Tests," *Chemosphere*, *33*, 2303–2320

Costan, G., M. Bermingham, C. Blaise, and J. F. Ferrard, 1993, "Potential Ecotoxicity Probe (PEEP): A Novel Index to Assess and Compare the Toxic Potential of Industrial Effluents," *Environ. Toxicol. Water Qual. J.*, *8*, 115–140

De Coen, W. M., C. R. Janssen, and G. Persoone, 1995, "Rapid Toxicity Screening of Sediment Pore Waters using Physiological and Biochemical Biomarkers of *Daphnia magna*," Second SETAC World Congress – Abstract Book, Society of Environmental Toxicology and Chemistry Press, Pensacola, USA, 197 pp

De Filipis, L. F., and F. Ziegler, 1993, "Effect of Sublethal Concentrations of Zinc, Cadmium, and Mercury on the Photosynthetic Carbon Reduction Cycle of *Euglena*," *J. Plant Physiol.*, *142*, 167–172

Denyer, S. P., 1989, "ATP Bioluminescence and Biocide Assessment: Effect of Bacteriostatic Levels of Biocide," in *ATP Luminescence Rapid Methods in Microbiology*, (P. E. Stanley, B. J. McCarthy, and R. Smither, Eds.), Society of Applied Bacteriology Technical Series 26, Blackwell Scientific Publishers, Oxford, UK, pp. 189–195

De Zwart, D., and W. Sloof, 1983, "The Microtox as an Alternative Assay in the Acute Toxicity Assessment of Water Pollutants," *Aquat. Toxicol.*, *4*, 129–138

Dutka, B. J., and J. F. Gorrie, 1989, "Assessment of Toxicant Activity in Sediments by the ECHA Biocide Biomonitor," *Environ. Pollut.*, *57*, 1–7

Dutka, B. J., and K. K. Kwan, 1981, "Comparison of Three Microbial Toxicity Screening Tests with the Microtox Test," *Bull. Environ. Contam. Toxicol.*, *27*, 753–757

Dutka, B. J., and K. K. Kwan, 1982, "Application of Four Bacterial Screening Procedures to Assess Changes in the Toxicity of Chemicals in Mixtures," *Environ. Pollut.*, *29*, 125–134

Dutka, B. J., and K. K. Kwan, 1984, "Studies on a Synthetic Activated Sludge Toxicity Screening Procedure with Comparison to Three Bacterial Toxicity Tests," in *Toxicity Screening Procedures Using Bacterial Systems* (D. Liu, and B. J. Dutka, Eds.), Marcel Dekker, New York, USA, pp. 125–138

Dutka, B. J., K. K. Kwan, S. Rao, A. Jurkovic, and R. McInnis, 1991, "Use of Bioassays to Evaluate River Water and Sediment Quality," *Environ. Toxicol. Water Qual. Int. J.*, 6, 309–327

Dutka, B. J., N. Nyholm, and J. Peterson, 1983, "Comparison of Several Microbiological Toxicity Screening Systems," *Water Res.*, 17, 1363–1368

Espiritu, Q. E., 1994, "Development of an Enzymatic Toxicity Test with Selected Phyllopod Species (*Crustacea: Anostraca and Cladocera*), *PhD thesis*, University of Ghent, Belgium, 279 pp

Espiritu, E. Q., C. R. Janssen, and G. Persoone, 1995, "Cyst-based Toxicity Tests: VII. Evaluation of the 1 Hour Enzymatic Inhibition Test (Fluotox) with *Artemia* Nauplii," *Environ. Toxicol. Water Qual.*, 10, 25–34

Felkner, I. C., B. Worthy, T. Christison, C. Chaisson, J. Kurtz, and P. J. Wyatt, 1989, Laser- Microbe Bioassay System," *Aquatic Toxicology and Hazard Assessment*, Vol. 12 (U. M. Cowgill, and L. R. Williams, Eds.), American Society for Testing and Materials, Philadelphia, USA, pp. 95–103

Gala, W. R., and J. P. Geisy, "Flow Cytometric Techniques to Assess Toxicity to Algae," in *Aquatic Toxicology and Risk Assessment*, Vol. 13 (W. Landis, W. H. Vander Schalie, Eds.), ASTM, Philadelphia, USA, pp. 237–246

Giddings, J. M., A. J. Stewart, R. V. O'Niel, and R. H. Gardner, 1983, "An Efficient Algal Bioassay Based on Short-Term Photosynthetic Response," in *Aquatic Toxicology and Hazard Assessment*, (R. C. Bahner, and D. J. Hansen, Eds.), American Society for Testing and Materials, Philadelphia, USA, p. 480

Greene, J. C., W. E. Miller, M. K. De Bacon, M. A. Long, and C. Bartels, 185, "A Comparison of Three Microbial Assay Procedures for Measuring Toxicity of Chemical Residues," *Arch. Environ. Contam. Toxicol.*, 14, 659–667

Hayes, K. R., W. S. Douglas, Y. Terrell, J. Fischer, L. A. Lyons, and L. J. Briggs, 1993, "Predictive Ability of the *Daphnia magna* IQ Toxicity Test for Ten Diverse Water Treatment Additives," *Bull. Environ. Contam. Toxicol.*, 51, 252–260

Horning, W. B., and C. I. Weber, 1985, "Short-Term Methods for Estimating the Chronic Toxicity of Effluents and Receiving Waters to Freshwater Organisms," US EPA, Report EPA/600/4–85/014, p. 161

Huggett, R. J., R. A. Kimerle, P. M. Merhle, and H. L. Bergman, 1992, *Biomarkers—Biochemical, Physiological, and Histological Markers of Antropogenic Stress*, Lewis Publishers, Boca Raton, USA, 347 pp

ISO (International Organization for Standardization), 1995, *Water Quality—Pseudomonas putida Growth Inhibition Test*, ISO/TC 147/DC 5, ISO, Paris, France

ISO (International Organization for Standardization), 1987, *Algal Growth Inhibition Test*, Draft ISO Standard ISO/DIS 10253.2, ISO, Paris, France

Janssen, C. R., E. Q. Espiritu, and G. Persoone, 1993, "Evaluation of the New "Enzymatic Inhibition" Criterion for Rapid Toxicity Testing with *Daphnia magna*," in *Progress in Standardization of Aquatic Toxicity Tests* (A. M. V. M. Soares, and P. Calow, Eds.), Lewis Publishers, Boca Raton, USA

Janssen, C. R., and G. Persoone, 1993, "Rapid Toxicity Screening Tests for Aquatic Biota: I. Methodology and Experiments with *Daphnia magna*," *Environ. Toxicol. Chem.*, *12*, 711–717

Johnson, B. T., 1993, "Activated Mutatox Assay for the Detection of Genotoxic Substances," *Environ. Toxicol. Water Qual. Int. J.*, *8*, 103–113

Johnson, I., 1995, "Identification of Screnning, Lethal and Sublethal Toxicity Tests for Assessing Effluent Toxicity," National Rivers Authority U.K., R&D Note 389

Juchelka, C. M., and T. W. Snell, 1994, "Using Rotifer Ingestion Rates for Rapid Toxicity Assessment," *Archiv. Environ. Contam. Toxicol.*, *26*, 549–554

Juchelka, C. M., and T. W. Snell, 1995, "Rapid Toxicity Assessment Using Ingestion Rate of Cladocerans and Ciliates," *Arch. Environ. Contam. Toxicol.*, *28*, 508–512

Keddy, C. J., J. C. Greene, and M. A. Bonnell,1995, Review of Whole-Organism Bioassays: Soil, Freshwater Sediment and Freshwater Assessment in Canada," *Ecotox. Environ. Saf.*, *30*, 221–251

King, E. F., and B. J. Dutka, 1986, "Respirometric Techniques," in *Toxicity Testing Using Microorganisms* (G. Bitton, and B. J. Dutka, Eds.), CRC Press, Boca Raton, USA

Krieg, N. R., J. P. Tomelty, and J. S. Wells, 1967, "Inhibition of Flagellar Coordination in *Spirillum Volutans*," *J. Bacteriol.*, *94*, 1431–1436

Kwan, K. K., 1992 (3), "Direct Toxicity Assessment of Solid Phase Samples Using the Toxi-Chromotest Kit," *Environ. Toxicol. Water Qual.*, *8*, 223–230

Kwan, K. K., and B. J. Dutka, 1990, "Simple Two-Step Sediment Extraction Procedure for Use in Genotoxicity and Toxicity Bioassays," *Tox. Assess. Int. J.*, *5*, 395–404

Kwan, K. K., B. J. Dutka, S. S. Roa, and D. Liu, 1990, "Mutatox Test: A New Test for Monitoring Environemental Genotoxic Agents," *Environ. Pollut.*, *65*, 323–404

Kwan, K. K., and B. J. Dutka, 1992a, "Evaluation of Toxi-Chromotest Direct Sediment Toxicity Testing Procedure and Microtox Solid-Phase Testing Procedure," *Bull. Environ. Contam. Toxicol.*, *49*, 656–662

Kwan, K. K., and B. J. Dutka, 1992b, "A Novel Bioassay Approach Direct Application of the Toxi-Chromotest and the SOS Chromotest to Sediments," *Environ. Toxicol. Water Qual.*, *7*, 49–60

Latif, M., G. Persoone, C. Janssen, , W. De Coen, and K. Svardal, 1995, "Cost-Effective Toxicity Testing of Waste Waters in Austria with Conventional and Cost-Effective Bioassays, *Ecotox. Environ. Saf.*, *32*, 139–146

Lesback, M. E., A. D. Andersen, C. M. De Graeve, and H. C. Bergmon, 1981, "Comparison of Bacterial Luminescence and Fish Bioassay Results for Fossil-Fuel Process Water and Phenolic Constituents," in *Aquatic Toxiclogy and Hazard Assessment*, (D. R. Branson, and K. L. Dickson, Eds.), American Society for Testing and Materials, Philadelphia, USA

Lewis, M.A., 1993, "Freshwater Primary Producers," in *Handbook of Ecotoxicology* (P. Calow, Ed.), Blackwell Scientific Publ, Oxford, UK, pp. 28–51

Lindgren, A., M. Sjöström, and S. Wold, 1996, "QSAR Modelling of the Toxicity of Some Technical Non-Ionic Surfactants Towards Fairy Shrimp," *Quant. Struct. Activ. Relat.*, submitted

Liu, D., 1981, "A Rapid Biochemical Test for Measuring Chemical Toxicity," *Bull. Environ. Contam. Toxicol.*, *26*, 145–149

Liu, D., and K. Thompson, 1983, "Toxicity Assessment of Chlorobenzenes Using Bacteria," *Bull. Environ. Contam. Toxicol.*, *31*, 105–110

Lukavsky, J., 1992, "The Evaluation of Algal Growth Potential and Toxicity of Water by Miniaturized Growth Bioassay," *Wat. Res.*, *26*, 1409–1413

Lukavsky, J., J. Komárek, A. Lukavská, Ludvík, and J. Pokorny, 1986, "Metabolic Activity and Cell Structure of Immobilized Algal Cells (*Chlorella, Scenedesmus*)," *Arch. Hydrobiol. Suppl.*, *73* (2), 261–276

Maltby, L., and P. Calow, 1989, "The Application of Bioassays in the Resolution of Environmental Problems: Past, Present and Future," in *Environmental Bioassay Techniques and their Application* (M. Munawar, G. Dixon, C. I. Mayfield, T. Reynoldson, and M. H. Sadar, Eds.), Hydrobiologia, Vol. 188/189, Kluwer Academic Publishers, Dordrecht, The Netherlands, pp. 65–76

Maron, D. M., and B. N. Ames, 1983, "Revised Methods for the *Salmonella* Mutagenicity Test," *Mutat. Res.*, *113*, 173–215

Mayfield, C. I., 1993, "Microbial Systems," in *Handbook of Ecotoxicology* (P. Calow, Ed.), Blackwell Scientific Publications, Oxford, UK, pp. 9–28

Moffat, B. D., and T. W. Snell, 1995, "Rapid Toxicity Assessment Using Esterase Biomarkers in *Brachionus Plicatilis* (Rotifera)," *Ecotox. Environ. Saf.*, *9*, 171–178

OECD (Organization for Economic Cooperation and Development), 1984a, "*Daphnia* Spp., Acute Immobilization and Reproduction Test," *OECD Guideline for Testing Chemicals*, No. 202, OECD, Geneva, Switzerland

OECD (Organization for Economic Cooperation and Development), 1984b, "Algal Growth Inhibition Test," *OECD Guideline for Testing Chemicals*, No. 201, OECD, Geneva, Switzerland

Packard, T. T., P. G. Garfield, and R. Martinez, 1983, "Respiration and Respiratory Enzyme Activity in Aerobic and Anaerobic Cultures of the Marine Denitrifying Bacterium *Pseudomonas perfectomarinus*," *Deep Sea Res.*, *30*, 227

Parker, C. E., and E. J. Jr. Pribyl, 1984, "Assessment of Bacterial ATP Response as a Measurement of Aquatic Toxicity," in *Toxicity Screening Procedures Using Bacterial Systems* (D. Liu, and B. J. Dutka, Eds.), Marcel Dekker, New York, USA, pp. 283–293

Pascoe, D., and R. W. Edwards, 1989, "Single Species Toxicity Tests," in *Aquatic Ecotoxicology: Fundamental Concepts and Methodologies*, Vol. II (A. Boudou, and F. Ribeyre, Eds.), CRC Press, Boca Raton, USA, pp. 93–126

Peakall, D., 1992, *Animal Biomarkers as Pollution Indicators*, Chapman and Hall, London, UK, p. 291

Persoone, G., 1991, "Cyst-based Toxicity Tests: I. A Promising New Tool for Rapid and Cost-effective Toxicity Screening of Chemicals and Effluents," *Z. Angew. Zool.*, *78*, 235–241

Persoone, G., 1992, "Cyst-based Toxicity Tests: VI. Toxkits and Fluotox Tests as Cost-Effective Tools for Routine Toxicity Screening," in *Biologische Testverfahren*, (K. G. Steinhäuser, and P. D. Hansen, Eds.), Gustav Fischer Verlag, Stuttgart, Germany, pp. 563–576

Persoone, G., 1996a, "Development and First Validation of a "Culture Free" Algal Microbiotest: The Algaltoxkit," in *Microscale Toxicology, Advances, Techniques and Practice*, (P. G. Wells, K. Lee, and C. Blaise, Eds.), CRC Press, Boca Raton, USA, in press

Persoone, G., 1996b, "Development and Validation of Toxkit Microbiotests with Invertebrates, in Particular Crustaceans," in *Microscale Toxicology, Advances, Techniques and Practice*," (P. G. Wells, K. Lee, and C. Blaise, Eds.), CRC Press, Boca Raton, USA, in press

Persoone, G., C. Blaise, T. W. Snell, C. Janssen, and Van Steertegem, 1992/1993, "Cyst-Based Toxicity Tests: II. Report on an International Intercalibration Exercise with Three Cost-Effective Toxkits," *Z. Angew. Zool.*, *79*, 17–36

Persoone, G., M. P. Goyvaerts, C. R. Janssen, W. De Coen, and M. Van Steertegem, 1993, "Cost-Effective Acute Hazard Monitoring of Polluted Waters and Waste Dumps with the Aid of Toxkits," Commission of the European Communities, Contract ACE 89/BE2/D3., 600 pp

Persoone, G., and C. R. Janssen, 1993, "Freshwater Invertebrate Toxicity Tests," in *Handbook of Ecotoxicology*, (P. Calow, Ed.), Blackwell Scientific Publ, Oxford, UK, pp. 51–65

Persoone, G., C. R. Janssen, and W. De Coen, 1995, "Cyst-Based Toxicity Tests: Comparison of the Sensitivity of the Acute *Daphnia magna* Test and Two Crustacean Microbiotests for Chemicals and Wastes," *Chemosphere*, *29*, 2701–2710

Peterson, H. G., F. P. Healey, and R. Wagemann, 1984, "Metal Toxicity to Algae: A Highly pH Dependent Phenomenon," *Can. J. Fish. Aquat. Sci.*, *41*, 974–979

Peterson, H. G., and F. P. Healey, 1985, "Comparative pH Dependent Metal Inhibition on Nutrient Uptake by *Scenedesmus Quadricauda* (Chlorophyceae)," *J. Phycol.*, *21*, 217–222

Quillardet, P., and M. Hofnung, 1993, "The SOS Chromotest – A Review," *Mutat. Res.*, *297*, 235–279

Qureshi, A. A., K. W. Flood, S. R. Thompson, S. M. Janhurst, C. S. Innis, and D. A. Rokosh, 1982, "Comparison of a Luminescent Bacterial Test with Other Bioassays for Determining Toxicity of Pure Compounds and Complex Effluents," in *Aquatic Toxicity and Hazard Assessment*, Fifth Conference, (J. G. Pearson, R. B. Foster, and W. E. Bishop, Eds.), American Society for Testing and Materials, Philadelphia, USA, pp. 179–195

Radetski, C. M., J. M. Ferard, and C. Blaise, 1995, "A Semi-Static Microplate Based Phytotoxicity Test," *Environ. Toxicol. Chem.*, *14*, 299–302

Reinhartz, A., I. Lampert, M. Hersberg, and F. Fish, 1987, "A New Short-Term Sensitive Bacterial Assay Kit for the Detection of Toxicants," *Tox. Assess. Int. Q.*, *2*, 193–206

Reteuna, C., P. Vasser, and R. Cabridenc, 1989, "Performances of Three Bacterial Assays in Toxicity Assessment," in *Environmental Bioassay Tecniques and their Application* (M. Munawar, G. Dixon, C. I. Mayfield, T. Reynoldson, and H. Sadar, Eds.), Kluwer Academic Publishers, Dordrecht, The Netherlands, pp. 149–153

Ribo, J. M., and K. L. E. Kaiser, 1983, "Effects of Selected Chemicals to Luminscent Bacteria and their Correlation with Acute and Sublethal Effects on other Organisms," *Chemosphere*, *12*, 1412–1442

Ribo, J. M., and K. L. E. Kaiser, 1987, "*Photobacterium phosphoreum* Toxicity Bioassay. I. Test Procedures and Applications," *Tox. Assess.*, *2*, 305–323

Rodrigue, D., K. Mailhiot, T. P. Hynes, and L. J. Wilson, 1995, "Aquatic Effects Monitoring in the Mining Industry: Review of Appropriate Technologies," in *Proc. Sudbury '95—Mining and the Environment* (T. P. Hynes, and M. C. Blanchette, Eds.), Canmet, Ottawa, Canada, pp. 813–819

Samuelson, G., and G. Oquist, 1977, "A Method for Studying Photosynthetic Capacities of Unicellular Algae Based on *in vivo* Chlorophyll Fluorescence," *Physiol. Plant, 40,* 315–319

Snell, T. W., and C. R. Janssen, 1995, "Rotifers in Ecotoxicology: A Review," *Hydrobiologia, 313–314,* 231–247

Snell, T. W., and C. R. Janssen, 1996, "Microscale Toxity Testing with Rotifers," in *Microscale Toxicology, Advances, Techniques and Practice,* (P. G. Wells, K. Lee, and C. Blaise, Eds.), CRC Publishers, Boca Raton, USA, in press

Snell, T. W., and G. Persoone, 1989a, "Acute Toxicity Bioassays Using Rotifers. II. A Freshwater Test with *Brachionus rubens,*" *Aquat. Toxicol., 14,* 81–92

Snell, T. W., and G. Persoone, 1989b, "Acute Toxicity Bioassays Using Rotifers. I. A Test for Brackfish and Marine Environments with *Brachionus plicatilis,*" *Aquat. Toxicol., 14,* 65–80

Snell, T. W., B. D. Moffat, C. Janssen, and G. Persoone, 1991a, "Acute Toxicity Tests Using Rotifers. IV. Effects of Cyst Age, Temperature and Salinity on the Sensitivity of *Brachionus calyciflorus,*" *Ecotox. Environ. Saf., 21,* 308–317

Snell, T. W., B. D. Moffat, C. Janssen, and G. Persoone, 1991, "Acute Toxicity Tests Using Rotifers: III. Effects of Temperature, Strain and Exposure Time on the Sensitivity of *Brachionus plicatilis,*" *Environ. Tox. Water Qual., 6,* 63–75

St-Laurent, D., C. Blaise, P. McQuarrie, R. Scroggins, and B. Trottier, 1992, "Comparative Assessment of Herbicide Phytotoxicity to *Selenastrum Capricornutum* Using Microplate and Flask Bioassay Procedures," *Environ. Toxicol. Water Qual., 7,* 35–48

Stanley, P. E., B. J. McCarthy, and R. Smther, 1989, "*ATP Luminescence Rapid Methods in Microbiology,*" Soc. Appl. Bacteriology Technical Series 26, Blackwell Scientific Publications, Oxford, UK

Tebo, L. B., 1985, "Technical Considerations Related to the Regulatory Use of Multispecies Toxicity Tests," in *Multispecies Toxicity Testing* (J. Cairns, Jr., Ed.), Pergamon Press, New York, USA, pp. 19–26

Thellen, C., C. Blaise, Y. Roy, and C. Hickey, Round Robin Testing with the *Selenastrum capricornutum* Microplate Toxicity Assay," *Develop. Hydrobiol., 188/189,* 259–268

Thomulka, K. W., and D. J. McGee, 1992, "Evaluation of Organic Compounds in Water Using *Photobacterium phosphoreum* and *Vibrio harveyi* Bioassays," *Fresenius Environ. Bull., 1,* 815–820

Toussaint, M. W., T. R. Shedd, W. H. Van der Schalie, and G. R. Leather, 1995, "A Comparison of Standard Acute Toxicity Tests with Rapid Screening Toxicity Tests," *Environ. Toxicol. Chem., 14,* 907–915

Trevors, J. T., 1986, "Bacterial Growth and Activity as Indicators of Toxicity," in *Toxicity Testing Using Microorganisms* (G. Bitton, and B. J. Dutka, Eds.), CRC Press, Boca Raton, FL, USA

Trevors, J. T., 1984, "A Method for Assessing the Effect of Pollutants on Electron Transport System (ETS) Activity in Soil and Sediment," in *Toxicity Screening Procedures Using Bacterial Systems* (D. Liu, and B. J. Dutka, Eds.), Marcel Dekker, New York, USA, pp. 163–173

Turbak, S. C., S. B. Olsen, and G. A. McFeters, 1986, "Comparison of Algal Assay Systems for Detecting Water Borne Herbicides and Metals," *Water. Res., 20,* 91–96

Van der Wielen, C., G. Persoone, M. P. Goyvaerts, B. Neven, and D. Quaghebeur, 1993, "Toxicity of the Effluents of Three Pharmaceutical Companies, as Assessed with a Battery of Tests," *Tribune de l'Eau*, *46*, 19–28

Vangheluwe, M., C. R. Janssen, and G. Persoone, 1995, "Sediment Toxicity Screening with Cost-Effective Microbiotests and Conventional Assays: A Comparative Study," Second SETAC World Congress—Abstract book, Society of Environmental Toxicology and Chemistry Press, Pensacola, USA, 23 pp

Vasseur, P., J. F. Ferard, J. Vial, and G. Larbraight, 1984, "Comparison de Tests Microtox et daphnie pour l'évaluation de la toxicité aigue d'effluents industriels," *Environ. Pollut. Ser.*, *34*, 225–235

Van Steertegem, M., and G. Persoone, 1993, "Cyst-Based Toxicity Tests: V. Development and Critical Evaluation of Standardized Toxicity Tests with the Brine Shrimp *Artemia* (Anostraca, Crustacea)," in *Progress in Standardization of Aquatic Toxicity Tests* (A. M. V. M. Soares, and P. Calow, Eds.), Lewis Publishers, Boca Raton, USA

Versteeg, D. J., 1990, "Comparison of Short- and Long-Term Toxicity Results for the Green Algae *Selenastrum Capricornutum*," in *Plants for Toxicity Assessment* (G. Wang, J. W. Gorsuch, and W. R. Lower, Eds.), American Society for Testing and Materials, Philadelphia, USA, pp. 40–48

Vojtísek, V., and V. Jirkú, 1983, "Immobilized Cells of Microorganisms," *Folia Microbiol.*, *28*, 309–340

Willemsen, A., M. A. Vaal, and D. de Zwart, 1995, "Microbiotests as Tools for Environmental Monitoring," National Institute of Public Health and Environmental Planning (RIVM), The Netherlands, Report No 9, 607042005, 39 pp

Wren, M. J., and D. McCaroll, 1990, "A Simple and Sensitive Bioassay for the Detection of Toxic Materials Using a Unicellular Green Alga," *Environ. Pollut.*, *64*, 87–91

Xu, H., and B. J. Dutka, 1987, "ATP-TOX System: A New, Rapid, Sensitive Bacterial Toxicity Screening System Based on the Determation of ATP," *Tox. Assess.*, *2*, 149–166

Xu, H., B. J. Dutka, and K. K. Kwan, 1987, "Genotoxicity Studies on Sediments Using A Modified SOS Chromotest," *Tox. Assess.*, *2*, 79–8

Xu, H., B. J. Dutka, and K. Schurr, 1989, "Microtitration SOS Chromotest: A New Approach in Genotoxicity Testing," *Tox. Assess.*, *4*, 105–114

27

Legislative Perspective in Ecological Risk Assessment

Jan Ahlers (*Berlin, Germany*)

and

Robert Diderich (*Paris, France*)

27.1 SUMMARY

The environmental risk assessment of a chemical substance involves

- *exposure assessment* leading to predicted environmental concentrations (**PEC**) of a chemical substance in different environmental compartments from releases due to its

Ecotoxicology, Edited by Gerrit Schüürmann and Bernd Markert.
ISBN 0-471-17644-3   © 1998 John Wiley & Sons, Inc. and Spektrum Akademischer Verlag.

production, processing, use and disposal. A PEC can be estimated based on measured concentrations in the environment or on mathematical modeling;

- *effects assessment*: data obtained from acute or long-term toxicity tests are used to extrapolate concentrations with expectedly no adverse effects on organisms or ecosystems (Predicted No Effect Concentration, **PNEC**);

- *risk characterization*: for each compartment (water, sediment, soil and atmosphere) the PEC, either based on modeling or on measured concentrations, is compared with the PNEC. If PEC > PNEC an attempt should be made to revise data of exposure and/or effects in an iterative process to conduct a refined risk characterization. In case the PEC remains larger than the PNEC, risk reduction measures have to be considered.

An assessment for Trichloroacetic acid (TCA) is presented to illustrate this procedure.

The advantages and drawbacks of the risk assessment described are discussed and proposals for further amendments especially for the compartments sediment and soil as well as for the marine environment are given. Further development for a more realistic estimation of releases into the environment is also necessary. Moreover it is suggested to allow more flexibility and expert judgement to meet the very heterogeneous data situation for existing chemicals. Sublethal effects and results of chronic studies have to be included in the future. Special attention should be given to persistence and bioaccumulation.

27.2 INTRODUCTION

As a consequence of the various chemicals acts, which came into force in industrial nations during the last two decades, the necessity arose to interpret the data which became available. Depending on the nature of the data to be investigated such interpretations are called:

- *hazard identification* (identification of the adverse effects which a substance has an inherent capacity to cause),

- *dose-response or concentration–effect assessment* (estimation of the relationship between dose or level of exposure to a substance and the incidence and severity of an effect),

- *exposure assessment* (estimation of the concentrations/doses to which human populations (workers, consumers and man exposed indirectly via the environment) or environmental compartments (surface water, sediment, soil, air) are or might be exposed),

- *risk characterization* (estimation of the incidence and severity of the adverse effects likely to occur in a human population or environmental compartment due to actual or predicted exposure to a substance).

The whole procedure starting with hazard identification is called risk assessment.

Applications of hazard identification are the various classification schemes or the derivation of environmental quality objectives. The exposure and the effect assessments already represent a more complex view of the available data and are the basis for the final risk characterization. The total risk assessment therefore represents the most comprehensive interpretation of the available data. In contrast to traditional approaches, which looked at each compartment separately, in a contemporary risk assessment the exposure of a chemical during its whole life cycle (production, use, disposal) to all environmental compartments is considered and the results of these exposure estimations are compared to (eco) toxicological data derived from organisms representative of these compartments.

The aim to protect man and the environment from the risks occurring from the production, use and disposal of chemical substances is approximately achieved by comparing the concentration in the environmental compartments with the concentrations at which no effects on organisms or ecological systems are expected to occur. The reliability to derive such exposure and effect concentrations on the basis of a limited data set will be discussed below. Human beings as well as ecosystems in the aquatic, terrestrial or air compartment are to be protected. For the environment the protection goals for practical reasons are at the moment limited to aquatic ecosystems which include sediment, terrestrial ecosystems, as well as the atmosphere. In addition to the specific compartments, effects which are relevant to the food chain (secondary poisoning) are considered as well as the microbiological activity of sewage treatment systems. The latter is evaluated because the functioning of waste water treatment plants has an important effect on the exposure of the aquatic environment.

Whereas in the past most legislative work was based mainly on effect data (e.g., classification and labelling, derivation of quality objectives, establishment of threshold values, most toxicological assessments) the necessity to compare these data to measured or calculated environmental concentrations was only recently introduced in legislation. For new chemicals, where the risk assessment is generally performed on the basis of premarketing data, several generic exposure models and scenarios have been developed in order to perform the necessary estimations. For existing chemicals, which have been on the market for a long time already, monitoring data can often be used additionally. On the other hand, when assessing the exposure of the environment to existing chemicals, it has to be borne in mind that large quantities of the substances have possibly been released over decades into the environment, and that accumulative processes may have already resulted in background concentrations in the environment which have to be taken into account.

27.3 RISK ASSESSMENT WITHIN THE EU

Within the legal framework of the European Union for new chemicals a certain base-set of data has to be submitted on which an assessment is performed. With increasing tonnage, however, more data become available. For existing chemicals risk assessments at the moment must only be performed if the substance is produced at a level of more than 1000 t/a and if it is on a priority list, thus indicating a potential risk. The producer is obliged to provide all the available data on effects, environmental fate and exposure, and at least a minimal set corresponding to the base-set for new chemicals and including mainly acute aquatic toxicity tests, a test on biodegradation, physicochemical data as well as some information on exposure. Therefore the data basis of a risk assessment will vary considerably from chemical to chemical.

In 1992 the German Federal Environmental Agency published a concept for the risk assessment of existing chemicals (Ahlers et al. 1992; Ahlers et al. 1993) on the basis of a comparison of exposure and effects, which reflected the experience of several years of assessment of existing chemicals. It's aim was to reflect uncertainties on the effect side with safety or assessment factors and on the exposure side with worst-case assumptions. As, in contrast to new chemicals, for existing chemicals the quality and quantity of data are largely heterogeneous, this concept demanded from the assessor a considerable amount of expert judgement and asked for flexibility to meet the different data situations in order to achieve a comparable assessment, regardless of whether only few data were available or a comprehensive data set could be used.

In 1993 after the publication of a new directive for the notification of new substances the EU member countries harmonized their risk assessment methods for notified new substances and common technical guidance documents were published (EC 1993). The following year, corresponding guidance documents were elaborated for existing substances (EC 1994) taking into account further methodological developments in that area as well as existing national assessment schemes (Ahlers et al. 1994).

Both documents have been combined by now (EC 1996). Similar concepts have been published recently by other institutions or organisations such as the US-EPA (Nabholz 1991) or the OECD (OECD 1995).

The procedure for environmental risk assessment is described in Fig. 27.1. Basically it consists of comparing the concentration in an environmental compartment with the highest concentration at which no effects on organisms or ecological systems representative of that compartment are expected to occur. Therefore after the initial step of compiling and validating all available information, comprehensive exposure and effect assessments have to be performed, before in the risk characterization step both results are compared. Afterwards it has to be decided whether there is no immediate concern, further testing or improvement of data is necessary or whether a risk reduction strategy has to be established. In this case the European Council Regulation on the Evaluation and Control of Existing Substances demands the consideration of possible substitution products and the performance of a cost-benefit analysis.

27.3.1 Environmental Exposure Assessment

Comprehensive monitoring data in the compartments water, sediment, soil and air would be the best basis for the determination of the environmental exposure of a chemical. However, such data are rarely available and usually only for

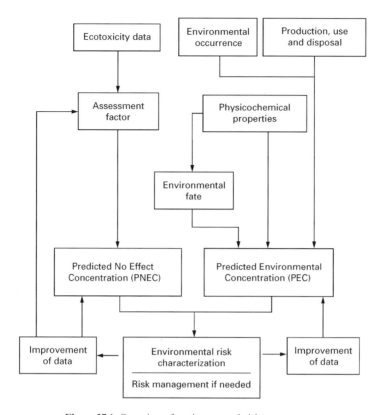

Figure 27.1 Overview of environmental risk assessment.

certain parts of the environment, such as main rivers, some lakes, marine estuaries or the atmosphere in industrial regions. Moreover, the environmental exposure may change rapidly due to alterations in the production (production rate, production technique, treatment of waste water or exhaust air). Therefore it is necessary to estimate the quantities of a chemical released to particular environmental compartments using all available data on production, use, disposal, its physical properties as well as information on environmental fate, including transport and degradation, in order to calculate *Predicted Environmental Concentrations* (*PEC*) as a function of time and space for every compartment (Fig. 27.1).

Releases

While the mathematical models predicting the distribution of a substance in the environment on a local or regional level can be validated and many validated models are already available, the situation is totally different regarding the release of a substance in the different compartments during the different stages of its life.

These release factors are extremely difficult to estimate accurately. They can vary from less than 1% (e.g., release of intermediates to air during their transformation in chemical synthesis) to 100% (e.g., release of detergents to waste water during their use).

For the development of the Guidance Documents for new substances, the most often indicated intended uses for the already notified substances were statistically evaluated. For ten types of use (e.g., textile dyes, paints and varnishes, etc.), all available data from literature and research reports were gathered to allow the prediction of the quantitative release of a substance. The specific properties of the given substance (e.g., fixation degree of a dye onto the cloth) are incorporated into these scenarios. These generic scenarios are meant to be representative of a certain operation and therefore applicable to every site (industrial or other) where this operation is performed. Since the operational conditions (e.g., water consumption) vary from site to site, these are chosen to represent a sufficient "worst case situation," the ultimate goal being to calculate a PEC which would be reached at less than 10% of the total number of sites.

These "use category documents" or "industry category documents" were also formally adopted for the corresponding Guidance Documents for existing substances in 1994. By that time it became clear that it was necessary to develop specific release scenarios for much more use categories. For a transition period it was agreed to use "worst case" release factors, which are based on the physico-chemical properties of a substance, allowing to perform an initial exposure assessment. The risk characterization based on this initial exposure assessment will identify those chemicals for which a more realistic exposure assessment is necessary (see Section 27.3.3).

Local PEC

Local environmental concentrations are those measured or estimated in the immediate vicinity of an industrial site. In Figure 27.2, all the relevant pathways and degradation processes are described which are used for risk assessment in the EU (EC 1994).

Elimination Processes Before Entering Surface Waters. In sewage treatment plants a substance can be eliminated by aerobic/anaerobic biodegradation (route 4), volatilization (route 3) and adsorption and precipitation (route 5). The elimination factor can either be determined from influent and effluent concentrations or approximately estimated by mathematical models using the physical properties of the substance in combination with data from laboratory biodegradation tests.

The most realistic simple models used to simulate waste water treatment plants are multimedia fugacity models (Struijs et al. 1991; Cowan et al. 1993). They permit to establish a complete mass balance, so that the concentration in raw sewage sludge can be estimated.

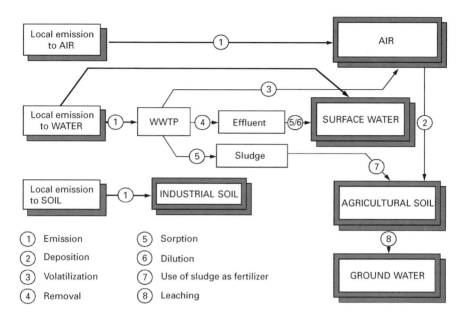

Figure 27.2 Schematic representation of the emission routes considered in the estimation of exposure concentrations at a local scale.

Dilution Processes Upon Entering Surface Waters (route 6). After biological treatment, the waste water is released to the surface water, either a river, a lake or the sea, where it gets diluted. For a river, the low flow is used, preferably the 10-*percentile* based upon daily measurements over one year. To choose a representative dilution factor for a given industrial branch it would be necessary to know the locations of all the sites belonging to that branch and the corresponding river flows. The gathering of this data and the statistical evaluation has yet to be done. A default dilution factor of 10 is used whenever more specific data is missing.

Emission to Air, Water and Soil. The local atmospheric concentration at ground level of a chemical released through stacks or vents (route 1) is estimated with a Gaussian plume model (OECD 1992b; Van Jaarsfeld 1990). The wet and dry atmospheric deposition (route 2) due to local emission to air is averaged over a radius of 1000 m from the source.

It is assumed that, at this distance, agriculture exists and the atmospheric deposition is used as an input into agricultural soil, together with the "contaminated" sewage sludge which is used as a fertilizer (route 7).

To calculate the local concentration in surface water, degradation, volatilization and sedimentation can be neglected. However, the sorption to suspended matter is considered as well as dilution. The diminished concentration of the substance in surface water and its concentration in sediment can be calculated using the organic carbon adsorption coefficient K_{oc}. For soil exposure of the geosphere, the K_{oc} value reflects the potential of a chemical to accumulate or to contaminate the ground water (route 8).

Regional PEC

The regional PEC is an expression of the possible background concentration in a region or small country after long-term use of a substance, once steady state distribution in the different compartments has been reached. Its estimation is especially important for persistent or accumulating substances as well as for substances which are indirectly released to a given compartment (e.g., diffusely to soil through atmospheric deposition). For existing chemicals, the regional concentration can be compared with that measured outside the immediate vicinity of an industrial site or a large town. For notified new substances, which are on the market for a short period of time only, regional concentrations can only be estimated by mathematical models and represent a kind of forecast for future background concentrations. The regional concentrations, whether calculated or based on measured data, are added to the local concentrations. In most cases though, the regional concentrations are negligible compared to the local ones. Multimedia fugacity models (Mackay et al. 1992; Van de Meent 1993) are used for the regional PEC estimation. The processes taken into consideration are partitioning between compartments, removal by biotic or abiotic degradation as well as influx to and efflux from the system.

For risk assessment in the European Union a regional standard environment has been defined, which is highly industrialised, relatively small (200×200 km) but densely populated (20 million inhabitants). It is assumed that 10% of European production takes place within this area, i.e., 10% of the total estimated emission is used as input for the region (EC 1994).

The characteristics of this "standard region" (e.g., area covered by surface water, degree of connection to biological waste water treatment) are European averages, although these parameters can significantly differ from one country to another. The aim of the regulation is to identify those substances which have to be regulated throughout the European Union.

Use of Monitoring Data

Whenever possible monitoring data for air, water, sediment and/or soil should be used in addition to the calculated PECs in the exposure assessment. These data have to be evaluated for their adequacy and representativeness. It has to be ascertained whether the data are the result of sporadic studies or whether they are detected at the same site over a certain period of time. Data from a prolonged programme, including seasonal fluctuations, are of special interest for the exposure assessment. The 90-percentile values are of highest preference. In addition, the measured data should be allocated to a local or regional scale in order to derive the nature of the environmental concentration estimated. This allows a comparison with the corresponding calculated PEC to determine which

PEC should be used in the risk characterization. If there is no spatial proximity between the sampling site and point sources of emission, the data represent a background concentration that has to be added to the calculated local PEC. Otherwise they reflect the releases into the environment through point sources, where, however, the background concentration is already included.

After the determination of a calculated PEC and selection of qualified monitoring data, calculated and measured concentrations have to be compared. Analysis and critical discussion of divergences are important steps for developing an environmental risk assessment for an existing chemical.

27.3.2 Environmental Effects Assessment

Introduction

The first step in the effects assessment process is evaluation of the available data. This step is of importance, especially for existing chemicals, as tests will have often been carried out using non-standard organisms and/or non-standardized methods. Due to the quantity and quality of valid data, appropriate assessment factors have to be derived in the next step. Finally, the Predicted No-Effect Concentration (PNEC) is calculated by dividing the lowest LC/EC_{50} values of acute tests or NOECs from long-term studies by an appropriate assessment factor (Figure 27.3).

Aquatic Compartment

Assessment factors are necessary to span the discrepancies between the aim to protect structure and function of natural ecosystems and the reality that the assessment can usually be based only on a few monospecies laboratory tests of varying quality. The following uncertainties have to be considered and should be reflected by the assessment factors (Ahlers et al. 1994):

- variances of the test results,
- differing sensitivities of each individual of one species (biological variance)
- differing sensitivities of species,
- small number of species tested compared to complexity of an ecosystem
- undiscovered sublethal effects,
- the possibility of additive or even synergistic effects due to the presence of several chemicals in environmental compartments, whereas in a laboratory test normally only a single substance is studied,
- extrapolation of acute toxicity (LC/EC_{50}) to chronic toxicity (NOEC),
- extrapolation of laboratory data to ecosystems.

These aspects have to be considered when a numerical value is derived for an assessment factor. However, it has to be kept in mind that it is extremely difficult

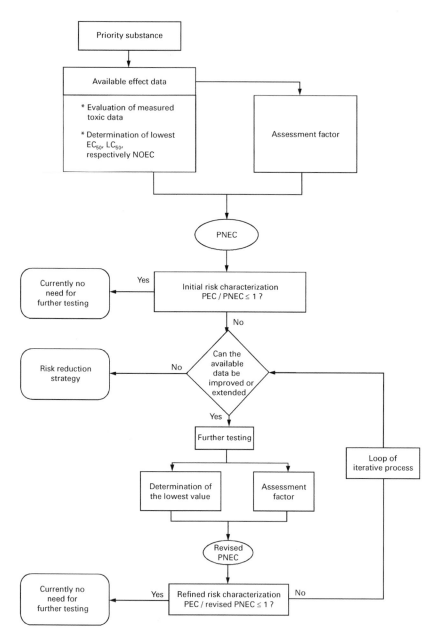

Figure 27.3 General procedure of the concentration effects assessment.

to quantify the above-mentioned uncertainties. The concept of safety factors rather represents a pragmatic approach to approximately evaluate the hazards caused by environmental chemicals.

A detailed assessment of the environmental risk is often only feasible for the water compartment. The quantity and quality of available data vary considerably, especially for existing chemicals, ranging from less than base-set information to very well studied chemicals. As pointed out above, the size of the assessment factor depends on the confidence with which a PNEC can be derived from the available data. This confidence increases if data are available on the toxicity to organisms at a number of trophic levels, and for taxonomic groups and various feeding strategies. Thus lower assessment factors can be used if organisms at several trophic levels have been studied and if long-term investigations are available in addition to acute ones. The EC regulation for the assessment of existing chemicals requires, as a minimum for the aquatic environment, acute toxicity data on fish, daphnids and algae. On this basis an assessment factor of 1,000 is considered to be appropriate to calculate a PNEC from the lowest valid LC/EC_{50} values obtained. If, in addition, long-term toxicity studies have been performed, considerably lower assessment factors can be applied to the NOEC derived from these tests. They range from 100 for one long-term study to 10 for three, although it seems questionable whether a factor of 10 is sufficient to meet all the remaining uncertainties in extrapolating from laboratory tests to the environment. It has to be mentioned that standard as well as non-standard organisms can be used in environmental risk assessment. However, the lower factors applicable if long-term studies are available should only be used if the most sensitive species of the acute tests have been tested in the long-term studies. Assessment factors resulting from field studies or from investigations using model ecosystems are derived on a case by case basis.

In special cases, deviations from these assessment factors are possible. However, these deviations should be substantiated and should be applied only in exceptional cases, e.g. when data from a wide selection of species are available covering at least three trophic levels and/or additional taxonomic groups other than those represented by the base-set species (EC 1994).

Another example are substances which are not continually released but are subject to intermittent release. In this case the exposure may be only of short duration and therefore only short-term effects need to be considered. Thus an assessment factor of 100 may be sufficient, which allows extrapolation from short-term laboratory tests to short-term effects in ecosystems. However, a case-by-case decision should be taken, especially in the case of substances with a high bioaccumulation potential.

Sediment

So far, very few tests for sediment organisms have been conducted. In the absence of any ecotoxicological data for sediment-dwelling organisms, the PNEC may be calculated provisionally using the equilibrium partitioning method with the PNEC for aquatic organisms and the sediment/water partitioning coefficient $K_{p(sed)}$ (OECD 1992a):

$$PNEC_{sed} = K_{p(sed)} \times PNEC_{aqu}$$

In the partitioning method it is assumed that sediment-dwelling and aquatic organisms are equally sensitive to the chemical, that the concentration in the pore water can be determined by the sediment water partition coefficient and that uptake of the chemical only occurs via the water phase. It is obvious that these assumptions can only be a first approximation. Especially for substances with a high octanol-water partition coefficient the ingestion of sediment may become the main uptake route, leading to an underestimation of the total uptake. As a first approach, an additional factor of 10 is applied to the PEC for highly accumulating substances.

Terrestrial Compartment

A risk assessment for the terrestrial compartment should be performed if release of the substance (application of sewage sludge in agriculture, direct application of the chemical, wet or dry deposition from the atmosphere) leads to relevant exposure in the soil. At the moment there is no assessment strategy available for effects on soil function like filtration, buffering capacity and metabolic capacity. Therefore the assessment has to be based solely on direct effects on soil organisms. In contrast to the aquatic compartment, however, the amount of toxicity data is usually very limited. In some cases acute tests with earthworms or plants are available or tests with microorganisms. Long-term tests are rarely found. Another problem arises from the generally reduced bioavailability of the chemicals. Furthermore, this bioavailability is largely influenced by the soil properties, e.g., content of organic matter. Therefore a standard soil has been defined with a content of 3.4% organic matter. Tests performed with different kinds of soil have to be normalized using relationships which describe the bioavailability of chemicals in soil.

The same assessment factors as for the aquatic system should be used for the terrestrial system according to the type of investigation (acute or long-term toxicity tests) and the number of trophic levels tested. The $PNEC_{soil}$ is calculated on the basis of the lowest effect value measured. If short-term tests with a producer, consumer and/or with a decomposer are available the test result is divided by a factor of 1000 to calculate the $PNEC_{soil}$. If only one terrestrial test is available, the risk assessment should be performed both on the basis of this test and on the basis of the aquatic toxicity data, in order to give an indication of the risk for soil organisms. For reasons of precaution, the highest PEC/PNEC ratio determines whether further tests are necessary.

In many cases, however, due to the lack of data only an approximation using the equilibrium partitioning method described above applied to soil can be performed. Again problems arise with substances which have a high accumulation potential.

Air

For a risk assessment of the air compartment biotic and abiotic effects have to be considered. However, almost no toxicological data or even internationally accepted test guidelines on animal species other than mammals are available. Therefore a quantitative effect assessment cannot be performed for this compartment and consequently a risk characterization is not possible. Instead, in a qualitative assessment effect concentrations observed should be described and compared to the PEC.

For the evaluation of an abiotic effect the following physicochemical parameters have to be considered:

- global warming
- ozone depletion in the stratosphere
- ozone formation in the troposphere
- acidification.

Secondary Poisoning

Secondary poisoning is the result of the transfer of chemicals via the food chain to different trophic levels (biomagnification) resulting in toxic effects in the higher levels in the trophic chain. The following scheme (Figure 27.4) shows the procedure in the assessment of secondary poisoning for the route water/fish/ fish-eating mammal or fish-eating bird proposed for risk assessment within the EU. In the first step it has to be examined whether there is an indication of a bioaccumulation potential. In the aquatic compartment this can be done by measuring the bioconcentration factor under equilibrium conditions or via uptake and depuration kinetics. In the absence of such measurements a log K_{ow} of higher than 3 can also be interpreted as an indication of a bioaccumulation potential. However, for several groups of chemicals (detergents, metal ions, substances with high molecular weight) this approach is not appropriate. If there are indications of a bioaccumulation potential as a second precondition for an assessment it has to be checked, if the chemical is to be classified on the basis of its mammalian toxicity as Very Toxic (T+), Toxic (T), or Harmful (Xn) with one of the risk phrases R47 (may cause birth effects), R48 (danger of serious damage to health by prolonged exposure), R60 (may impair fertility), R61 (may cause harm to the unborn child), R62 (possible risk of impaired fertility), R63 (possible risk of harm to the unborn child), R64 (may cause harm to breast-fed babies). In this case a PEC_{oral} is calculated from the PEC_{water} and the BCF. This value should be lower than the $PNEC_{oral}$, which can be derived from the No Observed Effect Level (NOEL) in dietary toxicity tests with animals representative of fish-eating birds or mammals.

27.3.3 Risk Characterization

As soon as all available information has been evaluated and PECs as well as PNECs have been derived, an initial risk characterization can be performed by comparing the PECs with the PNECs for the relevant compartments: aquatic ecosystem, terrestrial ecosystem, atmosphere, top predators, microorganisms in

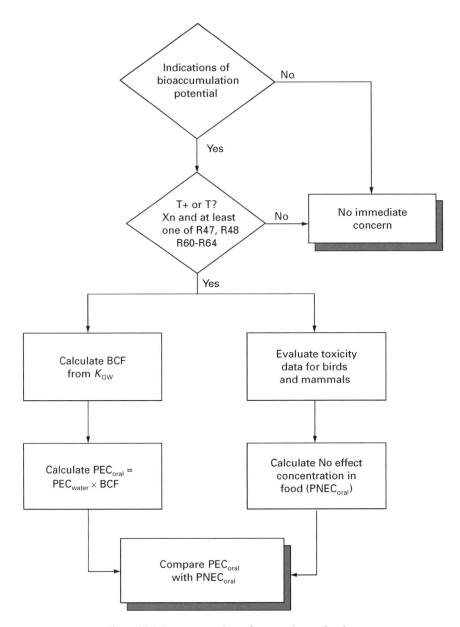

Figure 27.4 Assessment scheme for secondary poisoning.

sewage treatment plants (Figure 27.5). If in this initial assessment the PEC/ PNEC ratio is not greater than 1, no further information is required unless the production volume or the use pattern change considerably. However, if one or more PEC/PNEC ratios are greater than 1 an environmental risk is apparent.

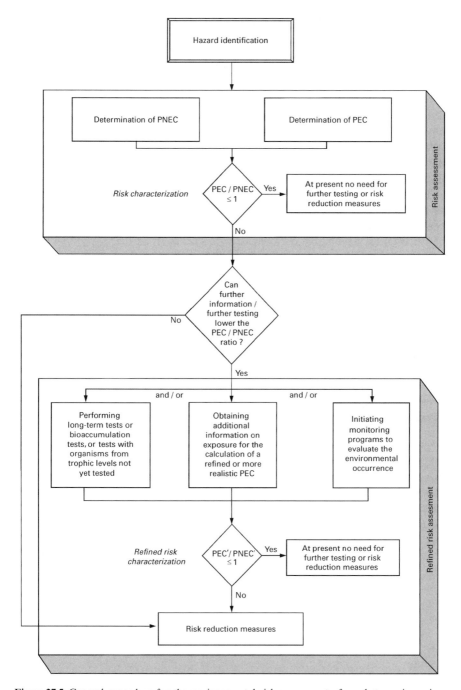

Figure 27.5 General procedure for the environmental risk assessment of a substance in a given compartment.

In this case it has to be decided whether it is possible to obtain more information on exposure and/or effects (e.g. by performing long-term tests or tests with organisms from trophic levels not yet considered). In addition, more information on exposure helps to avoid worst-case assumptions for calculating a refined PEC. In some cases, measurement of environmental concentrations becomes necessary. With this additional information, the new refined PEC and/or PNEC can differ from the initial values. Therefore, whenever new and relevant data are obtained, PEC and PNEC have to be compared again in a refined risk characterization. This procedure is repeated till the PEC/PNEC ratio is no longer greater than 1 (no need for further action) or till the data cannot be refined further to decrease the ratio below 1.

This iterative process has precautionary aspects, especially in the initial phase, as data gaps are filled in by worst-case assumptions or high assessment factors. It also saves resources and takes animal welfare into account, as only a minimum of tests has to be performed. Possible hazards are identified by the preliminary assessment at an early stage and a specific testing plan can be elaborated for maximum additional information at low expense.

For substances with a low water solubility, where no acute effects can be observed at the solubility limit, long-term effects cannot be ruled out, especially in the case of high lipophilicity. Therefore, a long-term study should be performed if for these substances the log $K_{OW} > 3$ (or BCF > 100) and PEC > 1/100th of the water solubility.

27.3.4 Example of a Risk Assessment of a Chemical Substance According to the EU Scheme

As pointed out above, especially for existing chemicals, the heterogenous data set often makes a risk assessment rather difficult. Generalizations or schematic procedures are rarely possible. A great deal of expert judgement is necessary in exposure assessment as well as in effect assessment. To illustrate these problems as well as possible solutions and to give a more practical description of a risk assessment according to the procedure discussed above, we present an assessment of trichloroacetic acid (TCA). It is an excerpt from a complete initial risk assessment of TCA prepared by the German authorities for the OECD programme on existing substances. For shortness sake, only one single use of the substance is reported here, and the assessment is limited to the aquatic compartment. The complete assessment and detailed description of the test results can be found in the "OECD SIDS initial assessment report" (OECD 1994).

Identity

Name: Trichloroacetic acid,TCA
CAS-Nr.: 76-03-9
Structural Formula: CCl_3COOH

Exposure

TCA is used as an auxiliary in textile finishing (ca 110 t/a). According to the German producer, TCA is included in dyes as an alkali-liberating auxiliary for fixing reactive dyes on fibres for printing processes. In the usual formulations, the relative ratio to the actual dye is 1:1. Its average concentration in the formulations is 40 g/kg. Ca. 2.9 g TCA are necessary to completely dye 100 g textile.

Environmental Fate

TCA is highly soluble in water (930 g/L). The pure solid compound has a vapour pressure of 120 Pa at 50°C. In aqueous solution, TCA dissociates ($p_{Ka} = 0.7$) and volatilization from water is not to be expected to a high degree. The measured $\log K_{OW}$ amounts to -0.27 (probably dissociated form). The $\log K_{OW}$ of undissociated TCA was calculated to be 1.33. Bioaccumulation factors of 0.4–1.0 and < 1.7 have been determined experimentally (OECD GL 305C). Bioaccumulation in fish is therefore considered to be low.

Tests on ready biodegradability of TCA in the aquatic medium are not available. Nevertheless, several test results on inherent biodegradability are available. In two Zahn-Wellens tests with industrial activated sludge, TCA proved to be non-biodegradable (ca. 5–10% DOC removal after 27 days). In a modified MITI-II-test (with 100 mg/L instead of 30 mg/L test substance; 100 mg/L inoculum) biodegradation rates of 0–46% in 28 days were recorded.

On the other hand, in a modified OECD confirmatory test, no significant biodegradation was observed; even after increasing the retention time to 48 hours, only a maximum of 10% biodegradation with regard to Cl^- liberation could be achieved.

In aerated lagoons receiving the effluents from mills producing wood pulp the removal was 22% in one treatment plant (age 3 years; hydraulic retention time 2.5 days) and 99% in a second (age 10 years; hydraulic retention time 5 days).

Despite the many negative biodegradation results in laboratory tests, a significant elimination of TCA in waste water treatment plants can be expected, based on monitoring data. Indeed, the elimination of TCA was monitored in 5 domestic waste water treatment plants (WWTPs) in Switzerland and ranged from 24 to 90%. An average removal rate of 66% was recorded (average influent concentration: 1.4 µg/L; effluent conc.: 0.43 µg/L).

Exposure Estimation/PEC Calculations

Local concentration due to the use of auxiliary in textile dyes

According to the use category document on textile dyes (EC 1993), the colouring capacities of dye-houses vary within a wide range. An average weight of processed goods of 3000 kg/day is assumed, implying a daily use of 86 kg TCA for dying the whole cloth. During the steaming process for fixation, TCA is decomposed to sodium bicarbonate, soda and chloroform. In a laboratory simulation test performed by the producer, TCA decomposed to 96.5%. With this decomposition rate a daily release rate of 3 kg is estimated.

The default volumes of water used during wet processing of textile material (including pretreatment such as scouring, desizing, mercerizing, bleaching, etc.) is ca. 250 $m^3/1000$ kg textile. For those textile processing units which do not perform all the pretreatment processes, lower waste water volumes would occur and subsequently higher concentrations of TCA.

With the waste water volumes indicated above, and the decomposition rate of 96.5% a concentration in the raw waste water of 4 mg/L is calculated.

Most textile processing sites are rejecting their waste water into domestic waste water treatment plants (ca. 93% in Germany). Assuming a dilution factor of 5 upon entering the WWTP, an influent concentration of ca 0.8 mg/L is calculated.

Elimination during biological treatment is assumed to be 66% (see above). With an elimination rate of 66% in WWTPs and assuming a further dilution factor of 10 upon entering the surface water, a PEC_{local} of ca. 27 µg/L is calculated. Distribution immediately upon entering the surface water cannot be expected.

For estimating the successive dilution steps, other data are available: in a recent survey of textile processing units by ETAD, the 20-percentile of the water flow of 25 waste water treatment plants receiving waste water from textile processing units is 5000 m^3/d. The 20-percentile of the water flow of 18 rivers receiving waste water from these is 45,000 m^3/d.

With a release of 3 kg/d TCA, an elimination in WWTPs of 66% and a dilution in a total of 50.000 m^3/d (5,000 + 45,000 m^3/d), the resulting PEC_{local} is 20 µg/L.

The above scenario assumes that the whole cloth is dyed. On the other hand, print colour is usually applied only to specific areas by screen printing systems to achieve the planned design. For comparison, the PEC will be calculated for an average use of 30 kg dye per day, i.e. 30 kg TCA. With a decomposition rate of 96.5%, 1.05 kg TCA are rejected per day into the waste water. Eliminated in WWTPs at a rate of 66% and diluted in 50,000 m^3/d, the resulting PEC_{local} is 7 µg/L.

In conclusion, the expected PEC_{local} in surface waters will be in the range 7 to 27 µg/L.

Effects on Aquatic Organisms

Unless otherwise stated, the tests were performed with the neutralized acid or with the sodium salt (the mass conversion from the TCA-Na-salt to the free acid was not performed; the concentrations marked "*" are related to the TCA-Na-salt).

a) toxicity to fish

Leuciscus idus	48h LC_{50}	> 10,000 mg/L
Poecilia reticulata	48h LC_{50}	9,160 mg/L
Alburnus alburnus	96h LC_{50}	9,300 mg/L
Pimephales promelas	96h LC_{50}	2,000 mg/L

Furthermore, the results from a long-term test are available:

Cyprinus carpio	63d LOEC	7 mg/L*

The weight loss was ca. 10% compared to the controls. No other behavioral or clinical changes were observed. Histopathological changes were muscular atrophy and hyaline degeneration as well as cell necrosis in the gills.

b) toxicity to invertebrates

Daphnia magna	24h EC_{50}	8,370 mg/L
	24h EC_{50}	> 10,000 mg/L*
	48h EC_{50}	2,000 mg/L
Nitocra spinipes	48h EC_{50}	4,800 mg/L

The EC_{50} values are consistently high. The lowest acute toxicity in a valid test was recorded as 2,000 mg/L. No long-term test results are available. Sublethal effects were determined with Dragonfly nymphs (several different species: *Somatochlora cingulata, Aeschna umbrosa* and species of genera *Aeschna* and *Basiaeschna*); significant effects were recorded in the oxygen consumption rate at 0.01 mg/L and 0.1 mg/L and the ammonia excretion rate at 0.01 mg/L.

c) toxicity to algae

Scenedesmus quadricauda	7d TT	200 mg/L*
Effect: growth inhibition (biomass); TT = toxicity threshold		
Ankistrodesmus minutissimus	14d EC_{50}	98 mg/L*
Chlorella pyrenoidosa	14d EC_{50}	0.3 mg/L*
	14d NOEC	0.01 mg/L*
Chlorella mucosa	14d EC_{50}	0.46 mg/L*
	14d NOEC	0.01 mg/L*
Chlorococcum sp.	14d EC_{50}	1.2 mg/L*
Dictospaerium pulchellum	14d EC_{50}	7 mg/L*
Scenedesmus acutus	14d EC_{50}	8.8 mg/L*
Effect: growth inhibition (biomass)		
mixed culture of Chlorococcales	24h EC_{10}	> 1,000 mg/L*
Effect: inhibition of oxygen production		

Algae are very sensitive to TCA. Only results from non-standardized tests are available. The strongest effects were recorded in a 14-day test with the species *Chlorella pyrenoidosa* and *Chlorella mucosa*. Although the recorded effects are based on biomass after the exponential growth phase, an analysis of the growth curves reported in the publication shows that the effect concentrations derived from biomass during the exponential growth phase are of the same order as those reported after 14 days.

d) Toxicity to microorganisms

Several tests with protozoa, cyanobacteria as well as other bacteria have been performed. The lowest value was a 14d EC_{50} of 5 mg/L obtained with a cyanobacterium.

Determination of $PNEC_{aqua}$

The lowest aquatic effect concentrations were determined with algae (14d-NOEC = 0.01 mg/L, NaTCA = 0.0086 mg/L TCA) and invertebrates (significant effects at < 0.01 mg/L). The acute sublethal effects recorded with the dragonfly nymphs underline the high sensibility of certain organisms to TCA. On the other hand, as this type of recorded effect does not fit the usual assessment schemes, the effect data from algae are used for the assessment here.

According to the EU Technical Guidance Document for the risk assessment of existing substances, a safety factor of $F = 50$ applied to the lowest NOEC would be appropriate, as long-term tests are available only for species from two trophic levels. However, as the toxicity to daphnids is approximately four orders of magnitude lower than to algae, a long-term test with daphnids would probably not result in a lower NOEC than that applied so far. Therefore an assessment factor of 10 is justified in this case.

$$PNEC_{aqua} = 8.6 \ \mu g/L/10 = 0.9 \ \mu g/L$$

Risk Characterization

With a PEC of 7–27 µg/L and a PNEC of 0.9 ug/L a PEC/PNEC ratio of 8–30 is derived. Based on the present data configuration, therefore, a risk for the aquatic compartment has to be expected.

Recommendations

Before recommending emission reductions, it should be checked whether the database can be improved. As the assessment factor cannot be lowered, it is only possible to revise the PEC. Therefore it is proposed to monitor the releases analytically in order to substitute some worst-case assumptions in the exposure assessment.

27.3.5 Comparison with Other Countries

A recent survey of risk assessment schemes throughout OECD countries (OECD 1995) has shown that in most of these countries comparable methods are used to identify those substances presenting a risk to the environment. The principle of comparing concentrations occurring or expected in the environment with concentrations known or expected to have no effect upon ecosystems is almost universal. The way effects are measured is comparable throughout OECD countries. Extrapolation from laboratory data to real situations (single species to other related species, short term toxicity data to long term data, etc.) is common to all schemes. Differences still exist for the extrapolation factors or safety factors (e.g. 100–1000 if acute toxicity data are available for three species from three trophic levels).

A more qualitative procedure is used in Japan (Environmental Agency Japan 1995) as defined in the Japanese Chemical Substances Control Law. The determining factor for regulating a substance is its inherent potential to cause pollution that may affect human health. Possible damage to the ecosystems is so far not considered directly for limiting the use of a substance within the scope of this law. The criteria for limiting the use of a substance are persistence, bioaccumulation and chronic toxicity to humans. For existing chemicals which are suspected to have a high possibility of persistence in the environment, monitoring programmes are initiated in surface waters. Monitoring in wildlife is conducted in a third step to verify the persistence of those chemicals needing special attention.

In Canada, while the risks assessment procedure is similar to that in the EU, the stringency of risk reduction measures is also dependent on bioaccumulation and persistency (Government of Canada/Environment Canada 1995). Depending on whether or not all the criteria are fulfilled, the use of the substance is limited till virtual elimination from the environment or life-cycle management is initiated to prevent or minimize release into the environment.

As the aim of the different countries is the same, i.e. the protection of man and the environment from adverse effects of chemical substances, the harmonization of risk assessment procedures throughout the OECD countries would favour the mutual acceptance of risk assessments among member countries and thereby speed up the process of assessing the huge number of existing chemicals which still have not undergone the procedure. It is clearly recognized among member countries and within the OECD Existing Chemicals Programme that this harmonization is a long-term goal.

27.4 DISCUSSION AND PERSPECTIVES

In Section 27.3 a description of the procedure of environmental risk assessment in the EU is presented. These assessment schemes represent a practical approach to obtaining an answer to the question as to whether there is a risk to the environment from the large number of chemicals which have been on the market in large quantities and for many years, or from those new chemicals which are to be placed on the market. As discussed by Klöpffer (1994b) such a single compound assessment, although it represents considerable progress compared to the older practice of judging organic pollutants by sum parameters, is limited by several problems, such as

- great number of existing substances (can be solved by priority-setting, e.g., Ahlers et al. 1991),
- complex mixtures consisting of many individual compounds (e.g. chlorinated paraffins, crude oil, gasoline) which have different physicochemical properties and thus a different distribution/degradation behaviour and different effects,
- synergistic effects (such effects have been described especially for lipophilic compounds due to interactions with the cell membrane of organisms by Witte et al. 1995)

Klöpffer (1994) concludes that therefore even a comprehensive single compound assessment cannot give the full truth and suggests combining such an assessment with a Life Cycle Assessment of products or ecobalances.

In addition it is obvious that many details of the assessment procedures have not yet been verified scientifically and some areas can only be treated in a qualitative manner.

Some methodological solutions, which are not validated due to lack of scientific evidence, represent a compromise between all the parties involved in order to keep the scheme practical. Therefore on the basis of scientific developments and practical experience, future amendments are possible and necessary. In this chapter some experience achieved in Germany on the basis of about 100 environmental risk assessments of existing chemicals performed in accordance with the criteria given in chapter 2 are discussed. Moreover the advantages and drawbacks of this procedure are discussed and some proposals for future amendments are presented.

27.4.1 Availability of Data

The reliability of the results of a risk assessment depends mainly on the quality and quantity of data. If, for example, a large number of acute and chronic tests have been performed on a wide variety of species covering several trophic levels, an expert will be able to estimate approximately a concentration where probably no effects on organisms or ecosystems will occur, especially if field tests or tests on artificial ecosystems have been performed, regardless of the assessment schemes applied. If, on the other hand, the release of a chemical to the environment during its whole life cycle is known (production, processing, use, disposal) and the distribution between the environmental compartments can be calculated from its physical properties and, in addition, monitoring data are available, then the PEC can be reasonably well estimated.

Such situations, however, occur very rarely. With decreasing quantity and quality of data the PEC and PNEC values can only be determined with large uncertainties and this has to be reflected by higher assessment factors and worst-case assumptions.

It is obvious that the largely different regulatory requirements for new and existing chemicals may hinder innovation, as for industry it is often much cheaper to continue the production of an existing chemical instead of placing a less hazardous new one on the market. The differences between new and existing chemicals could be diminished gradually. The need for further testing of new chemicals at production levels of 100 or 1000 t/a could be made dependent on the results of the initial risk assessment, similar to the procedure for existing substances. The experience gained when assessing new and existing chemicals has shown that, especially for a realistic exposure assessment, the base-set requirements are insufficient. A questionnaire should be developed, which, depending on the use pattern of a chemical, should be able to procure the necessary data.

One of the main uncertainties when estimating the elimination of a substance by biodegradation in a waste water treatment plant, in surface water or in soil, is the biodegradation rate in these compartments. Results from simulation tests are very rarely available and for some compartments no internationally agreed test guidelines have been made available so far.

For most exposure assessments the results from screening tests (like those according to the OECD guidelines on ready or inherent biodegradability) have therefore to be extrapolated into degradation rate constants, although these tests were not designed to provide this kind of information.

One of the major challenges of the years to come is to develop test guidelines for biodegradation simulation in different compartments as well as an iterative testing strategy. Compared to the aquatic compartment the data situation for sediment or soil assessment is usually much worse, in most cases it does not allow a quantitative assessment. As the long-time use of high production volume chemicals in most cases has lead to an exposure of soil and/or sediment to chemicals with a high accumulation potential, a minimal data set should include at least results of tests with plants, soil invertebrates and possibly soil micro-organisms. For exposure estimation more knowledge should be available on distribution and bioavailability of the chemical in soil. Tests for sediment assessment should be developed or standardised to enable a quantitative effect assessment.

For the air compartment the data available so far enable an assessment of abiotic effects. In addition by using physicochemical data in connection with generic models a preliminary exposure assessment is possible. As the concentrations of industrial chemicals in the air compartment are usually low due to dilution, chronic plant aeration tests should be developed in order to enable a quantitative risk assessment.

27.4.2 Assessment Scheme for the Aquatic Compartment

As discussed above a risk assessment for the aquatic compartment is—compared to the other compartments—in most cases possible. For existing chemicals a drawback of the concept to be applied in the EU is that it had to be elaborated in accordance with the concept for new chemicals. However, the data set building the basis for the assessment normally differs considerably between both groups of chemicals. For new chemicals there are either base set data available (acute studies on fish, daphnids, algae performed according to OECD guidelines and GLP) or in higher tonnages long-term studies with the same species. These very precise data lead to assessment factors of 1,000, 50 or 10 respectively and a testing strategy is limited in these tests. For existing chemicals on the other hand we are confronted with a very heterogeneous data situation (see the example above). The minimal data are three aquatic tests. However, they are not necessarily conducted according to OECD Guidelines or GLP. Non-standard organisms may be used and exposure time may vary. In addition other information may be available, e.g., acute tests with other species, long-term or chronic studies describing endpoints other than just lethality and often covering more sensitive life stages. During the last few years several alternative test methods—often using cell cultures—have been examined which describe sublethal endpoints (e.g. Ahlers et al. 1988; Ahlers et al. 1991). In addition, in some cases tests in model ecosystems or even field studies have been performed. Therefore it is evident that for existing chemicals a continuum of information from less than base-set up to a very comprehensive data situation exists. This information should not be reduced to just a few assessment factors. Instead, it would be advisable to allow the expert to choose an appropriate factor according to the complete data set. Otherwise important information remains unused. The above mentioned assessment factors should in such a scheme only be applied as a framework. For choosing an individual assessment factor the range of variations should ideally be considered, if several tests have been performed. Neither using the lowest effect concentration nor calculating the mean value are totally satisfactory. Moreover, the meaning of a test or its sensitivity should be taken into account when an assessment factor is derived. Clearly, a 72h NOEC for algae has to be assessed differently than a 21d NOEC for daphnids, a 28d NOEC for fish, an early life-stage test or a full life-cycle fish test and thus should have an influence on the assessment factor applied.

When further testing is necessary, the whole data situation should be reflected and an appropriate test should be chosen. For example, one criterion could be to choose a longer exposure time when chemicals with a high log K_{ow} are tested, as under the normal exposure time a thermodynamic equilibrium may not yet have been achieved.

Flexible responses to the PEC/PNEC ratios are also necessary depending on the kind of emission data used in the exposure assessment. If it is based mainly on generic scenarios or regional PECs the first step would be to gather information in order to be able to perform a site-specific exposure assessment for all sites concerned before revising the PNEC.

Bioavailability, different exposure scenarios, different modes of action, synergistic or antagonistic effects of other substances, all mitigate against a single concept being right for all chemical substances. No matter whether the exposure is estimated by a model or by measurement, consideration needs to be given as to how the exposure concentrations vary with time and distance in relation to the source. Undue regulatory attention could be paid to low volume, intermittent, point source discharges to the possible detriment of much higher volume discharges of a more diffuse nature (Brown 1994).

For the assessment of sediment the procedure described in the section *Sediment* on the basis of the equilibrium partitioning method is not satisfactory, especially as the uptake pathway via ingestion is not sufficiently taken into account. Moreover, the biotransformation process for substances adsorbed onto sediment may be quite different to that in surface water (Rönnpagel et al. 1995). Therefore Ahlf (1995) proposes a tiered approach including a test battery for sediment assessment.

27.4.3 Assessment Scheme for the Terrestrial Compartment

If there is a significant exposure to the terrestrial compartment an assessment on the basis of the pore-water concentrations in combination with the aquatic toxicity does not seem to be sufficient for high production volume chemicals.

For exposure assessment in the terrestrial compartment more information is necessary, especially on distribution, degradation and on the influence of adsorption upon availability to biodegradation and terrestrial organisms.

In the case of exposure at least tests with soil invertebrates (preferably earthworm reprotoxicity), plants (phytotoxicity) and soil microorganisms (dehydrogenase activity) should be available to derive an appropriate PNEC which can be compared with the PEC_{soil}. If the ratio PEC/PNEC is greater than 1 further tests are necessary. Römbke et al. (1996) suggest a testing strategy consisting mainly of long-term or chronic tests (e.g. reproduction) and of the estimation of bioaccumulation, which should be performed separately with plants and animals due to large physiological differences in uptake mechanisms. Whereas in level I the above-mentioned information should be available, in level II a collembole reproduction as well as a staphylinide generation test are suggested. Additional tests on species or trophic levels which are most sensitive according to level I might be performed in level II. However, so far only a few long-term tests are available and not much experience could be gained with them. If PEC/PNEC is still above 1, tests using model ecosystems could be performed as discussed by Morgan and Knacker (1994) in order to decide whether risk reduction measures for the terrestrial compartment are necessary.

27.4.4 Influence of Persistence and Bioaccumulation Potential

The influence of persistent substances is principally considered in the exposure assessment. However, a precise quantification is difficult for several reasons.

One problem is the lack of knowledge about how to extrapolate from laboratory biodegradation tests to the natural environment. This lack of conceptual framework is especially critical if the substance is not readily biodegradable and thus it or its transformation products may persist under natural conditions. Moreover, the question of persistent transformation products is especially important in cases where abiotic degradation is the main elimination process and thus mineralization cannot be expected.

The bioaccumulation potential is considered at several places in the assessment scheme too. Based on the PEC_{aqua} it is taken to estimate PEC_{oral} and is used for the assessment of secondary poisoning. Moreover, the fact that bioaccumulation leads to high concentrations in biota usually has an influence on toxicity. However, often the time before a steady-state is reached is relatively long compared to the exposure time in a laboratory test, especially for substances with a high bioaccumulation potential (e.g. DDT). In these cases neither acute tests nor the long-term studies normally applied are sufficient. Often only chronic tests, preferably studies over more than one generation, may be able to indicate the hazards. Klöpffer (1994a) points out that persistence is the central criterion of environmental hazard assessment of anthropogenic, organic chemicals. Special attention should be given if a substance is not only persistent, but also able to accumulate.

Cowan et al. (1995) suggest including a quantitative assessment of the potential for, and consequences of, persistence and bioaccumulation when conducting a quantitative environmental risk assessment. In this process it has to be evaluated if increased environmental concentrations occur due to differences between the rates of addition and loss. If this is the case, the possibility of impact on the ecosystem in the future exists. While such calculations are taken into account for soil and rivers, no corresponding methods exist for the marine environment. Due to high dilution of the chemical the conventional risk assessments on the basis of the PEC/PNEC ratio are not appropriate for protecting marine life in the future as the sea is a sink for many chemicals. Thus a prediction of addition and loss should be included in an assessment as well as an estimation of concentrations in marine biota.

Another possibility to give persistence and bioaccumulation greater consideration would be to enhance the assessment factor if a chemical shows a high bioaccumulation potential (Ahlers et al. 1993), thus taking the additional uncertainties discussed above into account. In addition, when further testing is required, the test design should be in accordance with the time necessary for reaching the steady-state level, i.e. the higher the $\log K_{OW}$ the longer the exposure time that should be applied in a long-term or chronic test. In such cases new standard test systems should also be considered to be able to design a testing strategy that can more realistically mimic the actual dose of the substance in the environment.

27.5 REFERENCES

Ahlers, J., M. Benzing, A. Gies, W. Pauli, and E. Rösick, 1988, "Yeast as a Unicellular Model System in Ecotoxicology and Xenobiochemistry," *Chemosphere, 17*, 1603–1615

Ahlers, J., I. Cascorbi, M. Foret, A. Gies, M. Köhler, W. Pauli, and E. Rösick, 1991, "Interaction with Functional Membrane Proteins – A Common Mechanism of Toxicity for Lipophilic Environmental Chemicals?," *Comp. Biochem. Physiol. C 100*, 111–113

Ahlers, J., R. Diderich, U. Klaschka, and B. Schwarz-Schulz, 1994, "Environmental Risk Assessment of Existing Chemicals," *Environ. Sci. Pollut. Res., 1*, 117–123

Ahlers, J., W. Koch, A. Lange, A. Marschner, and G. Welter, 1992, "Bewertung der Umweltgefährlichkeit von Alten Stoffen nach dem Chemikaliengesetz (ChemG)," Chemikaliengesetz-Heft 10, Texte 19/92, Umweltbundesamt, Berlin, Germany

Ahlers, J., W. Koch, A. Marschner, and G. Welter, 1993, "Environmental Hazard Assessment of Existing Chemicals," *Sci. Total Environ. Suppl.*, 1587–1596

Ahlers, J., A. W. Lange, and G. Welter, 1991, "Bewertung der Umweltgefährlichkeit von Altstoffen nach dem Chemikaliengesetz; Teil 1: Auswahl gefährlicher Stoffe durch das Umweltbundesamt," UWSF-Z. *Umweltchem. Ökotox., 3*, 104–106

Ahlf, W., 1995, "Ökotoxikologische Sedimentbewertung – Sedimenttoxizität, Biotest, Testkombination," UWSF-Z. *Umweltchem. Ökotox., 7*, 84–91

Brown, D., 1994, Conference on a Practical Look at Regulatory Harmonisation for New and Existing Chemicals in the EC, Abstracts, London, UK

Cowan, C. E., R. J. Larson, T. C. J. Feijtel, and R. A. Rapaport, 1993, "An Improved Model for Predicting the Fate of Consumer Product Chemicals in Wastewater Plants," *Water Res., 27*, 561–573

Cowan, C. E., D. J. Versteeg, R. J. Larson, and P. J. Kloepper-Sams, 1995, "Integrated Approach for Environmental Assessment of New and Existing Substances," *Reg. Toxicol. Pharmacol., 21*, 3–31

EC (European Commission), 1996, Technical Guidance Documents in Support of Directive 93/67/EEC on Risk Assessment of New Notified Substances and Regulations (EC) N° 1488/94 on Risk Assessment of Existing Substances, Brussels, Belgium

EC (European Commission), 1994, Risk Assessment of Existing Substances: Technical Guidance Document, XI/919/94-EN, Brussels, Belgium

EC (European Commission), 1993, Risk Assessment of Notified New Substances: Technical Guidance Document, Brussels, Belgium

Environmental Agency Japan, 1995, Chemicals in the Environment: Report on Environmental Survey and Wildlife Monitoring of Chemicals in F. Y. 1992 and 1993

Government of Canada/Environment Canada, 1995, Toxic Substances Management Policy

Klöpffer, W., 1994a, "Environmental Hazard Assessment of Chemicals and Products; Part II: Persistence and Degradability of Organic Chemicals," *Environ. Sci. Pollut. Res., 1*, 108–116

Klöpffer, W., 1994b, "Environmental Hazard Assessment of Chemicals and Products, Part III: The Limits to Single Compound Assessment," *Environ. Sci. Pollut. Res., 1*, 179–184

Mackay, D., S. Paterson, and W. Y. Shin, 1992, "Generic Models for Evaluating the Regional Fate of Chemicals," *Chemosphere*, *24*, 695–717

Morgan, E., and T. Knacker, 1994, "The Role of Laboratory Terrestrial Model Ecosystems in the Testing of Potentially Harmful Substances," *Ecotoxicology*, *3*, 213–233

Nabholz, J. V., 1991, "Environmental Hazard and Risk Assessment Under the United States Toxic Substances Control Act," in *QSAR in Environmental Toxicology–IV* (J. L. M. Hermens and A. Opperhuizen, Eds.), Elsevier, New York, USA, pp. 649–665

OECD, 1992a, "Report of the Workshop on Effects Assessment in Sediment," Copenhagen, Denmark, 13–15 May, 1992

OECD, 1992b, Screening Assessment Model System (SAMS), Version 1.1, Paris, France

OECD, 1994, SIDS Initial Assessment Report: Trichloroacetic acid, CAS-Nr. 76-03-9 Draft

OECD, 1995, "Draft report of the OECD Workshop on Environmental Hazard/Risk Asessment," London, UK, 24–27 May 1994

Römbke, J., C. Bauer, and A. Marschner, 1996, "Entwicklung einer Teststrategie zur Bewertung des Umweltgefährlichkeitspotentials von Umweltchemikalien im Kompartiment Boden," *UWSF-Z. Umweltchem. Ökotox.*, *8*, 158–166

Rönnpagel, K., W. Liß, and W. Ahlf, 1995, "Microbial Bioassays to Assess the Toxicity of Solid Associated Contaminants," *Ecotoxicol Environ. Saf.*, *31*, 99–103

Struijs, J., J. Stoltenkamp, and D. A. Van de Meent, 1991, "A Spreadsheet-Based Model to Predict the Fate of Xenobiotics in a Municipal Wastewater Treatment Plant," *Water Res.*, *25*, 891–900

Van Jaarsveld, J. A., 1990, "An Operational Atmospheric Transport Model for Priority Substances; Specifications and Instructions for Use," National Institute of Public Health and Environmental Protection (RIVM), The Netherlands, Report No. 222501 002

Van de Meent, 1993, Simpleboy a Generic Multimedia Fate Evaluation Model, National Institute of Public Health and Environmental Protection (RIVM), The Netherlands, No. 672720 001

Witte, I., H. Jacobi, and Juhl-Strauss, 1995, "Correlation of Synergistic Cytotoxic Effects of Environmental Chemicals in Human Fibroblasts with their Lipophilicity," *Chemosphere*, *31*, 4041–4049

Index

ENVIRONMENTAL SCIENCE AND TECHNOLOGY

A Wiley-Interscience Series of Texts and Monographs

Edited by JERALD L. SCHNOOR, *University of Iowa*
ALEXANDER ZEHNDER, *Swiss Federal Institute for Water Resources and Water Pollution Control*

PHYSIOCHEMICAL PROCESSES FOR WATER QUALITY CONTROL
Walter J. Weber. Jr., Editor

pH AND pION CONTROL IN PROCESS AND WASTE STREAMS
F. G. Shinskey

AQUATIC POLLUTION: An Introductory Text
Edward A. Laws

INDOOR AIR POLLUTION: Characterization, Prediction, and Control
Richard A. Wadden and Peter A. Scheff

PRINCIPLES OF ANIMAL EXTRAPOLATION
Edward J. Calabrese

SYSTEMS ECOLOGY: An Introduction
Howard T. Odum

INTEGRATED MANAGEMENT OF INSECT PESTS OF POME AND STONE FRUITS
B. A. Croft and S. C. Hoyt, Editors

WATER RESOURCES: Distribution, Use and Management
John R. Mather

ECOGENETICS: Genetic Variation in Susceptibility to Environmental Agents
Edwards J. Calabrese

GROUNDWATER POLLUTION MICROBIOLOGY
Gabriel Bitton and Charles P. Gerba, Editors

CHEMISTRY AND ECOTOXICOLOGY OF POLLUTION
Des W. Connell and Gregory J. Miller

SALINITY TOLERANCE IN PLANTS: Strategies for Crop Improvement
Richard C. Staples and Gary H. Toenniessen, Editors

ECOLOGY, IMPACT ASSESSMENT, AND ENVIRONMENTAL PLANNING
Walter E. Westman

CHEMICAL PROCESSES IN LAKES
Werner Stumm, Editor

INTEGRATED PEST MANAGEMENT IN PINE-BARK BEETLE ECOSYSTEMS
William E. Waters, Ronald W. Stark, and David L. Wood, Editors

PALEOCLIMATE ANALYSIS AND MODELING
Alan D. Hecht, Editor

BLACK CARBON IN THE ENVIRONMENT: Properties and Distribution
E. D. Goldberg

GROUND WATER QUALITY
C. H. Ward, W. Giger, and P. L. McCarty, Editors

TOXIC SUSCEPTIBILITY: Male/Female Differences
Edward J. Calabrese

ENERGY AND RESOURCE QUALITY: The Ecology of the Economic Process
Charles A. S. Hall, Cutler J. Cleveland, and Robert Kaufmann

ENVIRONMENTAL SCIENCE AND TECHNOLOGY
List of Titles (*Continued*)